Surface Science

Surface Science

Foundations of Catalysis and Nanoscience

Third Edition

KURT W. KOLASINSKI

Department of Chemistry, West Chester University, West Chester, PA, USA

A John Wiley & Sons, Ltd., Publication

Library of Congress Cataloging-in-Publication Data

Kolasinski, Kurt W.
    Surface science [electronic resource] : foundations of catalysis and nanoscience / Kurt W. Kolasinski. – 3rd ed.
    1 online resource.
    Includes bibliographical references and index.
    Description based on print version record and CIP data provided by publisher; resource not viewed.
    ISBN 978-1-118-30860-8 (MobiPocket) – ISBN 978-1-118-30861-5 (ePub) – ISBN 978-1-119-94178-1
    (Adobe PDF) – ISBN 978-0-470-66556-5 (hardback) (print)
    1. Surface chemistry. 2. Surfaces (Physics) 3. Catalysis. 4. Nanoscience. I. Title.
    QD506
    541'.33 – dc23

                                                                          2012001518

A catalogue record for this book is available from the British Library.

HB ISBN: 9781119990369
PB ISBN: 9781119990352

Set in 10/12pt Times-Roman by Laserwords Private Limited, Chennai, India
Printed and bound in Malaysia by Vivar Printing Sdn Bhd

Instructors can access PowerPoint files of the illustrations presented within this text, for teaching, at: http://booksupport.wiley.com

1   2012

To Kirsti and Annika

Contents

Acknowledgements

A significant number of readers, starting with Wayne Goodman, have asked for worked solutions to the exercises. After years of puttering about, these have now been included. Thank you for your patience, persistence, comments and pointing out items of concern. Those individuals who have pointed out errors in the first two editions include Scott Anderson, Eric Borguet, Maggie Dudley, Laura Ford, Soon-Ku Hong, Weixin Huang, Bruce Koel, Lynne Koker, Qixiu Li, David Mills and Pat Thiel. I would also like to thank David Benoit, George Darling, Kristy DeWitt and David Mills for help with answers to and data for exercises. A special thanks to Eckart Hasselbrink for a critical reading of Chapter 8, and Fred Monson for proofreading several chapters. I would particularly like to acknowledge Pallab Bhattacharya, Mike Bowker, George Darling, Istvan Daruka, Gerhard Ertl, Andrew Hodgson, Jonas Johansson, Tim Jones, Jeppe Lauritsen, Volker Lehmann, Peter Maitlis, Katrien Marent, Zetian Mi, TC Shen, Joseph Stroscio, Hajime Takano, Sachiko Usui, Brigitte Vögele, David Walton and Anja Wellner for providing original figures. Figures generated by me were drawn with the aid of Igor Pro, Canvas and CrystalMaker. The heroic efforts of Yukio Ogata in securing the *sumi nagashi* images were remarkable. Hari Manoharan provided the quantum corral image on the cover of the first edition. Flemming Besenbacher provided the image of the BRIM™ catalyst on the cover of the second edition. Tony Heinz provided the image of graphene on the cover of the third edition. Finally, sorry Czesław, but this edition took a whole lot of Dead Can Dance to wrap up.

Kurt W. Kolasinski
West Chester

October 2011

Introduction

When I was an undergraduate in Pittsburgh determined to learn about surface science, John Yates pushed a copy of Robert and McKee's *Chemistry of the Metal-Gas Interface* [1] into my hands, and said "Read this". It was very good advice, and this book is a good starting point for surface chemistry. But since the early 1980s, the field of surface science has changed dramatically. Binnig and Rohrer [2, 3] discovered the scanning tunnelling microscope (STM) in 1983 [3]. By 1986, they had been awarded the Nobel Prize in Physics and surface science was changed indelibly. Thereafter, it was possible to image almost routinely surfaces and surface bound species with atomic-scale resolution. Not long afterward, Eigler and Schweizer [4] demonstrated that matter could be manipulated on an atom by atom basis. The tremendous infrastructure of instrumentation, ideas and understanding that has been amassed in surface science is evident in the translation of the 2004 discovery of Novoselov and Geim [5] of graphene into a body of influential work recognized by the 2010 Nobel Prize in Physics.

With the inexorable march of smaller, faster, cheaper, better in the semiconductor device industry, technology was marching closer and closer to surfaces. The STM has allowed us to visualize quantum mechanics as never before. As an example, two images of a Si(100) surface are shown in Fig. I.1. In one case, Fig. I.1(a), a bonding state is imaged. In the other, Fig. I.1(b) an antibonding state is shown. Just as expected, the antibonding state exhibits a node between the atoms whereas the bonding state exhibits enhanced electron density between the atoms.

The STM ushered in the age of nanoscience; however, surface science has always been about nanoscience, even when it was not phrased that way. Catalysis has been the traditional realm of surface chemistry, and 2007 was a great year for surface science as celebrated by the awarding of the Nobel Prize in Chemistry to Gerhard Ertl "for his studies of chemical processes on solid surface". While it was Irving Langmuir's work – Nobel Prize in Chemistry, 1932 – that established the basis for understanding surface reactivity, it was not until the work of Gerhard Ertl that surface chemistry emerged from its black box, and that we were able to understand the dynamics of surface reactions on a truly molecular level.

Of course, these are not the only scientists to have contributed to the growth of understanding in surface science, nor even the only Nobel Prize winners. In the pages that follow, you will be introduced to many more scientists and, hopefully, to many more insights developed by all of them. This book is an attempt, from the point of view of a dynamicist, to approach surface science as the underpinning science of both heterogeneous catalysis and nanotechnology.

Surface Science: Foundations of Catalysis and Nanoscience, Third Edition. Kurt W. Kolasinski.
© 2012 John Wiley & Sons, Ltd. Published 2012 by John Wiley & Sons, Ltd.

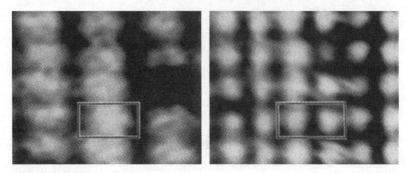

Figure I.1 *Bonding and antibonding electronic states on the Si(100) surface as imaged by STM. Reproduced with permission from R.J. Hamers, P. Avouris and F. Bozso, Phys. Rev. Lett. 59 (1987) 2071. ©1987 by the American Physical Society.*

I.1 Heterogeneous catalysis

One of the great motivations for studying chemical reactions on surfaces is the will to understand heterogeneous catalytic reactions. Heterogeneous catalysis is the basis of the chemical industry. Heterogeneous catalysis is involved in literally billions of dollars worth of economic activity. Neither the chemical industry nor civilization would exist as we know them today if it were not for the successful implementation of heterogeneous catalysis. At the beginning of the 20th century, the human condition was fundamentally changed by the transformation of nitrogen on nanoscale, potassium promoted, iron catalysts to ammonia and ultimately fertilizer. Undoubtedly, catalysts are the most successful implementation of nanotechnology,

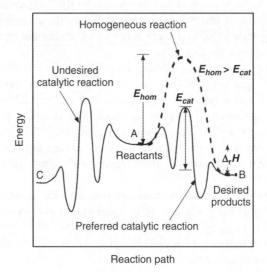

Figure I.2 *Activation energies and their relationship to an active and selective catalyst, which transforms A, the reactant, into B, the desired product, rather than C, the undesired product. E_{hom}, activation barrier for the homogeneous reaction; E_{cat}, activation barrier with use of a catalyst; $\Delta_r H$, change in enthalpy of reactants compared with products.*

not only contributing towards roughly 1/3 of the material GDP of the US economy [6], but also supporting an additional 3.2 billion people beyond what the Earth could otherwise sustain [7]. One aim of this book is to understand why catalytic activity occurs, and how we can control it.

First we should define what we mean by catalysis and a catalyst. The term catalysis (from the Greek λνσιζ and κατα, roughly "wholly loosening") was coined by Berzelius in 1836 [8]. Armstrong proposed the word catalyst in 1885. A catalyst is an active chemical spectator. It takes part in a reaction but is not consumed. A catalyst produces its effect by changing activation barriers as shown in Fig. I.2. As noted by Ostwald , who was awarded the Nobel Prize in Chemistry in 1909 primarily for this contribution, a catalyst speeds up a reaction; however, it does not change the properties of the equilibrated state. It does so by lowering the height of an activation barrier. Remember that whereas the kinetics of a reaction is determined by the relative heights of activation barriers (in combination with Arrhenius pre-exponential factors), the equilibrium constant is determined by the Gibbs energy of the initial state relative to the final state.

Nonetheless, the acceleration of reactions is not the only key factor in catalytic activity. If catalysts only accelerated reactions, they would not be nearly as important or as effective as they actually are. Catalysts can be designed not only to accelerate reactions: the best of them can also perform this task *selectively*. In other words, it is important for catalysts to speed up the right reactions, not simply every reaction. This is also illustrated in Fig. I.2, wherein the activation barrier for the desired product B is decreased more than the barrier for the undesired product C.

I.2 Why surfaces?

Heterogeneous reactions occur in systems in which two or more phases are present, for instance, solids and liquids, or gases and solids. The reactions occur at the interface between these phases. The interfaces are where the two phases and reactants meet, where charge exchange occurs. Liquid/solid and gas/solid interfaces are of particular interest because the surface of a solid gives us a place to deposit and immobilize a catalytic substance. By immobilizing the catalyst, we can ensure that it is not washed away and lost in the stream of products that are made. Very often catalysts take the form of nanoparticles (the active agent) attached to the surfaces of high surface area porous solids (the substrate).

However, surfaces are of particular interest not only because they are where phases meet, and because they give us a place to put catalysts. The surface of a solid is inherently different than the rest of the solid (the bulk) because its bonding is different. Therefore, we should expect the chemistry of the surface to be unique. Surface atoms simply cannot satisfy their bonding requirements in the same way as bulk atoms. Therefore, surface atoms will always have a propensity to react in some way, either with each other or with foreign atoms, to satisfy their bonding requirements.

I.3 Where are heterogeneous reactions important?

To illustrate a variety of topics in heterogeneous catalysis, I will make reference to a list of catalytic reactions that I label the (unofficial) Industrial Chemistry Hall of Fame. These reactions are selected not only because they demonstrate a variety of important chemical concepts, but also because they have also been of particular importance both historically and politically.

I.3.1 Haber-Bosch process

$$N_2 + 3H_2 \rightarrow 2NH_3$$

Nitrogen fertilizers underpin modern agriculture [7]. The inexpensive production of fertilizers would not be possible without the Haber-Bosch process. Ammonia synthesis is almost exclusively performed over an alkali metal promoted Fe catalyst invented by Haber, optimized by Mittasch and commercialized by Bosch. The establishment of the Haber-Bosch process is a fascinating story [7]. Ostwald (who misinterpreted his results), Nernst (who thought yields were intolerably low and abandoned further work), and Le Châtelier (who abandoned his work after an explosion in his lab), all could have discovered the secret of ammonia synthesis but did not. Technical innovations such as lower pressure reforming and synthesis, better catalysts and integrated process designs have reduced the energy consumption per ton of fixed nitrogen from 120 GJ to roughly 30 GJ, which is only slightly above the thermodynamic limit. This represents an enormous cost and energy usage reduction since over 130 million metric tons (MMt) of NH_3 are produced each year.

Ammonia synthesis is a structure sensitive reaction. Already a number of questions arise. Why an Fe catalyst? Why is the reaction run at high pressure and temperature? What do we mean by promoted, and why does an alkali metal act as a promoter? What is a structure sensitive reaction? What is the reforming reaction used to produce hydrogen, and how is it catalyzed? By the end of this book all of the answers should be clear.

I.3.2   Fischer-Tropsch chemistry

$$H_2 + CO \rightarrow \text{methanol or liquid fuels or other hydrocarbons (HC) and oxygenates}$$

Fischer-Tropsch chemistry transforms synthesis gas ($H_2 + CO$, also called syngas) into useful fuels and intermediate chemicals. It is the chemistry, at least in part, that makes synthetic oils that last 8000 km instead of 5000 km. It is the basis of the synthetic fuels industry, and has been important in sustaining economies that were shut off from crude oil, two examples of which were Germany in the 1930s and 1940s and, more recently, South Africa. It represents a method of transforming either natural gas or coal into more useful chemical intermediates and fuels. Interest in Fischer-Tropsch chemistry is rising again, not only because of the discovery of new and improved capture from old sources of natural gas, but also because biomass may also be used to produce synthesis gas, which is then converted to diesel or synthetic crude oil [9].

Fischer-Tropsch reactions are often carried out over Fe or Co catalysts. However, while Fischer-Tropsch is a darling of research labs, industrialists often shy away from it because selectivity is a major concern. A nonselective process is a costly one, and numerous products are possible in FT synthesis while only a select few are desired for any particular application.

I.3.3   Three-way catalyst

$$NO_x, CO \text{ and } HC \rightarrow H_2O + CO_2 + N_2$$

Catalysis is not always about creating the right molecule. It can equally well be important to destroy the right molecules. Increasing automobile use translates into increasing necessity to reduce automotive pollution. The catalytic conversion of noxious exhaust gasses to more benign chemicals has made a massive contribution to the reduction of automotive pollution. The three-way catalyst is composed of Pt, Rh and Pd. Pb rapidly poisons the catalyst. How does this poisoning (loss of reactivity) occur?

I.4   Semiconductor processing and nanotechnology

The above is the traditional realm of heterogeneous catalytic chemistry. However, modern surface science is composed of other areas as well, and has become particularly important to the world of micro- and nanotechnology [10–12]. Critical dimensions in microprocessors dropped below 100 nm in 2004 and now stand at 32 nm. The thickness of insulating oxide layers is now only 4–5 atomic layers. Obviously, there is a need to understand materials properties and chemical reactivity at the molecular level if semiconductor processing is to continue to advance to even smaller dimensions. It has already been established that surface cleanliness is one of the major factors affecting device reliability. Eventually, however, the engineers will run out of "room at the bottom". Furthermore, as length scales shrink, the effects of quantum mechanics inevitably become of paramount importance. This has led to the thought that a whole new device world may exist, which is ruled by quantum mechanical effects. Devices such as a single electron transistor have been built. Continued fabrication and study of such devices requires an understanding of atomic Legos® – the construction of structures on an atom-by-atom basis.

Figure I.3 shows images of some devices and structures that have been crafted at surfaces. Not only electronic devices are of interest. Microelectromechanical and nanoelectromechanical systems (MEMS and NEMS) are attracting increasing interest. The first commercial example is the accelerometer, which triggers airbags in cars and lets your iPhone™ know whether it should present its display in landscape or portrait mode. These structures are made by a series of surface etching and growth reactions.

The ultimate control of growth and etching would be to perform these one atom at a time. Figure I.4 demonstrates how H atoms can be removed one by one from a Si surface. The uncovered atoms are subsequently covered with oxygen, then etched. In Fig. I.4(b) we see a structure built out of Xe atoms. There are numerous ways to create structures at surfaces. We will investigate several of these in which the architect must actively pattern the substrate. We will also investigate self-assembled structures, that is, structures that form spontaneously without the need to push around the atoms or molecules that compose the structure.

I.5 Other areas of relevance

Surface science touches on a vast array of applications and basic science. The fields of corrosion, adhesion and tribology are all closely related to interfacial properties. The importance of heterogeneous processes in atmospheric and interstellar chemistry has been realized [13]. Virtually all of the molecular hydrogen that exists in the interstellar medium had to be formed on the surfaces of grains and dust particles. The role of surface chemistry in the formation of the over 100 other molecules that have been detected in outer space remains an active area of research [14–16]. Many electrochemical reactions occur heterogeneously. Our understanding of charge transfer at interfaces and the effects of surface structure and adsorbed species remain in a rudimentary but improving state [17–21].

I.6 Structure of the book

The aim of this book is to provide an understanding of chemical transformations and the formation of structures at surfaces. To do these we need to (i) assemble the appropriate vocabulary, and (ii) gain a familiarity with an arsenal of tools and a set of principles that guide our thinking, aid interpretation and enhance prediction. Chapter 1 introduces us to the structure (geometric, electronic and vibrational) of surfaces and adsorbates. This gives us a picture of what surfaces look like, and how they compare to molecules and bulk materials. Chapter 2 introduces the techniques with which we look at surfaces. We quickly learn that surfaces present some unique experimental difficulties. This chapter might be skipped in a first introduction to surface science. However, some of the techniques are themselves methods for surface modification. In addition, a deeper insight into surface processes is gained by understanding the manner in which data are obtained. Finally, a proper reading of the literature cannot be made without an appreciation of the capabilities and limitations of the experimental techniques.

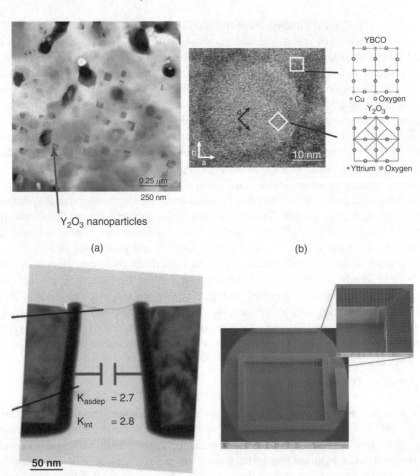

Figure I.3 *Examples of devices and structures that are made by means of surface reactions, etching and growth. (a) Transmission electron micrograph of yttria (Y_2O_3) nanocrystals in an yttrium barium copper oxide (YBCO) matrix. (b) Yttria nanocrystal embedded in YBCO layer of a second generation high temperature (high T_c) superconductor. Panels (a) and (b) reproduced from M. W. Rupich et al., IEEE Transactions on Applied Superconductivity 15, 2611. Copyright (2003), with permission from the IEEE. (c) An advanced CMOS device incorporating a low dielectric constant (low k) insulating layer. Reproduced from T. Torfs, V. Leonov, R. J. M. Vullers, Sensors and Transducers Journal, 80, 1230. Copyright (2007), with permission from the International Frequency Sensor Association (http://www.sensorsportal.com). (d) Micromachined thermoelectric generator fabricated on a silicon rim.*

After these foundations have been set in the first two chapters, the next two chapters elucidate dynamical, thermodynamic and kinetic principles concentrating on the gas/solid interface. These principles allow us to understand how and why chemical transformations occur at surfaces. They deliver the mental tools required to interpret the data encountered at liquid interfaces (Chapter 5) as well as in catalysis (Chapter 6), and growth and etching (Chapter 7) studies. Finally, in Chapter 8, we end with a chapter that resides squarely

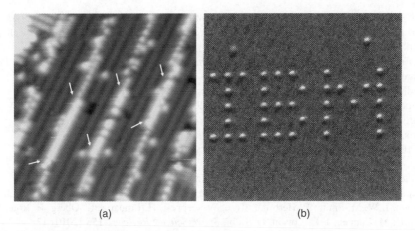

Figure I.4 *Examples of surface manipulation with atomic-scale resolution. (a) Nanolithography can be performed on a hydrogen-terminated silicon surface using a scanning tunnelling microscope (STM) tip to remove H atoms one at a time from the surface. (b) Individual Xe atoms can be moved with precision by an STM tip to write on surfaces. Panel (a) reproduced with permission from T.-C. Shen, C. Wang, G. C. Abeln, J. R. Tucker, J. W. Lyding, P. Avouris and R. E. Walkup,* Science *268 (1995) 1590. © 1995 American Association for the Advancement of Science. Panel (b) reproduced with permission from D. M. Eigler and E. K. Schweizer,* Nature *344 (2000) 524. © 2000 Macmillan Magazines Ltd.*

at the frontier of our knowledge: an investigation of the interfacial process probed and exited by photons, electrons and proximal probes.

Each chapter builds from simple principles to more advanced ones. Each chapter is sprinkled with Advanced Topics. The Advanced Topics serve two purposes. First, they provide material beyond the introductory level and can be skipped so as not to interrupt the flow of the introductory material. Second, they highlight some frontier areas. The frontiers are often too complex to explain in depth at the introductory level; nonetheless, they are included to provide a taste of the exciting possibilities of what can be done with surface science. Each chapter is also accompanied by exercises. The exercises act not only to demonstrate concepts arising in the text, but also as extensions to the text. They truly are an integral part of the whole and their solutions comprise the last eight chapters of this book. The exercises are not meant to be mere problems with answers to look up. Rather, they are intended to be exercises in problem solving applying the material in the text. The solutions, therefore, not only highlight and extend the material covered in the first eight chapters, they also detail methods of problem solving and the melding of concepts with mathematics to develop answers. Additional exercises can be found at the website that supports this book http://courses.wcupa.edu/kkolasinski/surfacescience/.

References

[1]   M. W. Roberts, C. S. McKee, *Chemistry of the Metal-Gas Interface*, Clarendon Press, Oxford, 1978.
[2]   G. Binnig, H. Rohrer, *Rev. Mod. Phys.*, **71** (1999) S324.
[3]   G. Binnig, H. Rohrer, C. Gerber, E. Weibel, *Phys. Rev. Lett.*, **49** (1982) 57.
[4]   D. M. Eigler, E. K. Schweizer, *Nature (London)*, **344** (1990) 524.
[5]   K. S. Novoselov, A. K. Geim, S. V. Morozov, D. Jiang, Y. Zhang, S. V. Dubonos, I. V. Grigorieva, A. A. Firsov, *Science*, **306** (2004) 666.

[6]   M. E. Davis, D. Tilley, *National Science Foundation Workshop on Future Directions in Catalysis: Structures that Function at the Nanoscale*, National Science Foundation, Washington, DC, 2003; http://www.cheme.caltech.edu/nsfcatworkshop/

[7]   V. Smil, *Enriching the Earth: Fritz Haber, Carl Bosch, and the Transformation of World Food Production*, MIT Press, Cambridge, MA, 2001.

[8]   K. J. Laidler, *The World of Physical Chemistry*, Oxford University Press, Oxford, 1993.

[9]   D. A. Simonetti, J. A. Dumesic, *Catal. Rev.*, **51** (2009) 441.

[10] P. Moriarty, *Rep. Prog. Phys.*, **64** (2001) 297.

[11] P. Avouris, Z. H. Chen, V. Perebeinos, *Nature Nanotech.*, **2** (2007) 605.

[12] G. Timp (Ed.), *Nanotechnology*, Springer Verlag, New York, 1999.

[13] E. Herbst, *Chem. Soc. Rev.*, **30** (2001) 168.

[14] D. J. Burke, W. A. Brown, *Phys. Chem. Chem. Phys.*, **12** (2010) 5947.

[15] L. Hornekaer, A. Baurichter, V. V. Petrunin, D. Field, A. C. Luntz, *Science*, **302** (2003) 1943.

[16] V. Wakelam, I. W. M. Smith, E. Herbst, J. Troe, W. Geppert, H. Linnartz, K. Oberg, E. Roueff, M. Agundez, P. Pernot, H. M. Cuppen, J. C. Loison, D. Talbi, *Space Science Reviews*, **156** (2010) 13.

[17] S. Fletcher, *J. Solid State Electrochem.*, **14** (2010) 705.

[18] N. S. Lewis, *Inorg. Chem.*, **44** (2005) 6900.

[19] [19] M. G. Walter, E. L. Warren, J. R. McKone, S. W. Boettcher, Q. Mi, E. A. Santori, N. S. Lewis, *Chem. Rev.*, **110** (2010) 6446.

[20] D. M. Adams, L. Brus, C. E. D. Chidsey, S. Creager, C. Creutz, C. R. Kagan, P. V. Kamat, M. Lieberman, S. Lindsay, R. A. Marcus, R. M. Metzger, M. E. Michel-Beyerle, J. R. Miller, M. D. Newton, D. R. Rolison, O. Sankey, K. S. Schanze, J. Yardley, X. Y. Zhu, *J. Phys. Chem. B*, **107** (2003) 6668.

[21] D. H. Waldeck, H. J. Yue, *Curr. Opin. Solid State Mater. Sci.*, **9** (2005) 28.

1

Surface and Adsorbate Structure

We begin with some order of magnitude estimates and rules of thumb that will be justified in the remainder of this book. These estimates and rules introduce and underpin many of the most important concepts in surface science. The atom density in a solid surface is roughly 10^{15} cm^{-2} (10^{19} m^{-2}). The Hertz-Knudsen equation

$$Z_w = \frac{p}{(2\pi m k_B T)^{1/2}} \tag{1.0.1}$$

relates the flux of molecules striking a surface, Z_w, to the pressure (or, equivalently, the number density). Combining these two, we find that if the probability that a molecule stays on the surface after it strikes it (known as the sticking coefficient s) is unity, then it takes roughly 1 s for a surface to become covered with a film one molecule thick (a monolayer) when the pressure is 1×10^{-6} Torr. The process of molecules sticking to a surface is called adsorption. If we heat up the surface with a linear temperature ramp, the molecules will eventually leave the surface (desorb) in a well-defined thermal desorption peak, and the rate of desorption at the top of this peak is roughly one monolayer per second. When molecules adsorb via chemical interactions, they tend to stick to well-defined sites on the surface. An essential difference between surface kinetics and kinetics in other phases is that we need to keep track of the number of empty sites. Creating new surface area is energetically costly and creates a region that is different from the bulk material. Size dependent effects lie at the root of nanoscience, and two of the primary causes of size dependence are quantum confinement and the overwhelming of bulk properties by the contributions from surfaces.

We need to understand the structure of clean and adsorbate-covered surfaces and use this as a foundation for understanding surface chemical processes. We will use our knowledge of surface structure to develop a new strand of chemical intuition that will allow us to know when we can apply things that we have learned from reaction dynamics in other phases and when we need to develop something completely different to understand reactivity in the adsorbed phase.

What do we mean by surface structure? There are two inseparable aspects to structure: electronic structure and geometric structure. The two aspects of structure are inherently coupled and we should never forget this point. Nonetheless, it is pedagogically helpful to separate these two aspects when we attack them experimentally and in the ways that we conceive of them.

When we speak of structure in surface science we can further subdivide the discussion into that of the clean surface, the surface in the presence of an adsorbate (substrate structure) and that of the adsorbate

Surface Science: Foundations of Catalysis and Nanoscience, Third Edition. Kurt W. Kolasinski.
© 2012 John Wiley & Sons, Ltd. Published 2012 by John Wiley & Sons, Ltd.

(adsorbate structure or overlayer structure). That is, we frequently refer to the structure of the first few layers of the substrate with and without an adsorbed layer on top of it. We can in addition speak of the structure of the adsorbed layer itself. Adsorbate structure not only refers to how the adsorbed molecules are bound with respect to the substrate atoms but also how they are bound with respect to one another.

1.1   Clean surface structure

1.1.1   Ideal flat surfaces

Most of the discussion here centres on transition metal and semiconductor surfaces. First we consider the type of surface we obtain by truncating the bulk structure of a perfect crystal. The most important crystallographic structures of metals are the face-centred cubic (fcc), body-centred cubic (bcc) and hexagonal close-packed (hcp) structures. Many transition metals of interest in catalysis take up fcc structures under normal conditions. Notable exceptions are Fe, Mo and W, which assume bcc structures and Co and Ru, which assume hcp structures. The most important structure for elemental (group IV: C, Si, Ge) semiconductors is the diamond lattice whereas compound semiconductors from groups III and V (III-V compounds, e.g. GaAs and InP) assume the related zincblende structure.

A perfect crystal can be cut along any arbitrary angle. The directions in a lattice are indicated by the Miller indices. Miller indices are related to the positions of the atoms in the lattice. Directions are uniquely defined by a set of three (fcc, bcc and diamond) or four (hcp) rational numbers and are denoted by enclosing these numbers in square brackets, e.g. [100]. hcp surfaces can also be defined by three unique indices and both notations are encountered as shown in Fig. 1.3. A plane of atoms is uniquely defined by the direction that is normal to the plane. To distinguish a plane from a direction, a plane is denoted by enclosing the numbers associated with the defining direction in parentheses, e.g. (100). The set of all related planes with permutations of indices, e.g. (100), (010), (001) etc, is denoted by curly brackets such as {001}.

The most important planes to learn by heart are the low index planes. Low index planes can be thought of as the basic building blocks of surface structure as they represent some of the simplest and flattest of the fundamental planes. The low-index planes in the fcc system, e.g. (100), (110) and (111), are shown in Fig. 1.1. The low-index planes of bcc symmetry are displayed in Fig. 1.2, and the more complex structures of the hcp symmetry are shown in Fig. 1.3.

The ideal structures shown in Fig. 1.1 demonstrate several interesting properties. Note that these surfaces are not perfectly isotropic. We can pick out several high-symmetry sites on any of these surfaces that are geometrically unique. On the (100) surface we can identify sites of one-fold (on top of and at the centre of

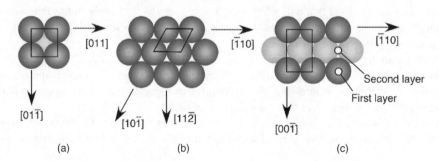

(a)   (b)   (c)

Figure 1.1   *Hard sphere representations of face-centred cubic (fcc) low index planes: (a) fcc(100); (b) fcc(111); (c) fcc(110).*

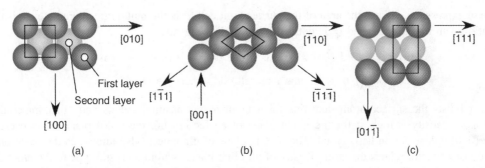

Figure 1.2 *Hard sphere representations of body-centred cubic (bcc) low index planes: (a) bcc(100); (b) bcc(110); (c) bcc(211).*

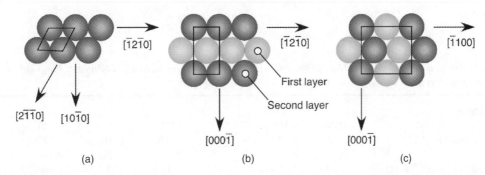

Figure 1.3 *Hard sphere representations of hexagonal close-packed (hcp) low index planes: (a) hcp(001) = (0001); (b) hcp(10$\bar{1}$0) = hcp(100); (c) hcp(11$\bar{2}$0) = hcp(110).*

one atom), two-fold (bridging two atoms) or four-fold co-ordination (in the hollow between four atoms). The co-ordination number is equal to the number of surface atoms bound directly to the adsorbate. The (111) surface has one-fold, two-fold and three-fold co-ordinated sites. Among others, the (110) presents two different types of two-fold sites: a long bridge site between two atoms on adjacent rows and a short bridge site between two atoms in the same row. As one might expect based on the results of co-ordination chemistry, the multitude of sites on these surfaces leads to heterogeneity in the interactions of molecules with the surfaces. This is important in our discussions of adsorbate structure and surface chemistry.

A very useful number is the surface atom density, σ_0. Nicholas [1] has shown that there is a simple relationship between σ_0 and the Miller indices hkl,

$$\sigma_0 = \frac{1}{A_{hkl}} = \frac{4}{Qa^2(h^2 + k^2 + l^2)^{1/2}} \quad \text{for fcc and bcc} \qquad (1.1.1)$$

and

$$\sigma_0 = \frac{1}{A_{hkl}} = \frac{2}{a^2[4r^2(h^2 + hk + k^2) + 3l^2]^{1/2}} \quad \text{for hcp} \qquad (1.1.2)$$

In these expressions, A_{hkl} is the area of the surface unit cell, a is the bulk lattice parameter, r is the hcp axial ratio given in Table 1.1 and Q is defined by the following rules:

$$\text{bcc} : Q = 2 \text{ if } (h + k + l) \text{ is even, } Q = 4 \text{ if } (h + k + l) \text{ is odd}$$

$$\text{fcc} : Q = 1 \text{ if } h, k, \text{ and } l \text{ are all odd, otherwise } Q = 2.$$

Table 1.1 lists the surface atom densities for a number of transition metals and other materials. The surface atom density is highest for the (111) plane of an fcc crystal, the (100) plane for a bcc crystal, and the (0001) plane for an hcp crystal. The (0001) plane of graphite is also known as the basal plane. A simple constant factor relates the atom density of all other planes within a crystal type to the atom density of the densest plane. Therefore, the atom density of the (111) plane along with the relative packing factor is listed for the fcc, bcc and hcp crystal types. Similarly for the diamond and zincblende lattices, the area of the surface units cell in terms of the bulk lattice constant a is

$$A_{100} = a^2/2 \tag{1.1.3}$$

$$A_{111} = (a^2/2) \sin 120° \tag{1.1.4}$$

Table 1.1 *Surface atom densities. Data taken from [2] except Si values from [3]. Diamond, Ge, GaAs and graphite calculated from lattice constants*

fcc structure

Plane	(100)	(110)	(111)	(210)	(211)	(221)	
Density relative to (111)	0.866	0.612	1.000	0.387	0.354	0.289	
Metal	Al	Rh	Ir	Ni	Pd	Pt	Cu
Density of (111)/cm^{-2} × 10^{-15}	1.415	1.599	1.574	1.864	1.534	1.503	1.772
Metal	Ag	Au					
Density of (111)/cm^{-2} × 10^{-15}	1.387	1.394					

bcc structure

Plane	(100)	(110)	(111)	(210)	(211)	(221)	
Density relative to (110)	0.707	1.000	0.409	0.316	0.578	0.236	
Metal	V	Nb	Ta	Cr	Mo	W	Fe
Density of (100)/cm^{-2} × 10^{-15}	1.547	1.303	1.299	1.693	1.434	1.416	1.729

hcp structure

Plane	(0001)	(10$\bar{1}$0)	(10$\bar{1}$1)	(10$\bar{1}$2)	(11$\bar{2}$2)	(11$\bar{2}$2)	
Density relative to (0001)	1.000	$\dfrac{3}{2r}$	$\dfrac{\sqrt{3}}{(4r^3 + 3)^1}$	$\dfrac{\sqrt{3}}{(4r^3 + 12)}$	$\dfrac{1}{r}$	$\dfrac{1}{2(r^3 + 1)^1}$	
Metal	Zr	Hf	Re	Ru	Os	Co	Zn
Density of (0001)/cm^{-2} × 10^{-15}	1.110	1.130	1.514	1.582	1.546	1.830	1.630
Axial ratio $r = c/a$	1.59	1.59	1.61	1.58	1.58	1.62	1.86
Metal	Cd						
Density of (0001)/cm^{-2} × 10^{-15}	1.308						
Axial ratio $r = c/a$	1.89						

	Diamond lattice			Zincblende		Graphite	
Element	C	Si	Ge	GaAs		C	
Areal Density/cm^{-2} × 10^{-15}						basal plane	
(100)	1.57	0.6782	0.627	0.626		3.845	
(111)	1.82	0.7839	0.724	0.723			

because the length of a side of the surface unit cell is given by $(\sqrt{2}/2)a$, the (100) unit cell is rectangular and the included angle in the (111) unit cell is 120°, just as in the fcc unit cells shown in Fig. 1.1. Therefore, the surface atom density ratio is

$$\sigma_{100}/\sigma_{111} = A_{111}/A_{100} = 0.866. \qquad (1.1.5)$$

Note that this is exactly the same factor as found in the table above for fcc metals.

The formation of a surface from a bulk crystal is a stressful event. Bonds must be broken and the surface atoms no longer have their full complement of co-ordination partners. Therefore, the surface atoms find themselves in a higher energy configuration compared to being buried in the bulk and they must relax. Even on flat surfaces, such as the low-index planes, the top layers of a crystal react to the formation of a surface by changes in their bonding geometry. For flat surfaces, the changes in bond lengths and bond angles usually only amount to a few per cent. These changes are known as relaxations. Relaxations can extend several layers into the bulk. The near surface region, which has a structure different from that of the bulk, is called the selvage. This is our first indication that bonding at surfaces is inherently different from that in the bulk both because of changes in co-ordination and because of changes in structure. On metal surfaces, the force (stress) experienced by surface atoms leads to a contraction of the interatomic distances in the surface layer.

1.1.2 High index and vicinal planes

Surface structure can be made more complex either by cutting a crystal along a higher index plane or by the introduction of defects. High index planes (surfaces with h, k, or $l > 1$) often have open structures that can expose second and even third layer atoms. The fcc(110) surface shows how this can occur even for a low index plane. High index planes often have large unit cells that encompass many surface atoms. An assortment of defects is shown in Fig. 1.4.

One of the most straightforward types of defect at surfaces is that introduced by cutting the crystal at an angle slightly off of the perfect [hkl] direction. A small miscut angle leads to vicinal surfaces. Vicinal surfaces are close to but not flat low index planes. The effect of a small miscut angle is demonstrated in Fig. 1.4. Because of the small miscut angle, the surface cannot maintain the perfect (hkl) structure over long distances. Atoms must come in whole units and in order to stay as close to a low index structure as possible while still maintaining the macroscopic surface orientation, step-like discontinuities are introduced into the surface structure. On the microscopic scale, a vicinal surface is composed of a series of terraces and steps. Therefore, vicinal surfaces are also known as stepped surfaces.

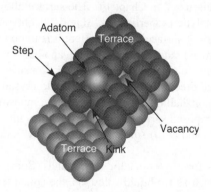

Figure 1.4 *Hard-sphere representation of a variety of defect structures that can occur on single-crystal surfaces.*

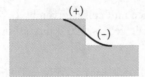

Figure 1.5 *Smoluchowski smoothing: the electrons at a step attempt to smooth out the discontinuity of the step.*

Stepped surfaces have an additional type of heterogeneity compared to flat surfaces, which has a direct effect on their properties [4]. They are composed of terraces of low index planes with the same types of symmetry as normal low index planes. In addition, they have steps. The structure of step atoms must be different from that of terrace atoms because of the different bonding that they exhibit. Step atoms generally relax more than terrace atoms. The effect of steps on electronic structure is illustrated in Fig. 1.5. The electrons of the solid react to the presence of the step and attempt to minimize the energy of the defect. They do this by spreading out in a way that makes the discontinuity at the step less abrupt. This process is known as Smoluchowski smoothing [5]. Since the electronic structure of steps differs from that of terraces, we expect that their chemical reactivity is different as well. Note that the top and the bottom of a step are different and this has implications, for instance, for diffusion of adsorbates over steps. It is often the case that diffusion in one direction is significantly easier than in the other. Furthermore, we expect that diffusion on the terraces may differ significantly from diffusion across steps (see § 3.2)

1.1.3   Faceted surfaces

Not all surfaces are stable. The formation of a surface is always endothermic, see Chapter 5. However, the formation of a larger surface area of low energy (low index) planes is sometimes favoured over the formation of a single layer of a high energy (high index) plane. Many high index planes are known to facet at equilibrium. Faceting is the spontaneous formation of arrays of low index planes separated by steps. Numerous systems exhibit ordered arrays of low index facets. These have been catalogued by Shchukin and Bimberg [6] and include vicinal surfaces of Si(111), GaAs(100), Pt(100), high index planes of Si(211) and low index planes of TaC(110).

1.1.4   Bimetallic surfaces

A surface composed of a mixture of two metals often exhibits unique properties. The catalytic behaviour of Au+Ni surfaces, for example, is discussed in Chapter 6. The surface alloy of $Pt_3Sn(111)$ has also attracted interest because of its unusual catalytic properties [7]. Materials containing mixtures of metals introduce several new twists into a discussion of surface structure. Here it is important to make a distinction between an alloy – a solid solution of one metal randomly dissolved in another – and an intermetallic compound – a mixture with a definite and uniform stoichiometry and unit cell. Consider a single crystal composed of two metals that form a true intermetallic compound. An ideal single crystal sample would exhibit a surface structure much like that of a monometallic single crystal. The composition of the surface would depend on the bulk composition and the exposed surface plane. Several examples of this type have been observed [8], e.g. $Cu_3Au(100)$; (100), (111) and (110) surfaces of Ni_3Al; (110) and (111) surfaces of NiAl as well as $TiPt_3(100)$.

Not all combinations of metals form intermetallic compounds. Some metals have limited solubility in other metals. In a solid solution, just as for liquid solutions, the solute tends to distribute randomly in the solvent. Furthermore, the solubility of a given metal may be different in the bulk than it is at the surface.

In other words, if the surface energy of one component of an alloy is significantly lower than that of the other component, the low surface energy species preferentially segregates to the surface. This leads to enrichment in the surface concentration as compared to the bulk concentration. Most alloys show some degree of segregation and enrichment of one component at the surface or in subsurface sites. The factors that lead to segregation are much the same as those that we encounter in Chapter 7 when we investigate growth processes. For a binary alloy AB, the relative strengths of the A−A, A−B and B−B interactions as well as the relative sizes of A and B determine whether alloy formation is exothermic or endothermic. These relative values determine whether segregation occurs. If alloy formation is strongly exothermic, i.e. the A−B interaction is stronger than either A−A or B−B interactions, then there is little tendency toward segregation. The relative atomic sizes are important for determining whether lattice strain influences the energetics of segregation. In summary, surface segregation is expected unless alloy formation is highly exothermic and there is good matching of the atomic radii.

If a bimetallic surface is made not from a bulk sample but instead from the deposition of one metal on top of another, the surface structure depends sensitively on the conditions under which deposition occurs. In particular, the structure depends on whether the deposition process is kinetically or thermodynamically controlled. These issues are dealt with in Chapter 7.

From these considerations, we can conceive of at least four configurations that arise from the deposition of one metal on top of another, as shown in Fig. 1.6. The formation of an intermetallic compound with a definite composition is illustrated in Fig. 1.6(a). The surface is rather uniform and behaves much as a pure metal surface with properties that are characteristic of the alloy and in all likelihood intermediate between the properties of either one of the constituent pure metals. If one metal is miscible in the other, it can incorporate itself into the surface after deposition. A random array of incorporated atoms, Fig. 1.6(b) is expected to dope the substrate. Doping means that the added atom changes the electronic character of the substrate metal. Thereby, the chemical reactivity and other properties such as magnetism, work function, etc, may also change. Whether the doping effect is long range or short range is greatly debated. An immiscible metal segregates into structures that minimize the surface area. Often step edges present sites with a high binding energy, which can lead to the structure shown in Fig. 1.6(c). If the binding of adatoms to themselves is stronger than to the substrate atoms, we expect the formation of islands to occur as shown in Fig. 1.6(d). In Chapter 6 we explore the implications of these different structures on catalytic reactivity.

1.1.5 Oxide and compound semiconductor surfaces

Oxides and compound semiconductors are covalent solids that exhibit a range of interesting properties. Most oxides can assume a number of structural forms. Metals often have more than one oxidation state, which means that they are able to form oxides of different stoichiometries or even mixed valence compounds in which more than one valence state is present. Different crystal structures have different optical and electronic properties. Defects as well as substitutional or interstitial impurities (dopants) also modify the optical and electronic properties as well as the chemistry of oxides and semiconductors. Because of the multifarious crystal structures exhibited by oxides and the strong role played by ubiquitous defects, it is more difficult to present general patterns in structure and reactivity [9].

Many oxides are insulators with band gaps in excess of 6 eV, such as SiO_2 (silica, quartz, glass, 35 different crystalline forms), Al_2O_3 (alumina, corundum or α-alumina as well as γ-alumina, which has a defect spinel structure, and several others) and MgO (magnesia with a NaCl structure). All of these are commonly used as substrates for supported catalysts. γ-alumina is particularly important for catalysis. It can take up water into its structure [10]; thus, the H content in γ-alumina may fall anywhere within the range $0 < n < 0.6$ for $Al_2O_3 \cdot n(H_2O)$. α-alumina doped with metal impurities is responsible for sapphire (Cr, Fe, or Ti doped) and ruby (Cr doped).

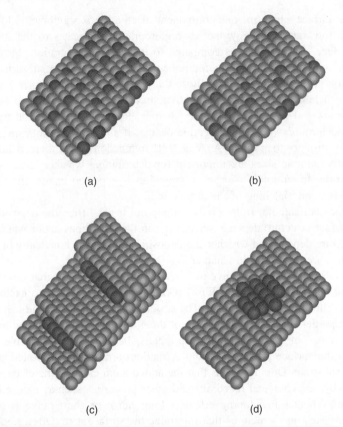

(a)   (b)

(c)   (d)

Figure 1.6   *Four limiting cases of the structure of a bimetallic surface prepared by metal-on-metal adsorption: (a) the formation of an intermetallic compound with a definite stoichiometry; (b) random absorption of a miscible metal; (c) segregation of an immiscible metal to the step edges; (d) segregation of an immiscible metal into islands.*

Other oxides are conductors or at least semiconductors with band gaps below 4 eV such as cerium oxides (ceria, CeO_2 with a fluorite structure but also Ce_2O_3), In_2O_3 (bixbyite type cubic structure and an indirect band gap of just 0.9–1.1 eV), SnO_2 (rutile, 3.6 eV direct band gap), and ZnO (wurzite, 3.37 eV direct band gap). The ability of ceria to form two different valence states allows its surface and near surface region to exist in a range of stoichiometries between CeO_2 and Ce_2O_3. As we shall see later, when used as an additive to the catalyst support for automotive catalysis, this allows ceria to act as a sponge that absorbs or releases oxygen atoms based on the reactive conditions. By substituting approximately 10% Sn into indium oxide, the resulting material, indium tin oxide (ITO), combines two seemingly mutually exclusive properties: a thin film of ITO is both transparent and electrically conductive. This makes ITO very important in display and photovoltaic technology. The properties of semiconducting oxides are responsive not only to oxygen content and metal atom substitution. Because they are semiconductors, their electrical and optical properties are also strongly size dependent when formed into nanotubes, nanowires and nanoparticles due to the effects of quantum confinement. In other words, when the size of the nanostructure becomes small enough for the electronic wavefunctions to sense the walls that contain them, the wavefunctions, and therefore also the properties of the electronic states, become size dependent.

The covalent nature of oxides means that certain cleavage planes are much more favoured to form surfaces. Certain planes exhibit a low surface energy, which means they are more stable and more likely to

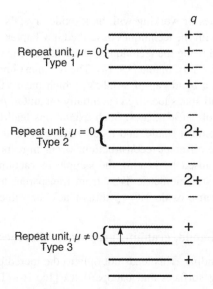

Figure 1.7 *Three types of planes formed by ionic crystals. q, ionic charge in layer; μ electric dipole moment associated with the lattice repeat unit.*

be observed because the crystal naturally breaks along these planes. This is in contrast to metals. Metallic bonding is much less directional. Consequently, many planes have similar surface energies and it is easier to make samples of crystalline metals that exhibit a variety of different planes. Another consequence of directional bonding is that the surfaces of covalent solids are much more susceptible to reconstruction.

The stability of covalent surfaces is described by Tasker's rules [11]. The surfaces of ionic or partly ionic crystals can be classified according to three types, as shown in Fig. 1.7. Type 1 consists of neutral planes containing a counterbalancing stoichiometry of anions and cations. Type 2 is made up of symmetrically arranged layers of positive and negative charge, which when taken as a repeat unit contain no net charge and no net dipole moment perpendicular to the unit cell. Type 3 exhibits a surface charge and there is a dipole moment perpendicular to the unit cell. Types 1 and 2 represent stable surfaces that exhibit a good balance of charge at the surface. Polar surfaces of Type 3 have very large surface energies and are unstable. These surfaces cannot exist unless they are stabilized by extensive reconstruction, faceting or the adsorption of counterbalancing charge.

The rocksalt (NaCl) structure is common among binary oxides of the general formula MX such as MgO and NiO. Both the {001} and {110} are stoichiometric planes of Type 1 with equal numbers of M^+ and X^- ions. Both are low energy planes and the {001} planes are the natural cleavage planes for rocksalt crystals. The {111} planes are of Type 3 and, therefore, are only expected to be observed in a bulk crystal if stabilized by reconstruction or adsorption, though they might be observed in a potentially metastable thin film.

The fluorite structure is common for MX_2 oxides such as CeO_2. Here the {001} are Type 3 planes. While the {110} are Type 1 planes, the {111} planes with exposed O^{2-} anions at their surfaces are of Type 2, have the lowest energy, and are the natural cleavage planes of a fluorite crystal. {111} planes that are terminated by metal cations are of Type 3 and are not normally observed.

The zincblende structure is characteristic of compound semiconductors such as GaAs and ZnS. While not oxides they are partially ionic. The {110} planes are the natural cleavage planes. They are neutral and of Type 1 containing the same number of anions and cations. Both the {001} and {111} planes contain either Ga or As atoms, that is, they are Type 3 planes which are unstable with respect to reconstruction. Crystals that expose {001} or {111} planes are composed of alternate layers of anions and cations.

One way to avoid the difficulties of working with bulk oxide crystals is to grow thin films of oxide on top of metal single crystals [12] with the techniques described in Chapter 7. Using this approach, a range of materials can then be investigated using the techniques of surface science to probe structure and reactivity.

An oxide of particular interest is TiO_2, titania [13, 14]. The two most important crystalline forms of titania are rutile and anatase. Rutile has a band gap of $\sim 3.2\,eV$, which means that it absorbs in the near UV by excitation from filled valence band states localized essentially on an $O(2p)$ orbital to the empty conduction band states primarily composed of $Ti(3d)$ states. Such excitations lead to very interesting photocatalytic properties. The (110) plane is the most stable, and it can assume several different reconstructions that are discussed in the next section. The different reconstructions occur in response to the oxygen content of the crystal. The release of oxygen caused by heating the sample in vacuum drastically changes the optical properties, changing a perfectly stoichiometric TiO_2 from transparent to blue to black with progressive loss of O atoms. A substoichiometric oxide TiO_{2-x} turns black for value of $x = 0.01$.

1.1.6   The carbon family: Diamond, graphite, graphene, fullerenes and carbon nanotubes

Carbon has unusually flexible bonding. This leads not only to the incredible richness of organic chemistry, but also to a range of interesting surfaces and nanoparticles [15, 16]. The richness of carbon chemistry is related to carbon's ability – in the vernacular of valence bond theory – to participate in three different types of orbital hybridization: *sp, sp^2* and *sp^3*. *sp* hybridization is associated with the ability to form triple bonds and is associated with linear structures in molecules. It is the least important for the understanding of carbonaceous solids. *sp^2* hybridization is associated with the formation of double bonds and planar structures. Below we see that this is the hybridization that leads to the formation of graphite, graphene, fullerenes and carbon nanotubes. *sp^3* hybridization is associated with tetrahedrally bound C atoms and forms the basis of C bound in the diamond lattice. The diamond lattice, Fig. 1.8(a), is shared by Si and Ge in their most stable allotropes. It is also geometrically the same as the zincblende structure common to the III-V family of semiconductors, e.g. GaAs and related compounds, with the important distinction that in the zincblende structure there are two chemically distinct atoms that alternately occupy lattice sites according to which plane is viewed. In general, the surfaces of semiconductors in the diamond or zincblende structure reconstruct to form surfaces with periodicities different than those of bulk crystals as are discussed further in § 1.2.

Graphite is a layered three-dimensional solid composed of six-membered rings of C atoms with *sp^2* hybridization. The honeycomb lattice of the basal plane of graphite is shown in Fig. 1.8(b). This is by far the most stable plane of graphite because it minimizes the number of reactive sites at unsaturated C atoms. The C atoms in any one layer are strongly covalently bonded to one another but the layers interact weakly through van der Waals forces. Because of this weak interplane interaction, the surface is easily deformed in an out-of-plane direction and C atoms in graphite or a graphene sheet can be pulled up by the approach of a molecule or atom as it attempts to adsorb [17, 18]. In their lowest energy form, the layers stack in an ABAB fashion. A single crystal of graphite would, of course, have nearly perfect order. However, the most commonly encountered form of ordered graphite is highly ordered pyrolytic graphite (HOPG), which contains many stacking faults and rotational domains in a macroscopic surface. The rotational domains are platelets of ordered hexagonal rings that lie in the same plane but which are rotated with respect to one another. Defects are formed where the domains meet. The domains can be as large as $1-10\,\mu m$ across.

Graphene is the name given to a single (or few) layer sample of carbon atoms bound in a hexagonal array. Graphite results from stacking together a large number of these layers. Carbon nanotubes (CNT) are made by rolling up graphene sheets such that the edge atoms can bind to one another and form a seamless tube with either one (single-wall nanotube, SWNT) or several (multiwall nanotube MWNT) layers. As can be seen from the structure depicted in Fig. 1.8(b), the manner in which the graphene sheet is rolled is very

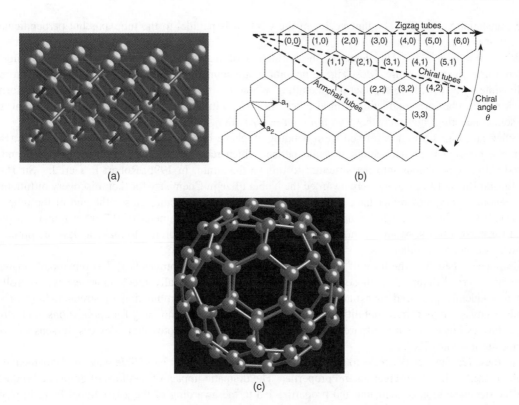

Figure 1.8 *(a) A ball and stick model of the diamond lattice with the unreconstructed (111) surface shown at the top. (b) The six-membered ring structure that defines the basal plane of graphite or a graphene sheet. a_1 and a_2 are the surface lattice vectors. When rolled to form a seamless structure, the resulting nanotube has either a zigzag, chiral or armchair configuration depending on the value of the chiral angle. (c) The structure of the fundamental fullerene, that of C_{60}.*

important for determining the structure of the nanotube. A CNT can be characterized by its chiral vector **C** given by

$$\mathbf{C} = n\mathbf{a}_1 + m\mathbf{a}_2 \tag{1.1.6}$$

where n and m are integers, and $\mathbf{a}_1$ and $\mathbf{a}_2$ are the unit vectors of graphene. The hexagonal symmetry of graphene means that the range of m is limited by $0 \leq |m| \leq n$. The values of n and m determine not only the relative arrangement of the hexagons along the walls of the nanotube, but also the nature of the electronic structure and the diameter. The diameter of a SWNT is given by

$$d = (a/\pi)\sqrt{n^2 + mn + m^2} \tag{1.1.7}$$

where the magnitude of the surface lattice vector $a = |\mathbf{a}_1| = |\mathbf{a}_2| = 0.246$ nm. The chiral angle, which can assume values of $0 \leq \theta \leq 30°$, is related to n and m by

$$\theta = \sin^{-1}\left(\frac{m}{2}\sqrt{\frac{3}{n^2 + mn + m^2}}\right). \tag{1.1.8}$$

The translation vector **T** is the unit vector of the CNT. It is parallel to the tube axis but perpendicular to the chiral vector.

When $m = 0$, $\theta = 0°$. This leads to the formation of nanotubes in which the hexagonal rings are ordered in a zigzag structure. Most zigzag nanotubes are semiconducting, except for those with an index n divisible by 3, which are conducting (metallic). When $n = m$, $q = 30°$ and armchair tubes are formed. All armchair tubes are conducting. Chiral tubes are formed in between these extremes. Most chiral tubes are semiconducting; however, those with $n - m = 3q$ with q an integer are conducting.

Fullerenes, Fig. 1.8(c), are formed from bending graphene into spherical or near spherical cages. However, the bending introduces strain into the lattice, which requires the rearrangement of bonds and the introduction of pentagons into the structure to relieve the strain. In 1996, Robert F. Curl Jr., Sir Harold Kroto, and Richard E. Smalley were awarded the Nobel Prize in Chemistry for their discovery of fullerenes. If a nanotube is capped rather than open, then a hemi-fullerene structure closes the end of the tube. The buckyball C_{60} was the first fullerene to be identified, and with a diameter of 0.7 nm it remains a poster child for nanoparticles. Atoms or molecules can be stuffed inside fullerenes to make a class of compounds known as endohedral fullerenes.

Graphene, whether in the form of layers [19] or rolled up into nanotubes [16], has remarkable materials properties very different than those of graphite. Layers are potentially much more easily, controllably and reproducibly produced than nanotubes. Graphene's unusual combination of strength, high thermal conductivity and high carrier mobility makes it particularly appealing not only for applications to electronic logic devices but also as a transparent electrode for displays and photovoltaic devices, in sensors and in composite materials [20].

For these reasons, ready access to graphene layers has ushered in an incredible wave of fundamental and applied research into their creation and properties. The dramatic impact of methods of graphene production and the demonstration of their unusual properties led to the awarding of the 2010 Nobel Prize in physics to Geim and Novoselov for their discoveries with regard to these materials.

Graphene is usually produced by one of four methods. The original method due to Novoselov et al. [21] was mechanical exfoliation. In this extremely complex method, a piece of adhesive tape is attached to a bulk sample of graphite and then ripped off. The layer that remains on the tape is thin but not a single layer. When pressed on to a substrate, a single layer can be transferred. The reproducibility and ability to form large layers improves with practice and films as large as $100 \, \mu m^2$ can be produced in this fashion. Chemical exfoliation works by intercalating a chemical species between the graphene layers. This causes lift off of graphene layers when properly treated. The method can be used industrially to form large quantities of graphene from graphite ore. Organic molecules or graphene oxide can be used as precursors from which they can be transformed by chemical reactions into graphene [22]. Of most interest for electronics applications are chemical vapour deposition methods for the production of graphene. These, along with methods for the production of CNTs with catalytic growth, are discussed in more detail in Chapter 7.

Graphene's unusual materials properties can be traced to its combination of covalent bonding with extended band structure. The hexagonal structure is composed of sp^2 hybridized C atoms. Single bonds with σ character hold these atoms together through localized covalent bonds directed in the plane of carbon atoms. In addition, an extended network of orbitals extending above and below the plane is formed with π symmetry. This network results from the highest occupied molecular orbitals (HOMO) overlapping to form the valance band while the lowest unoccupied molecular orbitals (LUMO) form the conduction band. These π bonds are shared equally by the carbon atoms, effectively giving them a 1 1/3 bond order each; but, more importantly, the electrons are all paired into bonding interactions within one carbon layer. Therefore, interactions within a plane are very strong; whereas out-of-plane interactions are quite weak. As a result, graphite is held together only by van der Waals interactions between the layers, and in-plane values of properties such as the electrical or thermal conductivity are much different than the out-of-plane values.

The "extra" 1/3 of a bond of each C atom can be removed by saturating it with H atoms. This produces a material known as graphane [23, 24]. Once fully saturated, the layer can take up 1/3 of a monolayer of H atoms, that is, one H atom for every 3 carbon atoms. This forces the layer to pucker and changes the electronic structure from that of a semiconductor/semimetal in which the conduction and valence bands meet at one Dirac point in k-space to an insulator with a fully developed band gap.

1.1.7 Porous solids

It is standard usage to refer to a solid substance by adding a suffix to its chemical notation to denote its phase. Hence crystalline Si is c-Si, amorphous Si is a-Si, alumina in the β structure is β-Al$_2$O$_3$ and in the gamma structure γ-Al$_2$O$_3$. In the spirit of this convention I denote a nanocrystalline substance as, e.g., nc-Si and a porous solid as, e.g., por-Si. This general notation avoids several possible ambiguities compared to some notation that is used in the literature. For instance, p-Si or pSi is sometimes used for porous silicon but also sometimes for p-doped Si.

A number of parameters are used to characterize porous solids. They are classified according to their mean pore size, which for a circular pore is measured by the pore diameter. The International Union of Pure and Applied Chemistry (IUPAC) recommendations [25] define samples with free diameters <2 nm as microporous, between 2 and 50 nm as mesoporous, and > 50 nm as macroporous. The term nanoporous is currently in vogue but undefined. The porosity ε is a measure of the relative amount of empty space in the material,

$$\varepsilon = V_p/V, \tag{1.1.9}$$

where V_p is the pore volume and V is the apparent volume occupied by the material. We need to distinguish between the exterior surface and the interior surface of the material. If we think of a thin porous film on top of a substrate, the exterior surface is the upper surface of the film (taking into account any and all roughness), whereas the interior surface comprises the surface of all the pore walls that extend into the interior of the film. The specific area of a porous material, a_p, is the accessible area of the solid (the sum of exterior and interior areas) per unit mass.

Porous solids can be produced in several ways. They can be grown from a molecular beam incident at a high angle in a technique known as glancing angle deposition (GLAD) [26]. Porous solids are often produced by etching an originally nonporous solid, for instance Si [27], Ge, SiC, GaAs, InP, GaP and GaN [28] as well as Ta$_2$O$_5$ [29], TiO$_2$ [30], WO$_3$ [31], ZrO$_2$ [32] and Al$_2$O$_3$ [33]. Mesoporous silicas and transition metal oxides (also called molecular sieves) are extremely versatile and can exhibit pore sizes from 2–50 nm and specific areas up to 1500 m^2 g^{-1}. They can be synthesized using a liquid crystal templating method that allows for control over the pore size [34, 35]. A similar templating strategy has also been used to produce mesoporous Ge [36] and porous Au [37]. By using a template of latex or silica microspheres (or, more generally, colloidal crystals), this range can be extended into the macroporous regime for a broad range of materials, including inorganic oxides, polymers, metals, carbon and semiconductors [38]. This method involves creating a mixture of the template and the target material. The template self-assembles into a regular structure and the target fills the space around the template. Subsequently, the template is removed by combustion, etching, some other chemical reaction or dissolution, leaving behind the target material in, generally, a powder of the porous material.

Porous materials are extremely interesting for the optical, electronic and magnetic properties [27, 39] as well as membranes [40, 41]. A much more in-depth discussion of their use in catalysis can be found in Thomas and Thomas [42]. Porous oxides such as Al$_2$O$_3$, SiO$_2$ and MgO are commonly used as (relatively) inert, high surface areas substrates into which metal clusters are inserted for use as heterogeneous catalysts. These materials feature large pores so that reactants can easily access the catalyst particles and products can easily leave them. Zeolites are microporous (and, more recently, mesoporous) aluminosilicates, aluminium phosphates (ALPO), metal aluminium phosphates (MeALPO) and silicon aluminium phosphates (SAPO) that are used in catalysis, particularly in the refining and reforming of hydrocarbons and petrochemicals. Other atoms can be substituted into the cages and channels of zeolites, enhancing their chemical versatility. Zeolites can exhibit both acidic and basic sites on their surfaces that can participate in surface chemistry. Microporous zeolites constrain the flow of large molecules; therefore, they can be used for shape-selective catalysis. In other words, only certain molecules can pass through them so only certain molecules can act as reactants or leave as products.

Another class of microporous materials with pores smaller than 2 nm is that of metal-organic frameworks (MOF) [43–45]. MOFs are crystalline solids that are assembled by the connection of metal ions or clusters through molecular bridges. Control over the void space between the metal clusters is easily obtained by controlling the size of the molecular bridges that link them. One implemented strategy, termed 'reticular synthesis' incorporates the geometric requirements for a target framework and transforms starting materials into such a framework. As synthesized, MOFs generally have their pores filled with solvent molecules because they are synthesized by techniques common to solution phase organic chemistry. However, the solvent molecules can usually be removed to expose free pores. Closely related are covalent organic frameworks (COF) which are composed entirely of light elements including H, B, C, N and O [46].

1.2   Reconstruction and adsorbate structure

1.2.1   Implications of surface heterogeneity for adsorbates

As suggested above, the natural heterogeneity of solid surfaces has several important ramifications for adsorbates. Simply from a consideration of electron density, we see that low index planes, let alone vicinal surfaces, are not completely flat. Undulations exist in the surface electron density that reflect the symmetry of the surface atom arrangement as well as the presence of defects such as steps, missing atoms or impurities. The ability of different regions of the surface to exchange electrons with adsorbates, and thereby form chemical bonds, is strongly influenced by the co-ordination numbers of the various sites on the surface. More fundamentally, the ability of various surface sites to enter into bonding is related to the symmetry, nature and energy of the electronic states found at these sites. It is a poor approximation to think of the surface atoms of transitions metals as having unsaturated valences (dangling bonds) waiting to interact with adsorbates. The electronic states at transition metal surfaces are extended, delocalized states that correlate poorly with unoccupied or partially occupied orbitals centred on a single metal atom. The concept of dangling bonds, however, is highly appropriate for covalent solids such as semiconductors. Non-transition metals, such as aluminium, can also exhibit highly localized surface electronic states.

The heterogeneity of low index planes presents an adsorbate with a more or less regular array of sites. Similarly, the strength of the interaction varies in a more or less regular fashion that is related to the underlying periodicity of the surface atoms and the electronic states associated with them. These undulations are known as corrugation. Corrugation can refer to either geometric or electronic structure. A corrugation of zero corresponds to a completely flat surface. A high corrugation corresponds to a mountainous topology.

Since the sites at a surface exhibit different strengths of interaction with adsorbates, and since these sites are present in ordered arrays, we expect adsorbates to bind in well-defined sites. Interactions between adsorbates can enhance the order of the overlayer; indeed, these interactions can also lead to a range of phase transitions in the overlayer [47]. We discuss the bonding of adsorbates extensively in Chapter 3 and adsorbate–adsorbate interactions in Chapter 4. The symmetry of overlayers of adsorbates may sometimes be related to the symmetry of the underlying surface. We distinguish three structural regimes: random, commensurate and incommensurate. Random adsorption corresponds to the lack of two-dimensional order in the overlayer, even though the adsorbates may occupy (one or more) well-defined adsorption sites. Commensurate structures are formed when the overlayer structure corresponds to the structure of the substrate in some rational fashion. Incommensurate structures are formed when the overlayer exhibits two-dimensional order; however, the periodicity of the overlayer is not related in a simple fashion to the periodicity of the substrate.

A more precise and quantitative discussion of the relationship of overlayer-to-substrate structure is discussed in § 2.5, which deals with low energy electron diffraction (LEED). The surface obtained by projecting one of the low index planes from the bulk unit cell onto the surface is called the (1×1) surface in Wood's notation because the surface unit cell is the same size as the bulk unit cell. Several examples of ordered overlayers and the corresponding Wood's notation are given in Fig. 1.9. Notice that n and m are proportional to the length of the vectors that define the parallelogram of the overlayer unit cell compared with the length of the unit vectors that define the clean surface unit cell. Therefore, the product nm is proportional to the area of the unit cell. A (2×2) unit cell is $2 \times$ larger than a (2×1) and $4 \times$ larger than a (1×1) unit cell area.

1.2.2 Clean surface reconstructions

In most instances, the low index planes of metals are stabilized by simple relaxations. Sometimes relaxation of the selvage is not sufficient to stabilize the surface as is the case for Au(111) and Pt(100). To minimize the surface energy, the surface atoms reorganize the bonding among themselves. This leads to surfaces – called reconstructions – with periodicities that differ from the structure of the bulk-terminated surface. For semiconductors and polar surfaces it is the rule rather than the exception that the surfaces reconstruct. This can be traced back to the presence of dangling bonds on covalent surfaces whereas metal

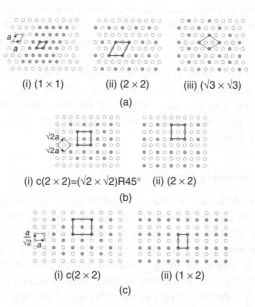

Figure 1.9 *Some commonly observed adsorbate structures on low index face-centred cubic (fcc) planes. (a) fcc(111), (i) (1×1), (ii) (2×2), (iii) $(\sqrt{3} \times \sqrt{3})$. (b) (i) fcc(100)–c(2 × 2), (ii) fcc(110)–c(2 × 2). (c) (i) fcc(100)–(2 × 2), (ii) fcc(110)–(1 × 2).*

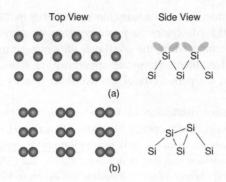

Figure 1.10 *The Si(100)–(2 × 1) reconstruction: (a) unreconstructed clean Si(100)–(1 × 1); (b) reconstructed clean Si(100)–(2 × 1).*

electrons tend to occupy delocalized states. Delocalized electrons adjust more easily to relaxations and conform readily to the geometric structure of low index planes. Dangling bonds are high-energy entities and solids react in extreme ways to minimize their number. The step atoms on vicinal surfaces are also associated with localized electronic states, even on metals. In many cases, vicinal surfaces are not sufficiently stabilized by simple relaxations and they therefore undergo faceting as mentioned above.

One of the most important and interesting reconstructions is that of the Si(100) surface, shown in Fig. 1.10, which is the plane most commonly used in integrated circuits. The Si(100)–(2 × 1) reconstruction completely eliminates all dangling bonds from the original Si(100)–(1 × 1) surface. The complex Si(111)–(7 × 7) reconstruction reduces the number of dangling bonds from 49 to just 19.

Another exemplary set of reconstructions is displayed by $TiO_{2-x}(110)$. As Fig. 1.11(a) shows, the expected structure with the same structure as the (110) face of the bulk unit cell is found on stoichiometric TiO_2 and for surfaces for which x is very small (corresponding to blue crystals). The images are dominated by 5-fold co-ordinate Ti atoms in the surface. Oxygen ions are not usually imaged by scanning tunnelling microscopy (STM), and the reasons why different atoms image differently are discussed in Chapter 2. The $(n \times m)$ nomenclature indicates the size of the reconstruction unit cell compared to the unreconstructed (1×1) unit cell. As the crystal loses more O atoms, it changes as shown in Figs 1.11 (a)–(d). The structure first transforms to a (1×3) pattern, which only exists over a narrow range of x. This is followed by two different (1×2) reconstructions. The first is an added row reconstruction. The second (1×2) structure is really an $(n \times 2)$ where n is a large but variable number for any particular surface (i.e. the structure often lacks good order in the [001] direction).

1.2.3   Adsorbate induced reconstructions

An essential tenant of thermodynamics is that at equilibrium a system possesses the lowest possible chemical potential, and that all phases present in the system have the same chemical potential. This is true in the real world in the absence of kinetic or dynamic constraints or, equivalently, in the limit of sufficiently high temperature and sufficiently long time so that any and all activation barriers can be overcome. This tenant must also hold for the gas phase/adsorbed phase/substrate system and has several interesting consequences. We have already mentioned that adsorbates can form ordered structures, viz. Figs 1.4, 1.10, 2.2 and 2.5. This may appear contrary to the influence of entropy, but if an ordered array of sites is to be maximally filled, then the adsorbates must also assume an ordered structure. The only constraint on the system is that chemisorption must be sufficiently exothermic to overcome the unfavourable entropy factors.

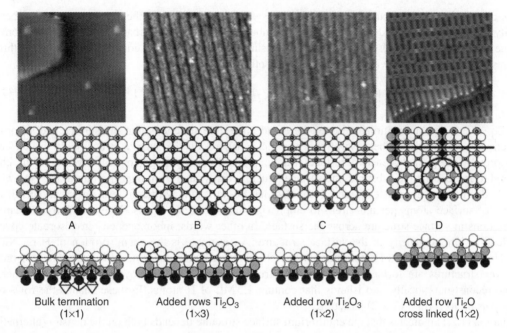

A   B   C   D

Bulk termination
(1×1)

Added rows Ti$_2$O$_3$
(1×3)

Added row Ti$_2$O$_3$
(1×2)

Added row Ti$_2$O
cross linked (1×2)

Figure 1.11   *The surface structures observed on TiO$_2$(110) as a function of increasing bulk reduction of the crystal. Upper panels are scanning tunnelling microscope images with 20 nm scan size. The lower two panels display the proposed surface structures. Reproduced from M. Bowker,* Curr. Opin. Solid State Mater. Sci. *10, 153. (2006) with permission from Elsevier.*

Adsorbates not only can assume ordered structures, they can also induce reconstruction of the substrate. One way to get rid of dangling bonds is to involve them in bonding. The Si–H bond is strong and non-polar. H atom adsorption represents a perfect method of capping the dangling bonds of Si surfaces. H atom adsorption is found to lift the reconstruction of Si surface, that is, the clean reconstructed surfaces are transformed to a new structure by the adsorption of H atoms. By adsorbing one H atom per surface Si atom, the Si(100)–(2 × 1) asymmetric dimer structure is changed into a symmetric dimer Si(100)–(2 × 1):H structure. The Si(111) surface takes on the bulk-terminated (1 × 1) structure in the presence of chemisorbed hydrogen.

Adsorbate induced reconstructions can have a dramatic effect on the kinetics of reactions on reconstructed surfaces. Of particular importance is the reconstruction of Pt(110). The clean surface is reconstructed into a (1 × 2) missing row structure, a rather common type of reconstruction. However, CO adsorption leads to a lifting of the reconstruction. Adsorbate induced reconstruction of a metal surface is associated with the formation of strong chemisorption bonds.

The surface does not present a static template of adsorption sites to an adsorbate. Somorjai [48] has collected an extensive list of clean surface and adsorbate-induced reconstructions. When an adsorbate binds to a surface, particularly if the chemisorption interaction is strong, we need to consider whether the surface is stable versus reconstruction [49]. For sufficiently strong interactions and high enough adsorbate concentrations we may have to consider whether the surface is stable versus the formation of a new solid chemical compound, such as the formation of an oxide layer, or the formation of a volatile compound, as in the etching of Si by halogen compounds or atomic hydrogen. Here we concentrate on interactions that lead to reconstruction.

The chemical potential of an adsorbate/substrate system is dependent on the temperature T, the chemical identity of the substrate S and adsorbate A, the number density of the adsorbates σ_A and surface atoms σ_S, and the structures that the adsorbate and surface assume. The gas phase is coupled, in turn, to σ_A through the pressure. Thus, we can write the chemical potential as

$$\mu(T,\sigma_S,\sigma_A) = \sum_i \sigma_{S,\,i}\mu_i(T) + \sum_j \sigma_{A,\,j}\mu_j(T). \tag{1.2.1}$$

In Eq. (1.2.1) the summations are over the i and j possible structures that the surface and adsorbates, respectively, can assume. The adsorption energy can depend on the surface structure. If the difference in adsorption energy between two surface structures is sufficiently large so as to overcome the energy required to reconstruct the substrate, the surface structure can switch from one structure to the next when a critical adsorbate coverage is exceeded. Note also that Eq. (1.2.1) is written in terms of areal densities (the number of surface atoms per unit area) to emphasize that variations in adsorbate concentration can lead to variations in surface structure across the surface. In other words, inhomogeneity in adsorbate coverage may lead to inhomogeneity in the surface structure. An example is the chemisorption of H on Ni(110) [50]. Up to a coverage of 1 H atom per surface Ni atom ($\equiv$ 1 monolayer or 1 ML), a variety of ordered overlayer structures are formed on the unreconstructed surface. As the coverage increases further, the surface reconstructs locally into islands that contain 1.5 ML of H atoms. In these islands the rows of Ni atoms pair up to form a (1×2) structure.

Equation (1.2.1) indicates that the equilibrium surface structure depends both on the density (alternatively called coverage) of adsorbates and the temperature. The surface temperature is important in two ways. The chemical potential of each surface structure depends on temperature. Therefore, the most stable surface structure can change as a function of temperature. Secondly, the equilibrium adsorbate coverage is a function of both pressure and temperature. Because of this coupling of adsorbate coverage to temperature and pressure, we expect that the equilibrium surface structure can change as a function of these two variables as well as the identity of the adsorbate [49]. This can have important consequences for working catalysts because surface reactivity can change with surface structure.

An example of the restructuring of a surface and the dependence on adsorbate coverage and temperature is the H/Si(100) system [51–54], Fig. 1.12. The clean Si(100) surface reconstructs into a (2×1) structure

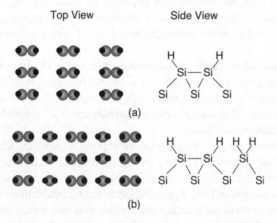

Figure 1.12 *The adsorption of H on to Si(100): (a) Si(100)–(2 × 1):H, θ (H) = 1 ML; (b) Si(100)–(3 × 1):H, θ (H) = 1.33 ML. Note: the structures obtained from the adsorption of hydrogen atoms on to Si(100) are a function of the hydrogen coverage, θ (H). 1 ML = 1 monolayer as defined by 1 hydrogen atom per surface silicon atom.*

accompanied by the formation of Si dimers on the surface. The dimers are buckled at low temperature but the rocking motion of the dimer is a low frequency vibration, which means that at room temperature the average position of the dimers appears symmetric. When H adsorbs on the dimer in the monohydride structure (1 H atom per surface Si atom), the dimer becomes symmetric, the dimer bond expands and much of the strain in the subsurface layers is relaxed. Further increasing the H coverage by exposure of the surface to atomic H leads to the formation of dihydride units (SiH_2). These only form in appreciable numbers if the temperature during adsorption is below $\sim$400 K. If the surface is exposed to H at room temperature, trihydride units (SiH_3) can also form. These are a precursor to etching by the formation of SiH_4, which desorbs from the surface. Northrup [55] has shown how these changes are related to the chemical potential and lateral interactions. Neighbouring dihydride units experience repulsive interactions and are unable to assume their ideal positions. This lowers the stability of the fully covered dihydride surface such that some spontaneous formation and desorption of SiH_4 is to be expected.

The LEED pattern of a Si(100) surface exposed to large doses of H at room temperature or below has a (1 × 1) symmetry, but it should always be kept in mind that diffraction techniques are very sensitive to order and insensitive to disorder. The (1 × 1) diffraction pattern has been incorrectly interpreted as a complete coverage of the surface with dihydrides with the Si atoms assuming the ideal bulk termination. Instead, the surface is rough, disordered and composed of a mixture of SiH, SiH_2 and SiH_3 units. The (1 × 1) pattern arises from the ordered subsurface layers, which are also probed by LEED. If the surface is exposed to H atoms at $\sim$380 K, a (3 × 1) LEED pattern is observed. This surface is composed of an ordered structure comprised of alternating SiH and SiH_2 units. Thus, there are 3 H atoms for every 2 Si atoms. Heating the (3 × 1) surface above $\sim$600 K leads to rapid decomposition of the SiH_2 units. The hydrogen desorbs from the surface as H_2 and the surface reverts to the monohydride (2 × 1) structure.

The reconstructed and non-reconstructed Pt surfaces are shown in Fig. 1.13. Of the clean low index planes, only the Pt(111) surface is stable versus reconstruction. The Pt(100) surface reconstructs into a quasi-hexagonal (hex) phase, which is $\sim$40 kJ mol^{-1} more stable than the (1 × 1) surface. The Pt(110) reconstructs into a (1 × 2) missing row structure. These reconstructions lead to dramatic changes in the chemical reactivity, which can lead to spatiotemporal pattern formation during CO oxidation [56], see Chapter. 6. The clean surface reconstructions can be reversibly lifted by the presence of certain adsorbates including CO and NO. This is driven by the large difference in adsorption energy between the two reconstructions. For CO the values are 155 and 113 kJ mol^{-1} on the (1 × 1) and hex phases, respectively, just large enough to overcome the energetic cost of reconstruction.

The surface structure not only affects the heat of adsorption, but it can also dramatically change the probability of dissociative chemisorption. O_2 dissociates with a probability of 0.3 on the Pt(100)–(1 × 1) surface. However, this probability drops to $\sim$10^{-4} − 10^{-3} on the Pt(100) hex phase. We investigate the implications of these changes further in Chapter 6. In Chapter 3 we discuss the dynamical factors that affect the dissociation probability.

1.2.4 Islands

A flux of gas molecules strikes a surface at random positions. If no attractive or repulsive interactions exist between the adsorbed molecules (lateral interactions), the distribution of adsorbates on the surface would also be random. However, if the surface temperature is high enough to allow for diffusion (§ 3.2), then the presence of lateral interactions can lead to non-random distributions of the adsorbates. In particular, the adsorbates can coalesce into regions of locally high concentration separated by low concentration or even bare regions. The regions of high coverage are known as islands. Since the coverage in islands is higher than in the surrounding regions, then according to Eq. (1.2.1), the substrate beneath the island

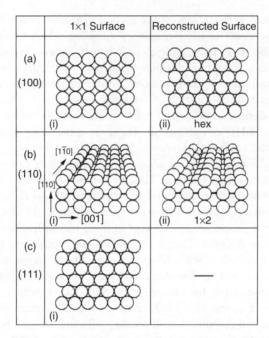

Figure 1.13 *Reconstructed and unreconstructed surfaces for the three low index planes of Pt: (a) (100) plane, (i) (1 × 1) unreconstructed surface, (ii) quasi-hexagonal (hex) reconstructed surface; (b) (110) plane, (i) (1 × 1) unreconstructed surface, (ii) missing row (1 × 2) reconstructed surface; (c) (111) plane, (i) (1 × 1) unreconstructed surface (Pt(111) is stable against reconstruction). Reproduced from R. Imbihl and G. Ertl, Chem. Rev.* **95**, *697. © (1995), with permission from the American Chemical Society.*

might reconstruct whereas regions outside of the islands might not. In some cases, such as H/Ni(110), it is the capacity of a reconstructed region to accommodate a higher coverage than an unreconstructed region that drives the formation of islands. In subsequent chapters, we shall see that islands can have important implications for surface kinetics (Chapter 4), in particular for spatiotemporal pattern formation (§ 6.8) as well as the growth of self assembled monolayers (Chapter 5) and thin films (Chapter 7).

1.2.5   Chiral surfaces

One of the great challenges in heterogeneous catalysis is to develop catalysts for asymmetric synthesis, that is, catalysts that accelerate the rate of reaction for only one of a pair of entantiomers. Chiral chromatography essentially works because entantiomers (molecules with the same chemical formula but with mirror image structures) have ever so slightly different rates of adsorption and desorption onto and off of the surfaces present in the chromatographic column. There are several different ways by which chiral recognition and chemical interaction dependent on chirality can be engineered into surfaces. One is to form a porous solid, the pores or porewalls of which are chiral. A number of chiral silicate zeolites have now been recognized [57]. Significant progress has been made in the construction of chiral metal-organic frameworks [58]. Porous solids such as por-Si, which are not inherently chiral, can be used to immobilize enzymes or chiral molecules [59]. The interactions of these adsorbed species can then be used for chiral surface processing. Indeed, chiral chromatography columns work by means of chiral selectors adsorbed on otherwise achiral surfaces.

Purely two-dimensional structures can also exhibit chiral interactions. The kink sites on vicinal metal surfaces can possess a chiral structure as was first recognized by Gellman and co-workers [60], who proposed a system for naming left-handed and highhanded kinks $(hkl)^S$ and $(hkl)^R$, respectively. Attard et al. [61] were the first to observe that naturally chiral metal surfaces can exhibit entantiospecific chemical reactivity, namely, in the electro-oxidation of glucose on Pt(643)^S. More recently, Greber et al. [62] have shown that D-cysteine binds 140 meV stronger than L-cysteine to the Au(17 11 9)^S kinks on a vicinal Au(111) surface.

Following King and co-workers [63], we use the low index or singular planes that contain the highest symmetry elements of each crystal class as the starting point for a description of chiral surfaces. For fcc and bcc crystals these are the {100}, {110}, and {111} planes, which contain 4-fold, 2-fold and 3-fold rotational axes, respectively. In an hcp system, the {0001} contains a 3-fold rotational axes, while the {10$\bar{1}$0} and {11$\bar{2}$0} planes present two symmetrically distinct 2-fold axes. These planes also contain mirror planes. Any plane that contains a mirror plane as a symmetry element is achiral. Any chiral surface is described by three inequivalent, non-zero Miller indices, such as the (643) and (17 11 9) examples given above.

The (R, S) naming convention depends on decomposing the kink into the three singular microfacets and assigning an absolute chirality (R or S) based on the density of atoms in the singular surfaces. King and co-workers have developed a more general (D, L) naming convention that does not depend on the presence of kinks. Instead, they consider a stereographic projection of the crystal symmetry. They define the surface chirality within a single stereographic triangle, Fig. 1.14, by the sense in which one visits the high-symmetry poles at its vertices: proceeding in order of increasing symmetry – {110}, {111}, {100} – a clockwise sequence implies D and an anticlockwise route L. For fcc surfaces, the 'D' label is matched with 'R' and the 'L' label is matched with 'S'. This system has the advantage that a general (hkl) surface of a primitive cubic crystal has the same chirality label, regardless of whether the lattice is simple cubic, fcc or bcc.

King et al. note that the one-dimensional close-packed chain of atoms is the only extended structural feature shared by all three common metallic crystal forms. They use this construct to define a surface

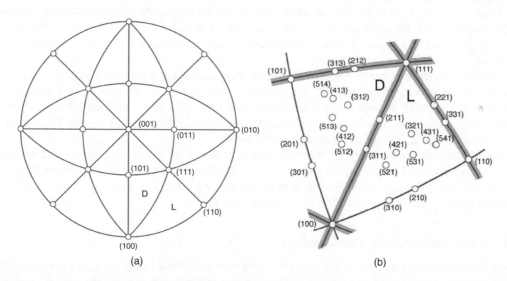

Figure 1.14 *Stereogram projected on (001) showing the mirror zones relevant to primitive fcc, bcc and simple cubic crystal surfaces. Note that right-handed axes are used. The chiralities of two triangles are labelled D and L according to symmetry-based convention of Pratt et al. [63]. Reproduced from S.J. Pratt, S.J. Jenkins, D.A. King, Surf. Sci., 585, L159. © (2005), with permission from Elsevier.*

containing two or more such chain directions as "flat", a surface containing precisely one such direction as "stepped", and a surface containing no such directions as "kinked". Their analysis allows them to draw the following conclusions regarding the symmetry and structure of fcc surfaces: (i) that all chiral surfaces are kinked; (ii) that kinked surfaces with one zero index are achiral; (iii) that surfaces having two equivalent and non-zero Miller indices are stepped; (iv) that the {110} surface is neither more nor less stepped than the most symmetric stepped surface; and (v) that only the {111} and {001} surfaces are truly flat. Therefore on fcc crystals, all chiral surfaces are kinked but not all kinked surfaces are chiral.

Similarly for bcc surfaces: (i) all chiral surfaces are kinked, with the exception of those that are merely stepped; (ii) stepped surfaces are those with three non-zero indices satisfying $|h| - |k| - |l| = 0$, or an equivalent permutation; (iii) all stepped surfaces with three inequivalent indices are chiral and devoid of kinks; (iv) kinked surfaces with one or more zero indices or at least two equivalent indices are achiral; (v) the {111} and {001} surfaces are, in fact, symmetric kinked surfaces; (vi) the {211} surface is a symmetric stepped surface analogous to fcc{110}; and (vii) only the {110} surface is truly flat. On bcc crystal, there are kinked chiral surfaces; however, there are also stepped chiral surfaces that do not have kinks. This is a new class of chiral surfaces that may potentially be more stable than kinked surfaces.

1.3   Band structure of solids

1.3.1   Bulk electronic states

A bulk solid contains numerous electrons. The electrons are classified as being either valence or core electrons. Valence electrons are the least strongly bound electrons and have the highest values of the principle quantum number. Valence electrons form delocalized electronic states that are characterized by three-dimensional wavefunctions known as Bloch waves. The energy of Bloch waves depends on the wavevector, **k**, of the electronic state. The wavevector describes the momentum of the electron in a particular state. Since momentum is a vector, it is characterized by both its magnitude and direction. In other words, the energy of an electron depends not only on the magnitude of its momentum, but also on the direction in which the electron moves. The realm of all possible values of **k** is known as k-space. k-space is the world of the solid described in momentum space as opposed to the more familiar world of *xyz* co-ordinates, known as real space.

Because of the great number of electrons in a solid, there are a large number of electronic states. These states overlap to form continuous bands of electronic states and the dependence of the energy on the momentum is known as the electronic band structure of the solid. Two bands are always formed: the valence band and the conduction band. The energetic positions of these two bands and their occupation determine the electrical and optical properties of the solid.

The core electrons are the electrons with the lowest values of the principal quantum number and the highest binding energy. These electrons are strongly localized near the nuclei and they do not form bands. Core electrons are not easily moved from their positions near the nuclei and, therefore, they do not directly participate either in electrical conduction or chemical bonding.

1.3.2   Metals, semiconductors and insulators

The simplest definition of metals, semiconductors and insulators is found in Fig. 1.15. In a metal the valence band and the conduction band overlap. There is no energy gap between these bands, and the conduction band is not fully occupied. The energy of the highest occupied electronic state (at 0 K) is known as the Fermi energy, E_F. At 0 K, all states below the Fermi level (a hypothetical energy level at energy $= E_F$) are occupied and all states above it are empty. Because the conduction band is not fully occupied, electrons are

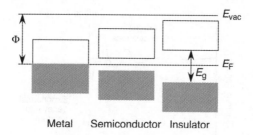

Figure 1.15 *Fermi energies, vacuum energies and work functions in a metal, a semiconductor and an insulator. The presence and size of a gap between electronic states at the Fermi energy, E_F, determines whether a material is a metal, semiconductor or insulator. E_g, band gap; Φ work function equal to the difference between E_F and the vacuum energy, E_{vac}.*

readily excited from occupied to unoccupied states and the electrons in the conduction band are, therefore, quite mobile. Excitation of an electron from an occupied state to an unoccupied state leads to an excited electronic configuration in which an electron occupies an excited state and an absent electron 'occupies' the original electronic state. This absent electron is known as a hole. The hole is a pseudo-particle that acts something like the mirror image of an electron. A hole is effectively a positively charged particle that can be characterized according to its effective mass and its mobility analogous to electrons. Electron excitation always creates a hole. Therefore, we often speak of electron-hole ($e^- - h^+$) pair formation. As these particles possess opposite charges, they can interact with one another. Creation of electron-hole pairs in the conduction band of metals represents an important class of electronic excitation with a continuous energy spectrum that starts at zero energy.

Figure 1.15 illustrates another important property of materials. The vacuum energy, E_{vac}, is defined as the energy of a material and an electron at infinite separation. The difference between E_{vac} and E_F is known as the work function, Φ,

$$e\Phi = E_{vac} - E_F. \tag{1.3.1}$$

and at 0 K it represents the minimum energy required to remove one electron from the material to infinity (e is the elementary charge). For ideal semiconductors and insulators, the actual minimum ionization energy is greater than Φ because there are no states at E_F. Instead, the highest energy electrons reside at the top of the valence band. Not apparent in Fig. 1.15 is that the work function is sensitively dependent on the crystallographic orientation of the surface, the presence of surface defects in particular steps [5, 64] and, of course, the presence of chemical impurities adsorbed on the surface.

The reason for the dependence of the work function on surface properties can be traced back to the electron distribution at the surface, Fig. 1.16. The electron density does not end abruptly at the surface. Instead, it oscillates near the surface before decaying slowly into the vacuum (Friedel oscillations). This distribution of electrons creates an electrostatic dipole layer at the surface. The surface dipole contribution D is equal to the difference between the electrostatic potential energy far into the vacuum and the mean potential deep in the bulk,

$$D = V(\infty) - V(-\infty). \tag{1.3.2}$$

If we reference the electrostatic potential with respect to the mean potential in the bulk, i.e. $V(-\infty) = 0$, then the work function can be written

$$e\Phi = D - E_F. \tag{1.3.3}$$

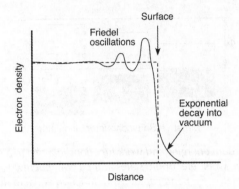

Figure 1.16   *Friedel oscillations: the electron density near the surface oscillates before decaying exponentially into the vacuum.*

Therefore, Φ is determined both by a surface term D and a bulk term E_F. With this definition of the work function, changes in D due to surface structure and adsorbates are responsible for changes in Φ because surface properties cannot affect E_F. More commonly, the work function is given in units of eV and Eq. (1.3.3) is written as

$$\Phi = -e\chi - \frac{E_F}{e} \qquad (1.3.4)$$

where χ is the surface potential in volts (the electrostatic potential step resulting from the inhomogeneous charge distribution at the surface) and E_F is in joules. On metal surfaces χ is negative.

The occupation of electronic states is governed by Fermi-Dirac statistics. All particles with non-integer spin such as electrons, which have a spin of $1/2$, obey the Pauli exclusion principle. This means that no more than two electrons can occupy any given electronic state. The energy of electrons in a solid depends on the availability of electronic states and the temperature. Metals sometimes exhibit regions of k-space in which no electronic states are allowed. These forbidden regions of k-space are known as partial bandgaps. At finite temperature, electrons are not confined only to states at or below the Fermi level. At equilibrium they form a reservoir with an energy equal to the chemical potential, μ. The probability of occupying allowed energy states depends on the energy of the state, E, and the temperature according to the Fermi-Dirac distribution

$$f(E) = \frac{1}{\exp[(E-\mu)/k_B T] + 1}. \qquad (1.3.5)$$

The Fermi-Dirac distribution for several temperatures is drawn in Fig. 1.17. At $T = 0\,\text{K}$, the Fermi-Dirac function is a step function, that is $f(E) = 1$ for $E < \mu$ and $f(E) = 0$ for $E > 0$. At this temperature the Fermi energy is identical to the chemical potential

$$\mu(T = 0\ K) \equiv E_F \qquad (1.3.6)$$

More generally, the chemical potential is defined as the energy at which $f(E) = 0.5$. It can be shown, see for instance Elliot [65], that

$$\mu(T) \approx E_F\left[1 - \frac{\pi^2}{12}\left(\frac{k_B T}{E_F}\right)^2\right]. \qquad (1.3.7)$$

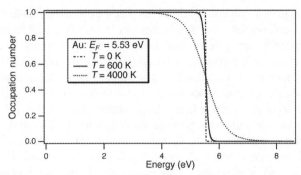

Figure 1.17   *Fermi-Dirac distribution for gold at three different temperatures, T, E_F, Fermi energy.*

A direct consequence of Fermi-Dirac statistics is that the Fermi energy is not zero at 0 K. In fact, E_F is a material dependent property that depends on the electron density, ρ, according to

$$E_F = \frac{\hbar^2}{2m_e}(3\pi^2\rho)^{2/3}. \tag{1.3.8}$$

E_F is on the order of several eV and E_F/k_B, known as the Fermi temperature, T_F, is on the order of several thousand Kelvin. Consequently, the chemical potential and E_F are virtually identical unless the temperature is extremely high.

Semiconductors exhibit a complete bandgap between the valence and conduction bands. The energy of the conduction band minimum is E_C and the energy at the valence band maximum is E_V. The magnitude of the bandgap is the difference between these two

$$E_g = E_C - E_V \tag{1.3.9}$$

The Fermi level of a pure semiconductor, known as an intrinsic semiconductor, lies in the bandgap. The exact position depends on the temperature according to

$$E_F = E_i = \frac{E_C + E_V}{2} + \frac{k_B T}{2}\ln\left(\frac{N_V}{N_C}\right) \tag{1.3.10}$$

where N_V and N_C are the effective densities of states of the valence and conduction bands, respectively.

The densities of states can be calculated from material specific constants and the temperature as shown, for instance, by Sze [66]. Equation (1.3.10) shows that the Fermi energy of an *intrinsic* semiconductor lies near the middle of the gap. As we shall see below, this is not true for the more common case of a doped (extrinsic) semiconductor.

In the bulk of a perfect semiconductor there are no electrons at the Fermi level, even though it is energetically allowed, because there are no allowed electronic states at this energy. An equivalent statement is that in the bulk of an ideal semiconductor, the density of states is zero at E_F. In any real semiconductor, defects (structural irregularities or impurities) introduce a non-zero density of states into the bandgap. These states are known as gap states. At absolute zero the valence band is completely filled, and the conduction band is completely empty. At any finite temperature, some number of electrons is thermally excited into the conduction band. This number is known as the intrinsic carrier density, and is given by

$$n_i = \sqrt{N_C N_V}\exp\left(-E_g/2k_B T\right) \tag{1.3.11}$$

The presence of a bandgap in a semiconductor means that the electrical conductivity of a semiconductor is low. The bandgap also increases the minimum energy of electron-hole pair formation from zero to $\geq E_g$. What distinguishes a semiconductor from an insulator is that E_g in a semiconductor is sufficiently small that either thermal excitations or the presence of impurities can promote electrons into the conduction band or holes into the valence band. Electrically active impurities are known as dopants. There are two classes of dopant. If a valence III atom, such as B, is substituted for a Si atom in an otherwise perfect Si crystal, the B atom accepts an electron from the Si lattice. This effectively donates a hole into the valence band. The hole is mobile and leads to increased conductivity. B in Si is a *p*-type dopant because it introduces a *positive* charge carrier into the Si band structure. B introduces acceptor states into the Si and the concentration of acceptor atoms is denoted N_A. On the other hand, if a valence V atom, such as P, is doped into Si, the P atom effectively donates an electron into the conduction band. Because the resulting charge carrier is *negative*, P in Si is known as an *n*-type dopant. Analogously, the concentration of donors is N_D.

The position of the Fermi energy in a doped semiconductor depends on the concentration and type of dopants. E_F is pushed upward by *n*-type dopants according to

$$E_F = E_i + k_B T \ln \left(\frac{N_D}{n_i} \right). \tag{1.3.12}$$

In a *p*-type material, E_F is pulled toward the valence band

$$E_F = E_i - k_B T \ln \left(\frac{N_A}{n_i} \right). \tag{1.3.13}$$

Equations (1.3.12) and (1.3.13) hold as long as the dopant density is large compared to the n_i and density of electrically active dopants of the opposite type.

An insulator has a large band gap. The division between a semiconductor and an insulator is somewhat arbitrary. Traditionally, a material with a bandgap larger than ~3 eV has been considered to be an insulator. The push of high technology and the desire to fabricate semiconductor devices that operate at high temperatures have expanded this rule of thumb. Hence diamond with a band gap of ~5.5 eV now represents the upper limit of wide bandgap semiconductors.

Graphite represents one other important class of material. It does have electronic states up to the Fermi energy. However, the conduction and valence band edges, which correlate with π^* antibonding and π bonding bands formed from p_z-like orbitals, only overlap in a small region of k space. Therefore, the density of states at E_F is minimal and graphite is considered a semi-metal.

1.3.3   Energy levels at metal interfaces

An interface is generally distinguished from a surface as it is thought of as the boundary between two materials (or phases) in intimate contact. At equilibrium, the chemical potential must be uniform throughout a sample. This means that the Fermi levels of two materials, which are both at equilibrium and in electrical contact, must be the same.

Figure 1.18 demonstrates what occurs when two bulk metals are brought together to form an interface. In Fig. 1.18(a) we see that two isolated metals share a common vacuum level but have different Fermi levels, E_F^L and E_F^R, as determined by their work functions, Φ_L and Φ_R. L and R refer to the left-hand and right-hand metals, respectively. When the two metals are connected electrically, electrons flow from the low work function metal to the high work function metal, L $\rightarrow$ R, until the Fermi levels become equal. Consequently, metal L is slightly depleted of electrons and metal R has an excess. In other words, a dipole

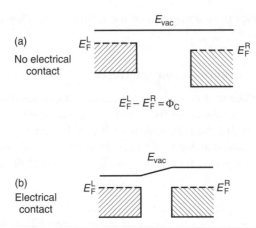

Figure 1.18 *Electronic bands of metals (a) before and (b) after electrical contact. The Fermi energies of the two metals align at equilibrium when electrical contact is made. E_{vac}, vacuum energy; E_F, Fermi energy; subscripts L and R refer to left-hand and right-hand metal, respectively.*

develops between the two metals and with this is associated a small potential drop, the contact potential

$$\Delta\varphi = \Phi_R - \Phi_L, \tag{1.3.14}$$

and an electrical field. The presence of the electric field is evident in Fig. 1.18(b) by the sloping vacuum level. Figure 1.18 demonstrates that the bulk work functions of the two metals remain constant. Since the Fermi energies must be equal at equilibrium, the vacuum level shifts in response.

In metals, screening of free charges by valence electrons is efficient. Screening is the process by which the electrons surrounding a charge (or charge distribution such as a dipole) are polarized (redistributed) to lower the energy of the system. Screening is not very effective in insulators because the electrons are not as free to move. Therefore, when the two metals are actually brought into contact, the width of this dipole layer is only a few angstroms. The energetic separation between E_F and E_{vac} is constant after the first two or three atomic layers. At the surface, however, this separation is not constant. For the interface of two metals this means that the position of E_{vac} does not change abruptly but continuously.

The presence of a dipole layer at the surface has other implications for clean and adsorbate covered surfaces. The difference between E_F and E_{vac} in the bulk is a material specific property. In order to remove an electron from a metal, the electron must pass through the surface and into the vacuum. Therefore, if two samples of the same metal have different dipole layers at the surface, they exhibit different work functions. In Fig. 1.5 we have seen that Smoluchowski smoothing at steps introduces dipoles into the surface. A linear relationship between step density and work function decrease has been observed [64]. Similarly, the geometric structure of the surface determines the details of the electronic structure at the surface. Thus, the work function is found to depend both on the crystallographic orientation of the surface as well as the presence of surface reconstructions.

The presence of adsorbates on the surface of a solid can introduce two distinct dipolar contributions to the work function. The first occurs if there is charge transfer between the adsorbate and the surface. When an electropositive adsorbate such as an alkali metal forms a chemical bond with a transition metal surface, the alkali metal tends to donate charge density into the metal and decreases the work function. An electronegative adsorbate, such as O, S or halogens, withdraws charge and increases the work function. The second contribution arises when a molecular adsorbate has an intrinsic dipole. Whether this contribution increases or decreases the work function depends on the relative orientation of the molecular dipole with respect to the surface.

1.3.4   Energy levels at metal-semiconductor interfaces

The metal-semiconductor interface is of great technological importance not only because of the role it plays in electronic devices [66]. Many of the concepts developed here are directly applicable to charge transfer at the electrolyte-semiconductor interface as well [67, 68]. It is somewhat more complicated than the metal-metal interface, but our understanding of it can be built up from the principles that we have learned above. The situation is illustrated in Fig. 1.19. Again, the Fermi levels of the metal and the semiconductor

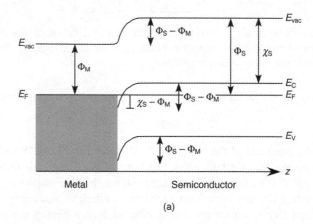

(a)

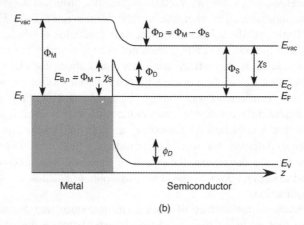

(b)

Figure 1.19   *Band bending in an n-type semiconductor at a heterojunction with a metal. (a) Ohmic contact* ($\Phi_S > \Phi_M$). *(b) Blocking contact (Schottky barrier,* $\Phi_S < \Phi_M$). *The energy of the bands is plotted as a function of distance z in a direction normal to the surface.* Φ_S, Φ_M *work function of the semiconductor and of the metal, respectively;* E_{vac}, *vacuum energy;* E_C, *energy of the conduction band minimum;* E_F, *Fermi energy;* E_V *energy of the valence band maximum;* E_g *band gap. Reproduced from S. Elliott,* The Physics and Chemistry of Solids, *John Wiley, New York.* © *(1998), with permission from John Wiley & Sons, Ltd.*

must be aligned at equilibrium. Equalization of E_F is accomplished by charge transfer. The direction of charge transfer depends on the relative work functions of the metal (M) and the semiconductor (S). The case of $\Phi_S > \Phi_M$ for an n-type semiconductor is depicted in Fig. 1.19(a), whereas $\Phi_S < \Phi_M$ is depicted in Fig. 1.19(b). Screening in a semiconductor is much less effective, which results in a near-surface region of charge density different from that of the bulk with a width of several hundred angstroms. This is called a space-charge region. In Fig. 1.19(a) the metal has donated charge to the semiconductor space-charge region. The enhanced charge density in the space-charge region corresponds to an accumulation layer. In Fig. 1.19(b) charge transfer has occurred in the opposite direction. Because the electron density in this region is lower than in the bulk, this type of space-charge region is known as a depletion layer. The shape of the space-charge region has a strong influence on carrier transport in semiconductors and the electrical properties of the interface. Figure 1.19(a) corresponds to an ohmic contact whereas Fig. 1.19(b) demonstrates the formation of a Schottky barrier.

In the construction of Fig. 1.19 we have introduced the electron affinity of the semiconductor, χ_S. This quantity as well as the band gap remain constant throughout the semiconductor. Importantly, the positions of the band edges at the surface remain constant. Thus, we see that the differences $E_{vac} - E_V$ and $E_{vac} - E_C$ are constant whereas the positions relative to E_F vary continuously throughout the space-charge region. The continuous change in E_V and E_C is called band bending.

E_{vac}, E_C and E_V all shift downward by the same amount in Fig. 1.19(a). In Fig. 1.19(b) the shifts are all upward but again E_{vac}, E_C and E_V all shift by the same amount. The potential at the surface is the magnitude of the band bending and is given by

$$eU_{surf} = E_{vac}^{surf} - E_{vac}^{bulk} \tag{1.3.15}$$

In Eq. (1.3.15) E_{vac} was chosen but the shifts in E_{vac}, E_C and E_V are all the same, hence any one of these three could be used in Eq. (1.3.15). The doping density determines the magnitude of the band bending and the depletion layer width, d. For an n-type semiconductor, the value is

$$U_{surf} = -\frac{eN_D d^2}{2\varepsilon\varepsilon_0} \tag{1.3.16}$$

where ε and ε_0 are the permittivities of the semiconductor and free space. In a p-type semiconductor, the sign is reversed and N_A is substituted for N_D. From Fig. 1.19 it is further apparent that

$$U_{surf} = \Phi_S - \Phi_M \quad \text{(Ohmic as in Fig. 1.19(a))} \tag{1.3.17}$$

and

$$U_{surf} = \Phi_M - \chi_S \quad \text{(Schottky as in Fig. 1.19(b))} \tag{1.3.18}$$

for n-type Ohmic and Schottky contacts, respectively. U_{surf} in the case of a Schottky contact is also known as the Schottky barrier height.

Of great importance for both device applications and for electrochemistry is that the extent of band bending can be changed by the application of an external bias. When a voltage U is applied across a semiconductor junction (either metal-semiconductor or electrolyte-semiconductor) it is not the chemical potential that is constant throughout the junction region, but the electrochemical potential, $\overline{\mu}$

$$\overline{\mu} = \mu - eU, \tag{1.3.19}$$

as shown in Fig. 1.20. In forward bias, electrons flow from the semiconductor to the metal and the barrier is reduced by eU. In reverse bias, little current flows from the metal into the semiconductor as the barrier

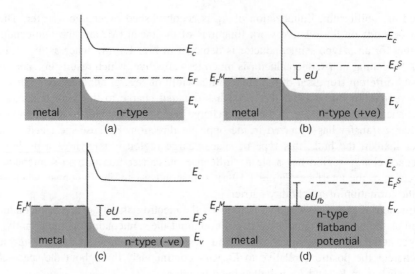

Figure 1.20   *The electrochemical potential and the effects of an applied voltage on a metal/semiconductor interface. (a) No applied bias. (b) Forward bias. (c) Reverse bias. (d) Biased at the flatband potential, U_{fb}. E_C, energy of the conduction band minimum; E_F, Fermi energy; E_V, energy of the valence band maximum; superscripts M and S refer to the metal and the semiconductor, respectively.*

height is increased by eU. The potential at which $\mu - eU = 0$ is known as the flatband potential U_{FB}, and therefore

$$U_{FB} = \mu/e, \qquad (1.3.20)$$

from which the Fermi level of the semiconductor can be determined.

1.3.5   Surface electronic states

All atoms in the bulk of a pure metal or elemental semiconductor are equivalent. The atoms at the surface are, by definition, different because they do not possess their full complement of bonding partners. Therefore surface atoms can be thought of as a type of impurity. Just as impurities in the bulk can have localized electronic states associated with them, so too can surface atoms. We need to distinguish two types of electronic states associated with surface atoms. An electronic state that is associated with the surface can either overlap in k-space with bulk states or it can exist in a bandgap. An overlapping state is known as a surface resonance. A surface resonance exists primarily at the surface. Nonetheless, it penetrates into the bulk and interacts strongly with the bulk electronic states. A true surface state is strongly localized at the surface and because it exists in a band gap, it does not interact strongly with bulk states. A surface state or resonance can either be associated with surface atoms of the solid (intrinsic surface state or resonance) or with adsorbates (extrinsic surface state or resonance). Structural defects can also give rise to surface states and resonances. Surface states and surface resonances are illustrated in Fig. 1.21.

Surface states play a defining role in determining the surface band structure of semiconductors. In effect, they can take the place of the metal overlayer, and the reasoning we used in § 1.3.4 can be used to describe band bending in the presence of surface states. If the electron distribution in a semiconductor were uniform all the way to the surface, the bands would be flat. The presence of surface states means that the surface may possess a greater or lesser electron density relative to the bulk. This non-uniform electron distribution

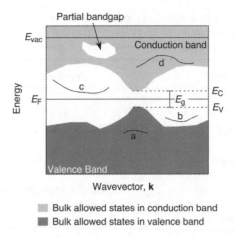

Figure 1.21 *The band structure of a semiconductor, including an occupied surface resonance (a), an occupied surface state (b), a normally unoccupied surface state (c), and a normally unoccupied surface resonance (d). E_{vac}, vacuum energy; E_F Fermi energy; E_V, energy of the valence band maximum; E_C, the conduction band minimum.*

again leads to a space-charge region and band bending. Surface states can either act as donor or acceptor states. Surface states can have a strong influence on the electrical properties of devices – especially on the behaviour of Schottky barriers [65, 66, 69].

The distinction between a resonance and a state may seem somewhat arbitrary. The difference is particularly obvious in normally unoccupied or empty states that exist above the Fermi level. Empty states can be populated by the absorption of photons with energies less than the work function. The strong interaction between bulk states and a resonance results in a short lifetime for electrons excited into the resonance. A surface state exhibits a much longer lifetime. These lifetimes have been measured directly by two photon photoemission, § 2.6.3. Lifetimes depend on the specific system. For example, on Si(111)–(2 × 1) the π^* normally unoccupied surface state has a lifetime of $\sim$200 ps [70].

If in a metal an energy gap exists somewhere between E_F and E_{vac}, an electron excited into this gap experiences an attractive force associated with its image potential. The result is a series of bound states, image potential states [71, 72], that are the surface analogue of Rydberg states in atomic and molecular systems. To a first approximation [72], the energies of these states form a series whose energy is given by

$$E_n = -\frac{R_\infty}{16 n^2} = -\frac{0.85 \text{ ev}}{n^2} \tag{1.3.21}$$

converging on the energy of the vacuum level, E_{vac}. These states are bound in the z direction (normal to the surface) but are free-electron-like in the plane of the surface. As n increases, the wavefunction of the image states overlaps less and less with the bulk leading to progressively less interaction between the two. The result is that the lifetime of the state increases with n. This has been measured directly by Höfer et al. [73] who have found that above Cu(100) the lifetime varies from 40 ± 6 to 300 ± 15 fs as n changes from 1 to 3. Similarly, the lifetime can be increased by the introduction of a spacer layer between the image state and the metal surface. Wolf, Knoesel and Hertel [74] have shown that the presence of a physisorbed Xe layer increases the $n = 1$ lifetime on Cu(111) from 10 ± 3 fs to 50 ± 10 fs.

1.3.6 Size effects in nanoscale systems

So far, we have confined our discussion to extended, essentially semi-infinite solids. We have seen that the surface acts differently than the bulk. What happens if we shrink our sample from a semi-infinite solid to

Table 1.2   *Clusters can be roughly categorized on the basis of their size. The properties of clusters depend on size. For clusters of different composition the boundaries might occur as slightly different values. The diameter is calculated on the assumption of a close packed structure with the atoms having the size of a Na atom*

	Small	Medium	Large
Atoms (N)	2–20	20–500	500–10^6
Diameter/nm	≤ 1.1	1.1–3.3	3.3–100
Surface/Bulk (N_s/N)	Not separable	0.9–0.5	$\leq 0.5 \sim 0.2$ for $N = 3\,000$ $\ll 1$ for $N \geq 10^5$
Electronic states	Approaching discrete	Approaching bands	Moving toward bulk behaviour
Size dependence	No simple, smooth dependence of properties on size and shape	Size dependent properties that vary smoothly	Quantum size effects may still be important but properties approaching bulk values

a finite cluster of atoms? Table 1.2 lists several characteristics of clusters and classifies them arbitrarily as small, medium and large. We can see that the relative number of surface atoms N_s, compared to the total number of atoms in the cluster N changes as a function of cluster size. Therefore, the relative importance of the surface in determining the properties of the cluster changes as a function of size. Because surface atoms act differently than bulk atoms, this also means that the properties of the cluster changes as a function of size, i.e. as a function of the number of atoms in the cluster. Small clusters are close to molecular in their behaviour with electronic states that are discrete or approaching discrete behaviour. Their properties do not change regularly with decreasing size. Large clusters have smoothly varying properties that approach bulk values. Different properties (vapour pressure, ionization potential, polarizability, etc) may change at different rates, with some having already attained bulk values and others still exhibiting size dependent behaviour. Medium sized clusters have electronic states that approach the band like states of large clusters, but which are still distinctly different from the bulk states. The properties of a cluster generally vary in a smooth manner as a function of size.

What separates clusters, also called nanoparticles, from bulk materials is that their properties are size dependent. A glass of pure water has a well-defined melting point of 273 K. The amount of energy required to remove a molecule of water from the surface of liquid water (the desorption energy) is also well defined. Half a glass of water also has the same melting point and desorption energy. This is not true for water clusters. The melting point of a water cluster $(H_2O)_N$ with $N = 10$ is not the same as for $N = 100$. Similarly, the amount of energy required to desorb a water molecule from $(H_2O)_{10}$ is not the same as for $(H_2O)_{100}$. This regime of size dependent properties is the hallmark of nanoscience, which challenges our conception of what a thing is, because a very small nanoglass of water does not behave the same as a macroscopic glass of water.

One reason for size dependent properties is the aforementioned change in the relative number of surface atoms to bulk atoms. Stated more generally, when surface contributions begin to dominate bulk contributions the cluster's properties become size dependent.

A second reason is quantum confinement. Quantum confinement can be defined as a phenomenon that occurs when a particle senses the size of the container it is in, because the container is smaller than the natural extent of the particle's wavefunction. The pertinent size of a particle is given by its de Broglie wavelength λ,

$$\lambda = h/p, \tag{1.3.22}$$

where h is the Planck constant and $p = mv$ is the linear momentum. When the container that confines a particle approaches the size of its de Broglie wavelength, the properties of the particle become dependent on the size of the container. For example, an excited electron in a solid can form a bound electron-hole pair known as an exciton. The radius of the exciton is well-defined. When the size of the particle approaches the size of the exciton, the electronic structure becomes size dependent. For Si, the onset of size dependent electronic structure is below $\sim 5\,\text{nm}$.

Perhaps the easiest example to consider is that of a particle in a box. If the box has a length L and the walls are impenetrable (the potential energy goes to infinity at $x = 0$ and $x = L$), the energy levels of the box are given by

$$E_n = \frac{n^2 h^2}{8\,mL^2}, \quad n = 1, 2, 3, \dots \tag{1.3.23}$$

The important things to draw from Eq. (1.3.23) are (i) the energy levels are quantized (n is an integer not a continuum variable), and (ii) the spacing between levels ΔE is inversely proportional to particle mass and the length of the box squared

$$\Delta E \propto \frac{1}{mL^2}. \tag{1.3.24}$$

When thermal excitations are greater than the spacing between levels, that is $k_B T \gg \Delta E$, quantum effects are smeared out and systems behave in the continuum fashion familiar to classical mechanics. Systems for which quantum effects are important are those for which $\Delta E \gtrsim k_B T$. Large values of ΔE are favoured by small m and small L. Quantum confinement is, therefore, most likely to affect light particles confined to small boxes, especially at low temperature. Sufficiently small particles exhibit quantum effects even at room temperature and above.

1.4 The vibrations of solids

1.4.1 Bulk systems

The vibrations of a solid are much more complex than the vibrations of small molecules. This arises from the many-body nature of the interactions of atoms in a solid. Analogous to electronic states, the vibrations of a solid depend not only on the movements of atoms but also the direction in which the atoms vibrate. The multitudinous vibrations of the solid overlap to form a phonon band structure, which describes the energy of phonons as a function of a wavevector. Partial bandgaps can be found in the phonon band structure. Within the Debye model [65], there is a maximum frequency for the phonons of a solid, the Debye frequency. The Debye frequency is a measure of the rigidity of the lattice. Typical values of the Debye frequency are $14.3\,\text{meV}$ ($115\,\text{cm}^{-1}$) and $34.5\,\text{meV}$ ($278\,\text{cm}^{-1}$) for Au and W, respectively, and $55.5\,\text{meV}$ ($448\,\text{cm}^{-1}$) for Si. The most rigid lattice is that of diamond which has a Debye frequency of $192\,\text{meV}$ ($1550\,\text{cm}^{-1}$). Note that the Debye frequency is the highest frequency obtainable within the model but that real crystals may exhibit slightly higher values. For instance, the highest longitudinal optical phonon mode of Si lies at $520\,\text{cm}^{-1}$.

From the familiar harmonic oscillator model, the energy of a vibrational normal mode in an isolated molecule is given by

$$E_v = (v + \tfrac{1}{2})\hbar\,\omega_0 \tag{1.4.1}$$

where v is the vibrational quantum number and ω_0 is the fundamental radial frequency of the oscillator.

In a 3D crystal with harmonic vibrations, this relationship needs to be modified in two important ways. First, we note that a crystal is composed of N primitive cells containing p atoms. Since each atom has three translational degrees of freedom, a total of $3pN$ vibrational degrees of freedom exists in the solid. The solutions of the wave equations for vibrational motion in a periodic solid can be decomposed in $3p$ branches. Three of these modes correspond to acoustic modes. The remaining $3(p-1)$ branches correspond to optical modes. Optical modes can be excited by the electric field of an electromagnetic wave if the excitation leads to a change in dipole moment. Acoustic and optical modes are further designated as being either transverse or longitudinal. Transverse modes represent vibrations in which the displacement is perpendicular to the direction of propagation. The vibrational displacement of longitudinal modes is parallel to the direction of propagation.

The second important modification arises from the 3D structure of the solid. Whereas the vibrations of a molecule in free space do not depend on the direction of vibration, this is not the case for a vibration in an ordered lattice. This is encapsulated in the use of the wavevector of the vibration. The wavevector is given by

$$|\mathbf{k}| = \frac{2\pi}{\lambda} \tag{1.4.2}$$

where λ is the wavelength and $\mathbf{k}$ is a vector parallel to the direction of propagation. Consistent with this definition, we introduce the radial frequency of the vibration, ω_k,

$$\omega_k = 2\pi \upsilon_k. \tag{1.4.3}$$

We now write the energy of a vibration of wavevector $\mathbf{k}$ in the pth branch as

$$E(\mathbf{k}, p) = (n(\mathbf{k}, p) + {}^1\!/\!_2)\hbar\,\omega_k(p). \tag{1.4.4}$$

The vibrational state of the solid is then represented by specifying the excitation numbers, $n(\mathbf{k},p)$, for each of the $3pN$ normal modes. The total vibrational energy is thus a sum over all of the excited vibrational modes

$$E = \sum_{\mathbf{k},p} E(\mathbf{k},p). \tag{1.4.5}$$

By direct analogy to the quantized electromagnetic field, it is conventional to describe the vibrations of a solid in terms of particle-like entities (phonons) that represent quantized elastic waves. Just as for diatomic molecules, Eqs (1.4.4) and (1.4.5) show that the vibrational energy of a solid is non-vanishing at 0 K as a result of zero point motion.

The mean vibrational energy (see Exercise 1.5) is given by

$$\langle E \rangle = {}^1\!/\!_2 \hbar\omega_k(p) + \frac{\hbar\omega_k(p)}{\exp(\hbar\omega_k(p)/k_\mathrm{B}T) - 1}. \tag{1.4.6}$$

Thus by comparison to Eq. (1.4.4), it is confirmed that the mean phonon occupation number is

$$n(\mathbf{k}, p) \equiv \langle n(\omega_k(p), T) \rangle = \frac{1}{\exp(\hbar\omega_k(p)/k_\mathrm{B}T) - 1} \tag{1.4.7}$$

and that it follows the Planck distribution law. This is expected for bosons, that is, particles of zero (more generally, integer) spin. Accordingly, the number of phonons in any given state is unlimited and determined solely by the temperature.

Again in analogy to electronic states, there are phonon modes that are characteristic of and confined to the surface. Surface phonon modes have energies that are well defined in k-space and sometimes exist in the partial bandgaps of the bulk phonons. Surface phonon modes that exist in bandgaps are true surface phonons, whereas those that overlap with bulk phonon in k-space are surface phonon resonances. Furthermore, since the surface atoms are under-co-ordinated compared to the bulk, the surface Debye frequency is routinely much lower than the bulk Debye frequency. In particular, the vibrational amplitude perpendicular to the surface is much larger for surface than for bulk atoms. Depending on the specific material, the root mean square vibrational amplitudes at the surface are commonly 1.4–2.6 times larger on surfaces than in the bulk.

1.4.2 Nanoscale systems

Just as the electronic properties of a system become size dependent in the nanoscale regime, so too do the vibrational properties. This is most conveniently probed through the vibrational Raman spectrum. The peak position as well as the peak shape depend not only on the size of the crystallite probed [75] but also its shape [76, 77]. There is still no comprehensive theory of the lattice dynamics of nanostructures. The theoretical approaches can be classified as either continuum models [78] or lattice dynamical models [77].

The Raman spectrum probes the optical phonons of the sample. For a large sample only the $q = 0$ phonons at the centre of the Brillouin zone are probed in the first order spectrum, where q is the momentum of the phonon in units of $2\pi/a$, and a is the lattice constant. As the sample size is reduced, the $q = 0$ selection rule is relaxed. This leads to a shift in the peak centre and a broadening of the peak, both of which are dependent on the phonon dispersion, that is, on the function $\omega(q)$ which details how the phonon frequency depends on momentum. For spherical nanocrystals of diameter L, the intensity of first-order Raman spectrum $I(\omega, L)$ is given by

$$I(\omega, L) = \int \frac{\exp(-q^2 L^2/2\alpha)}{[\omega(q) - \omega]^2 + (\Gamma_0/2)^2} d^3q \tag{1.4.8}$$

For a sample of nanoparticles [79] with a size distribution described by the function $\varphi(L)$, the experimentally observed Raman spectrum is described by a convolution of the intensity function of Eq. (1.4.8) with $\varphi(L)$ according to

$$I(\omega) = \int \varphi(L) I'(\omega, L) \, \mathrm{d}L. \tag{1.4.9}$$

1.5 Summary of important concepts

- Ideal flat surfaces are composed of regular arrays of atoms with an areal density on the order of 10^{15} cm^{-2} (10^{19} m^{-2}).
- Surfaces expose a variety of potential high-symmetry binding sites of different co-ordination numbers.
- Real surfaces always have a number of defects (steps, kinks, missing atoms, etc).
- Clean surfaces can exhibit either relaxations or reconstructions.
- Relaxations are modest changes in bond lengths and angles.
- Reconstructions are changes in the periodicity of the surface compared to the bulk-terminated structure.
- Adsorbates can form ordered or random structures, and may either distribute themselves homogeneously over the surface or in islands.
- Adsorbates can also lead to changes in the surface structure of the substrate inducing either a lifting of the clean surface reconstruction or the formation of an entirely new surface reconstruction.
- The occupation of electronic states is defined by the Fermi-Dirac distribution, Eq. (1.3.5).
- Surface electronic states exist in a bulk band gap.

- A surface resonance has a wavefunction that is concentrated at the surface, but it does not exist in a band gap and, therefore, interacts strongly with bulk states.
- Vibrations in solids are quantized and form bands analogous to the band structure of electronic states.
- The mean phonon occupation number follows the Planck distribution law, Eq. (1.4.7).
- Properties of matter become size dependent as the contributions of surface atoms start to outweigh those of bulk atoms and as quantum confinement effects set in.
- Both the electronic structure and the phonon spectrum change as particle size approaches the nanoscale.

1.6   Frontiers and challenges

- Excited electronic states. As we will see in the chapter on stimulated (nonthermal) chemistry, excited states play a central role and their description at surfaces is fraught with uncertainty.
- Plasmons. This particular type of surface electronic excitation is implicated in a range of phenomena including hot electron formation; nanoantenna performance, including the surface enhanced Raman effect; sub-wavelength non-linear and near-field optical effects.
- Graphene and the pursuit of other one atomic layer thick materials.
- The surface science of chirality.
- Characterization and explanation of the electronic, optical, geometric, magnetic and chemical properties of nanoscale features on surfaces.
- Characterization and description of defects at solid surfaces.
- The relationship between surface structure, stoichiometry and surface reactivity of oxide surfaces.
- Electronic structure of strongly correlated materials, such as high temperature superconductors, graphene and topological insulators.
- Controlling spins in nanostructures [80]. Is spintronics an academic curiosity or can practical devices be constructed?

1.7   Further reading

S. Elliott, *The Physics and Chemistry of Solids* (John Wiley & Sons Ltd, Chichester, 1998).

C. Kittel, *Introduction to Solid State Physics*, 8th ed. (John Wiley & Sons, Inc., New York, 2004).

Walter A. Harrison, *Solid State Theory* (Dover Publications, Inc., New York, 1979).

H. Lüth, *Solid Surfaces, Interfaces and Thin Films*, 4th ed. (Springer-Verlag, Berlin, 2001).

S. Roy Morrison, *Electrochemistry at Semiconductor and Oxidized Metal Electrodes* (Plenum, New York, 1980).

Herbert Over, *Crystallographic study of interaction between adspecies on metal surfaces*, Prog. Surf. Sci. **58** (1998) 249.

M. W. Roberts and C. S. McKee, *Chemistry of the Metal-Gas Interface* (Clarendon Press, Oxford, 1978).

G. A. Somorjai, *Introduction to Surface Chemistry and Catalysis*, 2nd ed. (John Wiley & Sons, Inc., New York, 2010).

S. M. Sze and K. K. Ng, *Physics of Semiconductor Devices*, 3rd ed. (John Wiley & Sons, Inc., New York, 2006).

S. Titmuss, A. Wander, and D. A. King, *Reconstruction of clean and adsorbate-covered metal surfaces*, Chem. Rev. **96**, (1996) 1291.

A. Zangwill, *Physics at Surfaces* (Cambridge University Press, Cambridge, 1988).

1.8   Exercises

1.1   (a) Determine the surface atom density of Ag(221). (b) For the basal (cleavage) plane of graphite, determine the surface unit cell length, a, the included angle between sides of the unit cell, γ, and the density of surface C atoms, σ_0, given that the C–C nearest neighbour distance is 1.415 Å. A representation of the graphite surface is given in Fig. 1.22.

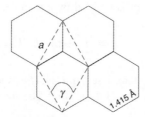

Figure 1.22   *A schematic drawing of the structure of the basal plane of graphite.*

1.2   The Pt(111) surface has a surface atom density of $\sigma_0 = 1.503 \times 10^{15}$ cm^{-2}. (a) Calculate the diameter of a Pt atom. (b) Calculate the atom density of the Pt(100) surface.

1.3   (a) Derive a general expression for the step density, ρ_{step}, of an fcc crystal with single-atom-height steps induced by a miscut angle χ from the ideal surface plane.

(b) Make a plot of step density versus miscut angle for the Pt(111) surface, which has a surface atom density of $\sigma_0 = 1.503 \times 10^{15}$ cm^{-2}.

1.4   Discuss why surface reconstructions occur. On what types of surface are reconstructions most likely to occur and where are they least likely to occur?

1.5   The Fermi energies of Cs, Ag and Al are 1.59, 5.49 and 11.7 eV, respectively [65]. Calculate the density of the Fermi electron gas in each of these metals as well as the Fermi temperature. Calculate the difference between the chemical potential and the Fermi energy for each of these metals at their respective melting points.

1.6   A clean Ag(111) surface has a work function of 4.7 eV. As a submonolayer coverage of Ba is dosed on to the surface, the work function drops and reaches a minimum of 2.35 eV [81]. Calculate the surface potential associated with the clean and Ba covered surfaces and explain the effect of Ba adsorption on the work function of Ag(111).

1.7   The work function of clean Al(111) is 4.24 eV [82] (a free electron *sp* metal), clean Ag(111) is 4.7 eV [81] (a coinage metal), and for bulk polycrystalline Cs is 2.14 eV (an alkali metal). Use the value of the Fermi energy given in Exercise 1.1 to calculate the surface potential for these three different types of metals.

1.8   The magnitude of an electric dipole μ is

$$\mu = 2qR \tag{1.8.1}$$

for charges $+q$ and $-q$ separated by a distance $2R$. The work function change $\Delta\Phi$ expressed in V associated with adsorption of a species with charge q, located at distance R from the surface (hence its image charge is at a distance $2R$ from the adsorbate and $-R$ below the plane of the surface) at a coverage σ is given by the Helmholtz equation

$$\Delta\Phi = \sigma\mu/2\varepsilon_0 = \sigma qR/\varepsilon_0, \tag{1.8.2}$$

when the dipole moment μ and coverage are expressed in SI units.

Calculate the work function changes expected for 0.1 ML of either peroxo (O_2^{2-}) or superoxo (O_2^-) species bound on Pd(111) given that their bond distance is 2 Å. The measured work function change is only on the order of 1 eV. Explain the difference between your estimates and the measured value.

1.9   The work function of Pt(111) is 5.93 eV. A Ru film has a work function of 4.71 eV. If Ru islands are deposited on a Pt(111) surface, in which direction does electron transfer occur?

1.10   Redraw Fig. 1.13 for a *p*-type semiconductor [65, 67].

1.11   Given that the partition function, *q*, is defined by a summation over all states according to

$$q = \sum_{i=1}^{\infty} \exp(-E_i/k_B T) \tag{1.8.3}$$

where E_i is the energy of the *i*th state. Use Eq. (1.4.4) to show that the mean vibrational energy of a solid at equilibrium is given by Eq. (1.4.6). Hint: The mean energy is given by

$$\langle E \rangle = k_B T^2 \frac{\partial(\ln q)}{\partial T} \tag{1.8.4}$$

1.12   The Debye temperature,

$$\theta_D = \hbar \omega_D / k_B, \tag{1.8.5}$$

is more commonly tabulated and determined than is the Debye frequency because of its relationship to the thermodynamic properties of solids.

(a) Calculate the Debye frequencies of the elemental solids listed in Table 1.3 in Hz, meV and cm^{-1}.

(b) Calculate the mean phonon occupation number at the Debye frequency and room temperature for each of these materials at 100, 300 and 1000 K.

1.13   The Debye model (see Table 1.3) can be used to calculate the mean square displacement of an oscillator in a solid. In the high-temperature limit this is given by

$$\langle u^2 \rangle = \frac{3N_A \hbar^2 T}{M k_B \theta_D^2} \tag{1.8.6}$$

Table 1.3   *Debye temperatures, θ_D, for selected elements; see Exercises 1.12, 1.13 and 1.14*

	Ag	Au	diamond	graphite	Pt	Si	W
θ_D (K)	225	165	2230	760	240	645	400

(a) Compare the root-mean-square displacements of Pt at 300 K to that at its melting point (2045 K). What is the fractional displacement of the metal atoms relative to the interatomic distance at the melting temperature?

(b) Compare this to the root-mean-square displacement of the C atoms at the surface of diamond at the same two temperatures.

1.14   The surface Debye temperature of Pt(100) is 110 K. Take the definition of melting to be the point at which the fractional displacement relative to the lattice constant is equal to ∼8.3% (Lindemann criterion [83, 84]). What is the surface melting temperature of Pt(100)? What is the implication of a surface that melts at a lower temperature than the bulk?

1.15   The bulk terminated Si(100)–(1 × 1) surface has two dangling bonds per surface atom and is, therefore, unstable toward reconstruction. Approximate the dangling bonds as effectively being half-filled sp^3 orbitals. The driving force of reconstruction is the removal of dangling bonds.

(a) The stable room temperature surface reconstructs into a (2 × 1) unit cell in which the surface atoms move closer to each other in one direction but the distance is not changed in the perpendicular direction. Discuss how the loss of one dangling bond on each Si atom leads to the formation of a (2×1) unit cell. Hint: The nearest neighbour surface Si atoms are called dimers.

(b) This leaves one dangling bond per surface atom. Describe the nature of the interaction of these dangling bonds that leads to (i) symmetric dimers and (ii) tilted dimers.

(c) Predict the effect of hydrogen adsorption on the symmetry of these two types of dimers [51, 52]. Hint: Consider first the types of bonds that sp^3 orbitals can make. Second, two equivalent dangling bonds represent two degenerate electronic states.

1.16 Describe the features a, b, c, and d in Fig. 1.21.

1.17 What is the significance of a band gap? What differentiates a partial band gap from a full band gap?

1.18 What are E_g, E_F, E_C, E_V and E_{vac} as shown, for instance, in Fig. 1.21?

References

[1] J. F. Nicholas, *An Atlas of Models of Crystal Structures*, Gordon and Breach, New York, 1961.

[2] M. W. Roberts, C. S. McKee, *Chemistry of the Metal-Gas Interface*, Clarendon Press, Oxford, 1978.

[3] H.-J. Gossmann, L. C. Feldman, *Phys. Rev. B*, **32** (1985) 6.

[4] H.-C. Jeong, E. D. Williams, *Surf. Sci. Rep.*, **34** (1999) 171.

[5] R. Smoluchowski, *Phys. Rev.*, **60** (1941) 661.

[6] V. A. Shchukin, D. Bimberg, *Rev. Mod. Phys.*, **71** (1999) 1125.

[7] R. M. Watwe, R. D. Cortright, M. Mavrikakis, J. K. Nørskov, J. A. Dumesic, *J. Chem. Phys.*, **114** (2001) 4663.

[8] C. T. Campbell, *Annu. Rev. Phys. Chem.*, **41** (1990) 775.

[9] D. P. Woodruff (Ed.), *The Chemical Physics of Solid Surfaces and Heterogeneous Catalysis: Oxide Surfaces, Vol. 9*, Elsevier, Amsterdam, 2001.

[10] K. Sohlberg, S. J. Pennycook, S. T. Pantelides, *J. Am. Chem. Soc.*, **121** (1999) 7493.

[11] P. W. Tasker, *J. Phys. C*, **12** (1979) 4977.

[12] H. J. Freund, *Faraday Discuss.*, **114** (1999) 1.

[13] U. Diebold, *Surf. Sci. Rep.*, **48** (2003) 53.

[14] M. Bowker, *Curr. Opin. Solid State Mater. Sci.*, **10** (2006) 153.

[15] G. L. Hornyak, J. Dutta, H. F. Tibblas, A. K. Rao, *Introduction to Nanoscience*, CRC Press, Boca Raton, FL, 2008.

[16] P. J. F. Harris, *Carbon Nanotube Science: Synthesis, Properties and Applications*, Cambridge University Press, Cambridge, 2011.

[17] P. Cabrera Sanfelix, S. Holloway, K. W. Kolasinski, G. R. Darling, *Surf. Sci.*, **532–535** (2003) 166.

[18] T. Zecho, A. Güttler, X. W. Sha, B. Jackson, J. Küppers, *J. Chem. Phys.*, **117** (2002) 8486.

[19] A. H. Castro Neto, F. Guinea, N. M. R. Peres, K. S. Novoselov, A. K. Geim, *Rev. Mod. Phys.*, **81** (2009) 109.

[20] A. K. Geim, *Science*, **324** (2009) 1530.

[21] K. S. Novoselov, A. K. Geim, S. V. Morozov, D. Jiang, Y. Zhang, S. V. Dubonos, I. V. Grigorieva, A. A. Firsov, *Science*, **306** (2004) 666.

[22] M. J. Allen, V. C. Tung, R. B. Kaner, *Chem. Rev.*, **110** (2010) 132.

[23] D. C. Elias, R. R. Nair, T. M. G. Mohiuddin, S. V. Morozov, P. Blake, M. P. Halsall, A. C. Ferrari, D. W. Boukhvalov, M. I. Katsnelson, A. K. Geim, K. S. Novoselov, *Science*, **323** (2009) 610.

[24] D. W. Boukhvalov, M. I. Katsnelson, A. I. Lichtenstein, *Phys. Rev. B*, **77** (2008) 035427.

[25] L. B. McCusker, F. Liebau, G. Engelhardt, *Pure Appl. Chem.*, **73** (2001) 381.

[26] J. J. Steele, M. J. Brett, *J. Mater. Sci.*, **18** (2007) 367.

[27] A. G. Cullis, L. T. Canham, P. D. J. Calcott, *J. Appl. Phys.*, **82** (1997) 909.

[28] S. Langa, J. Carstensen, M. Christophersen, K. Steen, S. Frey, I. M. Tiginyanu, H. Föll, *J. Electrochem. Soc.*, **152** (2005) C525.

[29] I. V. Sieber, P. Schmuki, *J. Electrochem. Soc.*, **152** (2005) C639.

[30] R. Beranek, H. Hildebrand, P. Schmuki, *Electrochem. Solid State Lett.*, **6** (2003) B12.

[31]   H. Tsuchiya, J. M. Macak, I. Sieber, L. Taveira, A. Ghicov, K. Sirotna, P. Schmuki, *Electrochem. Commun.*, **7** (2005) 295.

[32]   H. Tsuchiya, J. M. Macak, L. Taveira, P. Schmuki, *Chem. Phys. Lett.*, **410** (2005) 188.

[33]   H. Masuda, K. Fukuda, *Science*, **268** (1995) 1466.

[34]   X. He, D. Antonelli, *Angew. Chem., Int. Ed. Engl.*, **41** (2002) 214.

[35]   M. A. Carreon, V. V. Guliants, *Eur. J. Inorg. Chem.*, **2005** (2005) 27.

[36]   G. S. Armatas, M. G. Kanatzidis, *Science*, **313** (2006) 817.

[37]   G. W. Nyce, J. R. Hayes, A. V. Hamza, J. H. Satcher, Jr., *J. Phys. Chem. C*, **19** (2007) 344.

[38]   O. D. Velev, A. M. Lenhoff, *Curr. Opin. Colloids Interface Sci.*, **5** (2000) 56.

[39]   R. B. Wehrspohn, H. S. Kitzerow, *Phys. J.*, **4** (2005) 35.

[40]   M. Ulbricht, *Polymer*, **47** (2006) 2217.

[41]   C. Dekker, *Nature Nanotech.*, **2** (2007) 209.

[42]   J. M. Thomas, W. J. Thomas, *Principles and Practice of Heterogeneous Catalysis*, VCH, Weinheim, 1996.

[43]   N. W. Ockwig, O. Delgado-Friedrichs, M. O'Keeffe, O. M. Yaghi, *Acc. Chem. Res.*, **38** (2005) 176.

[44]   J. L. C. Rowsell, O. M. Yaghi, *Microporous and Mesoporous Materials*, **73** (2004) 3.

[45]   J. L. C. Rowsell, O. M. Yaghi, *Angew. Chem., Int. Ed. Engl.*, **44** (2005) 4670.

[46]   A. P. Cote, A. I. Benin, N. W. Ockwig, M. O'Keeffe, A. J. Matzger, O. M. Yaghi, *Science*, **310** (2005) 1166.

[47]   A. Patrykiejew, S. Sokołowski, K. Binder, *Surf. Sci. Rep.*, **37** (2000) 207.

[48]   G. A. Somorjai, *Introduction to Surface Chemistry and Catalysis*, John Wiley & Sons,Inc., New York, 1994.

[49]   G. A. Somorjai, *Annu. Rev. Phys. Chem.*, **45** (1994) 721.

[50]   G. Ertl, *Langmuir*, **3** (1987) 4.

[51]   K. W. Kolasinski, *Internat. J. Mod. Phys. B*, **9** (1995) 2753.

[52]   R. Becker, R. Wolkow, Semiconductor surfaces: Silicon, in *Scanning Tunneling Microscopy* (Eds: J. A. Stroscio, W. J. Kaiser), Academic Press, Boston, 1993, p. 149.

[53]   Y. J. Chabal, K. Raghavachari, *Phys. Rev. Lett.*, **54** (1985) 1055.

[54]   J. J. Boland, *Adv. Phys.*, **42** (1993) 129.

[55]   J. E. Northrup, *Phys. Rev. B*, **44** (1991) 1419.

[56]   R. Imbihl, G. Ertl, *Chem. Rev.*, **95** (1995) 697.

[57]   D. Avnir, C. Dryzun, Y. Mastai, A. Shvalb, *J. Mater. Chem.*, **19** (2009) 2062.

[58]   B. Kesanli, W. B. Lin, *Coord. Chem. Rev.*, **246** (2003) 305.

[59]   L. A. DeLouise, B. L. Miller, *Anal. Chem.*, **76** (2004) 6915.

[60]   C. F. McFadden, P. S. Cremer, A. J. Gellman, *Langmuir*, **12** (1996) 2483.

[61]   G. A. Attard, A. Ahmadi, J. Feliu, A. Rodes, E. Herrero, S. Blais, G. Jerkiewicz, *J. Phys. Chem. B*, **103** (1999) 1381.

[62]   T. Greber, Z. Sljivancanin, R. Schillinger, J. Wider, B. Hammer, *Phys. Rev. Lett.*, **96** (2006) 056103.

[63]   S. J. Pratt, S. J. Jenkins, D. A. King, *Surf. Sci.*, **585** (2005) L159.

[64]   K. Besocke, B. Krahl-Urban, H. Wagner, *Surf. Sci.*, **68** (1977) 39.

[65]   S. Elliott, *The Physics and Chemistry of Solids*, John Wiley & Sons Ltd, Chichester, 1998.

[66]   S. M. Sze, *Physics of Semiconductor Devices*, 2nd ed., John Wiley & Sons, Inc., New York, 1981.

[67]   S. R. Morrison, *Electrochemistry at Semiconductor and Oxidized Metal Electrodes*, Plenum, New York, 1980.

[68]   H. Gerischer, *Electrochim. Acta*, **35** (1990) 1677.

[69]   H. Lüth, *Surfaces and Interfaces of Solid Materials*, 3rd ed., Springer-Verlag, Berlin, 1995.

[70]   N. J. Halas, J. Bokor, *Phys. Rev. Lett.*, **62** (1989) 1679.

[71]   R. Haight, *Surf. Sci. Rep.*, **21** (1995) 275.

[72]   W. Steinmann, T. Fauster, Two-photon photoelectron spectroscopy of electronic states at metal surfaces, in *Laser Spectroscopy and Photochemistry on Metal Surfaces. Part I* (Eds: H.-L. Dai, W. Ho), World Scientific, Singapore, 1995, p. 184.

[73]   U. Höfer, I. L. Shumay, C. Reuß, U. Thomann, W. Wallauer, T. Fauster, *Science*, **277** (1997) 1480.

[74]   M. Wolf, E. Knoesel, T. Hertel, *Phys. Rev. B*, **54** (1996) 5295.

[75]   H. Richter, Z. P. Wang, L. Ley, *Solid State Commun.*, **39** (1981) 625.

[76]   I. H. Campbell, P. M. Fauchet, *Solid State Commun.*, **58** (1986) 739.

[77] S. P. Hepplestone, G. P. Srivastava, *Nanotechnology*, **17** (2006) 3288.

[78] M. A. Stroscio, M. Dutta, *Phonons in Nanostructures*, Cambridge University Press, Cambridge, 2001.

[79] M. N. Islam, S. Kumar, *Appl. Phys. Lett.*, **78** (2001) 715.

[80] R. Wiesendanger, *Curr. Opin. Solid State Mater. Sci.*, **15** (2011) 1.

[81] O. M. N. D. Teodoro, J. Los, A. M. C. Moutinho, *J. Vac. Sci. Technol. A*, **20** (2002) 1379.

[82] A. Hohlfeld, M. Sunjic, K. Horn, *J. Vac. Sci. Technol. A*, **5** (1987) 679.

[83] J. J. Gilvarry, *Phys. Rev.*, **102** (1956) 308.

[84] F. A. Lindemann, *Phys. Z.*, **11** (1910) 609.

2

Experimental Probes and Techniques

In Chapter 1 we were introduced to the structural, electronic and vibrational properties of solids and their surfaces. In this chapter, we investigate the techniques used to probe these properties. The emphasis here is to delve into the physical basis behind these techniques as well as the information that we can hope to gain from them. We do not emphasize the instrumental side of these techniques. For more information on the instrumentation of surface science, the reader is referred to the books of Vickerman [1], Ertl and Küppers [2], Woodruff and Delchar [3], Alford, Feldman and Mayer [4] as well as the other texts found in Further Reading. These books also introduce a number of surface sensitive techniques that are not covered in this chapter. Experimental surface science is an instrumentally intensive experimental science and many examples of best practise and the tricks of the trade have been compiled by Yates [5].

2.1 Ultrahigh vacuum

2.1.1 The need for UHV

The surface science approach to interfacial science now extends to all types of interfaces including the gas/solid but also liquid/solid, gas/liquid, etc interfaces. The surface science approach involves probing interfaces at the molecular level and developing molecular descriptions of reactions and properties. This requires maintaining a well-characterized interface of known concentration and structure. Many of the techniques used to probe surfaces also involve electron spectroscopy or electron beams. Both the impingement rate and mean free path depend on pressure and, as we shall show presently, both require ultrahigh vacuum (UHV), that is, pressure at and below the 10^{-9} mbar (or Torr) range, so that their values obtain experimentally desirable values.

We start by calculating the flux incident on a surface using the Hertz-Knudsen equation

$$Z_W = \frac{p}{\left(2\pi m k_B T\right)^{1/2}},$$ (2.1.1)

which at 300 K, for a gas with the mass of N_2, and assuming that there are 1×10^{15} surface atoms cm^{-2} in a monolayer (ML) reduces to

$$Z_W = \left(3.8 \times 10^5 \text{ ML s}^{-1}\right) p,$$ (2.1.2)

Surface Science: Foundations of Catalysis and Nanoscience, Third Edition. Kurt W. Kolasinski.
© 2012 John Wiley & Sons, Ltd. Published 2012 by John Wiley & Sons, Ltd.

with the pressure p given in torr. Thus, at a pressure of 1×10^{-6} torr, the number of molecules that hit the surface per second is roughly equal to the equivalent of one monolayer. In other words, on average in 1 s every surface atom is struck by at least one molecule from the gas phase. If each of these incident molecules stuck to the surface and became adsorbed, the surface would be completely covered in only 1 s. Therefore, if we want to keep the surface clean, or to keep a specifically covered surface unaltered by further adsorption, for the time it takes to perform an experiment, say $1\,\text{h} = 3600$ s, then the pressure has to be kept below 1×10^{-6} torr$/3600 \approx 3 \times 10^{-10}$ torr. The need for UHV to keep a surface unaltered by background adsorption is obvious.

The mean free path is the distance that a particle travels on average between collisions. In an ideal gas with mean velocity $\bar{c}$ and collision frequency Z, its value is given by

$$\lambda = \frac{\bar{c}}{Z} = \frac{k_{\mathrm{B}}T}{2^{1/2}\sigma p} \tag{2.1.3}$$

The collision cross section of N_2 is $\sigma = 0.43\,\text{nm}^2$. Therefore, at a pressure of $p = 1\,\text{atm} = 101\,325\,\text{Pa} = 760\,\text{torr}$, the mean free path is just 70 nm; whereas at 1×10^{-10} torr $= 1.3 \times 10^{-13}$ atm $= 1.3 \times 10^{-8}$ Pa, the mean free path is over 500 km. In electron spectroscopy, an electron must transit from the sample to a detector without scattering from any background gas over a flight path on the order of 1 m. Therefore, electron spectroscopy generally must be performed at a pressure below 7×10^{-3} Pa (5×10^{-5} torr), such that $\lambda \geq 1\,\text{m}$. In practise, even lower pressures are often required so that detector noise from electron multipliers is reduced to acceptable levels.

2.1.2 Attaining UHV

Technological advances in both pressure gauges and vacuum pumps made the achievement of UHV a (more or less) routine occurrence in the late 1950s and early 1960s. Shortly after this time is when the era of modern surface science began. The pumping speed S is defined by

$$S = Q/p, \tag{2.1.4}$$

where Q is the throughput in units such as Pa m^3 s^{-1} or torr l s^{-1} and p is the pressure in corresponding units. Different types of pumps have different pumping speeds, different pressure regions over which they maintain their pumping speed and different ultimate pressures. The ultimate pressure is the lowest pressure attainable by the pump. Equation (2.1.4) can also be used to calculate the pressure attained by a pump with pumping speed S for a chamber that has a leak rate (intentional or otherwise) given by Q. The ideal gas equation in the form of $pV = Nk_{\mathrm{B}}T$ can be used to show that 1 torr l s$^{-1} = 3.2 \times 10^{19}$ molecules s^{-1} at standard temperature and pressure, using

$$Q = k_{\mathrm{B}}T\frac{\mathrm{d}N}{\mathrm{d}t}. \tag{2.1.5}$$

Vacuum pumps are classified as either compression or entrapment pumps. Compression pumps are further categorized as either displacement pumps, which can move vast quantities of gas but have ultimate pressures in the rough vacuum range (atmospheric pressure to 0.1 Pa or 1 mtorr), or mass transfer pumps, which have much lower throughput but ultimate pressures in the UHV range. Rotary vane, roots blower and piston pumps are all displacement pumps. They are often used as rough pumps that back mass transfer pumps. In other words, the rough pump (or backing pump) is placed on the exhaust of the mass transfer pump. Oil diffusion and turbomolecular pumps are mass transfer pumps. Both of these types of pump require a backing pump to reduce the pressure on their outlet so that the pressure at their inlet can be as low as possible. They generally cannot be used at pressures above 10^{-5} torr. The ratio of the outlet pressure to the inlet pressure is the compression ratio of the pump. The compression ratio is often in the range of $10^6 - 10^9$ for diffusion and turbo pumps. It is molecular mass specific and depends on the type of pump.

Entrapment pumps include cryopumps, ion pumps, sorption pumps and titanium sublimation pumps. Cryopumps are surfaces cooled with liquid N_2 or liquid He. They are often combined with a Ti sublimation pump. Ti is particularly reactive toward most gases such that they are irreversibly adsorbed when they strike a fresh Ti surface. This process of irreversible adsorption is called gettering. Based on the pressure to which they are exposed, the Ti layer (the getter layer, for instance, on the surface of the cryopump) has to be periodically refreshed by subliming a layer of Ti from solid electrodes that are heated by passing current through them. Ion pumps ionize gases that enter the pump with the aid of a high electric field. These positive ions are then embedded in a negatively charged getter layer. Sorption pumps are composed of cryogenically cooled (usually with l-N_2) zeolites, which have very high surface areas. None of these pumps let the gases out of the back end of the pump and, because of this, all of these pumps need to be periodically regenerated as discussed above.

The type of gases pumped, the ultimate pressure and throughput required, as well as cost all factor into the decision of what type and combination of pumps are used on a UHV chamber. Diffusion pumps tend to have larger throughputs than turbo pumps but they tend to backstream oil (releasing oil into the chamber) and pump lighter molecules better than heavy ones. Turbo pumps, which operate much like a jet engine, pump heavier molecules better than light ones, particularly H_2 and He. They can be cycled between on and off relatively rapidly compared to diffusion pumps, which work with the use of jets of boiling hot oil and must cool down and warm up to turn off and back on. Reactive gases require special diffusion pump oils and are best not used in conjunction with cryopumps and sorption pumps. Ion pumps are able to achieve pressures below 2×10^{-11} torr when combined with a l-N_2 cooled Ti sublimation pump to increase their H_2 pumping capacity and are not susceptible to backstreaming. H_2, CH_4, and noble gases, however, are not pumped well by ion pumps. While H_2O is pumped exceedingly well by cryopumps, low boiling point gases such as H_2 and He are not. If ion pumps are subjected to too heavy a load of a gas, they tend to heat up and are susceptible to burping, the release of gas in pressure bursts. This can also happen to diffusion pumps or oil lubricated rotary vane displacement pumps when they are exposed to reactive gases such as silanes.

To attain the ultimate pressure of diffusion, turbo and ion pumps, the vacuum chamber must be baked out to reduce the desorption rate from the chamber walls. Because of the need for low outgassing rates and the ability to withstand a bake out temperature in the range of $125-150°C$, only special materials can be used in a UHV chamber. Stainless steel and nonporous glass are the fundamental materials of UHV chambers. Teflon and Macor are among the few polymers that can be used. Other metals such as W, Mo, Ta and OFHC Cu (oxygen free, high conductivity copper) are also suitable. The composition of the vacuum depends on the history of the chamber (particularly the baking conditions and what gases it has been exposed to) as well as the pumps. H_2, He, H_2O, CH_4, CO and CO_2 are the most commonly observed gases. Any time a filament is turned on (for instance, in a pressure gauge, electron gun or H atom source) all of these gases (apart from He) are observed to desorb from the surface of the filament either because they were adsorbed to begin with or because they diffuse out of the bulk of the filament and desorb after reaching the surface.

2.2 Light and electron sources

Many analytical and experimental techniques require the use of beams of electrons or photons. Here we discuss a few of these sources because their characteristics give us some insight into what is available and what sort of information can be gained from them. Different regions of the electromagnetic spectrum correspond to different types of characteristic excitations and it is useful to categorize the spectrum according to the energy (or equivalently the wavelength) of the photon. The infrared region lies at a wavelength of 700 nm to 1 mm (1.77 eV–1.24 meV), though this range is more commonly designated in wavenumbers (14 300 cm^{-1} to 10 cm^{-1}). The mid-IR range (several hundred to 4000 cm^{-1}) is commonly associated with vibrational excitations. The visible is at 400 nm $\leq \lambda \leq$ 700 nm, 1.8–3.1 eV and corresponds to low lying valance electron excitations. Ultraviolet (UV: 200 nm $\leq \lambda \leq$ 400 nm, 3.1–6.2 eV), vacuum ultraviolet (VUV: 100 nm $\leq \lambda \leq$ 200 nm, 6.2–12.4 eV) and extreme ultraviolet (XUV: 10 nm $\leq \lambda \leq$ 100 nm, 12.4–124 eV) light excite progressively deeper valence levels, while x-rays (0.01 $\leq \lambda \leq$ 100 nm, 125–125,000 eV) excite core level electrons.

Several materials problems can complicate the integration of light sources with surface science and UHV chambers. Visible light is easily transported through air by the use of conventional optical materials such as glass, quartz, sapphire and metals. IR light is subject to absorption by atmospheric H_2O and CO_2 so these may need to be excluded from the beam path. Light in the VUV and XUV regions can only be effectively transmitted in vacuum due to absorption by oxygen and nitrogen. VUV and XUV radiation interact strongly with solid materials [6]. Metals are no longer highly reflective, necessitating the use of specially designed multilayer mirrors [7]. The very best quartz cannot be used as an optical material below 175 nm. CaF_2 extends to 125 nm for use as a window and MgF_2 to 122 nm. The highest band gap material, LiF, can be used as a window down to 104.5 nm.

Electrons of the corresponding energies are capable of producing similar excitations but, as we shall see below and discussed particularly in the section on electron energy loss spectroscopy, there are differences in how electrons and photons induce excitations. In particular, one must remember that the electron is a massive particle with a linear momentum mv, charge $-e$ and a spin of $\frac{1}{2}$. The photon is massless, neutral and has angular momentum of 1. A useful conversion factor is to recall that 1 cm^{-1} corresponds to 8.065 meV.

2.2.1   Types of lasers

Table 2.1 tries to convey a sense of the range of properties that are encountered with commercially-available lasers. Lasers operate over a great range of temporal pulse durations (pulse widths), spectral bandwidth (linewidth), wavelengths and fluences. Since lasers can either be run in pulsed or continuous wave (cw) mode, two other important characteristics are the peak power (power while the pulse is on) and the time averaged power (J s^{-1} = W). More important for many practical purposes are the peak or time averaged irradiance (power per unit area) or fluence (energy per unit area).

2.2.2   Atomic lamps

Atomic emission lamps excited by any of a variety of discharges [6] are widely used as sources of VUV and XUV light. Commonly used emission lines are those of H (10.2 eV), and He (21.1 and 40.82 eV). The He lamp in particular has been a workhorse for UV photoemission experiments. Such lamps can deliver photon fluxes on the order of 10^{13} photons cm^{-2} s^{-1}. Different rare gases can be used to deliver other photon energies. These sources are continuous wave, i.e. not pulsed, sources that require windowless operation; hence, they affect the pressure in the surface science chamber. Furthermore, useful photon fluxes are, for most purposes, only available at discrete lines since the continuum radiation between these lines tends to be produced at much lower intensity. Xe flash lamps can be operated to provide μs pulses of light in the range of 170–3000 nm, with time-averaged powers exceeding 100 mW.

Table 2.1 *Types of lasers and their characteristics including typical wavelengths, pulse durations, pulse energy or power, and repetition rates*

Laser material	λ/nm	$h\nu$/eV	Characteristics
Solid state			
Semiconductor laser diode	IR–visible ~0.4–20 μm		Usually cw but can be pulsed, wavelength depends on material, GaN for short λ, AlGaAs 630–900 nm, InGaAsP 1000–2100 nm, used in telecommunications, optical discs
Nd^{3+}: YAG (1st harmonic)	1064	1.16	cw or pulsed, ~10 ns pulses, most common, 150 ps versions (and shorter) available, 10–50 Hz rep rate, 1 J to many J pulse energies. Nd^{3+} can also be put in other crystalline media such as YLF(1047 and 1053 nm) or YVO_4 (1064 nm)
Nd^{3+}: YAG (2nd harmonic)	532	2.33	
Nd^{3+}: YAG (3rd harmonic)	355	3.49	
Nd^{3+}: YAG (4th harmonic)	266	4.66	
Nd^{3+}: glass	1062 or 1054	2.33	~10 ps, used to make terawatt systems for inertial confinement fusion studies
Ruby ($Cr:Al_2O_3$ in sapphire)	694	1.79	~10 ns
Ti:sapphire	700–1000	1.77–1.24	fs to cw; 1 Hz, kHz, 82 MHz
Alexandrite (Cr^{3+} doped $BeAl_2O_4$)	700–820	1.77–1.51	Tattoo removal
Liquid			
Dye laser	300–1000	4.13–1.24	Rep rate and pulse length depend on pump laser; fs, ps, ns up to cw
Gas			
CO_2	10 600 (10.6 μm)	0.12	Long (many μs), irregular pulses, cw or pulsed at high rep rates, line tuneable, few W to >1 kW
Kr ion	647	1.92	cw, line tuneable, 0.1–100 W
HeNe	632.8	1.96	cw, 0.5–35 W
	543.5	2.28	
Ar ion	514.5	2.41	cw, line tuneable, Ar and Kr ion laser (or versions with both present) are commonly used in the entertainment industry for light shows
	488	2.54	
HeCd	441.6, 325	2.81, 3.82	cw, 1–100 mW
ArF excimer	193	6.42	~20 ns, 1–>1000 Hz, several W to over 1 kW, 100 mJ to >1 J
KrF excimer	248	5.00	30–34 ns
XeCl excimer	308	4.02	22–29 ns
XeF excimer	351	3.53	12 ns
F_2	154	8.05	1–several kHz rep rate, 1–20 W, 10–50 mJ pulse energies, 10 ns
N_2	337	3.68	1–3.5 ns, 0.1–1+ mJ, 1–20 Hz rep rate

(continued overleaf)

Table 2.1   *(continued)*

Laser material	λ/nm	hν/eV	Characteristics
Exotic			
Free electron laser	1 nm–THz		Wide variety of characteristics depending on machine characteristics
X-ray lasers			Lasing medium is a highly ionized plasma producing light at one wavelength
High harmonic generation	UV–X-ray		High powered laser pumps a gas to produce light at wavelengths that are harmonics (up to over 150) of the pump wavelength. Capable of producing attosecond pulses

2.2.3   Synchrotrons

Accelerating electrons emit radiation. The recognition of this was a primary objection to the planetary model of negatively charged electrons orbiting a positive nucleus and led Bohr to his *ad hoc* proposition for the quantization of angular momentum. Parasitic emission of light was observed as a complication in the construction and operation of high-energy electron accelerators. However, it was soon realized that this could be exploited as a light source and the first implementation was at General Electric in 1947 [7]. Third generation synchrotron sources operate continuously between 10–1000 eV (and beyond). They can deliver up to 10^{13} photons s^{-1} into a bandwidth of 0.01%. Different types of pulse trains are possible but typically high repetition rates (1–500 MHz) lead to low pulse energies in ~100 ps pulses [8]. Therefore, synchrotrons are ideal for applications that require high average photon fluxes of incoherent radiation but they generally cannot reach the high power densities required to study nonlinear phenomena. Fourth generation sources, incorporating energy recoverable linear accelerators and free electron lasers, promise to push pulse lengths into the sub-ps regime [9].

Synchrotrons have been instrumental in the development of a number of surface science techniques aimed at elucidating the electronic and geometric structure of surfaces and adsorbates, particularly in the VUV, XUV and x-ray regions. These techniques include not only photoemission, but also photoelectron diffraction and x-ray diffraction.

2.2.4   Free electron laser (FEL)

A free electron laser consists of an electron beam propagating through a periodic magnetic field [10–12]. This transverse oscillating magnetic field is known as the undulator. Lasing occurs because the wiggler and the radiation combine to produce a beat wave that is synchronized with the electrons. An FEL is continuously tuneable, capable of high peak and average powers, and can produce a range of pulse widths and patterns.

While designs vary and specifications vary, a "typical" FEL is a sub-picosecond (or longer), tuneable light source covering the range from the UV to the mid-infrared, with pulse energies of up to several to several hundred μJ, and at repetition rates up to 75 MHz. Not all parameters can be satisfied simultaneously but average powers in excess of 10 kW have been demonstrated in the infrared [11]. Because of the high peak power, the wavelength range could potentially be extended by conventional and high harmonic generation. The major advantage of an FEL over a synchrotron is that it can produce coherent light many orders of magnitude brighter than the incoherent synchrotron radiation [11]. FELs will soon be extended into shorter wavelength regions even up to the x-ray region.

2.2.5 Electron guns

A source of a beam of electrons with controllable energy is generally referred to as an electron gun, which usually is composed of a hot wire (the cathode) through which current is passed such that electrons are emitted from it by thermionic emission. Alternatively, the primary source of electrons might be a photocathode, that is, a cathode (a semiconductor or metal) that emits electrons as a result of photoemission. The advantage of a photocathode is that the primary source of electrons can be pulsed whereas thermionic emission is a continuous process. After the electrons are produced at the cathode they pass through a series of lenses that focus and accelerate the beam to a well-defined energy. The energy resolution of the beam is mainly defined by the spread of energies in the electrons that leave the cathode; however, space charge broadening caused by the repulsion between like charged particles packed into too small a volume can also lead to a broadening of the electron energy distribution. Space charge broadening also constrains the upper limit of current density that can be delivered by an electron gun. The maximum fluxes of electron guns are generally much lower than photon sources but their construction is relatively cheap and easy. In addition, they are broadly tuneable over very large energy ranges. Electrons usually interact much more strongly with matter than photons of the same energy; therefore, fewer electrons than photons are required for measurable signals or effects to be registered. This compensates somewhat for the fact that electron guns tend to have lower intensities than lamps or, even more so, lasers.

An extremely exciting development in the area of electron guns is the construction of new ultrafast pulsed high intensity sources [13]. This new generation of electron guns is capable of producing subpicosecond pulses with sufficient numbers of electrons to facilitate surface science experimentation. Because of the previously mentioned constraint on the mean free paths, electron guns can only be used effectively in high vacuum.

2.3 Molecular beams

Molecular beams are extremely important in dynamical studies at surfaces as well as in growth of materials by molecular beam epitaxy. There are two types of molecular beams (Fig. 2.1), Knudsen beams (or thermal beams) and supersonic jets. Here we take a quick look at the characteristics of both as well as how they are formed.

2.3.1 Knudsen molecular beams

The basic equation of gas dynamics is the Hertz-Knudsen Equation, Eq. (2.1.1). It defines the flux incident on any unit of surface area. Imagine now that instead of thinking of gas incident on a surface, we drill a hole of diameter D in a thin wall of the vacuum chamber. If the pressure in the chamber, which we call the stagnation pressure p_0 with corresponding number density ρ_0, is not too high, and the wall is not too thick, then when molecules emerge out the other side of the hole, they do not experience any collisions that change their properties from what they were on the other side of the hole. The beam that emanates from the hole has the same characteristics as those of a thermal gas. For instance, the temperature of the gas in the oven, the stagnation temperature T_0, is the same as the temperature of the gas in the beam and only one temperature suffices to define the characteristics of the gas in the beam. The beam differs from a thermal distribution only in that we have defined the direction in which the gas molecules travel to some degree by selecting out only those molecules from the half-space in front of the hole. If we call the direction pointing out the back of the hole the x direction, then only molecules with a positive value of the x component of velocity can make it through the hole. (Usually we call the direction along the surface normal the z direction; however, in this case it is going to define the direction of propagation of the molecule beam and this direction is conventionally called the x direction in the literature.)

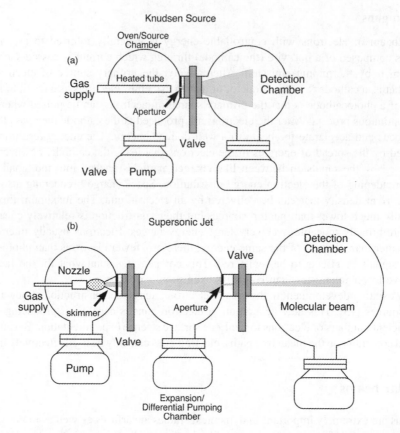

Figure 2.1   *Knudsen and supersonic molecular beam machines. Both beam sources require vacuum chambers for their production and implementation in surface science.*

Assuming the gas in the chamber (alternatively, we can replace the chamber with a nozzle, that is a tube closed all but for a small hole in or near one end) is at equilibrium then its translational and internal energy distributions are defined by Maxwell-Boltzmann distributions and its flux is defined by the Hertz-Knudsen equation. Furthermore, since the gas incident on the hole has a flux that varies with the angle θ from the surface normal according to a cosine distribution [14], the beam emanating from the hole also is described by a cosine distribution. This means that the beam expands out of the hole and as it propagates along x, the molecules spread out with a flux that falls off as $\cos \theta$. Since the beam is expanding, it initially has a cross sectional area and pressure equal to $\frac{1}{4}\pi d^2$ and p_0, respectively; however, as the cross sectional area expands, the effective pressure (or, more appropriately, the number density) in the beam drops. A beam of this type has properties defined by equilibrium thermal distributions at the temperature of the nozzle or oven from which the gas originates. It is known as an effusive beam or Knudsen beam [15] and the apparatus (beam machine) that makes such a beam is a Knudsen source.

2.3.2   Free Jets

We now change the geometry and conditions that create the beam to those shown in Fig. 2.2. The nozzle is a tube with a length much longer than its diameter. The diameter of the hole D is much smaller than that of the tube. The tube can be heated or cooled to temperature T_0. In some cases the end of the tube has a

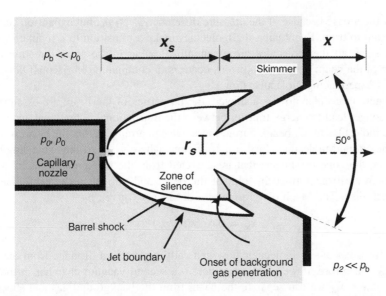

Figure 2.2 *A supersonic expansion and the formation of a supersonic jet. The nozzle-skimmer distance is x_S. The skimmer radius is r_S. The nozzle diameter is D. The pressure and gas density in the nozzle are p_0 and ρ_0. The pressure in the expansion chamber is p_b. Behind the skimmer the differentially pumped buffer chamber is at a pressure p_2. The distance along the centreline from the skimmer to the point of measurement is x.*

valve that can be rapidly opened and closed to form a pulsed jet, that is, one made up of pulses rather than running continuously. Pulse lengths can vary from the millisecond to the $\sim 100\,\mu$s regime. An alternative method of creating pulses is to place a chopper after the nozzle. A chopper is a disk with small slots cut in it and the combination of a high speed chopper and a pulsed nozzle can reduce pulse lengths to $\sim 10\,\mu$s.

The advantage of pulsed nozzles is that the nozzle flow rate is proportional to the pressure and nozzle diameter according to

$$N_{\text{flow}} \propto p_0 D^2 / T_0^{1/2}. \tag{2.3.1}$$

The number of molecules that flow through the nozzle is given by the product of flow rate and the duty cycle of the nozzle. The duty cycle is equal to the number of pulses per second, the repetition rate, multiplied by the pulse length in seconds (duty cycle = time open/total time = repetition rate × pulse length = $R_{\text{rep}} \times t_{\text{p}}$). Therefore, the pumping speed S required to keep the chamber into which the beam expands at some given pressure p_b scales according to

$$S \propto \frac{p_0 D^2}{p_b T_0^{1/2}} R_{\text{rep}}\, t_{\text{p}}. \tag{2.3.2}$$

For experimental reasons, we generally want to keep p_b as low as possible. For financial reasons, we generally want to keep S as low as possible. Therein lies a conflict especially because, as we shall soon see, several other interesting properties of the beam are dependent on the $p_0 D$ product and for most applications we would like to run the molecule beam with the maximum value of $p_0 D$.

If we crank up the pressure p_0 behind the nozzle – more accurately, as we increase the product $p_0 D$ – we eventually reach a point at which we can no longer neglect the effect of collisions. These collisions occur within a flow since only a subset of molecules is selected to go through the nozzle. This changes the properties of the molecules in the flow. Again, the molecules expand after leaving the nozzle. The beam accelerates out of the nozzle because of the pressure difference $p_0 \gg p_b$, but expansion causes the effective pressure in the beam to drop. Eventually, the molecules make a transition to a regime in which collisions no longer occur, since all the molecules are essentially flowing in the same direction and at the same speed. The region of the beam in which collisions occur is in continuum (or viscous) flow. Once collisions cease, the beam is experiencing molecular flow.

At the point where molecular flow is achieved, the properties of the beam are essentially frozen in as collisions can no longer lead to energy transfer between the translational, vibrational and rotational degrees of freedom of the molecules in the beam. Furthermore, we can arrange it that the beam is not influenced by collisions with background molecules. First, for conditions where $p_0 \gg p_b$ the expanding beam essentially clears out a region, the zone of silence, that is separated from the background gas by a series of shock waves. The shock wave front formed in front of the beam and normal to the centreline of the beam is known as the Mach disk. The Mach disk forms as a distance x_M given by

$$x_M = 0.67D (p_0/p_b)^{1/2}. \tag{2.3.3}$$

The transition from viscous to molecular flow generally occurs as a distance from the nozzle of about $x_{transition} < 0.1\ x_M$. Therefore, if we place an orifice to a second vacuum chamber in the beam path at a distance $x_{transition} \leq x \leq x_M$ we can separate the beam from the background gas and shock waves, ensure that the expansion has achieved its full effect and avoid any interference from skimming off the centre of the beam. The device used to select out the central portion of the beam is called a skimmer, which has a sloping conical shape. The walls are angled so as to avoid the formation of shock waves that would interfere with the expansion.

The beam now expands into a region of negligible pressure $p_2 \ll p_b$ in which no collisions occur either within the jet or with background molecules. We have formed a free jet. If one is used, the chopper is usually placed in this second chamber. For maximum beam intensity, the skimmer is placed at

$$x_S/D = 0.125[1/Kn (p_0/p_b)]^{1/3}, \tag{2.3.4}$$

where Kn is the Knudsen number

$$Kn = \lambda/D \tag{2.3.5}$$

and λ is the mean free path, Eq. (2.1.3), at the conditions in the stagnation chamber.

2.3.3   Comparison of Knudsen and supersonic beams

An important parameter with which we characterize a molecular beam is the Mach number M

$$M = V/a, \tag{2.3.6}$$

where V is the velocity of molecules in the beam and a is the speed of sound. For an ideal gas the speed of sound is given by

$$a = (\gamma RT/W)^{1/2}, \tag{2.3.7}$$

where $\gamma = C_p/C_v$ is the ratio of heat capacities at constant pressure and constant volume, T is the temperature of the gas and W is the mean molar mass. The Mach number is also the ratio of the parallel speed to the random speed at the centreline of the beam. The Mach number of an equilibrium gas contained in a chamber is $M = 0$, since there is no net flow and the speed distribution of such molecules is described by a Maxwell distribution. Here we are interested in probing the properties of beams for which $M > 1$, that is for supersonic beams.

Sonic and subsonic beams, such as Knudsen beams, do not change their mean speed and Mach number as they expand. This is not true for supersonic beams for which V and M increase throughout the viscous flow regime. V and M stop changing once the transition to molecular flow is achieved. The terminal value of the Mach number depends on the expansion conditions and can be approximated by

$$M_\infty = A\,Kn^{-((1-\gamma)/\gamma)}, \tag{2.3.8}$$

where A is a constant on the order of 1. More elaborate expressions for M_∞ are given by Miller [16].

Note that since $E_{K,\text{flow}} = \frac{1}{2}mV^2$, an increasing V corresponds to an increasing kinetic energy. If the kinetic energy of the beam is increasing and energy is conserved, where does the "extra" energy come from? The answer to this is found in another extremely interesting property of supersonic jets: *the translational, vibrational and rotational degrees of freedom are no longer described by a single temperature*.

Imagine an atomic beam. In the volume of the stagnation chamber gas molecules can translate in three dimensions and each dimension has a mean energy of $\frac{1}{2}k_B T$. We now force the molecules through the nozzle and let them expand. Atoms that are flowing in the beam collide with each other. Slow atoms get rear ended by fast atoms, both of which have a much higher collision cross section than molecules moving at the mean velocity of the beam. Collisions speed up the slow molecules and slow down the fast molecules. Furthermore, atoms moving transverse to the beam no matter what their speeds are more likely to collide than molecules moving parallel to the beam. They are either scattered out of the beam or else have their transverse velocity transferred into parallel velocity. The result is that the beam is focussed in the forward direction and all of the molecules move with almost the same velocity parallel to the beam axis. In a hard expansion, virtually all of the energy is removed from the two transverse directions and transferred into the parallel velocity component. Hence, while the total kinetic energy is conserved, the formerly random kinetic energy is converted into kinetic energy that is almost exclusively directed along the parallel component in the direction of the flow. Furthermore, the spread in velocities is substantially reduced, and the resultant supersonic jet possesses translational energy, which is very narrowly distributed about a very large stream velocity V. The velocity distribution follows a Maxwell distribution; however, the temperature of the distribution is very narrow and can easily be $T_{\text{trans}} < 10$ K. Supersonic expansion leads to substantial cooling during the expansion.

Now consider a molecular beam and recall that translational energy is equilibrated in just a few collisions whereas rotational energy requires tens of collisions and vibrational energy – depending mainly on the frequency of the vibration – requires tens to hundreds to thousands of collisions to equilibrate. In theory, both the rotational and vibrational distributions are also cooled during expansion. Rotational cooling is generally very effective and again $T_{\text{rot}} < 10$ K can be achieved. The rotational distributions are very close to Boltzmann but often exhibit a slight tail of overpopulation at high rotational energies. The exception is hydrogen. Because of its very large rotational constant it is difficult to cool rotationally. Vibrations act much differently. Very low wavenumber vibrations ($<200\,\text{cm}^{-1}$) can be cooled but high wavenumber vibrations are almost unaffected by expansion. They simply require too many collisions for redistribution of their energy. The energy that is removed from rotations (and low frequency vibrations) is transferred to kinetic energy along the flow direction. Therefore, supersonic molecular beams are generally characterized by high translational energy along the flow direction, extremely low translational and rotational temperatures but with vibrational temperatures close to T_0.

The maximum or terminal velocity of a molecular beam is given by assuming complete cooling of all degrees of freedom into the stream velocity,

$$V_\infty = \left(\frac{2RT_0}{W_{\text{avg}}} \frac{\gamma}{\gamma - 1} \right)^{1/2}, \tag{2.3.9}$$

where W_{avg} is the molar average of the molar masses in the beam. This result shows us that if we make a mixture of a light gas with a small amount of heavy gas, all of the molecules in the beam have the same velocity. Such a beam is called a seeded beam. Note that while the *velocity* is the same, the *kinetic energy* of the heavy gas is substantially higher than the kinetic energy of the light gas. If we seed, say, 1% N_2 in H_2 at 300 K, the velocity of the heavy gas (neglecting any change in heat capacity caused by varying the composition) is given by

$$E_{K,heavy} \approx \frac{W_{heavy}}{W_{avg}} \frac{\gamma}{\gamma - 1} R T_0. \tag{2.3.10}$$

This corresponds to $E_K(N_2) = 1.3\,eV$. Seeded beams can be used to produce extremely high kinetic energies with very low translational temperatures. If a light gas is seeded into a heavy gas, deceleration occurs, a process called antiseeding. It should be noted that concentration of the heavy species along the centreline occurs during seeding. This focussing effect scales roughly with the mass; hence a nominal 1% N_2 seeded beam in H_2 has approximately 14% N_2 along the centreline. As the concentration of the heavy species increases, the velocity of the heavy species starts to lag behind the prediction of Eq. (2.3.10). This is known as velocity slip. Velocity slip is less likely to occur at high values of $p_0 D$.

Along with the transformation of the energy distribution of the molecules within the beam, the most important property of a supersonic molecular beam is the enhancement of the intensity compared to a Knudsen beam. The intensity of molecules of mass m along the centreline of a thermal beam is

$$I_{Knudsen} = \frac{\rho_0 \pi r_S^2}{4} \left(\frac{8 k_B T_0}{\pi m} \right)^{1/2} \frac{1}{\pi x^2}. \tag{2.3.11}$$

Note that the intensity falls as $1/x^2$. The speed ratio is defined as

$$w = \left(\frac{m u^2}{2 k_B T} \right)^{1/2}, \tag{2.3.12}$$

where the most probably velocity of a thermal distribution is given by

$$u^2 = \frac{2 k_B T}{m} \tag{2.3.13}$$

and, therefore, $w = 1$ for a thermal beam and $w > 1$ for a supersonic expansion. Assuming that the expansion is sufficiently hard such that $w > 4$ and positioning the skimmer before the occurrence of the Mach disk, the intensity of a supersonic beam is

$$I_{supersonic} = \rho_S \pi r_S^2 u_S \frac{1}{\pi x^2} \left(\frac{x_S}{x_1} \right)^2 \left(\tfrac{1}{2} \gamma M_S^2 + \tfrac{3}{2} \right), \tag{2.3.14}$$

where x_1 is the distance at which the transition to molecular flow occurs. It can be shown that the ratio of intensities at any axial distance is given by

$$\frac{I_{supersonic}}{I_{Knudsen}} \simeq \left(\frac{\pi C_p}{k_B} \right)^{1/2} \gamma M_{eff}^2. \tag{2.3.15}$$

Thus for N_2 expanded to an effective Mach number $M_{eff} = (x_S/x_1) M_1 = 10$, which is a mild expansion, an ideal supersonic beam is approximately 470 times as intense as a Knudsen beam expanded at the same pressure and temperature. At $M_{eff} = 20$, the intensity ratio rises to 1860 [17].

For further information on supersonic jets and molecular beams in general, the reader is referred to the seminal review article of Anderson, Andres and Fenn [17] as well as the definitive 2-volume set edited by Scoles [18, 19].

2.4   Scanning probe techniques

Scanning tunnelling microscopy (STM) and the slew of scanning probe microscopy (SPM) techniques that arose in its wake represent a monumental breakthrough in surface science. Binnig and Rohrer were awarded the Nobel Prize for the discovery of STM [20] in 1987. The STM is not only a tool for detailed investigation of surface structure but also for the manipulations of atoms and molecules at surfaces [21, 22].

The basis of all scanning probe or proximal probe techniques is that a sharp tip is brought close to a surface. A measurement is then made of some property that depends on the distance between the tip and the surface. A variation on this theme is near-field scanning optical microscopy (NSOM), in which a small-diameter optical fibre is brought close to the surface. The diameter of the fibre is less than the wavelength of the light that is directed down the optical fibre. By working in the distance regime before the effects of diffraction have caused the light to diverge significantly (the near-field regime), objects can be imaged with a resolution far below the wavelength of the light.

Numerous scanning probe techniques exist [23]. These techniques rely on: the measurement of current; van der Waals, chemical, or magnetic forces; capacitance; phonons or photons when a probe is brought close to a surface and then scanned across it. We focus on three techniques that are widely used and illustrate the most salient aspects of scanning probe microscopy: scanning tunnelling microscopy, atomic force microscopy (AFM) and near-field scanning optical microscopy. All of these techniques are extremely versatile and can be operated under ultrahigh vacuum (UHV), at atmospheric pressure and even in solution [24].

2.4.1   Scanning tunnelling microscopy (STM)

The basic principle of scanning tunnelling microscopy is presented in Fig. 2.3. A sharp conductive tip, usually a W or Pt/Ir wire, is brought within a few nanometres or less of a conducting surface. A voltage difference is then applied between the tip and the surface. A measurement of the current at constant voltage (constant voltage imaging mode) or of the voltage at constant current (constant current imaging mode) is then made while the tip is scanned across the surface. As always, we set up a co-ordinate system in which the x and y axes lie in the plane of the surface and the z axis is directed away from the surface. Measurements of these kinds lead to the images found in Figs 2.4–2.6.

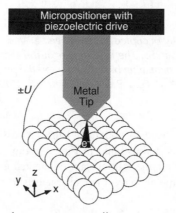

Figure 2.3   *Schematic drawing of a scanning tunnelling microscope tip interacting with a surface.*

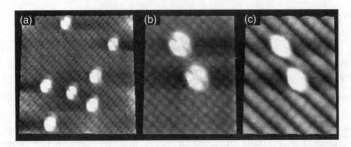

Figure 2.4 *An STM image of occupied states on a Si(100)–(2 × 1) surface nearly completely covered with adsorbed H atoms. The uncapped Si dangling bonds (sites where H is not adsorbed) appear as lobes above the plane of the H-terminated sites. The rows of the (2 × 1) reconstruction are clearly visible in the H-terminated regions. Reproduced from J. J. Boland, Phys. Rev. Lett. 65 (1990) 3325. © 1990, with permission from the American Physical Society.*

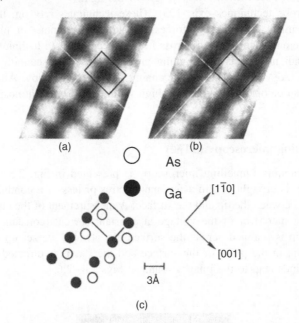

Figure 2.5 *Constant current STM images of the clean GaAs(110) surface. (a) The normally unoccupied states imaged at V = +1.9 V. (b) The normally occupied states imaged at V = −1.9 V. (c) Schematic representation of the positions of the Ga (•) and As (o) atoms. The rectangle is at the same position in (a), (b) and (c). This is an unusual example of chemically specific imaging based simply on the polarity of the tip. Reproduced from R. M. Feenstra, J. A. Stroscio, J. Tersoff and A. P. Fein, Phys. Rev. Lett. 58 (1987) 1192. ©1987, with permission from the American Physical Society.*

To interpret the STM images we need to understand the processes that control the flow of electrons between the tip and the surface. This is illustrated in Fig. 2.7. Fig. 2.7(a) represents a situation in which two metals are brought close to one another but are not connected electrically. The Fermi energies of the left-hand, E_F^L, and right-hand, E_F^R, metals have their characteristic values and as before (here the factor of e is included in the work function)

$$\Phi = E_{vac} - E_F. \qquad (2.4.1)$$

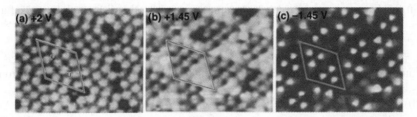

Figure 2.6 *Constant current STM images of the Si(111)–(7 × 7) surface. Notice how the apparent surface structure changes with voltage even though the surface atom positions do not change. This illustrates that STM images electronic states (chosen by the voltage) and not atoms directly. Reproduced from R. J. Hamers, R. M. Tromp and J. E. Demuth, Phys. Rev. Lett. 56 (1986) 1972. © 1986, with permission from the American Physical Society.*

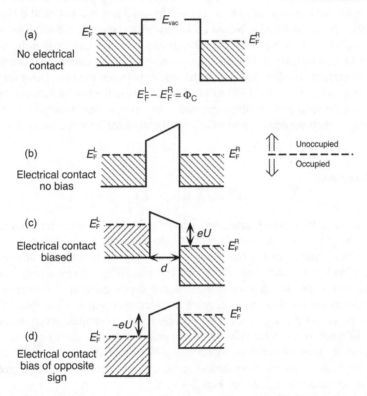

Figure 2.7 *Illustrations of the Fermi and vacuum level positions for two metals separated by distance d. (a) Isolated metals. (b) After electrical contact, in the absence of an applied bias. (c) Biasing shifts the relative positions of the Fermi levels and makes available unoccupied states in an energy window eU into which electrons can tunnel. (d) The direction of tunnelling is switched compared to the previous case simply by changing the sign of the applied bias. Adapted from J. Tersoff, N. D. Lang, Theory of scanning tunneling microscopy, in Scanning Tunneling Microscopy (Eds J. A. Stroscio, W. J. Kaiser), Academic Press, Boston, 1993, p. 1. © 1993, with permission from Academic Press.*

The offset in the Fermi energies (contact potential) is equal to the difference in work functions of the two metals

$$\Phi_c = E_F^L - E_F^R. \tag{2.4.2}$$

The solid line in the drawing represents the potential barrier that separates the electrons of the two metals.

In Fig. 2.7(b) the metals are brought closer together and are connected electrically. The Fermi energies of the two metals line up and the vacuum level shifts. The potential between the two metals is no longer constant; therefore, an electric field exists in the vacuum between the metals. Since the two electrodes are both metals, all states up to E_F are full and those above it are empty (ignoring thermal effects). Therefore, no current can flow between the two metals because electrons must always flow from occupied states to empty states, and in Fig. 2.7(b) the alignment of the Fermi levels means that the occupied states of the metals are also aligned.

Figures 2.7(c) and (d) demonstrate that we can not only turn on a current between the two metals but also that we can *control the direction of flow* of current by adjusting the potential difference between the two. In Fig. 2.7(c) the right-hand metal is biased positively with respect to the left-hand metal. This lowers E_F^R by an energy eU, where e is the elemental charge and U is the potential difference. The occupied states of the left-hand metal now lie at the same energy as some of the unoccupied states of the right-hand metal. Classically, no current can flow because of the barrier between the two. However, if the two metals are brought sufficiently close, the wavefunctions of the electrons in the two metals overlap and by the laws of quantum mechanics, tunnelling of electrons can occur. In a simple one-dimensional approximation [25], the tunnelling current, I, then depends exponentially on the distance between the two metals according to

$$I \propto e^{-2\kappa d} \tag{2.4.3}$$

where d is the distance and

$$\kappa^2 = \frac{2m}{\hbar^2}(eU_B - E). \tag{2.4.4}$$

E is the energy of the state from which tunnelling occurs and eU_B is the barrier height, which is approximately the vacuum level. κ is on the order of 1 Å^{-1}, hence, a change in separation of just 1 Å leads to an order of magnitude change in the tunnelling current. Typical tunnelling currents are on the order of nanoamps or less and the extreme sensitivity of the tunnelling current on distance translates into sub-angstrom spatial resolution. In addition to the distance dependence, it is important to remember the requirement that electrons always flow from occupied to unoccupied states. Therefore, STM images always represent a convolution of the density of states of occupied and unoccupied electronic states between the tip and the surface. In other words, *STM does not image atoms, it images electronic states*. However, since the density of electronic states is correlated with the positions of the nuclei, STM images are always correlated with the positions of atoms. Nonetheless, great care must be taken in interpreting STM images, as we shall see as we examine the images in Figs 2.4–2.6.

By comparing Figs 2.7(c) and 2.7(d), we see that the types of states that are imaged depend on the sign of the applied voltage. Electrons always flow toward positive voltage and from occupied states to empty states. Therefore we can control the direction of current flow simply by appropriately setting the voltage. Let us assume that the right-hand metal is the tip and the left-hand metal in Fig. 2.7 is the sample. Control of the voltage not only changes the direction of current flow, but it also changes whether the tip images the occupied or the unoccupied states of the sample.

STM images result from the convolution of tip and sample electronic structure as well as the distance between them. Unfortunately, the bumps in an image are not labelled with atomic symbols. For instance, an oxygen atom does not always look the same. It may appear as a protrusion at one voltage and as a depression at another. On one substrate it may be imaged at one voltage whereas a different voltage is required on a different substrate. The lack of chemical specificity in STM images represents one difficulty with the technique. Under special circumstances images can be associated with specific atoms, as in Fig. 2.6 and some other cases [26, 27]. Functional group identification has also been possible in several instances [28, 29].

2.4.2 Scanning tunnelling spectroscopy (STS)

The scanning tunnelling microscope has the ability to deliver more information than just topography. Equations (2.4.3) and (2.4.4) show that the image depends on the voltage on the tip and not just the topography. Control of the voltage determines the electronic states from which tunnelling occurs. Therefore, the STM can be used to measure the electronic structure with atomic resolution. This mode of data acquisition is known as scanning tunnelling spectroscopy (STS) [30].

The tunnelling current at bias voltage U can be shown [30] to depend on the surface density of states at energy E, $\rho_s(E)$, and the transmission probability through the barrier, $D(E)$, according to

$$I \propto \int_0^{eU} \rho_s(E)D(E) \, dE. \tag{2.4.5}$$

The differential conductance is given by

$$\sigma = \frac{dI}{dU}. \tag{2.4.6}$$

Both σ and the total conductivity I/U depend to first order on $D(E)$. Therefore, the normalized differential conductance depends only on the density of states and the tunnelling voltage

$$\frac{dI/dU}{I/U} = \frac{\rho_s(eU)}{(1/eU)\int_0^{eU} \rho_s(eU) \, dE}. \tag{2.4.7}$$

Simultaneous measurements of I versus U and position on the surface are sometimes called current imaging tunnelling spectroscopy (CITS) and provide images of the surface as well as the local electronic states at each position in the scan. This is illustrated in Fig. 2.8.

Figure 2.8 demonstrates the ability of STS to perform atom-resolved spectroscopy. The electronic structure of a Si(111)–(7 × 7) surface depends on which atom is probed. Furthermore, as the right-hand side of Fig. 2.8 shows, STS has the ability to identify how the chemisorption of an atom changes the electronic structure of the surface.

2.4.3 Atomic force microscopy (AFM)

Many materials are not conductors and, hence, cannot be imaged by STM. Soft interfaces and biological molecules may be sensitive to electron bombardment, which complicates STM analysis. These are some of the motivations to seek alternative approaches to SPM. These efforts were culminated in 1986 with the development of atomic force microscopy (AFM) by Binnig, Quate and Gerber [31].

If a tip is brought near a surface, it experiences attractive or repulsive interactions with the surface. These interactions may be of the van der Waals type or of a chemical, magnetic or electrostatic nature, for example. These forces and their ranges are outlined in Table 2.2. The tip is attached to a non-rigid cantilever and the force experienced by the tip causes the cantilever to bend. The deflection of the cantilever can be detected by reflection of a laser beam off the cantilever. AFM can image conducting as well as

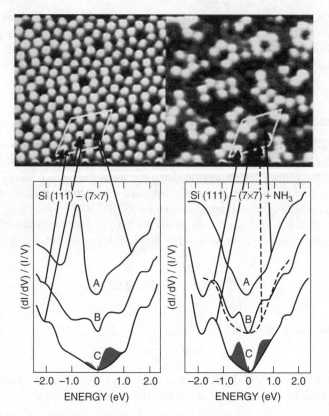

Figure 2.8   *Left-hand side: Topography of the unoccupied states of the clean (7 × 7) surface (top panels) and atom resolved tunnelling spectra (bottom panels). The curves represent spectra acquired over different sites in the reconstructed surface (Curve A: restatom, Curve B: corner adatom, Curve C: middle adatom). Negative energies correspond to occupied states, positive to empty states. Right-hand side: Same types of images and spectra obtained after exposure of a Si(111)–(7 × 7) surface to NH_3. The different sites exhibit different reactivities with respect to NH_3 adsorption with the restatoms being the most reactive and the middle adatoms being the least reactive. Reproduced from R. Becker and R. Wolkow, Semiconductor surfaces: Silicon, in* Scanning Tunnelling Microscopy *(Eds J. A. Stroscio, W. J. Kaiser), Academic Press, Boston, 1993, p. 193. © 1993, with permission from Academic Press.*

Table 2.2   *Interaction forces appropriate to scanning force microscopy and their ranges. Values taken from Takano et al. [32]*

Force	Range (nm)
Electrostatic	100
Double layer in electrolyte	100
van der Waals	10
Surface-induced solvent ordering	5
Hydrogen bonding	0.2
Contact	0.1

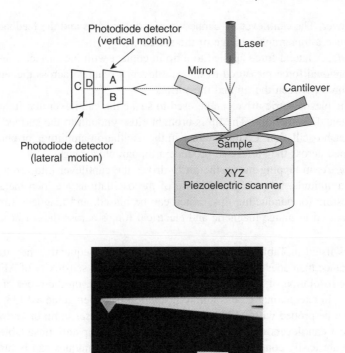

Figure 2.9 *Principal components for an optical lever type AFM. Detection of the reflected laser beam with a quadrant, position sensitive photodiode facilitates the simultaneous detection of bending and torsion of the cantilever. A scanning electron micrograph of a typical AFM cantilever and tip is shown in the lower panel. Reproduced from H. Takano, J. R. Kenseth, S.-S. Wong, J. C. O'Brien, M. D. Porter,* Chem. Rev. *99 (1999) 2845. © 1999, with permission from the American Chemical Society.*

insulating surfaces. In addition, it can harness a number of different forces to provide for the deflection of the cantilever. Just as in STM, a piezoelectric micropositioner and some type of electrical feedback mechanism for the micropositioner are integral parts of the microscope design.

As the piezoelectric drive moves a cantilever of the type shown in Fig. 2.9 in free space, the motion of the tip on the cantilever exactly follows the motion of the drive. As the tip approaches a surface it experiences any one of a number of forces. The action of a force on the tip leads to a displacement of the tip in addition to the motion of the piezoelectric drive, Δz. If the force constant, k_N, of the cantilever is known, the displacement can be translated into a force according to

$$F_N = k_N \Delta z. \tag{2.4.8}$$

The force curve is obtained by a measurement of the force as a function of the distance of the tip from the surface. The sensitivity of AFM is on the order of $10^{-13} - 10^{-8}$ N, which allows for direct measurements of even van der Waals interactions.

There are several modes in which images can be acquired.

- *Contact mode.* The tip is advanced toward the surface until physical contact is made. The feedback circuitry is then set to maintain a constant imaging parameter. The imaging parameter is often taken to be the force, which can be measured according to Eq. (2.4.8) by the deflection of a laser beam off the back of the cantilever. The cantilever is scanned in two dimensions and the feedback is simultaneously measured to produce a topographic image of the surface.

- *Friction force mode.* A lateral force applied to a tip in contact with the surface causes a twisting of the cantilever. The torsional force measured as the cantilever is scanned across the surface can be related to the frictional force between the tip and the surface.
- *Tapping mode.* The piezoelectric drive can be used to shake the cantilever at a frequency resonant with one of its fundamental oscillations. The tip is brought close enough to the surface such that it touches at the bottom of each oscillation. Changes either in the oscillation amplitude or phase can be measured as the tip is scanned across the surface to produce a topographic image.
- *Non-contact mode.* As in tapping mode, the piezo drives the cantilever into resonant or near-resonant oscillations. The amplitude, frequency and phase of the oscillations are then measured. This mode is particularly interesting for conducting tips, which can be biased, and magnetic tips. This transduction mode can then be used to image magnetic and electrical forces across the surface.

Many of the forces listed in Table 2.2 are quite long range. Consequently, they tend to fall off much more slowly with distance than does tunnelling current. Therefore, the sensitivity of AFM is much less than STM. As a result, the resolution of AFM is generally less than STM, on the order of several nanometres or tens of nanometres. In exceptional cases, atomic resolution has been achieved [33, 34].

Chemical forces can be probed with AFM as well. For these purposes, a tip of known chemical composition is required. Most cantilevers are made of Si_3N_4. Bare Si_3N_4 tips are unsuitable for chemical force sensing because they are easily contaminated. However, several techniques can be used to transform the tip into one with a known chemical composition. The tip can be coated with Au onto which molecules or colloidal particles can be transferred. Alkanethiol molecules and the techniques of self-assembly, treated in Chapter 5, can be used to transfer molecules terminated with a variety of functional groups to the tip. Attachment of silica spheres can also be accomplished. Silica is a versatile substrate onto which a variety of organic and biomolecules can be tethered. Takano et al. [32] have reviewed the extensive literature on tip modification.

Lieber and co-workers [35] have pursued a promising variant of tip modification. They have grown carbon nanotubes on a cantilever through a catalytic process. A nanotube tip provides two advantages for AFM. The first is the known chemical composition of the tip. The second is that carbon nanotubes have high aspect ratios. This can allow for increased resolution, particularly in cases in which deep narrow crevices are present on the surface. Conventional AFM tips have a radius of 30–50 nm, a nanotube tip with a radius of only 9 nm expresses much lower convolution effects.

Once a tip has been modified in a controlled manner, the interactions of the probe molecule with surfaces or molecules immobilized on a surface can be studied. This opens up a huge range of possibilities, particularly in biochemical and biomedical applications. Since the forces measured with a chemically modified tip can be chemically specific, imaging under the influence of chemical forces can be used for compositional mapping of the substrate. By grafting one protein onto the tip and another onto a substrate it is possible to study in great detail the interaction of single binding pairs. Similarly, DNA and antigen-antibody interactions can be studied [32].

2.4.4   Near-field scanning optical microscopy (NSOM)

Optical microscopy has a long history of providing insight into chemical, physical and biological phenomena. There is, as well, a familiarity and comfort with visual data since it is a human response to feel as though we understand something when we can see it – sometimes, no matter what we see. Hence, wide-ranging efforts have pushed the boundaries of optical microscopy to ever-greater resolution. Conventional far field microscopy is limited to a resolution of roughly $\lambda/2$, where λ is the wavelength of the light used for imaging. This resolution limit can be exceeded by over an order of magnitude by using so-called near field techniques [36]. Furthermore, the use of pulsed light sources and the relative ease of performing studies as a function of wavelength mean that optical microscopic techniques lend themselves readily to time-resolved and spectrally-resolved studies that can investigate dynamical behaviour as well as provide chemical identification. A range of coherent nonlinear optical imaging techinques [37] can lead to single molecule detection sensitivities, and breaking the diffraction limit. These techniques, such as stimulated emission depletion spectroscopy (STED), are grouped together under the term RESLOFT, which stands for reversible saturable optically linear fluorescence transitions [38].

In a series of papers written between 1928 and 1932, Synge [39–41] proposed how an aperture smaller than the wavelength of light could be used to obtain resolution far smaller than the wavelength. When light passes through such an aperture it is diffracted and the light propagates away from the aperture in a highly divergent fashion. Not only does the light diverge, but also diffraction leads to an undulating transverse intensity distribution. In the far field, a distance large compared to the wavelength of light and the size of the aperture, diffraction leads to the familiar Airy disk pattern in intensity. In the near field region, not only is the light divergent, but also the intensity distribution is evolving, as shown in Fig. 2.10. By working close enough to the aperture, resolution close to the size of the aperture and far below the wavelength of the light can be obtained.

The proposals of Synge were far too technically demanding to be experimentally realized in the 1930s. It was not until 1972 that the feasibility of subwavelength resolution with microwave radiation was demonstrated [42]. The advent of a device based on this effect capable of routine imaging, however, had to await several other discoveries. The widespread availability of high quality optical fibres, visible lasers and the AFM greatly aided the development of practical instruments for near field scanning optical microscopy (NSOM, also known as SNOM). Development followed from advances made by Pohl and co-workers at IBM Zürich [43] and Lewis et al. at Cornell University [44].

Figure 2.11 demonstrates a variety of operating modes that can be employed to perform NSOM experiments. There are several features that most NSOM instruments have in common. They require a high brightness light source. A laser is particularly well suited for this because of its monochromaticity, low

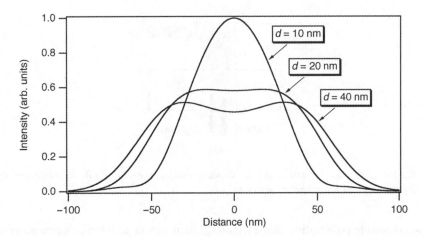

Figure 2.10 *Near-field intensity distributions are shown for 532 nm light that has passed through an aperture with a diameter of 100 nm at distances d from the aperture of 10, 20 and 40 nm. The distributions are normalized such that they all have the same integrated intensity.*

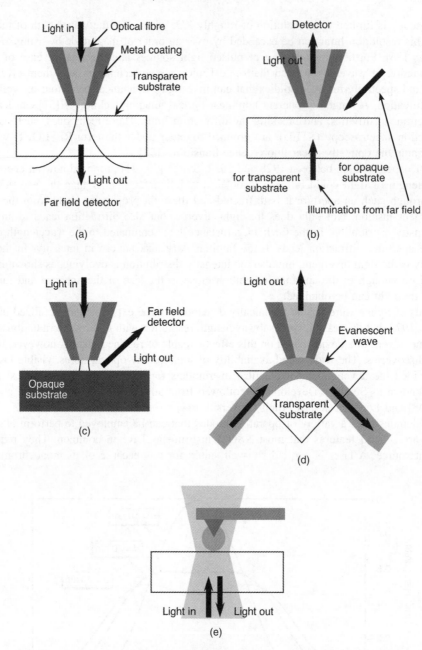

Figure 2.11 *NSOM data can be collected in various modes including (a) illumination; (b) collection; (c) reflection; (d) photon tunnelling; and (e) apertureless.*

divergence and favourable polarization characteristics. Lasers can in addition be operated either with continuous wave or pulsed radiation. The properties of laser light are particularly well suited for coupling into an optical fibre. The optical fibre plays two roles in NSOM. First, the end of an optical fibre can readily be tapered to minuscule dimensions. After tapering, the fibre is then coated with a metallic film, generally either Al or Ag. In this way, apertures with diameters of ∼100 nm can be made routinely. The optical fibre also acts as a light pipe that transports the photons from the laser to the aperture. Finally, the fibre can be easily incorporated into an AFM head. The AFM head is used to regulate the aperture-to-sample separation. Depending on the instrument, either the fibre or the substrate can be scanned to allow for 2D imaging. The instrumentation required to scan the substrate and to obtain NSOM images is similar to that used in AFM. Indeed, often the microscope is constructed such that NSOM and AFM images are collected simultaneously.

There are several principles that can be used to collect NSOM data:

- *Illumination mode.* The tip is used to illuminate the sample. A transparent substrate is used so that photons can be collected below the sample. The light may either be transmitted photons from the tip or fluorescence excited by the tip illumination.
- *Collection mode.* In collection mode, the fibre is used to gather photons and transport them to the detector. The sample is illuminated from the far field either from below or from the side.
- *Reflection mode.* The sample is irradiated through the optical fibre and the photons reflected by the sample are collected in the far field. This is useful for opaque samples. Using polarization tricks or wavelength separation, the aperture used for illumination can also be used to collect photons.
- *Photon tunnelling microscopy.* Total internal reflection can be used to pass light through a transparent sample. When the light reflects from the interface, an evanescent wave penetrates up to several hundred nanometres beyond the interface. If an aperture is brought into contact with the evanescent wave it propagates into the fibre and can then be transported to the detector. This technique is sometimes called scanning tunnelling optical microscopy (STOM) or photon scanning tunnelling microscopy (PSTM).
- *Apertureless near-field microscopy.* If an AFM tip is brought close to a surface and is irradiated by a far-field light source, a field enhancement occurs in a localized region between the tip and the sample. This can be used to image the surface if sophisticated filtering techniques are used to extract the signal from the large background of conventionally scattered radiation. Alternatively, a fluorescent molecule attached to the end of the tip can be excited by a laser and its fluorescence can be used to probe the sample.

Because of the small aperture diameter, typically $80-100\,nm$, the optical fibre only delivers $10^{-4} - 10^{-7}$ of the light coupled into it onto the sample. This corresponds to a few tens of nanowatts of light for typical input powers. While this may appear to be minute, it still amounts to $\sim 100\,W\,cm^{-2}$. The ultimate resolution of NSOM is roughly $12\,nm$. In practice, however, a resolution of $50\,nm$ is obtained. A number of alternative implementations of NSOM are being investigated that may be able to significantly improve the resolution [45].

Because NSOM images photons, it allows not only for nanoscale imaging but also for spectroscopic studies at high resolution without physical contact and without electron bombardment. Thus NSOM has the potential to measure the position, orientation and chemical identity of adsorbates on not only solid but also liquid surfaces. Betzig and Chichester demonstrated single molecule detection in 1993 [46]. With pulsed lasers, extremely high temporal resolution can be attained as well. This facilitates dynamical studies of molecular motion such as rotational reorientation and diffusion. The measurement of the time decay of fluorescence provides direct access to excited state lifetimes.

2.5   Low energy electron diffraction (LEED)

The most standard (and potentially most quantitative) method of surface structure determination is provided by the diffraction of low energy electrons [2] as illustrated in Fig. 2.12. In 1927, Davidson and Germer observed the angular distributions of electrons scattered from Ni and explained their data in terms of

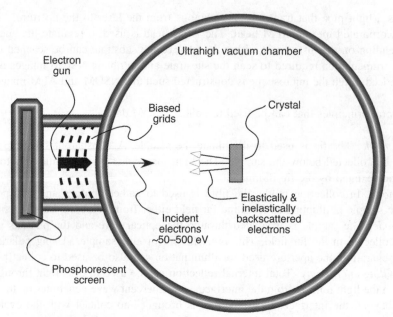

Figure 2.12   *Schematic drawing of a LEED chamber.*

the diffraction of the electrons from crystallites in their sample. This was one of the first experimental demonstrations of the wavelike behaviour of massive particles as predicted by de Broglie in 1921:

$$\lambda = \frac{h}{m_e v} \tag{2.5.1}$$

in which wavelength λ is inversely proportional to the linear momentum (product of mass and velocity). Davidson, who had begun his work on low energy electron diffraction in 1922, and G.P. Thomson, who studied high-energy electron diffraction from thin films, were awarded the 1937 Nobel Prize in Physics "for their discovery of the interference phenomena arising when crystals are exposed to electronic beams".

Electron scattering is surface sensitive if the electrons have the right energy. This is illustrated by the data in Fig. 2.13. The inelastic mean free path of electrons in a solid depends on the energy of the electrons in a manner that does not depend too strongly on the chemical identity of the solid [47]. Therefore, the curve in Fig. 2.13 is known as the universal curve and is given by the empirical expression

$$\lambda_M = c_0 \, E^{-2} + c_1 \, (lE)^{1/2}. \tag{2.5.2}$$

The constants depend on the class of material with $c_0 = 538$ and $c_1 = 0.41$ for elemental solids and $c_0 = 2170$ and $c_1 = 0.72$ for inorganic compounds. Here the inelastic mean free path λ_M (IMFP) is given in monolayers, E is the kinetic energy of the electron in eV, and l is the monolayer thickness in nanometres (typically 2–3 Å) calculated from

$$l^3 = \left(M / \rho n N_A\right) \times 10^{24} \tag{2.5.3}$$

with M = molar mass in g mol^{-1}, n the number of atoms in the molecule, and ρ the bulk density in kg m^{-3}. Multiplying λ_M by l gives the IMFP in nanometres. The universal curve is only a rough approximation. As suggested by Powell and implemented computationally by Penn [48], the IMFP (or the related electron stopping power SP) can be calculated from optical data. Powell and Jablonski provide a comprehensive discussion and collection of accurate IMPF values for electron energies from 50–10 000 eV [49]. Accurate

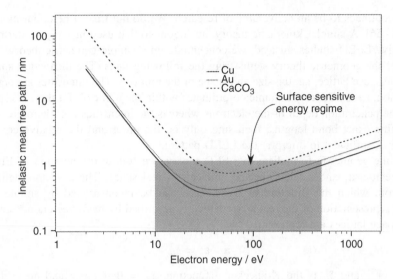

Figure 2.13 *The universal curve of electron mean free path* λ *in solid matter. Calculated from the data of Seah and Dench [45].*

values of SPs have now been calculated for over 40 elemental solids [50, 51]. For any material, an estimate of λ_M in Å good to about 10% in the range of 50–2000 eV is given by

$$\lambda_M = \frac{E}{E_p^2 \left[\beta^* \ln(\gamma E) - C/E + D^*/E^2 \right]}. \qquad (2.5.4)$$

The definitions of the parameters can be found in the original work of Tanuma et al. [52]. Here the incident electron energy E is given in eV, and $E_p = 28.8(N_v \rho/M)^{1/2}$ is the free-electron plasmon energy in eV (N_v is the number of valence electrons per atom or molecule and ρ is the density in g cm^{-3}). This emphasizes the relationship of the electron IMFP to not only the electron energy, but also the electron density of the material, and the mechanism by which inelastic scattering occurs. A web-based database of IMFP values is maintained by NIST.

The proportion of elastic scattering and the scattering mechanisms of electrons from surfaces depends strongly on the incident energy [53]. At very low energies, scattering of electrons is very weak; indeed, for a perfect crystal at 0 K an electron at the Fermi energy has an infinite mean free path. However, as the energy of the electron increases, different inelastic scattering events become possible. A few eV above E_F, phonons (collective lattice vibrations) can be excited. In the range above about 10–20 eV, plasmons (collective electron vibrations) can be excited. For incident energies between roughly 20 and 500 eV, electrons interact strongly with matter and their mean free path drops to ~5–10 Å. LEED measures only elastically scattered electrons; therefore, most of the electrons detected in LEED are scattered from the surface of the sample. Layers deeper than 3 or 4 atoms below the surface play virtually no role in the detected signal. To a first approximation, the patterns observed in LEED can be understood by considering only the symmetry of the surface of the sample.

The same property of electron scattering that makes it surface sensitive also makes diffraction of low energy electrons at a surface more complicated than x-ray scattering. Low energy electrons interact strongly with matter. This means that they have a short inelastic mean free path. This is good for surface sensitivity since no more than 3–4 layers contribute to the scattering. On the other hand, a complete description of the positions and intensities of diffraction spots as a function of the incident energy requires a dynamical theory that accounts for multiple scattering from all of the layers that contribute to scattering. Development of a full theory continues to be an active area of research. We do not treat it here, but more can be found elsewhere [2, 54–56]. A simpler kinematic theory, analogous to that used in x-ray scattering, is of limited use for quantitative LEED studies. Instead, we concentrate on a simple geometric theory.

The success of the geometric theory stems from the following facts. The diffraction spot positions are determined by the space lattice, i.e. the size and shape of the unit cell. The intensities are determined by the diffraction function, i.e. by the exact atomic co-ordinates within the unit cell. Further, the spot positions do not depend on the penetration depth of the electrons whereas the intensities do. Therefore, if we disregard trying to determine exact bond lengths, focussing only on the order and the relative coverage, then the geometric theory is sufficient to interpret the LEED pattern.

In presenting the geometric formalism of LEED theory, we follow the notation of Ertl and Küppers [2]. The basis vectors $\mathbf{a}_1$ and $\mathbf{a}_2$ describe the unit cell in real space. These vectors define the smallest parallelogram from which the structure of the surface can be reconstructed by simple translations. A reciprocal space representation of the real space lattice is described by basis vectors $\mathbf{a}_1{}^*$ and $\mathbf{a}_2{}^*$. The real and reciprocal space lattices are represented by

$$\mathbf{a}_i \cdot \mathbf{a}_j^* = \delta_{ij} \tag{2.5.5}$$

where $i, j = 1$ or 2 and δ_{ij} is the Kronecker δ function. $\delta_{ij} = 0$ if $i \neq j$ and $\delta_{ij} = 1$ if $i = j$. This means that $\mathbf{a}_i^* \perp \mathbf{a}_j$ for $i \neq j$. Introducing γ and γ^*, the angles between ($\mathbf{a}_1$ and $\mathbf{a}_2$) and ($\mathbf{a}_1{}^*$ and $\mathbf{a}_2{}^*$), respectively, we have

$$\mathbf{a}_1^* = \frac{1}{\mathbf{a}_1 \sin \gamma} \tag{2.5.6}$$

$$\mathbf{a}_2^* = \frac{1}{\mathbf{a}_2 \sin \gamma} \tag{2.5.7}$$

$$\sin \gamma = \sin \gamma^*. \tag{2.5.8}$$

The inverse relationship between real and reciprocal space means that a long vector in real space corresponds to a short vector in reciprocal space. Equation (2.5.8) follows from the relationship $\gamma^* = \pi - \gamma$.

The need for the reciprocal space description is made evident by Fig. 2.14. This figure shows that an image of the diffracted electrons corresponds to a reciprocal space image of the lattice from which the electrons diffracted. Hence, by uncovering the relationship between a reciprocal space image and the real space lattice, we can use LEED patterns to investigate the surface structure.

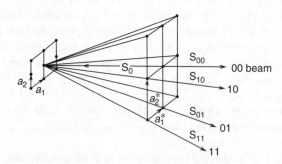

Figure 2.14   *The principle of diffraction pattern formation in a LEED experiment. The incident electron beam approaches along s_0. The specular beam exits along s_{00}. Reproduced from G. Ertl, J. Küppers,* Low Energy Electrons and Surface Chemistry, *2nd ed., VCH, Weinheim. © 1985, with permission from John Wiley & Sons, Ltd.*

Advanced Topic: LEED structure determination

We represent a surface overlayer by the basis vectors $\mathbf{b}_1$ and $\mathbf{b}_2$ and the corresponding reciprocal lattice vectors are given stars. The substrate and overlayer lattices are related by

$$\mathbf{b}_1 = m_{11}\mathbf{a}_1 + m_{12}\mathbf{a}_2 \tag{2.5.9}$$

$$\mathbf{b}_2 = m_{21}\mathbf{a}_1 + m_{22}\mathbf{a}_2 \tag{2.5.10}$$

$$\mathbf{b}_1^* = m_{11}^*\mathbf{a}_1^* + m_{12}^*\mathbf{a}_2^* \tag{2.5.11}$$

$$\mathbf{b}_2^* = m_{21}^*\mathbf{a}_1^* + m_{22}^*\mathbf{a}_2^*. \tag{2.5.12}$$

This can be written in matrix notation as

$$\mathbf{b} = \mathbf{M} \cdot \mathbf{a} \rightarrow \begin{pmatrix} \mathbf{b}_1 \\ \mathbf{b}_2 \end{pmatrix} = \begin{pmatrix} m_{11} & m_{12} \\ m_{21} & m_{22} \end{pmatrix} \cdot \begin{pmatrix} \mathbf{a}_1 \\ \mathbf{a}_2 \end{pmatrix}. \tag{2.5.13}$$

A similar relationship holds for the reciprocal space representation

$$\mathbf{b}^* = \mathbf{M}^* \cdot \mathbf{a}^* \rightarrow \begin{pmatrix} \mathbf{b}_1^* \\ \mathbf{b}_2^* \end{pmatrix} = \begin{pmatrix} m_{11}^* & m_{12}^* \\ m_{21}^* & m_{22}^* \end{pmatrix} \cdot \begin{pmatrix} \mathbf{a}_1^* \\ \mathbf{a}_2^* \end{pmatrix}. \tag{2.5.14}$$

It can be shown that $\mathbf{M}^*$ is related to $\mathbf{M}$ (and vice versa) by

$$\mathbf{M} = \frac{1}{\det \mathbf{M}^*} \begin{pmatrix} m_{22}^* & -m_{12}^* \\ -m_{21}^* & m_{11}^* \end{pmatrix} \tag{2.5.15}$$

$$\mathbf{M}^* = \frac{1}{\det \mathbf{M}} \begin{pmatrix} m_{22} & -m_{12} \\ -m_{21} & m_{11} \end{pmatrix} \tag{2.5.16}$$

The determinant of $\mathbf{M}^*$ is defined as

$$\det \mathbf{M}^* = m_{11}^* m_{22}^* - m_{21}^* m_{12}^*, \tag{2.5.17}$$

and analogously for the determinant of $\mathbf{M}$. Experimentally, the challenge is to determine the elements of $\mathbf{M}$ from the diffraction pattern measured on the LEED screen.

The diffraction condition from a one-dimensional lattice of periodicity a leads to constructive interference at angles φ when

$$a \sin \varphi = n\lambda \tag{2.5.18}$$

for an electron with a wavelength λ incident at normal incidence. n is an integer denoting the diffraction order. The wavelength of the electron is given by the de Broglie relationship, Eq. (2.5.1). The Bragg condition of Eq. (2.5.18) needs to be generalized to two dimensions. This leads to the Laue conditions

$$\mathbf{a}_1 \cdot (\mathbf{s} - \mathbf{s}_0) = h_1 \lambda \tag{2.5.19}$$

$$\mathbf{a}_2 \cdot (\mathbf{s} - \mathbf{s}_0) = h_2 \lambda \tag{2.5.20}$$

where $\mathbf{s}_0$ defines the direction of the incident beam (generally along the surface normal) and $\mathbf{s}$ defines the direction of the diffracted beam intensity maxima. h_1 and h_2 are integers and they are used to identify the diffraction reflexes that appear in the LEED pattern. The specular reflex at (00) is used as the origin and arises from electrons that are elastically scattered without diffraction.

The determination of a real space structure from a reciprocal space image may seem rather esoteric. The relationship between the two is shown in Fig. 2.15. By way of an example of pattern analysis, we will see that the simple geometric theory can lead to a rapid determination of surface structure as well as other properties of the adsorbed layer. An important concept to keep in mind is that while every ordered structure produces one and only one diffraction pattern for a given electron energy, a diffraction pattern can correspond to more than one real space structure. For example, a (4×2) array of occupied sites leads to exactly the same LEED pattern as a (4×2) array of vacancies. In addition, the superposition of symmetry related diffraction patterns from rotational domains of adsorbates leads to the observation of a composite pattern. For example, two (2×1) domains rotated by $90°$ with respect to each other lead to a pattern with apparent (2×2) symmetry. This is the case for most Si(100) surfaces that exhibit the (2×1) reconstruction.

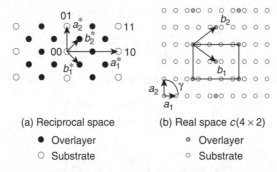

(a) Reciprocal space   (b) Real space $c(4 \times 2)$

● Overlayer   ⊙ Overlayer

○ Substrate   ○ Substrate

Figure 2.15   *Real space and reciprocal space patterns. (a) Reciprocal lattice (LEED pattern) composed of substrate (normal) spots ○ and overlayer (extra) spots ●. (b) Real lattice of the substrate (○) and overlayer (⊙). The solid line delineates the c(4 × 2) cell and the arrows depict the unit vectors.*

To begin a structure determination, we note that the position of the (00) spot does not change with electron energy (λ). This leads to easy identification of the specular reflex, as it is the only reflex that does not move when the electron energy is changed. There is no need to determine the absolute distance between spots. By referencing the positions of overlayer related diffraction spots to those of the substrate, we can strip the analysis of a variety of experimental parameters.

The surface unit cell is the smallest possible cell that reproduces the ordered structure by means of translation alone. For instance, in Fig. 2.15 whereas the $c(4 \times 2)$ cell is a natural selection to relate the overlayer structure to the substrate lattice, the unit vectors $\mathbf{b}_1$ and $\mathbf{b}_2$ actually define a unit cell with a smaller area that is rotated with respect to substrate lattice. There are five types of two-dimensional Bravais lattices from which all ordered surface structures could be built. These are shown in Fig. 2.16.

Two obvious notations emerge to describe the relative symmetry of surface layers. The most general notation uses the full matrix **M**. This notation was proposed by Park and Madden [57]. If the angle between $\mathbf{a}_1$ and $\mathbf{a}_2$ is the same as the angle between $\mathbf{b}_1$ and $\mathbf{b}_2$, the simpler form of notation that we have already encountered above and in Chapter 1 can be used, for example, Si(100)–(2×1). We express the overlayer structure in terms of $n = |\mathbf{b}_1/\mathbf{a}_1|$, $m = |\mathbf{b}_2/\mathbf{a}_2|$ and any angle of rotation, θ, between the two lattices in the form $(n \times m)$ R $\theta°$. If $\theta = 0°$, it is excluded from the notation. In addition a letter "c" must be added if the overlayer corresponds to the centred rectangular lattice of Fig. 2.16(c). Those who are sticklers for detail sometimes add the letter "p" for primitive, but it is usually omitted. Most observed overlayer structures can be expressed in these terms, known as Wood's notation [58].

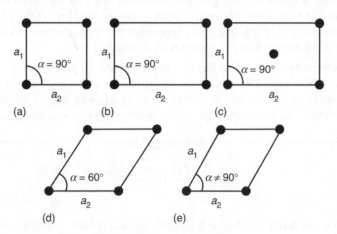

Figure 2.16 *The five types of surface Bravais lattices. (a) square, $a_1 = a_2$, $\alpha = 90°$. (b) primitive rectangular, $a_1 \neq a_2$, $\alpha = 90°$. (c) centred rectangular, $a_1 \neq a_2$, $\alpha = 90°$. (d) hexagonal, $a_1 = a_2$, $\alpha = 60°$. (e) oblique, $a_1 \neq a_2$, $\alpha \neq 90°$.*

If all of the elements of **M** are integers, the overlayer forms a simple structure. A simple structure is commensurate with the substrate, and all of the adsorbates occupy identical adsorption sites. If the elements are rational numbers, the overlayer forms a coincidence lattice, which is an incommensurate structure. If the elements of **M** are irrational, no common periodicity exists between the overlayer and substrate lattices. This structure type is known as an incoherent structure. The overlayer is also incommensurate with the substrate. With increasingly large values of the elements of **M**, the distinction between coincidence lattices and incoherent structures is lost.

A number of other factors affect the appearance of the diffraction pattern, namely the size of the spots and the contrast between the reflex maxima and the surrounding background. Spots are observed in the LEED pattern only if the surface is ordered in two dimensions. Streaking or broadening of spots in one direction is indicative of the loss of order in one dimension. A disordered region results in a diffuse background.

The width of a reflex is related to the monochromaticity of the electron beam and the size of the ordered region. The width of a reflex can be expressed in terms of the angular divergence of the beam, $\delta\varphi$. It can be shown [2] that in the absence of any instrumental broadening,

$$\delta\varphi = \frac{\lambda}{2d \cos\varphi}, \tag{2.5.21}$$

where d is the diameter of the ordered region, e.g. an island. Thus, the larger the ordered region, the sharper the spot in the LEED pattern.

A spread in the electron wavelength affects the pattern in two ways. Equation (2.5.21) shows that $\delta\varphi$ is directly proportional to any uncertainty in λ. In addition, the diffraction pattern represents a reciprocal space image of the surface over the coherence length of the electron beam. If the electron gun were a monochromatic point source, the coherence length would be the size of the incident electron beam. Since it is not, the radiation is not completely in phase across the diameter of the beam. The amount that the radiation can be out of phase and still give coherent scattering puts a limit on the coherence length. For conventional electron beams the coherence length $\leq 10\,\mathrm{nm}$. In other words, LEED is not sensitive to disorder on length scales $> 10\,\mathrm{nm}$, and again it is important to recall that diffraction techniques are proficient at finding order even when it is mixed with disorder.

Increasing surface temperature leads to an increase in the vibrational motion of the surface layer, which results in increased diffuse scattering (increased background), and a decrease in the intensity at the reflex maximum. Similar to x-ray scattering, the decrease in intensity can be described by an effective Debye-Waller factor [59]. The intensity drops exponentially with temperature according to

$$I = I_0 \exp(-2M) \tag{2.5.22}$$

where

$$2M = \frac{12h^2}{mk_B} \left(\frac{\cos\varphi}{\lambda}\right) \frac{T_s}{\theta_D^2}. \tag{2.5.23}$$

In Eq. (2.5.23), θ_D is the surface Debye temperature (assuming that scattering only occurs from the surface layer) and T_s is the surface temperature. Since the inelastic mean free path changes with electron energy, the value of θ_D changes with electron energy, converging on the bulk value for high energies. If the temperature becomes high enough, an overlayer of adsorbates can become delocalized due to diffusion, that is, it can enter a two-dimensional gas phase. Such a transition from an ordered phase to a delocalized phase would be evident in the LEED pattern as it gradually transformed from a sharp pattern to a diffuse pattern.

2.6   Electron spectroscopy

We treat here three types of electron spectroscopy in which an excitation leads to the ejection of an electron from the solid. The energy and possibly angular distribution of the ejected electrons are analyzed. The three most prolific forms of electron spectroscopy are illustrated in Fig. 2.17. X-rays can be used to excite photoemission. This technique is call x-ray photoelectron spectroscopy (XPS) or, especially in the older and analytical literature, electron spectroscopy for chemical analysis (ESCA). Ultraviolet light, especially that of rare gas discharge lamps, can induce photoemission as well: ultraviolet photoelectron spectroscopy (UPS). Auger electron spectroscopy (AES) involves excitation with either x-rays or, more usually, with electrons of similar energy. All of these techniques, as well as LEED and other aspects of the interactions of low energy electrons with surfaces are treated in the classic text by Ertl and Küppers [2].

2.6.1   X-ray photoelectron spectroscopy (XPS)

Deep core electrons have binding energies corresponding to the energies of photons that lie in the x-ray region. When a solid absorbs a photon with an energy in excess of the binding energy of an electron, a photoelectron is emitted and the kinetic energy of the photoelectron is related to the energy of the photon. Deep core electrons do not participate in bonding and their energies are characteristic of the atom from which they originate. To a first approximation, the energy of core electrons does not depend on the

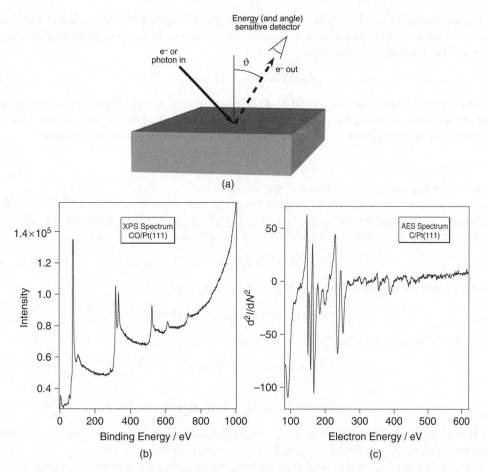

Figure 2.17 *(a) Schematic representation of electron spectroscopy. (b) A sample XPS spectrum of CO/Pt(111). (c) A sample AES spectrum of CO/Pt(111).*

environment of the atom. Therefore, XPS is particularly useful for elemental analysis. Not only can XPS identify the composition of a sample, but it can also be used to determine the composition *quantitatively*. The modern application of XPS owes much of its development to Siegbahn and co-workers in Uppsala, Sweden. Kai Siegbahn, whose father Manne became a Nobel laureate in 1924, was awarded the 1981 Nobel Prize for his contributions to the development of XPS/ESCA.

When the energies of core levels are investigated in more detail it is found that small but easily detected shifts do occur. These shifts, known as chemical shifts, depend on the bonding environment around the atom, in particular, on the oxidation state of the atom. To understand chemical shifts, we first need to understand the relationship between the photon energy, $h\nu$, and the electron binding energy, E_B, which is given by the Einstein equation

$$E_B = h\nu - E_K \tag{2.6.1}$$

where E_K is the electron kinetic energy, and the binding energy is referenced to the vacuum level. Because reproducibly clean surfaces of gold are easily prepared and maintained, the photoemission peaks of Au are conventionally used as a standard to calibrate the energy scale.

The binding energy of an electron is equal to the energy difference between the initial and final states of the atom. That is, the binding energy is equal to the difference in energies between the atom with n electrons and the ion with $n-1$ electrons

$$E_B(k) = E_f(n-1) - E_i(n) \tag{2.6.2}$$

where $E_f(n-1)$ is the final state energy and $E_i(n)$ is the initial state energy. If there were no rearrangement of all of the spectator electrons, the binding energy would exactly equal the negative of the orbital energy of the initial state of the electron $-\varepsilon_k$. This approximation is known as Koopmans' theorem:

$$E_B(k) = -\varepsilon_k \tag{2.6.3}$$

in which the binding energy is referenced to the vacuum level.

Koopmans' theorem is based on an effective one electron approximation in which the initial state wavefunction can be described by a product of the single-particle wavefunction of the electron to be removed, φ_j, and the $n-1$ electron wavefunction of the resulting ion, $\Psi_j(n-1)$,

$$\psi_i(n) = \varphi_j \Psi_j(n-1). \tag{2.6.4}$$

Within this picture the final state wavefunction of the ion is simply

$$\psi_f(n-1) = \Psi_j(n-1). \quad \text{(one electron)} \tag{2.6.5}$$

Consequently the energies required for Eq. (2.6.2) are

$$E_i(n) = \langle \psi_i(n)|H|\psi_i(n)\rangle \tag{2.6.6}$$

and

$$E_f(n-1) = \langle \Psi_j(n-1)|H|\Psi_j(n-1)\rangle \tag{2.6.7}$$

Substitution of Eqs (2.6.6) and (2.6.7) into Eq. (2.6.2) demonstrates that only one peak is expected in the photoelectron spectrum since both $E_i(n)$ and $E_f(n-1)$ are single valued.

The electrons are not frozen. The final state achieved by removing one electron corresponds to an ionic state in which a hole exists in place of the ejected photoelectron. This does not correspond to the ground state of the final ionic state. The remaining electrons relax and, thereby, lower the energy of the final state. This type of relaxation occurs regardless of the phase (gas, liquid, or solid) in which the atom exists. If the atom is located in a solid, then in addition to this atomic relaxation, there can be extra-atomic relaxation. In other words, not only the electrons in the ionized atom can relax but also the electrons on neighbouring atoms can relax in response to the ionization event.

These relaxation phenomena lead us to reconsider our description of the final state wavefunction, which we now write in terms of the eigenstates of the ion,

$$\psi_f(n-1) = u_k \Phi_{jl}(n-1) \tag{2.6.8}$$

where u_k is the wavefunction of the excited electron with momentum k and Φ_{jl} are the wavefunctions of the l ionic states that have a hole in the jth orbital. The relaxation of the ion core in the final state means that the wavefunction $\Psi_j(n-1)$ is not an eigenstate of the ion. Therefore, there is not a unique $\Psi_j(n-1)$ that corresponds to $\Phi_{jl}(n-1)$. Instead, the $\Psi_j(n-1)$ must be projected onto the true eigenstates of the ion. This results in the formation of one or more possible final state wavefunctions such that the final state energy required for Eq. (2.6.2) is

$$E_{fl}(n-1) = \langle \Phi_{jl}(n-1)|H|\Phi_{jl}(n-1)\rangle, \tag{2.6.9}$$

which has between 1 and l solutions. That there is more than one solution to Eq. (2.6.9) means that more than one peak appears in the spectrum.

The one electron or Hartree-Fock picture of Eq. (2.6.3) also neglects relativistic and electron correlation effects. Both of these tend to increase the electron binding energy; hence, a more accurate expression of the binding energy is

$$E_B(k) = -\varepsilon_k - \delta\varepsilon_{\text{relax}} + \delta\varepsilon_{\text{rel}} + \delta\varepsilon_{\text{corr}}. \qquad (2.6.10)$$

Recall that the energy of an electronic state is determined not only by the electronic configuration but also by angular momentum coupling. In the LS coupling scheme, the total angular momentum is given by the vector sum $j = l + s$. Any state with an orbital angular momentum $l > 0$ and one unpaired electron is split into two states, a doublet, corresponding to $j = l \pm \frac{1}{2}$. More complex multiplet splittings in the initial state arise from states with higher total spins. Two conventions are encountered for the naming of electronic states. The first is the familiar convention for electronic states nL_j (n = principle quantum number and $L = s, p, d, f, \ldots$ corresponding to $l = 0, 1, 2, 3, \ldots$). In x-ray spectroscopy, a nomenclature based on the shell is commonly used. The shell derives its name from n according to K, L, M, N, ... for $n = 1, 2, 3, 4, \ldots$. A subscript further designates the subshell. The numbering starts at 1 for the lowest (l,j) state and continues in unit steps up to the highest (l,j) state. Thus the shell corresponding to $l = 2$ has levels L_1, L_2 and L_3 corresponding to $2s$, $2p_{1/2}$, $2p_{3/2}$.

From the above discussion it is clear that both initial state and final state effects influence the binding energy. Initial state effects are caused by chemical bonding, which influences the electronic configuration in and around the atom. Thus, the energetic shift caused by initial state effects is known as a chemical shift, ΔE_b. To a first approximation, all core levels in an atom shift to the same extent. For most samples, it is a good approximation to assume that the chemical shift is completely due to initial state effects and that, in particular, the relaxation energy does not depend on the chemical environment. This is obviously an approximation, but it is a good general rule.

The chemical shift depends on the oxidation state of the atom. Generally, as the oxidation state increases, the binding energy increases. The greater the electron withdrawing power of the substituents bound to an atom, the higher the binding energy. This can be understood on the basis of simple electrostatics. The first ionization energy of an atom is always lower than the second ionization energy. Similarly, the higher the effective positive charge on the atom, the higher the binding energy of the photoelectron.

Most of the atomic relaxation results from rearrangement of outer shell electrons. Inner shell electrons of higher binding energy make a small contribution. The nature of a material's conductivity determines the nature of extra-atomic relaxation. In a conducting material such as a metal, valence electrons are free to move from one atom to a neighbour to screen the hole created by photoionization. In an insulator, the electrons do not possess such mobility. They react by being polarized by the core hole. Hence, the magnitude of the extra-atomic relaxation in metals (as much as $5-10$ eV) is greater than that of insulators.

Several other final state effects result in what are known as satellite features. Satellite features arise from multiplet splitting, shake-up events and vibrational fine structure. Multiplet splittings result from spin-spin interactions when unpaired electrons are present in the outer shells of the atom. The unpaired electron remaining in the ionized orbital interacts with any other unpaired electrons in the atom/molecule. The energy of the states formed depends on whether the spins are aligned parallel or anti-parallel to one another. In a shake-up event, the outgoing photoelectron excites a valence electron to a previously unoccupied state. This unoccupied state may be either a discrete electron state, such as a π^* or σ^* state, or, especially in metals, is better thought of as electron-hole pair formation. By energy conservation, the photoelectron must give up some of its kinetic energy in order to excite the shake-up transition; hence, shake-up features always lie on the high binding energy (lower kinetic energy) side of a direct photoemission transition. Occasionally, the valence electron is excited above the vacuum level. Such a double ionization event is known as shake-off. Shake-off events generally do not exhibit distinct peaks whereas shake-up transitions involving excitation to discrete states do. An excellent discussion of final state effects can be found in [60].

The width of photoemission peaks is determined to a large extent by the lifetime of the core hole (homogeneous broadening) and instrumental resolution. However, sample inhomogeneity and satellite features can also lead to peak broadening, especially if the instrumental resolution is insufficient to resolve the latter from the main photoemission feature. The instrumental resolution is determined not only by the resolution of the energy analyzer, but also by the wavelength spread in the incident x-ray beam. Through the Heisenberg uncertainty relation, the intrinsic peak width, Γ is inversely related the core hole lifetime, τ, by

$$\Gamma = h/\tau. \tag{2.6.11}$$

where h is the Planck constant. The lifetime generally decreases the deeper the core hole because core holes are filled by higher lying electrons, and the deeper the core hole, the more de -excitation channels there are that can fill it. Analogously, for a given energy level, the lifetime decreases (the linewidth increases) as the atomic number increases. Intrinsic lifetime broadening is an example of homogeneous broadening and has a Lorentzian lineshape.

2.6.1.1   Quantitative analysis

One of the great strengths of XPS is that it can be used not only for elemental analysis, but also for quantitative analysis [61]. XPS peak areas are proportional to the amount of material present because the photoionization cross section of core levels is virtually independent of the chemical environment surrounding the element. However, since the inelastic mean free path of electrons is limited, the detection volume is limited to a region near the surface. The distribution of the analyte in the near surface region also affects the measured intensity.

The area under an XPS peak must be measured after a suitable background subtraction has been performed. In order to relate the peak area to the atomic concentration, we need to consider the region that is sampled. X-rays interact weakly with matter and penetrate several μm into the bulk. The photoelectrons, on the other hand, have a comparatively short inelastic mean free path, as we have seen in Fig. 2.13 above. In addition, it depends on the kinetic energy of the electron ejected from the atom A, and the material through which it passes, hence, we denote the mean free path by $\lambda_M(E_A)$.

The measured intensity is compared to that of a known standard, and a ratioing procedure is used to eliminate the necessity of evaluating a number of instrumental factors. It must be borne in mind that the best standard is one that has exactly the same concentration and matrix, in short, one that is identical to the sample. This is, of course, impractical. Therefore, the proper choice of the standard is important and the approximations introduced by using one must be considered. Appropriate standards are the signal from a known coverage, as determined for example by LEED measurements, or that from a pure bulk sample. One should try to match both the concentration and the spatial distribution of the standard as closely as possible to those of the sample to be analyzed. Here, we consider only the most common case. For other distributions and factors affecting quantitative analysis, the reader is referred to Ertl and Küppers [2] and Powell and Jablonski [61].

Adsorbate A occupies only surface sites with a coverage $\boldsymbol{\theta_A}$ The coverage θ_A represents the covered fraction of B, the substrate. The uncovered portion is then $(1-\theta_A)$. The signal from A is not attenuated by electron scattering, hence

$$I_A = \theta_A I_A^0, \tag{2.6.12}$$

where I_A^0 is the signal from a known or standard coverage of A. Equation (2.6.12) can be used to determine the absolute coverage at an arbitrary coverage if the absolute coverage is known for I_A^0. If not, then Eq. (2.6.12) can only be used to determine a relative coverage compared to the coverage at I_A^0. The absolute coverage can be found at any arbitrary coverage if we use a ratio of XPS signals. The signal from B, given by the sum of the bare surface contribution and that of the covered portion, is

$$I_B = I_B^0 \left[1 - \theta_A + \theta_A \exp\left(-\frac{a_A \cos \vartheta}{\lambda_A} \right) \right] \tag{2.6.13}$$

where I_B^0 is the signal from a clean B substrate, λ_A is the inelastic mean free path, and ϑ the detection angle. Therefore,

$$\frac{I_A I_B^0}{I_B I_A^0} = \frac{\theta_A}{1 - \theta_A + \theta_A \exp\left(-\frac{a_A \cos \vartheta}{\lambda_A} \right)}. \tag{2.6.14}$$

2.6.2 Ultraviolet photoelectron spectroscopy (UPS)

The origins of photoelectron spectroscopy rest not with x-ray excitation but rather with photoemission caused by ultraviolet (UV) light. Hertz [62] and Hallwachs in the late 1880s discovered that negative charge was removed from a solid under the influence of UV irradiation. This is the photoelectric effect. Neither the electron nor the quantum nature of the photon were yet known at the time, and it was not until the landmark work of Einstein in 1905 [63] that it was recognized that light must be composed of photons with quantized energy $h\nu$. It was Einstein who proposed the linear relationship between the maximum kinetic energy of the photoelectrons and the photon energy,

$$E_{K,\max} = h\nu - \Phi, \tag{2.6.15}$$

where h is Planck's constant and Φ is the work function. Equation (2.6.15) is, of course, a special version of the more general Einstein equation, Eq. (2.6.1).

UV photons can excite photoemission from valence levels. In contrast to XPS, ultraviolet photoelectron spectroscopy with near UV light is difficult to use quantitatively. However, since valence electrons are involved in chemical bonding, UPS is particularly well suited to the study of bonding at surfaces. UPS readily provides measurements of the work function and band structure of the solid, the surface and adsorbed layers.

UPS as a method of studying band structure rather than simply work functions is a relatively new technique. This is largely because of instrumental reasons, for example, the need for UHV and strong sources of UV light. Eastman and Cashion [64] were the first to integrate a differentially pumped He lamp with a UHV chamber. A He lamp provides light at either 40.82 or 21.21 eV. Atomic emission lines are extremely sharp, which allows for a high ultimate resolution. The resolution is, in practice, limited by the energy resolution of the electron spectrometer. Other light sources covering the range of roughly 10–100 eV have found use, particularly synchrotrons and the laser based technique of high harmonic generation [65, 66]. Lasers have brought about the advent of multiphoton photoelectron spectroscopy [67–72], which is a particularly powerful technique that allows us to investigate the dynamics of electrons directly in the time domain.

Much of what we have learned above regarding XPS is applicable to UPS as well. Koopmans' theorem is again the starting point, in which the first approximation is that the negative of the orbital energy is equated to the measured binding energy. The binding energy, also called the ionization potential I_p, is the difference of the initial and final state energies. Relaxation effects are again important but tend to be smaller than those found in XPS. UPS can be performed with such high resolution that small shifts and satellites are easily detected. These satellites can in principle include vibrational structure. Normal resolution is insufficient for most vibrational structure but high-resolution studies have observed it [73]. A further difficulty with observing vibrational structure arises from the natural linewidth of adsorbed molecules. Chemisorbed species are strongly coupled to the substrate. The adsorbate–substrate interaction leads to excited state lifetimes that are generally on the order of 10^{-15} s. This translates into a natural linewidth of ~ 1 eV, which is much too broad to observe vibrational transitions. Physisorbed species are less strongly coupled and this has allowed for the observation of vibrational structure in exceptional cases. High-resolution studies can yield information not only on electron and hole lifetimes but also electron–phonon interactions and defect scattering [74].

Whereas XP spectra are conventionally referenced to the vacuum level, UP spectra are commonly referenced to the Fermi energy. Thus a binding energy of $E_B = 0$ corresponds to E_F. With this convention, the work function of the sample and, thereby, the absolute energy scale for an UP spectrum, is easily determined, but for experimental reasons, it is not as straightforward as Einstein's relationship suggests. The spectrometer work function, Φ_{sp}, must be accounted for when interpreting the measured electron kinetic energies, E_K, as shown in Fig. 2.18,

$$E_K = h\nu - E_B - \Phi_{sp}. \tag{2.6.16}$$

For excitation with a fixed photon energy $h\nu$, therefore, the maximum kinetic energy of photoelectrons is given by

$$E_{K,\max} = h\nu - \Phi_{sp}. \tag{2.6.17}$$

Meanwhile the minimum kinetic energy is given by

$$E_{K,\min} = \Phi - \Phi_{sp}. \tag{2.6.18}$$

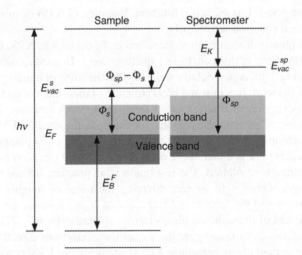

Figure 2.18   *The influence of the spectrometer work function, Φ_{sp}, on photoelectron spectra. Φ_s, work function of the sample; E_{vac}^{sp}, E_{vac}^{s} vacuum energies of the spectrometer and the sample, respectively; E_F, Fermi energy; E_K, electron kinetic energy; E_B^F, binding energy; h, Planck constant; ν, frequency of incident photon. Adapted from J. C. Vickerman, Surface Analysis: The Principal Techniques, John Wiley, Chichester, Sussex. © 1997 with permission from John Wiley & Sons, Ltd.*

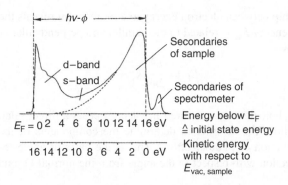

Figure 2.19 *A representative ultraviolet photoelectron spectrum. The relative intensities of primary and secondary electrons depend on instrumental factors. E_{vac}, vacuum energy of the sample; E_F, Fermi energy; h, Planck constant; v, frequency of the incident photon; Φ, sample work function. Reproduced from G. Ertl, J. Küppers, Low Energy Electrons and Surface Chemistry, 2nd ed., VCH, Weinheim. © 1985 with permission from John Wiley & Sons, Ltd.*

The width of the spectrum,

$$\Delta E = E_{K,\max} - E_{K,\min} = hv - \Phi \qquad (2.6.19)$$

can thus be used to determine the work function of the sample

$$\Phi = hv - \Delta E \qquad (2.6.20)$$

These relations are used to fix the energy scale as shown in Fig. 2.19. Figure 2.19 indicates the additional possibility of a signal arising from secondary electrons emanating from the spectrometer. Applying a voltage to the sample easily discriminates against the signal from secondaries.

2.6.2.1 *Angle-resolved ultraviolet photoemission (ARUPS)*

UPS can be used to interrogate the band structure of the solid as well as adsorbate levels. First, we note that the photon momentum is inconsequential in the photon energy range (10–100 eV) used in UPS. Therefore, the electron momentum is unchanged by photon absorption, that is $k_i \approx k_f$. Equivalently, we state that only Franck-Condon (vertical) transitions are observed in UPS.

To obtain a measure of the band structure of a solid we need a more in-depth understanding of the photoemission process [2, 75, 76]. The transition rate R between the initial state ψ_i and the final state ψ_f caused by a perturbation H' is given by Fermi's golden rule

$$R = \frac{2\pi}{\hbar} |\langle \psi_f | H' | \psi_i \rangle|^2 \delta(E_f - E_i - hv). \qquad (2.6.21)$$

$\delta(E_f - E_i - hv)$ is the Dirac delta function, it is unity when $E_f - E_i - hv = 0$, and zero elsewhere. The perturbation represents the interaction of the electromagnetic field of the photon with the atom/molecule to be excited. As electric dipole transitions are by far the most likely to be observed, we write

$$H' = \frac{e}{2mc}(\boldsymbol{A} \cdot \boldsymbol{P} + \boldsymbol{P} \cdot \boldsymbol{A}) \qquad (2.6.22)$$

where $\boldsymbol{P}$ is the momentum operator $-i\hbar\nabla$ and $\boldsymbol{A}$ is the vector potential of the photon's electric field.

Measuring the relationship between electron energy and momentum reveals the band structure of a solid. The photoelectron kinetic energy E_K is related to the parallel and perpendicular components of momentum, $k_{\parallel}$ and $k_{\perp}$, respectively, by

$$E_K = \frac{\hbar^2}{2m}(k_{\parallel}^2 + k_{\perp}^2). \tag{2.6.23}$$

The momentum of the photoelectron measured in vacuum bears a direct relationship to the momentum of the electrons in the solid [75]. In order to determine the components of the momentum, we need to measure the angular distribution of photoelectrons. This is the basis of angle-resolved UPS or ARUPS.

The differential cross section with respect to the angle from the normal Ω must be determined. This can be written (see [75])

$$\frac{d\sigma}{d\Omega} \propto |\langle \psi_f|\boldsymbol{P}|\psi_i\rangle \cdot \boldsymbol{A}_0|^2 \delta(E_f - E_i - h\upsilon). \tag{2.6.24}$$

Again we write the initial and final wavefunctions in a one-electron picture

$$\psi_i(n) = \phi_j \Psi_j(n-1) \tag{2.6.25}$$

$$\psi_f(n-1) = u_k \Phi_{jl}(n-1). \tag{2.6.26}$$

Accordingly, the differential photoionization cross section can be written

$$\frac{d\sigma}{d\Omega} \propto |\langle u_k|\boldsymbol{A} \cdot \boldsymbol{P}|\phi_i \Phi_{jl}|\psi_i\rangle|^2 \tag{2.6.27}$$

Similar to what we discussed above for XPS, satellite features are expected due to relaxation effects in addition to the potential for vibrational fine structure. The presence of the vector product $\boldsymbol{A} \cdot \boldsymbol{P}$ means that the differential cross section depends on the relative orientation between the polarization of the ultraviolet light and the transition dipole of the species to be excited. This property can be used to identify transitions and the adsorption geometry of molecular adsorbates.

A detailed interpretation of UP spectra is beyond the scope of this book but can be found in the references in Further Reading. Figure 2.20 shows the ability of UPS to identify adsorbate electronic states and the effects of the adsorbate on the substrate. The most complete understanding of angle-resolved photoemission can only be achieved with comparisons to calculations.

UPS is surface sensitive because the low energy of the photoelectrons leads to a short inelastic mean free path. It can be used to map out the band structure of solids as well as the electronic structure of adsorbates, and the effect of adsorption on the substrate's band structure. Chemisorption involves an exchange of electrons between the adsorbate and substrate. This leads to a change in work function, which is readily detected by the change in the width of the UP spectrum. Difference spectra, the difference between spectra of adsorbate-covered and clean surfaces, clearly show the changes in the substrate as well as adsorbate induced features.

Because of the strong coupling between adsorbate and substrate, the ionization energies measured in UPS are shifted compared to those observed in the gas phase. This is illustrated in Fig. 2.20 and embodied in the equation

$$E_{ad} = E_{gas} - (\Phi + \Delta\Phi) + E_{relax} + E_{bond\ shift}. \tag{2.6.28}$$

E_{ad} and E_{gas} are the ionization energies measured in the adsorbed and gas phases, respectively. E_{relax} accounts for final state relaxation processes and $E_{bond\ shift}$ is an initial state shift brought about by the adsorbate–substrate interaction. The remaining term accounts for changes in the work function; however, the exact form of this term remains controversial as is discussed elsewhere [2].

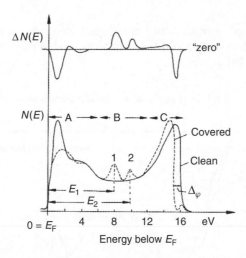

Figure 2.20   *Changes observed in ultraviolet photoemission spectra (lower panel) upon adsorption. The upper panel displays the difference spectrum. N(E)$_{covered}$ and N(E)$_{clean}$ are the count rates of photoelectrons from the adsorbate-covered and clean surface, respectively. ΔN(E) = N(E)$_{covered}$ − N(E)$_{clean}$. Reproduced from G. Ertl, J. Küppers,* Low Energy Electrons and Surface Chemistry, *2nd ed., VCH, Weinheim. © 1985 with permission from John Wiley & Sons, Ltd.*

For chemisorbed species, $E_{\text{bond shift}}$ is large. The orbital energies are pinned to the Fermi level of the substrate. The magnitudes of $E_{\text{bond shift}}$ and E_{relax} depend on the specific orbital and not all orbitals shift to the same extent. This can lead to overlaps in UP spectra from adsorbed species that do not occur in gas-phase spectra.

Physisorbed species exhibit $E_{\text{bond shift}} \approx 0$ as expected from the weak coupling to the substrate. The orbitals are pinned to the vacuum level. Küppers, Wandelt and Ertl [77] exploited this by using physisorbed Xe photoemission to demonstrate that the work function is defined locally not globally. For instance, steps and defects affect the surface dipole layer in their vicinity, and therefore exhibit a different work function than terraces. Similarly, a chemically heterogeneous surface, such as a bimetallic surface, can exhibit distinct photoemission peaks associated with Xe physisorption on the two distinct metal sites on the surface.

Advanced Topic: Multiphoton photoemission (MPPE)

An extremely potent extension of photoelectron spectroscopy, especially for dynamical studies, is multiphoton photoemission [78]. The most common version is two-photon photoemission (2PPE). Multiphoton photoemission is the surface analogue of multiphoton ionization (MPI), which has been used to great advantage in gas-phase spectroscopy. MPPE can be performed either with or without resonance enhancement. Resonance enhancement occurs when one (or more) photon(s) excites an electron into a real but normally unoccupied electronic state. Subsequently, further absorption of one (or more) photon(s) ionizes the electron. The sum of the photon energies has to be greater than the work function. Extremely high signal to noise can be achieved by using photons that individually do not have sufficient energy to cause photoemission. This can lead to essentially background free photoemission spectra. This is demonstrated in Fig. 2.21. The photoelectron signal is resonance enhanced due to the transient population of a normally unoccupied surface state. Figure 2.21 also demonstrates the power of polarization control to identify the nature of a transition. For instance, the diamond surface state is composed principally of sp^3 orbitals, which are aligned along the surface normal. Such normally unoccupied states are difficult to study by other spectroscopic techniques, though inverse photoemission has been profitably applied [79] as well.

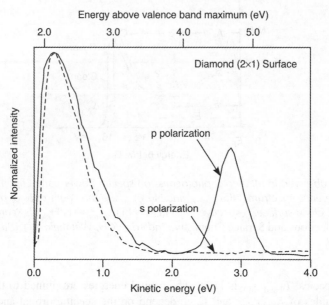

Figure 2.21 *Two-photon photoemission of the reconstructed C(111)–(2 × 1) diamond surface. The normally unoccupied surface state observed at ~3 eV is only observed with p-polarization due to selection rules. Reproduced from G.D. Kubiak, K.W. Kolasinski, Phys. Rev. B, 39, 1381. © 1989 with permission from the American Physical Society.*

Multiphoton events are an example of non-linear spectroscopy – a spectroscopy in which the signal is proportional to the light intensity raised to a power greater than one. Multiphoton transitions are extremely weak and, in general, they are only excited by high intensity light sources such as lasers. The use of lasers allows for high spectral and temporal resolution. High temporal resolution is required to study electron dynamics [80]. In this way, the lifetimes of surface states and image potential states have been directly measured [66], and the electron-electron and electron-phonon scattering processes that lead to thermalization of excited electrons have been probed [70, 72, 81]. Two-photon photoemission has also given us a direct glimpse of the motion of desorbing Cs atoms as they leave the surface [82].

2.6.3 Auger electron spectroscopy (AES)

Another workhorse technique of surface science, in particular because it can be easily incorporated into the same apparatus that is used for LEED, is Auger electron spectroscopy (AES). An Auger transition can be excited by photons, electrons or even by ion bombardment and is depicted in detail in Fig. 2.22. The phenomenon was first described by Auger in 1925 [83]. Conventionally, electron bombardment at energies in the 3–5 keV range is used for surface analytical AES. Auger transitions can, however, also be simultaneously observed during XPS measurements. Like XPS, AES can be used for quantitative elemental

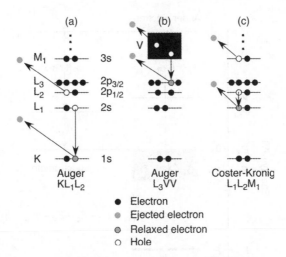

Figure 2.22   *A detailed depiction of Auger transitions involving (a) three core levels; (b) two core levels and the valence band; and (c) a Coster-Kronig transition in which the initial hole is filled from the same shell.*

analysis, and the measured signal is influenced by the inelastic mean free path of electrons, the atomic concentration and the concentration distribution.

Our first approximation to the photoemission process involved in XPS assumed a one-electron picture. Auger spectroscopy, in contrast, relies on the coupling between electrons. As shown in Fig. 2.22, after ionization of a core level, a higher lying (core or valence) electron can fill the resulting hole in a radiationless transition. This process leaves the atom in an excited state. The excited state energy is removed by the ejection of a second electron. Inspection of Fig. 2.22 makes clear that a convenient method of labelling Auger transitions is to use the x-ray notation of the levels involved. Thus the transition in Fig. 2.22 (a) is a KL_1L_2 transition and that in Fig. 2.22 (b) is a KL_1V transition in which the V stands for a valence band electron. In some cases, more than one final state is possible and the final state is added to the notation to distinguish between these possibilities, e.g. $KL_1L_3(^3P_1)$ as opposed to $KL_1L_3(^3P_2)$. Coster-Kronig transitions, Fig. 2.23(c), have particularly high rates and are, therefore, extremely important for determining the relative intensities of Auger transitions.

Auger spectroscopy can be used to detect virtually any element, apart from H and He. With increasing atomic number (Z), however, the probability of radiative relaxation of the core hole, i.e. x-ray fluorescence, increases. Furthermore, since the incident electron energy (the primary electron energy E_p) is often only 3 keV, the Auger transitions of interest in surface spectroscopy generally have energies below 2 keV. Thus, the primary excitation shell shifts upward as Z increases, e.g. K for Li, L for Na, M for potassium and N for ytterbium.

The energy of an Auger transition is difficult to calculate precisely as many electron effects and final state energies have to be considered. Fortunately for surface analytical purposes, the exact energy and lineshape of an Auger transition need only be considered in the highest resolution specialist applications. The low resolution required for quantitative surface analysis also means that the small energy differences between final state multiplets can be neglected. Auger transitions from different elements tend to be fairly widely spaced in energy and when they do overlap coincidentally there is often no easy way to use them for quantitative analysis.

The observed energy of an Auger transition is best approximated with reference to Fig. 2.23. Note that the energy scale is referenced to the Fermi level. The primary electron e_p^- must have sufficient energy to

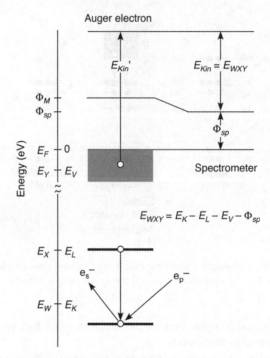

Figure 2.23   *Energy levels for a KLV Auger transition, including the influence of the spectrometer work function* Φ_{sp}. *Adapted from J.C. Vickerman,* Surface Analysis: The Principal Techniques, *John Wiley & Sons, Chichester.* © *1997 with permission from John Wiley & Sons, Ltd.*

ionize the core level of energy E_W. An electron at energy E_X fills the initial hole, and the energy liberated in this process is transferred to an electron at E_Y, which is then ejected into the vacuum. This electron must overcome the work function of the sample, Φ_e, to be released into the vacuum; hence the kinetic energy is reduced by this amount. However, the electron is detected by an analyzer with a work function Φ_{sp}. The vacuum level is constant and electrical contact between the sample and the analyser leads to alignment of their Fermi levels. Thus it is actually Φ_{sp} that must be subtracted from the energy of the electron to arrive at the final kinetic energy

$$E_{WXY} = E_K - E_L - E_V - \Phi_{sp}. \tag{2.6.29}$$

If the sample is an insulator, the Fermi levels do not automatically align. Sample charging can easily occur and special care must be taken in interpreting the spectra. The energy of the electrons created through the Auger process, the Auger electrons, is not dependent on the primary electron energy.

2.6.3.1   *Quantitative analysis*

The quantitative treatment of the intensity of Auger transitions is complicated not only by inelastic scattering but also by backscattering effects. For a complete discussion, see Ertl and Küppers [2] or Vickerman [1]. First, consider the geometry of excitation and detection (see Fig. 2.24). Some of the electrons from the primary excitation beam excite the emission of Auger electrons, and some are backscattered after elastic or inelastic scattering from the sample. Excitation occurs over a certain volume and the geometry of this excited region is determined by the composition of the sample (in particular the atomic number

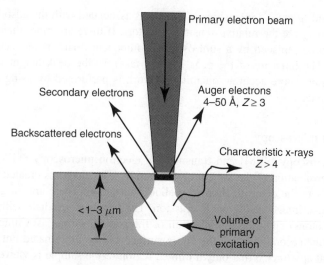

Figure 2.24 *The geometry of Auger electron spectroscopy. A primary electron beam excites the formation of Auger electrons as well as x-ray fluorescence. Backscattered and secondary electrons are also created in the process.*

of the components) and the primary electron beam energy. The volume of primary excitation takes on a teardrop shape for low atomic number samples and low primary electron beam energies. It develops into a more spherical shape for higher beam energies and higher Z. Auger electrons are created and propagate in all directions. However, only those Auger electrons that propagate back toward the detector, such as a cylindrical mirror analyzer (CMA) or, on older systems, a retarding field analyzer (RFA), and which escape the sample without experiencing inelastic scattering, are detected as Auger electrons. Therefore, the Auger electrons emanate from a shallow region whose cross section is defined by the cross section of the primary beam and the depth of which is determined by the inelastic mean free path of the Auger electrons at their specific kinetic energy. Inelastically scattered primary and Auger electrons excite the emission of other electrons from the sample (secondary electrons) with a continuous range of energies from E_p down to zero (more accurately, down to a low energy cut-off determined by the spectrometer detection system and stray fields). All of the Auger, backscattered and secondary electrons that are intercepted by the detector are detected. Each detector has a typical acceptance angle of electrons that it can detect. Electrons that leave the sample at too great an angle measured from the normal fall outside the acceptance angle.

Because of the presence of backscattered and secondary electrons, the signal acquired – the number of electrons as a function of the kinetic energy of the electrons $N(E)$ – is composed of peaks from the Auger electrons at well-defined energies along with a continuous background. The background has an intensity comparable to that of the Auger peaks and also has some smoothly varying structure. The signal $N(E)$ is called the integral signal or the integral spectrum. In many cases it is not easy to identify the peaks in the $N(E)$ spectrum; therefore, the differential signal $dN(E)/dE$ is calculated and plotted as a function of E. The energy of an Auger transition is labelled according to the most negative portion of a peak in the derivative spectrum.

Approximate compositional analysis can easily be obtained with the use of the measured relative sensitivity factors s_X, which can be found in the *Handbook of Auger Electron Spectroscopy* [84]. The mole fraction of component A in a binary mixture of A and B is given by

$$x_A = \frac{I_A/s_A}{I_A/s_A + I_B/s_B}. \tag{2.6.30}$$

In Eq. (2.6.30), I_A and I_B are the intensities of the peaks associated with the adsorbate and the substrate, respectively and s_A and s_B are the relative sensitivity factors. If there are more than two components then the denominator has to be replaced by a sum over all components. The mole fraction can be equated to the relative coverage. The intensity of the peak is often taken as the peak height when the data is taken in a differential mode, but more accurate quantitative work is performed by using the peak area for data taken in an integral mode.

2.6.4   Photoelectron microscopy

Scanning electron microscopy (SEM) and transmission electron microscopy (TEM) are well-established techniques for the determination of the structure of materials. The electrons created by photoemission also lend themselves readily to the determination of spatial information. There are, in general, three strategies that can be used to make images using photoemission. The first is to irradiate with a minute probe beam (either electrons or photons) and then to collect all of the photoelectrons as a function of the position of the scanned beam. In the second method, the entire surface can be illuminated but the photoelectrons are imaged through a small aperture so that only a limited region of the sample is viewed. The detector is then scanned across the sample to build up an image. In the third method, the entire sample is illuminated and the photoelectron distribution is imaged onto a position sensitive detector. A photoelectron microscope can be used not only to measure the compositional profile of a surface, but also the geometric profile of its magnetic structure [85].

Photoelectron microscopes are useful both for spectromicroscopy and microspectroscopy. In spectromicroscopy, micrographs (images) of electrons within a limited energy range are acquired. In other words, spectromicroscopy is energy-resolved microscopy. This is particularly important for compositional mapping. Microspectroscopy involves the acquisition of spectroscopic information, e.g. photoemission intensity versus kinetic energy, from a spatially resolved surface region. In this way the electronic structure of micro- or nano-scale structures can be measured, for example, as a function of size.

2.6.4.1   Profiling and xy mapping with XPS

XPS can also be used to garner information about the geometric distribution of an element in all three dimensions. An XPS signal becomes progressively more surface sensitive for large values of the take-off angle. For take-off angles normal to the surface, roughly the first 10 nm (20–40 atomic layers) of material are probed. If information for greater depths is required, XPS can be performed with simultaneous removal of the surface layers. Removal is generally achieved by sputtering, a process in which high-energy ions are used to bombard the surface and eject material from the sample. The measurement of atomic composition as a function of depth is known as depth profiling.

The spatial resolution of XPS in the *xy* plane is impeded by the difficulty of focussing x-rays. Using focussed x-rays, spatial resolutions down to ~50 μm have been achieved. Alternatively, imaging of the photoelectrons with a position sensitive detector has been used to achieve a resolution of ~10 μm. The pursuit of an XPS microscope remains an active area of research.

2.6.4.2   Depth profiling and xy mapping with AES

Depth profiling by the combination of AES and sputtering is more commonly employed than the combination with XPS. See Vickerman [1] and Hofmann [86] for a full discussion. Since high-energy electron beams can be focussed to spot sizes on the order of tens of nanometres, Auger spectroscopy lends itself to high-resolution surface microscopy more readily than does XPS. High-resolution electron microscopes can be used in two modes. They can be used as a normal scanning electron microscope (SEM) to image the topography of a sample. The advantage of a scanning Auger microscope is that images of elemental composition can be obtained. This allows for correlations to be drawn between topography and elemental composition.

2.6.4.3 *Photoemission electron microscope (PEEM)*

In conventional photoemission spectroscopy, the photon energy is chosen to be significantly greater than the work function of the sample. As we have learned, the work function of a heterogeneous surface is not uniform. Consider a surface that is partially covered with chemisorbed oxygen atoms. Oxygen causes a large increase in the surface dipole *in the region where it is adsorbed*. If measured by a technique that averages over a large area, this appears as a global change in the work function. However, if probed by a higher resolution technique, we find that the work function is only changed in the parts of the surface covered with oxygen. If we choose a photon energy that is just above the work function of the clean surface, photoemission is only observed from the clean surface regions. The oxygen-covered areas are dark, that is, they do not emit electrons.

This effect can be exploited to image the chemisorption of oxygen and other adsorbates that cause large changes in the work function. This technique is known as photoemission electron microscopy (PEEM) [85, 87]. It was originally conceived in the 1930s but it was not widely exploited in surface studies until Engel developed modern instruments in the 1980s. Under favourable circumstances, i.e. when adsorbates produce sufficiently diverse work function changes to be differentiated, PEEM is capable of submicron resolution coupled with moderate temporal resolution. This combination allows PEEM to image spatiotemporal pattern formation during surface reactions with ~200 nm resolution [88]. We speak more about this topic in Chapter 6.

2.7 Vibrational spectroscopy

Vibrational spectroscopies have long found an important place in the chemist's arsenal of analytical techniques and the same is true for adsorbed systems [89, 90]. Vibrational spectroscopy is extremely useful in identifying the types of bonds that are present in a sample. Shifts in vibrational frequencies can also be used to gain insight into subtle changes in bonding. A variety of spectroscopies can give information on vibrational motions: infrared absorption [91, 92], electron energy loss [23], Raman, sum frequency generation, inelastic neutron tunnelling [90], helium scattering [93], and inelastic electron scattering [94]. Ho has demonstrated that an STM can yield vibrational spectra with high spatial resolution [95]. NSOM can also be used to obtain spatially resolved infrared and Raman spectra. Here we shall concentrate on just two of these spectroscopies: infrared absorption and electron energy loss spectroscopy.

Most surface vibrational spectroscopy is performed on the fundamental transitions, that is, on the transition from $v = 0$ to $v = 1$. The harmonic oscillator approximation is usually appropriate at this low level of excitation. The energy levels of a harmonic oscillator are equally spaced, and given by

$$E_v = (v + \tfrac{1}{2})h\omega_0, \tag{2.7.1}$$

where v is the vibrational quantum number. The characteristic frequency of a harmonic oscillator is given by

$$\omega_0 = \frac{1}{2\pi}\sqrt{\frac{k}{\mu}} \tag{2.7.2}$$

where k is the force constant of the vibration (related to the bond strength), and μ is the reduced mass

$$\mu = \frac{m_1 m_2}{m_1 + m_2} \tag{2.7.3}$$

of the diatomic oscillators composed of atoms of mass m_1 and m_2. An N-atom polyatomic molecule has $3N-6$ (bent) or $3N-5$ (linear) vibrational modes in the gas phase. Additional modes are introduced by adsorption as the translational and rotational degrees of freedom are lost for localized adsorption. These are transformed into new modes – often denoted as frustrated translations and rotations. If we consider the CO molecule bound in an upright geometry, adsorption transforms the z translation into the vibration associated with the surface–CO bond. To a first approximation we can consider the various vibrational modes to be independent oscillators. The harmonic approximation breaks down for high levels of excitation because the oscillators are better described by a Morse potential than a harmonic potential. This is especially important for low frequency modes, such as the surface–adsorbate bond, because even moderate temperatures lead to significant populations in excited vibrational states. Anharmonicity not only results in a continuously decreasing level spacing with increasing values of v, but also in couplings between the various vibrational modes. In a real (anharmonic) oscillator, overtone transitions ($\Delta v > 1$) and combination bands (simultaneous excitation of more than one vibrational mode) can be observed.

Equations (2.7.1) and (2.7.2) are written in terms of radial frequency. More commonly, vibrational transitions are categorized in terms of inverse centimetres (cm^{-1}) in infrared spectroscopy or millielectron volts (meV) in electron energy loss spectroscopy. The inverse centimetre is the unit of wavenumber, which is denoted by $\bar{v}$ and is not to be confused with frequency (v) or angular frequency ($\omega = 2\pi v$). A useful conversion factor is $1\,meV = 8.065\,cm^{-1}$.

As shown in Fig. 2.25 the vibrational spectrum in the adsorbed phase differs greatly from that found in the gas phase. The loss of rotational fine structure is immediately obvious. The change in linewidth is also apparent. There are numerous mechanisms that lead to the increased linewidth in the adsorbed phase [92, 96–98]. Both homogeneous and inhomogeneous processes lead to line broadening. Inhomogeneous broadening is dominated by variations in the environment around the adsorbate. Under normal circumstances, inhomogeneous broadening is significant and leads to a Gaussian lineshape. Only the most well-ordered adsorbate structures on single crystal surfaces exhibit spectra that are not dominated by inhomogeneous broadening. Figure 2.25 also demonstrates that more than one frequency can be associated with the same vibration in the adsorbed phase, the C–O stretch in this case, even though there is a unique value in the gas phase. A prime example of this is the ideally terminated Si(111)–(1 × 1):H surface shown in Fig. 2.26, which Chabal and co-workers have exploited to study vibrational energy transfer processes in unparalleled detail [97]. Homogeneous broadening arises both from vibrational energy transfer and dephasing events. The study of vibrational energy transfer and dephasing in adsorbates was pioneered by Cavanagh, Heilweil, Stephenson and co-workers [99–103].

Vibrational modes are named according to the type of motion involved in the vibration. There are four characteristic types of vibration. *Stretches* change the bond length and are denoted with v (Greek nu not to be confused with the vibrational quantum number v). In-plane motions that change bond angles but not bond lengths are *bends*. These are denoted with δ and are sometimes further subdivided between *rocks*, *twists* and *wags*. Out-of-plane bends are assigned the symbol γ. *Torsions* (τ) change the angle between two planes containing the atoms involved in the motion.

Equation (2.7.2) shows that the fundamental vibrational frequency depends on the reduced mass. Isotopic substitution can therefore be used as an aid in identifying vibrational transitions. The value of k depends on the vibrational potential, which means that vibrational spectra are sensitive to changes in the vibrational potential introduced by changes in chemical bonding.

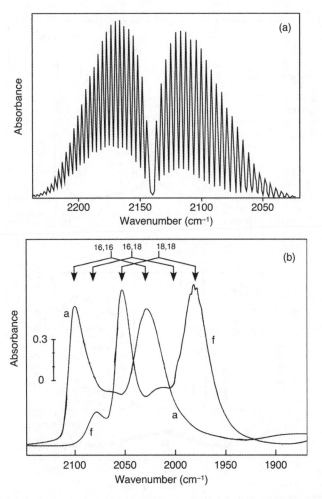

Figure 2.25 *The IR spectrum of (a) gas-phase CO versus that of (b) CO adsorbed on dispersed Rh clusters. The gas-phase spectrum exhibits rotational fine structure. The adsorbed CO forms a gem dicarbonyl species ($Rh(CO)_2$). Coupling between the two adsorbed CO molecules leads to two vibrational peaks. The effect of oxygen isotopic substitution is also evident. 16,16 refers to $Rh(C^{16}O)_2$, etc. Adapted from J.T. Yates, Jr., K. Kolasinski, J. Chem. Phys., 79, 1026. © 1983 with permission from the American Institute of Physics.*

2.7.1   IR spectroscopy

The application of IR spectroscopy to surfaces owes much of its early development to the work of Eichens [104, 105], Sheppard [106, 107] and Greenler [108, 109]. The absorption of infrared radiation by adsorbed species [91, 92] bears many similarities to IR absorption in other phases. The probability of a vibrational transition is proportional to the square of the transition dipole moment. The transition dipole moment is given by [91]

$$M_{\nu\nu'} = \int_{-\infty}^{\infty} \psi(\nu)\mu\psi(\nu') \, \mathrm{d}\tau \tag{2.7.4}$$

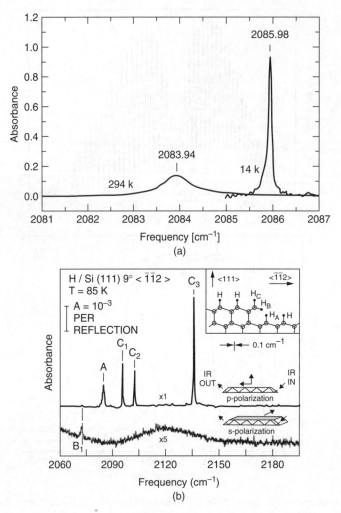

Figure 2.26   *The IR spectrum of H adsorbed on chemically prepared (a) flat and (b) stepped Si(111) surfaces. Part (a) reproduced from P. Jakob, Y.J. Chabal, K. Raghavachari, Chem. Phys. Lett., 187, 325. (c) 1991 with permission from Elsevier. Part (b) reproduced from P. Jakob, Y.J. Chabal, K. Raghavachari, S.B. Christman, Phys. Rev. B, 47, 6839. © 1993 with permission from the American Physical Society.*

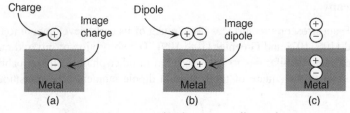

Figure 2.27   *Image dipole at a metallic surface.*

where ψ (v) and ψ (v') are the initial and final state vibrational wavefunctions, respectively, and μ is the dipole moment of the molecule. For the vibrational transition to have a non-zero probability, the integral in Eq. (2.7.4) must be non-zero along at least one of its components (x,y or z). The z-axis is conventionally chosen to lie along the surface normal. Group theory and symmetry arguments [90, 110] can be used to determine which modes are IR active, that is, capable of being observed in infrared spectroscopy. This symmetry constraint depends not only on the molecule but also on the symmetry of the adsorption site [111]. This property is useful for the determination of adsorbate structure.

The strength of an absorption feature depends not only on the transition dipole moment but also on the strength of the electric field associated with the infrared light incident on the sample and the orientation of the electric field vector with respect to the transition dipole. Therefore, we need to consider the electric field strength at the interface to understand what types of vibration can be observed. Semiconductors and insulators can support electric fields both perpendicular and parallel to their surfaces. A metal, however, can only support a perpendicular electric field. This is because of the image dipole effect illustrated in Fig. 2.27. Thus on a metal, only vibrations with a component along the surface normal are observed in IR absorption spectra. On semiconductors and insulators, the orientation of the adsorbate can be determined by measuring the intensity of spectral features as a function of the polarization angle of the incident light.

Infrared spectroscopy is performed in a number of distinct modes:

- *Transmission* (Fig. 2.28(a)): Only appropriate for transparent substrates and films that are sufficiently thin so as not to absorb too much of the incident light. Successfully employed in the study of dispersed metal supported catalysts, thin films on insulating and semiconducting substrates and adsorption onto the surfaces of porous solids.
- *Reflection* (Fig. 2.28(b)): Most appropriate for metal single crystals for which it is performed near grazing incidence. Known as reflection absorption infrared spectroscopy (RAIRS) or infrared reflection absorption spectroscopy (IRAS).
- *Diffuse reflectance* (Fig. 2.28(c)): Powders or rough surfaces scatter radiation diffusely rather than specularly. An absorption spectrum is measured by collecting all of the scattered radiation. Known as diffuse reflectance infrared Fourier transform spectroscopy (DRIFTS).
- *Internal reflection* (Fig. 2.28(d)): Infrared light can be bounced through a transparent substrate, particularly diamond, Si, Ge or ZnSe, and absorption can occur at each surface reflection. Known as attenuated total reflectance (ATR) or multiple internal reflection (MIR) spectroscopy. Can be used to interrogate the interface between any two phases – solid/solid, solid/liquid, solid/gas – as long as the substrate is transparent. Thus, studies of semiconductor/aqueous solution interfaces become possible, as required for *in situ* electrochemical studies, even though the liquid is opaque. In addition to adsorbed layers on the ATR element, this technique can also be used to study the surface of solids pressed against the semiconductor prism.

Other methods are discussed by Chabal [92].

Transmission is perhaps the most familiar form of infrared spectroscopy. In surface science, however, it is generally only used for high surface area samples and thin films, because a single adsorbed monolayer absorbs so little radiation. In transmission mode, the amount of absorbed radiation is characterized in terms of the transmittance

$$T = \frac{I}{I_0} \tag{2.7.5}$$

or absorbance

$$A = -\log T = \log \left(\frac{I_0}{I} \right) \tag{2.7.6}$$

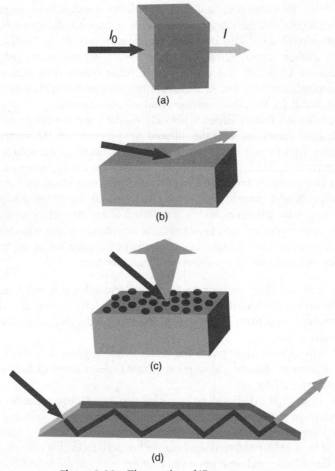

Figure 2.28   *The modes of IR spectroscopy.*

where I_0 is the incident intensity and I is the transmitted intensity. In the absence of reflection and scattering, the transmittance is given by

$$T = 10^{-acl} = \exp(-\alpha l) \qquad (2.7.7)$$

where a is the absorptivity, c the concentration and l the path length. The absorption coefficient, α, is related to the absorptivity by

$$\alpha = 2.3026 \, ac \qquad (2.7.8)$$

which is, in turn, related to the imaginary part, κ, of the complex refractive index, $\tilde{n}$, by

$$\alpha = \frac{4\pi\kappa}{\lambda_0} \qquad (2.7.9)$$

where λ_0 is the vacuum wavelength of the infrared light and κ is defined through

$$\tilde{n} = n + i\kappa. \qquad (2.7.10)$$

It follows that absorbance is linearly proportional to concentration and the path length

$$A = acl. \tag{2.7.11}$$

The absorption spectra collected by diffuse reflectance are expressed in terms of the Kubelka-Munk function

$$\frac{\alpha}{S} = \frac{(1 - R_\infty)^2}{2R_\infty} \tag{2.7.12}$$

where S is the scattering coefficient and R_∞ is the reflectivity of an (optically) infinitely thick sample. If the scattering coefficient is known (or assumed independent of wavelength) over the region of the spectrum, the Kubelka-Munk function directly transforms the measured reflectivity spectrum into the absorption spectrum.

In RAIRS and ATR, the reflection spectrum is measured for the clean substrate, $R_0(v)$, and the adsorbate-covered substrate, $R(v)$. The absorption spectrum is usually expressed in terms of the relative change in reflectivity

$$\frac{\Delta R(v)}{R_0(v)} = \frac{R_0(v) - R(v)}{R_0(v)} \tag{2.7.13}$$

The relative change in reflectivity is related to the absolute coverage (molecules per unit area), σ, and the adsorbate polarizability, α (v) (not to be confused with the absorption coefficient) according to [92]

$$\frac{\Delta R(v)}{R_0(v)} = \frac{8\pi^2 v}{c} F(\varphi)\sigma \ \text{Im} \ \alpha(v) \tag{2.7.14}$$

where φ is the angle of incidence and $F(\varphi)$ contains all the local field characteristics and the dielectric response of the system.

Equations (2.7.11) and (2.7.14) demonstrate that IR spectroscopy can be used to quantify the extent of adsorption. It is usually *assumed* that the absorbance and relative change in reflectivity are linearly proportional to adsorbate number density, as expected from Eqs (2.7.11) and (2.7.14). However, this is not always so [91]; in particular, deviations are to be expected if there are strong interactions between adsorbates (lateral interactions). While it is often a good approximation, especially at low coverage, the linear relationship between adsorbate coverage and infrared absorption needs to be confirmed by corroborating measures of coverage.

Surface infrared spectroscopy can only be performed if the substrate does not absorb strongly. Depending on the substrate, this leads to a cut-off in the low-frequency region of the spectrum, occurring at around $500-1000 \ \text{cm}^{-1}$. Because of this, infrared spectroscopy cannot be used to interrogate low-frequency vibrations. For example, the substrate–molecular adsorbate bond generally has a very low frequency. Surface phonons also have vibrational frequencies too low to be studied by IR spectroscopy.

IR spectroscopy has two other important characteristics. The first is that it is capable of very high spectral resolution ($<0.01 \ \text{cm}^{-1}$). This is much smaller than the linewidths observed for most adsorbates. High resolution studies, however, can lead to important information regarding vibrational energy transfer and dephasing. The second characteristic is that it can be used for high-pressure, *in situ* studies. In this case, any absorption from the gas phase must be subtracted. This characteristic allows for the study of working catalysts under realistic conditions.

2.7.2 Electron energy loss spectroscopy (EELS)

When electrons backscatter from a surface they can lose energy to the various degrees of freedom of the surface and adsorbed layer. Electron energy loss spectroscopy [23] relies upon the use of a monochromatic,

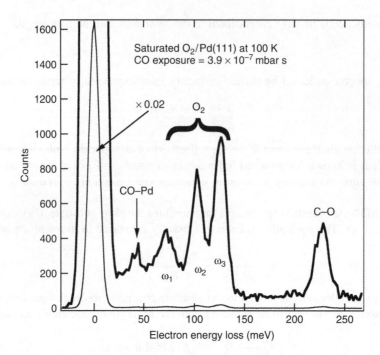

Figure 2.29   *The electron energy loss spectrum of co-adsorbed O_2 + CO on Pd(111). The species associated with ω_1, ω_2 and ω_3 are illustrated in Fig. 3.8. Adapted from K.W. Kolasinski, F. Cemič, A. de Meijere, E. Hasselbrink, Surf. Sci., 334, 19. © 1995 with permission from Elsevier.*

collimated beam of electrons, which is energy analyzed after it has scattered from the surface. The angular distribution of the scattered electrons contains further information. In principle, rotational, vibrational and electronic transitions can be observed in EELS. The study of vibrations by electron energy loss is often called high-resolution electron energy loss spectroscopy (HREELS) to differentiate it from the study of electronic transitions.

Figure 2.29 displays an EEL spectrum collected from a co-adsorbed layer of CO and O_2, which concentrates on the vibrational part of the spectrum. By far the largest peak in the spectrum is the elastic peak. The absolute energy of the elastically scattered electrons is unimportant as far as the position of peaks in the spectrum is concerned. The centre of the elastic peak, E_0, fixes the origin of the spectrum and is conventionally set equal to zero energy. The width of the elastic peak, conventionally expressed as a full width at half maximum (FWHM), is a measure of the spectral resolution.

The energy at which a peak occurs is

$$E = E_0 - \hbar\omega \tag{2.7.15}$$

where $\hbar\omega$ is the energy of the excited transition (generally in meV) and E is the detected energy. Thus, if we take $E_0 = 0$, the energy of a peak is the energy of the vibration excited by electron scattering.

2.7.2.1   *Three scattering mechanisms*

The proper interpretation of EEL spectra requires knowledge of how electrons scatter inelastically. The three mechanisms of scattering are dipole scattering, impact scattering and resonance scattering. Wave-particle duality lies behind these three scattering mechanisms.

In dipole scattering, the electric field associated with the charge of a moving electron interacts with the scatterer (adsorbate or surface phonon). The electron acts as a wave and the interaction is analogous to that of a photon in IR spectroscopy. On a metal, the same selection rule is followed, that is, only vibrations associated with a dipole moment change normal to a metal surface can be observed. Just as for photons, dipole scattered electrons emerge along the specular angle. The angle of incidence equals the angle of reflection, and both the elastically scattered and inelastically scattered electrons emerge along the same propagation direction. The analysis of the spectra associated with dipole scattering, therefore, follows the same methods as RAIRS. Whereas the energetic position of the loss peaks does not depend on E_0, the intensity of the peaks does. The intensity smoothly varies with E_0 roughly following $(E_0)^{1/2}$.

In impact scattering the electron acts like a particle. The electron bounces off the scatterer, experiencing a short-range interaction and exchanging momentum. Because of momentum exchange off-specular scattering ($\varphi_i \neq \varphi_f$) occurs. Thus, by measuring the angular distribution of the scattered electrons, impact and dipole scattering can be differentiated. The short-range interaction means that the selection rules valid for RAIRS are no longer appropriate for this scattering regime.

While we may attempt to classify an electron as either a wave or a particle, an electron is, after all, an electron. In resonance scattering [112], the electron interacts with the adsorbate as only an electron can. That is, the electron actually becomes trapped in a bound (or quasi-bound) electronic state. The trapping state can be a real excited state of the isolated ion. In electron scattering such a state is called a *Feshbach resonance*. Alternatively, a centrifugal barrier, which arises from the angular momentum associated with the electron-molecule scattering event, can trap an electron. This is known as a *shape resonance*. A shape resonance is a very short-lived state (on the order of a few femtoseconds). However, even a Feshbach resonance generally has a lifetime on the order of only $10^{-10} - 10^{-15}$ s. Resonance scattering is sensitively dependent on the incident electron energy, that is, resonances in the vibrational excitation probability are observed as a function of E_0. Resonances also have characteristic angular dependencies for both their excitation and decay. The angular distribution of the electrons scattered through a resonance provides information about the symmetry of the resonance, e.g. whether a σ or π symmetry state is involved. Because of the finite lifetime of the temporary negative ion state formed, the excitation of overtones often accompanies resonance scattering. In fact, this is one of the clearest signatures of this mechanism. It also is often the only way to study these higher lying vibrational states.

One of the great strengths of EELS is that it can interrogate the low frequency region ($0-800\,\mathrm{cm^{-1}}$) where IR cannot be used. Thus, EELS can be used to investigate both the substrate-adsorbate bond and phonons. The resolution of EELS is, however, significantly lower than that of IR spectroscopy. While it is relatively easy to obtain $10\,\mathrm{meV}$ resolution, it is only recently that resolution on the order of $1\,\mathrm{meV}$ ($\sim 8\,\mathrm{cm^{-1}}$) has been obtained. The use of an electron beam means that EELS is constrained to use in UHV. The intensity of loss features is dependent on the adsorbate coverage and, subject to the same caveats as for RAIRS, it can be used as a quantitative measure of adsorbate coverage.

2.8   Second harmonic and sum frequency generation

Second harmonic generation (SHG) and sum frequency generation (SFG) are two closely related techniques. They have found broad applications in surface and interface science ever since SHG was first used to probe adsorbate structure by Heinz, Tom and Shen [113], and SFG was first used to acquire vibrational spectra of adsorbed molecules by Zhu, Suhr and Shen [114]. The processes involved are illustrated in Fig. 2.30. Laser light illuminates an interface between two phases A and B, and excites a non-linear polarization response [115], Fig. 2.30(a). Two photons at the same frequency ω_1 can be mixed together to form a third photon at frequency $\omega = \omega_1 + \omega_1$. This is second harmonic generation, Fig. 2.30(b). Two photons with

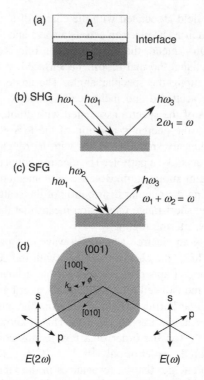

Figure 2.30 *(a) Two bulk materials joined by an interfacial region. (b) Two photons of the same frequency are mixed in second harmonic generation (SHG). (c) Two photons of different frequency are mixed in sum frequency generation (SFG). (d) The polarization components s and p are shown for a laser incident on a surface.*

distinct frequencies ω_1 and ω_2 and angles of incidence ϑ_1 and ϑ_2 can mix together to form a third photon at $\omega = \omega_1 + \omega_2$, which departs at ϑ_3. This is sum frequency generation, Fig. 2.30(c). The signal depends on the product of the intensities of the incident photons $I(\omega_1)$ and $I(\omega_2)$ according to [116]

$$I(\omega) \propto \left| e^\dagger(\omega) \cdot \chi^{(2)} : e(\omega_1) e(\omega_2) \right|^2 I(\omega_1) I(\omega_2). \tag{2.8.1}$$

Equation (2.8.1) is complex but can be understood in a straightforward conceptual manner with the aid of Fig. 2.30(d). The $e(\omega_i) \equiv F(\omega_i)\hat{e}(\omega_i)$ terms are the product of the transmission Fresnel factors and the unit polarization vectors of the optical electric fields $E(\omega_i)$. These terms introduce a dependence of the signal on the angles of incidence of the excitation lasers, the angle at which the signal appears, and the polarizations of the three optical fields.

Pulsed lasers are generally used to generate high intensities with low time averaged powers in order to maximize signals with minimal sample damage. The use of pulsed lasers in combination with the non-linear optical response described by Eq. (2.8.1) also opens up the possibility of introducing a time delay between photons at ω_1 and ω_2 to facilitate time-resolved measurements in SFG.

The second order non-linear response that determines the intensity of the SH or SF intensity is governed by the non-linear susceptibility $\chi^{(2)}$, which can be decomposed into contributions from the surface and the bulk

$$\chi^{(2)} = \chi_s^{(2)} + \chi_b^{(2)}. \tag{2.8.1}$$

The bulk term $\chi_b^{(2)}$ vanishes for electric dipole transitions for isotropic or centrosymmetric media as a result of the symmetry properties of the transition dipole moment. This means that there is no SH or SF response within the dipole approximation from vacuum or the bulk of a gas, a liquid or an elemental metal or semiconductor. This property makes SHG and SFG exquisitely surface and interface sensitive. It should be kept in mind, however, that for noncentrosymmetric media such as compound semiconductors or molecular solids, the bulk can contribute to the signal. For a well-ordered surface (or bulk when allowed) the orientation of the laser polarizations with respect to the surface crystallographic axes, ϕ in Fig. 2.30, also plays a role in the signal intensity [117].

If adsorbates are present at the interface, the second-order non-linear susceptibility contains contributions from both the substrate surface $\chi_{ss}^{(2)}$ and the adsorbates $\chi_{sa}^{(2)}$

$$\chi_s^{(2)} = \chi_{sa}^{(2)} + \chi_{ss}^{(2)} \tag{2.8.2}$$

$\chi^{(2)}$ is a third rank tensor with 27 elements. Many of these vanish due to symmetry constraints. The nonvanishing terms respond to the structure (orientation) of the adsorbate, and the polarization combination of the incident and detected polarizations. Polarization dependent SHG and SFG experiments can therefore be used to determine adsorbate structure.

Each term that contributes to $\chi_s^{(2)}$ can exhibit resonance behaviour at any of the frequencies (ω_1, ω_2 or ω) involved in the process if that frequency corresponds to a resonant transition in the vibrational or electronic structure of the adsorbate or surface. There is also a nonresonant response that contributes to the signal. One difficulty in the prediction of SH and SF spectra is that the resonant and nonresonant contributions add coherently, that is, their relative phases need to be taken into account, and the relative phase can have a frequency dependence.

In general, light detection in the visible is extremely efficient and much easier than in the infrared. Therefore, it is common in SHG and SFG to choose the frequencies such that the SH or SF signal occurs in the visible. Since its introduction by Shen [114], a very common combination is to choose an infrared frequency, which can be tuned through vibrational resonances of an adsorbate, and to mix it with a visible or near IR photon of fixed frequency to yield a visible photon. The fixed frequency pulse is called the up-conversion pulse. In time-resolved studies, the tuneable pulse is used to pump population into an excited state, and is called the pump pulse. The fixed frequency pulse (probe pulse) then up-converts the polarization field at the frequency of the first pulse into the photon that is detected. The signal intensity in such a pump-probe experiment is determined not only by the intensities of the pump and probe pulses, but also the lifetime of the excited state and the delay between the two pulses.

The versatility of SHG and SFG allow them to probe not only the UHV/solid interface [118] but also liquid [119, 120] and electrochemical interfaces [121]. It can be used to probe surface coverage [122], diffusion [123] and electron dynamics in real time [124]. Time-resolved vibrational studies [125] have led to a much greater fundamental understanding of energy transfer dynamics.

2.9 Other surface analytical techniques

A large fraction of research at surfaces throughout the 1970s and 1980s was devoted to developing an alphabet soup of surface sensitive analytical techniques. An extensive collection of these acronyms has been collected by Somorjai [126]. Experimentation in surface science requires the routine intervention of a multi-technique approach and, therefore, a familiarity with a large number of techniques is necessary for the surface scientist. For brevity, a number of these techniques have been left out, including synchrotron based techniques for structure determination [127], ion scattering [4], and He atom scattering [128].

2.10   Summary of important concepts

- Knudsen beams are molecular beams with thermal properties.
- Supersonic jets experience significant cooling during expansion, exhibit high translational energy and have significantly enhanced intensity compared to Knudsen beams.
- STM involves the tunnelling of electrons from occupied to unoccupied electronic states. The voltage between the tip and the surface determines the direction of current flow.
- STM images electronic states not atoms.
- AFM allows for atomic scale imaging on insulating surfaces and for direct measurements of intermolecular forces.
- NSOM extends optical spectroscopy to the nanoscale and even single molecule regime.
- Low energy ($\sim$20–500 eV) electrons penetrate only the first few atomic layers and can be used to investigate surface structure.
- The symmetry of LEED patterns is related to the periodicity of the substrate and adsorbate overlayer structure.
- XPS probes the electronic states associated with core levels and is particularly well suited to quantitative elemental analysis.
- UPS probes the electronic states associated with valence electrons, and is particularly well suited to the study of electronic changes associated with chemical bonding.
- AES is also used for quantitative elemental analysis.
- In IR spectroscopy at metal surfaces, a strict dipole selection rule means that only vibrations with a component along the surface normal can be observed.
- In EELS, electrons scatter through dipole, impact and/or resonance scattering mechanisms, and no strict selection rule can be assumed unless the mechanism is known.
- SHG and SFG can be interface sensitive even in the presence of a gas or liquid.
- SFG can be used not only to perform spectroscopy in the frequency domain, but also to perform pump-probe studies, which investigate dynamics directly in the time domain.

2.11   Frontiers and challenges

- For all forms of spectroscopy, higher energy resolution, lower detection sensitivity and higher data acquisition speed are the paths to the frontiers and the greatest challenges. However at surfaces, this comes with the bonus challenges of doing so with smaller spots sizes (higher spatial resolution) and less damage to the sample.
- Development of time-resolved microscopies such as ultrafast electron and x-ray diffraction techniques for four-dimensional electron microscopy [129] and dynamic transmission electron microscopy [130].
- Photoelectron spectroscopy with attosecond (10^{-18} s) resolution [131].
- Development of *in situ* photoelectron spectroscopy [132], x-ray and electron microscopies (i.e. scanning and transmission) that can be used to study interfaces at high pressures or in the presence of a liquid.
- Developing techniques for *routine* chemical analysis using scanning probe microscopes.
- Understanding SPM tip/surface or SPM tip/adsorbate interactions. How does tip size and shape affect AFM images? How does the electronic structure of the tip affect STM images and spectroscopic data? When and how do tip interactions influence SPM data?
- Building a complete IR spectroscopic database for adsorbates on well-defined single-crystal oxide surfaces comparable in quality and scope to that available for single-crystal metal surfaces.

2.12   Further reading

T. L. Alford, L. C. Feldman, and J. W. Mayer, *Fundamentals of Nanoscale Film Analysis*. (Springer Verlag, Berlin, 2007).

G. A. D Briggs and A. J Fisher, *STM experiment and atomistic modelling hand in hand: Individual molecules on semiconductor surfaces*, Surf. Sci. Rep. **33** (1999) 1.

F. Besenbacher, *Scanning tunnelling microscopy studies of metal surfaces*, Rep. Prog. Phys. **59** (1996) 1737.

Y. J. Chabal, *Surface infrared spectroscopy*, Surf. Sci. Rep. **8** (1988) 211.

R. C. Dunn, *Near-field scanning optical microscopy*, Chem. Rev. **99** (1999) 2891.

G. Ertl and J. Küppers, *Low Energy Electrons and Surface Chemistry* (VCH, Weinheim, 1985).

B. Feuerbacher and R. F. Willis, *Photoemission and electron states at clean surfaces*, J. Phys. C **9** (1976) 169.

A. S. Foster and W. A. Hofer, *Scanning Probe Microscopy: Atomic Scale Engineering by Forces and Currents*, Springer, New York, 2006.

K. Heinz, *LEED and DLEED as modern tools for quantitative surface structure determination*, Rep. Prog. Phys. **58** (1995) 637.

W. A. Hofer, A. S. Foster, and A. L. Shluger, *Theories of scanning probe microscopes at the atomic scale*, Rev. Mod. Phys. **75** (2003) 1287.

H. Ibach and D. L. Mills, *Electron Energy Loss Spectroscopy and Surface Vibrations* (Academic Press, New York, 1982).

J. W. Niemantsverdriet, *Spectroscopy in Catalysis*, 3rd ed. (Wiley-VCH, Weinheim, 2007).

A. Nilsson and L. G. M. Pettersson, *Chemical bonding on surfaces probed by X-ray emission spectroscopy and density functional theory*, Surf. Sci. Rep. **55** (2004) 49.

F. M. Mirabella, Jr. (Ed.), *Internal Reflection Spectroscopy: Theory and Applications* (Marcel Dekker, New York, 1992).

J. F. O'Hanlon, *A User's Guide to Vacuum Technology*, 3rd ed. (Wiley Interscience, New York, 2003).

E. W. Plummer and W. Eberhardt, *Angle-resolved photoemission as a tool for the study of surfaces*, Adv. Chem. Phys. **49** (1982) 533.

H. H. Rotermund, *Imaging of dynamic processes on surfaces by light*, Surf. Sci. Rep. **29** (1997) 265.

J. A. Stroscio and W. J. Kaiser (Eds), *Scanning Tunneling Microscopy. Methods of Experimental Physics* Vol. **27** (Academic Press, Boston, 1993).

H. Takano, J. R. Kenseth, S.-S. Wong, J. C. O'Brien, and M. D. Porter, *Chemical and biochemical analysis using scanning force microscopy*, Chem. Rev. **99** (1999) 2845.

S. O. Vansteenkiste, M. C. Davies, C. J. Roberts, S. J. B. Tendler, and P. M. Williams, *Scanning probe microscopy of biomedical interfaces*, Prog. Surf. Sci. **57** (1998) 95.

J.C. Vickerman, *Surface Analysis: The Principal Techniques*, 2nd ed. (John Wiley & Sons Ltd, Chichester, 2009).

D. P. Woodruff and T. A. Delchar, *Modern Techniques of Surface Science* (Cambridge University Press, Cambridge, 1994).

J. T. Yates, Jr. and T. E. Madey (Eds), *Vibrational Spectroscopy of Molecules on Surfaces* (Plenum Press, New York, 1987).

J. T. Yates, Jr., *Experimental Innovations in Surface Science: A Guide to Practical Laboratory Methods and Instruments*. (AIP Press (Springer-Verlag), New York, 1998).

2.13   Exercises

2.1   Derive Eq (2.1.2).

2.2   Show that the pressure p in a chamber of volume V, at initial pressure p_0, pumped with a pumping speed S changes as a function of time according to

$$p(t) = p_0 \exp(-t/\tau), \qquad (2.12.1)$$

where $\tau = V/S$ is the time constant of the chamber/pump combination.

2.3   (a) What is the gas flux striking a surface in air at 1 atm and 300 K?

(b) Calculate the pressure necessary to keep a $1 \, cm^2$ Pt(100) surface clean for 1 hr at 300 K, assuming a sticking coefficient of 1, no dissociation of the gas upon adsorption, and that "clean" means <0.01 ML of adsorbed impurities.

2.4   (a) Estimate the maximum exposure delivered per pulse (in units of ML/pulse) if N_2 is expanded through a chopped Knudsen source with a skimmer radius of $r_s = 50 \, \mu m$, an effective pressure of $p_0 = 20 \, Pa$ at the skimmer, a pulse length of 5 ms, and a skimmer to surface distance of $x = 15 \, cm$. Take the surface to be Si(111) with a density of surface atoms $\sigma_0 = 7.8 \times 10^{18} \, m^{-2}$.

(b) Using the same parameters, make an estimate for a hard expansion through a supersonic source with an effective Mach number of $M_{eff} = 20$.

2.5   Lieber and co-workers [133] used AFM to measure the adhesion force arising from the contact of CH_3 groups. The adhesion force was 1.0 nN. If the tip/surface contact area was $3.1 \, nm^2$ and the radius of a CH_3 group is $(0.2)^{1/2} \, nm$, calculate the interaction force resulting from the contact of two individual CH_3 groups.

2.6   In an STM image, does an adsorbate sitting on top of a surface always look like a raised bump compared to the substrate? Explain your answer.

2.7   Some adsorbates can be imaged in STM at low temperatures but disappear at higher temperatures even though they have not desorbed from the surface. Explain.

2.8   The dimer unit on a Si(100) surface has a bonding orbital just below E_F and an antibonding orbital just above E_F. Make a prediction about STM images that are taken at positive compared with negative voltages. Do the images look the same and, if not, how do they differ?

2.9   Use Eqs (2.5.19) and (2.5.20) to show that the diffraction reflexes appear at

$$\Delta s / \lambda = h_1 \mathbf{a}_1^* + h_2 \mathbf{a}_2^* \qquad (2.12.2)$$

Assume normal incidence of the incoming electrons and $\Delta \mathbf{s} = \mathbf{s} - \mathbf{s}_0$.

2.10   Describe how the spots in a LEED pattern would evolve if:

(a) incident molecules adsorbed randomly onto a surface forming an ordered overlayer only when one quarter of the substrate atoms are covered (a quarter of a monolayer);

(b) incident molecules formed ordered islands that continually grow in size until they reach a saturation coverage of a quarter of a monolayer;

(c) an overlayer is ordered in one direction but not in the orthogonal direction (either due to random adsorption or diffusion in the orthogonal direction).

2.11   Consider clean fcc(100), fcc(110) and fcc(111) surfaces (fcc = face-centred cubic). Draw the unit cell and include the primitive lattice vectors $\mathbf{a}_1$ and $\mathbf{a}_2$. Calculate the reciprocal lattice vectors $\mathbf{a}_1^*$ and $\mathbf{a}_2^*$ and draw the LEED pattern including the reciprocal lattice vectors.

2.12   For structures (a)–(i) in Fig. 2.31 determine the associated LEED patterns. Classify the structures in both Wood's and matrix notation.

2.13   Fractional coverage can be defined as the number of adsorbates divided by the number of surface atoms:

$$\theta = \frac{N_{ads}}{N_0}. \qquad (2.12.3)$$

For each of the structures in Exercise 2.12, calculate the coverage. Note any correlations between coverage and the LEED patterns.

2.14   Given LEED patterns (a)–(i) in Fig. 2.32 obtained from adsorbate-covered face-centred cubic (fcc) substrates, determine the surface structures. Substrate reflexes are marked ● while the additional adsorbate induced reflexes are marked ×. Assume no reconstruction of the surface.

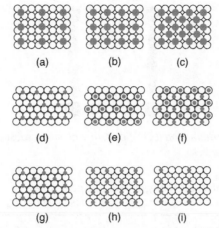

Figure 2.31   *Structures (a)–(i): see Exercise 2.11.*

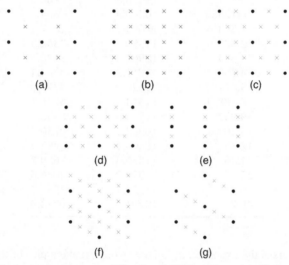

Figure 2.32   *Low-energy electron diffraction (LEED) patterns (a)–(i): see Exercise 2.13.*

2.15   When O atoms adsorb on Ag(110), a series of streaky LEED patterns are observed. The LEED patterns are elongated in one direction as shown in Fig. 2.33. Describe the structure of the overlayer and what might be causing this phenomenon. Refer directly to the structure of the overlayer with respect to the Ag(110) surface.

2.16   Determine all of the x-ray levels that are possible for the $n = 3$ shell.

2.17   Show that XPS can be made more surface sensitive by detecting the photoelectrons in the off normal direction.

2.18   Explain why Auger electron spectroscopy is surface sensitive. Are Auger peaks recorded at 90, 350 and 1500 eV equally surface sensitive?

2.19   The Auger data for the intensity of the Pt transition at 267 eV and the C transition at 272 eV found in Table 2.3 were taken for adsorption of CH_4 on Pt(111) at $T_s = 800$ K, $T_{gas} = 298$ K. At this temperature CH_4 dissociates and H_2 desorbs leaving C(a) on the surface. Use the data to make a

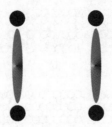

Figure 2.33 *Low-energy electron diffraction (LEED) pattern with sharp substrate spots (•) and streaky adsorbate overlayer related spots: see Exercise 2.15.*

Table 2.3 *Integrated Auger peak areas measured for Pt and C after dosing with CH₄*

$\varepsilon/10^{20}$ cm^{-2}	I_{Pt}	I_C
0	11350.8	166.2
0.1758	10248.4	664.7
1.593	11693.5	1871.3
2.477	11360.1	2382.5
3.454	10967.6	3178.5
4.290	11610.2	4464.4
5.071	9626.0	4277.8
6.188	11313.7	5876.1
9.957	11575.4	8358.7
14.20	10964.3	8730.4
17.35	10864.1	10292.9
21.49	11850.8	11689.7
25.65	12912.7	13086.6
30.51	11529.1	12753.0
48.75	10791.4	10713.9

plot of θ_{Pt} and θ_C versus the exposure, ε, in other words, make a plot of the uncovered and covered surface sites as a function of ε. The AES sensitivity factors are $s_{Pt} = 0.030$ and $s_C = 0.18$.

2.20 Consider the spectrum of adsorbed CO in Fig. 2.25. CO is adsorbed as a gem-dicarbonyl on the Rh atoms present on the Al_2O_3 substrate as shown in Fig. 2.34.

(a) If the CO molecules in the gem-dicarbonyl were independent oscillators, only one CO stretch peak would be observed. Explain why there are two peaks in the spectrum [134].

(b) Explain why substitution of ^{18}O for ^{16}O changes the positions of the bands.

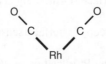

Figure 2.34 *CO adsorbed as a gem-dicarbonyl on rhodium atoms present on an Al_2O_3 substrate (substrate not shown).*

References

[1]   J. C. Vickerman, *Surface Analysis: The Principal Techniques*, 2nd Edition, John Wiley & Sons Ltd, Chichester, 2009.

[2]   G. Ertl, J. Küppers, *Low Energy Electrons and Surface Chemistry*, 2nd ed., VCH, Weinheim, 1985.

[3]   D. P. Woodruff, T. A. Delchar, *Modern Techniques of Surface Science*, 2nd ed., Cambridge University Press, Cambridge, 1994.

[4]   T. L. Alford, L. C. Feldman, J. W. Mayer, *Fundamentals of Nanoscale Film Analysis*, Springer Verlag, Berlin, 2007.

[5]   J. T. Yates, Jr., *Experimental Innovations in Surface Science: A Guide to Practical Laboratory Methods and Instruments*, AIP Press (Springer-Verlag), New York, 1998.

[6]   J. A. R. Samson, *Techniques of Vacuum Ultraviolet Spectroscopy*, John Wiley & Sons, Inc., New York, 1967.

[7]   C. Kunz (Ed.), *Synchrotron Radiation: Techniques and Applications, Topics in Current Physics, Vol. 10*, Springer-Verlag, Berlin, 1979.

[8]   R. H. Lipson, S. S. Dimov, P. Wang, Y. J. Shi, D. M. Mao, X. K. Hu, J. Vanstone, *Instrum. Sci. Technol.*, **28** (2000) 85.

[9]   T. Warwick, J. Arthur, H. A. Padmore, J. Stöhr (Eds), *Synchrotron Radiation Instrumentation, AIP Conf. Proc., Vol. 705*, American Institute of Physics, Melville, NY, 2004.

[10] M.-E. Couprie, J.-M. Ortéga, *Analysis*, **28** (2000) 725.

[11] P. G. O'Shea, H. P. Freund, *Science*, **292** (2001) 1853.

[12] J. Feldhaus, J. Arthur, J. B. Hastings, *J. Phys. B: At., Mol. Opt. Phys.*, **38** (2005) S799.

[13] B. J. Siwick, J. R. Dwyer, R. E. Jordan, R. J. D. Miller, *J. Appl. Phys.*, **92** (2002) 1643.

[14] M. Knudsen, *Ann. Phys.*, **353** (1916) 1113.

[15] M. Knudsen, *Ann. Phys.*, **333** (1909) 999.

[16] D. R. Miller (Ed.), *Free Jet Sources, Vol. 1*, Oxford University Press, New York, 1988.

[17] J. B. Anderson, R. P. Andres, J. B. Fenn, *Adv. Chem. Phys.*, **10** (1966) 275.

[18] G. Scoles (Ed.), *Atomic and Molecular Beam Methods, Vol. 1*, Oxford University Press, New York, 1988.

[19] G. Scoles (Ed.), *Atomic and Molecular Beam Methods, Vol. 2*, Oxford University Press, New York, 1992.

[20] G. Binnig, H. Rohrer, C. Gerber, E. Weibel, *Phys. Rev. Lett.*, **49** (1982) 57.

[21] D. M. Eigler, E. K. Schweizer, *Nature (London)*, **344** (1990) 524.

[22] P. Moriarty, *Rep. Prog. Phys.*, **64** (2001) 297.

[23] H. Ibach, D. L. Mills, *Electron Energy Loss Spectroscopy and Surface Vibrations*, Academic Press, New York, 1982.

[24] K. Itaya, *Prog. Surf. Sci.*, **58** (1998) 121.

[25] J. Tersoff, N. D. Lang, Theory of scanning tunneling microscopy, in *Scanning Tunneling Microscopy* (Eds: J. A. Stroscio, W. J. Kaiser), Academic Press, Boston, 1993, p. 1.

[26] Y. Sugimoto, P. Pou, M. Abe, P. Jelinek, R. P. Morita, Ó. Custance, *Nature (London)*, **446** (2007) 64.

[27] J. A. Stroscio, F. Tavazza, J. N. Crain, R. J. Celotta, A. M. Chaka, *Science*, **313** (2006) 948.

[28] D. M. Cyr, B. Venkataraman, G. W. Flynn, A. Black, G. M. Whitesides, *J. Phys. Chem.*, **100** (1996) 13747.

[29] B. Venkataraman, G. W. Flynn, J. L. Wilbur, J. P. Folkers, G. M. Whitesides, *J. Phys. Chem.*, **99** (1995) 8684.

[30] J. A. Stroscio, R. M. Feenstra, Methods of tunneling spectroscopy, in *Scanning Tunneling Microscopy* (Eds: J. A. Stroscio, W. J. Kaiser), Academic Press, Boston, 1993, p. 96.

[31] G. Binnig, C. F. Quate, C. Gerber, *Phys. Rev. Lett.*, **56** (1986) 930.

[32] H. Takano, J. R. Kenseth, S.-S. Wong, J. C. O'Brien, M. D. Porter, *Chem. Rev.*, **99** (1999) 2845.

[33] F. Ohnesorge, G. Binnig, *Science*, **260** (1993) 1451.

[34] F. J. Giessibl, *Science*, **267** (1995) 68.

[35] C. L. Cheung, J. H. Hafner, T. W. Odom, K. Kim, C. M. Lieber, *Appl. Phys. Lett.*, **76** (2000) 3136.

[36] J.-J. Greffet, R. Carminati, *Prog. Surf. Sci.*, **56** (1997) 133.

[37] W. Min, C. W. Freudiger, S. J. Lu, X. S. Xie, *Annual Review of Physical Chemistry, Vol 62*, **62** (2011) 507.

[38] S. W. Hell, *Science*, **316** (2007) 1153.

[39] E. H. Synge, *Philos. Mag.*, **6** (1928) 356.

[40]   E. H. Synge, *Philos. Mag.*, **11** (1931) 65.

[41]   E. H. Synge, *Philos. Mag.*, **13** (1932) 297.

[42]   E. A. Ash, G. Nicholls, *Nature (London)*, **237** (1972) 510.

[43]   D. W. Pohl, W. Denk, M. Lanz, *Appl. Phys. Lett.*, **44** (1984) 651.

[44]   A. Harootunian, E. Betzig, A. Lewis, M. Isaacson, *Appl. Phys. Lett.*, **49** (1986) 674.

[45]   R. C. Dunn, *Chem. Rev.*, **99** (1999) 2891.

[46]   E. Betzig, R. J. Chichester, *Science*, **262** (1993) 1422.

[47]   M. P. Seah, W. A. Dench, *Surf. Interface Anal.*, **1** (1979) 2.

[48]   D. R. Penn, *Phys. Rev. B*, **35** (1987) 482.

[49]   C. J. Powell, A. Jablonski, *J. Phys. Chem. Ref. Data*, **28** (1999) 19.

[50]   C. J. Powell, S. Tanuma, D. R. Penn, *Surf. Interface Anal.*, **43** (2011) 689.

[51]   S. Tanuma, C. J. Powell, D. R. Penn, *J. Appl. Phys.*, **103** (2008) 063707.

[52]   S. Tanuma, C. J. Powell, D. R. Penn, *Surf. Interface Anal.*, **21** (1994) 165.

[53]   G. Gergely, *Prog. Surf. Sci.*, **71** (2002) 31.

[54]   M. A. Van Hove, *Low-Energy Electron Diffraction: Experiment, Theory and Surface Structure*, Springer-Verlag, Berlin, 1986.

[55]   K. Heinz, *Rep. Prog. Phys.*, **58** (1995) 637.

[56]   K. Heinz, *Curr. Opin. Solid State Mater. Sci.*, **3** (1998) 434.

[57]   R. L. Park, H. H. Madden, Jr., *Surf. Sci.*, **11** (1968) 188.

[58]   E. A. Wood, *J. Appl. Phys.*, **35** (1964) 1306.

[59]   C. Kittel, *Introduction to Solid State Physics*, 6th ed., John Wiley & Sons, Inc., NY, 1986.

[60]   D. A. Shirley, Many-electron and final-state effects: Beyond the one-electron picture, in *Photoemission in Solids. I. General Principles* (Eds: M. Cardona, L. Ley), Springer-Verlag, Berlin, 1978, p. 165.

[61]   C. J. Powell, A. Jablonski, *J. Electron Spectrosc. Relat. Phenom.*, **178** (2010) 331.

[62]   H. R. Hertz, *Ann. Phys.*, **31** (1887) 983.

[63]   A. Einstein, *Ann. Phys.*, **17** (1905) 132.

[64]   D. E. Eastman, J. K. Cashion, *Phys. Rev. Lett.*, **27** (1971) 1520.

[65]   R. Haight, P. F. Seidler, *Appl. Phys. Lett.*, **65** (1994) 517.

[66]   R. Haight, *Surf. Sci. Rep.*, **21** (1995) 275.

[67]   J. H. Bechtel, W. L. Smith, N. Bloembergen, *Phys. Rev. B*, **15** (1977) 4557.

[68]   H. W. Rudolf, W. Steinmann, *Phys. Lett.*, **61A** (1977) 471.

[69]   H.-L. Dai, W. Ho (Ed.), *Laser Spectroscopy and Photochemistry on Metal Surfaces: Part 1, Vol. 5*, World Scientific, Singapore, 1995.

[70]   P. M. Echenique, R. Berndt, E. V. Chulkov, T. Fauster, A. Goldmann, U. Höfer, *Surf. Sci. Rep.*, **52** (2004) 219.

[71]   J. Gudde, U. Höfer, *Prog. Surf. Sci.*, **80** (2005) 49.

[72]   J. Gudde, W. Berthold, U. Höfer, *Chem. Rev.*, **106** (2006) 4261.

[73]   U. Höfer, E. Umbach, *J. Electron Spectrosc. Relat. Phenom.*, **54/55** (1990) 591.

[74]   R. Matzdorf, *Surf. Sci. Rep.*, **30** (1997) 153.

[75]   E. W. Plummer, W. Eberhardt, *Adv. Chem. Phys.*, **49** (1982) 533.

[76]   F. Reinert, S. Hufner, *New Journal of Physics*, **7** (2005) 97.

[77]   J. Küppers, K. Wandelt, G. Ertl, *Phys. Rev. Lett.*, **43** (1979) 928.

[78]   W. Steinmann, T. Fauster, Two-photon photoelectron spectroscopy of electronic states at metal surfaces, in *Laser Spectroscopy and Photochemistry on Metal Surfaces. Part I* (Eds: H.-L. Dai, W. Ho), World Scientific, Singapore, 1995, p. 184.

[79]   V. Dose, *Surf. Sci. Rep.*, **5** (1985) 337.

[80]   H. Petek, S. Ogawa, *Prog. Surf. Sci.*, **56** (1997) 239.

[81]   W. S. Fann, R. Storz, H. W. K. Tom, J. Bokor, *Phys. Rev. Lett.*, **68** (1992) 2834.

[82]   H. Petek, M. J. Weida, H. Nagano, S. Ogawa, *Science*, **288** (2000) 1402.

[83]   P. Auger, *J. Phys. Radium*, **6** (1925) 205.

[84]   K. D. Childs, B. A. Carlson, L. A. LaVanier, J. F. Moulder, D. F. Paul, W. F. Stickle, D. G. Watson, *Handbook of Auger Electron Spectroscopy: A Book of Reference Data for Identification and Interpretation in Auger Electron Spectroscopy*, 3rd ed., Physical Electronics, Inc., Eden Prairier, MN, 1995.

[85]   W. Swiech, G. H. Fecher, C. Ziethen, O. Schmidt, G. Schönhense, K. Grzelakowski, C. M. Schneider, R. Frömter, H. P. Oepen, J. Kirschner, *J. Electron Spectrosc. Relat. Phenom.*, **84** (1997) 171.

[86]   S. Hofmann, *Rep. Prog. Phys.*, **61** (1998) 827.

[87]   H. H. Rotermund, *Surf. Sci. Rep.*, **29** (1997) 265.

[88]   R. Imbihl, G. Ertl, *Chem. Rev.*, **95** (1995) 697.

[89]   J. T. Yates, Jr., T. E. Madey (Ed.), *Vibrational Spectroscopy of Molecules on Surfaces*, Plenum Press, New York, 1987.

[90]   M. Pemble, Vibrational spectroscopy from surfaces, in *Surface Analysis: The Principal Techniques* (Ed.: J. C. Vickerman), John Wiley & Sons Ltd, Chichester, 1997, p. 267.

[91]   B. E. Hayden, Reflection absorption infrared spectroscopy, in *Vibrational Spectroscopy of Molecules on Surfaces, Vol. 1* (Eds: J. T. Yates, Jr., T. E. Madey), Plenum Press, New York, 1987, p. 267.

[92]   Y. J. Chabal, *Surf. Sci. Rep.*, **8** (1988) 211.

[93]   A. P. Graham, J. P. Toennies, *Surf. Sci.*, **428** (1999) 1.

[94]   W. H. Weinberg, Inelastic electron tunneling spectroscopy of supported homogeneous cluster compounds, in *Vibrational Spectra and Structure, Vol. 11* (Ed.: J. R. Durig), Elsevier, Amsterdam, 1982, p. 1.

[95]   B. C. Stipe, M. A. Rezaei, W. Ho, *Science*, **280** (1998) 1732.

[96]   A. L. Harris, K. Kuhnke, M. Morin, P. Jakob, N. J. Levinos, Y. J. Chabal, *Faraday Discuss.*, **96** (1993) 217.

[97]   Y. J. Chabal, A. L. Harris, K. Raghavachari, J. C. Tully, *Internat. J. Mod. Phys. B*, **7** (1993) 1031.

[98]   M. Morin, P. Jakob, N. J. Levinos, Y. J. Chabal, A. L. Harris, *J. Chem. Phys.*, **96** (1992) 6203.

[99]   R. R. Cavanagh, J. D. Beckerle, M. P. Casassa, E. J. Heilweil, J. C. Stephenson, *Surf. Sci.*, **269/270** (1992) 113.

[100]   J. D. Beckerle, R. R. Cavanagh, M. P. Casassa, E. J. Heilweil, J. C. Stephenson, *J. Chem. Phys.*, **95** (1991) 5403.

[101]   J. D. Beckerle, M. P. Casassa, E. J. Heilweil, R. R. Cavanagh, J. C. Stephenson, *J. Electron Spectrosc. Relat. Phenom.*, **54/55** (1990) 17.

[102]   J. D. Beckerle, M. P. Casassa, R. R. Cavanagh, E. J. Heilweil, J. C. Stephenson, *J. Chem. Phys.*, **90** (1989) 4619.

[103]   E. J. Heilweil, M. P. Casassa, R. R. Cavanagh, J. C. Stephenson, *J. Chem. Phys.*, **81** (1984) 2856.

[104]   R. P. Eischens, S. A. Francis, W. A. Pliskin, *J. Phys. Chem.*, **60** (1956) 194.

[105]   R. P. Eischens, W. A. Pliskin, *Adv. Catal.*, **10** (1958) 1.

[106]   N. Sheppard, D. J. C. Yates, *Proc. R. Soc. London, A*, **238** (1956) 69.

[107]   N. Sheppard, T. T. Nguyen, The vibrational spectra of carbon monoxide chemisorbed on the surfaces of metal catalysts - A suggested scheme of interpretation, in *Advances in Infrared and Raman Spectroscopy, Vol. 5* (Eds: R. E. Hester, R. J. H. Clark), Heyden and Son, London, 1978.

[108]   R. G. Greenler, *J. Chem. Phys.*, **44** (1966) 310.

[109]   R. G. Greenler, R. R. Rahn, J. P. Schwartz, *J. Catal.*, **23** (1971) 42.

[110]   F. A. Cotton, *Chemical Applications of Group Theory*, 2nd ed., John Wiley & Sons, Inc., NY, 1971.

[111]   N. V. Richardson, N. Sheppard, Normal modes at surfaces, in *Vibrational Spectroscopy of Molecules on Surfaces, Vol. 1* (Eds: J. T. Yates, Jr., T. E. Madey), Plenum Press, New York, 1987, p. 267.

[112]   R. E. Palmer, P. J. Rous, *Rev. Mod. Phys.*, **64** (1992) 383.

[113]   T. F. Heinz, H. W. K. Tom, Y. R. Shen, *Phys. Rev. A*, **28** (1983) 1883.

[114]   X. D. Zhu, H. Suhr, Y. R. Shen, *Phys. Rev. B*, **35** (1987) 3047.

[115]   V. I. Gavrilenko, *Optics of Nanomaterials*, Pan Stanford World Scientific, Singapore, 2010.

[116]   M. B. Raschke, Y. R. Shen, *Curr. Opin. Solid State Mater. Sci.*, **8** (2004) 343.

[117]   T. A. Germer, K. W. Kolasinski, L. J. Richter, J. C. Stephenson, *Phys. Rev. B*, **55** (1997) 10694.

[118]   M. C. Downer, B. S. Mendoza, V. I. Gavrilenko, *Surf. Interface Anal.*, **31** (2001) 966.

[119]   A. J. Hopkins, C. L. McFearin, G. R. Richmond, *Curr. Opin. Solid State Mater. Sci.*, **9** (2005) 19.

[120]   G. L. Richmond, *Chem. Rev.*, **102** (2002) 2693.

[121]   F. Vidal, B. Busson, A. Tadjeddine, *Chem. Phys. Lett.*, **403** (2005) 324.

[122]   M. Dürr, U. Höfer, *Surf. Sci. Rep.*, **61** (2006) 465.

[123]   X. D. Zhu, T. Rasing, Y. R. Shen, *Phys. Rev. Lett.*, **61** (1988) 2883.

[124]   C. Voelkmann, M. Reichelt, T. Meier, S. W. Koch, U. Höfer, *Phys. Rev. Lett.*, **92** (2004) 127405.

[125]   H. Arnolds, M. Bonn, *Surf. Sci. Rep.*, **65** (2010) 45.

[126]   G. A. Somorjai, *Introduction to Surface Chemistry and Catalysis*, John Wiley & Sons, Inc., NY, 1994.

[127]   W. R. Flavell, Surface structure determination by interference techniques, in *Surface Analysis – The Principal Techniques* (Ed.: J. C. Vickerman), John Wiley & Sons Ltd, Chichester, 1997, p. 313.

[128]   A. P. Graham, *Surf. Sci. Rep.*, **49** (2003) 115.

[129]   A. H. Zewail, *Science*, **328** (2010) 187.

[130]   J. S. Kim, T. LaGrange, B. W. Reed, M. L. Taheri, M. R. Armstrong, W. E. King, N. D. Browning, G. H. Campbell, *Science*, **321** (2008) 1472.

[131]   A. L. Cavalieri, N. Müller, T. Uphues, V. S. Yakovlev, A. Baltuka, B. Horvath, B. Schmidt, L. Blümel, R. Holzwarth, S. Hendel, M. Drescher, U. Kleineberg, P. M. Echenique, R. Kienberger, F. Krausz, U. Heinzmann, *Nature (London)*, **449** (2007) 1029.

[132]   M. Salmeron, R. Schlögl, *Surf. Sci. Rep.*, **63** (2008) 169.

[133]   C. D. Frisbie, L. F. Rozsnyai, A. Noy, M. S. Wrighton, C. M. Lieber, *Science*, **265** (1994) 2071.

[134]   J. T. Yates, Jr., K. Kolasinski, *J. Chem. Phys.*, **79** (1983) 1026.

3

Chemisorption, Physisorption and Dynamics

In Chapter 1 we discussed the structure of adsorbates on surfaces and in Chapter 2 the methods we use to probe them. Now we turn to describing the interactions that hold adsorbates onto surfaces and the processes that get them there. In Chapter 4 we discuss the thermodynamics and kinetics of adsorption and the adsorbed phase. In this chapter, we investigate those aspects of energetics that control the motions and binding of atoms and molecules to surfaces. We concentrate mainly on the interactions of individual molecules with clean surface. The effects of higher coverages are dealt with in Chapter 4.

3.1 Types of interactions

When a molecule sticks to a surface, it can bind with either a chemical interaction (chemisorption) or a physical interaction (physisorption). Chemisorption involves the formation of a chemical bond between the adsorbate and the surface. Physisorption involves weaker interactions involving the polarization of the adsorbate and surface rather than electron transfer between them. Table 3.1 compares and contrasts several general aspects of chemisorption and physisorption.

While chemisorption and physisorption may seem like nicely distinguishable categories, there exists a more-or-less continuous spectrum of interaction strengths from one to the other. Nevertheless, it is a useful distinction to make in most cases. At 0 K all molecular motion apart from that associated with zero point energy ceases. This means that all chemical substances condense if the temperature is low enough. Another way to interpret this is that any chemical substance adsorbs (become either chemisorbed or physisorbed) if the temperature is low enough. Liquid nitrogen and liquid helium are the two most commonly used cryogenic coolants and, therefore, two important low temperatures to remember are the boiling points of N_2 (77 K) and He (4 K). Most substances stick to a surface held at l-N_2 temperature and virtually everything sticks at l-He temperature.

Chemisorption is highly directional, as are all chemical bonds. Therefore, adsorbates that are chemisorbed, (chemisorbates) stick at specific sites and they exhibit a binding interaction that depends strongly on their exact position and orientation with respect to the substrate. On metals, chemisorbed atoms tend to sit on sites of the highest co-ordination. For instance, O atoms on Pt(111) bind on the fcc three-fold hollow sites with a bond energy of $\sim$370 kJ mol^{-1} [1]. This is not surprising since a crystal of a fcc metal is constructed by stacking the atoms one layer on top of the next by placing the atoms in the sites of highest co-ordination. There are, of course, exceptions. H atoms usually bind on the

Surface Science: Foundations of Catalysis and Nanoscience, Third Edition. Kurt W. Kolasinski.
© 2012 John Wiley & Sons, Ltd. Published 2012 by John Wiley & Sons, Ltd.

Table 3.1   *A comparison between chemisorption and physisorption*

Chemisorption	Physisorption
electron exchange	polarization
chemical bond formation	van der Waals attractions
strong	weak
≥ 1 eV (100 kJ mol^{-1})	≤ 0.3 eV (30 kJ mol^{-1}), stable only at cryogenic temperatures
highly corrugated potential	less strongly directional
analogies with co-ordination chemistry	

sites of highest co-ordination, but can occupy two-fold sites on W(100) [2]. The diamond lattice of a semiconductor is composed of atoms stuck successively on the tetrahedrally directed dangling bonds of the surface. Thus, atoms tend to chemisorb on the highly localized dangling bonds of semiconductors in preference to sites of high co-ordination.

One important exception to the above rules is when an adsorbate forms such a strong bond with the substrate that it reacts to form a new compound. Oxygen forms strong bonds with a number of elements, for instance, Fe, Al and Si are highly susceptible to oxidation. Therefore, oxygen chemisorption can easily lead to formation not only of surface bound O atoms but also O bound in subsurface sites. The bonding of O atoms between and below the surface atoms of the substrate can be thought of as precursors to the formation of, e.g. Fe_2O_3, Al_2O_3 and SiO_2. H atoms, because of their small size, are also highly susceptible to the occupation of subsurface sites. Of particular interest is the H/Pd system [3, 4], in which H atoms can pack into the Pd lattice to a density greater than that of liquid H_2.

Physisorbed species (physisorbates) do not experience such strongly directional interactions. Therefore, they are more tenuously bound to specific sites and experience an attractive interaction with the surface that is much more uniform across the surface. In many cases, the interactions between physisorbates are as strong as or stronger than the interaction with the surface.

3.2   Binding sites and diffusion

The binding energy of an adsorbate depends on its position on the surface. This means not only that there are different binding sites at the surface but also that the sites are separated by energetic barriers, as shown in Fig. 3.1. To move across the surface, an adsorbate has to hop from one site to the next via pathways that traverse these barriers. Therefore, the barriers that separate the binding sites represent diffusion barriers. Figure 3.1(a) can be thought of as a one-dimensional (1D) potential energy surface (PES). Motion from one well to the next represents diffusion. Each well may also contain a number of bound vibrational states, which represent the vibrations of the adsorbate against the surface.

Consistent with Fig. 3.1, diffusion [5] is an activated process. For a uniform potential, we therefore expect diffusion to be governed by a simple Arrhenius form

$$D = D_0 \exp(-E_{\text{dif}}/RT) \tag{3.2.1}$$

where D is the diffusion coefficient, D_0 the diffusion pre-factor and E_{dif} is the activation energy for diffusion.

There are two extremes of temperature for which Eq. (3.1.1) does not describe diffusion. At very low temperatures for light adsorbates such as H and D, quantum effects can predominate [6]. In this

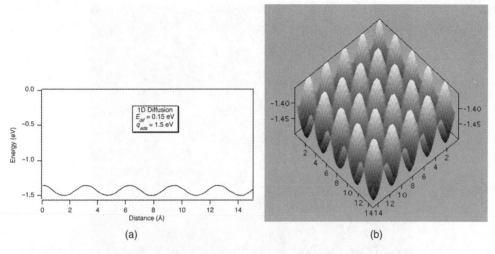

Figure 3.1 *The interaction potential of an adsorbate is corrugated as can be seen in these (a) 1D and (b) 2D representations of energy versus position on ideal defect-free surfaces.*

tunnelling regime, diffusion is independent of temperature. When the temperature is sufficiently high such that $RT_s \gg E_{dif}$, the adsorbate translates freely across the surface, performing a type of Brownian motion. In other words, it is bound in the z direction but it is not bound in x and y directions. This state of free 2D motion is known as a two-dimensional gas. The greater the corrugation in the adsorbate/surface potential, the larger the effective diffusion barrier experienced by the adsorbate and the higher T_s must be to form a 2D gas. Chemisorbates experience greater diffusion barriers than physisorbates. In most cases, the T_s required to form a 2D gas is sufficiently high to engender a significant rate of desorption.

For temperatures RT_s not $\gg E_{dif}$, we can interpret Eq. (3.1.1) directly. D is related to the hopping frequency, ν, the mean-square hopping length (related to the distance between sites) and the dimensionality of diffusion by

$$D = \nu d^2 / 2b. \tag{3.2.2}$$

For one-dimensional diffusion $b = 1$, whereas for uniform diffusion in a plane, $b = 2$. The root mean square distance, $\langle x^2 \rangle^{1/2}$, travelled by an adsorbate diffusing in 1D in a time t is given by

$$\langle x^2 \rangle^{1/2} = \sqrt{2Dt}. \qquad \text{(Uniform 1D Potential)} \tag{3.2.3}$$

Diffusion of a single particle in a uniform 1D potential is particularly simple; nonetheless, analysis of 1D diffusion and our knowledge of adsorbate interactions lead to several important conclusions. Figure 3.1 demonstrates that the adsorbate/surface potential is corrugated. This corrugation takes two forms: geometrical, as probed, e.g. by STM, and energetic, which results in diffusion barriers. Since different binding sites have different binding energies, they also exhibit different diffusion barriers. In a 1D potential, the highest barrier determines the overall rate of diffusion.

A surface is two-dimensional and this makes diffusion over a surface somewhat more complex. For a simple uniform 2D PES, we substitute $b = 2$ into Eq. (3.1.2) and arrive at

$$\langle x^2 \rangle^{1/2} = \sqrt{4Dt}. \qquad \text{(Uniform 2D Potential)} \tag{3.2.4}$$

However, the diffusion barrier is not always uniform across the surface, that is, E_{dif} depends on the direction in which the adsorbate diffuses. In this case, adsorbates diffuse anisotropically. This can lead to

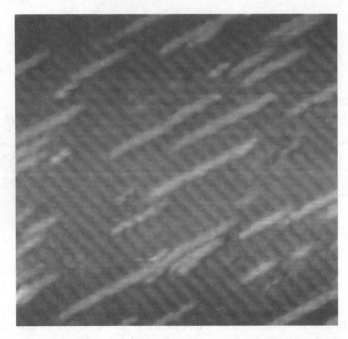

Figure 3.2   *Anisotropic growth of Si islands on Si, resulting from anisotropic diffusion. The lighter features in this scanning tunnelling microscope image are the Si atoms that have aggregated into islands on top of the Si(100) substrate. Reproduced from Y.W. Mo, R. Kariotis, B. S. Swartzentruber, M. B. Webb, M. G. Lagally, J. Vac. Sci. Technol., A, 8, 201. © (1990) with permission from AVS The Science & Technology Society.*

the formation of islands of adsorbates that do not have uniform shapes in both the x and y directions. For instance, on surfaces that exhibit rows and troughs, such as fcc(110) or the Si(100)–(2 × 1) reconstruction as shown in Fig. 3.2, diffusion along the rows is much easier than across the rows. If the diffusing adsorbates tend to stop when they meet each other, they form string-like islands that are much longer in the direction parallel to the row than perpendicular to it. Swartzentruber [7], for instance, has used an STM to track the motion of Si atoms deposited on a Si(100)–(2 × 1) surface. The atoms rapidly diffuse to form dimers (called addimers). The addimers also diffuse readily along the rows of dimers on the substrate. Diffusion in this direction can occur either directly over the dimers or between the dimers. Although the between-dimer site is the most energetically favourable, there is a barrier between the on-top and between-dimer positions. Therefore, addimers can diffuse for long periods up and down the rows before they hop into the more stable between-dimer sites. Defects have a great influence on diffusion. Defects can facilitate the transfer from on-top to between-dimer sites. In other cases, defects may act as repulsive walls or as sinks that trap adsorbates.

A typical defect, which we have already encountered, is a step. The electronic structure of steps differs from that of terraces. This leads to a significantly different binding potential at steps and likewise changes the diffusion barriers near steps. Frequently, as shown in Fig. 3.3, diffusion in the step-down direction – that is, from one terrace across a step to the terrace below – is more highly activated than is diffusion on the terrace. This additional barrier height is known as the Ehrlich-Schwoebel barrier, E_s [8, 9]. For the type of potential shown in Fig. 3.3, step-up diffusion is more highly activated than step-down or terrace diffusion. Consequently, only terrace and step-down diffusion are expected unless the temperature

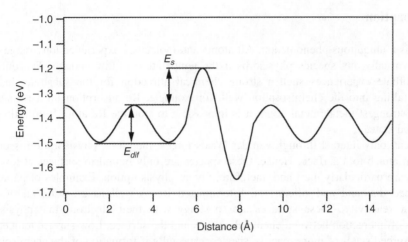

Figure 3.3 *A step changes the diffusion activation energy, E_{dif} [one-dimensional (1D) diffusion]. Step-up diffusion is often negligible because of the increased barrier; note also the increased binding strength at the bottom of the step – a feature that is often observed. E_s, Ehrlich-Schwoebel barrier.*

is high. Other types of defects as well as adsorbates also influence diffusion at surfaces [10]. The potentials shown in Fig. 3.1 are single-adsorbate potentials. Bowker and King [11, 12], and Reed and Ehrlich [13] have shown that lateral interactions play an important role in diffusion. Whereas D is constant in the absence of lateral interactions, repulsive interactions increase the value of D and attractive interactions decrease it. Thus, the study of diffusion profiles provides one means of measuring the strength of lateral interactions.

Molecules execute the type of diffusion described above. Atoms that do not interact too strongly with the substrate diffuse likewise. However, for strongly interacting atoms, particularly for metal atoms deposited on metal substrates, another diffusion mechanism can occur. This is known as the exchange mechanism of diffusion and is depicted in Fig. 3.4. This mechanism is particularly important for metal-on-metal growth systems.

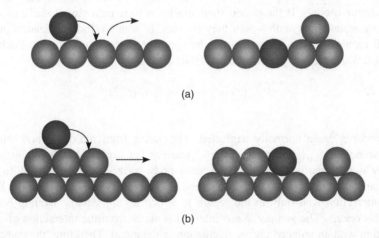

Figure 3.4 *The exchange mechanism of diffusion. Mass transport occurs via the replacement of one atom with another. This can happen either (a) on a terrace or (b) at a step.*

3.3   Physisorption

Physisorption is a ubiquitous phenomenon. All atoms and molecules experience long-range van der Waals forces. Thus, virtually any species physisorbs if the temperature is low enough. The only exception is when the adsorbate experiences such a strong chemical attraction for the substrate that it cannot be stopped from falling into the chemisorption well, for example, Re adsorption on a Re surface. For Re, which has the strongest metal–metal bond, it is impossible to stop the Re atom from falling directly into the chemisorbed state.

Rare gases can only interact through van der Waals forces; therefore, physisorption is the only way in which they can attach to a surface. Hence, these species are only bound to surfaces at low temperatures. Some surfaces are particularly inert and, therefore, favour physisorption. Examples of passivated surfaces are H-terminated diamond and silicon and S-terminated Re as well as graphite, all of which exhibit remarkably low reactivity. These surfaces are so stingy with their electrons that physisorption is the primary means of interaction between chemical species and the surface. Low surface temperature can also stabilize physisorbed states of more reactive species, especially if formation of the chemisorbed state is an activated process.

Van der Waals forces exist not only between adsorbates and substrates but also among adsorbates. Because the adsorbate–surface interaction is so weak in physisorption, lateral interactions are very important for physisorbed molecules and can be as strong as the adsorbate–substrate interaction. Furthermore, physisorption is less site-specific than chemisorption; therefore, at high coverage incommensurate structures, ordered structures that lack registry with the substrate, may be formed.

Advanced Topic: Theoretical Description of Physisorption

Zaremba and Kohn [14–16] laid the foundations for a theoretical description of physisorption in the mid 1970s. The total physisorption potential between an atom and a metal surface, $V_0(z)$, can be written in terms of a short-range repulsive potential, $V_{HF}(z)$, and a long-range van der Waals attraction, $V_{corr}(z)$. As is customary, z is chosen to be the atom–surface separation along the surface normal. The repulsive term arises from the overlap of (filled) atomic orbitals with the electronic states of the surface. The approaching atom experiences a Pauli repulsion when the electrons of the surface attempt to interact with the filled atomic orbitals. If the closed shell species is to remain closed shell, its orbitals must be orthogonal to any metal states with which they interact. It is this orthogonalization energy that gives rise to the Pauli repulsion. This interaction can be approximated using a Hartree-Fock treatment. The attractive van der Waals term is given by the Lifshitz potential

$$V_{corr}(z) = \frac{C}{(z - z_0)^3} + O(z^{-5}).$$ (3.3.1)

The terms of order z^{-5} are normally neglected. They arise from multipole and other higher order perturbations. Note that $V_{corr}(z)$ decreases as z^{-3} whereas the van der Waals interaction between two atoms decreases at z^{-6}, that is, because of the many-body surface, a physisorption interaction decays more slowly than does the van der Waals interaction between two isolated atoms. Equation (3.3.1) describes this interaction well only in the region in which no appreciable overlap between the atom and metal orbitals occurs. The van der Waals interaction arises from the interaction of an instantaneous dipole on the atom with an induced charge fluctuation in the metal. Therefore, the constant C is related to the polarizability of the atom, $\alpha(\omega)$, and the dielectric function of the metal, $\varepsilon(\omega)$

$$C = \frac{1}{4\pi} \int_0^\infty \alpha(i\omega) \frac{\varepsilon(i\omega) - 1}{\varepsilon(i\omega) + 1} \, d\omega. \tag{3.3.2}$$

z_0 is the position to which the Lifshitz potential is referenced. It is found from the weighted average of the centroid of the induced surface charge, $\overline{z}(i\omega)$,

$$z_0 = \frac{1}{2\pi C} \int_0^\infty \alpha(i\omega) \, \overline{z}(i\omega) \frac{\varepsilon(i\omega) - 1}{\varepsilon(i\omega) + 1} \, d\omega. \tag{3.3.3}$$

Equation (3.3.3) demonstrates that the position of the reference plane depends not only on the charge distribution but also on the dynamic screening properties of the surface.

The total potential can then be written

$$V_0(z) = V_{\text{HF}}(z) + V_{\text{corr}}(z). \tag{3.3.4}$$

Equation (3.3.4) succeeds in describing the physisorption potential because the equilibrium distance associated with a physisorption bond tends to be much larger than z_0. This condition ensures that the overlap of the metal and atom wavefunctions is sufficiently small to allow for the use of Eqs (3.3.2)–(3.3.4).

3.4 Non-dissociative chemisorption

3.4.1 Theoretical treatment of chemisorption

Haber suggested that adsorption was related to "unsaturated valence forces" in the surface of the substrate [17]. Langmuir [18, 19] subsequently formulated and confirmed the concept that chemisorption corresponds to the formation of a chemical bond between the adsorbate and the surface. This was one of his many contributions to surface science that garnered him the Nobel Prize in 1935. Theoretical approaches to the description of the adsorbate–surface bond have taken many approaches [14, 20]. All of these approaches need to contend with the problem of how an atom or molecule is attached to an essentially semi-infinite substrate. The standard methods of theoretical solid-state physics are well developed for dealing with the bulk of the substrate and these succeed in describing extended electronic states. However, the reduction of symmetry introduced by the surface leads to the formation of localized states and computational difficulties. Furthermore, the adsorbate represents a localized impurity in this reduced symmetry setting. Alternatively, one might approach the problem from the quantum chemical viewpoint. Quantum chemistry powerfully describes the properties of small molecules. One might then approach the surface as a cluster onto which an adsorbate is bound. The challenge is attempting to add a sufficient number of atoms to the cluster while still explicitly treating enough electronic wavefunctions. Obviously both approaches have their advantages and disadvantages.

Before we proceed to the specifics of chemisorption bond formation, a quick review of molecular orbital formation is in order. In Fig. 3.5(a), the familiar situation of molecular orbital (MO) formation from two atomic orbitals (AO) is presented. Recall that AO interactions depend both on the energy and symmetry of the states involved: the closer in energy the stronger the interactions and some symmetry combinations

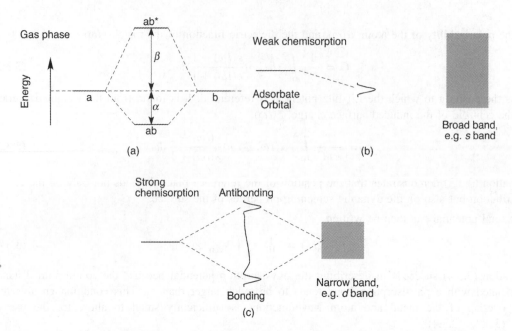

Figure 3.5   *Orbital interactions. (a) Gas phase. (b) Weak chemisorption. (c) Strong chemisorption. a, b, atomic orbitals; ab, ab*, bonding and antibonding molecular orbitals; α, β energy shifts of molecular orbitals with respect to the mean energy of a and b.*

are forbidden. Note also that the antibonding state is generally more antibonding than the bonding state is bonding, $\beta > \alpha$ in Fig. 3.5(a). Consequently, if ab and ab^* are both fully occupied, not only is the bond order zero, the overall interaction is repulsive. In the gas phase, an orbital must contain exactly zero, one or two electrons and the orbital energy is well defined (sharp) as long as the molecule remains bound. MOs extend over the entire molecule, though in some cases the electron density is localized about a small region of the molecule or even a single atom.

One of the first useful approaches applied to atomic absorption at surfaces was that of the Anderson-Grimley-Newns approach [14, 21–24]. The important point arising from this approach is that the types of electronic states that arise after chemisorption depend not only upon the electronic structure of the substrate and adsorbate but also the coupling strength between the adsorbate and the substrate. The adsorbate levels may end up either inside or outside the metal band. In the strong-coupling limit (Fig. 3.5(c)), in which the adsorbate level interacts strongly with a narrow band, e.g. the d band of a transition metal, the adsorbate and metal orbitals split into bonding and antibonding combinations, one below and one above the metal band. A weak continuous part extends between these split-off states. In the weak-coupling limit (Fig. 3.5(b)), in which the adsorbate level interacts with a broad band such as the s band of a transition metal, little of the metal density of states is projected onto the adsorbate. However, the adsorbate level is broadened into a Lorentzian shaped resonance centred on a narrow energy range.

The Grimley and Newns model based on the Anderson Hamiltonian is necessarily a local description of chemisorption because only the interaction of the adsorbate with its nearest neighbours is accounted for. Furthermore, electron correlation can, at best, only be included in an approximate fashion [25]. Consequently, other methods are required to get a quantitative understanding of chemisorption.

Figure 3.6 demonstrates the effect of surface proximity on the electronic states of an adsorbate, considered first in the weak chemisorption limit. As a molecule approaches a surface, its electronic states interact with

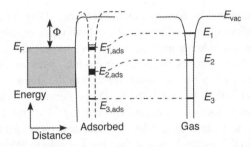

Figure 3.6 *Diagram of broadening and shifting of adsorbate levels as they approach a surface. E_F, Fermi energy; E_{vac}, vacuum energy; Φ work function of the surface material; E_1, E_2, E_3, energies of molecular orbitals 1, 2 and 3, respectively, of the molecule far from the surface; $E_{1,ads}$, $E_{2,ads}$, $E_{3,ads}$, energies of molecular orbitals 1, 2 and 3, respectively, of the adsorbed molecule; shaded area, occupied band (e.g. valence band).*

the electronic states of the metal. This broadens the MOs (spreads them out in energy) and it also lowers the energy of the MOs. Whereas the energetic ordering of MOs generally is not changed, the spacing between them may change. Furthermore, orbitals that were degenerate in the gas phase can have this degeneracy lifted by the presence of a surface. The reason why MOs experience a shift and broadening is that they interact with the electrons of the substrate. If there are no substrate electrons at the energy of an MO, as for the orbital of energy E_3 in Fig. 3.6, little interaction occurs and the MO remains sharp. This is particularly important for core levels, which lie below the valence band of the substrate and for MOs that fortuitously have energies that lie in a band gap of the substrate. As metals generally do not exhibit band gaps, the valence electronic states of adsorbates usually experience strong interactions, shifts and broadening. In the case of semiconductors and insulators where band gaps do exist, some adsorbate valence levels may show little interaction with the substrate if they fortuitously fall in a band gap.

Figure 3.6 is also instructive for the strong chemisorption limit. In this case, the bonding combination is shifted lower in energy than the initial adsorbate orbital. Broadening also occurs; indeed, the broadening is generally so severe that a continuous band connects the bonding and antibonding split-off states. In this limit, it is important to remember that while the final orbitals may be localized primarily on the adsorbate and correlate with specific initial adsorbate orbitals, the final orbitals are combination states that are composed of metal and adsorbate contributions. They also have a spatial extent that may include several substrate atoms.

The energy of an orbital with respect to the Fermi energy determines the occupation of that orbital. Just as for the substrate, all electronic states of the adsorbate that lie below E_F are filled, whereas those above E_F are empty. The broadening of electronic states has an interesting effect for states near E_F. When a state lies partially below and partially above E_F, as does the highest orbital in Fig. 3.6, it is partially filled. This is particularly important for the strong chemisorption limit because the bonding or antibonding combination often straddles E_F and is more or less occupied depending on its position relative to E_F. The interaction strength depends on the relative filling of the orbitals and their bonding or antibonding character. As we shall see shortly, the position of transition metal d bands relative to E_F is the most important factor in determining the filling of the bonding and antibonding combinations (for a given adsorbate orbital) and, therefore, the interaction strength.

The two most widely used theoretical approaches to chemisorption problems differ in how they set up the problem. In the repeated slab approach, a small slice of substrate is chosen. This is several atoms wide and several layers thick. To re-institute the symmetry properties that make bulk calculations more feasible, repeated boundary conditions are used, that is, a stack of slabs separated by a vacuum region is considered and the x and y directions fold back onto themselves. Alternatively, the cluster approach can be used to model the substrate. Increasingly, density functional theory (DFT) in combination with generalized gradient approximation corrections (GGA) is being used in both types of calculation. DFT has made great strides in recent years and has provided a method of treating larger and larger systems with increasing detail and accuracy. DFT methods attempt to model electronic states in terms of the electronic charge density with the GGA term accounting for the correction due to the shape of the electron distribution. An exchange-correlation functional is then tacked onto the charge density. Finding better expressions for the exchange-correlation functional remains an outstanding challenge for theoreticians. The awarding of the 1999 Nobel Prize in chemistry to Kohn marked the advances and successes of DFT theory.

One drawback of traditional DFT methods is that they are not able to calculate excited electronic states. Time-dependent DFT is, however, making rapid progress along these lines [26]. Configuration interaction (CI) calculations and other types of atomic/molecular orbital based strategies provide alternative methods to address excited states but the cost and time required to perform these calculations increase rapidly with the number of electrons, which make them difficult to perform on large systems. Currently, these methods are limited to 10–100 electrons if a high level of CI is included. This limits the usefulness of this approach for routine calculations with transition metal surfaces.

Whether a slab or cluster approach is more appropriate for a given system depends on the nature of the substrate. For semiconductors, for instance, bonding is highly localized. A cluster approach is therefore feasible, and clusters with on the order of 10 atoms can provide a useful model to study the interaction of, for instance, H with Si [27]. Metals require more atoms to be present for a fully developed band structure to evolve. The study of small metal clusters is itself an area of active research, largely because many properties as well as the reactivity of metals show cluster size dependence [28, 29]. While interesting in its own right, these small metal clusters do not necessarily represent a good model of larger metal aggregates and single crystal surfaces.

3.4.2   The Blyholder model of CO chemisorption on a metal

CO has long served as a model adsorbate [30] and a discussion of how CO binds with a surface illustrates a number of points that aid us in understanding non-dissociative molecular chemisorption. We need to consider the electronic structure of CO in the gas phase and how the electronic structure of CO is modified by the presence of a surface.

We start the discussion of CO chemisorption with the Blyholder model [31]. As in gas-phase or co-ordination chemistry, the frontier orbitals – the highest occupied molecular orbital (HOMO) and the lowest unoccupied molecular orbital (LUMO) – are assumed to have the greatest effect on adsorbate/surface chemistry. Figure 3.7 depicts the frontier orbitals of CO. The HOMO of CO is the 5σ MO. The LUMO is the $2\pi^*$. The former is roughly non-bonding with respect to the C–O bond, whereas the latter is antibonding with respect to C–O.

The alignment of the HOMO and LUMO with respect to the surface is also important in determining the bonding. The 5σ orbital is localized on the C end of the molecule. The $2\pi^*$ is symmetrically distributed along the molecular axis. The combination of the energetic and orientational aspects of the molecule/surface interaction can be summarized as follows. The 5σ orbital is completely occupied as it lies below the Fermi energy E_F. The $2\pi^*$ is partially occupied. The 5σ orbital interacts strongly with the metallic electronic states. Effectively, the electron density of the 5σ orbital is donated to the metal and new hybrid electronic states are formed (donation). These are predominately localized about the C end of the molecule; however, they also extend over several metal atoms. The $2\pi^*$ orbital accepts electron density from the metal through a process known as backdonation. Again, new hybrid electronic states are constructed which are primarily localized about the CO molecule but which also extend over several substrate atoms. The overlap of the 5σ orbital with the metal states is most favourable if the molecule is oriented with the C end toward

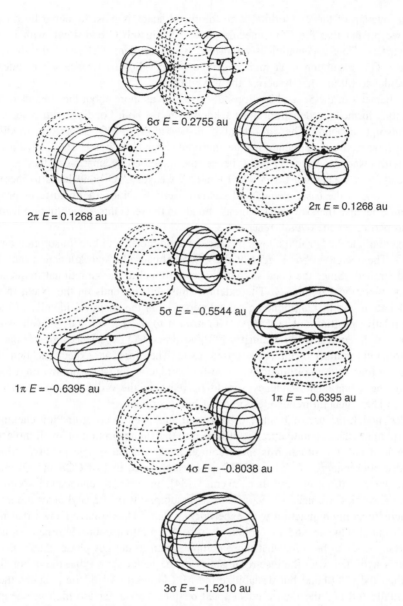

Figure 3.7 *The molecular orbitals of gas-phase CO. The wavefunction changes sign in going from the regions enclosed by solid lines to those enclosed by dashed lines. Energies, E, are given in atomic units (1 atomic unit = 27.21 eV). Orbitals with negative energies are occupied in the ground state of the neutral molecule. Reproduced from W. L. Jorgensen, L. Salem, The Organic Chemist's Book of Orbitals, Academic Press, New York. © (1973) with permission from Academic Press.*

the surface. The overlap of the $2\pi^*$ orbital with the metal states is most favoured by a linear geometry. Consequently, we predict that the CO molecule should chemisorb C end down with its axis along the normal to the surface. This expectation is confirmed by experiment. CO is nearly always bound in an upright geometry. One exception is at high coverage on the Ni(110) surface where lateral interactions cause CO molecules to tilt alternately across the rows [32].

The bonding character of the 5σ and $2\pi^*$ orbitals also tells us more about the chemisorption interaction. The hybrid orbitals formed by donation and backdonation are both bonding with respect to the M–CO bond (the chemisorption bond). The 5σ orbital is non-bonding in CO. Therefore, modification of this orbital does not have a strong influence on the intramolecular C–O bond. The $2\pi^*$, on the other hand, is antibonding with respect to the C–O bond, hence, the increased occupation of the $2\pi^*$ engendered by backdonation leads to a weakening of the C–O bond. A weaker C–O bond leads to increased reactivity of the CO. This is one of the basic elements of catalytic activity. Not only do surfaces provide a meeting place for reactants, but also they weaken (or break) bonds in the reactants, which in turn leads to a decrease in the activation barrier for the overall reaction.

The involvement of the $2\pi^*$ orbital allows us to understand the effect of chemisorption on the vibrational spectrum of CO. The decreased bond order of the CO bond caused by backdonation into the $2\pi^*$ results in a lower bond strength. Since the bond strength is proportional to the vibrational frequency, we observe a decrease in the vibrational frequency. The extent of this shift depends on the extent to which the $2\pi^*$ is filled. Backdonation into this antibonding orbital depends on the chemical identity of the metal, which determines the relative position of the d band with respect to the $2\pi^*$ orbital, as well as on the site that the CO occupies. As in co-ordination chemistry [33], adsorbed CO experiences greater backdonation with increased co-ordination number of the chemisorption site. Therefore, we expect backdonation to increase in the order on top (one-fold co-ordination) < two-fold bridge < four-fold hollow on a fcc(100) surface. Correspondingly the vibrational frequency should decrease in the order on top > two-fold bridge > four-fold hollow. These general trends are borne out by experiment.

The Blyholder model, the use of frontier orbitals and analogies to co-ordination chemistry, have been successful in explaining the qualitative trends observed in the chemisorption of small molecules. However, when looked at in detail, the situation is more complicated than the simple picture outlined above. For instance, the adsorption energy of CO on Ni(100) is nearly twice that of CO/Cu(100), nevertheless, the C–O stretch frequency differs by less than $25\,\mathrm{cm}^{-1}$ [34]. In order to understand specific systems and quantitative characteristics, we need to reconsider the assumptions of the Blyholder model.

Ab initio calculations in conjunction with experimental data [34, 35] have shown that not only the 5σ and $2\pi^*$ orbitals but also the 4σ and 1π orbitals must be taken into account. In other words, the formation of bonds at surface tends to be somewhat more complex than in the gas phase. Analysis of how the CO molecular orbitals hybridize with the electronic states of the metal shows that the σ bonding interactions slightly strengthen the C–O bond but destabilize the M–CO bond. All of the σ states that contribute to bonding are initially full and the lowest unoccupied σ state (6σ) is far too high in energy to contribute to the bonding. Therefore, the net effect of the σ states is to reorganize the electrons with respect to the gas-phase electron densities. They have little influence on the C–O bond energy but are repulsive with respect to the chemisorption bond rather than attractive as assumed in the Blyholder model. The π system, on the other hand, still acts as we expected in the Blyholder model. The π system is largely responsible for the M–CO bond and the backdonation into these hybrid orbitals is antibonding with respect to the C–O bond. Furthermore, the energetic position of the d bands with respect to the $2\pi^*$ orbital is still an important parameter in determining the orbital populations.

Several lessons are learned from CO adsorption. The fundamental principle of chemical bonding still prevails. Orbitals combine to form bonding and antibonding combinations. The relative occupation of bonding orbitals with respect to antibonding orbitals determines relative bond strengths. The filling of orbitals that strengthen one bond, for instance the M–CO bond, generally leads to a weakening of other bonds, e.g. the C–O bond. As a first approximation, we should consider the interaction of frontier molecular orbitals with the electronic states of the substrate near the Fermi energy. Any hybrid orbitals that fall below E_F are occupied and those above it are unoccupied. Adsorbate-associated orbitals are broadened by their interaction with the substrate. Therefore, they may straddle E_F and become partially occupied. To understand quantitatively the bonding in an adsorbate/substrate system, it may be necessary to consider other orbitals beyond the frontier orbitals.

3.4.3 Molecular oxygen chemisorption

Let us revisit the vibrational spectrum of O_2/Pd(111), Fig. 2.30, in the light of our understanding of the Blyholder model. The three distinct vibrational frequencies in the O–O stretching region are indications of three distinct molecular O_2 species with different bond orders. Decreasing vibrational frequency corresponds to increased occupation of the $\pi*$ related states and increasing M–O_2 binding. Vibrational spectroscopy alone does not allow us to unambiguously identify the geometry of the three O_2 species. However, with the aid of x-ray analysis and recourse to co-ordination chemistry, we are able to identify the three species as those shown in Fig. 3.8.

The interaction of O_2 with Pd is more complicated than that of CO with Pd. In part, this is caused by the increased reactivity of O_2 with Pd. CO does not dissociate at low temperature whereas O_2 partially dissociates when an O_2-covered surface is heated above $\sim$180 K. This can be related to the MO structure of O_2. The two $2\pi*$ orbitals are half-filled and degenerate in the gas phase. The degeneracy is lifted in the adsorbed phase because one orbital is oriented perpendicular and the other parallel to the surface. Both become available upon chemisorption for various amounts of backdonation. In a sense, the structures depicted in Fig. 3.8 are almost a snapshot of the dissociation of O_2 on Pd(111). It should come as no surprise to discover that the O_2 species that is most likely to dissociate is the ω_1 species, which is the O_2 species with the strongest M–O_2 bond and the weakest O–O bond. Nonetheless, other simple expectations about the O_2/Pd system are not borne out [36]. The most strongly bound species (ω_1) is the last to be populated when the surface is dosed at $T_s \approx 100$ K while the moderately bound, ω_2, is the first to be populated. Dissociation of O_2 does not occur until the temperature is raised. On the other hand, if adsorption is carried out at lower temperature, $T_s \approx 30$ K, a physisorbed molecular state is formed instead of the chemisorbed state [37]. This state has a vibrational frequency almost identical to that found in the gas phase, as expected for a van der Waals bound species that does not experience charge transfer into antibonding MOs from the surface.

The data for O_2/Pd(111) indicate that the predictions of simple equilibrium thermodynamics, i.e. that the system should attain the state of lowest free energy, are not borne out for the interaction of O_2 with a

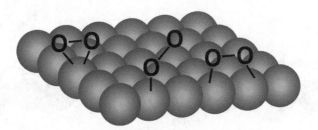

Figure 3.8 *O_2/Pd(111) adsorbate structure. The labelling of the three states (ω_1, ω_2 and ω_3) correlates with the loss peaks observed in the electron energy loss spectrum as shown in Fig. 2.30.*

low temperature Pd surface. Hence, activation barriers must separate the various molecular physisorbed, molecular chemisorbed and dissociatively adsorbed states. Reaction kinetics and dynamics control the state of the system and must prevent it from attaining the state of lowest free energy. This highlights the importance of always determining whether the system under consideration has attained equilibrium or merely a steady state. Reaction kinetics and dynamics are discussed thoroughly in the following chapters.

3.4.4   The binding of ethene

As a model of the binding of a polyatomic molecule to a surface, we consider ethene (ethylene). How would $H_2C=CH_2$ bind at a surface? In the gas phase, ethene is a planar molecule with a polarizable π-bond system connected with the $C=C$ double bond. At cryogenic temperature, H_2CCH_2 binds weakly to the surface, lying flat on the surface as depicted in Fig. 3.9(a). With a binding energy of $-73\,kJ\,mol^{-1}$ on Pt(111) [1] this species can be considered weakly chemisorbed. The adsorption geometry is determined by the π bonding interaction, which finds the strongest interaction when the H_2CCH_2 lies flat rather than end on.

A stronger interaction with the surface requires more extensive exchange of the π electrons with the surface. As π electrons are donated from H_2CCH_2 to the surface to form two σ bonds, the bond order drops to one between the C atoms. A planar geometry is no longer favourable for a H_2CCH_2 species. The adsorbate prefers to assume a structure closer to that of ethane. The C–C axis remains parallel to the surface, Fig. 3.9(b), but the H atoms bend up away from the surface so that the bonding about the C atoms can assume a structure consistent with sp^3 hybridization. This electronic rearrangement occurs when π-bonded C_2H_4/Pt(111) is heated above 52 K [38]. The transformation of the π-bonded planar species to the di-σ bound chemisorbed species is mirrored in the vibrational spectrum. The C–H stretches are IR inactive in the π-bonded species (no component along the normal) but not the chemisorbed species [38]. The C–C stretch, if it could be observed, would also shift to significantly lower frequency in the chemisorbed state.

On Pt(111), the di-σ species is stable up to $\sim$280 K with a binding energy of $-117\,kJ\,mol^{-1}$. Above this temperature, the molecule rearranges and at room temperature the stable configuration is that of ethylidyne ($C–CH_3$) as shown in Fig. 3.9(c) which requires the loss of one H atom per molecule via H_2 desorption. The migration of the H atom from one side of the molecule to the other is accompanied by the standing up of the molecule. The bond breaking and formation is activated, which is why this adsorption geometry is not formed at low temperature.

The three C_2H_4-derived species exhibit not only different binding geometries and energies but also different reactivities and roles in the hydrogenation of C_2H_4 [39]. Whereas we might not expect the

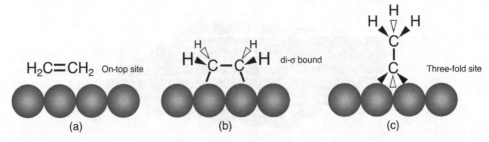

Figure 3.9   *The binding of ethene at a metal surface. (a) The weakly chemisorbed π-bonded C_2H_4. (b) The di σ-bonded chemisorbed state. (c) Ethylidyne.*

π-bonded species to be important at room temperature or above, high pressures ($\sim$1 atm) and the presence of H_2 in the gas phase can lead to appreciable coverages of this species. The coverage is determined by the dynamic balance between adsorption, desorption and reaction – topics that are dealt with in more detail in Chapters 4 and 6. The π-bonded species is more reactive than the di-σ bonded C_2H_4 toward hydrogenation while ethylidyne does nothing more than occupy sites that could otherwise be used for reaction. The catalytic reactivity, as discussed further in Chapter 6, depends sensitively on the strength of the adsorbate/surface interaction and the balance between a strong enough interaction to activate the adsorbate but not so strong that the adsorbate prefers to remain chemisorbed rather than reacting further.

3.5 Dissociative chemisorption: H_2 on a simple metal

The dissociation of H_2 on a surface is the prototypical example of dissociation at a surface [40–42]. The measurement of the dissociation probability (dissociative sticking coefficient) of H_2 and its theoretical explanation serve as the basis for a useful historical record of the advance, not only of our understanding of reaction dynamics at surfaces but also as a record of the ever more sophisticated experimental and theoretical techniques that have been brought to bear on surface reaction dynamics [43].

Figure 3.10 is an oversimplified but useful starting point to understand dissociative adsorption. The real-life situation is more complicated because each molecular orbital of the molecule interacts with the surface electronic states. Each combination of molecular orbital + surface orbital generates a bonding + antibonding pair. The positions and widths of each of these new hybrid orbitals change as the interaction of the molecule with the surface increases. Close accounting of orbital occupation must be made, not only to determine the filling of orbitals, but also to keep track of with respect to which bond an orbital is

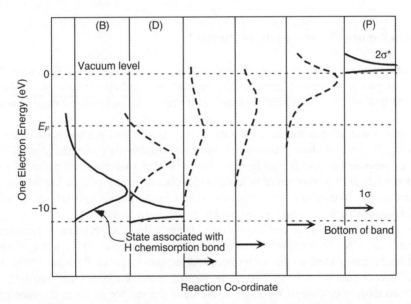

Figure 3.10 *Calculated changes in the electronic structure of the orbitals associated with H_2 as the molecule approaches a Mg surface. Moving to the left in the diagram represents motion toward the surface. Reproduced from J. K. Nørskov, A. Houmøller, P. K. Johansson, B. I. Lundqvist, Phys. Rev. Lett., 46, 257. © 1981, with permission from the American Physical Society.*

bonding or antibonding. For example, in the case of CO chemisorption, backdonation into the π system is bonding with respect to the M–CO bond but antibonding with respect to the C–O bond.

As a closed shell species approaches a surface, a repulsive interaction develops due to the Pauli repulsion of the electrons in the solid and molecule. A Pauli repulsion is the repulsive force that arises between filled electronic states because no more than two electrons can occupy an orbital as predicted by the Pauli exclusion principle. Filled states must, therefore, be orthogonal and orthogonalization costs energy. For a rare gas atom, this is the end of the story because excited states of the atom lie high in energy. For H_2, however, electron density can be donated from the metal into the $2\sigma^*$ antibonding orbital. Electron transfer leads to a weakening of the H–H bond and a strengthening of the adsorbate/surface interaction. Eventually, dissociation occurs if enough electron density can be transferred. Dissociation can be thought of as a concerted process in which the transfer of electron density leads to a gradual weakening of the H–H bond while M–H bonds form.

The $H_2 1\sigma$ and $2\sigma^*$ orbitals shift and broaden as they approach the surface. Electron transfer from the metal to the H_2 occurs because the $2\sigma^*$ drops in energy and broadens as H_2 approaches the surface. As it drops below E_F, electrons begin to populate the orbital and the H_2 bond grows progressively weaker while the M–H bonds become progressively stronger [44–46]. The process of charge transfer is illustrated in Fig. 3.10.

Because the MOs involved in bond formation have a definite geometry, the orientation of the molecule with respect to the surface must be correct to facilitate orbital overlap and charge transfer. Both the final state of the dissociated molecule (two adsorbed atoms) and the orientation of the $2\sigma^*$ orbital render a parallel approach of the H_2 the most favourable to charge transfer and dissociation. Furthermore, just as there is corrugation in a chemisorption potential, corrugation exists in the energetics of H_2 dissociation. In other words, there are specific sites on the surface that are more favourable to dissociation than are others. The corrugation in the H_2/surface interaction exhibits both energetic and geometric components [47].

3.6   What determines the reactivity of metals?

A number of new concepts have been introduced in the preceding sections. Therefore, this section is intended to act in part as a résumé, reinforcing concepts of bonding at surfaces. It also extends our understanding so that we can comprehend quantitative trends in adsorption energies and dissociation probabilities.

We already have some feeling for the reactivity of metals. Au and Ag are particularly attractive for jewellery not only because of their preciousness but also because they oxidize slowly. Fe is not very attractive for this purpose because it readily rusts. However, the oxidation behaviour of a metal is only one measure of reactivity. Pt is also prized in jewellery, particularly in Japan, and resists oxidation as well as other chemical attack. Nonetheless, it is one of the most useful metals in catalysis. So what makes Au so noble whereas metals such as Pt and Ni are highly catalytically active?

As is to be expected, surface structure plays a role in reactivity. The presence of structural defects can lead to enhanced reactivity. For instance, C–C and C–H bonds are more likely to be broken at step and kink sites [48] on Pt group metals than on terraces. Adsorption is generally more likely to be hindered by an energetic barrier (see § 3.9 below) on close packed metal surfaces – i.e. bcc(110), fcc(111) and hcp(001) – than on more open planes. We return to issues of the surface structure dependence of reactivity in Chapter 6. Putting aside these structural factors, we concentrate below on the electronic factors that differentiate one metal from another.

In this discussion we follow the theoretical work of Hammer and Nørskov [49–51], who investigated the chemisorption of H and O atoms and the dissociative chemisorption of H_2 molecules over a variety

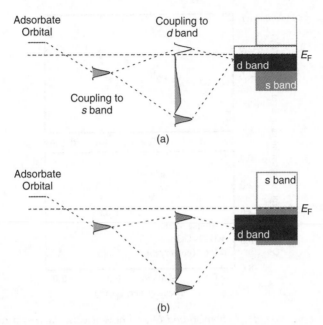

Figure 3.11 *The two-step conceptualization of chemisorption bond formation on transition metal surfaces. (a) Early transition metal. (b) Coinage metal.*

of metals. In the previous section, we considered a simple two-level picture of H_2 dissociation. We now look in more detail at how the molecular orbitals interact with the s and d bands of the metal and how this affects both binding energies and activation barriers.

We begin with the chemisorption of an atom, taking H as a representative example. The mixing of H and metal orbitals can be conceived of as a two-step process as depicted in Fig. 3.11. In the step 1, the hydrogen $1s$ orbital interacts with the s band of the metal. The s band of transition metals is very broad; thus, the interaction is of the weak chemisorption type. This leads to a bonding level far below the Fermi energy. Accordingly, the bonding level is filled and the overall interaction is attractive. The strength of this interaction is roughly the same for all transition metals. In step 2, we consider the interaction of the bonding level with the metal d band. This interaction leads to the formation of two levels. One is shifted to lower energies than the original bonding state. The second is positioned slightly higher in energy than the unperturbed metal d band. The low energy state is bonding with respect to adsorption while the high energy state is antibonding. The overall energetic effect of the resulting states depends on the coupling strength between the adsorbate-s band hybrid with the d band and the extent of filling of the antibonding state. The filling of the antibonding states is, in turn, determined by the position of the antibonding state with respect to the Fermi energy. As both the coupling strength and the Fermi energy depend on the identity of the metal, differences in the second step explain the differences in chemisorption bond strength between metals.

Moving from left to right across a row of transition metals, the centre of the d bands moves further below E_F. The antibonding state formed in step 2 follows the position of the d bands. For early transition metals (left-hand side), the antibonding state lies above E_F and is not filled. Consequently, both step 1 and step 2 are attractive and atomic chemisorption is strongly exothermic. For Cu, Ag and Au, however, the antibonding state lies below E_F and is filled. The resulting repulsion tends to cancel out the attraction

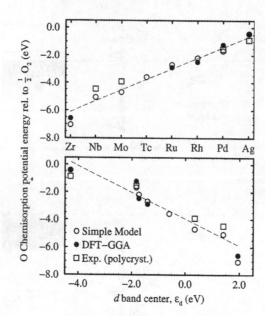

Figure 3.12   *The interaction strength of chemisorbed O and how it varies across a row of transition metals. In the upper panel, the good agreement between experimental and theoretical results is shown. In the lower panel, the linear relationship between interaction strength and the d band centre is demonstrated. Source of data for experimental results: I. Toyoshima, G. A. Somorjai, Catal. Rev – Sci. Eng., 19 (1979) 105. Reprinted from B. Hammer and J. K. Nørskov, Theoretical surface science and catalysis – Calculations and concepts, Adv. Catal., Vol. 45 (Eds B. C. Gates, H. Knözinger), Academic Press, Boston, p. 71. © 2000 with permission from Academic Press.*

from step 1. For both H and O on Cu, the overall interaction is marginally attractive, whereas on Au chemisorption is unstable. Thus, the trend across a row can be explained in terms E_d, the energy of the d band centre. This correlation is confirmed in Fig. 3.12 in which a linear relationship between E_d and the O atom binding energy is found.

To explain why Au is more noble than Cu, the magnitude of the coupling in step 2 needs to be considered. The orthogonalization energy between adsorbate and metal d orbitals, which is repulsive, increases with increasing coupling strength. This energy increases as the d orbitals become more extended. The $5d$ orbitals of Au are more extended than the $3d$ orbitals of Cu, which renders Au less reactive than Cu because of the higher energy cost of orthogonalization between the H $1s$ and Au $5d$ orbitals.

Hence, there are two important criteria influencing the strength of the chemisorption interaction: (i) the degree of filling of the antibonding adsorbate–metal d states (which correlates with E_d); and (ii) the strength of the coupling. The filling increases in going from left to right across a row of transition metals in the periodic chart and is complete for the coinage metals (Cu, Ag and Au). The coupling increases in going down a column in the periodic chart. It also increases in going to the right across a period.

The principles used to explain atomic adsorption can be extended to molecular adsorption as well. Indeed, in Chapter 6 we encounter examples (steam reforming of hydrocarbons over a Ni catalyst, Ru catalyst for ammonia synthesis, BRIM™ catalyst) in which the knowledge gained from first principles calculations has been used to fashion a modified working industrial catalyst. The two-step process is applicable, for instance, to CO. The result is a description of the bonding that is in accord with the Blyholder model. The 5σ derived states (bonding and antibonding combinations) are predominantly below the Fermi energy and, therefore, lead to a repulsive interaction. The $2\pi^*$ derived states lead to attractive interactions because the bonding combination lies below E_F while the antibonding combination is (at least partially) above E_F. Moving to the left in the periodic chart, the M–CO adsorption energy increases as the filling of the $2\pi^*$-metal antibonding combination rises further above E_F. However, the adsorption energy of C and O increases at a greater rate than that experienced by the molecular adsorbate. Therefore, a crossover from molecular to dissociative adsorption occurs. This happens from Co to Fe for the $3d$ transition, Ru to Mo for $4d$ and Re to W for $5d$. Similar trends are observed for N_2 and NO.

In the dissociative chemisorption of H_2 both the filled σ_g orbital and the unfilled σ_u^* MOs must be considered. The σ_g orbital acts much like the H $1s$ orbital did as explained above. The σ_u^* orbital also undergoes similar hybridization and it is the sum of both sets of interactions that determines whether an energy barrier stands in the way of dissociation. Again, the energetic positioning of the bonding and antibonding combinations with respect to the metal d band, the filling of the levels and the coupling strength are decisive. For dissociative adsorption, the strength of the σ_u^* interactions is the dominant factor that determines the height of the activation barrier. Similar trends appear for the chemisorption energy of H and the activation barrier height for H_2. Cu, Ag and Au exhibit substantial barriers (>0.5 eV). The barrier height, if present at all, decreases rapidly to the left. For instance for Ni, the close-packed Ni(111) surface has a barrier height of only 0.1 eV whereas other more open faces exhibit no barrier at all [52].

3.7   Atoms and molecules incident on a surface

Adsorption and desorption are the simplest and most ubiquitous surface reactions. All chemistry at surfaces involves adsorption and/or desorption steps at some point. Therefore, it is natural that we should start any discussion of reaction dynamics at surfaces with a thorough investigation of these two fundamental steps. Furthermore, we will find that a great deal of what we learn by the study of adsorption and desorption is transferable to the understanding of more complex surface processes.

3.7.1   Scattering channels

We now turn from a discussion of energetics and bonding interactions to a discussion of the dynamics of molecule/surface interactions. That is, we turn from looking at the adsorbed phase to how molecules approach and enter the adsorbed phase. A fundamental understanding of gas/solid dynamical interactions was greatly advanced by the introduction of molecular beam techniques into surface studies [53–60]. When these techniques were combined with laser spectroscopy in the groups of Ertl [61–63], Somorjai [64–66] Zare [67–69] and Kleyn, Luntz and Auerbach [70–72], we obtained our first glimpses of energy exchange between molecules and solid surfaces on a quantum state resolved level. Cavanagh and King [73, 74], Bernasek and co-worker [75–77], Haller and co-workers [78–83] and Kubiak, Sitz and Zare [84, 85] then extended such studies to molecular desorption and recombinative desorption. We are now in the position to address the question: what is the outcome of a generalized molecule/surface scattering event? To answer this question, as we shall see, we need to understand potential energy hypersurfaces and how to interpret them with the aid of microscopic reversibility.

We can classify these encounters in several ways, as illustrated in Fig. 3.13. Events in which no energy is exchanged correspond to elastic scattering. Although no energy is exchanged, momentum is transferred. For the molecular beam scattering depicted in Fig. 3.13, elastic scattering is characterized by scattering in which the angle of incidence is equal to the angle of reflection (specular scattering). The experimental and theoretical aspects of elastic scattering have received much attention in the literature [86, 87]. A special case of elastic scattering is diffraction. Diffraction is particularly important for light particles, as

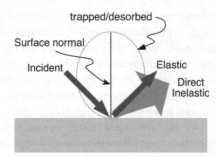

Figure 3.13   *Scattering channels for a molecule incident upon a surface.*

we have already seen for electrons, but it can also be important for heavier particles, particularly H_2 and He. Indeed, while LEED was important in demonstrating the wavelike nature of the electron, one of the first demonstrations of the wavelike nature of an atom was provided by the experiments of Stern who observed the diffraction of He from single crystal salt surfaces [88–90]. Even Ne can experience the effects of diffraction in scattering [91]. Diffractive scattering depends sensitively on the structure of the surface and He diffraction has been used as a powerful tool to examine the structure of surfaces [92] and may even find applications in microscopy [93]. A particle must lose energy to stick to a surface; therefore, an elastically scattered particle always leaves the surface and returns to the gas phase.

As far as reactions at surfaces are concerned, we are more interested in scattering events in which energy is exchanged. This is inelastic scattering. If a particle is to adsorb on a surface, it must lose a sufficient amount of energy. Therefore, we differentiate between two types of inelastic scattering. The first type is direct inelastic scattering in which the particle either gains or loses energy and is returned to the gas phase. A useful example of direct inelastic scattering is the scattering of He atoms, which can be used to investigate surface phonons, low frequency adsorbate vibrations and diffusion [94]. In Fig. 3.13, the direct inelastic channel has a broader angular distribution centred about a subspecular angle due to a range of momentum transfer events in which, on average, the scattered molecules have lost energy. Superspecular scattering is indicative of a net gain of energy by the scattered beam. Equating subspecular (superspecular) scattering with a net loss (gain) of energy from the beam to the surface is only strictly correct for a flat surface. In this case, the parallel component of momentum is conserved and only the normal component can change. Scattering from a corrugated surface can lead to coupling between parallel and normal momentum. Nonetheless, particularly if the incident kinetic energy is not too high, conservation of parallel momentum is often observed in direct inelastic scattering.

A special case of inelastic scattering is when the particle loses enough energy to be trapped in the adsorption well at the surface. This is known as trapping or sticking. Trapping is when a molecule is transferred from the gas phase into a state that is temporarily bound at the surface. If the molecule returns to the gas phase we call the overall process trapping/desorption. Sticking is when a molecule is transferred into a bound adsorbed state. The difference between trapping and sticking becomes most distinct in the discussion of precursor mediated adsorption (§ 3.7.5). The distinction between trapping and sticking is lost at high temperatures when surface residence times become very short even for chemisorbates. Because direct inelastic scattering and sticking lead to such different outcomes, we generally associate the term inelastic scattering only with the former. In contrast to elastic or direct inelastic scattering, molecules which leave the surface after trapping or sticking do so only after a residence time on the surface characteristic of the desorption kinetics. Desorption, regardless of whether it is normal thermal desorption or trapping/desorption, generally is associated with an angular distribution that is centred about the surface normal because the molecules have lost all memory of their initial conditions in the beam.

Molecular beam techniques are particularly well suited for the study of all forms of elastic and inelastic scattering including those that lead to reactions [95–97]. Supersonic molecular beams have found wide use in chemical dynamics [98, 99]. They allow the experimenter to control the angle of incidence and energy, as well as the flux of the impinging molecules. Molecular beams can be either continuous or pulsed. Pulsed molecular beams are particularly well suited to time-resolved studies of both dynamics and kinetics. A typical molecular beam apparatus is sketched in Fig. 3.25. For further details on molecular beam methods consult [95, 100–102].

As dynamicists studying surface reactivity, we are interested in the probability with which a particle adsorbs on the surface. From the above discussion, the sticking probability on a clean surface, or initial sticking coefficient, s_0, is readily defined as

$$s_0 = \underset{\theta \to 0}{Lim} \frac{N_{stick}}{N_{inc}} = \underset{\theta \to 0}{Lim} \frac{N_{stick}}{N_{el} + N_{in} + N_{stick}} \tag{3.7.1}$$

where N_{stick} is the number of particles that stick to the surface, N_{inc} is the total number incident on the surface, N_{el} is the number scattered elastically, and N_{in} is the number scattered inelastically but which do not stick. Values of s_0, even for simple molecules such as H_2 and O_2, can vary between 1 and $<10^{-10}$. This extreme range of sticking probabilities indicates that sticking is extremely sensitive to the exact form of the molecule–surface interaction potential. We investigate the mathematical treatment of the sticking coefficient in more detail in Chapter 4, in particular, how it changes with coverage. For now, we concentrate mainly on the dynamical factors that determine the sticking coefficient on a clean surface. In the following sections, we characterize the adsorption process by defining what we mean by non-activated and activated adsorption as well as direct and precursor mediated adsorption.

3.7.2 Non-activated adsorption

One of the most illustrative ways to conceptualize the binding of an adsorbate is in terms of a 1D potential as shown in Fig. 3.14. Such diagrams were introduced by Lennard-Jones in 1932 to aid in the understanding of H_2 adsorption on metal surfaces [103]; hence, the name Lennard-Jones diagrams. A molecule must lose energy to stick to a surface. Obviously, total energy is one of the decisive factors that determines whether a molecule can stick. The curve in Fig. 3.14(a) represents the energy of a molecule with zero kinetic energy as a function of the distance from the centre of mass of the molecule to the surface. We represent a molecule with some kinetic energy by a point above the curve. The distance between the point and the potential energy curve is equal to the kinetic energy.

Figure 3.14 represents a potential energy curve for non-activated adsorption. In non-activated adsorption, the molecule does not encounter an energetic barrier as it moves closer to the surface and adsorption is downhill all the way. From Fig. 3.14 we can draw several important conclusions. First, the less energy a molecule has, the easier it is for the molecule to roll downhill and stick in the chemisorption well. With increasing kinetic energy, the molecule has to lose progressively more kinetic energy when it strikes the surface. Therefore, the initial sticking coefficient decreases as the kinetic energy of the molecule increases. These intuitive expectations are true for direct non-activated adsorption except at extremely low energies. For physisorbed systems, as the velocity of the incident molecule approaches zero, the sticking coefficient actually tends to zero. This is a quantum mechanical effect, which has been observed for Ne incident on Si(100) [104]. This effect occurs only at exceptionally low energy, is not important for reactive systems, and will not be considered further.

Figure 3.14 depicts four hypothetical trajectories. In trajectory a, the molecule approaches the surface with a high kinetic energy. As it enters the chemisorption well, the total energy is constant but the distance between the trajectory and the potential energy curve, i.e. the kinetic energy, increases. The molecule is

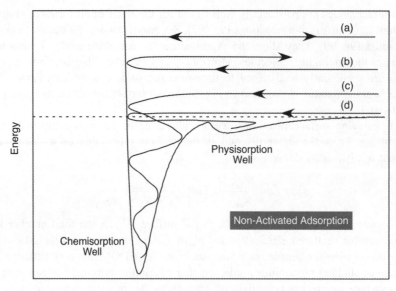

Figure 3.14 *A one-dimensional representation of non-activated adsorption: (a) elastic scattering trajectory; (b) direct inelastic scattering trajectory; (c) sticking event (chemisorption); (d) sticking event (physisorption).*

accelerated toward the surface by the attractive force of the chemisorption potential. The molecule then reflects from the repulsive back wall of the potential and retreats from the surface with the same total energy. This is an elastic scattering trajectory. Trajectory *b* approaches the surface, is reflected from the repulsive wall but, in this case, the molecule bounces off the surface with more kinetic energy than it had initially. This is a direct inelastic scattering trajectory in which the molecule gains energy. Trajectory *c* represents a sticking event. The molecule again approaches the surface and is scattered off the repulsive wall. Now, however, it loses sufficient energy on its first encounter with the wall that its total energy drops below the zero of the potential energy curve. The molecule then loses further energy and drops to the bottom of the chemisorption well. In trajectory *d*, the molecule sticks but in this case it falls into the physisorption well. After sticking in the physisorption well, a barrier must be overcome to enter the chemisorbed state.

Trajectory *c* poses an interesting question. How long does it take the molecule to reach the bottom of the chemisorption well once it has hit the surface? Note that for the molecule to stick, it need not lose *all* of its kinetic energy on the first bounce. It only has to lose *enough* energy to drop below the zero of the potential energy curve. Therefore, the molecule can retain some energy after the first encounter with the surface and it may take several bounces before the molecule fully equilibrates and falls to the bottom of the well. Indeed, from Fig. 3.14 we can easily imagine a trajectory that bounces off the back wall and only falls into the chemisorption well after it samples the physisorption well. As we shall see in § 3.7.5 and in Chapter 4, this result has important ramifications both for the dynamics and kinetics of adsorption.

Whereas the description of the adsorption of an atom onto a surface with a 1D potential energy curve may seem satisfactory, you should feel less confident about its ability to describe the interaction of an adsorbing molecule, especially for the case of dissociative adsorption. The interaction of a molecule is described by a multi-dimensional potential energy hypersurface (PES). A somewhat better description is given by a 2D representation in which the molecular centre of mass-surface separation, z, is plotted along

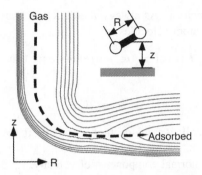

Figure 3.15   *2D PES for non-activated dissociative adsorption. z is the distance from the molecular centre of mass to the surface. R is the internuclear distance between the atoms of the molecule.*

one axis and the intramolecular bond distance of the (diatomic) molecule, R, is plotted along the other axis. Contours denote the change in energy along these two co-ordinates. Such a 2D PES for non-activated dissociative adsorption is depicted in Fig. 3.15. Far from the surface (large values of z), motion along the R axis corresponds to intramolecular vibration and motion along the z axis represents translation of the molecule with respect to the surface. At small values of z, the molecule is bound in an adsorption well. Motion along z now corresponds to the molecule–surface vibration.

In Fig. 3.15, the molecule dissociates along a non-activated pathway. On the surface, the chemisorption well is displaced to larger values of the interatomic separation because the minimum distance between the atoms in the dissociated state must be larger than in the molecule. If the diagram were continued to larger values of R, a periodic set of wells would be found, which corresponds to the two atoms bound at the preferred adsorption sites separated by increasing distances. The meaning of the R co-ordinate changes as a result of the transition from gaseous to adsorbed phase. In the gas phase, motion along R corresponds to the intramolecular vibration. In the adsorbed phase, R is still the distance between adsorbed atoms but no vibrational mode can be assigned directly to motion along this co-ordinate.

The dashed line in Fig. 3.15 represents the paths of minimum energy that connect the gas-phase molecule to the adsorbed phase. This can be thought of as the reaction co-ordinate. Drawing the energy as a function of the position along this path results in a 1D representation of energy versus position that is reminiscent of a Lennard-Jones diagram. Though some care must be taken in interpreting the energy profile along this path [105, 106], to a first approximation, it is convenient and instructive to visualize the reaction co-ordinate in this way.

3.7.3   Hard cube model

The effects of translational energy, a non-rigid surface, the relative masses of the impinging molecule and surface atom, and surface temperature upon direct non-activated adsorption are better understood by investigating so-called hard cube models [107–112]. The surface is modelled by a cube of mass m. The cube is confined to move along the direction normal to the surface with a velocity described by a 1D Maxwellian distribution at T_s. As the surface is assumed to be flat, the tangential momentum is conserved. The molecule with mass M and total kinetic energy E_K is assumed to be a rigid sphere, i.e. the internal degrees of freedom are unaffected by the collision. Its angle of incidence is ϑ_i. The attractive part of the potential is approximated by a square well of depth ε. The attractive well accelerates the molecule as it approaches the surface. Hence, the normal velocity of the molecule at impact is

$$u_n = -[2(E_K \cos^2 \vartheta_i + \varepsilon)/M]^{1/2}. \tag{3.7.2}$$

The surface atoms move with velocity v. As always, the direction away from the surface is positive. The probability distribution of surface atom velocities is

$$P(v)\,dv = (u_n - v)\exp(-a^2v^2)\,dv \tag{3.7.3}$$

where

$$a^2 = m/2k_BT_s. \tag{3.7.4}$$

The factor $(u_n - v)$ accounts for collisions being more probable if the cube is moving toward the atom than if it is moving away.

After collision, the parallel and normal components of velocity, v_p' and v_n' respectively, are calculated using the conservation of energy and linear momentum,

$$u_p' = u_p \tag{3.7.5}$$

and

$$u_n' = \frac{\mu - 1}{\mu + 1}u_n + \frac{2}{\mu + 1}v \tag{3.7.6}$$

where

$$\mu = M/m. \tag{3.7.7}$$

Trapping occurs if the molecule exchanges sufficient energy to drop below the zero of energy. Stated otherwise, trapping occurs if the normal velocity of the molecule after collision is below a critical velocity u_c' that can be contained in the well,

$$u_c' = (2\varepsilon/M)^{1/2}. \tag{3.7.8}$$

This condition occurs when the velocity of the surface cube is below a critical value of

$$v_c = \frac{1}{2}[(1 + \mu)(2\varepsilon/M)^{1/2} + (1 - \mu)u_n]. \tag{3.7.9}$$

The sticking coefficient is obtained by calculating the fraction of collisions for which the surface atom velocity is below v_c from Eqs (3.7.3) and (3.7.9)

$$s(u_n, \varepsilon, T_s) = \frac{\int_{-\infty}^{v_c}(u_n - v)\exp(-a^2v^2)dv}{\int_{-\infty}^{\infty}(u_n - v)\exp(-a^2v^2)dv}. \tag{3.7.10}$$

Integration yields

$$s(u_n, \varepsilon, T_s) = \frac{1}{2} + \frac{1}{2}\text{erf}(av_c) + \frac{\exp(-a^2v_c^2)}{2au_n\pi^{1/2}}. \tag{3.7.11}$$

where erf is the error function. The results of Eq. (3.7.11) are shown in Fig. 3.16 for parameters determined by Kleyn and co-workers [108] for NO/Ag(111). We see that s drops as expected for increasing molecular velocity. There is a weak dependence of s on T_s for the low velocities typical of thermal distributions; thus, the trapping probability falls slowly with increasing T_s. Furthermore, since the model involves an impulsive collision, s drops with increasing m because the collision becomes less inelastic. The effective mass of the surface cube depends on the point of impact as well as the initial velocity of the incident molecule.

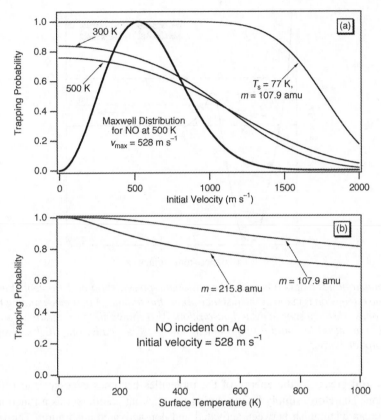

Figure 3.16 *The behaviour of the trapping probability as predicted by the hard cube model for NO incident on silver surface. (a) The effect of T_s and molecular velocity on s. A 500 K Maxwell velocity distribution is shown to convey the relative importance of various velocities to the thermally averaged trapping probability. (b) The dependence of s on the effective surface mass and T_s. The effective surface mass is chosen to be either one (107.9 amu) or two (215.8 amu) Ag atoms. Parameters determined by Kleyn and co-workers [108].*

3.7.4 Activated adsorption

Some molecules exhibit extremely low sticking coefficients that actually increase with increasing temperature, as first observed by Dew and Taylor for H_2 [113] and by Schmidt for N_2 [114]. To explain this, we need to consider the interaction of a molecule with a surface along a potential energy curve of the type shown in Fig. 3.17. In this case, the molecule encounters an activation barrier as it approaches the surface and, therefore, we call this activated adsorption. Trajectories on such a potential energy curve exhibit distinctly different behaviour than in the non-activated case. A low-energy molecule (trajectory *a*) reflects off the energetic barrier. In the absence of quantum mechanical tunnelling, a molecule must have an energy greater than or equal to the barrier height in order to have a chance to enter the chemisorption well (trajectory *b*). The sticking coefficient for activated adsorption is zero for low energy molecules. It then rises abruptly as the energy of the molecules approaches and exceeds the energy of the barrier. Classically, a step function is followed as the molecular energy exceeds the activation barrier. Quantum mechanically, tunnelling leads to a more gradual onset to barrier crossing. Other factors discussed below lead to a further rounding off of the step function behaviour. At sufficiently high energy, the sticking

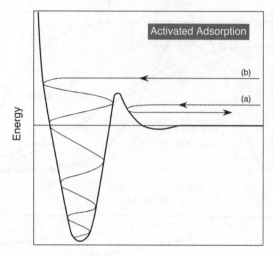

Distance from surface, *z*

Figure 3.17 *A Lennard-Jones diagram is a one-dimensional representation of the potential energy. In the case shown here, the one dimension is taken as the distance above the surface. A barrier separating the chemisorption well from the gas phase distinguishes activated adsorption. Also shown in the diagram are the energies of two hypothetical trajectories at (a) low and (b) high kinetic energy. Classically, only high energy trajectories can overcome the adsorption barrier.*

coefficient begins to decrease as the energy of the molecules becomes excessive, just as it did for non-activated adsorption. Therefore, simply by measuring the sticking coefficient as a function of the incident kinetic energy, we can distinguish between activated and non-activated adsorption. The latter statement is true for direct adsorption. In the next section, we discuss the complications introduced by the presence of an indirect (precursor mediated) path to the adsorbed phase.

Just as for non-activated adsorption, a 2D representation of the PES can be drawn for activated adsorption. Three examples are drawn in Fig. 3.18. These three PESs are distinguished by the position of the barrier. The barriers are characterized as early, middle or late based upon how closely the activated complex, which resides in the transition state (TS) located at the top of the barrier, resembles the gas phase molecule. The early barrier is in the entrance channel to adsorption and the interatomic distance in the TS is close to the equilibrium bond distance of the molecule. The late barrier is in the exit channel for adsorption and the TS has an elongated interatomic distance. As we shall see shortly, the position of the barrier and the shape of the PES are decisive in determining how a molecule best surmounts the activation barrier.

3.7.5   Direct versus precursor mediated adsorption

The possibility of transient mobility for an adsorbing molecule brings with it two distinct types of adsorption dynamics in non-activated adsorption. These two types of dynamics are known as direct adsorption and precursor mediated adsorption. The adsorption mechanism has implications not only for how molecular energy affects the sticking coefficient, but also for how adsorbate coverage affects the kinetics of adsorption. The effects on kinetics are discussed in Chapter 4. Table 3.2 presents a survey of some adsorption systems. A perusal of this table illustrates that a variety of adsorption dynamics are observed, and the mechanism of adsorption depends upon not only the chemical identity of the molecule and surface, but also on the presence of co-adsorbates, crystallographic plane, surface temperature and molecular energy.

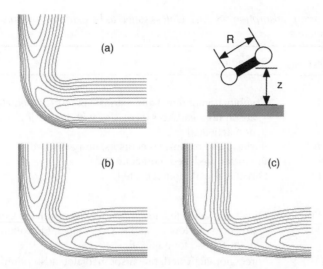

Figure 3.18 *Potential energy surfaces (PESs) for activated dissociative adsorption with (a) early, (b) middle and (c) late barriers. z, the distance from the molecular centre of mass to the surface (vertical axis); R, internuclear distance between the atoms of the (diatomic) molecule (horizontal axis). George Darling is thanked for providing these very fine model PESs.*

Direct adsorption corresponds to the case in which a molecule makes the decision to stick or scatter upon its first encounter with the surface. In the extreme limit, a molecule hits the surface, loses virtually all of its energy, and adsorbs at the site where it lands. In a less restrictive case, the molecule loses enough total energy to be bound, but it is still able to hop one or more sites away from the point of impact. For instance, the molecule might strike the surface at an on-top site before sliding into a bridge or hollow site.

We do not expect the energy of the gas molecule to have a strong effect on the sticking coefficient in direct non-activated adsorption. This is true at moderate energies; however, at extremes of high and low energy, the sticking coefficient is a function of energy. As already mentioned, the sticking coefficient drops at sufficiently high energies. At low molecular energies, two factors conspire to increase the sticking coefficient. One is that the molecule has little energy to lose. Second, a molecule may only be able to stick in one particular orientation or at one particular site. Consider, for example, NO on Pt(111), which must bind N end down. If a NO approaches O end first, it does not stick unless the molecule is able to reorient itself before it bounces off the surface. The slower the molecule, the greater the probability that the strong attraction of the surface for the N end pulls the molecule into the proper orientation before it strikes the surface. The process of molecular reorientation is known as steering. Molecules can be steered by attractive or repulsive forces, not only into the proper orientation, but also to the proper surface site.

Similarly, surface temperature does not have a strong influence on the sticking coefficient in direct non-activated adsorption, Fig. 3.16. At high temperatures, adsorbed molecules desorb after a short residence time on the surface. Eventually, the residence time becomes so short that it is difficult to distinguish between desorbing molecules and those that scatter directly [115].

In precursor mediated adsorption a molecule loses sufficient energy in the normal component of kinetic energy on the first bounce to stop it from returning to the gas phase. The molecule is trapped into a mobile precursor state. In some instances, the precursor may correlate with a stable adsorbed state, such as a physisorbed state. In other cases, the precursor may be a completely dynamical state. That is, the molecule enters the chemisorption well but a bottleneck in energy transfer stops it from falling immediately to

Table 3.2   *A survey of various adsorption systems. Unless stated to be extrinsic, precursor refers to the sticking behaviour on the clean surface*

Species	Surface	Properties	Reference
Atomic			
Xe	Pt(111)	Combination of direct and extrinsic precursor mediated adsorption leads to s increasing with increasing coverage, non-activated	[107]
Kr	Pt(111)	Intrinsic and extrinsic precursors, non-activated	[251]
Cs	W	Precursor mediated, non-activated	[157]
Ir, Re, W, Pd	Ir(111)	Direct at $\theta = 0$, non-activated	[252]
Molecular			
CO	Ni(100)	Direct non-activated adsorption on clean surface. Extrinsic precursor for low E_K and low T_s. Direct adsorption possible at $\theta(CO) > 0$ for higher E_K	[107]
	Pt(111)		[62, 253]
CO_2	Cu(110)	Direct non-activated, precursor mediated adsorption. Below 91 K s increases with increasing coverage.	[254]
	Ni(111)	Precursor mediated, non-activated	[255]
N_2	Fe(111)	Direct into first molecular state, activated transfer into second molecular state, K co-adsorption lowers barrier to transfer	[256, 257]
NO	Pt(111)	Direct non-activated	[258]
O_2	Ag(111)	Activation barrier between physisorbed and chemisorbed molecular states	[259]
	Pt(111)	Precursor mediated switching to direct for high E_K	[260]
	Pt(100)	s_0 two orders of magnitude lower on HEX-R0.7° reconstruction compared to (1 × 1), precursor switches to direct for high E_K	[261]
n-alkanes	Pt(111)	Precursor mediated, non-activated, s increases with coverage for C_2H_6 and C_3H_8	[107]
Si_2H_6	Si(111)-(7 × 7)	Precursor mediated, non-activated	[262]
Dissociative			
H_2	Si(111)-(7 × 7)	Direct activated, large barrier on clean terraces, lower barrier at steps, lower barrier in neighbourhood of adsorbed H	[168, 169]
	Si(100)-(2 × 1)		
	Ni(997)	Precursor mediated	[107]
	Cu(100)	Direct activated, large barrier	[43]
	Cu(111)		
	Ge(100)-(2 × 1)	Direct activated, large barrier	[266]
	Pt(111)	Activated on terraces but non-activated at steps with possible involvement of precursor	[267]
	W(110)	Direct activated	[267]
	Re(0001)		
	Mo(110)	Direct non-activated	[267]
	Rh(111)		
	Ir(111)		
N_2	Fe(111)	Precursor mediated	[256]
	W(110)	Direct activated	[107]
	W(100)	Precursor mediated at low E_K, direct at high trans, extrinsic precursor	[107]
	W(310)	Non-activated	[267]

Table 3.2 *(Continued)*

Species	Surface	Properties	Reference
Dissociative			
O_2	Pt(111)	Precursor mediated, barrier between molecular chemisorbed state and dissociative well	[268]
	W(110)	Precursor mediated for low E_K, direct activated for higher E_K	[242]
	Ir(110)-(2 × 1)		[269, 270]
	Ge(100)-(2 × 1)		[270]
CO_2	Ni(100)	Direct activated	[107]
	Si(111)-(7 × 7)		[271]
CH_4	W(110)	Direct activated	[107]
	Ni(111)		
	Ir(110)-(2 × 1)		
n-alkanes, C_2–C_4	Ni(100)	Direct activated	[107]
	Ir(110)-(2 × 1)	Direct activated and precursor mediated, low T_s and E_K favour precursor, high T_s and E_K favour direct	[107]
CH_3OH	Pt(111)	Direct activated and precursor mediated, low T_s and E_K favour precursor, high T_s and E_K favour direct, precursor associated with defect sites	[272]
C_6H_6	Si(111)-(7 × 7)	Precursor mediated from physisorbed state to chemisorbed	[273]
SiH_4	Si(111)-(7 × 7)	Direct activated	[262]
Si_2H_6	Si(100)-(7 × 7)	Precursor mediated switching to direct for high E_K, dissociation suppressed by H(a)	[262, 274, 275]
	Si(100)-(2 × 1)		

the bottom of the well. The total energy of the trapped precursor molecule may either be less than that required for stable adsorption or it may be higher. In the latter case, the precursor corresponds to a metastable state. The necessary criterion for trapping is that the normal component of kinetic energy is too low for the molecule to leave the surface. The precursor molecule now starts to hop from one site to the next. At each hop, the molecule either desorbs into the gas phase, migrates to the next site, or becomes adsorbed. Desorption either occurs because of energy transfer from the surface to the precursor molecule, or by intramolecular energy transfer from one molecular degree of freedom into the normal component of kinetic energy. The latter process is particularly important for metastable precursors.

The dependence on energy and temperature of precursor mediated adsorption is more complex than for direct adsorption because multiple steps must be considered. The dynamics of trapping into the precursor state are the dynamics of direct adsorption; thus, this step has the same energy dependence as direct adsorption. The desorption rate out of the precursor state increases with increasing surface temperature. In most cases, this rate increases faster than the transfer rate from the precursor to the adsorbed phase; therefore, increased surface temperature tends to decrease the rate of adsorption in precursor mediated adsorption. This behaviour is not, however, exclusive. If the barrier between the precursor state and the chemisorbed state has its maximum above $E = 0$, increasing T_s increases the rate of adsorption. This is found, for example, for the dissociative adsorption of O_2 on Pd(111) and Pt(111). For surface temperatures near l-N_2 temperature, O_2 at room temperature adsorbs into a molecular chemisorbed state, and no dissociation is observed. However, if the surface temperature is raised above ~180 K, the dissociative sticking coefficient rises as there is now a competition between desorption out of the molecularly chemisorbed precursor and dissociation. In Chapter 4 we investigate the kinetics of precursor mediated adsorption in more detail.

A change in molecular or surface temperature can also lead to a change in the adsorption dynamics, c.f. Table 3.2. If the energy is low, the molecule may be trapped into the physisorbed state and never move on to the chemisorbed phase. As the temperature is increased, adsorption can change from either direct or precursor mediated adsorption into the physisorbed state to precursor mediated adsorption into the chemisorbed phase. As the temperature increases further, the precursor may no longer represent an efficient mechanism for adsorption because of the high rate of desorption out of this state. At this point, the adsorption dynamics become dominated by direct adsorption. Whether such a scenario occurs as well as the relative efficacy of direct versus precursor mediated adsorption are determined by the exact details of the molecule surface interaction. By the end of this chapter, these factors should be clear.

3.8   Microscopic reversibility in Ad/Desorption phenomena

At equilibrium, the number of molecules per unit area striking a surface (the incident molecules) must be equal to the number per unit area receding from the surface (the emitted molecules) (no net mass flow). Likewise, there can be no net energy or momentum transfer. It follows that the angular distributions of the incident and emitted molecules must be identical (no flow directionality). While these statements may seem simplistic, they are so profound that they and their consequences have been repeatedly discussed (and misinterpreted) for more than a century. Whereas any remaining questions regarding angular distributions were settled in the 1970s and 1980s, issues pertaining to energy transfer continue to remain an active area of research, particularly now that quantum-state-resolved measurements are accessible.

James Clerk Maxwell [116] was the first to attempt a description of the angular and velocity distributions of molecules emitted from a surface. Maxwell proceeded with trepidation because, as he admitted, knowledge of the nature of the gas–surface interface was essentially non-existent in 1878. He assumed that a fraction, essentially the sticking coefficient, of the molecules strike the surface, and then are emitted with angular and velocity distributions characteristic of equilibrium at the surface temperature. The remaining molecules scatter elastically from the surface. Maxwell's assumptions were well founded and traces of them still pervade the thinking of many scientists to this day. However, as we shall see, these assumptions violate the laws of thermodynamics. First, we know that molecules also scatter inelastically. Second, and more importantly, we know that the sticking coefficient is not a constant.

Knudsen [117] made the next step forward when he formulated his law on the angular distribution of emitted molecules. According to Knudsen, the angular distribution must be proportional to $\cos \vartheta$, where ϑ is the angle measured from the surface normal. Such a distribution is often called diffuse. That Knudsen's proposition should attain the status of a 'law' is remarkable because (1) it was incorrectly derived, and (2) experiments of the type he performed to prove it should exhibit deviations from a $\cos \vartheta$ distribution. Smoluchowski [118] pointed out these and other deficiencies. In particular, he noted that Knudsen assumed that the sticking coefficient was unity – an assumption that cannot be generally valid. Gaede [119] scoffed at Smoluchowski's more general derivation. Gaede used an argument based on the second law of thermodynamics and the impossibility of building a perpetual motion machine, which he used to conclude that $s = 1$, and that the $\cos \vartheta$ distribution must be observed. Millikan [120], in the course of his oil drop experiments, set the cat among the pigeons when he demonstrated that specular scattering occurs. His results were unmistakably confirmed by Stern's observation of diffraction in atomic scattering at surfaces [88–90]. Therefore, the sticking coefficient cannot, in general, be unity. Nonetheless, Millikan's attempt at an explanation [121] was at best a description of a special case if not an outright violation of the second law, as he introduced a type of direct inelastic scattering characterized by a $\cos \vartheta$ distribution and Maxwellian velocity distribution.

The situation did not improve until the landmark work of Clausing [122]. By this time, Langmuir's investigations of gas–surface interactions had brought about a revolution in our understanding of surface processes at the molecular level. Clausing drew heavily on this work. In 1916, Langmuir stated: [123] "Since evaporation and condensation are in general thermodynamically reversible phenomena, the mechanism of evaporation must be the exact reverse of that of condensation, even down to the smallest detail." This remains a guiding principle that we use throughout the remainder of this chapter.

To analyze and describe correctly the energy and angular distributions of adsorbed, desorbed and scattered molecules, we need to apply correctly two closely related principles. These are microscopic reversibility, originally introduced by Tolman [124], and detailed balance, first proposed by Fowler [125]. These principles are formal statements of the idea expressed by Langmuir. Microscopic reversibility refers to individual trajectories whereas detailed balance refers to the rates of forward and reverse processes. That is, microscopic reversibility pertains to individual particles whereas detailed balance pertains to averages over ensembles of particles, though appropriate averaging can be used with microscopic reversibility to make predictions about rates.

Microscopic reversibility states that if we reverse individual trajectories in space and time, they must follow exactly the same trajectory. In other words, the interaction of a molecule with a surface (the effective potential energy curves) is independent of the direction of propagation. The same potential energy surface (or set of curves if electronic excited states are involved) governs the interaction of a molecule–surface system regardless of whether the molecule is approaching the surface or leaving the surface. In the strictest sense, microscopic reversibility does not apply to molecule–surface interactions. This is because of the symmetry of the problem. As was first noted by Clausing [122], we cannot reverse the co-ordinates of all the atoms (as the molecules must always approach the surface from above) and a less restrictive form of reversibility, called reciprocity, is obeyed for molecule–surface interactions [126, 127]. This is largely a matter for purists, and is commonly glossed over in the literature. The essential consequence of microscopic reversibility, however, is not lost: gas–surface interactions exhibit time-reversal symmetry such that adsorption and desorption trajectories can be related to one another.

Although the trajectories of individual particles are reversible, this does not mean that molecule–surface interactions cannot lead to irreversible changes, nor does it mean that molecules have memory. The question of reversible and irreversible chemical reactions is beyond our present discussion which is limited to the formation of a simple adsorbed layer rather than considering systems in which the surface reacts with the incident molecules to form a surface film with a new chemical composition such as an oxide layer or alloy. The question of memory is related to microscopic reversibility. The equivalence of the Maxwell-Boltzmann distributions for the incident and receding fluxes occurs because, when averaged over all molecules, both have exactly the same characteristics. This does not mean that when a molecule hits a surface, resides there for some time and subsequently desorbs, it enters the gas phase with exactly the same energy that it had before it became adsorbed. The wonder of thermodynamics is that, even though no individual molecule returns to the gas phase with exactly the same energy and direction of propagation, when averaged over all molecules in the system, the energy and propagation directions do not vary in time for a system at equilibrium when averaged over a sufficiently long period of time. Fluctuations in equilibrium distributions do occur on short time scales [128] but this is beyond our current discussion.

On the basis of this new knowledge of surface processes, thermodynamic arguments and the recognition that several channels (desorption, inelastic and elastic scattering) return molecules to the gas phase, Clausing proposed three principles.

1. The second law of thermodynamics demands the validity of the cosine law for the angular distribution of the molecules that leave a surface.
2. The principle of detailed balance demands that for every direction and every velocity, the number of emitted molecules is equal to the number of incident molecules.
3. Every well-defined crystalline surface, which emits molecules, has a sticking coefficient that is different from one, and that is a function of the angle of incidence and velocity.

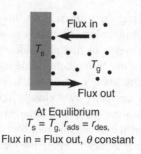

At Equilibrium
$T_s = T_g$, $r_{ads} = r_{des}$,
Flux in = Flux out, θ constant

Figure 3.19 *A gas in equilibrium with a solid. At equilibrium, the rates of forward and reverse reactions are equal, as are the temperatures of all phases. The balanced rates of adsorption and desorption lead to constant coverage. The balanced incident and exiting fluxes result in constant pressure and no net energy transfer between the gas and solid.*

Clausing's principles are essentially correct, though in some circumstance s is unity to a good approximation. We now follow their implications, use them to understand the dynamics of adsorption and desorption, and extend them to include the effects of molecular vibrations and rotations.

Consider, as in Fig. 3.19 a gas at equilibrium with a solid covered with an adsorbed layer. For a system at equilibrium, the temperature of the gas must be equal to the surface temperature, i.e. $T_g = T_s$. Recall that at equilibrium, chemistry does not stand still; rather, the rates of forward and reverse reactions exactly balance. In this case, the rates of adsorption and desorption are equal and, therefore, the coverage is constant.

Consider the case in which the sticking coefficient is not a function of energy. Take $s = 1$ as did Knudsen and Gaede (a similar argument can be made as long as s is a constant, as assumed by Maxwell). Since the gas-phase molecules are at equilibrium, by definition all degrees of freedom of the gas-phase molecules are in equilibrium, and these degrees of freedom–rotation, vibration and translation – are all described by one temperature, T_g. For $s = 1$, an equivalent way of stating that the rates of adsorption and desorption are equal is to state that the flux of molecules incident upon the surface is precisely equal to the flux of molecules returned to the gas phase by desorption, $r_{ads} = r_{des}$. In order to maintain $T_g = T_s$ = a constant, as is demanded by the state of equilibrium, not only must the fluxes of adsorbing and desorbing molecules be equal, but also the temperatures of the adsorbing and desorbing molecules must be equal. Furthermore, since the angular distribution of molecules incident upon the surface follows a cosine distribution based on simple geometric arguments, the angular distribution of the desorbing molecules must also follow a cosine distribution.

These results at first seem trivial. It appears obvious that the molecules, which are at equilibrium with a surface at T_s, should desorb with a temperature T_s and that since the surface is in equilibrium with the gas, $T_s = T_g$ = the temperature of the desorbed molecules. Nonetheless, this state of affairs is only a special case. It is not generally valid, nor is it normally observed.

We have learned that the sticking coefficient depends on energy. Therefore, we must consider the more general case of a sticking coefficient that is not constant. Consider the extreme case in which molecules below a certain energy stick with unit probability, but in which they do not stick at all above this energy, that is, $s = 1$ for $E \leq E_c$ and $s = 0$ for $E > E_c$, as shown in Fig. 3.20.

At equilibrium we must have $T_g = T_s$ and $r_{ads} = r_{des}$. The adsorbed molecules are in equilibrium with the surface. When they desorb, we intuitively expect that the desorbates leave the surface with a temperature equal to T_s. Consider, however, the implications of Fig. 3.20(b). From the form of the sticking coefficient as a function of energy, we know that only low-energy molecules stick to the surface. The high-energy

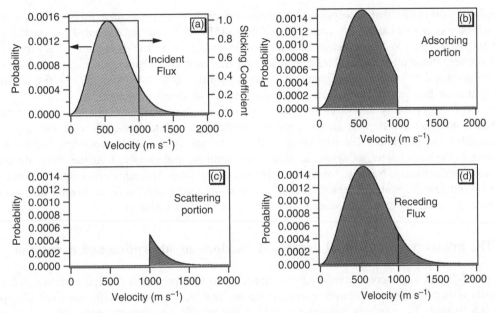

Figure 3.20 *Maxwell-Boltzmann distribution and a step function sticking coefficient. (a) Energy distribution for incident flux and the sticking function. (b) Energetic distribution of flux that adsorbs. (c) Energetic distribution of flux that does not adsorb. (d) Energetic distribution of total flux that leaves the surface.*

molecules reflect and return to the gas phase. By comparing Figs 3.20(a) and 3.20(b), we see that a Maxwell-Boltzmann distribution (often simply called a Boltzmann distribution or a Maxwellian distribution) at T_s has a higher mean energy than the mean energy of the molecules that stick because of the presence of the high-energy tail. Therefore, if the desorbed molecules entered the gas phase with a Boltzmann distribution at T_s, they would on average have a higher energy than the molecules that stick. The gas phase would gradually warm up, and the temperatures of the gas and the solid would no longer be equal. This violates our initial condition that the system is at equilibrium and because it spontaneously transports heat from a cool body to a warmer body – it could be used to create a perpetual motion machine. Hence, something must be wrong in our analysis.

The error is in the assumption of a Boltzmann distribution at T_s for the desorbed molecules. Figure 3.20(c) displays the true energy distribution of the desorbed molecules. The energy distribution of the desorbed molecules must *match exactly* that of the adsorbed molecules. Consequently, *whenever the sticking coefficient is a function of energy, the desorbed molecules will not leave the surface with a Boltzmann distribution at the surface temperature*. Indeed, although the adsorbed layer is in equilibrium with the surface, it will not, in general, desorb from the surface with an energy distribution that is described by a Maxwell-Boltzmann distribution.

Figures 3.20(b–d) describe how the system is able to maintain equilibrium. For a system in which the sticking coefficient is not unity, adsorption and desorption are not the only processes that occur. Scattering, which does not lead to sticking, must also be taken into account. Equilibration is not determined by adsorption and desorption alone: it is maintained by the sum of all dynamical processes. Therefore, the energy distribution of the incident flux must exactly equal the energy distribution of the sum of all fluxes that leave the surface. In other words, the energy distribution of the incident flux must equal the energy distribution of desorbed molecules plus scattered molecules, as is shown in Fig. 3.20(d).

Clausing's conclusion that desorption alone need not exhibit cosine angular or Maxwellian velocity distributions was not tested for 58 years. Then Comsa [55] reiterated and extended his arguments and within six months van Willigen measured non-cosine distributions in the desorption of H_2 from Fe, Pd and Ni surfaces. Soon after, molecular beam scattering was used to test these principles further [56, 129]. The adsorption and desorption experiments of Stickney, Cardillo and co-workers [57, 59, 60] led the way in demonstrating that adsorption and desorption are linked with detailed balance.

Microscopic reversibility allows us to understand the non-intuitive result that an adsorbed layer *at equilibrium* does not desorb with an energetic or angular distribution characteristic of our expectations of an equilibrium distribution. The root cause of this is the energy and angle dependence of the sticking coefficient. In the next chapter, we again encounter a non-intuitive result arising from the energy dependence of the sticking coefficient. Namely, we will see that deviations from Arrhenius behaviour are expected for adsorption and desorption at high temperatures. This deviation can be understood again once the importance of microscopic reversibility is appreciated within the context of transition state theory.

3.9   The influence of individual degrees of freedom on adsorption and desorption

The applicability of microscopic reversibility (reciprocity) to adsorption and desorption means that we can discuss these two reactions in nearly equivalent terms. That is, what pertains to one must also pertain to the other simply by reversing the sense in which the particle traverses the potential energy surface. As soon as we consider more than one degree of freedom, we must recognize that the potential energy curves shown above are simplifications of the true multidimensional potentials that govern the dynamics. Because the potential is multidimensional, the potential energy is actually described by a potential energy hypersurface (the "hyper" indicates that the "surface" contains more than two dimensions). Although I use the acronym PES, I use the term hypersurface when referring to the potential energy to avoid confusion with the physical surface at an interface.

3.9.1   Energy exchange

What if a gas is not in equilibrium with the solid? How long does it take the system to attain equilibrium? This leads us to a discussion of what is known as accommodation. By accommodation, we mean the process by which a gas at one temperature and a solid at a different temperature come into equilibrium. If a cold gas is exposed to a hot surface, on average the collisions are inelastic with the gas molecules gaining energy until the two phases attain the same temperature. In the older literature, and particularly in the applied and engineering literature, one encounters the concept of an accommodation coefficient introduced by Knudsen [117]. This is an attempt to quantify the efficiency of energy exchange between a solid and a gas. The thermal accommodation coefficient, α, is defined in terms of the energies of the incident, reflected and equilibrated molecules – E_i, E_r and E_s, respectively – as

$$\alpha = \frac{E_r - E_i}{E_s - E_i}.$$
(3.9.1)

If the molecules transfer no energy at all in a single encounter with the surface, $E_r = E_i$ and $\alpha = 0$. At the other extreme, the molecules equilibrate in one bounce and, therefore, $E_r = E_s$ and $\alpha = 1$. While the thermal accommodation coefficient is useful in engineering studies, it tells us little about the processes that lead to equilibration and is a poor quantity to use if we are to understand why certain gases equilibrate faster than others.

We need to realize that not all degrees of freedom are created equal. In virtually all circumstances (for large characteristic dimensions of length and not too high temperatures) we can consider only three distinct degrees of freedom – translation, rotation and vibration – neglect electronic excitations, and consider the translational degree of freedom to be a continuous variable. Rotations and vibrations are quantized and, in general, the spacing between rotational levels is significantly smaller than between vibrational levels. Energy exchange between collision partners depends on the relative spacing of the energy levels, the relative masses of the partners and the interaction potential. The closer together the spacing of energy levels and the more closely the energy levels of collision partners are to one another, the easier it is to transfer energy between them. Thus, translational energy levels, which are separated by infinitesimal increments, are most readily available for energy exchange in a gas–surface collision. Rotational levels are generally separated by energies small compared to $k_B T$ and small compared to the Debye frequency. Therefore, they should exchange energy readily but possibly more slowly than translations. Vibrational levels, except for rather heavy small molecules such as I_2 or large molecules such as long chain hydrocarbons, are generally separated by energies that are large compared to $k_B T$ and the Debye frequency. Hence, vibrations exchange energy at significantly slower rates. A full discussion of collisional energy transfer is found in Chapter 6 of Levine and Bernstein [130]. As a rule of thumb, translations equilibrate on the order of a few (10^0) collisions, whereas rotations require roughly $10^0 – 10^1$ collisions and vibrations, unless they have a very low frequency, require $10^4 – 10^6$ collisions. Molecular hydrogen, on account of its unusually small mass and large rotational and vibrational energy spacings, requires at least one order of magnitude more collisions to equilibrate.

The properties of the surface affect the rate of energy exchange. The mass of surface atoms plays a role as mentioned in the discussion of the hard cube model, as do the vibrational properties of the surface, that is, the phonon spectrum. For semiconductor and insulator surfaces, the only means of energy exchange (at normal temperatures) is via phonon excitations. For metals, however, low energy electron-hole pairs can be excited. These low energy excitations potentially play an important role in energy exchange at metal surfaces, especially for the excitation and de-excitation of the high frequency vibrational modes of adsorbing and desorbing molecules [131–136].

The existence of an attractive potential well is also important for energy exchange. The more attractive the potential energy surface, the greater the likelihood of energy exchange as demonstrated in the hard cube model. In part, this results from the acceleration that the molecule experiences as it approaches the surface. In addition, a strongly attractive potential energy surface may be able to reorient and distort the incident molecule. This can lead to enhanced energy exchange between the molecule and the surface as well as exchange of energy between different degrees of freedom within the molecule.

3.9.2 PES topography and the relative efficacy of energetic components

The shape of the PES defines the types of ad/desorption channels that are available: whether adsorption is activated or non-activated, direct or precursor mediated, molecular or dissociative. As can be seen in Table 3.2, the adsorption dynamics are determined by both the PES and the energy of the incident molecule. In the following sections, we discuss each degree of freedom individually. First, a few introductory remarks about PES topology are in order.

While it may seem obvious that vibrational excitation should facilitate dissociation, the degree to which vibrational energy can be converted into overcoming the activation barrier is determined by the exact shape of the PES. The topology of the PES also determines the relative efficacy of vibrational energy compared to translational and rotational energy [137]. To visualize the energetics of dissociative adsorption, we construct a two-dimensional PES such as those in Fig. 3.18. Starting from the upper left, motion of the molecule toward the surface corresponds to motion along y and the intramolecular vibration corresponds to motion along x. In the lower right, motion along y corresponds to the vibration of the chemisorbed atoms against the surface and motion along x corresponds to diffusion of the two atoms across the surface. Both the upper left and lower right can be probed spectroscopically by measurements performed at equilibrium. The difficulty arises in that the transition state governs the reaction dynamics. Ever since the invention of the concept [138, 139], one of the great preoccupations of chemical dynamics has been the direct observation of the transition state [130]. Great progress has been made in probing the transitions states of gas-phase reactions [140], as witnessed by the Nobel Prizes of 1986 and 1999 awarded to Herschbach, Lee and Polanyi, and Zewail, respectively, but direct observation of transition states at surfaces still remains elusive and (experimentally and computationally) challenging.

Three elbow potentials are represented in Fig 3.18. These are analogous to similar PESs drawn for gas-phase reactions and are similarly subject to the Polanyi rules developed from the study of gas-phase reactions [141–143]. For an early barrier, translational energy is most effective for overcoming the barrier and rotational and vibrational energy are ineffectual. For a very late barrier, vibrational energy is most effective. Rotational energy can also facilitate reaction through the mechanism discussed for H_2/Cu below, but it is less effective than vibrational energy. Translational energy plays a subordinate role on such a PES. For the middle barrier of 3.18(b), all of the molecular degrees of freedom play an important role in overcoming the barrier. Regardless of the barrier type, rotation always plays an important role because of its intimate relationship to the orientation of the molecule as it strikes the surface.

Hayden and Lamont [144] demonstrated that a higher dimensional PES is required to explain activated adsorption when they directly observed that sticking in the H_2/Cu system depends on internal energy and not just translational energy. The extensive experimental results of Rettner and Auerbach [43, 145–148] for both adsorption and desorption in the H_2/Cu system coupled with the trajectory calculations of Holloway and Darling [42, 149–154] have demonstrated the ability of state-resolved experiments to probe the PES of a surface reaction. In addition, this body of work has shown that even with the high dimensionality of surface reactions, a relatively simple PES and microscopic reversibility can explain the most intimate details of the formation and dissociation of H_2 on a metal surface.

The PES is also dependent on the surface co-ordinates. Binding energies and activation barriers depend on the position within the unit cell. Vibrational excitation and surface relaxation may also be important in ad/desorption dynamics. As a first approximation, the PES is usually conceived of in terms of a rigid surface. However, corrugation or the lack thereof must always be considered.

3.10   Translations, corrugation, surface atom motions

3.10.1   Effects on adsorption

Potential energy hypersurfaces can be thought of very much like topographical maps. We can use the intuition we have developed from our own experiences with geography and bobsledding to help us understand how molecules interact with surfaces. If you have ever taken a train from Berlin to Warsaw, you have experienced an approximation of the flat potential in Fig. 3.21(a). Once you have crossed the Oder River, the vast flat expanse of Wielkopolska gives one the impression that (if the tracks were perfectly frictionless) the engineer could turn off the engines, and the train would simple roll ahead until it reached Warsaw. Apart from friction, the only things that could stop the train are occasional defects in the tracks and planar landscape. Figure 3.21(b) represents a corrugated landscape. The consequences of corrugation are that there are preferred sites for an adsorbate to come to rest, and some directions are easier to move in than others. Corrugation also facilitates more extensive energy transfer between the adsorbate and the surface; for example, it leads to a coupling between normal and parallel momentum. Finally, as shown in Fig 3.21(c), not all regions are seismically inactive. Sometimes the surface must not be thought of as solely a rigid template, but also its motions must be considered. A surface, which is soft in comparison

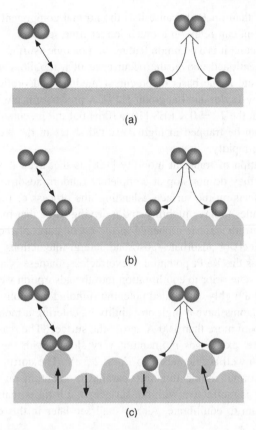

Figure 3.21   *The interaction of an ad/desorbing molecule with various models of solid surfaces. (a) A flat, rigid surface. (b) A rigid, corrugated surface. (c) A surface that is both corrugated and vibrationally active.*

to the force with which an adsorbate strikes it, is effective at dissipating energy, because energy can be transferred from the adsorbate to the phonons and vice versa.

Consider first the case of a rigid particle interacting with a flat, frictionless surface. This is represented in Fig. 3.21(a), and is roughly equivalent to the experience of a Xe atom incident upon a Pt(111) surface. The rare gas/Pt interaction potentials are extremely flat and non-directional as they are dominated by van der Waals forces [155]. Therefore, Xe has a low probability of stopping on the terraces of the Pt(111) surface [156]. The Xe atoms tend to roam across the flat planes until they encounter defects in the landscape. These defects take the form of either steps or previously adsorbed atoms (either Xe atoms or impurities).

Not all translational energy is the same. For Xe/Pt at low temperatures, the sticking coefficient is high; nevertheless, the atoms roam over large distances, as much as hundreds of angstroms, before coming to rest. Therefore, the important dynamical criterion for Xe sticking is not the loss of *all* translational energy upon the first encounter with the surface. Instead, *sufficient* energy must be lost from the normal component of translational energy, denoted $E_\perp$, or normal energy. The trapped but roaming Xe atom has a low value of $E_\perp$, but still has a significant parallel component of translational energy, $E_{||}$. The potential energy surface is conducive to efficient accommodation of $E_\perp$, but $E_{||}$ accommodates at a much slower rate. Xe adsorption is non-activated, so we expect that the sticking coefficient should decrease for sufficiently high kinetic energy. However, we can refine this statement and state that we expect the sticking coefficient of Xe to depend more strongly on $E_\perp$ than on $E_\parallel$ because it is the normal component that must be accommodated whereas the parallel component can relax on a much longer time scale.

While Xe/Pt is an extreme case, it is a common feature of molecule–surface PESs that $E_\perp$ accommodates more rapidly than $E_\parallel$. As mentioned above, the occurrence of a mobile state that can roam the surface searching for an adsorption site is the basis of precursor mediated adsorption. The concept of a mobile precursor was first proposed by Taylor and Langmuir [157]. A precursor is always bound above a well (even if this state does not appear in the $T_s = 0\,\mathrm{K}$ PES [105, 106]) but the precursor need not be accommodated into the well. A precursor can be trapped in highly excited states in the well if a dynamical bottleneck prevents it from equilibrating rapidly.

Another important implication of transient mobility [158] is that even though the Xe atoms strike the surface at random positions, they do not stop at completely random adsorption sites. The Xe atoms tend to form islands about the defects on the surface. Islanding, the process of forming patches of adsorbates, requires the adsorbing molecules to have mobility and a driving force that makes them stop at a preferred position. The driving force in this case is a change in the corrugation of the molecule–surface potential such as that presented by defects or adsorbate–adsorbate lateral interactions.

Most PESs are not as flat as the Xe/Pt potential. Nevertheless, flatness is not the only consideration for the formation of a long-lived state prior to equilibration into the adsorption well [159]. H is strongly bound on Cu(111), $q_\mathrm{ads} \approx 2.5\,\mathrm{eV}$, in a highly corrugated potential (binding at a hollow site ~0.5 eV stronger than on top site). Nonetheless, H atoms have a high probability of entering a mobile state that can live on the order of a ps or longer and roam more than 100 Å across the surface. The reason for this is that H is much lighter than Cu and, therefore, exchanges momentum very slowly with the phonons. As H approaches the surface, the chemisorption well accelerates it to $E_K \approx 2.5\,\mathrm{eV}$. The corrugation in the potential is very effective at converting the normal momentum into parallel momentum on the first bounce. Afterward, however, momentum transfer is ineffective with the surface, and it takes many collisions (on the order of tens of collisions) for the atom to equilibrate. As we shall see later in this chapter, such "hot" H atoms are highly reactive.

A PES with significant corrugation is depicted in Fig. 3.21(b). Corrugation leads to efficient exchange of energy between $E_\perp$ and $E_\parallel$. On a perfectly flat, frictionless potential, $E_\parallel$ never accommodates. Corrugation scrambles normal and parallel momentum, and can lead to more efficient energy transfer with phonons. This conceptualization of a corrugated but rigid surface can go a long way to explain many molecule–surface interactions. For instance, the dynamics of recombinative desorption of H_2 from Cu(100) can to a large extent be explained by visualizing a corrugated PES established by the symmetry of the Cu(100) surface and the motions of H atoms diffusing across it [42, 43, 47, 146, 149, 153, 160–162].

Figure 3.21(c) depicts a situation in which the surface plays a more active role in the adsorption and desorption dynamics. If a heavy molecule strikes a surface of comparatively light atoms, it can easily push the surface atoms around. This is a purely mechanical instance in which the lack of rigidity is important. However, we have seen in Chapter 1 that the presence of adsorbates can lead to reconstruction of surfaces. Adsorbate-induced reconstruction implies that surface atom motion can be inherent to the adsorption–desorption dynamics. This is the case for the H_2/Si system, in which surface atom motion during the adsorption–desorption process must be considered if we are to achieve an understanding of the dynamics. Although hydrogen is far too light to distort the surface mechanically, the Si atoms move under the influence of the forces accompanying the formation of chemical bonds. Surface atom excitations play a significant role in determining the fate of an incident hydrogen molecule because the height of the adsorption activation barrier is a function of the surface atom positions. Hence, it is not only the energy of the incident hydrogen molecules but also, more importantly, the energy (i.e. temperature) and positions of the surface Si atoms that determines to a large extent the sticking coefficient of H_2 on Si [163–169].

3.10.2 Connecting adsorption and desorption with microscopic reversibility

We have mentioned previously that adsorption trajectories are reversible. That is, if we simply run a movie of adsorption backwards then we obtain a movie that describes the reverse process, namely, desorption. We can use this fact and the discussion of adsorption in the previous section to make predictions about desorption.

When Xe adsorbs on Pt(111), it roams the surface until it eventually finds its final adsorption site. Therefore, when Xe desorbs, we predict that it is kicked out onto the terrace, diffuses over a long distance then, finally, departs the surface from some random site on the terrace. Low kinetic energy atoms are the most likely to stick. Therefore, microscopic reversibility demands that low energy atoms are also the most likely to be returned to the gas phase by desorption. The consequence is that, when averaged over the entire ensemble of desorbed atoms, the kinetic energy is lower than for an ensemble of molecules at the same temperature as the surface. In other words, for the desorbed molecules $T_{trans} < T_s$. This is known as translational cooling in desorption. It is a direct consequence of the process described in Fig. 3.20.

To better understand translational cooling, consider the following *Gedanken* experiment for a molecule that experiences non-activated adsorption. An adsorbed molecule has been excited all the way to the top of the chemisorption well, that is, to energy $E = 0$. We consider the desorption to be one dimensional. All modes other than the normal component of translational energy do not contribute to desorption. These spectator modes all have an equilibrium energy distribution at T_s and, within the 1D model, they do not couple to the reaction co-ordinate. The molecule has one final inelastic collision with the surface. Consider the surface to be effectively an oscillator with an energy distribution characterized by an equilibrium distribution at T_s as in the hard cube model. In the last encounter, the oscillator deposits all of its energy in the desorbing molecule. With no dynamical constraints on energy exchange, the surface oscillator transfers on average a distribution of energy characteristic of the equilibrium distribution at T_s into normal translation of the adsorbate. We call the amount of energy delivered $\langle kT \rangle$. The molecule then departs from the surface. The spectator modes all have an equilibrium distribution by definition in this model. The normal component of translational energy also has an equilibrium distribution. This description is what is demanded of a 1D model in which $s_0 = 1$, and is independent of kinetic energy.

Now consider a 1D model in which the oscillator finds it progressively more difficult to transfer larger amounts of energy. In this case, the surface oscillator delivers, on average, an amount of energy that is less than $\langle kT \rangle$. Whereas the spectator modes all have an equilibrium distribution of energy, the normal component of translational energy has less energy than expected for equilibrium at T_s. Translational cooling has been observed, for example, for desorption of NO/Pt(111) [65, 170], NO/Ge [171], Ar/Pt(111) [172] and Ar/H-covered W(100) [173]. It corresponds to the case in which s_0 decreases with increasing kinetic energy and is, therefore, a rather general phenomenon for systems that exhibit non-activated adsorption. Just as it is difficult for the surface to take away sufficient energy to allow increasingly fast molecules to stick, so too it is difficult for the surface to produce increasingly fast molecules in desorption.

In a real desorption system, the molecules do not all desorb from the $E = 0$ level. Molecules from a range of levels near the top of the chemisorption well experience one last encounter with the surface, then desorb. The inability of the surface to deliver $\langle kT \rangle$ leads to a subthermal energy distribution in the desorption co-ordinate. The levels near the top of the well are rapidly depleted by desorption. However, the surface is unable to maintain an equilibrium population distribution in these levels because of the inefficiency of energy transfer at high energy. As we shall see in Chapter 4, this effect leads not only to translational cooling in desorption, but also it represents a dynamical factor that causes non-Arrhenius behaviour of the desorption rate constant at high temperature [174–176].

As we have seen in § 3.5, H_2 adsorption on Cu is activated. In a simple one-dimensional Lennard-Jones type potential, we considered only the effect of energy along the reaction co-ordinate, which to a first approximation we take as the normal component of translational energy. Indeed, translational energy plays the most important role in determining the sticking coefficient of H_2 on Cu: the sticking coefficient increases strongly with increasing H_2 normal energy [43, 177]. Therefore, microscopic reversibility demands that since fast H_2 molecules are the most likely to stick, they are also the most likely to desorb. This is the case, and it is found experimentally that the desorbed H_2 is very "hot": in the translational degree of freedom. Furthermore, since molecules directed along the surface normal stick better than those incident at grazing angles, the desorbed molecules tend to be focussed along the surface normal.

H_2 adsorption on Si is also activated [164, 177]. However, the desorbed molecules are not translationally hot and, therefore, translational energy is not the primary means of promoting adsorption. The reason for this can be found in the role of surface excitations. Dissociative adsorption of H_2 on Si is activated in the surface co-ordinates. Effectively, we need to excite the surface into preferential configurations for the H_2 to dissociate and stick as H atoms. Therefore, in desorption, the energy of the activation is not efficiently conveyed to the desorbing H_2 molecules. Instead, a large fraction of the activation energy remains in the Si surface. The result is that molecules leave the surface not too hot, but a vibrationally excited surface is left behind.

3.10.3   Normal energy scaling

The Lennard-Jones model is a 1D model. Only energy directed along the reaction co-ordinate is effective in overcoming the barrier. Energy directed along other dimensions – so-called spectator co-ordinates – have no influence on the reaction dynamics. In the Lennard-Jones model, this dimension is taken as a barrier normal to the surface in which only the normal component of kinetic energy is capable of overcoming the barrier. Essentially, there was no evidence contradicting the 1D nature of the adsorption barrier until the work of Bernasek and co-workers [76], and Kubiak, Sitz and Zare [84, 85]. Their observations of superthermal population of the first vibrationally excited state could only be accounted for by including additional dimensions in the adsorption dynamics.

In the absence of corrugation, there is no coupling between the parallel and perpendicular components of momentum. Thus, in any flat surface system, if the dissociation barrier is directed along the surface normal, only the component of kinetic energy directed along the normal is effective in overcoming the barrier. In most systems, kinetic energy is the most efficient at overcoming an activation barrier. Thus, even in multidimensional systems, our first-order approximation is to expect normal energy scaling for the adsorption probability. Normal energy scaling is defined as follows. The normal component of the kinetic energy (the normal energy) varies with the angle of incidence, ϑ_{in}, as

$$E_\perp = E_{in} \cos^2 \vartheta_{in}. \tag{3.10.1}$$

Since the amount of normal kinetic energy decreases as $\cos^2 \vartheta_{in}$ at constant E_{in}, the effective barrier height in a 1D system increases as $1/\cos^2 \vartheta_{in}$ as shown in Fig. 3.22(c). If the adsorption probability scales with the magnitude of normal energy such that it decreases with angle as $\cos^2 \vartheta_{in}$, we say that activated adsorption exhibits normal energy scaling. This can be contrasted with total energy scaling, for which the exponent of $\cos^n \vartheta_{in}$ is $n = 0$, i.e. the adsorption probability scales with total kinetic energy rather than the component along the normal.

Normal energy scaling is observed for the dissociative sticking of CH_4 on W(110) and Ni(111), and n-alkanes/Ni(100) whereas total energy scaling is observed for the dissociative sticking of N_2/ W(100) [178]. For H_2 dissociation, normal energy scaling is commonly observed with the exception of Pt(111), Ni(100) and Fe(110) surfaces for which the parallel component inhibits dissociation. The presence of corrugation is expected to lead to a scaling characterized by $0 \leq n \leq 2$ [179]. Moreover, interaction potentials are corrugated, so how can we explain the prevalence of normal energy scaling?

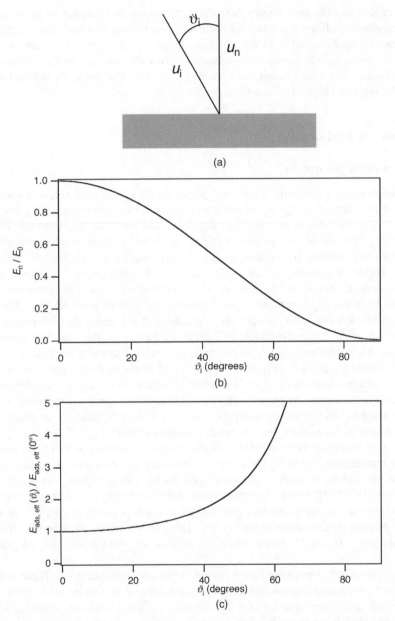

Figure 3.22 *A description of normal kinetic energy and the effective height of a one-dimensional barrier: (a) the molecule is incident upon the surface at angle ϑ_i from the normal with velocity u_i and normal velocity u_n; (b) the normal energy falls with increasing angle of incidence according to Eq. (3.10.1); (c) the decreasing normal energy of molecules incident at higher ϑ_i leads to an increasing effective barrier height if only the normal energy is capable of overcoming the barrier.*

Darling and Holloway [47] have shown how the combination of energetic and geometric corrugation combine counterbalancing effects that allow for normal energy scaling. Not only the energy of the activation barrier (energetic corrugation) but also the position of the barrier (geometric corrugation) varies with position across the surface. When they are combined in the right amounts, normal energy scaling results *because* the surface is corrugated rather than because it is flat. The reader is referred to the insightful discussion of Darling and Holloway for further details.

3.11   Rotations and adsorption

3.11.1   Non-activated adsorption

Our first impression might be that rotation should play a minor role in ad/desorption phenomena because it is, after all, the presence of energy in the normal component of kinetic energy that determines whether a molecule scatters or desorbs from a surface. Rotation, however, has two consequential effects upon chemical reactivity. The first is as a source of energy. If there is a way in which rotational energy can be channelled into the reaction co-ordinate, then rotational energy can be used to influence reactivity. One of the most important couplings is between rotational and kinetic energy. This channelling can occur either as rotation-to-translational energy transfer (R–T transfer) or in the opposite sense (T–R transfer). T–R transfer has been directly measured in, for example, the scattering of NO from Ag and Pt surfaces [62, 64, 69, 70, 180]. Rotation also changes the orientation of the molecule with respect to the surface. From the discussion of the Blyholder model in § 3.4.2, we have seen that the chemisorption potential is highly anisotropic. Molecules such as CO or NO bind in specific ways to a surface. A CO molecule on a transition metal surface chemisorbs only through the C end of the molecule and not through the O end. Therefore, a CO that approaches the surface O end first is not able to stick in its equilibrium chemisorption site unless it is able to rotate into the proper orientation. On the other hand, if a molecule is rotating rapidly, the surface might not be able to stop it and pull it into the proper orientation for sticking. Thus, a little rotation may be essential for sticking but too much enhances scattering.

The adsorption of molecular NO on Pt(111) is direct and non-activated with a sticking coefficient of ~0.85 for room temperature NO [170, 181]. At low coverage NO occupies an on-top site and is bound perpendicular to the surface through the N atom. NO sticks with a higher probability if it strikes the surface N end first [182, 183]. Just as for translations, not all rotations are the same. We can decompose rotational motion into two components: so-called helicopters, which rotate in the plane of the surface, and cartwheels, which rotate perpendicular to the surface. Think of the NO as an American football that is red on the N side and white on the O side. It prefers to stick to the surface with the red side down and the white side up.

Several competing effects determine how rotation influences ad/desorption. These are related to the competing effects of rotational-to-translational energy transfer and molecular orientation. A cartwheeling molecule can present the correct geometry for the chemisorbed state; however, as scattering measurements show [72, 180, 184–186], it can also efficiently couple its rotational energy into normal translational energy. This can give an added kick to the molecule and enhanced scattering. A helicoptering football never sticks red side down unless it is steered into an upright geometry. However, in-plane rotation does not couple at all to $E_\perp$ on a perfectly flat surface (think of the analogy to an oblong football) and, therefore, it allows a much softer landing at the surface with no prospect of enhanced desorption from rotational-to-translational energy transfer.

When looked at from the perspective of the chemisorbed molecule, it appears that only a cartwheeling motion can be excited in desorption. From the perspective of the surface, a Pt atom striking the normally bound NO from the bottom cannot excite in-plane rotations in an impulsive collision. Such a collision can only excite cartwheeling motion. Thus, we might predict that desorbed molecules should have a higher probability of being cartwheels. In adsorption, we might predict that helicopters would have a higher sticking coefficient because of the softer landing and the ability of the highly directional chemisorption well to steer the molecule along the incident trajectory. Our intuitive predictions appear to violate microscopic reversibility and can only be resolved by direct experimental observation.

First, we take the temperature of the desorbed molecules using laser spectroscopy. We find rotational cooling in desorption [63, 64, 73, 176, 187]. This has been observed for NO/Ru(001) [73], NO/Pt(111) [63, 64], OH/Pt(111) [188] and CO/Ni(111) [189] and appears to be a general phenomenon for non-activated adsorption systems. Analogously to the case of translational energy, this means that the sticking coefficient is a function of rotational energy and that it decreases for increasing rotational energy. Laser spectroscopy can also determine the degree of rotational alignment in a gas, i.e. the relative populations of the two components of rotational motion. Such an experiment has been performed for desorption of NO from Pt(111) [190]. For states with a low value of the rotational quantum number, J, there is no rotational alignment. However, as J increases there is a trend to a progressively greater number of in-plane rotors compared to out-of-plane rotors. In-plane rotation must, therefore, be more favourable for adsorption than out-of-plane rotation. As J increases, cartwheeling motion becomes progressively more unfavourable for adsorption.

The observation of rotational alignment in desorption has an important consequence for our understanding of ad/desorption dynamics. As we have stated above, a normally bound NO molecule cannot be excited into in-plane rotation when it is struck from below by a Pt atom. Nevertheless, in-plane rotation is excited as efficiently as – and at high J even more efficiently than – out-of-plane rotation. This is direct proof that desorption (and therefore adsorption) is not a direct process in which a molecule leaves from the bottom of the chemisorption well and pops out into the gas phase. Instead, the molecule gradually percolates up to the top of the chemisorption well after many collisions with the surface. After this multi-step process of climbing the ladder of vibrational states in the chemisorption well, the molecule leaves from the top of the chemisorption well. At the top of the well, the bonding interaction does not bear a strong resemblance to the well-oriented geometry at the bottom of the well. At the top of the well where all of the several NO-surface vibrational modes are multiply excited, the adsorbate geometry is not well-defined, and energy is easily transferred into both components of rotation. For the most highly excited rotational states, population is more readily built up in in-plane rotation. This is related to rotation-to-translation energy transfer. Since in-plane rotation couples weakly to normal translation, in-plane rotation does not enhance desorption, and the desorption rate has little dependence on the extent of in-plane rotational excitation. Out-of-plane rotation does couple to normal translation, therefore excitation of out-of-plane rotation enhances the desorption rate. The most highly excited out-of-plane rotation states therefore become progressively more depleted of population compared to the in-plane rotational states.

The implication for adsorption is that it, too, must be a multi-step process. A slow molecule rotating in-plane in a low J state is the most likely to stick. The molecules that stick do not fall directly to the bottom of the chemisorption well. The adsorbing molecules require many collisions with the surface until they finally lose their energy, attain the correct bonding geometry, and equilibrate with the surface. This may occur at the point of impact but we can also expect that the molecule may exhibit some mobility before attaining its final chemisorption site. The extent to which the molecule roams about the surface is determined by the relative rates of accommodation and the corrugation of the molecule−surface potential.

3.11.2 Activated adsorption

The role of rotation in activated adsorption is best illustrated by the case of H_2/Cu(100). Dissociative H_2 adsorption on Cu is activated as discussed in § 3.5. The overlap of the $2\sigma^*$ orbital with the metal orbitals is enhanced by a molecular orientation in which the internuclear axis is parallel to the surface [191]. We therefore predict rotational alignment in desorption with a preference for in-plane rotation, an expectation that is experimentally confirmed [148, 192] at low J.

The effect of rotational energy in H_2 dissociation has two distinct effects [146, 153]. For $J = 0$, the PES can re-orient the molecule as it approaches the surface and bring it into the best possible orientation for entering the transition state. As J increases, steering becomes less effective. Therefore, increasing rotational excitation is a hindrance to dissociative adsorption at low J. At high J, a portion of the rotational energy can be channelled into overcoming the activation barrier. This occurs because of the shape of the PES. Bond extension is required to reach the transition state. The extension of the H−H distance in the TS channels energy out of rotation and into the reaction co-ordinate. Hence, rotational excitation promotes dissociation at high J. Taking the two effects together, rotation hinders adsorption up to $J \sim 5$, whereupon the promotional effect of rotational energy becomes dominant. Similar effects should be observed in any system with an extended bond in the TS.

3.12   Vibrations and adsorption

For molecules that are not too large, the spacing between vibrational levels tends to be much greater than the energies of phonons and rotational states. Thus, vibrations tend not to couple efficiently to other degrees of freedom. Vibrations can couple to the electronic structure of a metal through electron-hole pair formation [132, 136, 193]. Because of the band gap found in semiconductors and insulators, electron-hole pair formation is not important at thermal energies for these materials. Because of the weak coupling of vibrations to other degrees of freedom, molecular vibrations are not of utmost importance for non-activated and, in particular, non-dissociative sticking of small molecules. This is not true for large molecules or for the sticking of clusters at surfaces. Large molecules and clusters have a plethora of low frequency vibrational modes that can effectively soak up translational energy. Their excitation upon collision with the surface can play an important role in the sticking dynamics.

For activated adsorption, vibrations can play a crucial role [42, 43]. For the dissociative adsorption of a diatomic molecule, the vibration is intimately related to the reaction co-ordinate. For H_2/Cu, H_2/Si and H_2/Pd, significant superthermal vibrational excitation in desorbed molecules has been measured [85, 194, 195]. These results demonstrate that vibrational excitation and the concomitant bond extension aid in reaching the transition state and overcoming the activation barrier. For any middle or late barrier PES, vibrational excitation is important in the ad/desorption dynamics.

An even more interesting case is that of methane dissociation on Ni surfaces, which exhibits vibrational mode selective chemistry [196]. In the dissociation of methane, it has been found by Utz and Beck and co-workers that not only is the presence of vibrational excitation important in dissociative adsorption, but also that the efficiency of channelling this excitation into dissociation depends on the specific vibrational mode that is excited. Vibrational excitation can be more efficient than translational energy at promoting dissociation. Furthermore, excitation of the symmetric ν_1 stretch is significantly more effective than excitation of the antisymmetric ν_3 mode. These observations prove that the statistical redistribution of energy within the reaction complex is significantly slower than the timescale for dissociation.

3.13   Competitive adsorption and collision induced processes

We now investigate what happens when a surface covered with one adsorbate is exposed to a different type of molecule. This is, of course, quite relevant to surface reactions as both types of adsorbates must usually be present on the surface for a reaction to ensue. The answer depends not only on the chemical identity of the surface and the adsorbing molecules, but also sometimes depends upon the order in which the molecules are exposed to the surface.

First consider the system $O_2 + CO$ on Pd(111) [36, 197]. When O_2 is exposed to a Pd(111) surface held at 100 K, three molecular chemisorbed states are formed. These three states have different binding energies to the surface. As the O_2–Pd bond strength increases the O–O bond is weakened and, therefore, the O–O stretch frequency is lowered. CO adsorbs in one molecular chemisorption state. Temperature programmed desorption (TPD) studies (§ 4.7) have shown that CO is bound more strongly than O_2.

Systems evolve to achieve the lowest Gibbs energy. Equilibrium is defined as the point of minimum Gibbs energy, and the state in which Gibbs energy no longer changes. However, the road to equilibrium is not always a direct one. As a first approximation, we neglect entropic effects. This should be justified at the low temperatures considered and for the two similar diatomic molecules in our system. CO should be able to displace O_2 from the surface since it is more strongly bound. Furthermore, we might expect that the least tightly bound O_2 species would be the first to be displaced followed by the intermediate, then the strongly bound O_2. That is, ω_3 before ω_2 before ω_1. The data in Fig. 3.23 show that while the first expectation is met, the second is not. CO can displace O_2 from the surface; however, it is the most weakly bound species, ω_3, which is actually the most able to compete with CO for adsorption sites. Furthermore, we see that CO can also induce the conversion of one type of O_2 into another. The reasons for this lie in the dynamics of competitive adsorption. CO and O_2 must compete for sites on the Pd surface. In this competition, we must consider three factors: (1) the energetics of chemisorption bond formation; (2) what types of sites the adsorbates prefer; and (3) the effect of the presence of one adsorbate on the energetics of a second adsorbate.

To understand the displacement of O_2 by CO on Pd(111), we have to consider not the relative energetics of the O_2 chemisorption states, but the relative energies of the O_2 states bound on the sites onto which

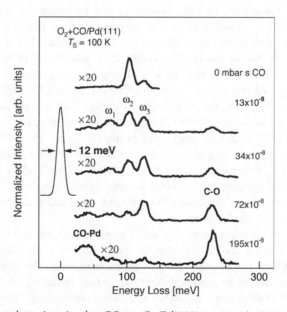

Figure 3.23 *Competitive adsorption in the $CO + O_2$/Pd(111) system is investigated by electron energy loss spectroscopy (EELS). The species associated with species ω_1, ω_2 and ω_3 are illustrated in Fig. 3.8. arb units, arbitrary units; T_s, surface temperature. Reproduced from K. W. Kolasinski, F. Cemič, A. de Meijere, E. Hasselbrink, Surf. Sci., 334, 19. © 1995, with permission from Elsevier.*

CO is most likely to adsorb. The results in Fig. 3.23 suggest that the most tightly bound O_2 species is bound to the same site onto which CO preferentially adsorbs. The ω_1 species is unable to compete with CO not only because the CO is more strongly bound, but also because the O_2 is not sufficiently strongly bound to establish a sizeable barrier to its displacement. ω_1–O_2 has two alternatives when CO pushes it out of its site. It can either convert to the ω_2 or ω_3 state or it can desorb from the surface. Quantitative analysis of the EEL spectrum reveals that both of these fates await displaced ω_1–O_2. The reason why CO is more strongly bound than O_2 is that it more effectively withdraws electrons from the surface. Thus with increasing CO coverage, the ability of Pd to donate electrons to adsorbates decreases. Since ω_3–O_2 requires the least amount of electron donation, and because it occupies a site different than that occupied by CO, it is able to survive to the highest coverages of CO co-adsorption. The details of CO displacement of chemisorbed O_2 depend sensitively on the PES. Whereas CO displaces O_2 from Pd(111) with near unit efficiency, on Pt(111) the efficiency is only 0.09 per adsorbing CO [198].

The question remains as to the mechanism by which one adsorbate displaces another. Rettner [199] has shown that a strongly adsorbing species can effectively transfer a portion of its heat of adsorption to a pre-adsorbed molecule, leading to desorption. Specifically H, N or O incident upon O_2 chemisorbed on Pt(111) facilitate efficacious displacement of the O_2. A similar mechanism is also likely in the displacement of O_2 by CO.

Somorjai and co-workers [200] have investigated a number of instances in which an ordered array of adsorbates is pushed aside by a co-adsorbate that is post-dosed to the surface. CO is able to compress pre-adsorbed S layers on both Re(0001) and Pt(111) surfaces, for example. The S atoms move into regions of locally higher coverage. In the process, sites are freed up that become available for CO chemisorption. If CO desorbs from the surface, as caused by an increase in the temperature, the S atoms move back into the ordered structure that they originally occupied. In other cases, adsorbing CO molecules can shepherd pre-adsorbed disordered molecules into an ordered phase. This has been observed for ethyne, ethylidyne, propylidyne, benzene, fluorobenzene, sodium, potassium and hydrogen on several metal surfaces. The common feature of all of these adsorbates is that they are relative electron donors compared to the electron accepting CO. In all of these cases, the pre-adsorbed molecule has a dipole associated with an orientation opposite to that of the CO dipole. Conversely, in co-adsorbed systems in which both adsorbates exhibit a dipole oriented in the same direction, disordering or segregation occurs. For instance in the $CO + O_2$/Pd(111) system no ordered structures are formed while in the $CO + S$/Pt(111) segregation occurs.

In Chapter 6 we study another important example of competitive adsorption. This occurs during the formation of self assembled monolayers (SAM) of alkanethiols from solution. With regard to SAM formation, an important aspect is that relatively inert surfaces must be used. The surface interacts weakly with the solvent but since it is present in a much greater concentration than the solute the physisorbed solvent covers the surface initially. The alkanethiol slowly chemisorbs to the surface by displacing the solvent molecules. When it binds to the surface, the bond is quite strong and, therefore, the alkanethiol easily pushes aside the physisorbed solvent molecules. Once bound, the alkanethiol cannot be removed from the surface by other adsorbates that bind less strongly.

In the above examples, the adsorption energy and lateral interactions were involved in displacement and shepherding of pre-adsorbed molecules. Another mechanism for displacement and dissociation of pre-adsorbed molecules utilizes the kinetic energy of an incident molecule. For moderately high incident kinetic energies, molecules that hit an adsorbed layer can impart sufficient momentum to cause chemical transformations [201, 202]. Two distinct processes need to be considered. The first is collision induced desorption in which the incident molecule transfers sufficient momentum to the adsorbed molecule such that the adsorbate departs from the surface. Ceyer and co-workers [203, 204] demonstrated that Ar incident on CH_4/Ni(111) induces desorption of CH_4 via a direct impulsive collision. The second process is collision induced dissociation [201, 203] in which the incident molecule fragments the adsorbate and the products remain, at least in part, on the surface. An impulsive collision transfers kinetic energy to the adsorbate, which then collides with the substrate. This mechanism leads to dissociation of CH_4 via a pathway that is equivalent to the dissociation of high translational energy CH_4 incident from the gas phase. The kinetic energy leads to a distortion of the molecule, which allows the molecule to overcome the dissociation barrier. For CH_4/Ni(111), collision induced desorption is always more probable than dissociation.

Advanced Topic: High Energy Collisions

Even higher energy collisions are also important in surface science. When considering incident energies on the order of a keV or larger, we usually are dealing with incident ions. At these translational energies, the differences between the dynamics of ions and molecules are largely inconsequential. High-energy ions efficiently penetrate the surface; therefore, we must consider collisions not only with the adsorbed layer and the substrate surface but also with the bulk. Energy can be lost by direct collisions between the ion and the nuclei of the target. Electronic excitations and charge exchange occur as well. All three of these energy exchange mechanisms are dependent on the translational energy and make different contributions as the ion slows. Electronic losses dominate at the highest energies but cause only small changes in the scattering angle. Elastic nuclear collisions produce large angle scattering, predominantly at low energies. Charge exchange is by far the least important relaxation pathway.

The nuclear scattering of ions from surfaces can be modelled quite successfully by classical hard-sphere kinematics. This allows for accurate predictions to be made about the scattering angles and has been exploited in the form of various ion scattering spectroscopies that can be used to probe the structure of surfaces and the selvedge [205]. The depth of penetration of ions depends on the relative masses of the projectile and the target nuclei, the ion energy and the angle of incidence. By adjusting these parameters and the ion dose, the depth and concentration of the implanted ions can also be controlled. Ion implantation has a range of technical applications in materials science, including the formation of buried interfaces, semiconductor doping and the modification of surface chemical properties [206].

High-energy particles cause a variety of radiation damage phenomena. When the projectile collides with a target nucleus in a primary collision, it transfers momentum, which leads to secondary collisions between the target and other substrate atoms. The atoms struck in the secondary collision go on to further generations of collisions. This process is known as a collision cascade. With the occurrence of so many collisions, it is unlikely that all of the atoms return to their equilibrium positions. Many defects are introduced into the lattice such as vacancies and the occupation of interstitial sites. Some atoms do not return to the substrate and are expelled into the gas phase. The process of removing substrate atoms by high-energy collisional processes is known as sputtering [207]. The number of atoms sputtered per incident ion, i.e. the sputter yield, is a function of the ion energy, angle of incidence, relative masses, the surface temperature and the ion flux. Sputter yields can range from 1–50 atoms per incident ion for typical mass combinations and collisions on the order of tens of keV. Sputtering occurs not only because of collision cascades. The energy transfer processes that decelerate the ion lead to a variety of thermal and electronic excitations that also contribute to the ejection of substrate ions. Sputtering combined with post-irradiation annealing to remove lattice defects is a commonly used method to prepare surfaces free of adsorbed impurities.

3.14 Classification of reaction mechanisms

Mass transport to, from and on the surface plays an essential role in surface reactions. When molecules are transferred between phases, it takes a finite time for them to equilibrate. Furthermore, we know that not every collision between a molecule and a surface leads to sticking. These characteristics affect the course of reactions at surfaces.

3.14.1   Langmuir-Hinshelwood mechanism

The most common surface reaction mechanism is one in which both reactants are adsorbed on the surface where they collide and form products. This is known as the Langmuir-Hinshelwood (LH) mechanism [208]. Adsorption, desorption and surface diffusion play essential roles in the LH mechanism. While it might be expected that the reaction rate should depend on the surface coverage of both species, the rate law may be complex and depend on the reaction conditions. Ultimately, the rate law can only be properly interpreted when the complete reaction mechanism is understood. Nonetheless, determination of the rate law is an important component in determining the reaction mechanism.

The dynamics of a LH reaction involves a convolution of the dynamics of adsorption, desorption and diffusion. Thus, in a sense, a convolution of all the dynamics we have studied up to this point is involved in LH dynamics. The interplay between reaction dynamics and the rate of catalytic reactions are discussed thoroughly in Chapter 4.

In a multi-step reaction mechanism, generally one reaction is a bottleneck and, therefore, determines the overall rate. This is known as the rate determining step (RDS) and it is defined more accurately in § 6.4.3. The dynamics of the RDS is the most important dynamics for any given LH reaction system. The RDS can be any one of a number of different types of surface reaction, e.g. adsorption, adsorbate decomposition, diffusion of an adsorbate to a reactive site or desorption of a product. Under normal conditions, the rate of ammonia synthesis in the Haber-Bosch process is determined by the rate of N_2 adsorption (see Chapter 6). The decomposition of Si_2H_6 on Si or Ge depends on the presence of free sites. If performed at low temperature, the reaction is self-limiting. However, at high temperatures recombinative desorption of H_2 liberates free sites. Thus, the decomposition rate is limited by the desorption rate of H_2.

The identity of the RDS generally depends on the reaction conditions: the pressure of the reactants in the gas phase, surface temperature and coverage. Consider the reaction of O_2 and CO to form CO_2 on platinum group metals [209]. The catalytic formation of CO_2 requires the reaction of adsorbed O atoms with adsorbed CO molecules. However, in the section on competitive adsorption, we have seen that whereas CO can adsorb on O_2-covered surfaces (and similarly for O-covered surfaces), O_2 cannot adsorb on CO-covered surfaces. Thus at high CO coverage, the rate is limited by O_2 dissociation and increases in the CO pressure are found to *inhibit* reaction. However, at low CO coverage, O_2 dissociation occurs rapidly. The reaction rate in this case depends on the coverage of both CO and O_2 and the rate increases with increases of O_2 and CO pressures. The extreme richness of CO oxidation kinetics is due to changes in the reaction dynamics and can lead to rate oscillations and formation of standing and travelling concentration gradients on the surface. These phenomena are known as spatiotemporal pattern formation and are explored further in Chapter 6.

The energy distribution of the products of a LH reaction informs us about the dissipation of energy during catalytic reactions. The oxidation of CO [52, 210] and H_2 [211, 212] on platinum group metals to form CO_2 and H_2O, respectively, are both highly exothermic reactions and energy deposition into the reaction product has been studied in both cases. As can be seen in Fig. 3.24, much of the exothermicity of the reaction is dissipated into the surface upon dissociative adsorption of O_2. In the case of CO_2, the product leaves the surface with high levels of rotational, vibrational and translational energy, carrying away ~80% of the energy it attains when it reaches the transition state. In contrast, H_2O leaves the surface with an energy that is much more characteristic of the surface temperature.

These energy distributions tell us something about the last interactions of the nascent product with the surface. H_2O essentially leaves the surface as if it were desorbed thermally. Thus, we can surmise that the

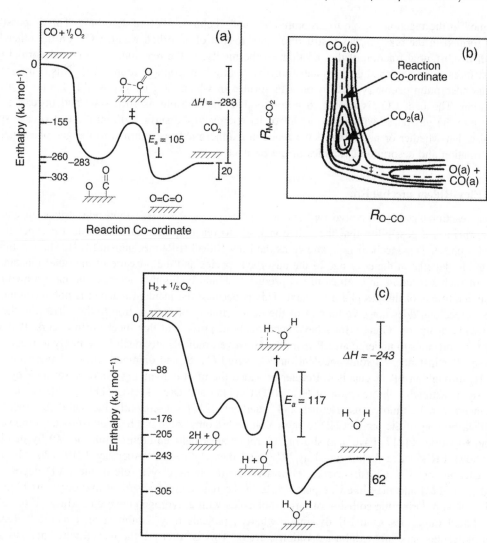

Figure 3.24 *The oxidation of CO to form CO_2 and H_2 to form H_2O on Pt(111) follow Langmuir-Hinshelwood mechanisms. The changes in energy along the reaction pathway are as follows. (a) Enthalpy changes associated with $CO + O_2$ reaction. The transition state ‡ is a stretched and bent CO_2 entity. (b) A two-dimensional potential energy surface of the $CO + O_2$ reaction, portraying the energetic changes as a function of the CO_2–surface distance [r(M–CO_2)] and the forming OC bond length [r(O–CO)]. (c) Enthalpy changes associated with the $H_2 + O_2$ reaction. Two intermediates (2H + O, and H + OH) are formed during the reaction. The transition state † reached prior to the formation of $H_2O(a)$ is also bent and stretched; however † resembles $H_2O(a)$ more closely than ‡ resembles $CO_2(a)$. Parts (a) and (b) Reproduced with permission from G. Ertl, Ber. Bunsenges. Phys. Chem. 86 (1982) 425. ©1982 Wiley VCH. Values for (c) taken from M. P. D'Evelyn and R. J. Madix, Surf. Sci. Rep. 3 (1983) 413.*

H_2O formed in the reaction is able to accommodate and chemisorb before desorbing. This is possible if the H_2O formed in the reaction has a structure close to that of adsorbed water. CO_2, on the other hand, carries off a substantial fraction of the reaction exothermicity. In the reaction, a CO bound normal to the surface diffuses to an adsorbed O atom and forms CO_2 in a bent configuration [213]. This contrasts with the normal adsorption geometry of CO_2 on a Pt surface in which CO_2 is very weakly bound in a linear configuration. The bent CO_2 is unable to accommodate into the physisorbed state and instead desorbs promptly into the gas phase without losing all of its energy to the surface. From these two case studies we surmise that whether or not some of the reaction exothermicity is returned to the gas phase depends on the formation and desorption dynamics of the product.

3.14.2   Eley-Rideal mechanism

A surface reaction need not involve two surface species. If a gas-phase molecule strikes an adsorbed molecule, there is a possibility that the collision leads to reaction and that the product escapes directly into the gas phase. This mechanism is known as the Eley-Rideal (ER) mechanism [214]. The reaction rate is expected to depend on the coverage of the adsorbed species and the pressure of the other reactant. The products of such a reaction, in contrast to LH products, should be highly energetic and have a memory of the initial conditions of the gas phase reactant. This is because the incident reactant is not accommodated with the surface and does not give up part of the exothermicity of the reaction to the surface in the form of its heat of adsorption. Great efforts have been made to prove that this mechanism occurs. It is rather unlikely for molecules to undergo an ER reaction. However, radicals are much more likely to react by this mechanism. For instance, H atoms incident on H covered Cu, Si and diamond surface have been shown to form H_2 through an ER mechanism. Rettner provided the most convincing evidence for an Eley-Rideal reaction by investigating the reaction of H(g) + D/Cu(111) and D(g) + H/Cu(111) [215]. While kinetic measurements can be ambiguous, Rettner and Auerbach [216] subsequently measured the energy and angular distributions of the product HD to show that both atoms could not be chemisorbed on the surface. Yates and co-workers [217] have also shown that removal of halogen atoms from Si(100) by incident H atoms follows ER kinetics. Lykke and Kay [218] and then Rettner and Auerbach [219, 220] observed a similar reaction for H + Cl/Au surfaces. H abstraction reactions play a role in the CVD deposition of diamond [221, 222] and may likewise play a role in the reactions of adsorbed hydrocarbons [223–225] and in ion pickup during the collision of large molecules with a hydrogen covered surface [226, 227].

Why should molecules shun ER dynamics whereas radicals may exhibit such dynamics? Reactions between molecules generally exhibit activation barriers. This is, after all, the reason why chemists search for catalysts that can lower the barriers to reactions. Weinberg [228] has applied transition state theory to argue quantitatively that whenever a barrier to reaction is present, LH dynamics is preferred to ER dynamics. ER dynamics requires the special case that a reaction is both barrierless and exothermic and these are exactly the types of reaction that radical species are prone to partake in. Shalashilin, Jackson and Persson [229–234] have shown that even when these conditions are fulfilled, as in the case of H incident upon H(a), the probability of ER dynamics is small. It is much more likely that the reaction in this case proceeds via an intermediate type of dynamics, which is explained in the next section.

3.14.3   Hot atom mechanism

The previous two mechanisms represent the extremes in equilibration or lack thereof of the reactants. Harris and Kasemo [235, 236] examined what might happen if one of the reactants were adsorbed while the other was not yet fully accommodated to the surface. This is the hot precursor or hot atom (HA) mechanism. Such a mechanism is quite interesting dynamically and leads to complex kinetics. Hints of this type of mechanism have been observed for O atoms incident on CO/Pt(111) to form CO_2 [237]. Similarly, when O_2 and CO are adsorbed onto Pt(111) at 100 K or below and then heated to $\sim$150 K, the O_2(a) dissociates and it appears that some of the liberated O atoms are able to react with CO before they accommodate with the surface [238]. The dynamics of such a process is not general but highly sensitive to the interaction potential as a similar reaction is not observed on Pd(111) under the same conditions [197].

The clearest indication of hot atom reactions has been found for H atoms incident on H-covered Cu, Si, Ni, Al and Pt surfaces. The classical trajectories studies on realistic PESs of Shalashilin, Jackson and Persson have elucidated the dynamics of the reaction mechanisms in great detail, in particular for the system H + D/Cu(111) and its isotopic analogues. In all of these systems, the reaction to form H_2 is barrierless and highly exothermic. These are necessary conditions for the occurrence of the ER mechanism; nonetheless, the ER reaction is quite improbable. For a surface covered with 0.5 ML of D, the probability of a direct ER reaction with an incident H atom is only $\sim$0.04. Much more likely is that the incident atom is deflected and scoots along the surface. The hot H atom exchanges energy very inefficiently with the surface but scatters inelastically from D(a) with much greater efficiency, losing about 0.1 eV per collision. A competitive process now sets in. The hot H atom can either play billiards with the D(a), eventually losing enough energy to stick (probability $\sim$0.5 of total incident atoms) or it can react with D(a) to form HD that desorbs from the surface ($\sim$0.4). Occasionally ($\sim$0.02) one of the D(a) atoms that has suffered a collision with the hot H atom goes on to react with another adsorbed D atom before it equilibrates with the surface (displacement reaction). Incident H atoms are particularly well suited to participate in hot atom reactions because they are highly reactive (and, therefore can partake in barrierless exothermic reactions) and they exchange energy slowly with the surface (which increases the lifetime of the hot precursor state). The kinetics of the reactions induced by incident H atoms are rather complex because they involve a convolution of scattering, sticking, ER reaction, HA reaction and displacement reaction. A model which successfully treats these kinetics has been developed by Küppers and co-workers [239, 240].

3.15 Measurement of sticking coefficients

Three types of sticking coefficients are encountered and measured. There is the initial sticking coefficient s_0, which is the sticking coefficient at zero coverage. An integral sticking coefficient is obtained by dividing the total coverage by the total exposure. The differential or instantaneous sticking coefficient is the sticking coefficient at a specific coverage and is the quantity that is properly used in rate equations. The differential sticking coefficient (hereafter as before, the sticking coefficient) is defined by

$$s(\sigma) = \frac{d\sigma}{d\varepsilon} = \lim_{\Delta \to 0} \frac{\Delta\sigma}{\Delta\varepsilon} \tag{3.15.1}$$

Thus, one method of determining $s(\sigma)$ is to take the derivative of an uptake curve (a plot of coverage versus exposure). Some combination of TPD, XPS or vibrational spectroscopy, perhaps supported by LEED measurements, is used to measure the coverage. For extremely low coverages, STM can be used to count adsorbates. s_0 is then determined by extrapolation to $\sigma = 0$. Alternatively, s_0 can be calculated from $\Delta\sigma/\Delta\varepsilon$ in the limit of vanishing σ.

The method of King and Wells [241] is a particularly useful variant of sticking coefficient measurement for reasonably large sticking coefficients. This utilizes a molecular beam in conjunction with the pressure changes that occur when the surface is exposed to the beam. The method is illustrated in Fig. 3.25. The vacuum chamber has a base pressure of p_b when the beam is off. When the beam enters the chamber but is blocked from hitting the crystal by a movable shutter upon which no adsorption occurs, the pressure rises to p_0. When the shutter is removed, the pressure, $p(t)$, changes as a function of time as molecules

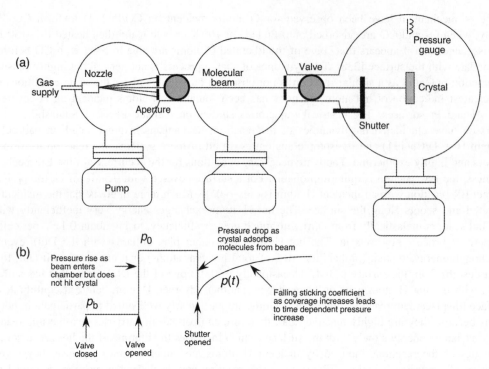

Figure 3.25   *The King and Wells method of sticking coefficient determination: (a) apparatus, (b) pressure curve.*

adsorb onto and eventually saturate the surface. For a chamber with constant pumping speed, the sticking coefficient is

$$s(t) = \frac{p_0 - p(t)}{p_0 - p_b}. \tag{3.15.2}$$

Assuming that molecules do not diffuse out of the beam irradiated area, a subsequent measure of the total coverage at the end of the experiment can be used to calculate the absolute sticking coefficient. Thereby a measurement of $p(t)$ can be transformed into a measurement of $s(\theta)$.

As we can anticipate from the discussion of adsorption dynamics above, the magnitude of the sticking coefficient depends on the details of the PES that describes the particular molecule/surface interaction. Its magnitude is specific to a given molecule–surface system, i.e. s_0 of CH_4 not only is very different on Ni compared to Pt but also is different for Ni(111) compared to Ni(110). In general, s_0 depends on molecular parameters such as angle of incidence, kinetic energy and the rovibrational state. It also depends on the T_s, the orientation of the molecule upon impact and the point of impact in the surface unit cell. The point of impact in the surface unit cell is also referred to as the surface impact parameter. If adsorption is non-activated, we expect s_0 is close to one and that it exhibits a weak dependence on molecular parameters. Whether T_s is important in determining s_0 depends on whether sticking is direct or precursor mediated.

For activated adsorption, the s_0 can be very small and depends strongly on some, if not all, of these parameters. If s_0 were determined by a single activation barrier, we would expect a thermally averaged sticking coefficient, i.e. s_0 measured in a system at equilibrium with $T_s = T_{gas}$, to exhibit a simple Arrhenius behaviour. This approximation is observed in some cases, particularly if a small range of energy or surface

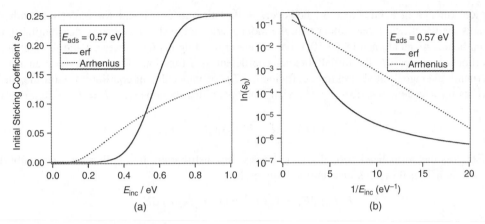

Figure 3.26 *Arrhenius behaviour (dotted line) compared to the sigmoidal form (solid line, erf = error function) for sticking in a model activated adsorption system with an adsorption barrier $E_{ads} = 0.57\,eV$. (a) The initial sticking coefficient s_0 is plotted versus incident energy E_{inc}. (b) An Arrhenius plot of ln s_0 vs $1/E_{inc}$.*

temperature is considered. It is best to think of the activation energy derived from such an analysis as an apparent activation energy – a value of E_a that describes the behaviour of the system over a certain temperature range but not a value that can be simply related to a particular feature on a PES. Furthermore, an experiment using a monoenergetic molecular beam for a system exhibiting only one classical barrier would reveal the step-like threshold behaviour show in Fig. 3.26. More likely, a range of barriers is present in the system because different values of the activation barrier are encountered for different vibrational states as well as for different molecular orientations and surface impact parameters. Rather than exhibiting threshold behaviour, the sticking coefficient as a function of kinetic energy often displays a sigmoidal form, as shown in Fig. 3.26.

Rettner et al. [242] demonstrated that a sigmoidal shape is expected to describe s_0 when the system has a Gaussian distribution of barrier heights centred about a mean value E_0. Michelsen et al. [146] later determined that the sticking coefficient data of numerous molecular beam experiments was best fitted by a function containing an error function of the form

$$s_0(E_K, \vartheta, v, J, T_s) = \frac{A(v,J)}{2}\left[1 + \mathrm{erf}\left(\frac{E_n - E_0(v,J)}{W(v,J,T_s)}\right)\right] \tag{3.15.3}$$

in which s_0 depends on the normal energy E_n through the kinetic energy E_K and the angle of incidence ϑ as well as T_s and the rovibrational quantum numbers v and J. $A(v,J)$ is the saturation sticking coefficient. $E_0(v,J)$ is the point of inflection found at the centre of the sigmoidal curve and is close to but not identically equal to the dissociation threshold energy. The width of the barrier distribution W depends on v, J and T_s and is empirically defined in terms of the range of barriers. As presented above, Eq. (3.15.3) has been introduced solely because it fits the form of the data but without much theoretical basis. A theoretical justification for this form has been presented by Luntz [243]. On this basis, W is defined as the sum of two terms: one representing the distribution of barriers and a second representing a distribution in energy transfer between the molecule and the surface. Nonetheless, debate still exists over how to interpret the parameters in the error function fit.

Using a molecular beam and, for instance, the method of King and Wells, the dependence of s_0 on translational, vibrational and rotational energy as well as angle of incidence and surface temperature can be determined as long as these variables can be varied independently. What is required for the modelling of reaction kinetics in a bulb reactor, however, is the sticking coefficient for molecules with thermal energy and angular distributions, rather than the monoenergetic translational energy distribution found in a molecular beam. As mentioned in § 3.10.3, normal energy scaling is often observed and for such a system, an experiment can measure the initial sticking coefficient as a function of normal kinetic energy $s_0(E_n)$. The thermally averaged initial sticking coefficient $s_0(T)$ for molecules in equilibrium with the surface, in other words for a system described by Maxwell-Boltzmann distributions with $T = T_s = T_{gas}$, is given by

$$s_0(T) = \int_0^\infty F(E_n)\, s_0(E_n)\, dE_n \qquad (3.15.4)$$

where $F(E_n)$ is the normalized normal kinetic energy flux distribution. In a bulb reactor at equilibrium, in which there is no net flow, this distribution is given by

$$F(E_n)\, dE_n = (kT)^{-1} \exp(-E_n/k_B T)\, dE_n. \qquad (3.15.5)$$

Equation (3.15.5) represents the fraction of molecules striking the surface with normal kinetic energy in the range of E_n to $E_n + dE_n$. The result of a calculation involving Eqs. (3.15.4) and (3.15.5) yields an initial sticking coefficient thermally averaged to take proper account of the translational energy and, implicitly, the angle of incidence dependence of the sticking coefficient. An even more complete and accurate calculation of the thermally averaged initial sticking coefficient would also average over the distribution functions for rotational and vibrational levels as well as the surface temperature. In particular, these averages would have to be made if excited rotational and vibrational states are present in the reactive mixture and if the sticking coefficient is found to vary strongly on the rovibrational state. Quite frequently, excited vibrational states play a very important role whereas the variation of sticking with rotational states is much smoother. Therefore, Eq. (3.15.4) must be extended by summing up the contribution of each vibrational state according to

$$s_0(T) = \sum_v f_v(T) \int_0^\infty F(E_n)\, s_0^v(E_n)\, dE_n. \qquad (3.15.6)$$

In Eq. (3.15.6), the fractional population in vibrational state $v, f_v(T)$, is summed over all (populated) vibrational states and $s_0^v(E_n)$ represents the functional dependence of the initial sticking coefficient on E_n for each vibrational state v. $f_v(T)$ is a normalized distribution.

3.16   Summary of important concepts

- Physisorption is a weak adsorption interaction in which polarization (dispersion) forces such as van der Waals interactions hold the adsorbate on the surface.
- Chemisorption is a strong adsorption interaction in which orbital overlap (sharing of electrons) leads to chemical bond formation.
- Binding sites at surfaces are separated by energy barriers. Therefore, diffusion on surfaces is an activated process.
- The chemisorption bond is formed by hybridization of substrate electronic states with the molecular orbitals of the adsorbate.
- As a first approximation, the interaction of frontier molecular orbitals with the substrate should be considered to understand chemisorption bonding and adsorbate structure.
- On transition metals, chemisorption bond formation can be considered a two-step process. In step 1, the frontier orbitals are broadened and shifted by the interaction with the s band. In step 2, bonding and antibonding hybrids are formed by the interaction with the d band.

- The strength of the chemisorption bond depends on the position of the hybrid orbitals with respect to E_F.
- The strength of chemisorption correlates with the energy of the d band centre. The lower the d band relative to E_F, the weaker the bond. Therefore, transition metals to the left of a row bind simple adsorbates more strongly than those on the right.
- In general, a strengthening of adsorbate–surface bonding leads to a weakening of intramolecular bonds in the adsorbate.
- Sufficiently strong chemisorption can lead to the scission of intramolecular bonds in the adsorbate (dissociative chemisorption).
- Adsorption can either be a non-activated or activated process.
- Dissociative chemisorption is most commonly associated with activated adsorption. The height of the activation barrier depends on the molecular orientation and the impact position within the unit cell.
- For non-activated adsorption, the sticking coefficient tends to one for low energy molecules but decreases for very high energy molecules.
- For activated adsorption, sticking can only occur if the incident molecule has sufficient energy to overcome the adsorption barrier. Molecules with energy far in excess of the barrier height may have difficulty sticking as they cannot follow the minimum energy path.
- Adsorption occurs on a multidimensional potential energy hypersurface (PES) and the effect on the sticking coefficient of placing energy in any particular degree of freedom depends on the shape of the PES.
- Adsorption can either be direct or precursor mediated.
- Adsorption and desorption are connected by microscopic reversibility.
- In any system for which the sticking coefficient is a function of energy, the desorbed molecules do not have an energy distribution corresponding to an equilibrium distribution at the surface temperature.
- Corrugation is the variation of barrier heights across the surface.
- Whereas initial sticking coefficient values for activated adsorption may exhibit Arrhenius behaviour over some range of temperature, a more general expectation is that they follow the sigmoidal form of Eq. (3.15.3).

3.17 Frontiers and challenges

- What is the role of nonadiabatic excitations in adsorption and desorption dynamics, reactivity and vibrational energy exchange?
- Understanding the dynamics of O_2 dissociative adsorption. What is the role of spin?
- The adsorption and desorption of H_2. Or as it is stated in more technologically relevant terms: Hydrogen storage. Hydrogen storage schemes always involve adsorption and desorption of H_2 (usually dissociative) and may also involve the permeation of adsorbed H atoms into the bulk of a material. Of particular interest is the understanding of H_2 adsorption and desorption in systems that do not involve a (relatively expensive) metal centre [244].
- The role of coverage effects in adsorption dynamics, for instance in the sticking of H_2 on Si.
- Understanding the nature of precursor states and the transition from precursor mediated to direct adsorption. What is going on in the energy transfer dynamics that differentiates between the two mechanisms?
- What functional form describes the initial sticking coefficient as a function of incident energy? What is the equivalent of the Arrhenius equation for sticking, and how is it justified theoretically?
- Can we extrapolate our detailed knowledge of the ad/desorption dynamics of diatomics to polyatomics?

3.18   Further reading

G. Antczak and G. Ehrlich, *Jump processes in surface diffusion*, Surf. Sci. Rep. **62**, (2007) 39.

J. A. Barker and D. J. Auerbach, *Gas-surface interaction and dynamics: Thermal energy atomic and molecular beam studies*, Surf. Sci. Rep. **4** (1985) 1.

G. P. Brivio and M. I. Trioni, *The adiabatic molecule-metal surface interaction: Theoretical approaches*, Rev. Mod. Phys. **71** (1999) 231.

G. Comsa and R. David, *Dynamical parameters of desorbing molecules*, Surf. Sci. Rep. **5** (1985) 145.

M. C. Desjonquères and D. Spanjaard, *Concepts in Surface Physics*, 2nd ed. (Springer-Verlag, Berlin, 2002).

Douglas J. Doren and John C. Tully, *Dynamics of precursor-mediated chemisorption*, J. Chem. Phys. **94** (1991) 8428.

Axel Groß, *Reactions at surfaces studied by ab initio dynamics calculations*, Surf. Sci. Rep. **32** (1998) 291.

J. Harris, *On the adsorption and desorption of H_2 at metal surfaces*, Appl. Phys. A: Mater. Sci. Process. **47** (1988) 63.

M. Dürr and U. Höfer, *Dissociative adsorption of molecular hydrogen on silicon surfaces*, Surf. Sci. Rep. **61** (2006) 465.

E. Hasselbrink and B. I. Lundqvist (Eds), *Handbook of Surface Science: Dynamics, Vol. 3* (Elsevier, Amsterdam, 2008).

A. Hodgson, *State resolved desorption measurements as a probe of surface reactions*, Prog. Surf. Sci. **63** (2000) 1.

D. A. King and D. P. Woodruff (Eds), *The Chemical Physics of Solid Surfaces and Heterogeneous Catalysis: Adsorption at Solid Surfaces*, Vol. 2 (Elsevier, Amsterdam, 1983).

C.-L. Kao and R. J. Madix, *The adsorption dynamics of small alkanes on (111) surfaces of platinum group metals*, Surf. Sci. **557** (2004) 215.

C. T. Rettner and M. N. R. Ashfold (Eds), *Dynamics of Gas-Surface Interactions* (The Royal Society of Chemistry, Cambridge, 1991).

C. T. Rettner, H. A. Michelsen, and D. J. Auerbach, *From quantum-state-specific dynamics to reaction rates: The dominant role of translational energy in promoting the dissociation of D_2 on Cu(111) under equilibrium conditions*, Faraday Discuss. Chem. Soc. **96** (1993) 17.

A. M. Wodtke, J. C. Tully, and D. J. Auerbach, *Electronically non-adiabatic interactions of molecules at metal surfaces: Can we trust the Born-Oppenheimer approximation for surface chemistry?* Int. Rev. Phys. Chem. **23** (2004) 513.

3.19   Exercises

3.1   Draw the structure of the fcc(100), fcc(111) and fcc(110) surfaces (top view). Indicate the unit cell and identify all possible surface adsorption sites expected for chemisorption.

3.2   Given that the mean lifetime of an adsorbate is

$$\tau = \frac{1}{A} \exp(E_{des}/RT_s) \qquad (3.19.1)$$

where A is the pre-exponential factor for desorption and E_{des} is the desorption activation energy, show that the mean random walk distance travelled by an adsorbate is

$$\langle x^2 \rangle^{1/2} = \left(\frac{4D_0}{A} \exp\left(\frac{E_{des} - E_{dif}}{RT} \right) \right)^{1/2}. \qquad (3.19.2)$$

3.3   When pyridine adsorbs on various metal surfaces, it changes its orientation as a function of coverage. Describe the bonding interactions that pyridine can experience and how this affects the orientation of adsorbed pyridine [245].

3.4   Cyclopentene, C_5H_8, is chemisorbed very weakly on Ag(111). Given that the double bond in C_5H_8 leads to a dipole that is oriented as shown in the Fig. 3.27, suggest a configuration for the molecule bound at low coverage on a stepped Ag surface with (111) terraces [246].

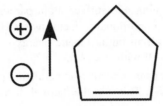

Figure 3.27 *The dipole associated with cyclopentene.*

3.5 When CO binds in sites of progressively higher co-ordination number (on-top $\rightarrow$ two-fold bridge $\rightarrow$ three-fold hollow $\rightarrow$ four-fold hollow), both the π and σ contributions to bonding increase in magnitude [247].
   (a) Predict the trends that are expected in the CO stretching frequency and chemisorption bond energy with change of site.
   (b) When the π bonding interaction with the surface is weak, which adsorption site is preferred?

3.6 CO bound to Pt(111) submerged in 0.1 M $HClO_4$ exhibits an FTIR peak associated with a linearly bond on-top species at $2070\,cm^{-1}$ [248]. 0.6 ML of Ru is deposited on the Pt(111) electrode to form islands of Ru. When CO is adsorbed on the resulting surface the peak at $2070\,cm^{-1}$ shifts by $-10\,cm^{-1}$, and decreases in intensity while a new peak appears at $1999\,cm^{-1}$. The new peak is shifted by $+6\,cm^{-1}$ compared to the peak associated with CO bound in an on-top site on a clean Ru electrode. Interpret the data as to where and how the CO is bound.

3.7 The amount of energy, δE, transferred in the collision of a molecule with a chain of atoms in the limit of a fast, impulsive collision (that is, a collision that is fast compared to the time that it takes the struck atom to recoil and transfer energy to the chain) is given by the Baule formula,

$$\delta E = \frac{4\mu}{(1+\mu)^2}(E_i + q_{ads}) \tag{3.19.3}$$

where $\mu = M/m$, M = the mass of the molecule, m is the mass of one chain atom, E_i is the initial kinetic energy of the molecule before it is accelerated by q_{ads} (the depth of the attractive well, effectively the heat of adsorption). Estimate the energy transfer for H_2, CH_4 and O_2 incident upon copper or platinum chains. Take the incident energy to be
   (a) the mean kinetic energy at 300 K;
   (b) $E_K = 1.0\,eV$.
Take the well depths to be 20 meV, 50 meV and 200 meV for H_2, CH_4 and O_2, respectively.

3.8 When a molecule strikes a surface it loses on average an amount of energy $\langle \Delta E \rangle$ given by [159]

$$\langle \Delta E \rangle = -\gamma \alpha_{dsp} E \tag{3.19.4}$$

where γ is a constant characteristic of the potential energy surface, α_{dsp} is a constant that depends on the collision partners and E is the kinetic energy upon collision. For H/Cu(111), $\alpha_{dsp} = 0.0024$, $\gamma = 4.0$ and the binding energy chemisorbed H is 2.5 eV.
   (a) For an H atom with an initial $E_K = 0.1\,eV$, 10 Å away from the surface, calculate the energy transfer on the first bounce.
   (b) Assuming the same amount of energy transfer on each subsequent collision, how many collisions are required for the H atom to reach the bottom of the well?
   (c) Given that α_{dsp} changes from one molecule to the next, in a manner analogous to the Baule formula, we write

$$\alpha_{dsp} = k\frac{4\mu}{(1+\mu)^2} \tag{3.19.5}$$

where μ is calculated assuming one surface atom participates in the collision, calculate α_{dsp} for CO assuming the same proportionality factor as for H. Then make a rough estimate of the number of collisions CO with an initial kinetic energy of 0.1 eV requires to reach the bottom of a 1.2 eV chemisorption well with $\gamma = 4.0$.

3.9   Classically, a chemical reaction cannot occur if the collision partners do not have sufficient energy to overcome the activation barrier. This and the thermal distribution of energy are the basis of the Arrhenius formulation of reaction rate constants. For an atom, the thermal energy is distributed over the translational degrees of freedom. The velocity distribution is governed by Maxwell distribution

$$f(v) = 4\pi \left(\frac{M}{2\pi RT} \right)^{3/2} v^2 \exp \left(\frac{-Mv^2}{2RT} \right),$$ (3.19.6)

where M is the molar mass and v the speed. Assuming that there is no steric requirement for sticking, i.e. that energy is the only determining factor, calculate the sticking coefficient of an atomic gas held at

(a) 300 K and

(b) 1000 K

for adsorption activation barriers of $E_{ads} = 0$, 0.1, 0.5 and 1.0 eV.

3.10   A real molecule has quantized rotational and vibration energy levels. The Maxwell-Boltzmann distribution law describes the occupation of these levels. The distribution among rotational levels is given by

$$N_{vJ} = N_v \frac{hc}{k_B T} (2J + 1) \exp \left(\frac{-E_{rot}}{k_B T} \right),$$ (3.19.7)

where N_{vJ} is the number of molecules in the rotational state with quantum numbers v and J and N_v is the total number of molecules in the vibrational state v. The energy of rigid rotor levels is given by

$$E_{rot} = hcB_v J (J + 1)$$ (3.19.8)

where B_v is the rotational constant of the appropriate vibrational state. The vibrational population is distributed according to

$$N_v = N \exp \left(\frac{-hcG_0(v)}{k_B T} \right)$$ (3.19.9)

where N is the total number of molecules and $G_0(v)$ is the wavenumber of the vibrational level v above the ground vibrational level. At thermal equilibrium the mean energy is distributed according to

$$\langle E \rangle = \langle E_{trans} \rangle + \langle E_{rot} \rangle + \langle E_{vib} \rangle$$ (3.19.10)

where for a diatomic molecule

$$\langle E_{trans} \rangle = 2k_B T$$ (3.19.11)

$$\langle E_{rot} \rangle = k_B T$$ (3.19.12)

$$\langle E_{vib} \rangle = \sum_{n>0} \frac{h\nu_n}{\exp(h\nu_n/k_B T) - 1}.$$ (3.19.13)

Note that Eq. (3.19.13) neglects the contribution of zero-point energy to the vibrational energy. Assume that the sticking coefficient exhibits Arrhenius behaviour when the temperature of the gas is varied. For the same barrier heights as in Exercise 3.9, calculate the sticking coefficient for molecules with mean total energies of 0.1, 0.5 and 1.0 eV. Use NO as the molecule and assume that only $v = 1$ contributes to the vibrational energy for which $G_0 = 1904\,\mathrm{cm}^{-1}$.

3.11 Consider an extremely late barrier in which translational energy plays no role, vibrational energy is 100% effective at overcoming the barrier and rotational energy is 50% efficient. Calculate the classical sticking coefficient of H_2 and D_2 as a function of rovibrational state for the first three vibrational levels and an adsorption barrier of 0.5 eV. Assume that zero point energy plays no role and that molecules can be described as rigid rotors. The vibrational energy spacings and rotational constants of H_2 and D_2 are given in Table 3.3.

Table 3.3   *Rotational and vibrational constants for H_2 and D_2*

v	H_2		D_2	
	B_v/cm^{-1}	$G_0(v)/\mathrm{cm}^{-1}$	B_v/cm^{-1}	$G_0(v)/\mathrm{cm}^{-1}$
0	59.3	0.0	29.9	0.0
1	56.4	4161.1	28.8	2994.0
2	53.5	8087.1	28.0	5868.8

3.12 Consider the adsorption of D_2. Assuming that normal translational energy is 100% effective and vibrational energy is 60% effective at overcoming the adsorption barrier, calculate the sticking coefficient of the first three vibrational levels as a function of normal translational energy. Neglect the effects of rotation.

3.13 The flux of molecules striking a surface follows a cosine distribution, $\cos \vartheta$ where ϑ is the angle from surface normal. If the perpendicular component of translational energy is effective at overcoming the adsorption barrier and the parallel component is not, the angular distribution of the flux that sticks is tightly constrained about the surface normal. The desorbing flux is similarly peaked about the surface normal. It is often observed that the desorbing flux can be described by a $\cos^n \vartheta$ distribution in which $n > 1$, the greater the value of n, the more peaked the distribution. The angular distribution of D_2 desorbing from Cu(100) has been measured by Comsa and David [249]. They measured the relationship given in Table 3.3 between the normalized desorption intensity, $N(\vartheta) / N(0°)$, and the desorption angle measured from the surface normal, ϑ: Determine n.

Table 3.4   *The angular distribution of D_2 thermally desorbed from Cu(100)*

ϑ	0°	5°	10°	15°	20°	25°	30°	35°	45°
$\dfrac{N(\vartheta)}{N(0°)}$	1.00	0.99	0.98	0.77	0.63	0.48	0.38	0.21	0.06

3.14 The sticking of molecular hydrogen on Si is highly activated in the surface co-ordinates. Bratu and Höfer [156, 166] have determined the sticking coefficient of H_2 on Si(111)–(7 × 7) as a function of surface temperature and have recorded the data given in Table 3.5. Using an Arrhenius formulation (Eq. 3.19.14), determine the pre-exponential factor and the activation barrier height.

3.15 If a chemical reaction proceeds with a single activation barrier, the rate constant should follow the Arrhenius expression. Accordingly for the sticking coefficient, s_0, we write

$$s_0 = A_s \exp(-E_a/RT),$$ (3.19.14)

Table 3.5   *Values for the initial sticking coefficient of H_2 on Si(111)–(7 × 7) as a function of surface temperature*

T_s / K	587	613	637	667	719	766	826	891	946	1000	1058
s_0	2.8×10^{-9}	6.5×10^{-9}	1.3×10^{-8}	2.0×10^{-8}	5.3×10^{-8}	1.7×10^{-7}	5.4×10^{-7}	1.3×10^{-6}	2.3×10^{-6}	2.7×10^{-6}	5.0×10^{-6}

where A is a constant, E_a is the adsorption activation energy, R is the gas constant and T is the temperature. However, if a distribution of barriers rather than a single barrier participates in the reaction, a sigmoidal form is followed. In the case of H_2 sticking on Cu(100), the sticking coefficient as a function of kinetic energy is found to follow

$$s_0(E_n) = \frac{A_s}{2}\left[1 + \mathrm{erf}\left(\frac{E_n - E_0}{W}\right)\right] \tag{3.19.15}$$

where E_0 is the mean position of a distribution of barriers that has a width W. Given the data in Table 3.6 of s_0 for $H_2(v = 0)$ versus E_n, determine E_0 and W. Make plots of s_0 vs E_n with different values of E_0 and W to observed the effects these have on the shape of the sticking curve.

Table 3.6   *Initial sticking coefficient, s_0, with normal kinetic energy, E_n. Source: Michelsen et al [146]*

$E_\perp$ / eV	0.1	0.2	0.3	0.4	0.5	0.6	0.7	0.8	0.9	1.0
s_0	4×10^{-6}	1×10^{-4}	0.0021	0.0167	0.0670	0.151	0.219	0.245	0.250	0.250

3.16   Classically we assign $1/2\,k_B T$ of energy to each active degree of freedom and, therefore, we assign a value of $3/2\,k_B T$ to the kinetic energy. This is true for a *volume* sample of a gas. For a *flux* of gas, such as that desorbing from a surface, the answer is different. Use the Maxwell velocity distribution to show that the equilibrium mean kinetic energy of a flux of gas emanating from (or passing through) a surface is

$$\langle E_{\mathrm{trans}}\rangle = 2k_B T_s. \tag{3.19.16}$$

The mean kinetic energy is defined by the moments of the velocity distribution according to

$$\langle E_{\mathrm{trans}}\rangle = \frac{1}{2}\frac{mM_3}{M_1} \tag{3.19.17}$$

where the moments are calculated according to

$$M_i = \int_0^\infty v^i f(v)\,dv. \tag{3.19.18}$$

3.17   For desorption from a rigid surface and in the absence of electron-hole pair formation or other electronic excitation, a desorbing molecule will not lose energy to the surface after it passes through the transition state. In the absence of a barrier the mean energy of the desorbed molecules is roughly equal to the mean thermal expectation value at the surface temperature, $\langle k_B T_s\rangle$. The excess energy above this value, as shown in Figure 3.28, is equal to the height of the adsorption activation energy. Therefore a measurement of the mean total energy of the desorbed molecules,

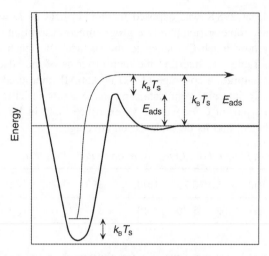

Figure 3.28   *Energy as a function of molecular distance from the surface, z · E$_{ads}$, adsorption activation energy; T$_s$, surface temperature; k$_B$, Boltzmann constant.*

$\langle E \rangle$, can be used to estimate the height of the adsorption barrier [163]. The mean total energy is given by

$$\langle E \rangle = \langle E_{trans} \rangle + \langle E_{rot} \rangle + \langle E_{vib} \rangle. \tag{3.19.19}$$

(a) Given the data for D_2/Cu(111) given in Table 3.7 [160], show that the above approximations hold and, therefore, to a first approximation we need not consider surface atom motions in the ad/desorption dynamics.

Table 3.7   *Data for D_2/Cu(111) and D_2/Si(100)–(2 × 1). Sources: Cu: Rettner, Michelsen and Auerbach [160] and Si: Kolasinski, Nessler, de Meijere and Hasselbrink [163]*

T_s (K)	$\langle E_{rot} \rangle$ (K)	T_{vib} (K)	$\langle E_{trans} \rangle$ (K)	E_{ads} (eV)
D_2/Cu(111): 925	1020	1820	3360	0.5
D_2/Si(100): 780	330	1700	960	0.8

(b) Given the data for D_2/Si(100)–(2 × 1) shown in Table 3.7 [163] show that the above approximations do not hold and, therefore, that the single static potential picture cannot be used to interpret the ad/desorption dynamics.

   In both cases assume that the vibrational distribution is thermal and is described by a temperature T_{vib}.

3.18   A Pt(997) is a stepped surface that contains 8 times more terrace atoms than step atoms. When CO is exposed to a Pt(997) surface held at $T_s = 11$ K, the ratio of CO molecules adsorbed at the atop terrace sites to those adsorbed in atop step sites is 3.6:1 [250]. Discuss what the expected coverage ratio is in terms of simple Langmuirian adsorption. Propose an explanation for why the Langmuirian result is not obtained.

3.19   CO from a bottle held at 298 K was exposed to Pt(111) held at $T_s = 80$ K. Thermal desorption detected by a mass spectrometer (temperature programmed desorption or TPD, see Chapter 4) was then used to quantify how much CO stuck to the surface after each defined exposure. The data obtained are given in Table 3.8. Interpret the data in terms of the adsorption dynamics of CO on Pt(111). Note that the saturation coverage of CO is 0.50 ML measured with respect to the Pt(111) surface atom density. To interpret your data, first make a plot of TPD peak area vs exposure and then convert this into a plot of CO coverage vs exposure (both in ML), from which you can infer the CO sticking coefficient and its dependence on coverage.

Table 3.8   *TPD peak areas as a function of CO exposure on to Pt(111) at 80 K*

Exposure / Langmuir	0.198	0.350	0.733	1.082	1.393	1.755	2.033	2.481	2.843	3.172
Integrated TPD peak area	1.46	2.62	5.70	8.48	11.3	13.8	16.8	15.9	16.1	16.6

3.20   Explain why multilayer absorption can occur for physisorption but not for chemisorption.

3.21   A metal single crystal sample is dosed with a Knudsen source, i.e. a nozzle that produces an equilibrium flux rather than a supersonic jet. The gas expands out of the nozzle with a cosine angular distribution and then intercepts the crystal. The flux intercepted by the crystal F_N depends on the temperature in the nozzle T_N, the distance of the crystal away from the nozzle d, the axial distance from the centre of the crystal x, the area of the hole in the nozzle A_N, and the pressure in the gas manifold behind the nozzle p_m according to

$$F_N = \frac{p_m A_N}{\sqrt{2\pi m k_B T_N}} \frac{\cos^4(\tan^{-1}(x/d))}{\pi d^2}.$$ (3.19.20)

The effective sticking coefficient is defined according to

$$\theta_{total} = s_{eff}\left(\frac{F_N t}{\sigma_0}\right),$$ (3.19.21)

where θ_{total} is the total coverage accumulated after a time t on the crystal, which has σ_0 surface atoms per unit area. The total coverage, however, is made up of contributions from adsorption directly from the nozzle as well as adsorption from background gas. Assuming that the background is made up from molecules that originate from the nozzle and miss the crystal (or do not stick on the first bounce), then the background pressure is proportional to the nozzle pressure, that is $p_{bkg} = c_0 p_m$, show that the true sticking coefficient of molecules emanating directly from the nozzle is given by

$$s_N = s_{eff} - s_{bkg}\left(\frac{c_0 \sqrt{T_N} \pi d^2}{A_N \sqrt{T} \cos^4(\tan^{-1}(x/d))}\right).$$ (3.19.22)

3.22   Consider the bonding of NO on Pt(111).
(a) Describe a likely binding geometry for adsorbed NO and justify your answer.
(b) The adsorption of NO on Pt(111) is non-activated. The sticking of NO is less likely for molecules that have high translational energy compared to low translational energy. Molecules with high rotational quantum number (high J states) also stick less effectively than low J states. Discuss the magnitude of the sticking coefficient and how it will change as the temperature of the NO is changed.

(c) Real Pt(111) surfaces often have a step density of about 1%. On these surfaces NO is found to dissociate at room temperature but only at very low coverages. Typically, the first 1% of a monolayer (ML) dissociates but after this all further adsorption is molecular. If the NO is dosed onto the clean surface at 90 K (a temperature at which there is no surface diffusion of NO), virtually no dissociation is observed. Discuss the dissociation of NO. Explain a possible dissociation mechanism.

3.23 When 0.25 L (L = Langmuir) of O_2 is dosed onto a Pd(111) surface at 30 K, a peak can be seen in the vibrational spectrum. The wavenumber of this peak is $1585 \, cm^{-1}$. This value is virtually the same as the vibrational frequency of a gas-phase O_2 molecule. When the crystal is warmed to above 50 K, the $1585 \, cm^{-1}$ peak disappears and a peak at $850 \, cm^{-1}$ appears. No O_2 is observed to desorb from the surface below 100 K. Interpret these results and describe what is occurring in the adsorbed layer.

References

[1] R. M. Watwe, R. D. Cortright, M. Mavirakis, J. K. Nørskov, J. A. Dumesic, *J. Chem. Phys.*, **114** (2001) 4663.
[2] K. Christmann, *Surf. Sci. Rep.*, **9** (1988) 1.
[3] B. D. Kay, C. H. F. Peden, D. W. Goodman, *Phys. Rev. B*, **34** (1986) 817.
[4] C. H. F. Peden, B. D. Kay, D. W. Goodman, *Surf. Sci.*, **175** (1986) 215.
[5] R. Gomer, *Rep. Prog. Phys.*, **53** (1990) 917.
[6] L. J. Lauhon, W. Ho, *Phys. Rev. Lett.*, **85** (2000) 4566.
[7] B. S. Swartzentruber, *Phys. Rev. Lett.*, **76** (1996) 459.
[8] G. Ehrlich, F. G. Hudda, *J. Chem. Phys.*, **44** (1966) 1039.
[9] R. L. Schwoebel, E. J. Shipsey, *J. Appl. Phys.*, **37** (1966) 3682.
[10] G. L. Kellogg, *Surf. Sci. Rep.*, **21** (1994) 1.
[11] M. Bowker, D. A. King, *Surf. Sci.*, **71** (1978) 583.
[12] M. Bowker, D. A. King, *Surf. Sci.*, **72** (1978) 208.
[13] D. A. Reed, G. Ehrlich, *Surf. Sci.*, **102** (1981) 588.
[14] G. P. Brivio, M. I. Trioni, *Rev. Mod. Phys.*, **71** (1999) 231.
[15] E. Zaremba, W. Kohn, *Phys. Rev. B*, **13** (1976) 2270.
[16] E. Zaremba, W. Kohn, *Phys. Rev. B*, **15** (1977) 1769.
[17] F. Haber, *Z. Elektrochem.*, **20** (1914) 521.
[18] I. Langmuir, *Phys. Rev.*, **8** (1916) 149.
[19] I. Langmuir, *J. Am. Chem. Soc.*, **40** (1918) 1361.
[20] M. C. Desjonquères, D. Spanjaard, *Concepts in Surface Physics*, 2nd ed., Springer-Verlag, Berlin, 2002.
[21] P. W. Anderson, *Phys. Rev.*, **124** (1961) 41.
[22] T. B. Grimley, *Proc. R. Soc. London, A*, **90** (1967) 751.
[23] T. B. Grimley, Theory of chemisorption, in *The Chemical Physics of Solid Surfaces and Heterogeneous Catalysis, Vol. 2* (Eds: D. A. King, D. P. Woodruff), Elsevier, Amsterdam, 1983, p. 333.
[24] D. M. Newns, *Phys. Rev.*, **178** (1969) 1123.
[25] W. Brenig, K. Schönhammer, *Z. Phys.*, **276** (1974) 201.
[26] I. Vasiliev, S. Öğüt, J. R. Chelikowsky, *Phys. Rev. Lett.*, **82** (1999) 1919.
[27] D. J. Doren, Kinetics and dynamics of hydrogen adsorption and desorption on silicon surfaces, in *Adv. Chem. Phys., Vol. 95* (Eds: I. Prigogine, S. A. Rice), John Wiley & Sons, Inc., NY, 1996, p. 1.
[28] H. Haberland (Ed.), *Clusters of Atoms and Molecules, Vol. 52*, Springer-Verlag, Berlin, 1993.
[29] H. Haberland (Ed.), *Clusters of Atoms and Molecules II, Vol. 56*, Springer-Verlag, Berlin, 1994.
[30] J. C. Campuzano, The adsorption of carbon monoxide by the transition metals, in *The Chemical Physics of Solid Surfaces and Heterogeneous Catalysis: Chemisorption Systems, Vol. 3A* (Eds: D. A. King, D. P. Woodruff), Elsevier, Amsterdam, 1990, p. 389.

[31]   G. Blyholder, *J. Phys. Chem.*, **68** (1964) 2772.

[32]   M. D. Alvey, M. J. Dresser, J. T. Yates, Jr., *Surf. Sci.*, **165** (1986) 447.

[33]   M. R. Albert, J. T. Yates, Jr., *The Surface Scientist's Guide to Organometallic Chemistry*, American Chemical Society, Washington, DC, 1987.

[34]   A. Föhlisch, N. Nyberg, P. Bennich, L. Triguero, J. Hasselström, O. Karis, L. G. M. Pettersson, A. Nilsson, *J. Chem. Phys.*, **112** (2000) 1946.

[35]   P. Hu, D. A. King, M.-H. Lee, M. C. Payne, *Chem. Phys. Lett.*, **246** (1995) 73.

[36]   K. W. Kolasinski, F. Cemič, E. Hasselbrink, *Chem. Phys. Lett.*, **219** (1994) 113.

[37]   R. Imbihl, J. E. Demuth, *Surf. Sci.*, **173** (1986) 395.

[38]   H. Steininger, H. Ibach, S. Lehwald, *Surf. Sci.*, **117** (1982) 685.

[39]   G. A. Somorjai, K. R. McCrea, Dynamics of reactions at surfaces, in *Adv. Catal., Vol. 45* (Eds: B. C. Gates, H. Knözinger), Academic Press, Boston, 2000, p. 386.

[40]   J. Harris, *Appl. Phys. A*, **47** (1988) 63.

[41]   A. E. DePristo, Dynamics of dissociative chemisorption, in *Dynamics of Gas-Surface Interactions* (Eds: C. T. Rettner, M. N. R. Ashfold), The Royal Society of Chemistry, Cambridge, 1991, p. 47.

[42]   G. R. Darling, S. Holloway, *Rep. Prog. Phys.*, **58** (1995) 1595.

[43]   H. A. Michelsen, C. T. Rettner, D. J. Auerbach, The adsorption of hydrogen at copper surfaces: A model system for the study of activated adsorption, in *Surface Reactions: Springer Series in Surface Sciences, Vol. 34* (Ed.: R. J. Madix), Springer-Verlag, Berlin, 1994, p. 185.

[44]   B. I. Lundqvist, J. K. Nørskov, H. Hjelmberg, *Surf. Sci.*, **80** (1979) 441.

[45]   P. Nordlander, S. Holloway, J. K. Nørskov, *Surf. Sci.*, **136** (1984) 59.

[46]   J. K. Nørskov, A. Houmøller, P. K. Johansson, B. I. Lundqvist, *Phys. Rev. Lett.*, **46** (1981) 257.

[47]   G. R. Darling, S. Holloway, *Surf. Sci.*, **304** (1994) L461.

[48]   G. A. Somorjai, *Adv. Catal.*, **26** (1977) 1.

[49]   B. Hammer, J. K. Nørskov, *Surf. Sci.*, **343** (1995) 211.

[50]   B. Hammer, J. K. Nørskov, *Nature (London)*, **376** (1995) 238.

[51]   B. Hammer, J. K. Nørskov, *Adv. Catal.*, **45** (2000) 71.

[52]   G. Ertl, Dynamics of reactions at surfaces, in *Adv. Catal., Vol. 45* (Eds: B. C. Gates, H. Knözinger), Academic Press, Boston, 2000, p. 1.

[53]   R. T. Brackmann, W. L. Fite, *J. Chem. Phys.*, **34** (1961) 1572.

[54]   J. N. Smith, Jr., W. L. Fite, *J. Chem. Phys.*, **37** (1962) 898.

[55]   G. Comsa, *J. Chem. Phys.*, **48** (1968) 3235.

[56]   R. L. Palmer, J. N. Smith, Jr., H. Saltsburg, D. R. O'Keefe, *J. Chem. Phys.*, **53** (1970) 1666.

[57]   A. E. Dabiri, T. J. Lee, R. E. Stickney, *Surf. Sci.*, **26** (1971) 522.

[58]   G. Marenco, A. Schutte, G. Scoles, F. Tommasine, *J. Vac. Sci. Technol.*, **9** (1971) 824.

[59]   M. Balooch, R. E. Stickney, *Surf. Sci.*, **44** (1974) 310.

[60]   M. J. Cardillo, M. Balooch, R. E. Stickney, *Surf. Sci.*, **50** (1975) 263.

[61]   F. Frenkel, J. Häger, W. Krieger, H. Walter, C. T. Campbell, G. Ertl, H. Kuipers, J. Segner, *Phys. Rev. Lett.*, **46** (1981) 152.

[62]   C. T. Campbell, G. Ertl, J. Segner, *Surf. Sci.*, **115** (1982) 309.

[63]   F. Frenkel, J. Häger, W. Krieger, H. Walter, G. Ertl, J. Segner, W. Vielhaber, *Chem. Phys. Lett.*, **90** (1982) 225.

[64]   M. Asscher, W. L. Guthrie, T.-H. Lin, G. A. Somorjai, *Phys. Rev. Lett.*, **49** (1982) 76.

[65]   W. L. Guthrie, T.-H. Lin, S. T. Ceyer, G. A. Somorjai, *J. Chem. Phys.*, **76** (1982) 6398.

[66]   M. Asscher, W. L. Guthrie, T.-H. Lin, G. A. Somorjai, *J. Chem. Phys.*, **78** (1983) 6992.

[67]   G. M. McClelland, G. D. Kubiak, H. G. Rennagel, R. N. Zare, *Phys. Rev. Lett.*, **46** (1981) 831.

[68]   J. E. Hurst, Jr., G. D. Kubiak, R. N. Zare, *Chem. Phys. Lett.*, **93** (1982) 235.

[69]   G. D. Kubiak, J. E. Hurst, Jr., H. G. Rennagel, G. M. McClelland, R. N. Zare, *J. Chem. Phys.*, **79** (1983) 5163.

[70]   A. W. Kleyn, A. C. Luntz, D. J. Auerbach, *Phys. Rev. Lett.*, **47** (1981) 1169.

[71]   A. C. Luntz, A. W. Kleyn, D. J. Auerbach, *J. Chem. Phys.*, **76** (1982) 737.

[72] A. C. Luntz, A. W. Kleyn, D. J. Auerbach, *Phys. Rev. B*, **25** (1982) 4273.

[73] R. R. Cavanagh, D. S. King, *Phys. Rev. Lett.*, **47** (1981) 1829.

[74] D. S. King, R. R. Cavanagh, *J. Chem. Phys.*, **76** (1982) 5634.

[75] S. L. Bernasek, S. R. Leone, *Chem. Phys. Lett.*, **84** (1981) 401.

[76] R. P. Thorman, S. L. Bernasek, *J. Chem. Phys.*, **74** (1981) 6498.

[77] L. S. Brown, S. L. Bernasek, *J. Phys. Chem.*, **82** (1985) 2110.

[78] D. A. Mantell, S. B. Ryali, B. L. Halpern, G. L. Haller, J. B. Fenn, *Chem. Phys. Lett.*, **81** (1981) 185.

[79] D. A. Mantell, S. B. Ryali, G. L. Haller, J. B. Fenn, *J. Chem. Phys.*, **78** (1983) 4250.

[80] D. A. Mantell, Y.-F. Maa, S. B. Ryali, G. L. Haller, J. B. Fenn, *J. Chem. Phys.*, **78** (1983) 6338.

[81] D. A. Mantell, K. Kunimori, S. B. Ryali, G. L. Haller, J. B. Fenn, *Surf. Sci.*, **172** (1986) 281.

[82] D. W. J. Kwong, N. DeLeon, G. L. Haller, *Chem. Phys. Lett.*, **144** (1988) 533.

[83] G. W. Coulston, G. L. Haller, *J. Chem. Phys.*, **95** (1991) 6932.

[84] G. D. Kubiak, G. O. Sitz, R. N. Zare, *J. Chem. Phys.*, **81** (1984) 6397.

[85] G. D. Kubiak, G. O. Sitz, R. N. Zare, *J. Chem. Phys.*, **83** (1985) 2538.

[86] U. Valbusa, General principles and methods, in *Atomic and Molecular Beam Methods, Vol. 2* (Ed.: G. Scoles), Oxford University Press, New York, 1992, p. 327.

[87] G. Boato, Elastic scattering of atoms, in *Atomic and Molecular Beam Methods, Vol. 2* (Ed.: G. Scoles), Oxford University Press, New York, 1992, p. 340.

[88] O. Stern, *Die Naturwissenschaften*, **21** (1929) 391.

[89] I. Estermann, R. Frisch, O. Stern, *Z. Phys.*, **73** (1931) 348.

[90] R. Frisch, O. Stern, *Z. Phys.*, **84** (1933) 430.

[91] H. Schlichting, D. Menzel, T. Brunner, W. Brenig, J. C. Tully, *Phys. Rev. Lett.*, **60** (1988) 2515.

[92] G. Comsa, B. Poelsema, Scattering from disordered surfaces, in *Atomic and Molecular Beam Methods, Vol. 2* (Ed.: G. Scoles), Oxford University Press, New York, 1992, p. 463.

[93] R. B. Doak, R. E. Grisenti, S. Rehbein, G. Schmahl, J. P. Toennies, C. Wöll, *Phys. Rev. Lett.*, **83** (1999) 4229.

[94] A. P. Graham, *Surf. Sci. Rep.*, **49** (2003) 115.

[95] J. A. Barker, D. J. Auerbach, *Surf. Sci. Rep.*, **4** (1985) 1.

[96] M. Asscher, G. A. Somorjai, Reactive scattering, in *Atomic and Molecular Beam Methods, Vol. 2* (Ed.: G. Scoles), Oxford University Press, New York, 1992, p. 488.

[97] D. J. Auerbach, Multiple-phonon inelastic scattering, in *Atomic and Molecular Beam Methods, Vol. 2* (Ed.: G. Scoles), Oxford University Press, New York, 1992, p. 444.

[98] P. Casavecchia, *Rep. Prog. Phys.*, **63** (2000) 355.

[99] Y. T. Lee, *Science*, **236** (1987) 793.

[100] J. B. Anderson, R. P. Andres, J. B. Fenn, *Adv. Chem. Phys.*, **10** (1966) 275.

[101] G. Scoles (Ed.), *Atomic and Molecular Beam Methods, Vol. 1*, Oxford University Press, New York, 1988.

[102] G. Scoles (Ed.), *Atomic and Molecular Beam Methods, Vol. 2*, Oxford University Press, New York, 1992.

[103] J. E. Lennard-Jones, *Trans. Faraday Soc.*, **28** (1932) 333.

[104] F. Shimizu, *Phys. Rev. Lett.*, **86** (2001) 987.

[105] D. J. Doren, J. C. Tully, *Langmuir*, **4** (1988) 256.

[106] D. J. Doren, J. C. Tully, *J. Chem. Phys.*, **94** (1991) 8428.

[107] C. R. Arumainayagam, M. C. McMaster, G. R. Schoofs, R. J. Madix, *Surf. Sci.*, **222** (1989) 213.

[108] E. W. Kuipers, M. G. Tenner, M. E. M. Spruit, A. W. Kleyn, *Surf. Sci.*, **205** (1988) 241.

[109] E. K. Grimmelmann, J. C. Tully, M. J. Cardillo, *J. Chem. Phys.*, **72** (1980) 1039.

[110] R. M. Logan, R. E. Stickney, *J. Chem. Phys.*, **44** (1966) 195.

[111] R. M. Logan, J. C. Keck, *J. Chem. Phys.*, **49** (1968) 860.

[112] A. W. Kleyn, Growth and etching of semiconductors, in *Handbook of Surface Science: Dynamics, Vol. 3* (Eds: E. Hasselbrink, I. Lundqvist), Elsevier, Amsterdam, 2008, p. 29.

[113] W. A. Dew, H. S. Taylor, *J. Phys. Chem.*, **31** (1927) 281.

[114] O. Schmidt, *Z. Phys. Chem.*, **133** (1928) 263.

[115] M. Head-Gordon, J. C. Tully, C. T. Rettner, C. B. Mullins, D. J. Auerbach, *J. Chem. Phys.*, **94** (1991) 1516.

[116] J. C. Maxwell, *Philos. Trans. R. Soc. London*, **170** (1879) 231.

[117]   M. Knudsen, *Ann. Phys.*, **339** (1911) 593.

[118]   M. V. Smoluchowski, *Ann. Phys.*, **33** (1910) 1559.

[119]   W. Gaede, *Ann. Phys.*, **41** (1913) 289.

[120]   R. A. Millikan, *Phys. Rev.*, **21** (1923) 217.

[121]   R. A. Millikan, *Phys. Rev.*, **22** (1923) 1.

[122]   P. Clausing, *Ann. Phys.*, **4** (1930) 36.

[123]   I. Langmuir, *J. Am. Chem. Soc.*, **38** (1916) 2221.

[124]   R. C. Tolman, *Proc. Natl. Acad. Sci. U. S. A.*, **11** (1925) 436.

[125]   R. H. Fowler, E. A. Milne, *Proc. Natl. Acad. Sci. U. S. A.*, **11** (1925) 400.

[126]   E. P. Wenaas, *J. Chem. Phys.*, **54** (1971) 376.

[127]   I. Kuscer, *Surf. Sci.*, **25** (1971) 225.

[128]   D. A. McQuarrie, *Statistical Thermodynamics*, University Science Books, Mill Valley, CA, 1973.

[129]   J. N. Smith, Jr., R. L. Palmer, *J. Chem. Phys.*, **56** (1972) 13.

[130]   R. D. Levine, R. B. Bernstein, *Molecular Reaction Dynamics and Chemical Reactivity*, Oxford University Press, New York, 1987.

[131]   S. Holloway, *Surf. Sci.*, **299/300** (1994) 656.

[132]   D. M. Newns, *Surf. Sci.*, **171** (1986) 600.

[133]   J. W. Gadzuk, *J. Vac. Sci. Technol., A*, **5** (1987) 492.

[134]   H. Metiu, J. W. Gadzuk, *J. Chem. Phys.*, **74** (1981) 2641.

[135]   J. C. Tully, *J. Electron Spectrosc. Relat. Phenom.*, **45** (1987) 381.

[136]   M. Persson, B. Hellsing, *Phys. Rev. Lett.*, **49** (1982) 662.

[137]   D. Halstead, S. Holloway, *J. Chem. Phys.*, **93** (1990) 2859.

[138]   H. Eyring, M. Polanyi, *Z. Phys. Chem.*, **B 12** (1931) 279.

[139]   J. C. Polanyi, *Science*, **236** (1987) 680.

[140]   J. C. Polanyi, A. H. Zewail, *Acc. Chem. Res.*, **28** (1995) 119.

[141]   J. C. Polanyi, W. H. Wong, *J. Chem. Phys.*, **51** (1969) 1439.

[142]   M. H. Mok, J. C. Polanyi, *J. Chem. Phys.*, **51** (1969) 1451.

[143]   J. C. Polanyi, *Acc. Chem. Res.*, **5** (1972) 161.

[144]   B. E. Hayden, C. L. A. Lamont, *Phys. Rev. Lett.*, **63** (1989) 1823.

[145]   C. T. Rettner, H. A. Michelsen, D. J. Auerbach, C. B. Mullins, *J. Chem. Phys.*, **94** (1991) 7499.

[146]   H. A. Michelsen, C. T. Rettner, D. J. Auerbach, R. N. Zare, *J. Chem. Phys.*, **98** (1993) 8294.

[147]   C. T. Rettner, H. A. Michelsen, D. J. Auerbach, *Faraday Discuss.*, **96** (1993) 17.

[148]   S. J. Gulding, A. M. Wodtke, H. Hou, C. T. Rettner, H. A. Michelsen, D. J. Auerbach, *J. Chem. Phys.*, **105** (1996) 9702.

[149]   G. R. Darling, S. Holloway, *Surf. Sci.*, **268** (1992) L305.

[150]   G. R. Darling, S. Holloway, *J. Chem. Phys.*, **97** (1992) 5182.

[151]   S. Holloway, G. R. Darling, *Surf. Rev. Lett.*, **1** (1994) 115.

[152]   G. R. Darling, S. Holloway, *J. Electron Spectrosc. Relat. Phenom.*, **64/65** (1993) 571.

[153]   G. R. Darling, S. Holloway, *J. Chem. Phys.*, **101** (1994) 3268.

[154]   G. R. Darling, Z. S. Wang, S. Holloway, *Phys. Chem. Chem. Phys.*, **2** (2000) 911.

[155]   D. Kulginov, M. Persson, C. T. Rettner, D. S. Bethune, *J. Phys. Chem.*, **100** (1996) 7919.

[156]   P. S. Weiss, D. M. Eigler, *Phys. Rev. Lett.*, **69** (1992) 2240.

[157]   J. B. Taylor, I. Langmuir, *Phys. Rev.*, **44** (1933) 423.

[158]   J. V. Barth, *Surf. Sci. Rep.*, **40** (2000) 75.

[159]   D. V. Shalashilin, B. Jackson, *J. Chem. Phys.*, **109** (1998) 2856.

[160]   C. T. Rettner, H. A. Michelsen, D. J. Auerbach, *J. Chem. Phys.*, **102** (1995) 4625.

[161]   C. T. Rettner, D. J. Auerbach, H. A. Michelsen, *Phys. Rev. Lett.*, **68** (1992) 1164.

[162]   H. A. Michelsen, C. T. Rettner, D. J. Auerbach, *Phys. Rev. Lett.*, **69** (1992) 2678.

[163]   K. W. Kolasinski, W. Nessler, A. de Meijere, E. Hasselbrink, *Phys. Rev. Lett.*, **72** (1994) 1356.

[164]   K. W. Kolasinski, *Internat. J. Mod. Phys. B*, **9** (1995) 2753.

[165]   P. Bratu, U. Höfer, *Phys. Rev. Lett.*, **74** (1995) 1625.

[166] P. Bratu, U. Höfer, in *Lasers in Surface Science*, unpublished conference proceedings, Trieste, Italy, 1994.

[167] U. Höfer, *Appl. Phys. A*, **63** (1996) 533.

[168] M. Dürr, M. B. Raschke, E. Pehlke, U. Höfer, *Phys. Rev. Lett.*, **86** (2001) 123.

[169] E. J. Buehler, J. J. Boland, *Science*, **290** (2000) 506.

[170] J. A. Serri, M. J. Cardillo, G. E. Becker, *J. Chem. Phys.*, **77** (1982) 2175.

[171] A. Mödl, T. Gritsch, F. Budde, T. J. Chuang, G. Ertl, *Phys. Rev. Lett.*, **57** (1986) 384.

[172] J. E. Hurst, Jr., L. Wharton, K. C. Janda, D. J. Auerbach, *J. Chem. Phys.*, **83** (1985) 1376.

[173] C. T. Rettner, E. K. Schweizer, C. B. Mullins, *J. Chem. Phys.*, **90** (1989) 3800.

[174] E. K. Grimmelmann, J. C. Tully, E. Helfand, *J. Chem. Phys.*, **74** (1981) 5300.

[175] J. C. Tully, *Surf. Sci.*, **111** (1981) 461.

[176] C. W. Muhlhausen, L. R. Williams, J. C. Tully, *J. Chem. Phys.*, **83** (1985) 2594.

[177] S. F. Bent, H. A. Michelsen, R. N. Zare, Hydrogen recombinative desorption dynamics, in *Laser Spectroscopy and Photochemistry on Metal Surfaces* (Eds: H. L. Dai, W. Ho), World Scientific, Singapore, 1995, p. 977.

[178] C. R. Arumainayagam, R. J. Madix, *Prog. Surf. Sci.*, **38** (1991) 1.

[179] L.-Q. Xia, M. E. Jones, N. Maity, J. R. Engstrom, *J. Chem. Phys.*, **103** (1995) 1691.

[180] D. C. Jacobs, K. W. Kolasinski, S. F. Shane, R. N. Zare, *J. Chem. Phys.*, **91** (1989) 3182.

[181] M. J. Cardillo, *Annu. Rev. Phys. Chem.*, **32** (1981) 331.

[182] E. W. Kuipers, M. G. Tenner, A. W. Kleyn, S. Stolte, *Phys. Rev. Lett.*, **62** (1989) 2152.

[183] E. W. Kuipers, M. G. Tenner, A. W. Kleyn, S. Stolte, *Nature (London)*, **334** (1988) 420.

[184] C. Haug, W. Brenig, T. Brunner, *Surf. Sci.*, **265** (1992) 56.

[185] A. W. Kleyn, A. C. Luntz, D. J. Auerbach, *Surf. Sci.*, **152/153** (1985) 99.

[186] T. F. Hanisco, C. Yan, A. C. Kummel, *J. Chem. Phys.*, **97** (1992) 1484.

[187] D. S. King, D. A. Mantell, R. R. Cavanagh, *J. Chem. Phys.*, **82** (1985) 1046.

[188] D. S. Y. Hsu, M. C. Lin, *J. Chem. Phys.*, **88** (1988) 432.

[189] M. A. Hines, R. N. Zare, *J. Chem. Phys.*, **98** (1993) 9134.

[190] D. C. Jacobs, K. W. Kolasinski, R. J. Madix, R. N. Zare, *J. Chem. Phys.*, **87** (1987) 5038.

[191] P. J. Feibelman, *Phys. Rev. Lett.*, **67** (1991) 461.

[192] D. Wetzig, R. Dopheide, M. Rutkowski, R. David, H. Zacharias, *Phys. Rev. Lett.*, **76** (1996) 463.

[193] H. Morawitz, *Phys. Rev. Lett.*, **58** (1987) 2778.

[194] K. W. Kolasinski, S. F. Shane, R. N. Zare, *J. Chem. Phys.*, **96** (1992) 3995.

[195] L. Schröter, R. David, H. Zacharias, *Surf. Sci.*, **258** (1991) 259.

[196] A. Utz, *Curr. Opin. Solid State Mater. Sci.*, (2009).

[197] K. W. Kolasinski, F. Cemič, A. de Meijere, E. Hasselbrink, *Surf. Sci.*, **334** (1995) 19.

[198] C. Åkerlund, I. Zorič, J. Hall, B. Kasemo, *Surf. Sci.*, **316** (1994) L1099.

[199] C. T. Rettner, J. Lee, *J. Chem. Phys.*, **101** (1994) 10185.

[200] G. A. Somorjai, *Chem. Rev.*, **96** (1996) 1223.

[201] J. D. Beckerle, A. D. Johnson, Q. Y. Yang, S. T. Ceyer, *J. Chem. Phys.*, **91** (1989) 5756.

[202] J. D. Beckerle, Q. Y. Yang, A. D. Johnson, S. T. Ceyer, *J. Chem. Phys.*, **86** (1987) 7236.

[203] J. D. Beckerle, A. D. Johnson, Q. Y. Yang, S. T. Ceyer, *J. Vac. Sci. Technol., A*, **6** (1988) 903.

[204] J. D. Beckerle, A. D. Johnson, S. T. Ceyer, *Phys. Rev. Lett.*, **62** (1989) 685.

[205] T. L. Alford, L. C. Feldman, J. W. Mayer, *Fundamentals of Nanoscale Film Analysis*, Springer Verlag, Berlin, 2007.

[206] P. D. Townsend, J. C. Kelley, N. E. W. Hartley, *Ion Implantation, Sputtering and Their Applications*, Academic Press, London, 1976.

[207] V. S. Smentkowski, *Prog. Surf. Sci.*, **64** (2000) 1.

[208] C. N. Hinshelwood, *The Kinetics of Chemical Change*, Clarendon Press, Oxford, 1940.

[209] T. Engel, G. Ertl, *Adv. Catal.*, **28** (1979) 1.

[210] T. Engel, G. Ertl, Oxidation of carbon monoxide, in *The Chemical Physics of Solid Surfaces and Heterogeneous Catalysis, Vol. 4* (Eds: D. A. King, D. P. Woodruff), Elsevier, Amsterdam, 1982, p. 73.

[211] S. T. Ceyer, W. L. Guthrie, T.-H. Lin, G. A. Somorjai, *J. Chem. Phys.*, **78** (1983) 6982.

[212] A. de Meijere, K. W. Kolasinski, E. Hasselbrink, *Faraday Discuss.*, **96** (1993) 265.

[213] A. Alavi, P. Hu, T. Deutsch, P. L. Silvestri, J. Hutter, *Phys. Rev. Lett.*, **80** (1998) 3650.

[214] D. D. Eley, E. K. Rideal, *Nature (London)*, **146** (1946) 401.

[215] C. T. Rettner, *Phys. Rev. Lett.*, **69** (1992) 383.

[216] C. T. Rettner, D. J. Auerbach, *Phys. Rev. Lett.*, **74** (1995) 4551.

[217] C. C. Cheng, S. R. Lucas, H. Gutleben, W. J. Choyke, J. T. Yates, Jr., *J. Am. Chem. Soc.*, **114** (1992) 1249.

[218] K. R. Lykke, B. D. Kay, *Proc. SPIE-Int. Soc. Opt. Eng.*, **1208** (1990) 18.

[219] C. T. Rettner, D. J. Auerbach, *Science*, **263** (1994) 365.

[220] C. T. Rettner, *J. Chem. Phys.*, **101** (1994) 1529.

[221] C. Lutterloh, A. Schenk, J. Biener, B. Winter, J. Küppers, *Surf. Sci.*, **316** (1994) L1039.

[222] D. D. Koleske, S. M. Gates, B. D. Thoms, J. N. Russell, Jr., J. E. Butler, *J. Chem. Phys.*, **102** (1995) 992.

[223] M. Xi, B. E. Bent, *J. Vac. Sci. Technol. B*, **10** (1992) 2440.

[224] M. Xi, B. E. Bent, *J. Phys. Chem.*, **97** (1993) 4167.

[225] L. H. Chua, R. B. Jackman, J. S. Foord, *Surf. Sci.*, **315** (1994) 69.

[226] E. W. Kuipers, A. Vardi, A. Danon, A. Amirav, *Phys. Rev. Lett.*, **66** (1991) 116.

[227] E. R. Williams, G. C. Jones, Jr., L. Fang, R. N. Zare, B. J. Garrison, D. W. Brenner, *J. Am. Chem. Soc.*, **114** (1992) 3207.

[228] W. H. Weinberg, Kinetics of surface reactions, in *Dynamics of Gas-Surface Interactions* (Eds: C. T. Rettner, M. N. R. Ashfold), The Royal Society of Chemistry, Cambridge, 1991, p. 171.

[229] B. Jackson, M. Persson, *Surf. Sci.*, **269/270** (1992) 195.

[230] M. Persson, B. Jackson, *J. Chem. Phys.*, **102** (1995) 1078.

[231] M. Persson, B. Jackson, *Chem. Phys. Lett.*, **237** (1995) 468.

[232] D. V. Shalashilin, B. Jackson, M. Persson, *Faraday Discuss.*, **110** (1998) 287.

[233] D. V. Shalashilin, B. Jackson, M. Persson, *J. Chem. Phys.*, **110** (1999) 11038.

[234] M. Persson, J. Strömquist, L. Bengtsson, B. Jackson, D. V. Shalashilin, B. Hammer, *J. Chem. Phys.*, **110** (1999) 2240.

[235] J. Harris, B. Kasemo, E. Törnqvist, *Surf. Sci.*, **105** (1981) L288.

[236] J. Harris, B. Kasemo, *Surf. Sci.*, **105** (1981) L281.

[237] C. B. Mullins, C. T. Rettner, D. J. Auerbach, *J. Chem. Phys.*, **95** (1991) 8649.

[238] T. Matsushima, *Surf. Sci.*, **123** (1982) L663.

[239] T. Kammler, D. Kolovos-Vellianitis, J. Küppers, *Surf. Sci.*, **460** (2000) 91.

[240] A. Dinger, C. Lutterloh, J. Küppers, *J. Chem. Phys.*, **114** (2001) 5338.

[241] D. A. King, M. G. Wells, *Surf. Sci.*, **29** (1972) 454.

[242] C. T. Rettner, L. A. DeLouise, D. J. Auerbach, *J. Chem. Phys.*, **85** (1986) 1131.

[243] A. C. Luntz, *J. Chem. Phys.*, **113** (2000) 6901.

[244] G. C. Welch, R. R. S. Juan, J. D. Masuda, D. W. Stephan, *Science*, **314** (2006) 1124.

[245] C. M. Mate, G. A. Somorjai, H. W. K. Tom, X. D. Zhu, Y. R. Shen, *J. Chem. Phys.*, **88** (1988) 441.

[246] M. D. Alvey, K. W. Kolasinski, J. T. Yates, Jr., M. Head-Gordon, *J. Chem. Phys.*, **85** (1986) 6093.

[247] A. Föhlisch, M. Nyberg, J. Hasselström, O. Karis, L. G. M. Pettersson, A. Nilsson, *Phys. Rev. Lett.*, **85** (2000) 3309.

[248] W. F. Lin, M. S. Zei, M. Eiswirth, G. Ertl, T. Iwasita, W. Vielstich, *J. Phys. Chem. B*, **103** (1999) 6969.

[249] G. Comsa, R. David, *Surf. Sci.*, **117** (1982) 77.

[250] J. Yoshinobu, N. Tsukahara, F. Yasui, K. Mukai, Y. Yamashita, *Phys. Rev. Lett.*, **90** (2003) 248301.

[251] A. F. Carlsson, R. J. Madix, *J. Chem. Phys.*, **114** (2001) 5304.

[252] S. C. Wang, G. Ehrlich, *J. Chem. Phys.*, **94** (1991) 4071.

[253] L. K. Verheij, J. Lux, A. B. Anton, B. Poelsema, G. Comsa, *Surf. Sci.*, **182** (1987) 390.

[254] S. Funk, B. Hokkanen, J. Wang, U. Burghaus, G. Bozzolo, J. E. Garces, *Surf. Sci.*, **600** (2006) 583.

[255] S. L. Tang, M. B. Lee, J. D. Beckerle, M. A. Hines, S. T. Ceyer, *J. Chem. Phys.*, **82** (1985) 2826.

[256] C. T. Rettner, H. Stein, *Phys. Rev. Lett.*, **59** (1987) 2768.

[257] L. J. Whitman, C. E. Bartosch, W. Ho, G. Strasser, M. Grunze, *Phys. Rev. Lett.*, **56** (1986) 1984.

[258] J. A. Serri, J. C. Tully, M. J. Cardillo, *J. Chem. Phys.*, **79** (1983) 1530.

[259] M. E. M. Spruit, E. W. Kuipers, F. H. Geuzebroek, A. W. Kleyn, *Surf. Sci.*, **215** (1989) 421.

[260] K.H. Allers, H. Pfnür, P. Feulner, D. Menzel, *Z. Phys. Chemie (Neue Folge)*, **197** (1996) 253.

[261] X.C. Guo, J. M. Bradley, A. Hopkinson, D. A. King, *Surf. Sci.*, **310** (1994) 163.

[262] S. M. Gates, *Surf. Sci.*, **195** (1988) 307.

[263] K. W. Kolasinski, *Internat. J. Mod. Phys. B*, **9** (1995) 2753.

[264] M. B. Raschke, U. Höfer, *Appl. Phys. B*, **68** (1999).

[265] E. J. Buehler, J. J. Boland, *Science*, **290** (2000) 506.

[266] L. B. Lewis, J. Segall, K. C. Janda, *J. Chem. Phys.*, **102** (1995) 7222.

[267] G. Ehrlich, Activated Chemisorption in *Chemistry and Physics of Solid Surfaces VII, Vol. 10* (Eds.: R. Vanselow, R. F. Howe), Springer-Verlag, New York, **1988**, pp. 1.

[268] C. T. Rettner, C. B. Mullins, *J. Chem. Phys.*, **94** (1991) 1626.

[269] R. W. Verhoef, D. Kelly, W. H. Weinberg, *Surf. Sci.*, **306** (1994) L513.

[270] C. B. Mullins, Y. Wang, W. H. Weinberg, *J. Vac. Sci. Technol., A*, **7** (1989) 2125.

[271] P. W. Lorraine, B. D. Thomas, R. A. Machonkin, W. Ho, *J. Chem. Phys.*, **96** (1992) 3285.

[272] L. Diekhöner, D. A. Butler, A. Baurichter, A. C. Luntz, *Surf. Sci.*, **409** (1998) 384.

[273] D. E. Brown, D. J. Moffatt, R. A. Wolkow, *Science*, **279** (1998) 542.

[274] S. K. Kulkarni, S. M. Gates, C. M. Greenlief, H. H. Sawin, *Surf. Sci.*, **239** (1990) 13.

[275] J. R. Engstrom, D. A. Hansen, M. J. Furjanic, L. Q. Xia, *J. Chem. Phys.*, **99** (1993) 4051.

4

Thermodynamics and Kinetics of Adsorption and Desorption

In this chapter we begin with a discussion of the thermodynamics of surface processes and move on to kinetics. Here we treat adsorption, desorption and the influence of lateral interactions. By concentrating on a statistical mechanical approach to kinetics, we see the importance of dynamics in surface processes. Further investigation of the thermodynamics of surfaces as it pertains to liquid interfaces and growth is presented in Chapters 5 and 7, respectively. Here we broaden the discussions presented in Chapter 3 by treating explicitly surface reactions at finite coverages. This is, of course, a necessary extension in order to handle the kinetics of surface chemical reactions, which are discussed in detail in Chapters 6 and 7.

4.1 Thermodynamics of Ad/Desorption

4.1.1 Binding energies and activation barriers

The energetics of the potential energy hypersurface is important for dynamics, thermodynamics and kinetics. We can use Lennard-Jones diagrams to define the relationships between a number of quantities. The simplest case is that of an atom approaching a surface along the z co-ordinate though, again, we can generalize this by considering z to represent the reaction co-ordinate. Figure 4.1(a) depicts the cases of non-activated adsorption and activated adsorption is shown in Fig. 4.1(b). As drawn, both potentials include a physisorption well. Note that the physisorption well is located further from the surface than the chemisorption well. This is consistent with the usual trend in chemistry that shorter bonds correspond to stronger bonds. At low temperatures, a species can be trapped in a physisorbed state even though a more strongly bound chemisorption state exists. Once an adsorbate settles into a physisorption well, it must overcome a small barrier to pass into the chemisorbed state.

At absolute zero for a classical system, there is no ambiguity in defining the heat released by adsorption, q_{ads}, the desorption activation energy, E_{des}, the adsorption activation energy, E_{ads}, and the adsorption bond binding energy (bond strength), $\varepsilon(M-A)$. In the case of non-dissociative, non-activated adsorption, $E_{ads} = 0$ and these relations are almost trivial

$$\varepsilon(M-A) = E_{des}, \text{(non-activated)} \tag{4.1.1}$$

Surface Science: Foundations of Catalysis and Nanoscience, Third Edition. Kurt W. Kolasinski.
© 2012 John Wiley & Sons, Ltd. Published 2012 by John Wiley & Sons, Ltd.

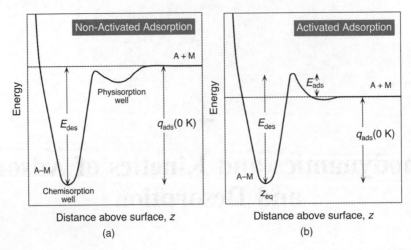

Figure 4.1 *One-dimensional potential energy curves for molecular adsorption: (a) nonactivated adsorption; (b) activated adsorption.* E_{ads}, E_{des}, *adsorption activation energy and desorption activation energy, respectively;* q_{ads}, *heat released by adsorption;* z_{eq}, *adsorbate-surface bond length.*

and

$$\varepsilon(M\text{--}A) = q_{ads}. \tag{4.1.2}$$

For a quantum mechanical system, the energy differences must be calculated from the appropriate zero-point energy levels, c.f. Fig. 4.5. $\varepsilon(M\text{--}A)$ is temperature independent, whereas, as we shall see in the next section, q_{ads} depends on temperature. For the moment these caveats need not concern us.

For activated adsorption, $E_{ads} > 0$ and the following relationships hold. q_{ads} is the difference between the bottom of the chemisorption well and the zero of energy, taken as the energy of the system when the adsorbate is infinitely far from the surface. E_{des} is the difference between the bottom of the chemisorption well and the top of the adsorption barrier. E_{ads} is the height of the activation barrier when approaching the surface from $z = \infty$. The defining relationships are now written,

$$E_{des} = E_{ads} + \varepsilon(M\text{--}A) \tag{4.1.3}$$

and

$$\varepsilon(M\text{--}A) = q_{ads} = E_{des} - E_{ads}. \quad \text{(activated)} \tag{4.1.4}$$

In dissociative adsorption, the intramolecular adsorbate bond with dissociation energy $\varepsilon(A\text{--}A)$ is also broken. Figure 4.2 depicts activated dissociative adsorption of a diatomic molecule A_2. The dissociation energy of the atomic fragments and heat of adsorption are then given by

$$\varepsilon(M\text{--}A) = \tfrac{1}{2}[E_{des} - E_{ads} + \varepsilon(A\text{--}A)]. \quad \text{(dissociative)} \tag{4.1.5}$$

and

$$q_{ads} = 2\varepsilon(M\text{--}A) - \varepsilon(A\text{--}A). \tag{4.1.6}$$

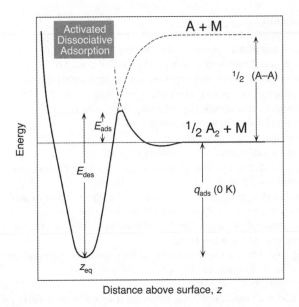

Figure 4.2   *Activated dissociative adsorption. E_{ads}, E_{des}, adsorption activation energy and desorption activation energy, respectively; q_{ads}, heat released by adsorption; z_{eq}, adsorbate–surface bond length.*

4.1.2   Thermodynamic quantities

Here we give a brief account of surface thermodynamics. More extensive discussions follow particularly in Ch 5, and can also be found elsewhere [1–4]. The most fundamental quantity in thermodynamics is the Gibbs energy, G. Any system relaxes to the state of lowest Gibbs energy *as long as no dynamical or kinetic constraints exist* that block it from reaching global equilibrium. Therefore, a spontaneous change is always accompanied by a decrease in Gibbs energy, i.e. $\Delta G < 0$ for all spontaneous processes. The relationship between the change in Gibbs energy, entropy and enthalpy is

$$\Delta G = \Delta H - T\,\Delta S. \tag{4.1.7}$$

Adsorption confines a gas to a surface, which results in an unfavourable entropy change, $\Delta_{ads}S < 0$, unless by some unusual process the substrate experiences an extraordinarily large positive entropy change that can compensate for this. Therefore, $\Delta_{ads}H$ must be negative (exothermic) for spontaneous adsorption to occur.

4.1.3   Some definitions

As always in thermodynamics, it is essential to make clear and consistent definitions of the symbols used in the mathematical treatment. Table 4.1 defines several of the symbols used here.

Two different definitions of σ_0 appear in the literature. The reader must always be attentive as to whether coverage is defined with respect to the number of sites or the number of surface atoms. These two definitions are equivalent in some instances. However, there are instances in which the saturation coverage is, say, one adsorbate for every two surface atoms. A saturated layer of adsorbates then has a coverage of 0.5 ML

Table 4.1   *Definition of symbols*

N_{ads}	Number of adsorbates (molecules or atoms, as appropriate)
N_0	Number of surface sites or atoms (as defined by context)
N_{exp}	Number of atoms/molecules exposed to (incident upon) the surface
A_s	Surface area (m^2 or, more commonly, cm^2)
σ	Areal density of adsorbates (adsorbates cm^{-2}), $\sigma = N_{ads}/A_s$
σ_0	Areal density of sites or surface atoms (cm^{-2})
σ_*	Areal density of empty sites (cm^{-2})
σ_{sat}	Areal density that completes a monolayer (cm^{-2})
θ	Coverage, fractional number of adsorbates (monolayers, ML), also sometimes called fractional coverage, $\theta \equiv \sigma/\sigma_0$
θ_{sat}	Saturation coverage, $\theta_{sat} \equiv \sigma_{sat}/\sigma_0$, where σ_0 is the number of surface atoms
δ	Relative coverage defined with respect to saturation, $\delta \equiv \sigma/\sigma_{sat} \equiv \theta/\theta_{sat}$
ε	Exposure, amount of gas incident on the surface, units of cm^{-2} or Langmuir
$\varepsilon(M-A)$	Binding energy of the M–A bond
q_{ads}	Heat released when a single particle adsorbs (positive for exothermic)
L	Langmuir, unit of exposure, $1\,L = 1 \times 10^{-6}$ torr $\times 1$ s
s	Sticking coefficient, $s = \sigma/\varepsilon$
	Integral sticking coefficient: total coverage divided by exposure, meaningful only if s is constant or as $\theta \rightarrow 0$ ML
	Instantaneous or differential sticking coefficient at coverage θ $s(\theta) = dN_{ads}/dN_{exp} = d\sigma/d\varepsilon$, evaluated at a specific value of θ
s_0	Initial sticking coefficient, sticking coefficient as $\theta \rightarrow 0$ ML

when defined with respect to the number of surface atoms, but 1.0 ML when defined with respect to sites. The former definition is more absolute, but the latter definition can lead to useful simplifications in kinetics calculations. The latter definition also has the operational simplification of not requiring any knowledge about the surface structure: a monolayer is simply defined as the number of molecules in the saturated layer. Two definitions of σ_0 result in two definitions of fractional coverage: θ is defined in respect to the number of surface atoms, and δ is used when fractional coverage is defined with respect to the number of sites. Unfortunately in the literature θ is used almost exclusively, regardless of whether it has been defined with respect to the number of surface atoms or the number of sites. This convention is also found in this text.

A clean surface is also something of a matter of definition. The normal sensitivity of many surface analytical techniques, apart from STM and a few others, is roughly 0.01 ML. Thus, we often consider a clean surface to be one that has ≤ 0.01 ML of impurities. Detection limits are a source of major experimental difficulty in surface science, and these difficulties must be kept in mind when results are analyzed.

4.1.4   The heat of adsorption

Returning to thermodynamics, we have two goals. First, we seek to define q_{ads} more precisely and understand its behaviour. Second, we would like to be able to relate the heat release measured in calorimetry [5] to thermodynamic parameters. The enthalpy is defined by

$$H = U + pV \tag{4.1.8}$$

where U is the internal energy, p the pressure and V the volume. For an ideal gas in molar units

$$H_g = U_g + p_g V_g = U_g + RT. \tag{4.1.9}$$

For the adsorbed gas, the pV term is negligible, thus

$$H_a = U_a. \tag{4.1.10}$$

The enthalpy change in going from the gas to the adsorbed phase is, therefore,

$$\Delta_{ads}H = H_a - H_g = U_a - U_g - RT. \tag{4.1.11}$$

In Figures 4.1 and 4.2, we have chosen the origin of energy such that the internal energy of the system is zero at infinite separation and $0\,K$. The internal energy depends on the sum of translational, rotational and vibrational energies of the gas (or adsorbate) – a quantity with obvious temperature dependence. We now make the following two identifications

$$-q_{ads} = U_a - U_g \tag{4.1.12}$$

and

$$q_c = RT. \tag{4.1.13}$$

From Eq. (4.1.12) we can appreciate that while q_{ads} is related to $\varepsilon(M-A)$, the equality of the two is valid only at $0\,K$. $\varepsilon(M-A)$ is essentially a single particle quantity, whereas q_{ads} involves a convolution over the thermal distributions of adsorbed and gas phase particles. q_c is the heat of compression arising from the transformation of a gas of finite volume into an adsorbed layer of essentially zero volume. By convention, the heat of adsorption (a positive quantity for exothermic adsorption) is often quoted in surface science rather than the adsorption enthalpy (a negative quantity for exothermic adsorption).

In general, the heat of adsorption is a coverage dependent quantity, hence

$$-\Delta_{ads}H(\theta) = q_{ads}(\theta) + q_c(\theta) = q_{st}(\theta) \tag{4.1.14}$$

where $q_{st}(\theta)$ is the isosteric heat of adsorption, $\Delta_{ads}H(\theta)$ is the differential adsorption enthalpy and $q_{ads}(\theta)$ is the differential heat of adsorption. At room temperature q_c is only $2.5\,kJ\,mol^{-1}$, hence in practice it is usually negligible.

The isosteric heat of adsorption is defined through the Clausius-Clapeyron equation,

$$q_{st}(\theta) = RT^2 \left(\frac{\partial \ln p}{\partial T} \right)_\theta = -R \left(\frac{\partial \ln p}{\partial (1/T)} \right)_\theta \tag{4.1.15}$$

where p is the equilibrium pressure that maintains a coverage θ at temperature T. It can be shown [5] that the heat measured in a single crystal adsorption calorimetry experiment is the isosteric heat of adsorption.

One final quantity of interest is the integral adsorption enthalpy. This represents the total enthalpy change (generally in molar units) recorded when the coverage changes from zero to some final value θ_f. The integral adsorption enthalpy is related to the heat of adsorption by

$$\Delta_{ads}H_{int} = \frac{\int_0^{\theta_f} -q_{ads}(\theta)\, d\theta}{\int_0^{\theta_f} d\theta}. \tag{4.1.16}$$

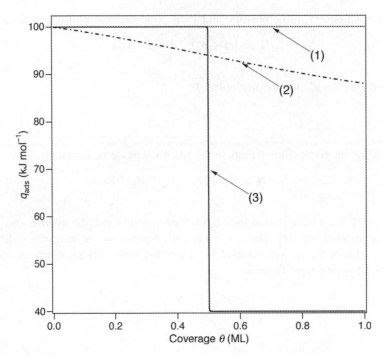

Figure 4.3   *Three different behaviours of the heat of adsorption, q_{ads}, as a function of coverage, θ. Case (1): the surface is composed of one and only one type of non-interacting site. Case (2): q_{ads} decreases linearly with θ. Case 3: the surface is composed of two types of sites with different binding energies that fill sequentially. As shown in §4.3, case 3 can also arise from strong lateral interactions.*

Figure 4.3 displays several possible scenarios for the dependence of q_{ads} on θ. If the surface has only one type of site, and all of these sites adsorb particles independently, then q_{ads} is a constant as shown in Fig. 4.3, curve (1) and $\Delta_{ads}H_{int} = -q_{ads}$ (in molar units). For a surface that has two independent adsorption sites with different characteristic adsorption energies that fill sequentially, a step-like behaviour such as that found in Fig. 4.3, curve (3) is observed. Chemisorption involves charge transfer, and the capacity of a surface to accept or donate charge is limited. Consequently, as more and more particles adsorb, the ability of the surface to bind additional adsorbates likely drops. Thus, q_{ads} drops with increasing θ. This type of behaviour is replicated in Fig. 4.3, curve (2). In addition, as the coverage increases the distance between adsorbates decreases. Lateral interactions become increasingly likely, and these influence q_{ads}, leading to changes as a function of θ, cf. §4.3.

4.2   Adsorption isotherms from thermodynamics

Consider adsorption onto a solid substrate in which adsorption occurs from an ideal monatomic gas in equilibrium with the solid. There are σ_0 equivalent sites on the surface, and not more than one adsorbate can bind on each. The adsorbates are non-interacting, thus the binding energy, ε, is independent of coverage.

The probability of having N_{ads} adsorbed atoms is given by the ratio

$$P(N_{ads}) = \frac{Q_{ads}\exp(N_{ads}\mu_{ads}/k_BT)}{\Xi} \tag{4.2.1}$$

where Q_{ads} is the canonical partition function, Ξ the grand canonical partition function, and μ_{ads} the chemical potential of the adsorbed atoms (see Exercise 4.1). This can be used to calculate the equilibrium fractional coverage of adsorbates with respect to the number of sites

$$\theta = \frac{\exp\left(\dfrac{\mu_{ads} + \varepsilon}{k_B T}\right)}{1 + \exp\left(\dfrac{\mu_{ads} + \varepsilon}{k_B T}\right)}. \tag{4.2.2}$$

At equilibrium, the chemical potential of all phases present must be equal. Therefore, we can calculate μ_{ads} by equating it to the chemical potential of monatomic ideal gas, which can be readily found [2]. Thus,

$$\mu_{ads} = k_B T \ln\left[\frac{p}{k_B T}\left(\frac{h^2}{2\pi m k_B T}\right)^{3/2}\right] \tag{4.2.3}$$

where p is the equilibrium pressure of the gas, m the mass of the atom and h the Planck constant. Substitution into Eq. (4.2.2) leads to

$$\theta = \frac{p}{p + p_0(T)} \tag{4.2.4}$$

where

$$p_0(T) = \left(\frac{2\pi m k_B T}{h^2}\right)^{3/2} k_B T \exp\left(\frac{-\varepsilon}{k_B T}\right) \tag{4.2.5}$$

is the pressure required to obtain an equilibrium coverage of $\theta = 0.5$ at temperature T.

Equation (4.2.5) describes the equilibrium coverage found on a surface as a function of the adsorbate binding energy, pressure and temperature. This equation is known as the Langmuir isotherm. In §4.6 the Langmuir isotherm is derived from kinetics. The thermodynamic interpretation of this isotherm is that the equilibrium coverage is determined by the chemical potential difference introduced by the adsorption energy between the adsorbed phase and the gas phase. The shape of the isotherm is given in Fig. 4.4. This general isotherm holds in many real chemisorption systems, at least at low coverage. Note that no *a priori* knowledge of the ad/desorption dynamics is required, nor are any details of the adsorption kinetics apart from the assumption of independent sites that adsorb no more than one adsorbate. Accordingly, no dynamical information can be obtained from isotherms. Dynamics, reflected in the kinetics of reaction, affects the time required to attain equilibrium, but it does not actually determine the shape of the isotherm.

There are numerous isotherms bearing names of many catalytic chemists [6]. An exposition of all of these is of little more than academic interest. The occurrence of other isotherms arises because of the breakdown in the assumptions of the Langmuir model, particularly when we consider physisorbed layers. The two most suspicious assumptions are that adsorption stops when N_0 sites on the surface are filled, that is that $\theta_{sat} = 1\,\mathrm{ML}$, and that the adsorbates are non-interacting.

Relaxation of the assumption of saturation at 1 ML leads to the Brunauer-Emmett-Teller (BET) isotherm [7]. This isotherm is used in countless thousands of measurements every year to determine the surface area of powders and porous solids, such as high surface area supported catalysts [8, 9]. The BET isotherm is obtained as above but with the addition that occupied sites can be filled with a second layer of binding energy of ε'. This leads to the isotherm equation [2]

$$\theta = \frac{pp_0(T)}{\left[p_0(T) + p - p\exp\left(\dfrac{\varepsilon' - \varepsilon}{k_B T}\right)\right]\left[p_0(T) - p\exp\left(\dfrac{\varepsilon' - \varepsilon}{k_B T}\right)\right]}. \tag{4.2.6}$$

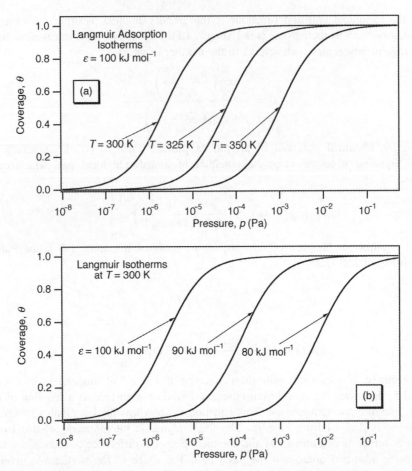

Figure 4.4   *Langmuir isotherms exhibit a dependence on the temperature and binding energy. (a) Constant heat of adsorption for various temperatures T. (b) Constant temperature for various adsorption energies ε.*

A more useful representation of the BET isotherm for practical determination of surface area is

$$\frac{p}{n^a(p^\circ - p)} = \frac{1}{n^a_m C} + \frac{C-1}{n^a_m C}\frac{p}{p^\circ}, \tag{4.2.7}$$

where n^a is the amount absorbed at the relative pressure p/p°, p° is the saturation vapour pressure at the temperature at which the experiment is performed (i.e. the pressure at which the adsorbate would simply condense), n^a_m is the monolayer capacity and C is a constant determined by the shape of the isotherm. The amount absorbed is often reported in units of moles adsorbed per unit mass of the porous material since at this point the surface area of the material is not known. A plot of $p/[n^a(p^\circ - p)]$ *vs* p/p° is generally linear over the range $0 \leq p/p^\circ \leq 0.3$. The slope $(C-1)/n^a_m C$ and intercept $1/(n^a_m C)$ are used to determine C and n^a_m. The BET surface area can then be determined from

$$A_s(\text{BET}) = n^a_m N_A a_m \tag{4.2.8}$$

by using a molecule with a known cross sectional area a_m. The adsorption of N_2 at 77 K, for which $a_m = 0.162$ nm, is recommended for probing mesoporous samples. Kr adsorption is recommended for samples with low specific surface areas of <2 m^2 g^{-1}. Hg porosimetry is recommended for macroporous materials.

4.3   Lateral interactions

Relaxation of the non-interacting adsorbate assumption leads to isotherms that are much more complex. The interactions are built into the isotherm equation with modified partition functions. Two of the most widely used approximations to treat effects of lateral interactions are the Bragg-Williams approximation and the quasi-chemical approximation. An in-depth discussion of these approximations can be found in Desjonquères and Spanjaard [2]. The inclusion of lateral interactions is necessitated by the observation not only of deviations from the Langmuir isotherm, but also of a range of 2D phase transitions [10].

Adsorbates interaction occur in four classes [11, 12]:

- Direct interactions due to wavefunction overlap. The formation of chemical bonds between adsorbates is an extreme case of attractive interaction. Brown and King, for example, have observed the formation of NO dimers on metal surfaces [13]. Usually, however, such interactions are repulsive due to Pauli repulsion.
- Indirect (substrate mediated) interactions. The binding of an adsorbate often shifts the d states of neighbouring transition metal atoms downwards. This causes a weaker metal atom–adsorbate interaction for subsequent adsorbates.
- Elastic interactions. A local distortion of the lattice in the vicinity of an adsorbate leads to a repulsive interaction with neighbouring adsorbates.
- Non-local electrostatic effects. Dipole–dipole (or higher multipole–multipole) and van der Waals forces can exist between adsorbates. Dipole interactions can be either repulsive or attractive depending on the relative orientation of the interacting dipoles.

Indirect interactions can arise either from electronic or structural changes induced by the presence of an adsorbate. Chemisorption is associated with charge transfer between the adsorbate and the surface. This means that the binding of an adsorbate to one site can lead to changes in the binding affinity of neighbouring sites. The changes can lead to either repulsive or attractive interactions; indeed, the sign of the interaction may change with distance from the adsorbate. For example, the surface phase diagram of H/Pd(100) is best fit by assuming a nearest-neighbour repulsion of -1.6 kJ mol^{-1}, a next-nearest neighbour attraction of $+1.2$ kJ mol^{-1} and a nonpairwise "trio" interaction of $+0.81$ kJ mol^{-1} [14]. The formation of a chemisorption bond is also accompanied by changes in the substrate atom positions. This may amount to a simple relaxation or a well-developed reconstruction of the substrate. In Chapter 7 we explore some of the consequences of the strain fields that arise during chemisorption, while adsorbate-induced reconstructions are explored in §1.2.2. A restructuring of the surface is often a local event which affects only those adsorbates that attempt to bind in the restructured region.

All mechanisms of lateral interactions display some type of distance dependence. In general, the importance of lateral interactions grows with decreasing distance between adsorbates. Hence, lateral interactions are important at high coverage. Nevertheless, even at low coverages, lateral interactions cannot always be neglected. In many adsorption systems, the formation of islands is observed even at low temperatures. The formation of islands is influenced by both lateral interactions and dynamics. Once an island is formed, the local coverage within the island is high compared to the globally averaged coverage. This circumstance can lead to invalidation of the assumption of non-interacting, isolated particles even for very low global coverages.

A straightforward model to treat the effects of lateral interactions can be formulated with the following assumption [15–17]:

- Adsorption occurs on a fixed number of equivalent sites.
- Interactions are limited to a nearest-neighbour interaction of energy ω, which is independent of coverage.
- Desorption is an equilibrium process.
- Adsorption is confined to just one layer on top of the surface.

With this set of approximations, the heat of adsorption is found to be

$$q_{ads}(\theta) = q_0 + \frac{1}{2}z\omega\left[1 - \frac{1 - 2\theta}{[1 - 4\theta(1 - \theta)(1 - \exp(\omega/k_B T))]^{1/2}}\right] \tag{4.3.1}$$

where q_0 is the heat of adsorption at zero coverage and z is the maximum possible number of nearest neighbours. In this convention, repulsive interactions correspond to negative values of ω. Equation (4.3.1) has two interesting limits. For $\omega/T \to 0$, that is, for temperatures sufficiently large to ensure a completely disordered layer, Eq. (4.3.1) reduces to

$$q_{ads}(\theta) = q_0 + z\omega\theta. \tag{4.3.2}$$

In the limit of $\omega/T \to -\infty$, the heat of absorption resembles a step function switching between the values q_0 and $q_0 + z\omega$ at $\theta = 0.5$ ML. These two extremes correspond to the behaviour depicted in Fig. 4.3 curves (2) and (3), respectively. Indeed, curve (2) was drawn using $q_0 = 100\,\text{kJ mol}^{-1}$, $\omega = -2\,\text{kJ mol}^{-1}$, $z = 6$ and $T = 1000\,\text{K}$; whereas curve (3) was calculated with $q_0 = 100\,\text{kJ mol}^{-1}$, $\omega = -10\,\text{kJ mol}^{-1}$, $z = 6$ and $T = 80\,\text{K}$. Such step function behaviour approximates what is observed for CO/Ru(001) [18]. Therefore, the behaviour of curve (3) is ambiguous as it can be caused either by sequential filling of two sites or strong repulsive interactions.

4.4   Rate of desorption

The kinetics of surface reactions has many similarities with kinetics in other areas of chemistry [19, 20]. However, several differences also exist, the most essential of which is that *reactions can only occur on a limited number of sites*. The possibility of sites with different reactivities is another complicating factor. We begin by treating the simplest of surface reactions – adsorption and desorption – and move on from there to consider chemical reactivity at surfaces.

The rate of a chemical process is described by the Polanyi-Wigner equation. In this case we are interested in the rate of desorption, r_{des}, which is the rate of change in the coverage as a function of time. In its most general form, the Polanyi-Wigner equation is written

$$r_{des} = -\frac{\partial \theta}{\partial t} = \nu_n \theta^n \exp(-E_{des}/RT_s) \tag{4.4.1}$$

where ν_n is the pre-exponential factor of the chemical process of order n, E_{des} is the desorption activation energy, and T_s is the surface temperature. As written in Eq. (4.4.1), the rate is in terms of monolayers per second. It could equally well be formulated in terms of molecules per unit area per second by substituting the absolute coverage $\sigma = \sigma_0\theta$, where θ is referred to the number of surface atoms.

4.4.1 First-order desorption

Consider a first-order process, i.e. atomic or non-associative molecular desorption. The absolute rate can be written in terms of the concentration term and the rate constant, k_{des},

$$r_{des} = \sigma_0 \theta k_{des} \tag{4.4.2}$$

for which the rate constant can be written in an Arrhenius form

$$k_{des} = A \exp(-E_{des}/RT_s). \tag{4.4.3}$$

Conventionally, the pre-exponential factor of a first-order process is denoted A.

Adsorption and desorption are often reversible processes. For a simple first-order process, we can express the lifetime of the adsorbate on the surface, τ, through the Frenkel Equation,

$$\tau = \frac{1}{k_{des}} = \frac{1}{A} \exp(E_{des}/RT_s) = \tau_0 \exp(E_{des}/RT_s) \tag{4.4.4}$$

Equation (4.4.4) is an important result to keep in mind. It states that the adsorbate lifetime on the surface is finite, and that it depends on the temperature and the desorption activation energy. As the temperature increases, the lifetime drops exponentially, but even at high temperature there is a finite lifetime. Obviously, in catalytic chemistry it is important for the surface lifetime of reactants to be long compared to the time it takes for adsorbates to diffuse across the surface and react.

Equation (4.4.1) is easily extended to higher order processes. The kinetics of desorption might at first appear to be completely equivalent to the kinetics of chemical reactions in other phases; however, important complications arise. Even for non-interacting adsorbates, we have seen that more than one binding state can be populated at the surface. Each one of these binding states has associated with it unique values of A and E_{des}. In this way, the desorption rate depends not only on the identity of the adsorbate, but also the binding state that it occupies. Surface diffusion between sites of different binding energy can lead to further complications, as is the case for NO/Pt(111) for which binding at the steps is significantly stronger than on the terraces [21].

Work, for instance by King [22, 23], Menzel [18], Schmidt [24] and co-workers has shown that lateral interactions can play an important role in the kinetics of desorption, particularly at high coverage. Lateral interactions affect the rate of desorption by making both A and E_{des} coverage dependent. Consequently, Eq. (4.4.3) is more generally written as

$$k_{des} = A(\theta) \exp(-E_{des}(\theta)/RT_s) \tag{4.4.5}$$

such that the coverage dependence of the kinetic parameters is accounted for. The Bragg-Williams and quasi-chemical approximations [2, 3] can be invoked to formulate expressions for the changes in A and E_{des} with coverage. Temperature programmed desorption (§4.7) and adsorption isotherms (§4.2) provide methods for determining the magnitude of such interactions.

Frequently A and E_{des} vary in concert. The relationship

$$\ln A(\theta) = E_{des}(\theta)/RT_\theta + c, \tag{4.4.6}$$

where T_θ is the isokinetic temperature [19], is followed more or less in these systems. Such a relationship between A and E_{des} is known as the compensation effect. A compensation effect can result from a number of sources such as (i) a heterogeneous surface that contains adsorption sites with a range of binding energies [25], (ii) lateral interactions [26], in particular if they are strong enough to give rise to coverage-dependent phase changes in the adsorbed layer [27], and (iii) adsorbate induced changes in the substrate structure [28]. More on the compensation effect follows in §4.4.3.

4.4.2   Transition state theory treatment of first-order desorption

We want to develop a deeper understanding of A and E_{des}. For the moment, we implicitly neglect the coverage dependence of the kinetic parameters. Direct atomic and simple molecular desorption are examples of one-step reactions, known as elementary reactions, of the type

$$A^* \rightleftharpoons A + {}^*$$

in which desorption occurs directly from one adsorbed phase into the gas phase. The reaction is in some ways equivalent to unimolecular dissociation. The rate of an elementary reaction is simply the product of the concentration and the rate constant, as in Eq. (4.2.4). The rate constant can be interpreted with thermodynamic or statistical mechanical methods. Transition state theory is the foundation for these formulations. In transition state theory, the reactants follow a multi-dimensional potential energy hypersurface as discussed in Chapter. 3. The reactants and products are separated by a transition state, which is at some position along the potential energy surface. The chemical entity at the transition state is called the activated complex. If an activation barrier separates the reactants and products, the transition state is located at the top of the barrier. If there is no maximum in PES, the definition of the transition state is somewhat arbitrary.

The main assumptions of conventional transition state theory (CTST) are [20]:

- Once the transition state is reached, the system carries on to produce the products.
- The energy distributions of the reactants follow Maxwell-Boltzmann distributions.
- The whole system need not be at equilibrium but the concentration of the activated complex can be calculated based on equilibrium theory.
- The motion along the reaction co-ordinate is separable from other motions of the activated complex.
- Motion is treated classically.

Extensions to CTST can be formulated that improve upon each of these assumptions, but most of these extensions need not concern us here. As shown by Tully and co-workers [29], however, it is essential to relax the first assumption. This is done by introducing the transmission coefficient [20], κ, which defines the probability with which an activated complex proceeds into the product channel. This is a number that is strictly ≤ 1. The transmission coefficient is a dynamical correction to CTST. We shall soon see the fundamental importance of κ.

First, we note that there are three equivalent ways to write an equilibrium constant. For the general reaction

$$a\text{A} + b\text{B} \rightleftharpoons c\text{C} + d\text{D}$$

the equilibrium constant can be written in terms of concentrations (more accurately activities), rate constants or molecular partition functions.

$$K = \frac{[C]^c[D]^d}{[A]^a[B]^b} = \frac{\overrightarrow{k}}{\overleftarrow{k}} = \frac{q_C^c q_D^d}{q_A^a q_B^b} \exp\left(\frac{-E_0}{RT}\right), \tag{4.4.7}$$

where the square brackets indicate concentrations, $\overrightarrow{k}$ and $\overleftarrow{k}$ represent the rate constants of the forward and reverse reactions, respectively, and the q_Xs are molecular partition functions per unit volume. The energy

E_0 is the molar energy change accompanying the conversion of reactants to products at 0 K, everything being in the appropriate standard state. The final expression follows from the CTST formulation of the rate constant, which for the forward reaction is

$$\vec{k}_{CTST} = \frac{k_B T}{h} \frac{q_{\ddagger}}{q_A q_B} \exp(-E_0/RT) \tag{4.4.8}$$

The molecular partition function is the product of the partition functions for all degrees of freedom

$$q_X = q_{trans} q_{rot} q_{vib} q_{elec} \tag{4.4.9}$$

The electronic partition function,

$$q_{elec} = \sum_i g_{ei} \exp(-\varepsilon_{elec,i}/k_B T) \tag{4.4.10}$$

where g_{ei} is the degeneracy of the electronic state of energy $\varepsilon_{elec,i}$, is usually unity because excited electronic states tend to be high in energy and the ground state is often a singlet. The translational partition function is

$$q_{trans} = \prod_i \frac{(2\pi m k_B T)^{1/2}}{h} \tag{4.4.11}$$

where m is the mass and i the dimensionality. For instance, for a system confined to two dimensions, $i = 2$ and $q_{trans} = (2\pi m k_B T)/h^2$. Eq. (4.4.11) has units of m^{-i}, and represents the partition function per unit volume in 3D or per unit area in 2D. The rotational partition function depends on whether the molecule is linear

$$q_{rot} = \frac{8\pi^2 I k_B T}{\sigma h^2} = \frac{k_B T}{\sigma h c B}. \tag{4.4.12}$$

or non-linear

$$q_{rot} = \frac{8\pi^2 (8\pi^3 I_A I_B I_C)^{1/2} (k_B T)^{3/2}}{\sigma h^3}. \tag{4.4.13}$$

σ is the symmetry number, which is an integer determined by the symmetry of reactants, transition state and products that is the number of indistinguishable orientations of the molecule. The rules for determining σ are given by Laidler [20]. For a heteronuclear diatomic molecule $\sigma = 1$ and for a homonuclear diatomic or symmetrical linear molecule $\sigma = 2$. The vibrational partition function is written as the product of the partition functions of all the ($3N - 5$ or $3N - 6$) normal modes of the molecule,

$$q_{vib} = \prod_i \frac{1}{1 - \exp(-h\nu_i/k_B T)} \tag{4.4.14}$$

where ν_i is the fundamental frequency of the ith oscillator. In the partition function of the transition state, the reaction co-ordinate is not bound, and this motion is not included in the partition function.

Laidler, Gladstone and Eyring [20, 30] used absolute rate theory to express the rate of desorption in terms of the partition functions. From the assumption of equilibrium between the transition state and the adsorbed phase, we write

$$K = \frac{\sigma_{\ddagger}}{\sigma_a} = \frac{q^{\ddagger}}{q_a} \exp\left(-\frac{E_{des}}{RT}\right) \tag{4.4.15}$$

where σ_a and $\sigma_{\ddagger}$ are the surface concentrations (per unit area) of the adsorbate and transition state, respectively. The precise definition of E_{des} is the activation energy at 0 K, that is, the energy required to elevate an adsorbed molecule in the lowest vibrational state to the lowest vibrational state of the activated complex. Note that in Eq. (4.4.15) the partition function of the surface has been omitted. The partition function of the surface is difficult to calculate, and is generally assumed to change little upon adsorption. Changes in the electronic and vibrational degrees of freedom of the substrate can occur upon adsorption. It has been argued that the electronic changes should not result in a significant change in the partition function [31]; therefore their neglect is justified. If the adsorbate does not significantly affect the phonon spectrum of the substrate, then the vibrational changes are also insignificant. However, in instances where adsorption leads to significant substrate rearrangement and phonon spectrum changes such as H/Si or H/W(100), the neglect of the surface partition function becomes tenuous and must be tested experimentally. One further caveat is that, as discussed by Menzel [32], a configurational partition function for the adsorbed layer may need to be included. This is required to explain experimental results in the CO/Ru(001) system [18, 33], and is in general important for adsorbed layers that can undergo structural phase transitions. This term can be neglected at low coverage and for non-interacting adsorbates.

The concentration of the activated complex is calculated from Eq. (4.4.15)

$$\sigma_{\ddagger} = \sigma_{a} \frac{q^{\ddagger}}{q_{a}} \exp\left(-\frac{E_{des}}{RT}\right). \tag{4.4.16}$$

The activated complex can be thought of as having one loose vibrational mode that corresponds to the motion leading to desorption. This is expressed in a factor of $k_{B}T/h\nu$, and when factored out of $q^{\ddagger}$ to give $q_{\ddagger}$, we have

$$\sigma_{\ddagger} = \sigma_{a} \frac{k_{B}T}{h\nu} \frac{q_{\ddagger}}{q_{a}} \exp\left(-\frac{E_{des}}{RT}\right). \tag{4.4.17}$$

This rearranges to

$$\nu\sigma_{\ddagger} = \sigma_{a} \frac{k_{B}T}{h} \frac{q_{\ddagger}}{q_{a}} \exp\left(-\frac{E_{des}}{RT}\right). \tag{4.4.18}$$

The left-hand term is the concentration of the activated complex multiplied by the frequency with which it leaves the transition state, which corresponds to the reaction rate. Recall, however, that we must correct the CTST result by introducing the transmission coefficient,

$$k_{des} = \kappa k_{CTST} \tag{4.4.19}$$

thus

$$r_{des} = \sigma_{a} \kappa \frac{k_{B}T}{h} \frac{q_{\ddagger}}{q_{a}} \exp\left(-\frac{E_{des}}{RT}\right). \tag{4.4.20}$$

According to Eq. (4.4.20) then, the pre-exponential factor in desorption is

$$A = \kappa \frac{k_{B}T}{h} \frac{q_{\ddagger}}{q_{a}}. \tag{4.4.21}$$

Note that although A has the units of frequency, it should not be confused with an "attempt" frequency. The frequency along the desorption co-ordinate was already subsumed into the rate in Eq. (4.4.20). The value of A tells us about the relative floppiness of the adsorbed phase compared to the transition state through the ratio of the partition functions, as well as the dynamical corrections to CTST through κ.

If we assume that $\kappa \approx 1$ and no change in the partition function between the adsorbed phase and the transition state ($q_{\ddagger}/q_{a} \approx 1$), we obtain $k_{B}T/h = 6.3 \times 10^{12}$ s^{-1} at room temperature whence the assumption that $A \sim 10^{13}\,\text{s}^{-1}$ comes. This is no more than a rough guess. For example, Ibach et al. [34] showed that for CO/Ni(100), $A \approx 6 \times 10^{16}\,\text{s}^{-1}$, and that it *increases* with coverage. Values of $\sim 10^{13}\,\text{s}^{-1}$ can only be expected if no dynamical corrections to CTST occur, the adsorbates are non-interacting, and no change in the vibrational, rotational and translational degrees of freedom occur. This seems highly unlikely and, indeed, for CO adsorption on metal surfaces at low coverage, A varies from $10^{13}-10^{17}\,\text{s}^{-1}$ [28]. Reasonable values of A can range from $10^{11}-10^{19}\,\text{s}^{-1}$. The value of $10^{13}\,\text{s}^{-1}$, however, does serve as a useful benchmark. Assuming for the moment that $\kappa \approx 1$, which is often true for simple atomic and non-dissociative molecular adsorption, values $>10^{13}\,\text{s}^{-1}$ indicate that $q_{\ddagger}/q_{\text{a}} > 1$. This means that the transition state is much "looser" than the adsorbed phase. A loose transition state in this sense means that it has degrees of freedom that are more easily excited by thermal energy than the adsorbed phase. An example of a loose transition state is a localized adsorbate that desorbs through an activated complex that is a two-dimensional gas. If $A < 10^{13}\,\text{s}^{-1}$, the transition state is constrained. A constrained transition state occurs if the molecule must take on a highly specific configuration in the activated complex. As a rule, pre-exponential values in excess of $10^{13}\,\text{s}^{-1}$ are found for non-activated, simple adsorption. Constrained transition states and low values of κ are generally associated with activated adsorption, as we shall see shortly.

4.4.3   Thermodynamic treatment of first-order desorption

Comparing Eqs. (4.4.7), (4.4.8) and (4.4.19) we write the rate constant for desorption as

$$k_{\text{des}} = \kappa \frac{k_{\text{B}}T}{h} K^{\ddagger} \tag{4.4.22}$$

where $K^{\ddagger}$ is the equilibrium constant for formation of the activated complex. Recalling that

$$\Delta G^{\circ} = -RT \ln K \tag{4.4.23}$$

we write

$$k_{\text{des}} = \kappa \frac{k_{\text{B}}T}{h} \exp\left(\frac{-\Delta^{\ddagger}G^{\circ}}{RT}\right). \tag{4.4.24}$$

$\Delta^{\ddagger}G^{\circ}$ is the standard Gibbs energy of activation and, as usual, it can be split into standard enthalpy and entropy of activation as $\Delta^{\ddagger}H^{\circ} - T\Delta^{\ddagger}S^{\circ}$,

$$k_{\text{des}} = \kappa \frac{k_{\text{B}}T}{h} \exp\left(\frac{\Delta^{\ddagger}S^{\circ}}{R}\right) \exp\left(\frac{-\Delta^{\ddagger}H^{\circ}}{RT}\right). \tag{4.4.25}$$

The standard enthalpy of activation is related to a general activation energy E_{a} by

$$E_{\text{a}} = \Delta^{\ddagger}H^{\circ} + RT. \tag{4.4.26}$$

Thus, for desorption we write

$$k_{\text{des}} = \kappa \frac{k_{\text{B}}T}{h} \exp\left(\frac{\Delta^{\ddagger}S^{\circ}}{R}\right) \exp\left(\frac{-(E_{\text{des}} - RT)}{RT}\right) = e\kappa \frac{k_{\text{B}}T}{h} \exp\left(\frac{\Delta^{\ddagger}S^{\circ}}{R}\right) \exp\left(\frac{-E_{\text{des}}}{RT}\right). \tag{4.4.27}$$

To define more precisely what we mean by the activation energy and how it relates to the PES, we turn to Fig. 4.5. First we note, as shown by Fowler and Guggenheim [35], that the activation energy, in this

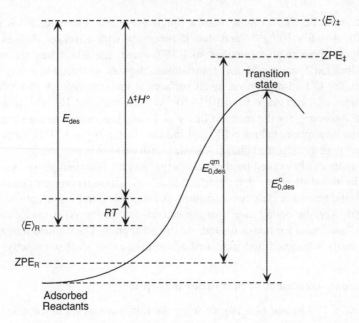

Figure 4.5   *The classical barrier height, $E_{0,\,des}^c$, is the energy difference from the bottom of the well to the top of the activation barrier. The quantum mechanical barrier, $E_{0,\,des}^{qm}$, is a similar difference defined between reactant and transition state zero-point energy levels ZPE_R and $ZPE_{\ddagger}$, respectively. E_{des} is defined as the difference between the mean energy of the reactants $\langle E \rangle_R$ and the mean energy of the molecules in the transition state $\langle E \rangle_{\ddagger}$. The standard enthalpy of activation $\Delta^{\ddagger} H^{\circ}$ is also shown.*

case E_{des}, is given by the difference between the mean energy of the reactants $\langle E \rangle_R$ and the mean energy of the molecules in the transition state $\langle E \rangle_{\ddagger}$

$$E_{des} = \langle E \rangle_R - \langle E \rangle_{\ddagger} \tag{4.4.28}$$

Since both $\langle E \rangle_R$ and $\langle E \rangle_{\ddagger}$ are temperature dependent, E_{des} is, in principle, also temperature dependent. The classical barrier height on the PES is $E_{0,\,des}^c$. E_{des} is not identical to $E_{0,\,des}^c$. As can be seen in Fig. 4.5, the two energies are identical at 0 K. At any other temperature, E_{des} and $E_{0,\,des}^c$ are different, though they have similar values.

To account for this expected temperature dependence, it is useful to introduce a more general mathematical definition of the activation energy of desorption

$$E_{des} = -R \frac{d \ln k}{d(1/T)} = RT^2 \frac{d \ln k}{dT}. \tag{4.4.29}$$

Frequently it is found that Eq. (4.4.29) obeys the form

$$E_{des} = E_{0,\,des}^{qm} + mRT. \tag{4.4.30}$$

Returning now to the Arrhenius formulation of the rate constant, we see that the pre-exponential factor is related to the entropy of activation

$$A = e\kappa \frac{k_B T}{h} \exp\left(\frac{\Delta^{\ddagger} S^{\circ}}{R}\right). \tag{4.4.31}$$

Within this framework and assuming $\kappa \approx 1$, we interpret pre-exponential factors of $\sim 10^{13}\,\text{s}^{-1}$ as being associated with $\Delta^{\ddagger}S^{\circ} = 0$. Larger pre-exponentials have $\Delta^{\ddagger}S^{\circ} > 0$ and smaller values correspond to $\Delta^{\ddagger}S^{\circ} < 0$. A greater entropy is associated with a greater number of accessible configurations for the system. The direct relationship to the partition functions should now be clear.

These results can also be used to gain further insight into compensation effects. Equation (4.4.24) shows us that the rate of desorption depends fundamentally on the standard Gibbs energy of activation. Equations (4.4.26) and (4.4.31) articulate how the activation energy and the pre-exponential factor depend on the standard enthalpy and entropy of activation, respectively. Therefore, A and E_{des} are fundamentally linked to $\Delta^{\ddagger}G^{\circ}$. The compensation effect arises because $\Delta^{\ddagger}H^{\circ}$ and $\Delta^{\ddagger}S^{\circ}$ vary strongly with coverage, but do this in such a way that $\Delta^{\ddagger}G^{\circ}$ is relatively constant in accord with

$$\Delta^{\ddagger}G^{\circ}(\theta) = \Delta^{\ddagger}H^{\circ}(\theta) - T\,\Delta^{\ddagger}S^{\circ}(\theta) \approx \text{constant.} \tag{4.4.32}$$

Those systems which exhibit a compensation effect are those in which either (1) an increase in $\Delta^{\ddagger}H^{\circ}$ is accompanied by an increase in $\Delta^{\ddagger}S^{\circ}$, or (2) a decrease in $\Delta^{\ddagger}H^{\circ}$ is accompanied by a decrease in $\Delta^{\ddagger}S^{\circ}$. Compensation effects are well known for reactions performed in a series of different solvents, or homologous reactions carried out with a series of different substituents introduced into one of the reactants [20]. Laidler has explained the solvent effect by suggesting that stronger binding between a solute molecule and the solvent decreases the enthalpy. This also simultaneously lowers the entropy by restricting the rotational and vibrational freedom of the solvent molecules. Transposing this argument to surfaces, higher adsorption energies tend to decrease the freedom of diffusion as well as frustrated translational and rotational modes. Regardless of whether the range of adsorption enthalpies is caused by surface heterogeneity or lateral interactions, the result is the same: $\Delta^{\ddagger}H^{\circ}$ and $\Delta^{\ddagger}S^{\circ}$ rise and fall in unison. In analogy to reactions in solution, we can think of an adsorbate at high coverage as being "solvated" by its neighbouring adsorbates. Attractive or repulsive lateral interactions increase with the number of neighbours, and the compensation effect observed in desorption can be considered in the same terms as those observed in solution.

4.4.4 Non-first-order desorption

In the absence of lateral interactions and for well-mixed adlayers, there is no ambiguity in using the Polanyi-Wigner equation to describe desorption. It is trivial to write down the rate law expected for simple desorption reactions. For evaporation, or desorption from any phase that has a constant coverage because it is being replenished by another state, zero-order kinetics governs desorption

$$A(\text{l}) \rightarrow A(\text{g}) \quad r_{\text{des}} = k_{\text{des}}. \tag{4.4.33}$$

For simple atomic and non-associative molecular desorption, first-order kinetics is expected.

$$A(\text{a}) \rightarrow A(\text{g}) \quad r_{\text{des}} = \theta\, k_{\text{des}}. \tag{4.4.34}$$

In the case of recombinative desorption such as $H + H \rightarrow H_2$, the bimolecular character of the reaction suggests second-order kinetics

$$A(\text{a}) + A(\text{a}) \rightarrow A_2(\text{g}) \quad r_{\text{des}} = \theta^2 k_{\text{des}}. \tag{4.4.35}$$

With the same caveats as for first-order desorption, a "normal" pre-exponential factor for second-order desorption can be calculated. This is on the order of $10^{-2}\,\text{cm}^2\,\text{s}^{-1}$ but can range from $10^{-4}\text{--}10^{-1}\,\text{cm}^2\,\text{s}^{-1}$.

The presence of lateral interactions not only make A and E_{des} coverage dependent, they can also change the effective reaction order. On Si(100) surfaces, the formation of dimers leads to a (2×1) reconstruction. These dimers are stabilized by what may be considered either a π bond or a Peierls interaction [36]. When a H atom binds to the dimer, this added stabilization is lost, leaving a dangling bond on the other end of the dimer to which the H atom has bonded. When a second H atom attempts to bind to the surface, it can bind either on the same dimer or on a neighbouring dimer that still enjoys its full stabilization. The former option is the lower energy proposition; therefore, H preferentially pairs up on surface dimers [37]. In a series of investigations, D'Evelyn and co-workers have shown that this pairing effect can lead to first-order desorption of H_2 from Si(100) surfaces, as well as for H_2 and HBr from Ge(100) [38–41]. First-order desorption occurs because the H_2 desorbs preferentially from doubly occupied dimers. The desorption rate is linearly dependent on the coverage of doubly occupied dimers. Since the concentration of doubly occupied dimers varies close to linearly with $\theta(H)$, desorption is effectively first-order in $\theta(H)$. Deviations from linearity are predicted at low $\theta(H)$ and Höfer, Li and Heinz [42] have observed such deviations. Consequently, the effective reaction order with respect to $\theta(H)$ of H_2 desorption from Si(100) is coverage dependent.

Reider, Höfer and Heinz [43] have shown that the desorption of H_2 from Si(111) follows non-integer desorption kinetics. For coverages below 0.2 ML, the effective reaction order is 1.4–1.7. This number has no absolute meaning but it demonstrates that desorption involves a more complex mechanism than implied by Eq. (4.4.35). The kinetics of H_2 desorption from Si(111) can be explained by the presence of two types of binding sites A and B, possibly the rest atoms and adatoms, with slightly different ($\sim$0.15 eV) binding energies. Under the assumption that desorption is a bimolecular reaction that only occurs from one of the sites, the desorption kinetics can be accurately modelled. A difference in the binding energies between the rest atom and adatom sites has been confirmed theoretically, which lends more support to this model [44].

The desorption of H_2 from Ag(111) is another illustrative example. Hodgson and co-workers [45] have shown that H can occupy either surface or subsurface sites. Both zero-order and half-order H_2 desorption kinetics are observed. Zero-order kinetics reigns when the desorbing phase has a constant concentration. In this case, the desorbing phase maintains a constant coverage because H diffusing out of the subsurface sites continuously replenishes it. Half-order kinetics is rather unusual. However, if it is assumed that H forms islands on the surface and that desorption occurs only from the perimeter of the islands, half-order kinetics is a direct consequence. This is consistent with LEED measurements on the system. Island formation arises from lateral interactions. This again highlights how lateral interactions can strongly affect the kinetics of surface reactions. The example of H_2/Ag(111) demonstrates that the reaction order only explicitly contains information regarding how the rate of reaction is affected by the supply of reactants, rather than containing detailed dynamical information. A reaction mechanism must be consistent with the measured kinetics. The kinetics deliver insight into the mechanism; however, it cannot unambiguously determine the mechanism. Only for an elementary step is the kinetic order unambiguously and uniquely defined by dynamics. For any given reaction order, there is an unlimited number of corresponding composite reaction mechanisms. For more information on the kinetic effects of lateral interactions, in particular their effect on surface reactions, see the review of Lombardo and Bell [46].

4.5   Kinetics of adsorption

4.5.1   CTST approach to adsorption kinetics

The mechanics used to determine the rate of adsorption are much the same as for desorption. First, we recognize that as a result of microscopic reversibility, there is one and only one transition state for adsorption and desorption. That is, the transition state is the same regardless of the direction from which it is approached. One subtlety in the kinetics of adsorption is that only a limited number of sites on the surface can react. In surface kinetics we must always keep track not only of the number of sites, but also whether these sites are indistinguishable and whether the number of sites occupied by an adsorbate is greater than, equal to, or less than one.

We start with a clean surface and again assume that an equilibrium distribution between the gas phase and the activated complex prevails. Thus

$$\frac{\sigma_{\ddagger}}{c_g\sigma_*} = \frac{q^{\ddagger}}{q_g q_*}\exp\left(\frac{-E_{ads}}{RT}\right) \tag{4.5.1}$$

where $\sigma^{\ddagger}$ and σ_* are the areal densities of the activated complex and empty sites, respectively, c_g is the number density of gas-phase molecules, and the qs are partition functions labelled accordingly. The activation energy is again strictly referenced to absolute zero.

Using the same procedure as for desorption, we identify one vibrational mode to extract from $q^{\ddagger}$, and equate $vc^{\ddagger}$ with the rate of adsorption, which yields

$$r_{ads} = c_g\sigma_*\frac{k_B T}{h}\frac{q_{\ddagger}}{q_g q_*}\exp\left(\frac{-E_{ads}}{RT}\right). \tag{4.5.2}$$

Equation (4.5.2) is valid only on the clean surface. In order to proceed and to characterize the rate of adsorption for increasing coverage, we need to know something about the adsorption dynamics.

4.5.2 Langmuirian adsorption: Non-dissociative adsorption

The name of Irving Langmuir is indelibly linked with surface science, so it should come as no surprise that the most useful starting point for the understanding of adsorption kinetics is the Langmuir model of adsorption [47, 48]. First, let us get an idea of some orders of magnitude. There are roughly 10^{15} surface atoms cm^{-2}. The flux of molecules attempting to stick on these atoms is given by the Hertz-Knudsen equation

$$Z_w = \frac{N_A p}{\sqrt{2\pi MRT}} = \frac{p}{\sqrt{2\pi mk_B T}} \tag{4.5.3}$$

where N_A is Avogadro's number, M is the molar mass (in kg mol^{-1}), m the mass of a particle (kg) and p the pressure (Pa). For $M = 28\,g\,mol^{-1}$ at standard ambient temperature and pressure (SATP $\equiv$ $p = 1\,atm$, $T = 298\,K$), $Z_w = 2.92 \times 10^{23}\,cm^{-2}\,s^{-1}$. That is, about a mole per second of molecules of molar mass 28 (e.g. N_2 and CO) hits the surface per square centimetre at SATP. This value of the impingement rate does not have much of an intuitive feel to it. If, on the other hand, we calculate the impingement rate for $M = 28\,g\,mol^{-1}$, at 1×10^{-6} torr we find $Z_w(T = 298\,K$, $p = 1 \times 10^{-6}$ torr, $M = 28\,g\,mol^{-1}) = 3.84 \times 10^{14}\,cm^{-2}\,s^{-1}$. Since surfaces have roughly 10^{15} atoms cm^{-2}, this exposure is roughly the equivalent of exposing each surface atom to one molecule from the gas phase. If all of these molecules had stuck, the coverage would be roughly 1 ML. Hence a convenient unit of exposure is the Langmuir: $1\,L \equiv 1 \times 10^{-6}$ torr $\times 1\,s = 1 \times 10^{-6}$ torr s. Langmuirs are defined in terms of torr rather than the SI unit Pa because, historically, they were the units of choice. Thus, a rule of thumb is that 1 L exposure leads to $\sim$1 ML coverage *if the sticking coefficient is unity and independent of coverage.*

The next step is to define the behaviour of the sticking coefficient, s, as a function of coverage. In the Langmuir model, we assume $s = 1$ on empty sites and $s = 0$ on filled sites. Therefore, for non-dissociative adsorption, the sticking coefficient is equal to the probability of striking an empty site

$$s = 1 - \sigma/\sigma_0 = 1 - \theta, \tag{4.5.4}$$

and varies with coverage as shown in Fig. 4.6. This behaviour indicates that the sticking coefficient decreases because of simple site blocking rather than because of any chemical or electronic effects. This model does not mean that the adsorbing molecule stick where it hits. This statement is too restrictive. It merely means that the adsorbing molecule *makes the decision* of whether it sticks or not on the first bounce. The observation of Langmuirian adsorption kinetics does not rule out the possibility of transient mobility after the first collision with the surface.

Further assumptions are (1) adsorption stops when all of the sites are full, that is, saturation occurs at 1 ML, (2) the surface is homogeneous thus containing only one type of site, and (3) the adsorbates are non-interacting. Consequently, the Langmuir model can be thought of as describing ideal adsorption – a basis from which we can model all adsorption and the deviations from ideal adsorption. We now have sufficient information to calculate the rate of adsorption and its dependence on pressure, temperature and surface coverage. The rate of change of the surface coverage is simply the impingement rate multiplied by the sticking coefficient

$$\frac{\partial \sigma}{\partial t} = r_{ads} = Z_w s = \frac{p}{\sqrt{2\pi m k_B T}}(1 - \theta) = c_A k_{ads}(1 - \theta). \tag{4.5.5}$$

where c_A is the number density of A in the gas phase, and the rate constant for adsorption $k_{ads} = (k_B T/2\pi m)^{1/2}$.

Several extensions to Langmuirian adsorption, which do not change its essential features, are quite useful. One is that we might want to define the saturation coverage to be θ_{max} rather than strictly one. This allows for absolute comparison between molecules that have different absolute coverages. Second, the initial sticking coefficient need not be exactly unity; thus, we give it the more general symbol s_0. Finally, we allow for the possibility that adsorption may be activated. The generalized adsorption rate is then

$$r_{ads} = Z_w s_0(1 - \theta) \exp\left(\frac{-E_{ads}}{RT}\right) = \frac{p}{\sqrt{2\pi m k_B T}} s_0(1 - \theta) \exp\left(\frac{-E_{ads}}{RT}\right). \tag{4.5.6}$$

The coverage at time t is given by integrating Eq. (4.5.7) (see also Exercises 4.3–4.6)

$$d\sigma = s(\sigma)d\varepsilon = s(\sigma)Z_w dt \tag{4.5.7}$$

where ε is the exposure. The coverage is linearly proportional to the exposure only if the sticking coefficient is constant as a function of coverage, which is often true at very low coverage, for metal on metal adsorption, or condensation onto multilayer films. Exposure, as we have mentioned above, is often expressed in the experimentally convenient units of pressure times time or Langmuirs. Adsorption within this model would lead to a growth in coverage versus exposure as shown in Fig. 4.7.

Extensions to the Langmuir model are essentially provisions for non-ideal behaviour. Values of $s_0 < 1$ indicate the importance of dynamical corrections to the sticking coefficient as discussed in Chapter 3. For instance, not every empty site on the surface might be equally capable of adsorbing molecules. It could also be indicative of steric constraints related to the orientation of the molecule when it strikes the surface. Lateral interactions could lead to island formation. Filled sites might not be totally inert toward sticking. Below we discuss the phenomenon of precursor mediated adsorption, in which sticking occurs even when a filled site is encountered by the incident molecule.

Comparing this result to our CTST result in Eq. (4.5.6) seems to bring little joy. First remember that the CTST result is valid only at $\theta = 0$. Next, assume for the moment that neither the adsorbate nor activated complex is localized. The activated complex is a 2D (atomic) gas, whereas the gas phase is three dimensional. The ratio of the partition functions is $h/(2\pi mk_BT)^{1/2}$, hence

$$r_{ads}^{CTST} = c_g \frac{k_BT}{h} \frac{h}{\sqrt{2\pi mk_BT}} \exp\left(\frac{-E_{ads}}{RT}\right). \tag{4.5.8}$$

Identifying $c_g k_B T$ as the pressure, the final CTST result is

$$r_{ads}^{CTST} = \frac{p}{\sqrt{2\pi mk_BT}} \exp\left(\frac{-E_{ads}}{RT}\right). \tag{4.5.9}$$

At $\theta = 0$, Eq. (4.5.6) reduces to

$$r_{ads} = \frac{p}{\sqrt{2\pi mk_BT}} s_0 \exp\left(\frac{-E_{ads}}{RT}\right). \tag{4.5.10}$$

These results agree exactly except for the factor of s_0. Recalling that the results of CTST need to be corrected for barrier re-crossings by the transmission coefficient, we make the identification

$$s_0 = \kappa \tag{4.5.11}$$

Thus, the study of sticking coefficients and their dependence on various experimental parameters is itself a study of the validity of CTST and its corrections. The implication of Eq. (4.5.11) has been further investigated by Tully and co-workers [29, 49, 50]. As we have seen in Chapter 3, the initial sticking coefficient s_0 is a function of energy (temperature) and not a constant. When we recall that κ and, therefore, s_0 are related to the Arrhenius pre-exponential factor A by Eq. (4.4.31), we see that A has an additional dependence on energy that was not envisioned in the original conception of A. At high temperatures, deviation from Arrhenius behaviour is expected as a consequence. That is, a plot of $\ln k_{des}$ vs $1/T$ yields straight lines at low and moderate temperatures but exhibit curvature to lower values for sufficiently high temperatures. This is essentially a consequence of microscopic reversibility. The dynamical correction to adsorption kinetics (s_0) equivalently applies to desorption kinetics (κ).

4.5.3 Langmuirian adsorption: Dissociative adsorption

With the same set of assumption as defined above, we can also treat dissociative adsorption. Assuming that dissociation requires two adjacent empty sites, the sticking probability varies with coverage as does the probability of finding two adjacent empty sites. This probability is given by $(1 - \theta)^2$ and, therefore,

$$s = s_0(1 - \theta)^2 \exp(-E_{ads}/RT) \tag{4.5.12}$$

Figures 4.6 and 4.7 show how s changes with σ and how σ increases with ε for Langmuirian dissociative adsorption.

Deviations from ideal behaviour are more common and pronounced for dissociative adsorption than for simple adsorption. As discussed in Chapter 3 regarding H_2 dissociation, the orientation of the molecule is often crucial in determining the sticking coefficient. The position in the unit cell where the molecule strikes is also important, that is, sticking is not uniform with regard to the point of impact. Often defects exhibit a much higher sticking coefficient than terrace sites. This can lead to a sticking coefficient that changes suddenly with coverage if the defect sites become decorated with immobile adsorbates. Adsorption can be self-poisoning. In other words, the adsorbate may lead to changes in the electronic structure that

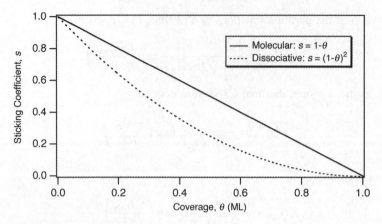

Figure 4.6 *Langmuir models (molecular and dissociative) of the sticking coefficient, s, as a function of coverage, θ.*

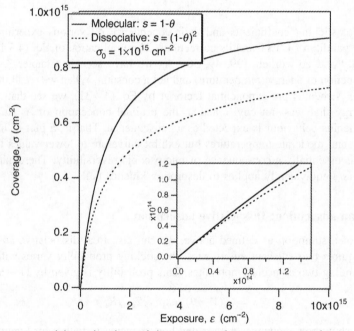

Figure 4.7 *Langmuir models (molecular and dissociative) of coverage, σ, as a function of exposure, ε.*

decrease the sticking coefficient at a more rapid rate than predicted by simple site blocking. The latter is sometimes modelled phenomenologically by assuming that an adsorbed molecule blocks more than one site (see Exercise 4.6), or that the number of sites blocked is coverage dependent. Such models are of limited mechanistic value because the root cause of such effects is generally electronic. Deviations from ideal behaviour are all the more likely for adsorption that is highly activated.

4.5.4 Dissociative Langmuirian adsorption with lateral interactions

Consider now a model in which all of the assumptions of Langmuirian adsorption are kept except that we allow for $s_0 \neq 1$, and pairwise lateral interactions characterized by the interaction energy ω. Dissociative adsorption requires two adjacent empty sites. King and Wells [51] have shown that the number of adjacent empty sites is given by

$$\theta_{OO} = 1 - \theta - \frac{2\theta(1 - \theta)}{[1 - 4\theta(1 - \theta)(1 - \exp(\omega/k_B T_s)]^{1/2} + 1} \tag{4.5.13}$$

The strength of the lateral interactions can be characterized by the term

$$B = 1 - \exp(\omega/k_B T_s). \tag{4.5.14}$$

Introducing

$$s = s_0 \theta_{req} \tag{4.5.15}$$

we define θ_{req} as the coverage dependence of the sites required for adsorption. For Langmuirian dissociative adsorption, $\theta_{req} = \theta_*^2 = (1 - \theta)^2$, whereas in the present model, $\theta_{req} = \theta_{OO}$.

In the limit of weak interactions or sufficiently high temperature to ensure no short-range order in the overlayer, $B = 0$. Substitution into Eqs. (4.5.13) and (4.5.15) yields

$$s = s_0(1 - \theta)^2 \tag{4.5.16}$$

as expected. For large repulsive interactions, $\omega \ll 0$ and $B = 1$. Therefore

$$\begin{aligned} s &= s_0(1 - 2\theta) \quad &\text{for } \theta < 0.5 \\ s &= 0 \quad &\text{for } \theta \geq 0.5 \end{aligned} \tag{4.5.17}$$

This dependence on θ arises from the adsorbates spreading out across the surface in an ordered array in which every other site is occupied. For large attractive interactions $\omega \gg 1$, $B \to \infty$ and, therefore,

$$s = s_0(1 - \theta). \tag{4.5.18}$$

The latter is the same result as for non-dissociative adsorption because the adsorbates coalesce into close-packed islands leaving the remainder of the surface completely bare. Intermediate values of ω result in the behaviour shown in Fig. 4.8.

4.5.5 Precursor mediated adsorption

Taylor and Langmuir [52] observed that the sticking coefficient of Cs on W did not follow Eq. (4.5.4) but rather, it followed the form found in Fig. 4.9. They suggested that adsorption is mediated by a precursor state through which the adsorbing atom passes on its way to the chemisorbed state. Precursor mediated adsorption has subsequently been observed in numerous systems, c.f. Table 3.2. Precursor states can be classified either as intrinsic or extrinsic. An intrinsic precursor state is one associated with the clean surface whereas an extrinsic precursor state is due to the presence of adsorbates. The explanation of how a precursor state manages to keep the sticking coefficient above the value predicted by the Langmuir model is that the incident molecule enters the precursor state and is mobile. Therefore, the adsorbing molecule

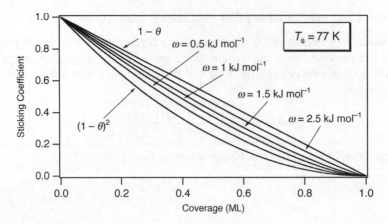

Figure 4.8 *The effect of lateral interactions on the dissociative sticking coefficient as a function of interaction strength, ω, and coverage θ at a fixed surface temperature $T_s = 77\,K$.*

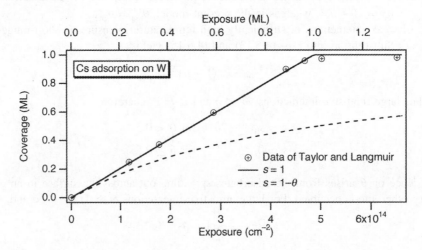

Figure 4.9 *Sticking of Cs on W. Replotted from the data of J. B. Taylor, I. Langmuir, Phys. Rev., 44 (1933) 423.*

can roam around the surface and hunt for an empty site. This greatly enhances the sticking probability because an adsorbed molecule has a certain probability to stick even if it collides with a filled site.

Kisliuk [53] was the first to provide a useful kinetic model of precursor mediated adsorption. The basis of the Kisliuk model is that the total rate of adsorption is viewed as a competitive process as outlined in Fig. 4.10. King and co-workers [23, 51] extended this by including the effects of lateral interactions, and considering the effects on both adsorption and desorption kinetics. Madix [54], Weinberg [55] and co-workers have presented modified Kisliuk models that account for a combination of direct and precursor mediated adsorption as well as an initial direct sticking coefficient that differs from the trapping probability into the precursor state.

The Kisliuk model has an intuitive formulation. Consider non-dissociative molecular adsorption on a finite number of equivalent sites. When a molecule strikes an empty site, it has a probability f_a of becoming adsorbed. Otherwise it has a probability f_m of migrating to a neighbouring site or f_d of desorbing. If a

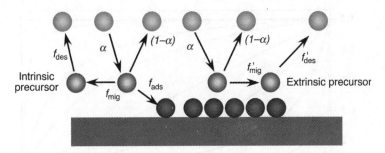

Figure 4.10 *The Kisliuk model of precursor mediated adsorption. Incident molecules trap into intrinsic or extrinsic precursors. Thereafter, sticking becomes a competitive process between desorption out of the precursor and transfer into the stable chemisorbed state. α is the probability to enter into the precursor state.*

molecule strikes an occupied site, it cannot chemisorb but it can migrate or desorb with probabilities, f'_m and f'_d, respectively. King introduced a trapping coefficient α that describes the probability of entering the precursor state. In our treatment α is independent of coverage and is the same for intrinsic and extrinsic precursors.

The incident molecule hops across the surface and makes a decision at each hop of whether to chemisorb, migrate or desorb. By summing up the probability over all possible hops, the sticking coefficient is calculated.

$$s = \alpha \left(1 + \frac{f_d}{f_a} \right)^{-1} \left[1 + K \left(\frac{1}{\theta_{req}} - 1 \right) \right]^{-1}, \tag{4.5.19}$$

where

$$K = \frac{f'_d}{f_a + f_d} \tag{4.5.20}$$

The initial sticking coefficient is

$$s_0 = \alpha (1 + f_d/f_a)^{-1}, \tag{4.5.21}$$

and

$$f_d/f_a = r_d/r_a, \tag{4.5.22}$$

where r_{des} and r_{ads} are the desorption rate and adsorption rate, respectively, from the precursor state. In other words, s_0 is determined by α and the competition between adsorption and desorption from the precursor state. As both adsorption and desorption are activated processes, we expect that sticking through a precursor state should have a temperature dependence. If $E_{des} > E_a$, increasing the surface temperature decreases the sticking coefficient. However if $E_{des} < E_a$, increasing T_s favours sticking.

The ratio of the s to s_0 is

$$\frac{s(\theta)}{s_0} = \left[1 + K \left(\frac{1}{\theta_{req}} - 1 \right) \right]^{-1} \tag{4.5.23}$$

As written, Eq. (4.5.23) can be used to describe either dissociative or non-dissociative adsorption so long as the correct form of θ_{req} is inserted. If a plot of $s(\theta)/s_0$ can be fitted by a unique value of K, the

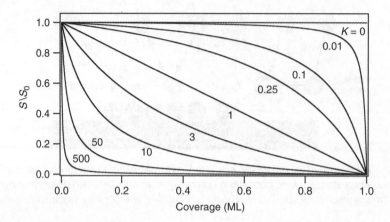

Figure 4.11   *The change of sticking coefficient, s, with coverage for precursor mediated adsorption is characterized by the parameter K. For K = 0, the sticking coefficient is constant, whereas for K = 1 it drops linearly with coverage as in Langmuirian adsorption. Large values of K decrease s relative to Langmuirian adsorption.*

precursor model describes the sticking behaviour. Determination of K can then provide some idea of the relative rates of sticking onto bare sites compared to the desorption rates from the intrinsic and extrinsic precursor states. It can also be used to measure the effects of lateral interactions in dissociative adsorption. Figure 4.11 shows how s depends on K. In addition, it demonstrates that a precursor can actually decrease sticking compared to Langmuirian adsorption if desorption out of the precursor is rapid.

4.6   Adsorption isotherms from kinetics

In §4.2 we introduced the adsorption isotherm, that is, an equation that describes the equilibrium coverage at constant temperature. There are various named isotherms that depend on the assumptions made to describe the rates of adsorption and desorption [6]. The basis of all isotherms is that we consider a system at equilibrium. Therefore, $T_s = T_g = T$, $r_{des} = r_{ads}$ and the coverage is constant, $d\theta/dt = 0$.

4.6.1   Langmuir isotherm

Using the set of assumptions posited in the Langmuir model of adsorption, we can derive the Langmuir isotherm. Equating the rates of adsorption, Eq. (4.5.5), and desorption, Eq. (4.4.2), yields

$$pk_{ads}(1 - \theta) = \theta \, k_{des}. \tag{4.6.1}$$

Both sides of Eq. (4.6.1) are written such that the rate is expressed in molecules per unit area per second, thus,

$$k_{ads} = (2\pi mk_B T)^{-1/2} e^{-E_{ads}/RT} \tag{4.6.2}$$

and

$$k_{des} = \sigma_0 A e^{-E_{des}/RT} \tag{4.6.3}$$

Note that the adsorption rate constant has been written so that it can be used with pressure rather than number density, as was done in Eq. (4.5.5). The exponential term allows us to account for the possibility of activated adsorption. Rearranging Eq. (4.6.1) yields

$$\frac{\theta}{1-\theta} = p\frac{k_{ads}}{k_{des}} = pK.$$ (4.6.4)

$K = k_{ads}/k_{des}$ is effectively an equilibrium constant. Purists will worry about it having units and how the standard state of the adsorbed layer is defined [1]; thus, it is more properly called the Langmuir isotherm constant. Rearranging we obtain the Langmuir isotherm describing the equilibrium coverage as a function of pressure

$$\theta = \frac{pK}{1+pK}.$$ (4.6.5)

Alternatively, concentration can be used in place of pressure, such that this expression can also be used to describe adsorption from any fluid (gas, liquid or solution) phase onto a solid substrate as long as the rate constants in Eqs (4.6.2) and (4.6.3) are written with the proper units.

Multiplying both sides of Eq. (4.6.1) by σ_0 and recalling that $\sigma = \theta\sigma_0$, we can derive (see Exercise 4.29) that

$$\sigma = (1 - \sigma/\sigma_0)pK\sigma_0 = (1 - \sigma/\sigma_0)pb$$ (4.6.6)

where

$$b = (2\pi m k_B T)^{-1/2}A^{-1}e^{-(E_{ads}-E_{des})/RT}.$$ (4.6.7)

The difference between the desorption and adsorption activation energies is equal in magnitude to the enthalpy of adsorption. Equation (4.6.6) can be rearranged to

$$\frac{1}{\sigma} = \frac{1}{bp} + \frac{1}{\sigma_0}.$$ (4.6.8)

A plot of $1/\sigma$ vs $1/p$ can be used to determine σ_0 and b. Since b is determined by the temperature and enthalpy of adsorption, we see that both of these quantities, as well as the pressure, are responsible for determining the equilibrium coverage σ. Note that in the limit of very low coverage, $\sigma \ll \sigma_0$ and $1/\sigma - 1/\sigma_0 \approx 1/\sigma$, hence,

$$\sigma = bp.$$ (4.6.9)

In other words, the coverage is linearly related to the pressure in the limit of low coverage. This relationship is favoured at high temperature, low pressure, and for weakly bound systems for which the equilibrium coverage is naturally small.

4.6.2 Classification of adsorption isotherms

The majority of adsorption isotherms can be grouped into one of six types [56] shown in Fig. 4.12. The measurement of adsorption isotherms is particularly important for the characterization of porous solids, in which case the physisorption of an inert gas is used. The Langmuir isotherm shown in Fig. 4.4 is a reversible Type I isotherm. A reversible isotherm exhibits no hysteresis when measured with increasing or decreasing pressure. The IUPAC recommends against calling this a Langmuir isotherm, and that recommendation is almost universally rejected in the literature. This isotherm reaches a limiting value (the monolayer coverage) as p/p° approaches 1. The low coverage linear region of this isotherm is called the Henry's law region. Type I isotherms are followed by microporous solids having relatively small external surfaces (e.g. activated

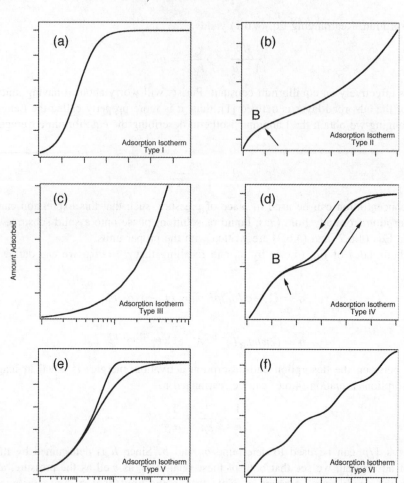

Figure 4.12　*The six types of physisorption isotherms in which coverage (either relative coverage θ or absolute coverage σ) is plotted against pressure, p, of a gaseous adsorbing species (or concentration, c, of a species dissolved in a liquid solution). The coverage is often expressed as specific coverage, that is, coverage per unit mass of the substrate.*

carbons, molecular sieve zeolites and certain porous oxides), for which the limiting uptake is governed by the accessible micropore volume rather than by the internal surface area. This is also the most commonly observed isotherm for chemisorption since chemisorption is strictly limited to the monolayer regime.

The reversible Type II isotherm is the normal form of isotherm obtained with a non-porous or macro-porous adsorbent. The Type II isotherm represents monolayer adsorption that is followed by unlimited multilayer condensation. Point B at the onset of the almost linear middle section of the isotherm is often taken to indicate the transition from the completion of the first monolayer to the formation of the first multilayer.

The reversible Type III isotherm is convex to the $p/p°$ axis over its entire range. Isotherms of this type are not common. They are observed when lateral interactions play an important role in determining the enthalpy of adsorption.

The Type IV isotherm is irreversible and exhibits a hysteresis loop, which is associated with capillary condensation taking place in mesopores. This isotherm also is characterized by limiting uptake over a range of high $p/p°$. The initial part of the Type IV isotherm is much like a Type II isotherm, and is attributed to monolayer–multilayer adsorption. Type IV isotherms are found for many mesoporous industrial adsorbents.

The Type V isotherm is uncommon. It is related to the Type III isotherm in that the adsorbent–adsorbate interaction is weak, but is obtained with certain porous adsorbents.

The Type VI isotherm represents stepwise multilayer adsorption on a uniform non-porous surface. The step height, the sharpness of which depends on the system and the temperature, represents the monolayer capacity for each physisorbed layer and, in the simplest case, remains nearly constant for two or three adsorbed layers. Amongst the best examples of Type VI isotherms are those obtained with Ar or Kr on graphitized carbon blacks at liquid nitrogen temperature.

4.6.3   Thermodynamic measurements via isotherms

The van't Hoff equation relates the equilibrium constant to the isosteric heat of adsorption

$$\left(\frac{\partial \ln K}{\partial T}\right)_\theta = \frac{-q_{st}}{RT^2}. \tag{4.6.10}$$

Substituting from Eq. (4.6.3) for K and using $\partial(1/T)\partial T = -1/T^2$ yields

$$\left(\frac{\partial \ln p}{\partial(1/T)}\right)_\theta = -\frac{q_{st}}{R}. \tag{4.6.11}$$

This is the same result as Eq. (4.1.15), which was derived from the Clausius-Clapeyron equation. Thus, a plot of $\ln p$ at constant coverage vs $1/T$ yields a line of slope $-q_{st}/R$. A series of such measurements as a function of θ leads to the functional form of the dependence of q_{st} on θ. Should the plot of $\ln p$ versus $1/T$ not yield a straight line, then one of the assumptions of the Langmuir model must be in error. One can try, for instance, to fit the measured curves by introducing modified sticking coefficients. Plots of q_{st} versus θ can be used to investigate lateral interactions by determining whether models such as the Bragg-Williams or quasi-chemical approximation can accurately predict the plots.

4.7   Temperature programmed desorption (TPD)

4.7.1   The basis of TPD

Temperature programmed desorption [17, 57–60] is a conceptually straightforward technique as illustrated in Fig. 4.13. A surface is exposed to a gas. Exposure can be performed by several means. The most uniform method is to backfill the chamber with the desired gas to some predetermined pressure for a measured length of time. Alternatively, a tube may be brought close to the sample and the gas is allowed to flow through it. This has the advantage of exposing the sample to relatively more gas than the rest of the chamber, but the disadvantage of supplying a highly non-uniform gas flux, which can lead to non-uniform coverage across the surface. An array doser [61, 62] attached to the end of the tube greatly improves the uniformity. In special cases, a molecular beam may be used to dose the surface. The latter is particularly relevant to dynamical studies in which a molecular beam is used so that the energetic characteristics of the molecules in the molecular beam can be systematically varied. The compendium of Yates [63] should be consulted for an explanation of a wide range of experimental techniques that can be used to improve the performance of TPD. Figure 4.13 demonstrates one of the most important methods of improving data

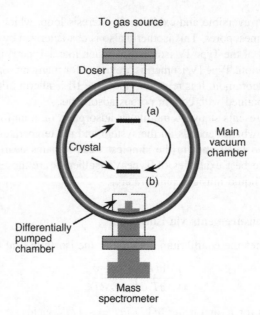

To gas source

Doser

(a)

Crystal

Main
vacuum
chamber

(b)

Differentially
pumped
chamber

Mass
spectrometer

Figure 4.13 *Temperature programmed desorption. The procedure begins by dosing the crystal at position (a). The crystal is then moved to (b), where mass-selective detection is performed with a mass spectrometer.*

quality in that the mass spectrometer is placed behind an aperture in a differentially pumped chamber. This reduces the background pressure in the detector and ensures that the detected species desorb directly from the sample rather than from the sample holder, chamber walls or any other surfaces. Automation of the technique [64] can further improve data quality by facilitating the averaging of many spectra.

After exposure, the sample is rotated to face the detector. Early flash desorption experiments, based on the method of Langmuir and Taylor [52] and revived by Ehrlich [65], merely detected the pressure rise caused by rapid increase in the sample temperature. Redhead [66] slowed down the heating rate, and showed how a thermal desorption spectrum could be used to determine surface kinetics. A quadrupole mass spectrometer (QMS) is employed in the modern technique. A QMS has the advantage of mass-selectively measuring the pressure rise. Therefore, it determines whether multiple products are formed, and discriminates against background gases.

Consider the situation in Fig. 4.13. Molecules desorb from the walls at a rate L while they are removed from the chamber of volume V by a vacuum pump operating at a pumping speed S. The vacuum chamber contains a gas of density c_g and molecules desorb from the sample of surface area A_s with a time dependent rate $A_s r_{ads}(t)$. The time dependence is important because a temperature ramp, usually linear in time, is applied to the sample according to

$$T_s = T_0 + \beta t \tag{4.7.1}$$

where T_s is the sample temperature, T_0 is the base temperature (the temperature at $t = 0$) and β is the heating rate. The change in the number of gas-phase molecules is given by

$$V \frac{dc_g}{dt} = A_s r_{des}(t) + L - c_g S \tag{4.7.2}$$

Assume that L and S are constant, that is, that the sample holder does not warm up leading to spurious desorption, and that the pressure rise that accompanies desorption is not large enough to cause displacement of molecules from the walls, or to affect the pumping speed. At the base temperature, which is low enough to ensure no desorption from the sample, the gas phase attains a steady state composition given by

$$V \frac{dc_g}{dt} = L - c_g S = 0 \qquad (4.7.3)$$

The steady-state solution is

$$c_{gs} = L/S \qquad (4.7.4)$$

which corresponds to a steady-state pressure of

$$p_s = k_B T_g c_g = k_B T_g L/S. \qquad (4.7.5)$$

The pressure change caused by desorption is

$$\Delta p = p - p_s. \qquad (4.7.6)$$

The relationship between the pressure change and the rate of desorption can be written

$$V \frac{d\Delta p}{dt} + S\Delta p = kT_g A_s r_{des}(t) \qquad (4.7.7)$$

In the limit of large pumping speed, the first term is negligible and

$$\Delta p = \frac{k_B T_g A_s}{S} r_{des}(t). \qquad (4.7.8)$$

In other words, the measured pressure change is directly proportional to desorption rate, and the position and shape of the desorption peak contains information about the kinetics parameters, including lateral interactions, that affect the rate.

4.7.2   Qualitative analysis of TPD spectra

Figure 4.14 displays thermal desorption spectra measured for cyclopentene adsorbed on Ag(221) [67]. These desorption spectra contain several features. Four different peaks, α_1–α_4, appear in the spectra with relative sizes that depend on the amount of cyclopentene exposed to the surface. The α_1 and α_2 peaks have a constant peak temperature while the α_3 and α_4 peaks shift with increasing coverage. The lowest temperature peaks do not exhibit saturation while the two highest temperature peaks do. Two obvious parameters that can be used to characterize the peaks are the peak area, A_p, and the temperature measured at the peak maximum, T_p (the peak temperature).

Integrating a desorption spectrum over all time

$$A_p \propto \frac{S}{A_s k_B T_g} \int_0^\infty \Delta p \, dt = \int_0^\infty r_{des}(t) \, dt = \sigma \qquad (4.7.9)$$

we see that the peak area (or sum of peak areas for a multiple peak spectrum) is directly proportional to the adsorbate coverage. Furthermore, if we integrate from $t = 0$ up to some intermediate time t_i, the included peak area is proportional to the amount desorbed. This allows us to determine the coverage relative to the total initial coverage at every point along the desorption curve. If an absolute coverage measurement

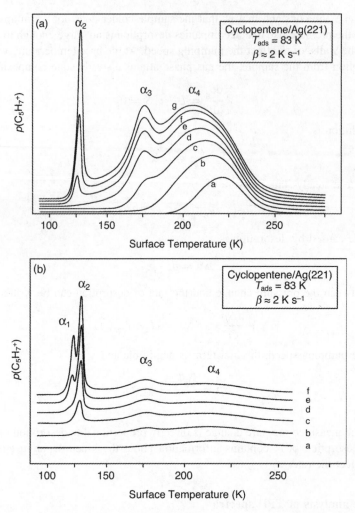

Figure 4.14   *Temperature programmed desorption (TPD) spectra for cyclopentene/Ag(221) at (a) low coverage and (b) high coverage. In parts (a) and (b), adsorption was carried out at a surface temperature of $T_{ads} = 83\,K$ and the heating rate was $2\,K\,s^{-1}$. T_{ads} is used to denote the surface temperature at the time of adsorption as distinguished from the surface temperature during the TPD experiment. The exposures (in units of $10^{14}\,cm^{-2}$) to cyclopentene in part (a) are: spectrum a, 1.2; spectrum b, 2.5; spectrum c, 3.1; spectrum d, 3.7; spectrum e, 4.9; spectrum f, 6.2; spectrum g, 6.3. The cyclopentene exposures in part (b) are: spectrum a, 7.4; spectrum b, 8.6; spectrum c, 9.3; spectrum d, 12.5; spectrum e, 15.6; spectrum f, 18.8. $\alpha_1 - \alpha_4$ refer to desorption states. Reproduced from M. D. Alvey, K. W. Kolasinski, J. T. Yates, Jr., M. Head-Gordon, J. Chem. Phys., 85 (1986) 6093. © (1986) with permission from the American Institute of Physics.*

is available for the initial coverage, for instance by XPS, the desorption curve can be used to calculate the absolute coverage at every point along the curve. Note, however, that for a multiple peak spectrum resulting from multiple adsorption sites, the area under each peak is not necessarily equal to initial coverage of the different adsorption sites. Interconversion between the sites can occur during the acquisition of the spectrum, as occurs, for example, in the H/Si system when di- and mono-hydride sites are both occupied.

Temperature programmed desorption is the beauty and beast of surface kinetics. Its beauty lies in its simplicity. TPD rapidly leads to an approximate picture of what is going on in surface kinetics. The beast lies in the accurate and unambiguous interpretation of the data. The sensitivity of TPD is generally on the order of 0.01 ML unless special measures are taken which can take this limit down by up to two orders of magnitude. As long as the pumping speed of the chamber in which desorption measurements are performed is sufficiently high, the desorption rate is directly proportional to the pressure rise in the chamber. In most modern surface science chambers this condition is easily met. However, under special situations it may still be of concern. For instance, the pumping speed of H_2 is comparatively low in turbomolecular pumped systems. Additionally, if the conductance between the sample and the pressure measurement device is constricted in some way or if spurious desorption from the walls or sample holder occur, interferences arise. Since these experimental concerns can be solved, TPD or its cousin, temperature programmed reaction spectrometry (TPRS), remains the technique of choice for initial investigations of surface kinetics. In one quick experiment, the identity of reaction products and the temperatures at which they appear can be determined.

Returning to the Polanyi-Wigner formulation of reaction kinetics, Eq. (4.4.1), we see that a direct measurement of reaction rate can lead to information on the kinetic parameters governing the reaction. Substituting a linear temperature ramp into Eq. (4.4.1), and assuming that ν_n and E_{des} are *independent of coverage*, yields

$$\frac{E_d}{RT_p^2} = \frac{n\nu^n}{\beta}\theta_p^{n-1}\exp\left(\frac{-E_d}{RT_p}\right). \tag{4.7.10}$$

For first-order desorption

$$\frac{E_d}{RT_p^2} = \frac{\nu}{\beta}\exp\left(\frac{-E_d}{RT_p}\right) \tag{4.7.11}$$

while for second-order desorption

$$\frac{E_d}{RT_p^2} = \frac{2\nu^2}{\beta}\theta_p\exp\left(\frac{-E_d}{RT_p}\right) \tag{4.7.12}$$

where θ_p is the coverage at T_p. Equations (4.7.11) and (4.7.12) show that T_p is independent of initial coverage, θ_i, for first-order desorption, and T_p shifts to lower temperatures as θ_i increases for second-order desorption. An increasing T_p with increasing coverage is indicative of an order $0 < n < 1$. Zero-order desorption strictly does not lead to a peak. Integration of the Polanyi-Wigner equation shows that

$$\theta_p = \theta_i/e \quad (n = 1) \tag{4.7.13}$$

and

$$\theta_p^{n-1} = n^{-1}\theta_0^{n-1} \quad (n > 1) \tag{4.7.14}$$

The peak shape can also be used to analyze desorption. Zero-order desorption from finite coverages leads to a series of peaks that all share the same leading edge. Eventually as the coverage drops during the desorption experiment, the desorption order changes to $n > 0$ and, therefore, a peak is observed. First-order desorption leads to asymmetric peaks. Second-order desorption leads to symmetric peaks. All of these generalization apply only if A and E_{des} are coverage independent.

Returning to the desorption spectra in Fig. 4.14, we can make several qualitative conclusions about the adsorption of cyclopentene on Ag(221). The large peaks at low temperature do not saturate, and the leading edges of the peaks as a function of increasing coverage overlap. These are clear indications of evaporation from a physisorbed layer. The occurrence of two peaks in the low temperature range, well-separated from the higher temperature peaks due to chemisorption, are indicative of a change in binding energy between the first (few) physisorbed layer(s) on top of the chemisorbed layer compared to the physisorbed layers on top of the first physisorbed layer. This frequently occurs. Menzel and co-workers [68] have made high-resolution TPD studies of the desorption of rare gases. They have shown that the desorption peaks from the first through fourth layers for Ar physisorbed on Ru(001) can be resolved. The binding energy eventually converges on the sublimation energy of the bulk material, and indicates a decreasing influence of the metal substrate on the binding of the physisorbate as the separation from the surface increases.

The two higher temperature peaks in Fig. 4.14 are due to the chemisorbed layer. Since the adsorption of cyclopentene is molecular, desorption should be first-order. However, the α_4 peak shifts to lower energy even before the α_3 peak appears. This is an indication that strong lateral interactions are present because a first-order desorption peak should not shift with increasing coverage. As shown by Adams [16] sufficiently strong lateral interactions can lead not only to shifting and broadening of TPD peaks, but also to the formation of separate peaks although only one binding state is occupied. Proof of the influence of lateral interactions for cyclopentene on Ag(221) would require an in-depth analysis [17] that is beyond the scope of this discussion, but it provides a likely explanation for the appearance of two peaks in the chemisorption region even though only one binding site is occupied. The review of Lombardo and Bell [46] should be consulted for more details on techniques for simulating TPD spectra and the effects of lateral interactions.

4.7.3 Quantitative analysis of TPD spectra

Figure 4.15 exhibits a variety of simulated TPD spectra which demonstrate how the shape and size of the desorption peak varies with reaction order, coverage, E_{des} and A. Note that independent of reaction order, the desorption rate at T_p from a full monolayer is on the order of $0.1 \, ML \, s^{-1}$. This rule of thumb works for a variety of adsorbates. Strongly bound adsorbates have large A factors and weakly bound ones have small A factors. At the end of the day, nature likes to desorb a maximum of $\sim 0.1 \, ML \, s^{-1}$ from a full layer for normal heating rates.

Redhead [66] showed that the activation energy for desorption for first-order desorption is related to the peak temperature by

$$E_{des} = RT_p(\ln(AT_p/\beta) - 3.46). \tag{4.7.15}$$

Equation (4.7.15) is often used with an assumed value of A. However, it should be used as no more than a rule of thumb. It is a good first approximation to obtain a feeling for the binding energies, but a final value can only be arrived at by a full quantitative analysis. The only circumstances under which Eq. (4.7.15) should be applied are when A is known from experiment and for peaks that are well resolved showing no indication for lateral interactions.

Numerous methods have been developed to analyze TPD spectra and these have been critically reviewed by de Jong and Niemantsverdriet [69] and Yates and co-workers [70]. The relatively quick technique of Falconer and Madix [71] provides a useful approximation to the kinetic parameters. Consider a desorption rate written as a general function of the (non-negative order) expression of the rate dependence on coverage, $g(\theta)$. This yields

$$r_{des} = \nu \, g(\theta) \, \exp(-E_{des}/RT_s). \tag{4.7.16}$$

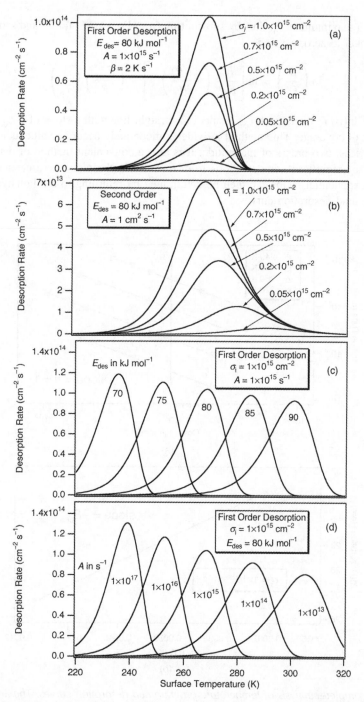

Figure 4.15 *Simulated thermal desorption spectra. In all cases the desorption activation energy, E_{des}, and pre-exponential factor, A, are independent of coverage. (a) First-order desorption for various coverages σ (note the asymmetric peak with constant T_p). (b) Second-order desorption for various coverages (note the symmetric peak with T_p that decreases with increasing coverage). (c) Effect of change in E_{des} in steps of 5 kJ mol^{-1}. (d) Effect of change in A in steps of 10^1. In parts (c) and (d) similar shifts in peak position are observed for first-order desorption in this temperature and coverage range. β, heating rate; σ_i, initial coverage.*

The rate reaches a maximum at T_p, thus by substituting from Eq. (4.7.1) for T, and setting the derivative with respect to t equal to zero, we find

$$\ln\left(\frac{\beta}{T_p^2}\right) = \ln\left[\frac{R\nu}{E_{des}}\left(\frac{dg(\theta)}{d\theta}\right)_{T_p}\right] - \left(\frac{E_{des}}{R}\right)\left(\frac{1}{T_p}\right) \tag{4.7.17}$$

Therefore, a plot of $\ln(\beta/T_p^2)$ versus $(1/T_p)$ results in a straight line with a slope of $-E_{des}/R$ provided that E_{des} is independent of coverage. This method provides a moderately accurate value as long as the heating rate is varied by at least two orders of magnitude. It is also a convenient method of determining whether lateral interactions are significant, as deviations from linearity in the plots are a clear indication of their presence. The pre-exponential factor can then be determined by returning to the Polanyi-Wigner equation and fitting the measured desorption curves.

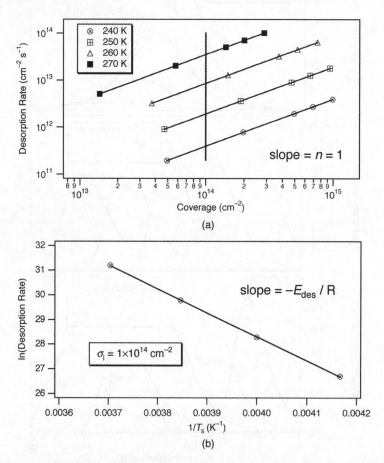

Figure 4.16 *The complete analysis of temperature programmed desorption curves. The analysis is performed for the spectra in Fig. 4.15(a) at the arbitrarily chosen coverage of $\sigma = 1 \times 10^{14}$ cm^{-2}. (a) Desorption rate plotted against coverage for various surface temperatures, T_s. (b) ln (desorption rate) plotted against the reciprocal of T_s at $\sigma = 1 \times 10^{14}$ cm^{-2}. The straight lines in (a) and (b) confirm that desorption is first order and that the kinetic parameters are coverage independent, respectively.*

The only fireproof method of determining the kinetics parameters is the complete analysis of TPD spectra. This method is involved but accurate. A family of desorption curves is measured as a function of initial coverage, θ_i or σ_i depending on whether an absolute coverage is known, as shown in Fig. 4.15. These are used to construct a family of $\sigma(t)$ curves via Eq. (4.7.9). Because of the known (linear) heating rate, this corresponds to a knowledge of $\sigma(T_s)$ as a function of σ_i. Then an arbitrary value of coverage, σ_1, is chosen that is contained in each of the desorption curves. The desorption rate at this coverage, $r_{ads}(\sigma_1)$, and the temperature at which this rate was obtained, T_1, are then read off from each of these desorption curves. As shown in Fig. 4.16, a plot of $\ln(r_{ads}(\sigma_1))$ versus $1/T_1$ is used to determine $E_{des}(\sigma_1)$ and $A(\sigma_1)$. This follows directly from the Polanyi-Wigner equation, restated as

$$\ln(r_{des}(\sigma)) = \ln(\nu_n(\sigma)\sigma^n) - \frac{E_{des}(\sigma)}{R}\frac{1}{T_s}. \qquad (4.7.18)$$

Equation (4.7.18) shows that the slope unambiguously determines $E_{des}(\sigma)$. By choosing a set of coverage values, the functional form of $E_{des}(\sigma)$ vs coverage can be determined. The pre-exponential factor and the reaction order are determined from the intercept. This can lead to ambiguity in the values of $A(\sigma)$ and n if only a relative rather than absolute coverage is known. Likewise the functional form of $A(\sigma)$ is determined by a series of intercepts determined for different coverages.

4.8   Summary of important concepts

- Whereas adsorption can be activated or non-activated, desorption is always activated. Correspondingly, adsorption is always exothermic.
- The heat of adsorption per molecule generally depends on the number of adsorbates. Changes in the heat of adsorption are associated with adsorption at different types of site, interaction between adsorbates (lateral interactions) and/or changes in electronic structure associated with adsorption.
- The equilibrium coverage is determined by the heat of adsorption, the temperature and the pressure in the gas phase. (adsorption isotherms)
- The compensation effect yields desorption rate constants that change little with coverage due to counterbalancing changes in the enthalpy and entropy of activation.
- The kinetics of adsorption and desorption share many aspects (Polanyi-Wigner and Arrhenius formulations) in common with kinetics in other phases. However, on surfaces there are a limited number of adsorption sites and, therefore, the concentration terms that are used in surface kinetics differ fundamentally from those used in other phases.
- Non-integer orders in desorption kinetics demonstrate that desorption is more complicated than an elementary reaction. More complex reaction orders can arise from strong lateral interactions, multiple binding sites and/or non-random adsorbate distributions.
- The kinetics of adsorption can only be calculated if a model is assumed to describe how the sticking coefficient changes with coverage.
- Within the Langmuir model of adsorption, adsorption only occurs if the incident molecule strikes an empty site, adsorption saturates at 1 ML and adsorption is random. The sticking coefficient then drops as $(1 - \theta)$ for molecular adsorption and $(1 - \theta)^2$ for dissociative adsorption requiring two adjacent empty sites.
- In precursor-mediated adsorption, the sticking coefficient does not follow $(1 - \theta)$. Adsorption occurs via a competitive process. Molecules enter a mobile state from which desorption competes with the search for an empty site.

- Adsorption isotherms can be derived by equating the rates of adsorption and desorption at equilibrium.
- Kinetic and thermodynamic parameters can be determined by the measurement of isotherms and TPD spectra.

4.9 Frontiers and challenges

- The thermodynamics of nanoparticles. Much of the thermodynamics of surfaces (more on this in Chapter 5) is couched in black box terms in which the dividing surface can be variously defined and interpreted. When the size of the particle becomes comparable to the uncertainty in how the interface is defined then the difficulty of the problem takes on new dimensions. For instance, what is the temperature of one nanoparticle?
- Microcalorimetry on single crystal surfaces for adsorption and reactions.
- Description of coverage effects on the heat of adsorption.
- Characterization and description of lateral interactions.

4.10 Further reading

M. Boudart and G. Djéga-Mariadassou, *Kinetics of Heterogeneous Catalytic Reactions* (Princeton University Press, Princeton, NJ, 1984).

M. C. Desjonquères and D. Spanjaard, *Concepts in Surface Physics*, 2nd ed. (Springer-Verlag, Berlin, 2002).

D. A. King, *Thermal desorption from metal surface: A review*, Surf. Sci. **47** (1975) 384.

D. A. King and D. P. Woodruff (Eds), *The Chemical Physics of Solid Surfaces and Heterogeneous Catalysis: Adsorption at Solid Surfaces*, Vol. 2 (Elsevier, Amsterdam, 1983).

K. J. Laidler, Chemical Kinetics, 3rd ed. (HarperCollins, New York, 1987).

T. E. Madey and J. T. Yates, Jr., *Desorption methods as probes of kinetics and bonding at surfaces*, Surf. Sci. **63** (1977) 203.

D. Menzel, Thermal desorption, in *Chemistry and Physics of Solid Surfaces IV*, Vol. 20, edited by R. Vanselow and R. Howe (Springer-Verlag, New York, 1982) p. 389.

M. W. Roberts and C. S. McKee, *Chemistry of the Metal-Gas Interface* (Clarendon Press, Oxford, 1978).

J. T. Yates, Jr., Thermal desorption of adsorbed species, in *Solid State Physics: Surfaces. Methods of Experimental Physics*, Vol. 22, edited by R. L. Park and M. G. Lagally (Academic Press, New York, 1985) p. 425.

4.11 Exercises

4.1 The canonical partition function appropriate for the Langmuir model is [2]

$$Q_{\text{ads}} = C_{N_0}^{N_{\text{ads}}} \exp\left(\frac{N_{\text{ads}}\varepsilon}{k_B T}\right) \tag{4.11.1}$$

and the grand canonical partition function is

$$\Xi = \sum_{N_{\text{ads}}}^{N_0} Q_{\text{ads}} \exp\left(\frac{N_{\text{ads}}\mu_{\text{ads}}}{k_B T}\right) \tag{4.11.2}$$

where $C_{N_0}^{N_{\text{ads}}}$ is the number of configurations and μ_{ads} is the chemical potential of the adsorbate phase. Derive Eq. (4.2.1) by calculating

$$\theta = \langle N_{\text{ads}} \rangle / N_0 \tag{4.11.3}$$

where $\langle N_{\text{ads}} \rangle$ is the average number of adsorbed atoms.

4.2 (a) If the sticking coefficient of NO is 0.85 on Pt(111) at 250 K, what is the initial rate of adsorption of NO on to Pt(111) at 250 K if the NO pressure is 1.00×10^{-6} Torr? (b) If the dissociative sticking coefficient of O_2 is constant at a value of 0.25, how many oxygen atoms will be adsorbed on a Pd(111) surface after it has been exposed to 1.00×10^{-5} Pa of O_2 for 1 s. The system is held at 100 K.

4.3 Consider non-dissociative adsorption. Assuming both adsorption and desorption are first-order processes, write down an expression for the coverage of an adsorbate as a function of time. The system is at a temperature T, there is only one component in the gas phase above the surface at pressure p and the saturation surface coverage is given by θ_{max}. Assume adsorption is non-activated and that the sticking is direct.

4.4 Consider sticking that requires an ensemble of two adjacent sites as is often true in dissociative adsorption. Assuming that there is no desorption, write down an expression for the coverage of an adsorbate as a function of time. The system is at a temperature T, there is only one component in the gas phase above the surface at pressure p and the saturation surface coverage is given by θ_{max}. Assume adsorption is non-activated and that the sticking is direct.

4.5 Consider sticking that requires an ensemble of three adjacent sites. Assuming that there is no desorption, write down an expression for the coverage of an adsorbate as a function of time. Compare the results to what is expected for adsorption that requires only one site or two sites for the case of methane sticking on a Pt(111) surface. Take $N_0 = 1.503 \times 10^{15}$ cm^{-2}, $\theta_{max} = 0.25$ ML, $T = 800$ K and the initial sticking coefficient is 2.5×10^{-7}. You will have to consider exposures up to 1×10^{22} cm^{-2}.

4.6 Consider sticking that requires an ensemble of only one surface atom but that poisons n sites. In other words, each adsorbing molecule appears to occupy n sites. Assuming that there is no desorption, write down an expression for the coverage of the adsorbate as a function of time. Compare the result for $n = 7$ to what is expected for adsorption that requires only one site for the case of methane sticking on a Pt(111) surface. Take $N_0 = 1.503 \times 10^{15}$ cm^{-2}, $\theta_{max} = 1$, $T = 800$ K and the initial sticking coefficient is 2.5×10^{-7}. You will have to consider exposures up to 1×10^{22} cm^{-2}.

4.7 Given that the kinetic parameters for diffusion are $D_0 = 5 \times 10^{-6}$ cm^2 s^{-1} and $E_{dif} = 20.5$ kJ mol^{-1} and those for desorption are $A = 10^{12}$ s^{-1} and $E_{des} = 110$ kJ mol^{-1} (first order) for CO on Ni(100), how far does the average CO molecule roam across the surface at $T_s = 480$ K (the top of the temperature-programmed desorption peak).

4.8 The experimentally measured [72] desorption prefactor for NO desorption from Pt(111) is $A = 10^{16 \pm 0.5}$ s^{-1} with a TPD peak temperature of 340 K and an initial sticking coefficient $s_0 = 0.9$. Take the partition function of the transition state as equal to the partition function of the free molecule and determine whether the experimental value is consistent with the transition state theory prediction. In the gas phase: $\tilde{v}(NO) = \omega_e = 1904$ cm^{-1}, $B_0 = 1.705$ cm^{-1}. In the adsorbed phase: $\tilde{v}(NO) = 1710$ cm^{-1}, $\tilde{v}(NO-Pt) = 306$ cm^{-1}, $\tilde{v}$(frust rot) $= 230$ cm^{-1}, $\tilde{v}$(frust trans) $= 60$ cm^{-1}.

4.9 Estimate the evaporative cooling rate for a hemispherical droplet of Si with a radius of 900 nm held at its melting point.

4.10 Show that for precursor mediated adsorption described by the Kisliuk model, a plot of $\ln[(\alpha/s_0) - 1]$ versus $1/T_s$ is linear with a slope of $-(E_{des} - E_{ads})/R$, where E_{des} and E_{ads} are the activation energies for desorption and adsorption out of the precursor state.

4.11 Consider precursor mediated adsorption through an equilibrated precursor state. The activation barrier to desorption out of the precursor is E_{des} and the activation barrier separating the precursor from the chemisorbed state is E_a. Prove mathematically that in precursor mediated adsorption, if $E_{des} > E_a$, increasing the surface temperature decreases the sticking coefficient and if $E_{des} < E_a$, increasing T_s favours sticking.

4.12   Consider the competitive absorption of two molecules A and B that adsorb non-dissociatively on the same sites on the surface. Assuming that both follow Langmuirian adsorption kinetics and that each site on the surface can only bind either one A or one B molecule, show that the equilibrium coverages of A and B are given by

$$\theta_A = \frac{K_A p_A}{1 + K_A p_A + K_B p_B} \qquad (4.11.4)$$

$$\theta_B = \frac{K_B p_B}{1 + K_A p_A + K_B p_B} \qquad (4.11.5)$$

4.13   Many different pressure units are encountered in surface science and conversions are inevitable. Show that

$$Z_w = 3.51 \times 10^{22}\,\text{cm}^{-2}\text{s}^{-1} \frac{p}{\sqrt{MT}} \qquad (4.11.6)$$

where p is given in torr, M in g mol^{-1} and T in Kelvin. Derive a similar expression that relates the SI units of pressure (Pa), molar mass (kg mol^{-1}) to the flux in m^{-2} s^{-1}.

4.14   Show that for dissociative adsorption within the Langmuir model, the isotherm equation is given by

$$\theta = \frac{(pK)^{1/2}}{1 + (pK)^{1/2}} \qquad (4.11.7)$$

4.15   An alternative representation of the Langmuir isotherm for a single component system is

$$c = \frac{\lambda_V p}{\lambda_p + p} \qquad (4.11.8)$$

where c is the surface coverage, λ_v is the Langmuir volume, λ_p is the Langmuir pressure and p the pressure. Note that the dimensions of λ_v set the dimension of c, which is often expressed in some form of concentration units (such as molecules adsorbed per gram of adsorbent) rather than areal density. Use Eq. (4.6.4) to derive an equation of the form of Eq. (4.11.8) and determine the meaning of λ_v and λ_p. Instead of fractional coverage explicitly use the areal density and saturation coverage σ_{sat}.

4.16   Methane is often found adsorbed on coal. Consider the reversible, non-dissociative molecular adsorption of CO_2 and CH_4 on coal. The parameters describing adsorption are given below

	λ_v/mol g^{-1}	λ_p/kPa
CO_2	0.0017	2100
CH_4	0.00068	4800

Calculate the equilibrium coverage for CO_2 and CH_4 on coal exposed to a total gas pressure of 3500 kPa for (a) pure CH_4, (b) pure CO_2, and (c) a 50/50 mixture of CH_4 and CO_2. Which species is bound more strongly and what is the implication of pumping CO_2 into a coal seam saturated with CH_4?

4.17 Given that A adsorbs in an on-top site on a fcc(111) lattice (fcc, face-centred cubic) with an initial heat of adsorption of $q_0 = 100\,\text{kJ}\,\text{mol}^{-1}$, calculate the heat of adsorption as a function of coverage for repulsive lateral interactions of 0, 2, 5 and $10\,\text{kJ}\,\text{mol}^{-1}$. Make a series of plots to demonstrate the effects of repulsive interactions at $T = 80\,\text{K}$ and $300\,\text{K}$.

4.18 On Pt(111) straight-chain alkanes exhibit non-activated molecular adsorption at low T_s. The desorption activation energy as determined by TPD is 18.6, 36.8, 42.4 and $54.0\,\text{kJ}\,\text{mol}^{-1}$ for CH_4, C_2H_6, C_3H_8 and C_4H_{10}, respectively [73]. Explain this trend.

4.19 Derive Eqs (4.5.16–18).

4.20 Write out expressions of Eq. (4.5.23) in the limits of
    (a) large desorption rate from the precursor;
    (b) large adsorption rate into the chemisorbed state;
    (c) large desorption rate from the chemisorbed state.

   Explain the answers for (b) and (c) with recourse to the value of s_0.

4.21 In recombinative desorption, the pre-exponential factor can be written [19]

$$A = D\overline{v} \tag{4.11.9}$$

where D is the mean molecular diameter and $\overline{v}$ is the mean speed of surface diffusion,

$$\overline{v} = \frac{\lambda}{\tau} \tag{4.11.10}$$

where λ is the mean hopping distance (the distance between sites) and τ is the mean time between hops,

$$\tau = \tau_0 \exp(E_{\text{dif}}/RT). \tag{4.11.11}$$

where τ_0 is the characteristic hopping time. Use this information and the expression for the desorption rate constant to show that a compensation effect could be observed in a system that exhibits a coverage dependent diffusion activation energy, E_{dif}.

4.22 When 0.5 L of H_2 is dosed onto Ir(110)–(2 × 1) at 100 K, it adsorbs dissociatively. Its TPD spectrum exhibits one peak at 400 K. When propane is dosed to saturation at 100 K on initially clean Ir(110)–(2 × 1), the H_2 TPD spectrum exhibits two peaks, one at 400 K and one at 550 K [74]. Interpret the origin of these two H_2 TPD peaks from C_3H_8 adsorption.

4.23 Show that for the simple adsorption reaction

$$A +\,^* \rightleftharpoons A^*$$

the following relationship holds:

$$\theta_* = \frac{1}{1 + (\theta_A/\theta_*)} \tag{4.11.12}$$

4.24 Using transition state theory and the lattice-gas approximation [75] the rate constants for adsorption and desorption can be written

$$k_a = (1 - \theta)k_a^0 \sum_i P_{0,i} \exp(-\varepsilon_i^*/k_B T) \tag{4.11.13}$$

$$k_d = \frac{(1 - \theta)}{\theta} k_d^0 \exp(\mu_a/k_B T) \sum_i P_{0,i} \exp(-\varepsilon_i^*/k_B T), \tag{4.11.14}$$

where k_a^0 and k_d^0 are the rate constants in the limit of low coverage, $P_{0,i}$ is the probability that a vacant site has the environment denoted by index i and ε_i^0 describes the lateral interactions in the activated state. The units of these rate constants are such that the net rate of coverage change can be written

$$\frac{d\theta}{dt} = k_a p - k_d \theta. \tag{4.11.15}$$

Show that Eqs (4.11.14) and (4.11.15) obey microscopic reversibility by showing that at equilibrium

$$\mu_a = \mu_d. \tag{4.11.16}$$

You will need to use the standard expression for the chemical potential of gas phase particles

$$\mu_g = \mu_g^\circ + k_B T \ln p. \tag{4.11.17}$$

4.25   The initial sticking coefficient at 500 K for CH_4 on a terrace site on Ni(111) is 2.1×10^{-9} whereas on a step site it is 2.8×10^{-7}. Assuming an Arrhenius form to describe the temperature dependence of the initial sticking coefficient and that the entropy of activation is the same for both terrace and step sites, calculate the difference in the barrier to dissociation between a terrace and step site.

4.26   The dissociative sticking coefficient of H_2 on Si(100) is $\sim 1 \times 10^{-11}$ at room temperature. There are 6.8×10^{18} Si atoms per m^2 on the Si(100) surface. Estimate the coverage of H atoms that results from exposing a Si(100) surface to 1×10^{-5} Pa of H_2 for 3 min. Justify the use of a constant sticking coefficient.

4.27   Determine the steady-state saturation coverage that results if atomic H is exposed to a surface. Assume that the sticking coefficient of H atoms on an empty site is s_{ad} and the abstraction probability is s_{ab} if the H atom strikes a filled site. Assume that one and only one H atom can adsorb per surface atom and that both s_{ad} and s_{ab} are independent of coverage. Can the coverage ever reach 1 ML (as defined by one adsorbate per surface atom)? Assume that T_s is low enough that desorption can be neglected.

4.28   Under the same assumptions found in Exercise 4.26, calculate how the H atom coverage θ changes with time.

4.29   Derive Eq. (4.6.6).

4.30   The equilibrium vapour pressure of water in Pa is given as a function of absolute temperature T by

$$\ln p = 23.195 - \frac{3.814 \times 10^3}{T - 46.29}.$$

(a) Assuming that the desorption activation energy is given by the enthalpy of vaporization of pure water $40.016 \, kJ \, mol^{-1}$, calculate the coverage of water expected on a surface exposed to the equilibrium vapour pressure of water at 298 K.

(b) What is the value of K in Eq. (4.6.4)?

References

[1]   S. Ĉerny, Energy and entropy of adsorption in *The Chemical Physics of Solid Surfaces and Heterogeneous Catalysis, Vol. 2* (Eds: D. A. King, D. P. Woodruff), Elsevier, Amsterdam, 1983, p. 1.

[2]   M. C. Desjonquères, D. Spanjaard, *Concepts in Surface Physics*, 2nd ed., Springer-Verlag, Berlin, 2002.

[3]   T. L. Hill, *An Introduction to Statistical Thermodynamics*, Dover Publications, New York, 1986.

[4]   A. W. Adamson, A. P. Gast, *Physical Chemistry of Surfaces*, 6th ed., John Wiley & Sons, New York, 1997.

[5]   Q. Ge, R. Kose, D. A. King, *Adv. Catal.*, **45** (2000) 207.

[6]   E. Swan, A. R. Urquhart, *J. Phys. Chem.*, **31** (1927) 251.

[7]   S. Brunauer, P. H. Emmett, E. Teller, *J. Am. Chem. Soc.*, **60** (1938) 309.

[8]   J. Rouquerol, F. Rouquerol, K. S. W. Sing, *Adsorption by Powders and Porous Solids: Principles, Methodology and Applications*, Academic Press, Boston, 1999.

[9]   J. Rouquerol, D. Avnir, C. W. Fairbridge, D. H. Everett, J. M. Haynes, N. Pernicone, J. D. F. Ramsay, K. S. W. Sing, K. K. Unger, *Pure Appl. Chem.*, **66** (1994) 1739.

[10] S. K. Sinha, *Ordering in Two Dimensions*, Elsevier, New York, 1980.

[11] T. L. Einstein, *Critical Reviews in Solid State and Materials Sciences*, **7** (1978) 261.

[12] B. Hammer, J. K. Nørskov, *Adv. Catal.*, **45** (2000) 71.

[13] W. A. Brown, D. A. King, *J. Phys. Chem. B*, **104** (2000) 2578.

[14] G. Ertl, *Langmuir*, **3** (1987) 4.

[15] J. S. Wang, *Proc. R. Soc. London, A*, **161** (1937) 127.

[16] D. L. Adams, *Surf. Sci.*, **42** (1974) 12.

[17] D. A. King, *Surf. Sci.*, **47** (1975) 384.

[18] H. Pfnür, P. Feulner, H. A. Engelhardt, D. Menzel, *Chem. Phys. Lett.*, **59** (1978) 481.

[19] M. Boudart, G. Djéga-Mariadassou, *Kinetics of Heterogeneous Catalytic Reactions*, Princeton University Press, Princeton, NJ, 1984.

[20] K. J. Laidler, *Chemical Kinetics*, HarperCollins, New York, 1987.

[21] J. A. Serri, J. C. Tully, M. J. Cardillo, *J. Chem. Phys.*, **79** (1983) 1530.

[22] C. G. Goymour, D. A. King, *J. Chem. Soc., Faraday Trans.*, **69** (1973) 749.

[23] A. Cassuto, D. A. King, *Surf. Sci.*, **102** (1981) 388.

[24] E. G. Seebauer, A. C. F. Kong, L. D. Schmidt, *Surf. Sci.*, **193** (1988) 417.

[25] B. Meng, W. H. Weinberg, *J. Chem. Phys.*, **100** (1994) 5280.

[26] J. W. Niemantsverdriet, K. Wandelt, *J. Vac. Sci. Technol., A*, **6** (1988) 757.

[27] P. J. Estrup, E. F. Greene, M. J. Cardillo, J. C. Tully, *J. Phys. Chem.*, **90** (1986) 4099.

[28] V. P. Zhdanov, *Surf. Sci. Rep.*, **12** (1991) 183.

[29] E. K. Grimmelmann, J. C. Tully, E. Helfand, *J. Chem. Phys.*, **74** (1981) 5300.

[30] K. J. Laidler, S. Glasstone, H. Eyring, *J. Chem. Phys.*, **8** (1940) 659.

[31] P. Stoltze, J. K. Nørskov, *Phys. Rev. Lett.*, **55** (1985) 2502.

[32] D. Menzel, Thermal Desorption in *Chemistry and Physics of Solid Surfaces IV, Vol. 20* (Eds: R. Vanselow, R. Howe), Springer-Verlag, New York, 1982, p. 389.

[33] H. Pfnür, D. Menzel, *J. Chem. Phys.*, **79** (1983) 2400.

[34] H. Ibach, W. Erley, H. Wagner, *Surf. Sci.*, **92** (1980) 29.

[35] R. H. Fowler, E. A. Guggenheim, *Statistical Thermodynamics*, Cambridge University Press, Cambridge, UK, 1939.

[36] K. W. Kolasinski, *Internat. J. Mod. Phys. B*, **9** (1995) 2753.

[37] J. J. Boland, *Adv. Phys.*, **42** (1993) 129.

[38] M. P. D'Evelyn, Y. L. Yang, L. F. Sutcu, *J. Chem. Phys.*, **96** (1992) 852.

[39] M. P. D'Evelyn, S. M. Cohen, E. Rouchouze, Y. L. Yang, *J. Chem. Phys.*, **98** (1993) 3560.

[40] Y. L. Yang, M. P. D'Evelyn, *J. Vac. Sci. Technol., A*, **11** (1993) 2200.

[41] M. P. D'Evelyn, Y. L. Yang, S. M. Cohen, *J. Chem. Phys.*, **101** (1994) 2463.

[42] U. Höfer, L. Li, T. F. Heinz, *Phys. Rev. B*, **45** (1992) 9485.

[43] G. A. Reider, U. Höfer, T. F. Heinz, *J. Chem. Phys.*, **94** (1991) 4080.

[44] K. Cho, E. Kaxiras, J. D. Joannopoulos, *Phys. Rev. Lett.*, **79** (1997) 5078.

[45] F. Healey, R. N. Carter, A. Hodgson, *Surf. Sci.*, **328** (1995) 67.

[46] S. J. Lombardo, A. T. Bell, *Surf. Sci. Rep.*, **13** (1991) 1.

[47] I. Langmuir, *J. Am. Chem. Soc.*, **38** (1916) 2221.

[48] I. Langmuir, *J. Am. Chem. Soc.*, **40** (1918) 1361.

[49] J. C. Tully, *Surf. Sci.*, **111** (1981) 461.

[50] C. W. Muhlhausen, L. R. Williams, J. C. Tully, *J. Chem. Phys.*, **83** (1985) 2594.

[51] D. A. King, M. G. Wells, *Proc. R. Soc. London, A*, **339** (1974) 245.

[52] J. B. Taylor, I. Langmuir, *Phys. Rev.*, **44** (1933) 423.

[53] P. Kisliuk, *J. Phys. Chem. Solids*, **3** (1957) 95.

[54] C. R. Arumainayagam, M. C. McMaster, R. J. Madix, *J. Phys. Chem.*, **95** (1991) 2461.

[55] H. C. Kang, C. B. Mullins, W. H. Weinberg, *J. Chem. Phys.*, **92** (1990) 1397.

[56] K. S. W. Sing, D. H. Everett, R. A. W. Haul, L. Moscou, R. A. Pierotti, J. Rouquérol, T. Siemieniewska, *Pure Appl. Chem.*, **57** (1985) 603.

[57] L. A. Pétermann, *Prog. Surf. Sci.*, **3** (1974) 1.

[58] T. E. Madey, J. T. Yates, Jr., *Surf. Sci.*, **63** (1977) 203.

[59] M. W. Roberts, C. S. McKee, *Chemistry of the Metal-Gas Interface*, Clarendon Press, Oxford, 1978.

[60] J. T. Yates, Jr., Thermal desorption of adsorbed species, in *Solid State Physics: Surfaces. Methods of Experimental Physics, Vol. 22* (Eds: R. L. Park, M. G. Lagally), Academic Press, New York, 1985, p. 425.

[61] C. T. Campbell, S. M. Valone, *J. Vac. Sci. Technol., A*, **3** (1985) 408.

[62] A. Winkler, J. T. Yates, Jr., *J. Vac. Sci. Technol., A*, **6** (1988) 2929.

[63] J. T. Yates, Jr., *Experimental Innovations in Surface Science: A Guide to Practical Laboratory Methods and Instruments*, AIP Press (Springer-Verlag), New York, 1998.

[64] S. Haegel, T. Zecho, S. Wehner, *Rev. Sci. Instrum.*, **81** (2010) 033904.

[65] G. Ehrlich, *Adv. Catal.*, **14** (1963) 255.

[66] P. A. Redhead, *Vacuum*, **12** (1962) 203.

[67] M. D. Alvey, K. W. Kolasinski, J. T. Yates, Jr., M. Head-Gordon, *J. Chem. Phys.*, **85** (1986) 6093.

[68] M. Head-Gordon, J. C. Tully, H. Schlichting, D. Menzel, *J. Chem. Phys.*, **95** (1991) 9266.

[69] A. M. de Jong, J. W. Niemantsverdriet, *Surf. Sci.*, **233** (1990) 355.

[70] J. B. Miller, H. R. Siddiqui, S. M. Gates, J. N. Russell, Jr., J. T. Yates, Jr., J. C. Tully, M. J. Cardillo, *J. Chem. Phys.*, **87** (1987) 6725.

[71] J. L. Falconer, R. J. Madix, *Surf. Sci.*, **48** (1975) 393.

[72] R. J. Gorte, L. D. Schmidt, J. L. Gland, *Surf. Sci.*, **109** (1981) 367.

[73] J. F. Weaver, A. F. Carlsson, R. J. Madix, *Surf. Sci. Rep.*, **50** (2003) 107.

[74] P. D. Szuromi, J. R. Engstrom, W. H. Weinberg, *J. Chem. Phys.*, **80** (1984) 508.

[75] V. P. Zhdanov, *J. Chem. Phys.*, **114** (2001) 4746.

5

Liquid Interfaces

Liquid interfaces are important in many biological and real world settings, in particular the interactions of water with other materials. Membrane and micelle formation as well as protein folding all involve bonding interactions with water molecules at their surfaces. The interaction of surfactants at interfaces, specifically the lowered surface tension accompanying adsorption, is one of the main factors determining the effectiveness of products such as soaps, detergents and lubricants. Here we concentrate mainly on the interactions of aqueous systems and the liquid/solid interface.

What is the nature of the liquid/solid interface and how does it differ from the gas/solid interface? We know what the bulk of the solid is like. In a crystal we have a system that exhibits long-range and short-range order. Bonding distances, angles and composition are all well defined and uniform. The bulk of a liquid is a much less ordered system. There is short-range order, nearest neighbour distances and angles have a high correlation, but no long-range order. Composition is constant, but the exact positions of neighbours are in flux and constantly changing even if, on average, they exhibit short-range order. Solutes exist surrounded by a characteristic solvent shell. A huge difference between gases and liquids is that liquids with a high dielectric constant support stable ionic species. Another defining difference is that while in an ideal gas we can neglect intermolecular interactions, in an ideal liquid, interactions are assumed to be present but *uniform* for all species. So what are the consequences of these differences, and what does the interface between these two phases look like?

5.1 Structure of the liquid/solid interface

Consider a physisorbed system. A physisorbed layer at a sufficiently high temperature can exist in a disordered liquid-like state in which the molecules are highly mobile and not particularly bound to any one site. At the liquid/solid interface this is known as non-specific adsorption (we will modify this definition shortly when we consider solutions). As mentioned in § 4.7.2, each successive layer in a multilayer physisorption system is bound slightly differently. The presence of the substrate perturbs the physisorbate slightly, and the first layer is the most strongly bound, the one most influenced by the crystallography of the substrate. The second layer senses the surface less but still acts differently than the same molecule in the liquid phase because it is interacting with other physisorbates that are strongly perturbed from the normal liquid state. For Ar physisorbed on Ru(001) Menzel and Schlichting [1] were able to resolve binding strength differences in the first four physisorbed layers. The binding energy (and perturbation to the

Surface Science: Foundations of Catalysis and Nanoscience, Third Edition. Kurt W. Kolasinski.
© 2012 John Wiley & Sons, Ltd. Published 2012 by John Wiley & Sons, Ltd.

adsorbate) decreases gradually in these first few layers. Eventually, for thick enough layers the molecules are unaffected by the presence of the solid, and they assume all the properties of the bulk liquid.

An ionic solute interacts strongly with the solvent as well as the solid. It might chemisorb with a well-defined adsorbate geometry as we expect in any chemisorption system. This is called specific adsorption. The ion exists with a solvent shell in the solution but if it is specifically adsorbed, it must shed this shell at least partially when it adsorbs to the surface. This only occurs if the formation of a chemisorption bond between the ion and the surface can overcome that portion of the energy of solvation between the solvent and the solute that is lost as a result of specific adsorption. This does not mean that a specifically adsorbed ion has lost all interaction with the solvent. However, it does mean that there is no solvent *intervening* between it and the surface. A non-specifically adsorbed ion is one that has not lost its solvent shell. In this case the solvent-solute interaction is stronger than the solute-surface interaction and only a physisorption interaction between the solvated ion (with intervening solvent molecules) and the surface occurs. These types of interaction are illustrated in Fig. 5.1, which demonstrates that the liquid/solid interface is very complex. Whereas the gas/solid interface is confined to just the adsorbed layer and the solid surface, the liquid/solid interface is composed not only of these, but also a region of space in which the liquid changes from the structure and composition of the adsorbed layer to the structure and composition of the bulk liquid.

The structure of the electrolyte/solid interface is complex and several models of it have developed over the years [3–5]. Figure 5.1 depicts the Stern layer, which is composed of the inner and outer Helmholtz layers, on top of which resides a diffuse Gouy layer. The development of this model began with the work of Helmholtz [6]. The inner Helmholtz layer is defined by the plane that contains a layer of fully or partially solvated ions. Whether ions are specifically or non-specifically adsorbed depends on the electric field strength and the chemical identity of the ions. The outer Helmholtz layer is defined by the plane of fully solvated ions that reside above the inner layer; together these form the electric double layer. Depending on the extent to which the charge of the first layer is compensated by excess charge at the solid surface, the outer layer is composed either of like charged ions (full compensation), counterions (no compensation) or a mixture of the two. Gouy [7] and Chapman [8] developed a continuum model in which a continuous charge distribution above the surface is contained in a diffuse mobile layer. The solvent in this layer is depicted as a continuous medium rather than as a molecular liquid. Stern [9] combined both of these concepts to create the model depicted in Fig. 5.1. The shape of the interface influences the formation of the Stern layer and a model calculation of the potential and ion distribution around a spherical surface was subsequently achieved by Debye and Hückel [10].

These examples lead us to see how there are two very different descriptions for the liquid/solid interface. One is atomistic much like what we have developed for the gas/solid interface. The other is phenomenological and more continuum-like (or, perhaps, more like a black box). We can divide the liquid/solid system into four regions. The bulk liquid, the bulk solid, the surface of the solid along with its adsorbates and a region just above the adsorbed layer that is different from the bulk liquid. Throughout this chapter, we try to develop both of these models to better understand how these four regions interact and influence the chemistry that occurs at the liquid/solid interface.

5.1.1 The structure of the water/solid interface

Water is the exception to almost every rule and its interactions with other forms of matter are extremely complex. If it were not of such paramount importance to our world and life itself, we might never study water and simply write it off as pathological. However, because of its unique importance, we specifically treat the water/solid interaction here.

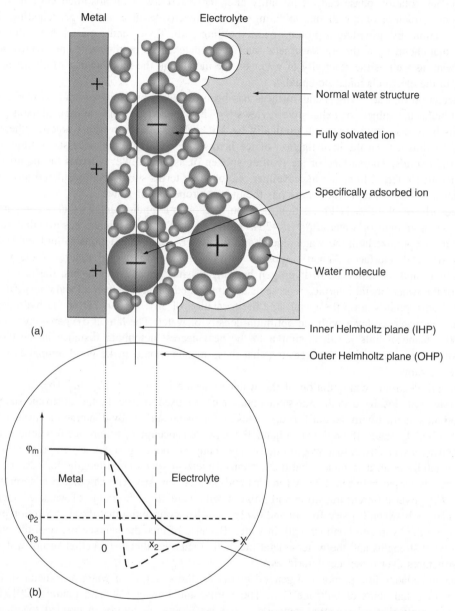

Figure 5.1 *A model of the liquid/solid interface with examples of specific and non-specific adsorption. (a) The structure of the Helmholtz layers and a pure solvent is also shown. (b) The potential drop across the interface in case of non-specific (– – –) and specific (- - -) ion adsorption. ϕ_m, ϕ_s and ϕ_2 are potentials inside the metal, in the electrolyte, and at the outer Helmholtz plane x_2, respectively. A linear potential drop across the electrochemical interface is generally assumed for non-specifically adsorbing ions, while in the case of specific adsorption, a steeper potential gradient (close to the metal surface) and an "overshooting" of the potential with respect to the solution value is assumed [2]. Reproduced from D. M. Kolb, Surf. Sci. 500, 722. © 2002 with permission from Elsevier.*

The aqueous solution phase can significantly alter the reactivity of an adsorbed system by changing the electronic structure of the metal, solvating charged reactive states, and/or participating directly in chemical reactions by providing a proton or ion shuttling path. Taylor and Neurock have shown [11], in particular, that chemistry at the aqueous/metal interface is determined by the steric and electrostatic effects of the solvent, as well as the reactivity of water on the metal, and the modification of chemical reactivity according to the electrochemical potential.

The interaction of water with metal surfaces has been reviewed by Hodgson and co-workers [12, 13] as well as being the subject of exhaustive reviews by Thiel and Madey [14] and Henderson [15]. These interactions have also been treated theoretically by Michaelides and co-workers among others [16, 17]. Much can be drawn from the investigation of ice layers but we should not necessarily think of ice-like layers as rigid solids. The surface of ice is often spoken of as liquid-like; whereas the liquid water/solid interface is often referred to as ice-like. Neither one of these terms should be considered too strictly and, again, it must be kept in mind that water has the unique properties of water.

Water adsorbs on close packed transition metal surfaces with a binding energy that is comparable to the hydrogen bond strength of water, $E_{\mathrm{HB}}^{\mathrm{water}} \approx 23\,\mathrm{kJ\,mol^{-1}}$ [12]. This equivalence ensures that stable water structures must optimize both the water–metal and the water–water interactions to find stable adsorption structures at metal interfaces. When combined with the highly directional nature of the water–water hydrogen bond and the influence of hydrogen bonding to subsequent water layers, the result is that the structure of the water–metal interface represents a complicated balance between local considerations, such as the local adsorption site on the metal, and the formation of an extended 2D or 3D hydrogen bonding network in the water ice layer with the optimum density of water. The lateral oxygen–oxygen separation in an (0001) plane of bulk ice Ih is similar to the next nearest neighbour distance in the close packed surfaces of the transition metals, suggesting that these metals should make good templates for growing crystalline ice films.

The ground electronic configuration of the water molecule is $1a_1^2\,2a_1^2\,1b_2^2\,3a_1^2\,1b_1^2$. In water ice, the electronic states are localized so that the same notation can be used to describe the electronic states. The $3a_1$ and $1b_1$ orbitals interact with the unfilled d_{z^2} states on the metal surface by donating electrons into unfilled or partially filled d_{z^2} states. If, on the other hand, the d_{z^2} state is occupied, a four-electron bond to the surface can form, whereby electrons that would otherwise occupy antibonding states are emptied into the Fermi level. Michaelides et al. determined that the optimized metal–oxygen bond lengths for adsorbed water on different metals varied between 2.25 Å (on Cu) and 3.02 Å (on Au). The calculated adsorption energies for water, E_{ads}, over different transition and noble metals were all in the range of weak interactions, with binding energies between 0.13 eV for Au and 0.42 eV for Rh, confirming that these adsorption energies are comparable to the hydrogen bond strength in water. Because of this, the balance between the water–water hydrogen-bond strengths and the water–metal interaction energies dictates whether or not water can form ordered structures over these metal surfaces.

Doering and Madey first proposed a general model for the structure of water at a metal surface on the basis of a truncated plane of bulk ice [18]. The Bernal and Fowler [19] and Pauling [20] (BFP) rules for bulk ice imply that a hexagonal network of oxygen atoms are hydrogen bonded together to create rings of six waters, each water maintaining a tetrahedral geometry and each O–O axis containing only one hydrogen. To preserve the tetrahedral bonding geometry of water, the oxygen atoms buckle to form "puckered" rings, with each oxygen vertically displaced from its nearest neighbour by ∼0.97 Å. Doering and Madey modified these rules to account for the preference of water to bind to a metal surface via its oxygen lone pair. The result is a structure where water forms two-dimensional islands of ice with a bilayer structure, half of the water molecules binding directly to the metal through the oxygen with the other half hydrogen bonding to those below. This model predicts ice structures where each of the waters in the upper half of the bilayer has one unsatisfied hydrogen bond, leaving OH dangling "H-up" toward the vacuum,

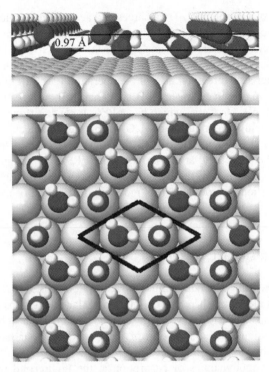

Figure 5.2 *Water bilayer structure for adsorption on closed packed metal surfaces. Reproduced from C. Clay, A. Hodgson,* Curr. Opin. Solid State Mater. Sci. *9, 11.* © *2005 with permission from Elsevier Science.*

as shown in Fig. 5.2. The molecules in the lower layer are close to flat. The energy difference between the H-up and the H-down structure is not great and the specific structure depends on the balance between adsorption and hydrogen bonding energies as determined by the identity of the metal. The structure of water near a Pt(111) surface is influenced by the polar anchoring of the first monolayer to the metal. This surface-induced ordering decays over a distance of ∼30 ML. It should also be emphasized that the structure of water near room temperature is not static and that while the O atoms tend to maintain their structure, there is significantly more randomness to the location of the H atoms.

The structure of a water layer is affected not only by the metal, but also by the presence of co-adsorbates, whether or not the water can dissociate, and any electric field. Water has an intrinsic dipole, and an electric field tends to align it. A positive electric field orients the water with the oxygen toward the surface and the H atoms away from it. When the field changes sign, the water molecules flip over. The presence of preadsorbed O atoms can lead to the dissociation of some water molecules and the formation of mixed $OH + H_2O$ layers. This occurs on Pt(111). On Ru(0001) either a metastable intact water layer or a partially dissociated layer can be formed. In the partially dissociated layer the water lies flat [12].

The ordering of the water molecules at an insulator surface responds to the pH due to an effect that is analogous to the effect of an electric field. The water dipoles flip by 180° as the pH is adjusted from low to high since the charging at the surface changes. For pure water interacting with a solid, the net charge on the solid determines how well ordered the water layer is, as well as its orientation. As the pH or potential is changed, the net surface charge changes from positive to negative. The isoelectric point (IEPS) or point of zero charge (PZC) is defined by the condition that the surface charge is neutral. In the colloid literature this is called the point of zero zeta potential. The charges and accompanying field act on the water to orient it. The dipoles are most randomly oriented when the interface is neutral. The dipole orientation reverses, at least for some of the water molecules, from one side of the isoelectric point to the other.

On quartz and mica surfaces, the water bound on the surface has an ice-like ordered structure rather than a disordered liquid water structure even at room temperature [21, 22]. Oxides such as quartz (SiO_2), TiO_2 and sapphire (Al_2O_3) are protonated at low pH such that their surfaces are covered by $-OH$ groups. Deprotonation occurs at high pH. On CaF_2, which has a PZC at pH 6.2, there is a change in orientation of water molecules above and below this point. On Al_2O_3 (sapphire) this occurs just above pH 8. The most ordered structures are found at low and high pH as the water molecules respond to increased hydrogen bonding and/or an enhanced electric field at the surface. The IEPS is important because the changes in the surface dipole cause changes not only in the water layer above it, and the dipoles associated with water, but also in the distribution of ions in solution. This modified distribution of ions can engender changes in competitive adsorption.

5.2   Surface energy and surface tension

Previously when we discussed the thermodynamics of adsorption at the gas/solid interface, we assumed that the gas was ideal. In an ideal gas, there are no intermolecular interactions and as long as the pressure is not too high, nor the temperature too low, all gases tend toward ideal behaviour. While it simplifies some thermodynamic expressions, the assumption of ideal gas behaviour has little significance for the discussion of adsorption at the gas/solid interface.

When we consider the thermodynamics of adsorption at the liquid/solid interface, we no longer can assume that intermolecular interactions in the liquid phase are negligible. Indeed, they have very strong effects on the energetics of the liquid/solid interface. An ideal gas is perfectly randomized. An ideal liquid/solution is also random, but we have already introduced the concept that the structure of the liquid is different near the interface in the Helmholtz layer. The composition of an ideal gas, liquid or solution is uniform everywhere. The composition and structure of a solution near the interface need not be uniform *even at equilibrium*. While in an ideal gas there are no intermolecular interactions, in an ideal solution there must be intermolecular interactions and these are assumed to be the same for all components. If the interactions are not equal for all components, then when adsorption occurs, we need to consider how the environment about the molecule changes in going from the liquid phase to the adsorbed phase. Even if a molecule has an exothermic adsorption enthalpy from the gas phase, it might not have an exothermic adsorption enthalpy from the solution phase if its enthalpy of solvation is high. Although entropy favours a uniform composition and a completely disordered system, the interfacial region may exhibit order and a composition differing from the bulk liquid if there are energetic factors that override the entropic factor.

5.2.1   Liquid surfaces

Gibbs [23] laid the foundation for the thermodynamics of fluid interfaces. To help us understand the interactions that occur at interfaces, we need to define and utilize surface energy (associated with energy required for surface creation) and surface tension (associated with the energy required for surface deformation). For a liquid, these two terms are synonymous, and we use them interchangeably. For largely historical reasons, the term surface tension is commonly used to refer to liquid surfaces. As we shall see in Chapter 7, this terminology is somewhat ambiguous for solids; therefore, we confine the use of surface tension to liquids.

Table 5.1 Selected values of surface tension taken from Adamson and Gast [27]

	Temperature/°C	γ/mN m^{-1}
Liquid-Vapour Interface		
Perfluoropentane	20	9.89
Heptane	20	20.14
Ethanol	20	22.39
Methanol	20	22.50
Benzene	30	27.56
	20	28.88
Water	25	72.13
	20	72.94
Hg	25	485.5
	20	486.5
Ag	1100	878.5
Cu	1357 (T_f)	1300
Pt	1772 (T_f)	1880

We begin by discussing liquid surfaces. Surface tension, γ, is defined by the relationship between the amount of work performed in enlarging a surface, δW^s, and the surface area created, dA. At constant temperature and pressure the amount of work done to expand reversibly a volume by an amount dV is

$$\delta W_{T,P} = p \, dV \qquad (5.2.1)$$

By direct analogy, the amount of work required to expand reversibly a surface is given by

$$\delta W_{T,P}^s = \gamma \, dA \qquad (5.2.2)$$

Hence, the surface tension is the surface analogue of pressure. Whereas three-dimensional pressure is perpendicular to the surface (the direction of expansion), the surface tension is parallel to the surface. Surface tension, unlike the pressure of an ideal gas, is a material dependent property. Surface tension scales with the force required to expand the surface area of a material. The surface energy of a material in the solid state is larger than the surface energy of that material in the liquid phase and tends to scale with the heat of sublimation. Typical values of surface energy range from 2300 mN m^{-1} for W(l) to 72.1 mN m^{-1} for H$_2$O(l) to 0.37 mN m^{-1} for He(l). Equivalent units are J m^{-2}. As confirmed in Table 5.1, there is a general tendency for surface energy to decrease with increasing temperature.

The change in Gibbs energy upon a change in surface area for a pure liquid is given by

$$dG = -S \, dT + V \, dp + \gamma \, dA \qquad (5.2.3)$$

At constant temperature and pressure, this reduces to

$$dG_{T,p} = \gamma \, dA, \qquad (5.2.4)$$

which shows that surface tension can also be thought of as the surface Gibbs energy per unit area of a pure liquid.

A system at equilibrium has attained a minimum value of the Gibbs energy. For a surface at constant T, p and composition, the condition for minimum Gibbs energy is found by minimizing the surface Gibbs energy according to

$$\int \gamma(\boldsymbol{m}) \, dA = \text{minimum} \tag{5.2.5}$$

In Eq. (5.2.5) we write the surface tension as a function of $\boldsymbol{m}$, a unit vector along the surface normal. This allows for any potential orientation dependence in the surface energy, which is essential for a solid as is shown in Chapter 7. For a material such as a liquid in which γ is isotropic, that is, in which the surface energy does not depend on the surface orientation, the minimum of the integral in Eq. (5.2.5) is obtained when the surface area of the material is minimized. Consequently, liquids and other isotropic materials contract into a sphere (in the absence of other forces such as gravity that can distort the particle shape) because a sphere has the minimum surface area.

Consider the case of adding a liquid onto the surface of a solid (or another liquid): we need to generalize Eq. (5.2.3) to include the presence of more than one material. The Gibbs energy is then

$$dG = -S \, dT + V \, dp + \sum_i \gamma_i \, dA_i + \sum_j \mu_j \, dn_j \tag{5.2.6}$$

for a system with i interfaces and n_j moles of j components of chemical potential μ_j. At constant T and p

$$dG_{T,p} = \sum_i \gamma_i \, dA_i + \sum_j \mu_j \, dn_j \tag{5.2.7}$$

and the equilibrium configuration of the system is obviously much more complex than a simple sphere. The system now attempts to maximize the areal fraction of low surface tension components while simultaneously maximizing the concentration of low chemical potential components. The system must balance the tendency toward minimum surface area with the drive to form substances with the lowest chemical potential.

5.2.2   Curved interfaces

Figure 5.3 depicts two commonly encountered curved interfaces involving a liquid. The Young-Laplace equation relates the radius of curvature r of a sphere and the surface tension γ to the pressure difference Δp between the two phases

$$\Delta p = 2\gamma/r. \tag{5.2.8}$$

Δp is also known as the Laplace pressure. When the capillary constant

$$a = (2\gamma/g\rho)^{1/2}, \tag{5.2.9}$$

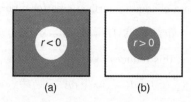

(a)   (b)

Figure 5.3   *(a) A spherical gas bubble of radius $r < 0$ suspended in a liquid. (b) A spherical liquid drop of radius $r > 0$ suspended in a gas. Note that there is a change in sign of r between the two cases.*

is large compared to the dimensions of the liquid, we can neglect the effects of gravity, and the shape of the interface is determined solely by surface energy. In Eq. (5.2.9), g is the gravitational acceleration and ρ is the density.

For a more general curved interface, $2/r$ is replaced by $1/(r_1 + r_2)$ where r_1 and r_2 are the two principle axes of curvature. Equation (5.2.8) tells us that in a sphere of uniform radius r, the pressure inside the bubble/droplet is also uniform; furthermore, the smaller the radius, the greater the pressure difference. Whereas the pressure inside a gas bubble of radius $r = 10$ mm in water is 140 Pa greater than the outside pressure, a similar bubble with $r = 10$ nm experiences a pressure of 1.4×10^7 Pa higher than the pressure in the liquid.

The Laplace pressure has an effect on the vapour pressure of liquids confined to drops. The vapour pressure found in thermodynamic tables refers to the vapour pressure measured above a planar surface. The elevated pressure experienced by molecules in a droplet causes them to evaporate more easily than the molecules contained behind a flat interface. To calculate the effect of the Laplace pressure on vapour pressure we start by calculating the molar Gibbs energy of a vapour at its normal vapour pressure p^*,

$$G_m^* = G_m^\circ + RT \ln(p^*/p^\circ). \tag{5.2.10}$$

G_m° is the standard state molar Gibbs energy and p° is the standard pressure (usually 1 bar). Compare this to the molar Gibbs energy at an arbitrary pressure p, that is, take p as the pressure inside the drop,

$$G_m = G_m^\circ + RT \ln(p/p^\circ). \tag{5.2.11}$$

The change in the molar Gibbs energy resulting from the curvature of the interface is given by

$$\Delta G_m = G_m - G_m^* = RT \ln(p/p^*). \tag{5.2.12}$$

This change in the Gibbs energy as a function of curvature, and hence also size, is known as the Gibbs-Thompson effect. Similarly, we can calculate the change in molar Gibbs energy as a function of the pressure increase brought about by the curved interface at constant temperature according to

$$\Delta G_m = \int dG = \int V_m \, dp - S_m \, dT = \int_0^{\Delta p} V_m \, dp = V_m \, \Delta p. \tag{5.2.13}$$

Here we use the usual definition of the change in Gibbs energy at constant composition, isothermal conditions, and have assumed that the change in the molar volume V_m is negligible. Substituting from Eq. (5.2.8) for Δp and equating the result with Eq. (5.2.12), we arrive at the Kelvin equation

$$RT \ln \frac{p}{p^*} = \frac{2\gamma V_m}{r} \quad \text{or} \quad p = p^* \exp\left(\frac{2\gamma V_m}{RTr}\right). \tag{5.2.14}$$

Recall that for a droplet, the radius of curvature r is positive and for a bubble r is negative. Therefore, for a droplet in contact with its vapour, the liquid has a higher vapour pressure than normal and the vapour pressure increases with decreasing size. The Kelvin equation is one of the first and most fundamental equations of nanoscience in that it demonstrates that the properties of matter can be size dependent, and that important deviations from size independent behaviour arise prominently in the nanoscale regime of $r < 100$ nm. For example, for water at 300 K, the change in vapour pressure is only $\sim 1\%$ for $r = 100$ nm but the deviation is $\sim 10\%$ at 10 nm and almost 300% at 1 nm. The Kelvin equation is also fundamental to Ostwald ripening, which is discussed in detail in Chapter 7. Ostwald ripening is a phenomenon in which large particles, because they are more stable (have a lower effective vapour pressure), grow at the expense of high vapour pressure, less stable smaller particles. The physics behind the Kelvin equation is also important for nucleation theory, which is also discussed in Chapter 7.

The change in sign of r when considering a bubble means that small bubbles are unstable. Only after bubbles have passed a certain critical radius will they continue to grow with a high probability, whereas small bubbles tend to collapse. One consequence of this is that superheating of liquids is possible if heterogeneous nucleation of bubbles – bubble formation at the surfaces of the container or dust/impurity particles – can be suppressed.

5.2.3   Capillary waves

To this point we have always considered liquid interfaces to be smooth. While they may be curved, we have not taken into account their fluctuations. When considered at a molecular level, the gas/liquid interface is neither sharp nor immobile. A liquid interface at equilibrium continually exchanges material, not only between the surface and bulk of the liquid but also between the surface and the gas phase. Global thermodynamic values are obtained by averaging sufficiently large samples over sufficiently long periods of time. However, at the molecular level, the gas/liquid interface is continually changing. Any snapshot of this interface has a tendency to deviate from a well defined plane or simple interfacial region, even though any deviation from a smooth profile of necessity (i) increases the surface area from its minimum value, and (ii) simultaneously raises the Gibbs energy of the system above its minimum value.

The pressure differences created at a curved interface support the formation of deformations in the liquid surface known as capillary waves, also called Goldstone fluctuations. In a liquid of thickness h, these waves are of the form [24, 25]

$$\psi = A \cos(kx - \omega t) \cosh k(z + h). \tag{5.2.15}$$

and the dispersion relationship between the wavevector k and the angular frequency ω is

$$\omega^2 = (\gamma k^3 / \rho) \tanh kh. \tag{5.2.16}$$

The amplitudes of these oscillations are small compared to their wavelengths. First described by Smoluchowski and Mandelstam, the debate continues as to whether they lead to a wavevector dependence of the surface tension, and what form this dependence takes [26, 27]. This capillary wave formalism is valid for atomic, molecular and even polymer liquids, so long as the thickness of the film is at least twice the radius of gyration of the polymer [28]. From the dispersion relation, we can derive the relaxation lifetime of capillary waves

$$\tau_D = \frac{\rho \Lambda}{8\pi^2 \eta}. \tag{5.2.17}$$

Silicon has a viscosity of $\eta = 0.7\,\text{mPa}\,\text{s}^{-1}$ just above its melting point at 1688 K [29]. Its relaxation time varies from $\tau_D \sim 3\,\text{ns}$ for a capillary wave of wavelength $\Lambda = 248\,\text{nm}$, 13 ns at $\Lambda = 532\,\text{nm}$, $\sim 25\,\text{ns}$ for $\Lambda = 800\,\text{nm}$, and 53 ns for $\Lambda = 1064\,\text{nm}$. Thus, capillary waves tend to *suppress the formation of the smallest structures* because short wavelengths correspond to high curvatures and high pressures. The lifetime increases with increasing wavelength.

When the liquid is confined to a region of space not too much larger than the maximum allowed wavelength, standing waves form. These must satisfy the boundary condition of joining up at the confining edge. For a rectangular enclosure, the wavevector is bounded by

$$k^2 = \pi^2 \left(\frac{m^2}{r_1^2} + \frac{n^2}{r_2^2} \right), \tag{5.2.18}$$

where m and n are integers and r_1 and r_2 are the side lengths. Since the relaxation time of capillary waves increases with increasing wavelength, the longest possible capillary wave (lowest harmonics of m and n) is the most probable to form unless some driving force is present to select a higher harmonic.

The formation of a nonplanar interface is influenced by the presence of a solute, which results in what is known as the Mullins–Sekerka instability [30]. A single-component melt solidifying at steady state maintains a planar solid–liquid interface as long as heat is removed through the solid to maintain local interfacial equilibrium. Tiller et al. [31] showed that a planar solidifying interface of a two-component solution is unstable above a certain critical concentration even at equilibrium. To explain these effects, Mullins and Sekerka calculated the time dependent amplitude of a sinusoidal perturbation applied to the interface. They considered alloy surfaces (equivalently dopants at high concentrations in semiconductors) and used a linear perturbation analysis of interface stability that includes the effects of capillary waves. Their equilibrium analysis has been extended to include nonequilibrium interface kinetics [32–35]. These instabilities can lead to supersaturation of ion implanted dopants in silicon as well as the formation of cellular structures [32, 36].

The properties of capillary waves are not only changed by the presence of a solute but also by the presence of a monolayer. This is most striking for the presence of capillary waves on water covered with a Langmuir film, that is, with a monomolecular layer of an amphiphile as discussed in §5.4. This case has been discussed thoroughly by Lucassen-Reynders and Lucassen [25]. The viscoelastic properties of the monolayer significantly change the energy dissipation dynamics of capillary waves. This is responsible for the calming effect of placing oil on water [25, 37, 38].

5.3 Liquid films

5.3.1 Liquid-on-solid films

Consider the case in which one material, B, is deposited on a second, A, and there is no mixing between the two components. The absence of mixing may occur because the two components only form compounds that have a higher chemical potential than either of the two pure components, or because of a large activation barrier to the formation of a compound or to penetration of the deposited component into the bulk of the substrate. As long as deposition is carried out at a sufficiently low temperature, B cannot overcome the activation barrier and it stays on the surface.

Consider a liquid placed on top of a non-deformable solid. In other words, we neglect small changes in γ that may arise in Eq. (5.2.7) because of liquid/solid interactions. As long as the interaction between the liquid and the substrate is not too strong, this is a good approximation; however, we return to this point later. In Fig. 5.4 we sketch a liquid on a substrate and use this to define the contact angle, ψ.

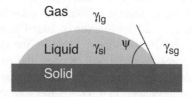

Figure 5.4   *The contact angle, ψ, is defined by the tangent to the gas/liquid interface at the point of intersection with the solid. $\psi < 90°$ is indicative of attractive liquid/solid interactions, whereas $\psi > 90°$ signifies repulsive interactions. γ_{lg}, γ_{sg} and γ_{sl} are the surface energies at the liquid/gas, substrate/gas and substrate/liquid interfaces, respectively.*

At equilibrium the forces on the three components (substrate, liquid, gas phase) must be balanced. The tangential components of the force exerted by the gas/substrate interface are equal and opposite to the forces exerted by the liquid/substrate interface and the gas/liquid interfaces. Balancing of the forces lead to the Young equation

$$\gamma_{lg} \cos \psi + \gamma_{sl} = \gamma_{sg} \qquad (5.3.1)$$

We can define two interesting limits to Eq. (5.3.1). When $\psi = 0$, the deposit spreads across the surface and coats it uniformly – it wets the surface. This is the familiar situation of water wetting (clean) glass. For $\psi > 0$, the deposit forms a droplet and does not wet the surface. This describes the nonwetting interaction of Hg with glass or water on siloxane-covered glass (such as the treatments that are applied to windshields to make rain roll off). In terms of the surface tensions, these two limits correspond to

$$\psi = 0 \Rightarrow \gamma_{sg} \geq \gamma_{sl} + \gamma_{lg} \qquad (5.3.2)$$

$$\psi > 0 \Rightarrow \gamma_{sg} < \gamma_{sl} + \gamma_{lg} \qquad (5.3.3)$$

Surface tension is a bulk property that does not give us an idea of the microscopic processes that lead to wetting or droplet formation. The key to wetting is the balance of the strengths of the A–A, A–B and B–B interactions. If the A–B interactions are strong compared to the B–B interactions, B would prefer to bond to the substrate rather than to itself. This leads to wetting. If, on the other hand, the B–B interactions are strong compared to the A–B interactions, B would prefer to stick to itself rather than substrate. In this case, B attempts to make a droplet to minimize the area of the B/A interface and maximize the number of B–B interactions. Accordingly, taking Φ_{AA} and Φ_{AB} as the energies of A–A and A–B bonds, respectively, we may define the energy difference

$$\Delta = \Phi_{AA} - \Phi_{AB} \qquad (5.3.4)$$

According to Eq. (5.3.4) we can restate the wetting condition as $\Delta < 0$. Equivalently, the non-wetting condition is $\Delta > 0$.

A glass surface is usually covered with Si–OH moieties except at high pH where these units deprotonate to form Si–O$^-$. In either case, these highly polar groups have a strong attractive interaction with polar molecules such as water. Hence water wets a glass surface. A greasy glass surface is coated with a hydrocarbon film. The nonpolar hydrocarbon interacts weakly with water and the interaction of hydrogen bonds between water molecules dominates over the water–hydrocarbon interaction. Hence, the water forms droplets to minimize the water–hydrocarbon interactions and maximize the water–water interactions.

As a thought experiment, consider a monolayer thick uniform film of water covering a hydrophobic substrate. We now let the system evolve to equilibrium. The water is attracted to itself and repelled by the substrate. In order to minimize the contact area between water and the substrate, the film spontaneously breaks up into 3D islands. Early on, the formation of islands is a random process. Some accrete material faster than others do, thus producing a broad distribution of island sizes.

The exact shape of the islands depends on the relative values of the substrate and liquid surface energies. In the limit of extremely repulsive liquid/substrate interactions, the equilibrium island shape would be a sphere. This is known as superhydrophobicity. For this example we assume the islands are hemispherical. The non-wetting nature of the interaction means that it costs more energy to create a liquid/solid interface than it does to create the vapour/liquid interface. Furthermore, since the vapour/liquid surface area increases more rapidly with island radius than the contact area, small islands are unstable with respect to large islands.

The relative instability of small islands with respect to large islands means that large islands grow at the expense of small islands, a result that we can anticipate on the basis of the Kelvin equation but which we see pertains to droplets at the liquid/solid as well as the gas/liquid interface. Given no mass transport limitations an equilibrium configuration of the system is attained in which a single large hemispherical island is formed. If, however, during island growth, the mean distance between islands exceeds the characteristic diffusion length of the liquid molecules, then the islands lose communication with one another and a quasi-equilibrium (metastable steady state) structure of relatively few large islands is attained. The diffusion rate, the amount of material deposited, and the rate of island nucleation determine the exact size distribution. As mentioned above, the process of large island growth at the expense of small islands is known as Ostwald ripening. Inherently, Ostwald ripening leads to a broad size distribution and no preferred or optimum size is achieved unless all the material is able to form just one big island.

The structure of the surface as well as the composition affect the hydrophobicity. Plants have figured this out and use it to create the lotus effect [39, 40]. Water balls up on lotus leaves, which leads to the self-cleaning nature of the leaves since the water droplets tend to pick up dust as they roll off. Appropriate roughness facilitates the trapping of air (composed mainly of nonpolar N_2 and O_2) below the water droplets, which leads to a hydrophobic interaction. To make ideally superhydrophobic surfaces with $\psi = 180°$ requires the combination of chemical and structural factors [41].

Now let us return to the balance of forces. The Young equation was derived from a balance of tangential forces. What about the forces normal to the substrate interface? The vertical component of the liquid/gas interface pulls the substrate upward and, if the substrate is deformable, the force must be balanced by an elastic deformation of the substrate. However, in a classic force balance picture, the three phases meet at the contact line and the stress – a force per unit area – diverges at a line because its area vanishes. The way out of this conundrum sounds rather simple once stated and formulated mathematically [42]: the vertical force arising from the liquid/gas interface may diverge; however, a singularity can be avoided if the stress in the substrate at the contact line diverges in an equal and opposite manner once the deformation has formed. By conservation of mass, the resulting upward bump in the surface along the contact line is surrounded on either side by dimples into the surface. The characteristic length scale of the deformation is given by $\gamma_{lg} \sin \psi / E_s$, where E_s is the Young's modulus of the substrate.

5.4 Langmuir films

When a small amount of an insoluble liquid is poured onto another liquid, it spreads out across the surface. The history of the observation of these films has been recounted by Laidler [43] and is summarized here. The water-calming effects of oil films on water were first noted by Pliny the Elder in 560 BC. The effects of these films later motivated Benjamin Franklin to investigate them. The truly modern era of the study of insoluble liquid films was ushered in by Agnes Pockels who pursued many studies in her kitchen. She communicated her results to Lord Rayleigh, and the fascination she awakened in him led him to study these films further. Rayleigh was the first to calculate that the films had to have a thickness of approximately one molecule. Langmuir took up the cause and greatly improved the equipment, known as a Langmuir trough, used to study these films [44]. Because of his extensive studies of molecular films on the surface of a liquid, these films are called Langmuir films.

A particularly interesting type of Langmuir film is formed when an amphiphile is poured onto the surface of water. Amphiphiles are molecules that are polar (hydrophillic) on one end and nonpolar (hydrophobic) on the other. Examples include stearic acid, $C_{17}H_{35}CO_2H$, alkanethiols, $CH_3(CH_2)_n SH$ and perfluoronated alkanethiols. The hydrophobic end is generally a long alkyl chain, either hydrogenated or perfluoronated. Many different polar end groups are suitable to form amphiphiles such as $-CH_2OH$, $-COOH$, $-CN$, $-CONH_2$, $-CH=NOH$, $-C_6H_4OH$, $-CH_2COCH_3$, $-NHCONH_2$, $-NHCOCH_3$. When such a molecule is deposited on the surface of water, the polar end is attracted to the water while the alkyl chain is repelled.

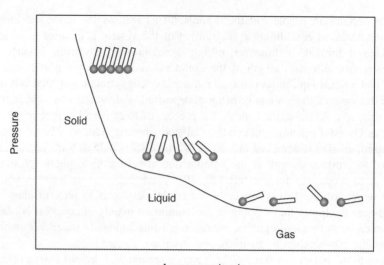

Figure 5.5 *Phase diagram of surface pressure versus area per molecule (amphiphile) in a Langmuir film. The (hydrophobic) tails should only be considered rigid in the solid-like phase in which tail–tail interactions lock them into an ordered structure.*

The structure of the film and the orientation of the molecules depend on the area available to each molecule. The area can be varied by containing the molecules and applying pressure to the film with the aid of a movable barrier. The Langmuir trough provides not only a containment area with a movable barrier but also a means to measure the applied pressure.

When a three-dimensional gas is subjected to increasing pressure it eventually condenses, first into a liquid and then into a solid. Each phase exhibits progressively less compressibility and changes in the extent of ordering. Similarly, a two-dimensional Langmuir film exhibits different phases based on the applied pressure. One way to think of the effect of pressure is that the application of pressure effectively changes the area available for each molecule.

The effects of pressure are illustrated in Fig. 5.5. When the area per molecule is large, a disordered gaseous phase is formed. A gas possesses neither short-range nor long-range order and there is little interaction between the molecules. As the area is decreased a phase transition occurs into a liquid-like phase. Liquids are less compressible than gases and exhibit short-range but no long-range order. In these fluid phases, the chain is not rigid and adds to the disorder in the films. Further reduction in the area compresses the film into a solid-like phase. Langmuir determined that the molecules are standing up straight because the area per molecule is independent of chain length. This has been confirmed by x-ray and polarized infrared absorption measurements. If the area is decreased even further, the film collapses. Molecules get pushed out of the first layer and shearing leads to bilayer formation. The order in the solid state means that the molecules all point with the polar end group in the water and the alkyl chains stand up away from the surface.

Various condensed phases have been observed [45]. Not all phase transitions are observed for all films. The types of condensed phases that can be formed depend sensitively on the intermolecular interactions within the film and between the film and the liquid. Unlike a solid surface, the liquid under the Langmuir film does not present an ordered lattice to the amphiphile; therefore, lateral interactions must be responsible for the formation of ordered solid-phase Langmuir films. Mobility is not a restriction, as it

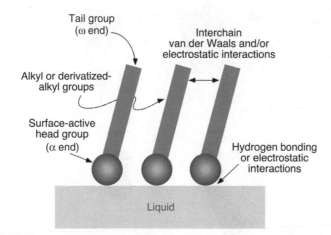

Figure 5.6   *A Langmuir film is a monomolecular layer of an amphiphile on a liquid. In this example, the polar head group of the amphiphile interacts with the liquid through hydrogen bonding or electrostatic interactions. By aligning and tilting, the tail groups maximize their attractive interactions.*

is in the formation of ordered layers by molecular beam epitaxy (MBE) or chemical vapour deposition (CVD, see Chapter 7), instead, the degree of ordering depends on the applied pressure and the strength of intermolecular interactions.

Figure 5.6 shows the interactions that occur within a Langmuir film. The head group of the amphiphile (the most polar end) is hydrogen bonded to the liquid (usually water). Hydrogen bonding is a relatively weak interaction of order $20\,\text{kJ}\,\text{mol}^{-1}$ (~ 0.2 eV) that can be equated to physisorption. In some instances, electrostatic forces rather than hydrogen bonding bind the head group to the polar liquid. The hydrophobic chains experience a repulsive interaction with the water. An upright configuration is thus favoured, but this interaction alone would not result in ordering. Ordering arises from non-polar chain–chain interactions. These interactions, caused by van der Waals forces, are maximized when neighbouring molecules align with one another. Although van der Waals forces are weak, the chains generally contain 8–20 or more carbon atoms and this contribution can amount to tens of $\text{kJ}\,\text{mol}^{-1}$. If the chain is not composed solely of methylene groups (CH_2) then hydrogen bonding may also play a role, e.g. for an ether (ROR) tail. Further impetus toward ordering can be given by the (polar or non-polar) interactions of end groups.

5.5   Langmuir-Blodgett films

Both physical (through the Kelvin equation) and chemical (through the contact angle) forces influence the formation of the liquid/solid interface. We can anticipate that the combination of these effects is particularly important at the nanoscale. We now investigate the formation of liquid/solid interfaces in more detail as well as considering the characteristics of molecular films at these interfaces.

5.5.1   Capillary condensation and meniscus formation

We start by considering the interaction of a vapour with confined spaces either in the form of a conical pore or in the crevice formed by the contact of two particles [5]. Consider first a conical pore with a perfectly wetting surface exposed to a vapour, Fig. 5.7(a). The most common case is a pore exposed to air

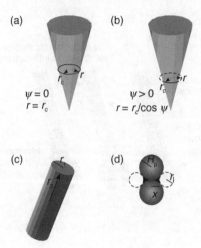

Figure 5.7   *(a) A conical pore with a hydrophilic surface induces the condensation of water. (b) Capillary condensation in a pore with a partially wetting surface. (c) A cylindrical pore with a width r$_1$ significantly smaller than the length. (d) An illustration of meniscus formation in the presence of two spherical particles of radius R$_p$.*

and the vapour in question is that of water. Water condenses on the hydrophilic surface, filling the pore up to the point at which a curved interface with radius of curvature r_c is formed. The concave interface can be treated as though a bubble of radius r_c is positioned atop the condensed water. The Kelvin equation allows us to calculate r_c once equilibrium has been reached,

$$RT \ln \frac{p}{p^*} = -\frac{2\gamma V_m}{r_c}. \tag{5.5.1}$$

The formation of a concave liquid/gas interface (the meniscus), brought about by the attractive water–surface interaction, reduces the vapour pressure of the liquid and drives the condensation of the liquid. The radius r_c corresponds to the capillary radius at the top of the meniscus. If the surface of the pore is not perfectly wetting but instead exhibits a contact angle $\psi > 0$ as shown in Fig. 5.7(b), then curvature of the interface is reduced such that $r = r_c/\cos \psi$. The level of the liquid is comparatively less than in the fully hydrophilic case because the vapour pressure reduction is less pronounced.

In a cylindrical pore of radius r_1, the sidewalls are straight which corresponds to an infinite radius of curvature $r_2 = \infty$. Hence, we need to make the substitution $1/r_1 + 1/r_2$ for $2/r_c$ in Eq. (5.5.1), which allows us to estimate the radius of pores that accommodate capillary condensation

$$r_1 = -\frac{\gamma V_m}{RT \ln(p/p^*)}. \tag{5.5.2}$$

This equation also holds approximately for any other irregular pore shape for which the length is much longer than the width.

The two particles with radius R_p depicted in Fig. 5.7(d) also induce the condensation of vapour and the formation of a meniscus with a radius of curvature r. The pressure associated with the presence of this liquid applies a force to the particles, which is known as the capillary force. In the case where the particles are not immersed in liquid and the liquid is present due to capillary condensation, $x \gg |r|$ and $1/r_1 + 1/r_2 = 1/x - 1/r \approx -1/r$. The effective pressure acting on the particles due to condensation is

$$\Delta p = \gamma/r \tag{5.5.3}$$

which acts over an area of πx^2 and, therefore, a capillary force of

$$F = \frac{\pi x^2 \gamma}{r}, \tag{5.5.4}$$

which can in turn be related by simple geometry to the particle radius as

$$F = 2\pi \gamma R_{\mathrm{p}}. \tag{5.5.5}$$

Irregularities in the surface profile of real particles tends to lower the capillary force compared to the value calculated with Eq. (5.5.5). The capillary force literally forces liquids into powders and porous materials. Powders and nanoparticles that are not kept dry will aggregate into larger masses that are held together by the capillary force. Another important consequence of the capillary force is that, as a liquid dries within a porous material, it retreats progressively into smaller and smaller pores. According to Eq. (5.5.3), the smaller the pore, the greater the pressure applied by the liquid to the material. Enormous pressures on the order of hundreds of atmospheres can result in material with nanoscale pores, see Exercise 5.16. As discussed in detail by Bellet and Canham [46], this can tear apart thin porous films and alter their structure. Similarly, the structure of powders, nanoparticles and nanowires can be changed as a result of the effects of drying. Cracking of porous films occurs below a certain critical thickness h_{c} that depends on the surface tension of the liquid γ_{L}, the surface tension and bulk modulus of the solid γ_{S} and E_{S}, respectively, as well as the porosity ε and the pore radius r_{p} according to

$$h_{\mathrm{c}} = \left(\frac{r_{\mathrm{p}}}{\gamma_{\mathrm{L}}}\right)^2 E_{\mathrm{S}}(1-\varepsilon)^3 \gamma_{\mathrm{S}}. \tag{5.5.6}$$

Drying stress can be avoided if the liquid is replaced by a supercritical fluid [47]. As the supercritical fluid is extracted from the pores, no meniscus forms because there is no vapour/liquid interface and, consequently, no capillary force.

A meniscus is also formed when a planar solid is pushed into a liquid. The balance of surface tension forces at the liquid/solid interface determines the profile of the meniscus. Figure 5.8 illustrates how the relative surface tensions of the liquid and solid lead to either concave or convex meniscuses. The surfaces of oxides tend to be covered with hydroxyl groups. These surfaces are polar and hydrophilic and, therefore, the

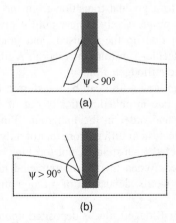

(a)

(b)

Figure 5.8 *Meniscus formation on a planar substrate. (a) A hydrophilic substrate in water. (b) A hydrophobic substrate in water. ψ, contact angle.*

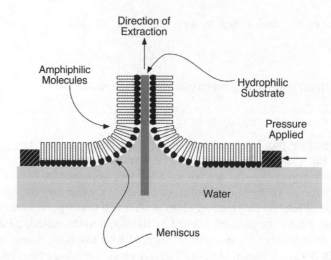

Figure 5.9   *The transfer of a Langmuir film onto a solid substrate via vertical deposition. In this example the substrate is hydrophilic and interacts attractively with the head group of the amphiphile. Movable barriers are required to maintain a constant surface pressure in the film to ensure uniform deposition.*

attractive forces associated with the hydroxyl groups lead to upward sloping meniscuses in water. Common hydrophilic substrates include Al_2O_3 (alumina, sapphire) Cr_2O_3, SnO_2, SiO_2 (glass, quartz), Au and Ag. A non-polar surface is not wet by water and forms a downward curving meniscus. Si is a particularly versatile substrate in that it can be made hydrophilic when an oxide layer is present or hydrophobic when the surface is H-terminated or coated with siloxanes. A Si surface normally has a thin (several nm) oxide layer on it, known as the native oxide. This surface is readily transformed into a H-terminated surface by dipping in acidic aqueous solutions containing fluoride.

5.5.2   Vertical deposition

Now consider the situation depicted in Fig. 5.9 in which a solid substrate interacts with a Langmuir film. The formation of a meniscus leads to a gradual transition from the surface of the liquid to the surface of the substrate. The liquid becomes progressively thinner until it gives way to the solid substrate. Thus, the molecules in the Langmuir film ride up the meniscus and gradually the binding switches from a molecule–liquid interaction to a molecule–substrate interaction. A film of amphiphiles transferred onto a solid substrate is known as a Langmuir-Blodgett film.

Figure 5.9 demonstrates in addition how a film can be transferred from the surface of the liquid to the surface of the substrate. If the substrate is pulled vertically out of the liquid and a movable boundary is used to maintain the pressure on and order in the Langmuir film, the film can be transferred to the substrate. The contact angle of a static system differs from that of a dynamic system in which the substrate is moving. As long as the velocity of the substrate is not too high, the general geometry of Fig. 5.9 is maintained and the film is transferred. When the film is retracted too rapidly, the system breaks down and transfer does not occur. In other words, deposition of a Langmuir-Blodgett film only occurs below a critical velocity.

Figure 5.9 depicts how a Langmuir-Blodgett film is deposited upon retraction of a hydrophilic surface; conversely, deposition during immersion can take place for a hydrophobic surface. The substrate need not be a "clean" substrate; that is, the substrate can be one upon which a Langmuir-Blodgett film has

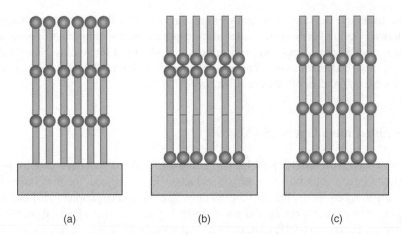

Figure 5.10   *X, Y and Z multilayer Langmuir-Blodgett (LB) films. (a) X-type films deposit only on downstrokes. (b) Y-type films are deposited successively on upstrokes and downstrokes. (c) Z-type films deposit only on upstrokes. Downstroke, insertion of the substrate into the LB trough; upstroke, retraction of the substrate from the LB trough.*

already been deposited. In this way, multilayer films can be built up that are composed of several layers of amphiphiles. Figure 5.10 displays several types of film that can be formed and defines the classifications of multilayer films. X multilayer films are built up from non-polar–non-polar interactions that bind the amphiphile to the surface. Subsequently, head–tail interactions bind one layer to the next. Such a film arises exclusively for amphiphiles in which the polarity does not differ greatly between the two termini. These are generally the least stable multilayer films. Y multilayer films contain amphiphiles in alternating head-tail–tail-head configurations. They are the most stable multilayer films and are most likely to occur for strongly polar head groups. Z multilayer films are an inverted version of X films; that is, the head group binds to the substrate and, thereafter, tail–head interactions hold neighbouring layers together. Z films may be more tightly bound to the substrate than X films, but the lack of tail–tail, head–head interactions between layers results in a much less stable film than found in Y films.

Langmuir-Blodgett films are usually weakly bound to the substrate by physisorption. Hence, these films are not stable with regard to washing in solvents. Exposure to air may gradually cause their destruction. However, particular combinations, such as RSH+Au, can lead to chemisorbed films.

5.5.3   Horizontal lifting (Shaefer's method)

Rather than approaching perpendicular to the surface, the substrate can be pressed onto a Langmuir film in a parallel geometry. This method of transferring Langmuir films onto a substrate is known as horizontal lifting or Shaefer's method. While Langmuir and Shaefer may have been the first to perform in-depth scientific studies on the horizontal deposition method, the technique has been used in Japan for over 800 years in the decorative arts [48, 49]. The Japanese technique of *sumi-nagashi* (meaning "ink flow") utilizes a horizontal deposition method to transfer marbled patterns onto paper. Legend has it that this art form was bestowed upon the Hiroba family by the god Kasuga in 1151. The family continues the tradition of making *sumi-nagashi* prints today. The technique consists of placing inks, generally black, red and indigo, on top of water. The inks are suspensions of organic pigments or graphite in protein; hence, they spread out over the surface of the water and do not mix with it. After completing the design with various brush strokes and by blowing on the film, the artist transfers the pattern by placing a sheet of paper on top of the film. Here the technique deviates from that of Langmuir and Shaefer in that paper is a porous material and the ink is drawn into the fibres of the material by capillary forces rather than remaining on the outer surface of the substrate. Sheets of *sumi-nagashi* are traditionally used for writing poetry but recently they have also found use as endpapers, book covers (in particular, this book's cover owes its design to an example of *sumi-nagashi*) and in other decorative uses.

5.6   Self assembled monolayers (SAMs)

Self assembled monolayers (SAMs) are molecular assemblies formed spontaneously by immersion of an appropriate substrate into a solution of an active surfactant in an organic solvent. Interest in SAMs has grown exponentially in recent years, but the first report by Zisman and co-workers [50] in 1946 went almost unnoticed by his contemporaries. By far the most studied system is that of an alkanethiol ($HS(CH_2)_nX$ with $X = CH_3$, CF_3, $CHCH_2$, CH_2OH, $COOH$, etc) interacting with a Au surface as first investigated by Nuzzo and Allara [51]. Alkanethiols are often abbreviated RSH or HSRX, where X represents the tail group. The head group is the sulfur end of the molecule, which is denoted the α end. The tail group is denoted as the ω end. Numerous systems form self-assembled monolayers as a result of adsorption either from solution or from the gas phase [52–54]. These include:

- Alkanethiols on Au, Ag, Cu and GaAs.
- Dialkyl sulfides (RSR') and disulfides (RS–SR') on Au.
- Organosilanes ($RSiCl_3$, $RSi(OCH_3)_3$, $RSi(NH_2)_3$) on SiO_2, Al_2O_3, glass, quartz, mica, GeO_2, ZnSe and Au.
- Alcohols (ROH) and amines (RNH_2) on Pt.
- Carboxylic acids (fatty acids, RCOOH) on Al_2O_3, CuO, AgO and Ag.
- Alcohols or terminal alkenes ($RC{=}CH_2$) on H-terminated Si.

Multifarious organic functionalities have been tethered to Si surfaces through the formation of Si–C bonds [55, 56]. The formation of Si–C bonds often requires activation by irradiation, a peroxide initiator, heating or an applied voltage. These conditions are harsher than those usually associated with the formation of SAMs, especially for the prototypical alkanethiol/Au system. This budding area of SAM research is of great interest because of the ability to pattern Si (and por-Si) surfaces as well as the technological implications of many Si-based materials.

The formation of an RSH SAM generally follows Langmuirian kinetics. The rate of adsorption is dependent on the concentration of the alkanethiol in solution, which is usually on the order of 10^{-3} mol l^{-1}. Jung and Campbell [57] have determined the sticking coefficient for C_2–C_{18} straight-chain alkanethiols. Langmuirian kinetics (see Chapter 4) were followed up to ~80% of the saturation coverage. The value of the initial sticking coefficient is of the order 10^{-6}–10^{-8}, increasing with increasing chain length. The initial sticking coefficient is on the order of unity for the adsorption of RSH from the gas phase. Clearly, the presence of a solvent hinders adsorption of the alkanethiol. It does so in two ways. First, solvent molecules (ethanol in the case studied by Jung and Campbell) adsorb on the surface and must be displaced by the incoming RSH. Second, a solvation shell surrounds a dissolved molecule. The alkanethiol must shed the solvent molecules that constitute the solvation shell before it can adsorb on the surface. The combination of these two processes leads to an adsorption barrier in the case of adsorption from the solution that is not present for adsorption from the gas phase.

A monolayer forms at room temperature in minutes to hours, though days may be required to obtain the most highly ordered, close-packed structure. A spectacular coincidence in the balance of intermolecular forces allows for formation of ordered structures at room temperature. Most RSH/Au SAMs are observed to disorder if heated above $\sim$380 K. Increased ordering of a completed monolayer can be achieved by cooling, particularly if a temperature below $\sim$250 K can be reached [58].

5.6.1 Thermodynamics of self-assembly

Here we follow Israelachvili [59], who discusses intermolecular forces, aggregation and self-assembly in some detail, in particular how they result in micelle, bilayer and membrane formation. Chemical equilibrium is defined by the state in which all components of the system have the same chemical potential. For an equilibrated solution composed of identical molecules drawn into aggregates composed of N molecules ($N = 1$ monomer, $N = 2$ dimer, $N = 3$ trimer, etc with an aggregate of N molecules being an N-mer), the chemical potential is expressed as

$$\mu = \mu_1^0 + k_B T \log X_1 = \mu_2^0 + \tfrac{1}{2} k_B T \log \tfrac{1}{2} X_2 = \mu_3^0 + \tfrac{1}{3} k_B T \log \tfrac{1}{3} X_3 = \dots, \tag{5.6.1}$$

where μ_N^0 is the standard part of the chemical potential, i.e. the mean interaction free energy per molecule, for aggregates of number N and X_N is the activity, which for low enough concentrations can be taken as the concentration. We can write this more succinctly as

$$\mu = \mu_N = \mu_N^0 + \frac{k_B T}{N} \log\left(\frac{X_N}{N}\right), N = 1, 2, 3, \dots \tag{5.6.2}$$

The equilibrium constant K of the reaction between N monomers and an aggregate of N molecules

$$N \, \text{monomer} \underset{k_N}{\overset{k_1}{\rightleftharpoons}} N\text{-mer}$$

can be written equivalently in terms of the chemical potential or a ratio of rate constants

$$K = \exp\left[-\frac{N(\mu_N^0 - \mu_1^0)}{k_B T} \right] = \frac{k_1}{k_N}. \tag{5.6.3}$$

The law of mass action can be used to write the rate of association

$$r_a = k_1 X_1^N \tag{5.6.4}$$

and the rate of dissociation

$$r_d = k_N (X_N / N) \tag{5.6.5}$$

from which we derive

$$X_N = N \left[X_1 \exp\left(\frac{\mu_1^0 - \mu_N^0}{k_B T} \right) \right]^N. \tag{5.6.6}$$

In Eq. (5.6.6), we have chosen the monomer as our reference state. The concentration, defined in terms of volume fraction or mole fraction, must satisfy the conservation relation that the total concentration C sums to unity

$$C = \sum_{N=1}^{\infty} X_N. \tag{5.6.7}$$

Note that this derivation is valid only so long as the aggregates do not begin to aggregate among themselves.

So when do molecules aggregate? If all aggregates experience the same interactions with the solvent, the value of μ_N^0 is constant for all values of N. Substitution of this constant value into Eq. (5.6.6) means that in this case

$$X_N = NX_1^N \tag{5.6.8}$$

and since $X_1 < 1$, all other concentrations must be lower yet, $X_N < X_1$, with the difference becoming increasingly large as N increases. If μ_N^0 increases as a function of N, the situation is even less favourable for aggregation. Therefore, the condition for aggregation is that μ_N^0 must be a decreasing function of N. If μ_N^0 is a complicated function of N, for example, one that exhibits a minimum at a certain value of N, then there is a size distribution of aggregates, the distribution function is controlled by the chemical potential through Eq. (5.6.6).

5.6.2   Amphiphiles and bonding interactions

For the formation of Langmuir films, it is essential to use amphiphiles to obtain ordered films. The same considerations and interactions found in Fig. 5.6 are also pertinent to the formation of SAMs because these interactions are what determine the chemical potential and provide the driving force for self-assembly. Nonetheless, there are two factors that differentiate Langmuir films from self-assembled monolayers. First, the head group is chemisorbed to the surface, with typical bond energies $>100\,\text{kJ}\,\text{mol}^{-1}$ ($>1\,\text{eV}$). Second, the substrate is flat with a defined arrangement of surface atoms. Therefore, a driving force toward ordering is provided by the directionality of chemisorption. While chemisorption may force the head groups to seek registry with the surface, the ordering of the chains and tails is brought about predominantly by the intermolecular interactions outlined above.

The substrate must be unreactive to everything but the molecule that is attempting to form a monolayer. Au and H-terminated Si are both rather inert surfaces. Many solvents only physisorb to these surfaces. The target molecule easily displaces the physisorbed molecules. The tenacious chemisorption of impurities would poison the formation of the SAM. Another important characteristic of the substrate is the spacing of the surface atoms. The Au(111) surface provides a virtually perfect template for alkanethiols. The S atoms are able to assume a close-packed structure, presumably by filling the three-fold hollow sites, with S–S distances of 4.99 Å. This distance is sufficient to ensure that the chains are not crowded, and have sufficient space to tilt and form an ordered structure that maximizes the chain–chain interactions.

5.6.3   Mechanism of SAM formation

The dynamics of SAM formation of alkanethiols on gold has been elucidated by Poirier and Pylant [60] and is depicted in Fig. 5.11. They investigated the interaction of various RSH molecules with a Au(111) surface under UHV conditions. Nonetheless, the observations they made seem to be generally applicable to the growth of SAMs from solution as well as to the process that occurs during the formation of ordered Langmuir films. The description below applies to studies carried out at room temperature. The major differences from adsorption out of solution are that solvent molecules need not be displaced from either the surface or the solvation shell.

The clean Au(111) is reconstructed into a herringbone structure. At low coverage, the adsorbed alkanethiol is present as a mobile two-dimensional gas. In this phase, the alkyl chain is oriented roughly parallel to the surface. As the coverage is increased, ordered islands of a solid phase begin to appear. The solid and 2D gas phases are in equilibrium and the appearance of the solid phase represents a first-order phase transition. The solid phase, denoted solid$_1$, is anchored to the surface through the S atom. The molecular axis is still parallel to the surface. The appearance of solid$_1$ leads to local changes in the Au(111)

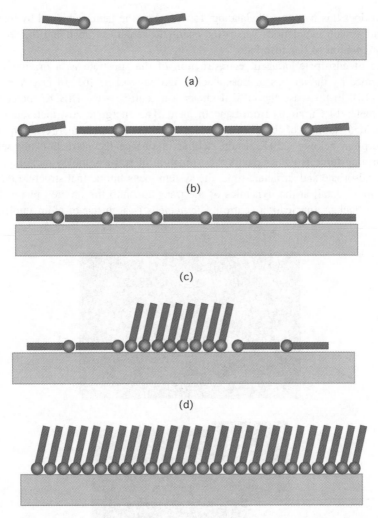

Figure 5.11 *Mechanism of self assembly: (a) disordered, mobile lattice gas; (b) lattice gas plus ordered islands; (c) saturated surface-aligned phase; (d) nucleation of solid-phase islands; (e) saturated solid phase. Self-assembly progresses in stages, as long as a sufficient supply of adsorbates is available, until the thermodynamically most favoured final state is reached. Not depicted in the figure is a reconstruction of the surface, which often accompanies the process.*

reconstruction. That is, the substrate changes its structure only under the nucleated islands. As the coverage is increased yet further, the surface becomes saturated with solid$_1$. Continued adsorption results in a second phase transition to a second solid phase (solid$_2$). In solid$_2$ the RSH stand upright and the same packing density is obtained as that achieved at saturation for a SAM deposited from solution. The driving forces of this phase transition are lateral interactions in the solid phase as well as the propensity of the alkanethiols to form as many Au–S bonds as possible.

This growth mechanism is a perfect illustration of the effects of Eq. (1.2.1). It demonstrates how multiple (substrate and adsorbate) phases can co-exist at the same temperature. It also demonstrates how a sequence of adsorbate structures can be populated as σ_A increases. The system as a whole needs to minimize its free energy, and it does this by counterbalancing, for instance, the gains achieved by increased adsorption and increased chain–chain interactions versus the energetic penalty of changing the reconstruction of the substrate or the orientation of the adsorbate.

The importance of adsorbate-induced reconstruction of the substrate as a driving force behind self-assembly is confirmed by the studies of Besenbacher and co-workers [61] of hexa-*tert*-butyldecacyclene (HtBDC) on Cu(110). In this case the critical nucleation centre is two HtBDC molecules that join and simultaneously expel ∼14 Cu atoms from beneath them. The energetic cost of removing these atoms is balanced by the added stability of the chemisorbed molecules. Thermal energy is required to aid in this process, and adsorption below ∼250 K results neither in surface reconstruction nor self assembly. The thermodynamic driving force is what favours the formation of self-assembled monolayers. Given enough of the appropriate adsorbate and sufficient time, the system locks into a final structure that is well ordered.

Let us look in more detail at the dynamics of the transition into the ordered phase. Growth of a new phase starts from a nucleus (nucleation centre). The nucleus is a cluster of atoms/molecules of a critical

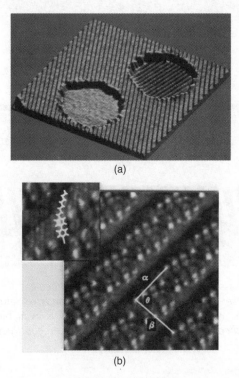

(a)

(b)

Figure 5.12 *Self-assembled rows of 4'octyl-4-biphenylcarbonitrile (8CB) molecules in corrals on graphite. (a) A 3000 × 3000 Å^2 constant-height scanning tunnelling microscope (STM) image. Several hours after deposition, ordered monolayer films have formed on the terrace and in one corral but not the other. Note how the rows in the corral are not aligned with the rows formed on the surrounding terrace. The second corral is filled with a disordered layer of 8CB molecules. (b) A 95 × 95 Å^2 constant-height image with molecular resolution. The position of one molecule is marked by the model overlay. Reproduced from D. L. Patrick, V. J Cee, T. P. Beebe Jr, Science 265, 231. © 1994 with permission from the American Association for the Advancement of Science.*

size with the appropriate orientation. The formation of this nucleus is a random process. The nucleus is assembled by statistical fluctuations that bring together the constituent units in the proper structure. Once the critical nucleus forms, the rest of the phase assembles around it.

Beebe and co-workers [62] have studied the formation of ordered domains of 4′-octyl-4-biphenylcarbonitrile (8CB) on graphite. To facilitate their study, they used specially prepared graphite substrates with circular etch pits. The one-monolayer-deep depressions act as molecular corrals that fence off the adsorbates inside from the surrounding terraces, allowing for detailed studies of isolated surface regions. 8CB self-assembles on graphite on a time scale of hours. Figure 5.12 exhibits an STM image of a graphite sample immersed in pure liquid 8CB at room temperature. Studies of such images reveal several interesting aspects of the dynamics of nucleation and self-assembly. Figure 5.12 illustrates three molecular corrals surrounded by regular graphite terraces. Two of the three corrals and the terraces are covered by regular arrays of 8CB. The third corral is filled with a disordered layer of 8CB. This corral appears empty because the disordered 8CB molecules are highly mobile. This image confirms the picture painted above: the adsorbate undergoes a phase transition from a disordered phase into the ordered phase. Furthermore, this phase transition is irreversible at room temperature. Once the ordered phase is formed, it does not spontaneously revert to the disordered phase.

The ordering of 8CB is much more rapid than the formation of nucleation centres. Nucleation occurs over a period of hours. Ordering, on the other hand, occurs too rapidly (tens of milliseconds) to be observed by the STM. Effectively, the disordered molecules wait around for a nucleation centre to form but once it appears, the remaining molecules snap into position. The formation of a nucleus is obviously difficult. By studying the nucleation rate in corrals with radii from roughly 50 to 1350 Å, Beebe and co-workers determined that nucleation is more likely to occur on terraces rather than steps. Quite possibly, the disorder brought about by the steps destabilizes the nucleation centres, making them less favourable sites for nucleation. This is in marked contrast to numerous metal-on-metal and semiconductor-on-semiconductor growth systems in which islands preferentially nucleate at defect sites.

Advanced Topic: Chemistry with Self Assembled Monolayers

The head groups in a SAM are strongly anchored and are, therefore, more tightly bound to the substrate than LB films. SAMs can be washed and rinsed with solvents. Chemical reactions with the tail or between chains can be performed without disrupting the monolayer. Once anchored to the surface with an ω-functional group exposed, the full tool chest of organic or organosilane chemistry can be brought into action to transform the termination of the SAM into a bewildering array of different molecular entities. ω-substitution of bulky groups has the tendency to reduce the order in the SAM.

These properties of SAMs make them ideal candidates for fundamental studies on molecular recognition. A SAM makes it possible to tether an ordered array of specific functional groups onto the surface. Interactions of this group with another molecule can then be studied with the advantage that the tethered molecule has a known orientation and is fixed in space. Prime and Whitesides, for example, have used SAMs to study specific protein binding interactions [63]. As discussed in Chapter 2, SAM formation on the tip of an AFM has been used to make tips of known chemical termination [64]. These layers can then be used to study chemical interactions between the tip-bound molecules and molecules immobilized on a substrate. This has proved a fruitful method of investigating the chemistry of a variety of biomolecules. The stability and uniform chemical structure of SAMs also make them ideal surfaces to study tribology, the science of friction, and the influence of interfacial layers on electrochemistry and charge transfer at surfaces.

5.7   Thermodynamics of liquid interfaces

As discussed at the beginning of this chapter, liquid interfaces are inherently more complex than the gas/solid interface. Because of this, liquid/solid, liquid/liquid and gas/liquid interfaces are more difficult to probe and understand with molecular detail. Hence a thermodynamic approach is useful for characterization of adsorbed layers and the formation of these interfaces.

5.7.1   The Gibbs model

Figure 5.13 illustrates the construction of a thermodynamic description of a two-phase system. Two homogeneous phases with volumes V^A and V^B are separated by an interfacial region of finite volume in which the structure and composition changes gradually from that of phase B to phase A as shown in Fig. 5.13(b). In the Guggenheim model, this interfacial region is treated explicitly. In the Gibbs model, the interface is represented by an infinitesimally thin plane that separates the two phases. The volume of a plane vanishes; therefore, the total volume is

$$V = V^A + V^B. \tag{5.7.1}$$

All other expressions for extensive properties of the system can be written in terms of the contributions from the three regions. Thus, n_j the number of moles of component j, U the internal energy and S entropy can be written as

$$n_j = n_j^A + n_j^B + n_j^S \tag{5.7.2}$$

$$U = U^A + U^B + U^S \tag{5.7.3}$$

$$S = S^A + S^B + S^S. \tag{5.7.4}$$

To determine values of the surface contributions we take advantage of the homogeneous phase values that are determined far from the interface. Writing the internal energy per unit volume of phases A and B and the concentrations of j in phases A and B as u^A and u^B, and c_j^A and c_j^B, respectively, the interfacial internal energy is

$$U^S = U - u^A V^A - u^B V^B \tag{5.7.5}$$

and the number of moles of component j at the interface is

$$n_j^S = n_j - c_j^A V^A - c_j^B V^B. \tag{5.7.6}$$

5.7.2   Surface excess

Having defined the quantity of molecules at the interface in Eq. (5.7.6), we also define a quantity that is similar to, but, because of the arbitrary definition of the ideal Gibbs interface, is not identical to the coverage. This is the surface excess

$$\Gamma_j = n_j^S / A. \tag{5.7.7}$$

We are using moles because we are specifically thinking of liquids and solutions, and concentrations are generally reported in moles per unit volume. Surface excess can also be defined in terms of molecules per unit area, as is common for coverage, instead of moles per unit area. Furthermore, there is nothing in the above discussion that could not be applied to the gas/solid interface. Note, however, that for the case of adsorbates confined to the surface of a solid (with no subsurface penetration) exposed to vacuum

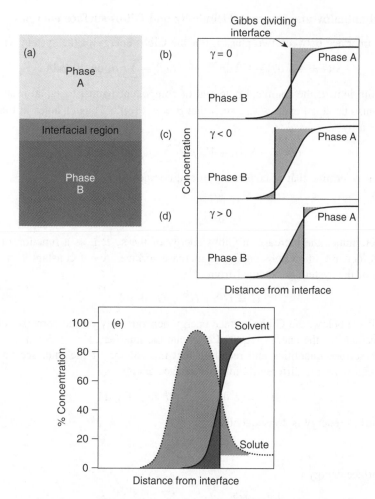

Figure 5.13 *The definition of the Gibbs model of an interface. (a) The system is composed of two homogeneous phases separated by an interfacial region. (b)–(d) In the interfacial region, the composition varies. The Gibbs surface is drawn at an arbitrary plane in the interfacial region, conventionally such that the surface excess of one component vanishes as shown in (b). (e) In a solution, the composition of a solute near the interface is generally not the same as in the bulk and the Gibbs surface is drawn such that the surface excess of the solvent vanishes.*

or gas, there are none of the uncertainties of what is meant by the extent of the solid and gas phases, and where the interface lies. For liquid interfaces, the position of the Gibbs dividing plane is arbitrary; however, conventionally it is drawn such that the differences in composition on either side of the plane balance, and $\Gamma = 0$ of the liquid (at the vapour/liquid interfaces) or solvent (for solution/solid interfaces).

Whereas the position of the Gibbs dividing plane and, therefore, G are arbitrary quantities, i.e. ones that depend on a choice, it can be shown that the relative absorption

$$\Gamma_j^{(1)} \equiv \Gamma_j^S - \Gamma_1^S \frac{c_j^A - c_j^B}{c_1^A - c_1^A} \tag{5.7.8}$$

is independent of the choice of the position of Gibbs surface. $\Gamma_j^{(1)}$ is the relative absorption of component j with respect to component 1 (usually taken to be the solvent) and is an experimentally measurable quantity.

5.7.3   Interfacial enthalpy and internal, Helmholtz and Gibbs surface energies

Consider a planar interface between two phases. For the Gibbs energy of the system we write

$$dG = -S\,dT + V^A\,dp^A + V^B\,dp^B + \sum \mu_j\,dn_j + \gamma\,dA. \tag{5.7.9}$$

Recall that at equilibrium μ_j the chemical potential of component j must be equal in all phases, and that the pressure must also be equal everywhere such that $p = p^A = p^B$. Thus changes in the Gibbs energy of the system are given by

$$dG = -S\,dT + V\,dp + \sum \mu_j\,dn_j + \gamma\,dA \tag{5.7.10}$$

from which we can determine that at constant T, p and composition

$$\left.\frac{\delta G}{\delta A}\right|_{T,p,n} = \gamma. \tag{5.7.11}$$

Surface tension determines the increase in Gibbs energy of the system as a function of area. Since γ is positive, it always increases the Gibbs energy to increase surface area at constant T, p and composition. The Gibbs energy of the system is obtained from

$$G = G^A + G^B + G^S + \gamma A \tag{5.7.12}$$

because as we will see below, the Gibbs surface energy depends only on the composition of the interface and we need to account for the energy required to create the interface.

Turning now to surface quantities and recalling that the volume of the interface vanishes within the Gibbs model, we can write the differential internal surface energy as

$$dU^S = T\,dS^S + \sum \mu_j\,dn_j^S + \gamma\,dA \tag{5.7.13}$$

and the internal surface energy is consequently

$$U^S = TS^S + \sum \mu_j n_j^S + \gamma A. \tag{5.7.14}$$

The Helmholtz surface energy is

$$F^S = U^S - TS^S = \sum \mu_j n_j^S + \gamma A. \tag{5.7.15}$$

The interfacial enthalpy is

$$H^S = U^S - \gamma A. \tag{5.7.16}$$

Finally, the Gibbs surface energy is

$$G^S = H^S - TS^S = F^S - \gamma A = \sum \mu_j n_j^S. \tag{5.7.17}$$

While in differential form it is

$$dG^S = -S^S\,dT + \sum \mu_j\,dn_j^S - A\,d\gamma. \tag{5.7.18}$$

From Eq. (5.7.17) we see that the surface Gibbs energy is always positive. In other words, the formation of a surface is energetically unfavourable. Dynamically this can be understood by considering that an increase of surface area by the addition of bulk atoms to the surface is accomplished by moving fully co-ordinated bulk atoms to surface sites of lower co-ordination. This process is energetically unfavourable, both because the atom must be moved to the surface and because of the less favourable bonding environment at the surface compared to the bulk.

5.7.4   Gibbs adsorption isotherm

It can be shown that at constant temperature

$$d\gamma = -\sum \Gamma_i \, d\mu_i. \tag{5.7.19}$$

Eq. (5.7.19) is known as the Gibbs adsorption isotherm and is valid only for liquid surfaces. For solid surfaces additional terms accounting for strain must be added. Specifically, for a two-component system composed of a solvent 1 and solute 2, the Gibbs adsorption isotherm is written

$$d\gamma = -\Gamma_1 \, d\mu_1 - \Gamma_2 \, d\mu_2 = -\Gamma_2 \, d\mu_2, \tag{5.7.20}$$

since we choose the Gibbs surface such that $\Gamma_1 = 0$. The chemical potential of the solute is a function of the activity of the solute a according to

$$\mu_2 = \mu_2^\circ + RT \ln a_2 \tag{5.7.21}$$

where $a_2 = \Gamma_2 c_2 / c^\circ$ is the activity of the solute with $c^\circ = 1$ mol dm^{-1}. Differentiating Eq. (5.7.21) with respect to a and inserting into Eq. (5.7.20) yields

$$\Gamma_2 = -\frac{1}{RT} \frac{d\gamma}{d \ln a_2} = -\frac{a_2}{RT} \frac{d\gamma}{da_2} \tag{5.7.22}$$

Equation (5.7.22) shows us that there are two types of molecules: those that are enriched at the interface compared to their bulk concentration ($\Gamma_2 > 0$), and those that avoid the interface ($\Gamma_2 < 0$). Molecules that are enriched at the surface are called surface active molecules or surfactants. This class of molecules, such as the amphiphilic molecules encountered above in connection with Langmuir and Langmuir-Blodgett films, lower the surface tension of the interface. A molecule that avoids the interface increases the surface tension when it is added to the solution. Equation (5.7.22) also provides us with a method of determining the surface excess by measuring surface tension as a function of concentration. This is particularly easy at low concentrations where activity can be replaced by concentration and then $\partial\gamma/\partial a_2 \approx \Delta\gamma/\Delta c_2$.

5.8   Electrified and charged interfaces

We return now to investigate in more detail the implications of the structure of charged interfaces, as depicted in Fig. 5.1. When a metal electrode is biased positively, the electrons in the Friedel oscillation are drawn back into the metal, while the nuclei remain fixed. The counterions move in closer to the surface, and the capacity of the Helmholtz layer increases. The changes in the electron distribution also change the dipole associated with the surface region.

5.8.1   Surface charge and potential

Biasing changes the effective charge on the surface and changes the coverage of adsorbed ions. The adsorption of anions is enhanced by positive charges at the surface and are suppressed when the surface is sufficiently biased to surpass some critical value of excess negative charge. Cations behave, of course, in just the opposite manner. Charge transfer and the associated dipole moment of molecular adsorbates are also changed with bias.

The approximation of a Helmholtz layer controlling the charge distribution at and near the metal interface pertains to the limit of not too dilute solutions. At low concentrations, as described in the Stern-Gouy model, beyond the Helmholtz layer the bulk electrolyte concentrations are only reached after an extended distance. In the second region, the ion distribution is governed by a Poisson-Boltzmann distribution [5], which is a long-range gradual sloping of the concentrations from the large excess near the electrode to the limiting bulk values.

Calculation of the charge distribution at and near the interface is an important but difficult task. Determine the value of the electric field as a function of distance, the surface charge, and capacitance of the double layer are required. All of these influence the chemical and physical characteristics of the interface as well as charge transfer and transport to and from the interface. The particulars obviously require detailed knowledge (or at least a detailed model) of the interfacial structure. The surface charge density is denoted σ^0, while that in the Stern layer is σ^i and that in the Gouy layer is σ^d. When a surface carries no net charge, it is at its point of zero charge (pzc). As discussed above, the pzc depends on the pH and the applied bias.

With a model of the charge density as a function of the distance from the surface, it is possible to calculate the surface potential ψ^0, the potential at the inner Helmholtz plane ψ^i, and the potential at the top of the Stern layer ψ^d. However, none of these potentials can be measured directly, so what is the potential that enters into equilibrium thermodynamics? The electrical potential energy difference between two points is equal to the electrical work required to move a unit charge from Point A to Point B. The potential is the work divided by the unit charge. The inner potential is defined by the work required to move a test charge from infinity to the inside of a phase. This is known as the Galvani potential φ and is the potential that enters into equilibrium electrochemistry. It is to be distinguished from the external potential, also known as the Volta potential, ψ and the surface potential χ, which are all related by

$$\varphi = \chi + \psi. \tag{5.8.1}$$

The Volta potential is defined by the work required to move a test charge from infinity (either in vacuum, an inert gas, or a solution) to a point close to the surface ($\sim 1\,\mu m$ away). This distance is sufficiently far away that chemical and image force interactions with the interface can be neglected. Volta potentials and Volta potential differences can be measured directly; however, since we cannot deconvolute the chemical work from electrical work in crossing between two different phases, we cannot unambiguously measure φ, χ nor changes in these. We can, nonetheless, measure changes in the differences of φ and χ.

To relate the Galvani potential to measurable quantities, we need to define the electrochemical potential $\overline{\mu}$ in terms of the chemical potential μ, the charge z_i on species i, and the Faraday constant F,

$$\overline{\mu}_i = \mu_i + z_i F \varphi. \tag{5.8.2}$$

The chemical potential is related to the temperature and activity as given in Eq. (5.7.21). The term $z_i F \varphi$ is effectively the work required to move one mole of electrons into a phase with a potential φ at its interior. To move charge from phase A to phase B requires the amount of work determined by

$$\Delta \overline{\mu}_i = \overline{\mu}_i^B - \overline{\mu}_i^A = \mu_i^B - \mu_i^A + z_i F (\varphi_B - \varphi_B). \tag{5.8.3}$$

At equilibrium the electrochemical potential is everywhere the same, thus $\Delta \overline{\mu}_i = 0$, which requires that

$$\Delta \varphi = \varphi_B - \varphi_A = \frac{\mu_i^B - \mu_i^A}{z_i F}. \tag{5.8.4}$$

Whereas the electrochemical potential is everywhere the same at equilibrium, the Galvani potential is not, that is, the Galvani potential difference $\Delta \varphi$ does not in general equal zero unless by coincidence the inner potentials of the two phases happen to be the same. A consequence of Eq. (5.8.4) is that when two dissimilar metals are brought into contact, as discussed in Eq. 1.3.3, a contact potential $\Delta \varphi$ develops because of the flow of charge to the interface between the two metals. Similarly, the flow of charge and the formation of dipole layers at the electrolyte/electrode interface also lead to a contact potential. The electrical field in the dipole layer contains contributions not only from the presence of ions but also due to the polarization of the solvent.

Consider the specific example of a gold electrode in contact with a solution containing Au^{3+} at equilibrium. The appropriate electrochemical reaction is

$$Au(s) \rightleftharpoons Au^{3+}(aq) + 3e^- \tag{5.8.5}$$

in which the electrons on the right-hand side are in the metal. From the stoichiometry of this reaction, we then write the relationship between chemical potentials of the components of the system as

$$\overline{\mu}_{Au} = \overline{\mu}_{Au^{3+}(aq)} + 3\overline{\mu}_{e^-} \tag{5.8.6}$$

We denote the Galvani potential in the solution as φ_S and that in the metal as φ_M. Assuming that the Au atoms are neutral in the metal such that $\overline{\mu}_{Au} = \mu_{Au}$, and taking the standard activity $a_0 = 1$, we use Eq. (5.8.2) to write

$$\mu^0_{Au} + RT \ln a_{Au} = \mu^0_{Au^{3+}} + RT \ln a_{Au^{3+}} + 3F\varphi_S + \mu^0_{e^-} + 3RT \ln a_{Au} - 3F\varphi_M. \tag{5.8.7}$$

By definition, the activity of Au atoms in the bulk is unity and since the effective concentration of electrons in the metal is constant, we can also take their activity to be one; therefore the Galvani potential difference is

$$\Delta\varphi = \varphi_M - \varphi_S = \Delta\varphi^0 + \left(\frac{RT}{3F}\right) \ln a_{Au^{3+}} \tag{5.8.8}$$

and the standard Galvani potential difference is

$$\Delta\varphi^0 = \frac{\mu^0_{Au^{3+}} + \mu^0_{e^-} - \mu^0_{Au}}{2F}. \tag{5.8.9}$$

The potential of one electrode in isolation is not measurable; however, when we introduce a reference electrode whose Galvani potential difference is constant, the potential E of our working electrode with respect to this reference electrode is measurable. Moreover,

$$E - E^0 = \Delta\varphi - \Delta\varphi^0, \tag{5.8.10}$$

can be determined experimentally. The value E^0 is the value of E measured at unit activity (and similarly for $\Delta\varphi^0$). Consequently, we are justified in writing the familiar Nernst equation

$$E = E^0 + \left(\frac{RT}{zF}\right) \ln a_{M^{z+}} \tag{5.8.11}$$

for a metal electrode M in contact with a solution containing its ions M^{z+}.

5.8.2 Relating work functions to the electrochemical series

As is discussed in Chapter 8, charge transfer can only occur between an adsorbate or solution species and a solid substrate if acceptor and donor energy levels exist at the appropriate energy in both systems. How do we tell where the energy levels of two disparate systems line up? For the gas/solid interface this is a rather straightforward exercise and is easily accomplished by photoelectron spectroscopy, which can measure directly the binding energy of any occupied orbital. The presence of a solution complicates

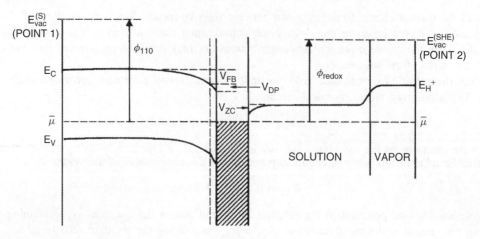

Figure 5.14 *An energy level diagram depicting the levels associated with the connection of a semiconductor electrode to a standard hydrogen electrode. Reproduced from H. Reiss, A. Heller, J. Phys. Chem., 89, 4207. ©1985 with permission from the American Chemical Society.*

the matter greatly as has been discussed, for instance, by Gerischer [65], Reiss [66, 67] and Trasatti [68, 69]. This issue is of particular importance when considering the results of *ab initio* calculations because without a method of aligning the well characterized but relative values of the electrochemical series to the absolute scale of work functions and the vacuum level, no quantitative comparison to experimental results can be made. This issue remains controversial and the uncertainty in the absolute scale is as much as ±0.2 eV.

Alignment of energy levels is particularly important in semiconductor photochemistry and electrochemistry. Therefore, following Reiss and Heller [66], Fig. 5.14 has been constructed. Prior to forming an electrical connection, the vacuum levels of the semiconductor, E_{vac}^{S}, and the standard hydrogen electrode (SHE), E_{vac}^{SHE}, are equal. The difference between the Fermi energy and the vacuum level determines the work function. Noting that the electrochemical potential $\bar{\mu}$ is equivalent to the Fermi energy and setting our absolute reference scale by setting $E_{vac}^{SHE} = 0$, we write

$$\bar{\mu} = \bar{\mu}_{redox} = -\Phi_{redox}/e, \tag{5.8.12}$$

where Φ_{redox} is the work function of the SHE. Once the two electrodes (the semiconductor and a Pt electrode both immersed in a 1 M solution of $HClO_4$ exposed to 1 atm of H_2 at 298 K) are brought into electrical contact, the vacuum levels shift such that the electrochemical potential is the same everywhere. There is band bending of the conduction band minimum E_c and the valence band maximum E_v in the semiconductor prior to immersion. The band bending is measured in terms of the flat band potential V_{FB}. V_{FB} is measured by finding the slope of a Mott-Schottky plot of capacitance versus applied potential. The situation is slightly less complicated if the semiconductor is replaced by a metal in that there is no band bending ($V_{FB} = 0$) in the latter case. A further change in potential in front of the semiconductor may occur and is denoted by the dipole layer potential V_{DP}. Similarly, the potential of zero charge V_{zc} represents the response of the Pt electrode to the Stern layer above it. The energy of an electron added to a H^+ ion to produce a H atom is denoted E_{H^+}.

From Fig. 5.14 we can determine that alignment of the Fermi levels leads to

$$E_{vac}^{S} - E_{vac}^{SHE} = V_{FB} + V_{DP} - V_{zc} \tag{5.8.13}$$

From our choice of an absolute reference, substituting for E_{vac}^S by using

$$\Phi_S/e = E_{vac}^S - \overline{\mu}, \tag{5.8.14}$$

and for $\overline{\mu}$ from Eq. (5.8.12), we obtain

$$\Phi_{redox}/e = V_{FB} + V_{DP} - V_{zc} - \Phi_S/e. \tag{5.8.15}$$

Similarly for the chemical potential

$$\overline{\mu}_{redox} = \overline{\mu} = V_{FB} + V_{DP} - V_{zc} - \Phi_S/e. \tag{5.8.16}$$

Equations (5.8.15) and (5.8.16) are both referenced to Point 2 in Fig. 5.14. For the specific case of InP(110) in contact with the SHE, Reiss and Heller [66] have found $\Phi_S = 5.78\,eV$, $V_{FB} = 1.05\,V$, $V_{DP} \approx 0$, and $V_{zc} = -0.3\,V$. Therefore,

$$\overline{\mu}_{redox} = \overline{\mu}_{H^+/H_2} = -4.43\,eV, \tag{5.8.17}$$

and the work function of the SHE is $\Phi_{SHE} = 4.43\,eV$. This value is in line with the recommendation from the IUPAC [68] of $4.44 \pm 0.02\,eV$. New attempts to unambiguously determine Φ_{SHE} continue to be developed. Attempts have been made to measure absolute reduction potentials by measuring the energy given off when an ion solvated in a cluster of water molecules in a supersonic molecular beam combines with an electron. The energy is measured by counting the number of water molecules and hydrogen atoms that desorb from the cluster in response to the energy released by ion neutralization. The product of the number of water molecules desorbed times the activation energy for desorption is then equal to the energy released by electron capture of the solvated ion [70, 71]. These direct experimental determinations suggest a value in the range of 4.29–4.45 eV based on two independent methods. Computational methods are becoming comprehensive enough to include solvation accurately and to deliver results with close to chemical accuracy. Truhlar's group among others has taken up this challenge and reports a low value of 4.28 ± 0.09 eV [72].

5.9 Summary of important concepts

- At the liquid/solid interface we need to consider not only the structure of the adsorbed layer but also that the near-surface region of the liquid has a structure and composition that can differ markedly from that of the bulk.
- Strongly solvated solution species do not lose their solvation shell and can become non-specifically adsorbed. A solvated species that loses at least part of its solvation shell and chemisorbs to the solid is specifically adsorbed.
- The structure of water near the surface depends on the substrate, the presence of specifically adsorbed ions and the presence of an electric field. Field effects and direct ion-water dipole interactions influence the orientation of the water molecules. Water dissociates on some metal surfaces to form mixed H + OH layers.
- By adjusting the potential and the pH, the net charge on the surface can be tuned from positive to neutral to negative.
- Surface energy and surface tension are synonymous for liquids.
- Surface energy can be thought of as being analogous to a two-dimensional pressure, compare Eqs (5.2.1) and (5.2.2), or else as the surface Gibbs energy per unit area of a pure liquid, Eq. (5.2.4).

- Curved interfaces act differently than planar interfaces. One consequence is that the vapour pressure changes, which leads to the instability of small droplets and capillary condensation within small pores and between small particles.
- Langmuir films are monomolecular films on the surface of a liquid.
- When these films are transferred onto a solid substrate, they are called Langmuir-Blodgett films.
- Self-assembled monolayers are ordered monolayer films that form spontaneously on a solid substrate. It is the chemical potential that provides the thermodynamic driving force for self-assembly.
- Whereas the chemisorption interaction between the head group and the surface accounts for the bulk of binding interaction in SAMs, it is the weak, predominantly noncovalent interactions between the chains and endgroups that lead to the order in the SAM.
- The Gibbs model of adsorption allows us to define the surface excess and calculate fundamental thermodynamics quantities at liquid interfaces.
- The Gibbs surface energy is always positive and creating additional surface area is always energetically unfavourable.
- The Stern-Guoy model describes the interface between an electrode and a solution in terms of a charged adsorbed layer, a layer of solvated ions above this (which together form the electric bilayer), followed by a diffuse and mobile continuous medium that extends into the remainder of the solution.
- The electrochemical potential and chemical potential are related by Eq. (5.8.2). The electrochemical and chemical potentials of neutral species are the same. At equilibrium, it is the electrochemical potential of all charged species that is the same throughout the system.
- The inner or Galvani potential is not everywhere the same even at equilibrium, and it is the Galvani potential that underlies the integrity of the Nernst equation.
- Relating electrochemical potentials to work functions and an absolute scale is a nontrivial but important exercise that allows us to understand how the electronic levels of electrodes and solution species are aligned with respect to one another.

5.10   Frontiers and challenges

- The structure of the liquid/solid interface and how it couples to reaction dynamics.
- Measurement of absolute half-cell potentials and the relationship of $E°$ values to band edges in real solutions.
- Quantum chemical theoretical description of heterogeneous electrochemistry (charge transfer and electrocatalysis) with chemical accuracy.
- Description and understanding of the interface between ionic liquids and solids, in particular, charge transfer in such systems and how they relate to batteries and other electrochemical devices.
- The science and technology of fuel cells and batteries. This includes a number of topics, e.g. charge transfer, mass transport, electrocatalysis and phase separations affected by membranes.
- The mechanism of self-assembly. How can we learn to control it, and how do systems perform error correction?

5.11   Further reading

H.-J. Butt, K. Graf, and M. Kappl, *Physics and Chemistry of Interfaces*, 2nd ed. (Wiley-VCH, Weinheim, 2006).

C. Clay and A. Hodgson, *Water and mixed OH/water adsorption at close packed metal surfaces*, Curr. Opin. Solid State Mater. Sci. **9** (2005) 11.

L. H. Dubois and R. G. Nuzzo, *Synthesis, structure and properties of model organic surfaces*, Annu. Rev. Phys. Chem. **43** (1992) 437.

M. A. Henderson, *The interaction of water with solid surfaces: Fundamental aspects revisited*, Surf. Sci. Rep. **46**, 1–308 (2002).

A. J. Hopkins, C. L. McFearin, and G. R. Richmond, *Investigations of the solid-aqueous interface with vibrational sum-frequency spectroscopy*, Curr. Opin. Solid State Mater. Sci. **9** (2005) 19.

J. N. Israelachvili, *Intermolecular and Surface Forces*, 3rd ed (London, Academic Press, 2011).

D. M. Kolb, *An atomistic view of electrochemistry*, Surf. Sci. **500** (2002) 722.

G. M. Nathanson, P. Davidovits, D. R. Worsnop, and C. E. Kolb, *Dynamics and kinetics at the gas-liquid interface*, J. Phys. Chem. **100** (1996) 13007.

G. L. Richmond, *Molecular bonding and interactions at aqueous surfaces as probed by vibrational sum frequency spectroscopy*, Chem. Rev. **102** (2002) 2693.

F. Schreiber, *Structure and growth of self-assembling monolayers*, Prog. Surf. Sci. **65** (2000) 151.

D. K. Schwartz, *Langmuir-Blodgett film structure*, Surf. Sci. Rep. **27** (1997) 24.

J. S. Spendelow, P. K. Babu, and A. Wieckowski, *Electrocatalytic oxidation of carbon monoxide and methanol on platinum surfaces decorated with ruthenium*, Curr. Opin. Solid State Mater. Sci. **9** (2005) 37.

C. D. Taylor and M. Neurock, *Theoretical insights into the structure and reactivity of the aqueous/metal interface*, Curr. Opin. Solid State Mater. Sci. **9**, 49–65 (2005).

P. A. Thiel and T. E. Madey, *The interaction of water with solid-surface: Fundamental aspects*, Surf. Sci. Rep. **7**, 211 (1987).

A. Ulman, *An Introduction to Ultrathin Organic Films from Langmuir-Blodgett to Self-Assembly* (Academic Press, Boston, 1991).

A. Ulman, *Formation and structure of self-assembled monolayers*, Chem. Rev. **96** (1996) 1533.

5.12   Exercises

5.1   Derive the Young Equation, Eq. (5.3.1).

5.2   Consider a hemispherical liquid island of radius r with surface energies $\gamma_{sl} > \gamma_{lg}$ in equilibrium with its vapour. Calculate the island surface energy as a function of r and demonstrate that small islands are unstable with respect to large islands. Assume the substrate to be rigid and that the island energy is composed only of island–substrate and island–vapour terms.

5.3   Consider the dynamics of deposition of X, Y and Z multilayer films. For each case, determine whether deposition occurs on the downstroke (insertion of substrate into the LB trough), upstroke (retraction) or in both directions. Discuss the reasons for these dependencies.

5.4   The sticking coefficient is defined as

$$s = r_{ads}/Z_w \tag{5.12.1}$$

and represents the probability of a successful adsorption event. The collision frequency in solution is given by

$$Z_w = c_{sol}\left(\frac{k_B T}{2\pi m}\right)^{1/2} \tag{5.12.2}$$

where c_{sol} is the concentration in molecules per cubic metre. The initial sticking coefficient of $CH_3(CH_2)_7SH$ on a gold film is 9×10^{-8} [57]. Assuming a constant sticking coefficient, which is valid only at low coverage, estimate the time required to achieve a coverage of 0.01 ML for adsorption from a 5×10^{-3} mol l^{-1} solution. Take the surface density of atoms to be 1×10^{19} m^{-2}.

5.5   Your lab partner has prepared two Si crystals but has not labelled them. One is H-terminated, the other is terminated with an oxide layer. Propose and explain an experiment you could perform in your kitchen that would distinguish the two.

5.6   Explain the observed trend that C_4 straight-chain amphiphiles generally do not form LB films or SAMs that exhibit a structure that is as well ordered as that of C_{12} straight-chain amphiphiles.

5.7   Describe what would occur during vertical deposition of a LB film if the barriers of the trough were stationary and a large surface area substrate were used.

5.8   After a 4 h exposure to pure, deoxygenated H_2O, a H-terminated Si(111) surface is found to have an oxygen atom coverage of 0.6 ML measured with respect to the number of Si atoms in the Si(111)−(1 × 1) layer. Estimate the sticking coefficient, (i) assuming that all of the oxygen is the result of dissociative H_2O adsorption and (ii) assuming that all of the oxygen results from the adsorption of OH^-.

5.9   Calculate the energy released when $H^+(aq)$ reacts with a dangling bond to form an Si–H bond on an otherwise hydrogen-terminated surface. You will need the following: The ionization potential of H atoms $IP(H) = 13.61\,eV = 1313\,kJ\,mol^{-1}$; the enthalpy of solvation of protons $\Delta_{solv}H(H^+) = -11.92\,eV = -1150\,kJ\,mol^{-1}$ and the Si–H bond strength $D(Si–H) = 3.05\,eV = 294\,kJ\,mol^{-1}$. Assume that the enthalpy of solvation of the Si dangling bond on an otherwise H-terminated surface is the same as the enthalpy of solvation of the Si–H unit that is formed.

5.10   In UHV surface science the "usual" first-order pre-exponential factor is $1 \times 10^{13}\,s^{-1}$. In electrochemistry the "usual" pre-exponential factor for electron transfer at an electrode surface is $1 \times 10^4\,m\,s^{-1}$. Explain how these values are derived.

5.11   When Si is placed in HF(aq) a dark current of about $0.4\,\mu A\,cm^{-2}$ is measured, i.e. a current that is measured in the absence of an applied potential and in the absence of illumination. Assuming that this dark current is due to an etching reaction, which is initiated by the absorption of F^- (aq) and that one electron is injected into the Si substrate for every one Si atom that is removed by etching, calculate the etch rate of unilluminated Si in HF(aq).

5.12   Derive the Nernst equation for the Fe^{2+}/Fe^{3+} redox couple.

5.13   The dipole moments of O(a) and OH(a) on Pt(111) are 0.035 and $0.05e$ Å, respectively [73]. Calculate the shift in the adsorption energy caused by the interaction of the electric field in the double layer with the dipoles when the surface is biased at 1 V relative to the point of zero charge.

5.14   Calculate the radius of curvature, and discuss capillary condensation in a conical pore that has surfaces with (i) $\psi = 90°$ and (ii) $\psi = 180°$.

5.15   Consider a material with cylindrical pores exposed to air at 25°C with a humidity of 85%. Into pores of what size will water condense?

5.16   Calculate the effective pressure resulting from capillary forces and the critical film thickness for a porous silicon film with a porosity $\varepsilon = 0.90$ when dried in air after rinsing in water or ethanol. The mean pore diameter is $r_p = 5\,nm$. $\gamma_{EtOH} = 22.75\,mN\,m^{-1}$, $\gamma_{water} = 71.99\,mN\,m^{-1}$, $\gamma_{Si} = 1000\,mN\,m^{-1}$, $E_{Si} = 1.62 \times 10^{11}\,N\,m^{-2}$.

5.17   Show that the relative absorption $\Gamma_j^{(1)}$ is independent of the position of the Gibbs surface.

5.18   Derive Eq. (5.7.19).

5.19   The Langmuir isotherm also holds for adsorption from solution onto a solid substrate. Recognizing that the pressure p of a gas is related to concentration c, it is straightforward to rewrite the isotherm as

$$\theta = \frac{Kc}{1 + Kc}. \tag{5.12.3}$$

The specific surface area a_s (area per unit mass) is used to characterize high surface area solids such as activated charcoal and porous materials. Therefore, coverage is also conveniently expressed in terms of quantities expressed per unit mass of the substrate. Thus, fractional coverage is equal to the ratio of the number of molecules adsorbed per gram of adsorbent N_s divided by the number of sites per gram of adsorbent (= the number of molecules required to form a monolayer per gram of adsorbent N_m). More conveniently yet, we convert number of molecules to moles (e.g. $n_s = N_s/N_A$) and

$$\theta = n_s/n_m. \tag{5.12.4}$$

Use Eq. (5.12.3) and (5.12.4) to show that

$$\frac{c}{n_s} = \frac{1}{n_m K} + \left(\frac{1}{n_m}\right)c. \tag{5.12.5}$$

References

[1]   H. Schlichting, D. Menzel, *Surf. Sci.*, **272** (1992) 27.
[2]   J. O. M. Bockris, A. K. N. Reddy, *Modern Electrochemistry*, 2nd ed., Plenum Press, New York, 1998.
[3]   H. Gerischer, Principles of electrochemistry, in *The CRC Handbook of Solid State Electrochemistry* (Eds: P. Gellings, H. Bouwmeester), CRC Press, Boca Raton, 1997, p. 9.
[4]   R. Guidelli, W. Schmickler, *Electrochim. Acta*, **45** (2000) 2317.
[5]   H.-J. Butt, K. Graf, M. Kappl, *Physics and Chemistry of Interfaces*, 2nd ed., Wiley-VCH, Weinheim, 2006.
[6]   H. von Helmholtz, *Wied. Ann.*, **7** (1879) 337.
[7]   G. Gouy, *J. Phys. (Paris)*, **9** (1910) 457.
[8]   D. C. Chapman, *Philos. Mag.*, **25** (1913) 475.
[9]   O. Stern, *Z. Elektrochem.*, **30** (1924) 508.
[10] P. Debye, E. Hückel, *Phys. Z.*, **24** (1923) 185.
[11] C. D. Taylor, M. Neurock, *Curr. Opin. Solid State Mater. Sci.*, **9** (2005) 49.
[12] C. Clay, A. Hodgson, *Curr. Opin. Solid State Mater. Sci.*, **9** (2005) 11.
[13] A. Hodgson, S. Haq, *Surf. Sci. Rep.*, **64** (2009) 381.
[14] P. A. Thiel, T. E. Madey, *Surf. Sci. Rep.*, **7** (1987) 211.
[15] M. A. Henderson, *Surf. Sci. Rep.*, **46** (2002) 1.
[16] A. Michaelides, P. Hu, *J. Am. Chem. Soc.*, **122** (2000) 9866.
[17] A. Michaelides, V. A. Ranea, P. L. de Andres, D. A. King, *Phys. Rev. Lett.*, **90** (2003) 216102.
[18] D. L. Doering, T. E. Madey, *Surf. Sci.*, **123** (1982) 305.
[19] J. D. Bernal, R. H. Fowler, *J. Chem. Phys.*, **1** (1933) 515.
[20] L. Pauling, *J. Am. Chem. Soc.*, **57** (1935) 2680.
[21] G. L. Richmond, *Chem. Rev.*, **102** (2002) 2693.
[22] A. J. Hopkins, C. L. McFearin, G. R. Richmond, *Curr. Opin. Solid State Mater. Sci.*, **9** (2005) 19.
[23] J. W. Gibbs, *Collected Works, Vol. 1, Thermodynamics*, Longmans, London, 1928.
[24] L. D. Landau, E. M. Lifshitz, *Fluid Mechanics*, Addison-Wesley, Reading, MA, 1959.
[25] E. H. Lucassen-Reynders, J. Lucassen, *Adv. Colloid Interface Sci.*, **2** (1970) 347.
[26] P. Tarazona, R. Checa, E. Chacon, *Phys. Rev. Lett.*, **99** (2007) 196101.
[27] A. ten Bosch, *Phys. Rev. E*, **73** (2006) 031605.
[28] Z. Jiang, H. Kim, X. Jiao, H. Lee, Y. J. Lee, Y. Byun, S. Song, D. Eom, C. Li, M. H. Rafailovich, L. B. Lurio, S. K. Sinha, *Phys. Rev. Lett.*, **98** (2007) 227801.
[29] H. Sasaki, E. Tokizaki, X. M. Huang, K. Terashima, S. Kimura, *Jpn. J. Appl. Phys.*, **34** (1995) 3432.
[30] W. W. Mullins, R. F. Sekerka, *J. Appl. Phys.*, **35** (1964) 444.
[31] W. A. Tiller, K. A. Jackson, R. W. Rutter, B. Chalmers, *Acta Metal.*, **1** (1953) 428.
[32] M. J. Aziz, J. Y. Tsao, M. O. Thompson, P. S. Peercy, C. W. White, *Phys. Rev. Lett.*, **56** (1986) 2489.
[33] D. E. Hoglund, M. O. Thompson, M. J. Aziz, *Phys. Rev. B*, **58** (1998) 189.
[34] P. Galenko, *Phys. Rev. E*, **76** (2007) 031606.

[35]   P. Galenko, *Phys. Rev. B*, **65** (2002) 144103.

[36]   A. G. Cullis, *J. Vac. Sci. Technol. B*, **1** (1983) 272.

[37]   P. Behroozi, K. Cordray, W. Griffin, F. Behroozi, *Am. J. Phys.*, **75** (2007) 407.

[38]   F. Zhu, R. C. Miao, C. L. Xu, Z. Z. Cao, *Am. J. Phys.*, **75** (2007) 896.

[39]   G. E. Fogg, *Nature (London)*, **154** (1944) 515.

[40]   A. B. D. Cassie, S. Baxter, *Nature (London)*, **155** (1945) 21.

[41]   L. Gao, T. J. McCarthy, *J. Am. Chem. Soc.*, **128** (2006) 9052.

[42]   E. R. Jerison, Y. Xu, L. A. Wilen, E. R. Dufresne, *Phys. Rev. Lett.*, **106** (2011) 186103.

[43]   K. J. Laidler, *The World of Physical Chemistry*, Oxford University Press, Oxford, 1993.

[44]   I. Langmuir, *J. Am. Chem. Soc.*, **39** (1917) 1848.

[45]   A. W. Adamson, A. P. Gast, *Physical Chemistry of Surfaces*, 6th ed., John Wiley & Sons, Inc., New York, 1997.

[46]   D. Bellet, L. Canham, *Adv. Mater.*, **10** (1998) 487.

[47]   G. W. Scherer, Drying stresses, in *Ultrastructure Processing of Advanced Materials* (Eds: D. R. Uhlmann, D. R. Uhlrich), John Wiley & Sons, Inc., New York, 1992, p. 179.

[48]   S. Usui, **1996**. The author is deeply grateful to Sachiko Usui of International Research Center for Japanese Studies, Kyoto, Japan for providing information on Sumi-nagashi.

[49]   S. Hughes, *Washi: The World of Japanese Paper*, Kodansha International, Tokyo, 1978.

[50]   W. C. Bigelow, D. L. Pickett, W. A. Zisman, *J. Colloid Interface Sci.*, **1** (1946) 513.

[51]   R. G. Nuzzo, D. L. Allara, *J. Am. Chem. Soc.*, **105** (1983) 4481.

[52]   A. Ulman, *An Introduction to Ultrathin Organic Films from Langmuir-Blodgett to Self-Assembly*, Academic Press, Boston, MA, 1991.

[53]   A. Ulman, *Chem. Rev.*, **96** (1996) 1533.

[54]   F. Schreiber, *Prog. Surf. Sci.*, **65** (2000) 151.

[55]   J. M. Buriak, *Chem. Commun.*, (1999) 1051.

[56]   R. Boukherroub, *Curr. Opin. Solid State Mater. Sci.*, **9** (2005) 66.

[57]   L. S. Jung, C. T. Campbell, *Phys. Rev. Lett.*, **84** (2000) 5164.

[58]   L. H. Dubois, R. G. Nuzzo, *Annu. Rev. Phys. Chem.*, **43** (1992) 437.

[59]   J. N. Israelachvili, *Intermolecular and Surface Forces*, 2nd ed., Academic Press, London, 1991.

[60]   G. E. Poirier, E. D. Pylant, *Science*, **272** (1996) 1145.

[61]   M. Schunack, L. Petersen, A. Kühnle, E. Lægsgaard, I. Stensgaard, I. Johannsen, F. Besenbacher, *Phys. Rev. Lett.*, **86** (2001) 456.

[62]   D. L. Patrick, V. J. Cee, T. P. Beebe, Jr., *Science*, **265** (1994) 231.

[63]   K. L. Prime, G. M. Whitesides, *Science*, **252** (1991) 1164.

[64]   H. Takano, J. R. Kenseth, S.-S. Wong, J. C. O'Brien, M. D. Porter, *Chem. Rev.*, **99** (1999) 2845.

[65]   H. Gerischer, W. Ekardt, *Appl. Phys. Lett.*, **43** (1983) 393.

[66]   H. Reiss, A. Heller, *J. Phys. Chem.*, **89** (1985) 4207.

[67]   H. Reiss, *J. Phys. Chem.*, **89** (1985) 3783.

[68]   S. Trasatti, *Pure Appl. Chem.*, **58** (1986) 955.

[69]   S. Trasatti, *Electrochim. Acta*, **36** (1991) 1659.

[70]   R. D. Leib, W. A. Donald, J. T. O'Brien, M. F. Bush, E. R. Williams, *J. Am. Chem. Soc.*, **129** (2007) 7716.

[71]   W. A. Donald, R. D. Leib, J. T. O'Brien, E. R. Williams, *Chemistry – A European Journal*, **15** (2009) 5926.

[72]   C. P. Kelly, C. J. Cramer, D. G. Truhlar, *J. Phys. Chem. B*, **110** (2006) 16066.

[73]   J. K. Nørskov, J. Rossmeisl, Á. Logadóttir, L. Lindqvist, J. R. Kitchin, T. Bligaard, H. Jónsson, *J. Phys. Chem. B*, **108** (2004) 17886.

6

Heterogeneous Catalysis

From the outset of UHV surface science studies, legitimate concerns have been raised regarding whether a pressure gap and/or a materials gap exist that would make UHV studies on model systems irrelevant for the high-pressure world of industrial catalysis. The pressure gap signifies the uncertainties in extrapolating kinetic data over as many as 10 orders of magnitude. One method to surmount this barrier has been to combine UHV methods with high-pressure reaction vessels [1]. The materials gap is posed by the uncertainties derived from using single crystals to model the highly inhomogeneous materials, often composed of small metal clusters on oxide substrates, that comprise industrial catalysts. These gaps have now been breached. The oxidation of CO over platinum group metals [2, 3] and the ammonia synthesis reaction [4, 5] are arguably the two best-understood heterogeneous catalytic reactions. Using kinetic parameters derived from UHV studies on model catalysts it is possible to model the reaction rates observed on high surface area catalysts at high pressures [5–7]. Even complex reactions involving the formation and conversion of hydrocarbons over metal catalysts can be understood across these gaps [6, 8, 9]. Hence, pressure and materials gaps exist only insofar as gaps exist in our knowledge of how to apply properly the lessons learned from UHV surface science studies. These gaps are not intrinsic barriers to the understanding of heterogeneous catalysis in terms of elementary reactions and fundamental principles of dynamics.

6.1 The prominence of heterogeneous reactions

Much of industrial chemistry involves transformation of carbon containing molecules. There are seven building block molecules, all of which are derived from petroleum, upon which the chemical industry relies: benzene, toluene, xylene, ethylene, propylene, 1,3-butadiene and methanol [10]. This is reflected in a list of materials made by industry. Table 6.1 lists the most important bulk chemicals ranked by production, as well as gasoline, the top three gases and the top two metals. Heterogeneous processes feature as a major role in all of these syntheses apart from ethylene, urea, N_2, and O_2 production. The contact process for H_2SO_4 synthesis is carried out with a supported liquid catalyst. Cl_2, NaOH and Al are all produced by heterogeneous electrochemical reactions. The catalysts are mainly composed of small metal particles supported on high surface area porous solids such as oxides or zeolites.

Table 6.1 reveals that the refining of petroleum and the production of petrochemicals are not possible at the present scale and cost without the use of catalysis [11]. The use of zeolites is particularly important in this area as both a catalyst and support. Zeolites are of interest as solid base catalysts for reactions such as

Surface Science: Foundations of Catalysis and Nanoscience, Third Edition. Kurt W. Kolasinski.
© 2012 John Wiley & Sons, Ltd. Published 2012 by John Wiley & Sons, Ltd.

Table 6.1   *US bulk chemical production in 2005 from C&ENews July 10, 2006. Aluminum, ammonia, phosphate rock and, pig iron data for 2010 from USGS Mineral Commodity Summaries 2011. Gasoline data from Energy Information Administration for 2009. Process information taken from* Ullmann's Encyclopedia of Industrial Chemistry *(Wiley-VCH, Weinheim, 2006)*

Material	Production/kMt	Process
1. H_2SO_4	36 520 (Europe 19 024; World 165 000)	Contact process. Catalytic oxidation of SO_2 over K_2SO_4 promoted V_2O_5 on a silica gel or zeolite carrier at 400–450 °C and 1–2 atm, followed by exposure to wet sulfuric acid.
2. Phosphate rock (P_2O_5, phosphoric acid)	26 100 (China 65 000; World 176 000)	Digestion of phosphate rock with H_2SO_4, HNO_3 or HCl
3. Ethylene (C_2H_4)	23 974 (Europe 21 600; World 75 000)	Thermal steam cracking of naptha or natural gas liquids at 750–950 °C, sometimes performed over zeolite catalysts to lower process temperature.
4. Propylene	15 333 (Europe 15 406)	By-product from ethylene and gasoline production (fluid catalytic cracking or FCC). Also produced intentionally from propane over Pt or Cr supported on Al_2O_3.
5. Ethylene dichloride	11 308 (Europe 6 646)	C_2H_4 + HCl + O_2 over copper chloride catalyst, $T > 200$ °C, with added alkali or alkaline earth metals or $AlCl_3$
6. Cl_2	10 175 (Europe 10 381)	Chlor-alkali process, anodic process in the electrolysis of NaCl. Originally at graphite anodes but new plants now exclusively use membrane process involving a Ti cathode (coated with Ru + oxide of Ti, Sn or Zr) or porous Ni coated steel (or Ni) cathode (with Ru activator).
7. NaOH	8 384 (Europe 10 588)	Chlor-alkali process, cathodic process in the electrolysis of NaCl.
8. NH_3	8 300 (China: 42 000, World: 131 000)	Haber-Bosch process, 30 nm Fe crystallites on Al_2O_3 promoted with K and Ca. SiO_2 added as a structural stabilizer, 400–500°C, 200–300 bar. Recently a more active but more expensive Ru catalyst has been developed.
9. Benzene	7 574 (Europe 7 908)	From catalytic cracking of petroleum. FCC process over solid acid catalysts (silica, alumina, zeolite) sometimes with added molybdena
10. Urea	5 801	Basaroff reactions of $NH_3 + CO_2$
11. Ethylbenzene	5 251 (Europe 4 276)	Ethylene + benzene over zeolite catalyst at $T < 289$ °C and pressures of $\sim$ 4 MPa
12. Styrene	5 042 (Europe 4 963)	Dehydrogenation of ethylbenzene in the vapour phase with steam over Fe_2O_3 catalyst promoted with Cr_2O_3/K_2CO_3 at $\sim$ 620 °C and as low pressure as practicable
Gasoline	379 800	FCC over solid acid catalysts (alumina, silica or zeolite) is performed to crack petroleum into smaller molecules. This is followed by naptha reforming at 1–3 MPa, $300 \le T \le 450$ °C, performed on Pt catalysts, with other metals, e.g. Re, as promoters.

Table 6.1   *(Continued)*

Gases	Production/Mm³	
1. N_2	26 448 (Europe 21 893)	Liquefaction of air
2. O_2	16 735 (Europe 26 128)	Liquefaction of air
3. H_2	13 989 (Europe 10 433)	Steam reforming of CH_4 (natural gas) or naptha over Ni catalyst supported on Al_2O_3, aluminosilicates, cement, and MgO, promoted with uranium or chromium oxides at T as high as 1 000 °C.

Metals		
1. Fe (pig iron)	29 000 (China: 600 000; World: 1 000 000)	Carbothermic reduction of iron oxides with coke/air mixture at $T \approx 2000$ °C.
2. Al	1 720 (China: 16 800; World: 41 400)	Hall-Heroult process. Electrolysis of bauxite (Al_2O_3) on carbon electrodes at 950–980 °C.

condensations, alkylations, cyclizations and isomerizations [12]. Further development of zeolites as solid base catalysts will require finding new materials that are less easily deactivated by CO_2 and water. Zeolites can also present acids sites or a combination of both acid and base sites. Their flexibility is further enhanced by the nature of their porosity and active sites. Their micropores can be tuned in a size range comparable to that of molecules. The nature of the active sites within these pores is controlled by substitution of transition metals and other elements into their walls [13]. A particularly popular type of zeolite known as ZSM-5 has the proper pore structure and active sites to convert biomass-derived molecules into aromatic hydrocarbons and olefins [10].

Since heterogeneous reactions occur at the gas/solid or liquid/solid interface, it is generally favourable to create maximum surface area in order to maximize the number of reactive sites. In other words, a general aim is to maximize the dispersion of our catalyst. Dispersion is the fraction of atoms in the catalyst that reside at the surface compared to the total number of atoms in the particle. The dispersion depends on the size and shape of the particle, but in general, dispersion increases with decreasing size. For example, if we consider spherical particles of radius r, mass m, density ρ and composed of N spherical atoms, we can calculate the volume of the particle and the atoms, as well as their surface areas and the number of surface atoms that cover the particle. Thereby, we can calculate (see Exercise 6.1) the relationship between the dispersion D and the radius of the particle

$$D = \left(\frac{48 M}{\pi \rho N_A} \right)^{1/3} \frac{1}{r}. \tag{6.1.1}$$

As expected, the smaller the particle, the greater the dispersion. For example, $D = 0.55$ for an $r = 1$ nm sphere but $D = 0.11$ for $r = 5$ nm. Whether or not the highest dispersion catalyst is actually the optimal catalyst depends on a number of factors that are considered in more detail below.

6.2   Measurement of surface kinetics and reaction mechanisms

The measurement of surface kinetics follows many of the same strategies as the measurement of kinetics in other phases [14], the basis of which always relates to the measurement of composition as a function of time. The difference is that in surface kinetics the measurements of the rate of consumption of gas-phase reactants and the appearance of gas-phase products are insufficient to determine the reaction mechanism and kinetics. This is because a complex set of surface reactions leads to quite complex reaction orders for the gas-phase species. These reaction orders often depend on temperature and gas composition and, therefore, cannot unambiguously be interpreted in terms of elementary steps. A true reaction mechanism cannot be determined without measurements of surface coverages, and an identification of which species are bound to the surface. For instance, in something as simple as a CO thermal desorption spectrum, a two-peak spectrum can be interpreted in at least three ways: (1) CO binds in two distinct binding sites, (2) strong lateral interactions account for the splitting of the desorption peak from a single binding site, or (3) CO partially dissociates, the low temperature peak is due to simple molecular desorption while the high temperature peak is due to recombinative desorption. This again highlights that a TPD peak area is not uniquely proportional to the initial coverage of any given binding site. A combination of TPD with, for example, vibrational spectroscopy on the layer before desorption and as a function of heating would deliver not only the absolute rate of the process, but also its proper interpretation in terms of a reaction mechanism.

Temperature programmed desorption applied to surface reactions is sometimes called temperature programmed reaction spectrometry (TPRS). An example is shown in Fig. 6.1 [15]. For the interpretation of a bimolecular reaction, this method requires four types of thermal desorption measurements. For the $CO + O_2$ reaction this would be a set of thermal desorption spectra conducted as a function of coverage for (1) pure CO, (2) pure O_2, (3) mixed $CO + O_2$ adlayers, and (4) pure CO_2 (the product). The interpretation of Fig. 6.1 is then straightforward. The low temperature O_2 desorption peaks are similar irrespective of the presence of CO. CO desorption is identical for both $CO + O_2$ and CO overlayers. The differences in the mixed phase are the formation of CO_2 and the suppression of a high-temperature O_2 desorption peak.

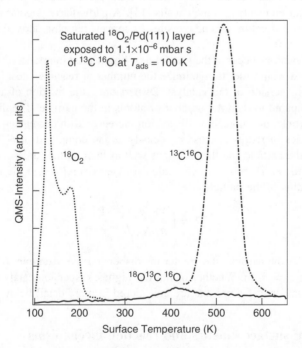

Figure 6.1 *A co-adsorbed layer of $^{18}O_2 + {}^{13}C^{16}O$ was prepared at $T_s = 100$ K on Pd(111). When heated, three products are observed in temperature programmed reaction spectrometry (TPRS): $^{18}O_2$, $^{13}C^{16}O$ and $^{13}C^{16}O^{18}O$. The CO_2 arises from the Langmuir-Hinshelwood reaction of $CO(a) + O(a)$. Reproduced from K. W. Kolasinski, F. Cemič, A. de Meijere, E. Hasselbrink, Surf. Sci., 334, 19. © 1995 with permission from Elsevier Science.*

The desorption of CO_2 arises from reaction-limited desorption caused by the Langmuir-Hinshelwood reaction of $CO(a) + O(a)$. This conclusion is arrived at as follows. EELS measurements [16] show that $O_2(a)$ dissociates on Pd(111) for $T_s \geq 180$ K, and above 250 K only atomic $O(a)$ is left on the surface. CO_2 is only formed above 300 K, therefore, the reaction must be between $CO(a)$ and $O(a)$ and not $CO(a)$ and $O_2(a)$. O_2 dissociation is a prerequisite for reaction, and the reaction proceeds between two chemisorbed species. The desorption of CO_2 is limited by its rate of formation, not its rate of desorption. Pure CO_2 would desorb from Pd(111) at <100 K, thus at ~ 400 K, the surface lifetime of adsorbed CO_2 is extremely short. As soon as the CO_2 is created by reaction, it desorbs from the surface. There is no high-temperature recombinative desorption of O_2 because $O(a)$ reacts completely with $CO(a)$, which is present in excess for the conditions in Fig. 6.1. The CO desorption peak matches that from a surface dosed only with CO because by the time the surface has reached 500 K, all of $O_2(a)$ and $O(a)$ have either desorbed or reacted, leaving the surface clean apart from $CO(a)$. Thus, CO oxidation follows a Langmuir-Hinshelwood mechanism described by the following reactions

$$CO + {}^* \rightleftharpoons CO^* \qquad (6.2.1)$$

$$O_2 + 2^* \rightleftharpoons 2O^* \qquad (6.2.2)$$

$$O^* + CO^* \rightleftharpoons CO_2 + 2^* \qquad (6.2.3)$$

where * represents an empty site.

Note that Fig. 6.1 utilized isotopic labelling. Isotopic labelling plays three roles in surface kinetics measurements. First, it is used to obtain better signal-to-noise ratios in the data. This is particularly important for CO and CO_2, which are often present in the background gases. Second, it is used to distinguish the products. N_2 and CO both appear at mass 28. Introduction of ${}^{18}O$, ${}^{15}N$ or ${}^{13}C$ into the reactants shifts the N_2 and CO mass spectral peaks accordingly when the isotopically labelled species is incorporated into the desorbed product. Isotopic labelling facilitates the identification of reaction pathways by explicating which bonds are breaking. When ${}^{18}O_2 + {}^{13}C{}^{16}O$ are dosed onto Pd(111) as in Fig. 6.1, the products observed are ${}^{18}O-{}^{18}O$, ${}^{18}O-{}^{13}C-{}^{16}O$ and ${}^{13}C{}^{16}O$. The lack of ${}^{18}O-{}^{16}O$ and ${}^{16}O-{}^{13}C-{}^{16}O$ demonstrate that CO adsorption is non-dissociative and that all of the atomic oxygen arises from the dissociation of O_2. Isotopic substitution can be coupled with vibrational spectroscopy [17] since many vibrational peaks shift sufficiently between isotopologues to facilitate spectral assignments.

While we know that $CO + O_2$ reaction occurs via a LH mechanism, we can ask whether the reaction has a greater propensity to occur at certain sites on the surface. Under the conditions used for the above $CO + O_2/$ Pd(111) reaction, we do not expect a great sensitivity to a particular type of site. However, this is not always the case. The concept of active sites was first proposed by Taylor [18], and has been the subject of numerous studies since. For instance, Somorjai [19] has shown that step and kink sites can be particularly active in breaking C–C and C–H bonds in hydrocarbons. Consequently, vicinal surfaces of Pt are much more reactive for these reactions than flat surfaces. Ertl and co-workers [20] have identified steps to be the active site in the dissociation of NO on Ru(001). This surface exhibits two different types of steps, which differ in geometric structure and reactivity. On one type of step, the O atom, tends to chemisorb and remain at the step. This deactivates the step, that is, this step is self-poisoned by its dissociation of NO. Oxygen atoms diffuse away from the other type of step leaving them clean and active for further dissociation. Hence, the first type of step is not an active site unless the temperature is sufficiently high to allow for the diffusion of $O(a)$ away from it. The kinetics of reactions that depend on a certain active site, therefore, depends on the concentration of these sites, and whether or not they become blocked during the course of the reaction. There need not be just one active site for a reaction. In a structure-sensitive reaction, pronounced differences exist between the reactivities of different active sites. In a structure-insensitive reaction, the difference in the reactivity of different types of active sites is negligible.

A particularly interesting case of active sites controlling a structure-sensitive reaction is that of the MoS_2 hydrodesulfurization (HDS) catalyst, the reactivity of which follows the CoMoS model of Topsøe and Clausen, and which has been elucidated by Besenbacher, Nørskov and co-workers [21, 22]. In the model, it is suggested that the active sites are the edges of supported single monolayer MoS_2 clusters with a size of 10–20 Å. Co promotes the reactivity of the MoS_2 clusters by replacing Mo at edge sites. DFT calculations confirm STM images that the edge atoms in MoS_2 clusters exhibit a localized electronic state along the edge of the cluster (the brim state). Whereas the basal plane, as represented by the interior of the MoS_2 clusters, is inert with respect to hydrogenation of and eventual cleavage of a C–S bond in a molecule such as thiophene, the brim state is particularly active. Adding Co to the cluster maintains the brim state. However, the shape of the cluster changes from a triangular cluster with Mo along the edge to a hexagonal cluster with alternate edges containing Mo and Co, as shown in Fig. 6.2. The industrial catalyst that has been developed from these studies has been called the BRIM™ catalyst.

The magic of catalysis is contained in a fine balancing act. The surface must be reactive enough to break the appropriate bonds and hold adsorbates on the surface, but not so reactive that it inactivates the products. This balancing act is broadly applicable to heterogeneous catalysis, and has been demonstrated directly, for instance, during the oxidation of CO on $RuO_2(110)$ surfaces [23]. When Ru(001) is exposed to stoichiometric mixtures of CO + O_2, the conversion probability to form CO_2 is extremely low. However, when CO oxidation is performed in a large excess of O_2, the reactivity is superior to that of Pd, which we have seen above is a very efficient CO oxidation catalyst. In excess O_2, a crystalline film of $RuO_2(110)$ grows. This surface reveals three types of surface atom: a Ru atom, a two-fold co-ordinated bridging O atom (O_{br}) and a three-fold co-ordinated O atom (O_{3f}). The binding site of CO as well as the transition state for this reaction have been determined [24]. CO binds, as expected, on the Ru atom. It reacts with the O_{br} atom to form CO_2. To form the TS on an O-covered Ru(001) surface, the Ru–O and Ru–CO bonds are significantly weakened. However, the O_{br}–CO TS is formed with little cost to the Ru–O_{br} bond

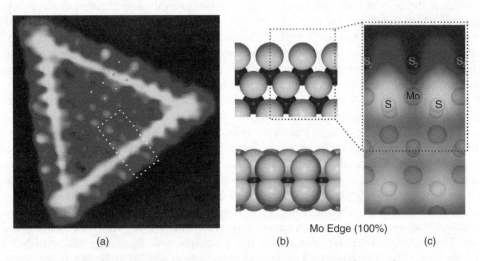

(a)   (b)   Mo Edge (100%)   (c)

Figure 6.2   *(a) Atom-resolved STM image of a CoMoS nanocluster. Size is 51 Å × 52 Å. Notice the very intense brim associated with the Co-substituted S edge (shorter edges). (b) Ball model of the proposed CoMoS structure. The CoMoS cluster is shown in top view exposing the unpromoted Mo edge and a Co-promoted S edge (Mo: dark, S: bright, Co: dark with spot). Also shown on the basal plane is a single Co inclusion. The Mo edge appears to be unaffected by Co and is shown in a side view ball model. The Co substituted S edge with a tetrahedral coordination of each edge Co is also shown. Reproduced from J. V. Lauritsen, R. T. Vang, F. Besenbacher, Catal. Today, 111, 34. © 2006 with permission from Elsevier.*

strength. This significantly reduces the activation energy, and leads to the site specificity. In this case the site specificity is not associated with a defect site. The O_{br} site is part of the ideal $RuO_2(110)$ surface. However, reaction with O_{br} is strongly favoured over O_{3f} or with O chemisorbed on Ru(001).

Molecular beam techniques have also been applied to the measurement of surface kinetics, and have been reviewed extensively by Madix and co-workers [25–27]. These studies are generally performed at isothermal conditions, and can involve one, two or even three molecular beams [28, 29]. The use of pulsed molecular beams and so-called molecular beam relaxation spectrometry (MBRS) allows for reaction products to be measured with moderate temporal resolution. Thus surface residence times can be established. This is particularly useful, for instance, to distinguish between LH and ER kinetics. Procedures for the analysis of data waveforms can be found elsewhere [25, 30–33].

6.3   Haber-Bosch process

Ammonia is an exceedingly useful and important chemical. Annual global production exceeds 131 million tons [34]. From it, nitrogen fertilizers are produced, as well as a host of other nitrogen-containing chemicals. The production of ammonia consumes more than 3×10^{17} J which represents $>1\%$ of global energy consumption [35], therefore, the discovery of a more efficient and lower temperature ammonia synthesis would have profound implications not only for fossil fuel consumption, but also the world-wide economy. Ammonia is synthesized according to the reaction

$$N_2 + 3H_2 \rightleftharpoons 2NH_3 \quad \Delta H_{298}^0 = -46.1 \, \text{kJ mol}^{-1}. \tag{6.3.1}$$

The reason why ammonia is so important industrially is virtually the same reason why Rxn (6.3.1) has to be performed catalytically: the triple bond in N_2 is so strong that N_2 is practically inert. Consequently, N_2 is a poor source of nitrogen whereas ammonia represents a usefully reactive source of nitrogen for industrial chemistry and agriculture. The truly magical aspect of catalysis is that the catalyst in the Haber-Bosch process is so effective at breaking the N≡N triple bond that virtually none of the energy consumed is used to break that bond. As we will see below, the activation energy of the catalyzed reaction is close to zero. Instead, the energy is required to compress and heat the gases, and to heat the catalyst so that the active sites remain clean.

From the above, we anticipate that N_2 dissociation is the rate determining step in ammonia synthesis. In other words, the dissociative sticking coefficient of N_2 is low, and an effective ammonia synthesis catalyst is one that efficiently dissociates N_2.

The ammonia synthesis reaction is conceptually simple. NH_3 is not only thermodynamically stable, it is by far the most stable nitrogen hydride. Therefore, selectivity is not an issue for the catalyst. The role of the catalyst is to produce efficiently and rapidly an equilibrium distribution of products and reactants. The key to the Haber-Bosch process is that the catalyst efficiently breaks the N≡N bond. This releases N atoms onto the surface where they collide with adsorbed H atoms and eventually form NH_3. The elementary steps of the reaction can be written [36]

$$N_2(g) + {}^* \rightleftharpoons N_2^* \tag{6.3.2}$$

$$N_2^* + {}^* \rightleftharpoons 2N^* \tag{6.3.3}$$

$$N^* + H^* \rightleftharpoons NH^* + {}^* \tag{6.3.4}$$

$$NH^* + H^* \rightleftharpoons NH_2^* + {}^* \tag{6.3.5}$$

$$NH_2^* + H^* \rightleftharpoons NH_3 * + {}^* \tag{6.3.6}$$

$$NH_3^* \rightleftharpoons NH_3(g) + {}^* \tag{6.3.7}$$

$$H_2(g) + 2^* \rightleftharpoons 2H^* \tag{6.3.8}$$

In principle, any one of the reactions could act as the rate determining step, if it has a rate that is significantly slower than all of the other elementary steps. Rxn (6.3.3) is the slow step under normal conditions with all of the subsequent reactions ensuing rapidly.

Iron is effective as a catalyst because it lowers the barrier to N_2 dissociation. It does this by gently coaxing the N≡N bond to break while simultaneously forming Fe–N chemisorption bonds. The activation barrier for N_2 dissociation is significantly lower than the N≡N bond strength, 941 kJ mol^{-1} (10.9 eV), precisely because these two processes occur simultaneously. Now we arrive at the second important property of Fe surfaces. They are able to break N≡N bonds, but in doing so they make an Fe–N bond that has just the right strength. If the M–N bond is extremely strong, as it is for early transition metals, then the N(a) is rendered inert because it forms a surface nitride, and does not react further. If, on the other hand, the M–N bond is too weak, then the surface residence time of N(a) becomes so short that N_2 may be formed instead of NH_3. This would lower the efficiency of the catalyst.

It would appear, then, that the activation energy for N_2 dissociation (E_{ads}) and heat of dissociative N_2 adsorption (q_{ads}) are linked, and vary systematically across the periodic chart. This is an expression of the Brønsted-Evans-Polanyi relation, which states that the activation energy and reaction energy are linearly related for an elementary reaction [37, 38]. The linearity of this relationship has been confirmed for N_2 dissociation on a variety of transition metals [39]. Furthermore, the relation holds for different classes of sites. The less reactive close-packed sites on Mo(110), Fe(111), Ru(001), Pd(111) and Cu(111) exhibit a linear relationship between E_{ads} and q_{ads}, while more reactive step sites on these surfaces exhibit a parallel linear relationship with correspondingly lower values of E_{ads}. Consequently, the trend of NH_3 synthesis reactivity, which peaks for Fe and Ru on single crystal surfaces, carries through to practical catalysts because the linear relationship is valid for ideal as well as defect sites. We look at this relationship between activation energy and enthalpy of the rate determining step in more detail in §6.10.

The Haber-Bosch process operates at 200–300 bar and 670–770 K over an Fe/K/CaO/Al$_2$O$_3$ catalyst. To make the catalyst, Fe$_3$O$_4$ is fused with a few percent of K$_2$O + CaO + Al$_2$O$_3$. The mixture is then reduced (activated) by annealing in a H$_2$/N$_2$ mixture at ~670 K so that metallic Fe particles form on the high surface area oxide substrate provided by Fe$_3$O$_4$/Al$_2$O$_3$. Al$_2$O$_3$ is the preferred substrate additive because it acts as a structural promoter, ensuring high dispersion of the Fe clusters and hindering their tendency to sinter into larger particles. The CaO may assist in this process. The K acts as an electronic promoter. We shall discuss below how promoters act. For now it is enough to know that promoters do exactly what the name implies: they promote the formation of the desired product.

Inspection of Rxn (6.3.1) reveals an exothermic reaction with a quite unfavourable entropy factor. Low temperatures and high pressure should therefore favour the forward reaction. High pressure is used in the industrial process, but the reason for the high temperature is not immediately obvious. As we shall soon see, the activation barrier for N_2 dissociation is small or even negative [40, 41]; thus, the high temperature is not required to activate nitrogen. High temperatures ensure rapid diffusion and reaction of adsorbed intermediates, and the rapid desorption of product NH_3 so that sufficient free surface sites are available to accept adsorbing N_2.

The composition of the industrial catalyst has changed little since its introduction. Haber first demonstrated the viability of the catalytic production of NH_3. But it was Mittasch who, in a demonstration of brute force combinatorial chemistry, performed over 6500 activity determinations on roughly 2500 different catalysts while developing the Haber-Bosch process [42]. One hundred tons of such a catalyst are required for a 1000 ton per day plant. By definition, a catalyst is not consumed in the reaction. Nonetheless, the

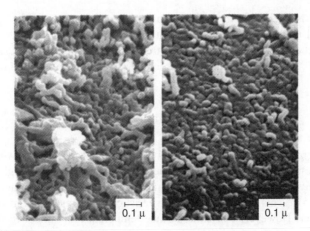

Figure 6.3 *The ammonia synthesis catalyst as revealed by high-resolution scanning electron microscopy. Reproduced with permission from G. Ertl, D. Prigge, R. Schloegl, M. Weiss,* J. Catal., *79 (1983) 359.* © *1983 Academic Press.*

lifetime of the catalyst is finite. The Haber-Bosch catalyst is particularly robust with roughly a 10-year lifetime. N_2 is provided by purified air, whereas H_2 is generated by the water gas shift reaction (§6.6).

The ammonia synthesis catalyst is a complex mixture of metals and oxides. What is the structure of the working catalyst, and how does this affect its activity? Ertl, Schlögl and co-workers have developed the following picture of the working catalyst [42]. As Fig. 6.3 demonstrates, the Haber-Bosch catalyst has a complex structure. The porous oxide substrate is covered with an inhomogeneous layer of metallic Fe clusters. Scanning Auger microscopy reveals that K preferentially segregates to the surface of the Fe clusters. The K is bound in a type of oxide that is not representative of a known bulk phase. This potassium oxide compound is chemisorbed on Fe and covers about a third of the surface. Oxygen enhances the thermal stability of K. Transmission electron microscopy unambiguously reveals that the Fe clusters are crystalline, and that they preferentially expose the (111) face. We now need to determine whether the geometric structure of the particles and the presence of K are critical for the performance of the catalyst. These questions are best answered by direct surface science experiments carried out on well-characterized model catalysts.

First, let us tackle the question of whether the presence of (111) crystallites is important. Heterogeneous catalytic reactions can be usefully classified as either structure-sensitive or structure-insensitive reactions [43]. Boudart and Djéga-Mariadassou [44] discuss a number of examples of both classes of reaction. Classically, the method of determining structural sensitivity is to measure the turnover number ($\equiv$ rate of product molecules formed per catalyst atom) of a reaction as a function of catalyst particle size. Such studies convolute numerous factors, and a definitive approach is to measure the rate of reaction as a function of the exposed crystal face on single-crystal model catalysts.

Somorjai and co-workers have taken precisely this approach [45, 46]. They prepared single crystal Fe samples of various orientations under standard UHV conditions, then followed the reaction of stoichiometric mixtures of N_2 and H_2. The use of a special reaction vessel allowed them to study the reaction at high pressure (20 bar) and a temperature of 773 K. The reaction rate depends strongly on the crystallographic orientation, varying by over two orders of magnitude between the three low index planes. The reaction rate increases in the order Fe(110) < Fe(100) < Fe(111). This demonstrates unequivocally the structure sensitivity of the reaction. The clean Fe(111) surface is the most active; nonetheless, it still makes a poor catalyst. The conversion efficiency of turning N_2 into NH_3 is only of order 10^{-6}.

If dissociation of N_2 is the rate determining step in ammonia synthesis, we expect to observe a low sticking coefficient, and that the sticking coefficient should increase in the order Fe(110) < Fe(100) < Fe(111). Bozso, Ertl and co-workers confirmed this trend, and measured a dissociative sticking coefficient that is of order 10^{-6} [40, 47]. Further studies [48] demonstrated that on Fe(111) the adsorption process is best described by

$$N_2(g) \underset{k_{-1}}{\overset{k_1}{\rightleftharpoons}} N_2(a) \underset{k_{-2}}{\overset{k_2}{\rightleftharpoons}} 2N(a), \tag{6.3.9}$$

and is actually characterized by a small negative activation barrier ($-0.034\,eV$) when the surface temperature is varied and the pre-exponential factor is assumed constant. Nonetheless, high vibrational and translational energies are effective at promoting adsorption into the molecular precursor and, therefore, dissociation of N_2 on Fe(111) [49–51]. This behaviour is characteristic of a direct adsorption process that proceeds over an activation barrier. However, when the surface temperature is varied, the rate of N_2 dissociation decreases. The negative apparent activation barrier indicates that dissociative N_2 adsorption does not occur via a simple elementary reaction. Consistent with Rxn (6.3.9), dissociation of N_2 incident upon the surface with thermal energies proceeds via (at least) a two-step sequence that involves a molecularly chemisorbed intermediate (precursor). The competition between desorption of N_2 (k_{-1}) and the dissociation of adsorbed N_2 (k_2) results in a negative apparent activation energy. The direct dissociative pathway is accessible only for N_2 molecules that have extremely high energies. Calculations show that even Rxn (6.3.9) is an oversimplification [52]. Four molecularly adsorbed states are found, and dissociation proceeds successively through these states. In addition, Mortensen et al. have shown that the sticking data can equally well be fitted by a small positive activation barrier ($+0.03\,eV$) if the temperature dependence of the pre-exponential is treated explicitly. Usually it is a good approximation to assume that A is constant; however, for such small barriers and large temperature ranges, the temperature dependence of A should be taken into account.

The behaviour of N_2 on Fe(111) shows certain parallels to what we have seen for O_2/Pd(111). In both cases, a direct transition from the gas-phase molecule to the dissociative adsorbed phase does not occur readily for molecules with thermal incident energies. Adsorption into a molecular chemisorbed state opens a dissociation pathway that has a significantly lower or even vanishing barrier. These two examples illustrate the multi-dimensional nature of the gas/molecule potential energy surface. Furthermore, they demonstrate how a surface can open new low activation reaction pathways by first accommodating a molecule into the adsorbed phase. Once chemisorbed onto the surface, the O_2 or N_2 molecules are able to traverse regions of configuration space that are not directly accessible from the gas phase. In doing so they can dissociate via a lower barrier path.

That the ammonia synthesis reaction is structure sensitive has practical implications for a practical catalyst. Our first impression of an effective catalyst might be that we would want a catalyst with a high dispersion. A dispersion of one indicates that all atoms reside on the cluster surface. In this way, no atoms would be "wasted" in the bulk below the surface. However, as the particle size changes, the geometric structure of the cluster changes. In particular, the number of sites of different co-ordinations is a function of the cluster size. If a particular type of site were required for catalysis, then the major concern would be to make a catalyst with a maximum of these sites rather than a maximum of surface atoms of all types. In the BRIM™ catalyst, for instance, each catalyst particle must have the required edge atoms. For ammonia synthesis on Fe catalysts – a structure-sensitive reaction – work in Somorjai's lab has demonstrated the importance of seven-fold co-ordinated (C_7) iron atoms [45]. The C_7 sites on iron are the active sites for ammonia synthesis.

If K acts to enhance the reactivity of the catalysts, it should accomplish this by increasing the dissociative sticking coefficient of N_2. This is found for Fe(111) and Fe(100) surfaces [53]. Additionally, in the presence of a sufficient amount of pre-adsorbed K, the two Fe surfaces no longer exhibit the structural sensitivity of the clean surface. K-covered Fe(111) and Fe(100) are roughly equally efficient at dissociating N_2. Thus, K not only acts to promote the sticking coefficient, but also it assists in making all exposed surfaces of the Fe particles equally reactive.

The pursuit of new ammonia synthesis catalysts remains an active area of scientific pursuit. A Ru based catalyst has been introduced. Enzymatic fixation of nitrogen occurs in nature at ambient temperatures. The active part of the enzyme is believed to be a $MoFe_7S_9$ cluster embedded among organic moieties. Fundamental theoretical studies have been carried out in Nørskov's group [41] to investigate whether an understanding of the enzymatic process may lead to an economically viable low-temperature synthetic route. Jacobsen [35] has shown that ternary nitrides (Fe_3Mo_3N, Co_3Mo_3N and Ni_2Mo_3N) can exhibit high activity for ammonia synthesis. A Cs promoted Co_3Mo_3N catalyst has an activity that exceeds that of a commercial catalyst by $\sim$30%. There is continuing interest in the surface chemistry of ammonia synthesis and the search for new catalyst formulations derived from fundamental understanding.

6.4 From microscopic kinetics to catalysis

Among the primary objectives of chemical studies at surfaces are the microscopic description of elementary steps in complex catalytic reactions, and the determination of kinetic parameters that describe the rates of overall processes. This has been achieved for several reactions. Two of the most important are the oxidation of CO by either O_2 or NO on Pt and the ammonia synthesis reaction. We start with the specific case of ammonia synthesis, then look at a more general method of performing kinetic analysis and defining the rate determining step.

6.4.1 Reaction kinetics

We seek to develop a general kinetic analysis that allows us to define and identify the rate determining step, and to analyze kinetic data. We start by summarizing a few important principles. Consider a generic reversible gas-phase reaction

$$R_1 + R_2 \rightleftharpoons P, \qquad (6.4.1)$$

which follows a mechanism composed of three elementary steps.

$$R_1 \underset{k_{-1}}{\overset{k_1}{\rightleftharpoons}} 2\,I_1 \qquad (6.4.2)$$

$$R_2 + I_1 \underset{k_{-2}}{\overset{k_2}{\rightleftharpoons}} I_2 \qquad (6.4.3)$$

$$I_1 + I_2 \underset{k_{-3}}{\overset{k_3}{\rightleftharpoons}} P \qquad (6.4.4)$$

We assume that all energy transfer processes are sufficiently rapid (quasi-equilibrated) that all of the kinetic relations can be written in their high-pressure limit. The sum of the three elementary steps is equal to the overall reaction.

As each step is reversible, we can write an equilibrium constant K_i for each individual step, and the overall equilibrium constant K_{eq} is equal to the product of all of the individual equilibrium constants. Furthermore, the equilibrium constant of an elementary reaction is equal to the ratio of the forward and reverse rate constants for that reaction. Hence we can write

$$K_{eq} = \frac{[P]}{[R_1][R_2]} = K_1 K_2 K_3 = \frac{k_1}{k_{-1}} \frac{k_2}{k_{-2}} \frac{k_3}{k_{-3}}. \qquad (6.4.5)$$

Using transition state theory, we can express an elementary rate constant as the product of a pre-factor $v^{\ddagger}$ (which might also include a standard state concentration) and $K_i^{\ddagger}$ the equilibrium constant for the formation of an activated complex from step i

$$k_i = v^{\ddagger} K_i^{\ddagger}. \tag{6.4.6}$$

For a system at steady state, the concentration of all reaction intermediates is constant. The net rate of a reaction step r_i is equal to the rate of the forward reaction $r_{i,f}$ minus the rate of the reverse reaction $r_{i,r}$; whereas the rate of change of concentration of any reactant or product must account for the net rate contributions of all reactions that consume or produce that species. For the reaction mechanism above, this allows us to write

$$\frac{dI_1}{dx} = 0 = 2r_1 - r_2 - r_3 \tag{6.4.7}$$

$$\frac{dI_2}{dx} = 0 = r_2 - r_3. \tag{6.4.8}$$

Note the factor of 2 that is included based on the stoichiometry of Rxn (6.4.2). From Eqs (6.4.7) and (6.4.8) we deduce that as steady state

$$r_1 = r_2 = r_3. \tag{6.4.9}$$

The rate of an elementary step is the product of a rate constant k_i and the appropriate concentration term that defines the order of the step. The rate constant is generally assumed to be related to an activation energy and a pre-exponential factor by the Arrhenius law.

6.4.2 Kinetic analysis using De Donder relations

The net rate of an elementary step

$$r_i = r_{i,f} - r_{i,r} \tag{6.4.10}$$

can also be written in terms of the affinity A_i

$$r_i = r_{i,f}[1 - \exp(-A_i/RT)]. \tag{6.4.11}$$

This is the De Donder relation [54, 55]. The affinity is the thermodynamic driving force for the elementary step. It is equal to the negative of the Gibbs energy change with respect to the extent of reaction. This is given by the difference between the Gibbs energies of the products and reactants at the reaction temperature and activity of each of the species involved,

$$A_i = -\sum_j v_{ii} G_j = -\sum_j v_{ij}[G_j^o + RT \ln a_j], \tag{6.4.12}$$

where v_{ij} are the stoichiometric coefficients for the reactants and products of step i (negative for reactants and positive for products), G_j^o are the standard Gibbs energies, and a_j are the activities. Because the equilibrium constant is also related to G_j^o by

$$K_{i\,eq} = \exp\left(-\sum_j v_{ij} G_j^o \bigg/ RT\right), \tag{6.4.13}$$

the affinity and equilibrium constant are related. We express this relationship with a dimensionless variable z_i called the reversibility of step i [54]

$$z_i = \exp(-A_i/RT) = \prod_j a_j^{v_{ij}}/K_{i\text{ eq}}. \tag{6.4.14}$$

As the reversibility approaches zero the step approaches irreversibility, while it is one for an equilibrated step. The total reversibility of the composite reaction is the product of the individual reversibilities and the total affinity is the sum of the affinities of the elementary steps. Hence for the three-step mechanism above

$$z_{\text{total}} = z_1 z_2 z_3 = \frac{a_P}{K_{\text{eq}} a_{R_1} a_{R_2}} \tag{6.4.15}$$

and

$$A_{\text{total}} = A_1 + A_2 + A_3 = -\ln(a_P/K_{\text{eq}} a_{R_1} a_{R_2}). \tag{6.4.16}$$

The conditions under which the reaction is run determine the parameters a_{R_1}, a_{R_2}, a_P and z_{total}. Using the De Donder relations we can write expressions for the net rates r_1, r_2 and r_3 in terms of the activities, equilibrium constants, and the reversibilities z_1, z_2 and z_{total}. Using the steady state condition and Eq. (6.4.9), we can determine the values of z_1 and z_2. It can be shown [54] that only the three forward rate constants k_1, k_2 and k_3 as well as $K_{\text{eq}}, K_{1\text{eq}}$ and $K_{2\text{eq}}$, need to be specified in order to completely determine the kinetics. The equilibrium constants are generally readily available based on thermodynamic data. This leaves experimental determination of the three rate constants for a complete description of the kinetics.

All of the above has been derived for a gas-phase reaction. It is also all applicable to surface reactions. However, one must also keep in mind the single greatest differentiating factor in catalysis for surface reactions: one must count sites. For instance, if the above reaction scheme were to be extended to a surface with I_1, I_2 and R_2 present as adsorbed species (assuming that R_1 dissociatively adsorbs, and that it and P are so weakly bound as to have negligible coverage) then we would also have to keep track of the number of empty sites θ_* according to

$$\theta_* = 1 - \theta_{I_1} - \theta_{I_2} - \theta_{R_2}. \tag{6.4.17}$$

6.4.3 Definition of the rate determining step (RDS)

The colloquial definition of a rate determining step is that it is the slowest step in a reaction. However, for the three-step reaction above at steady state, Eq. (6.4.9), we see that the net rate of each step is the same. Certainly it is not the net rate that determines the RDS. A better definition of the RDS has long been discussed, for instance by Boudart and Tamaru [56], Dumesic [54], and Campbell [57]. In simple reaction mechanisms, identification of the RDS is often easy; however, even for simple reactions the identification of an RDS can be complicated if, for instance, a slow step precedes a fast step in a consecutive reaction. For generality and for analysis of complex reactions a more quantitative definition of the rate determining step is required. The identification of the rate determining step is important in catalysis because if we can identify the RDS, we have identified the elementary step that must be addressed by the catalyst to accelerate the rate of reaction.

Campbell [57] has defined a method to quantifiably identify the RDS by introducing a parameter he defines as the degree of rate control for step i

$$X_{\text{rc, i}} = \frac{k_i}{R} \frac{\delta R}{\delta k_i} \tag{6.4.18}$$

where the overall rate of reaction is R, and the partial derivative of R with respect to the elementary rate constant k_i, $\delta R / \delta k_i$, is taken while holding all other rate constants k_j constant. The concept behind $X_{rc,i}$ is simple: Change the forward and reverse rate constant of step i by a small amount that still leads to a measurable change in R, say 1%. Since both forward and reverse rate constants are changed equally, the equilibrium constant of that step is not affected. Subsequent to this change, calculate the change in the overall rate R. The step that makes the greatest difference – that is, the step that changes R the most – is the step with the greatest $X_{rc,i}$ value. This is the rate determining step.

For any reaction mechanism that is composed of a series of consecutive steps and which has a single RDS, $X_{rc,i} = 1$ for the RDS, and is zero for all other steps. Any step with $X_{rc,i} \geq 0.95$ acts as a rate determining step. Some complex mechanisms do not have a single RDS. There are then several rate limiting steps with positive values of $X_{rc,i}$. If a step has a negative value of $X_{rc,i}$ it is an inhibition step. This method is extremely general. The only drawback is that in order to implement it efficiently, the rate equations must be handled using finite difference or other numerical methods. On the other hand, such kinetic analysis is routine with desktop computers so this does not present a serious obstacle to implementation.

6.4.4   Microkinetic analysis of ammonia synthesis

Here we describe in detail the kinetic model of ammonia synthesis developed by Stoltze and Nørskov [5, 7]. The model combines the experimental results from UHV single-crystal studies and quantum mechanical calculations. With no adjustable parameters, the model predicts rates in good agreement with high-pressure measurements made over an industrial catalyst. The results show unequivocally that the pressure gap can be overcome. First, the model begins with a set of elementary steps as proposed by Ertl [36] and described in Rxns (6.3.2)–(6.3.8).

An essential difference between surface kinetics and kinetics in other phases is that the reactions occur with a limited number of surface sites. An empty surface site is denoted * and a site occupied by species X is denoted X*. The number of surface sites is included explicitly within the model, and the total of occupied + unoccupied sites is constant. As indicated by the double arrows, each step is assumed reversible, and all reactions – apart from the rate determining step – are treated as equilibria. Further assumptions in the model are that the gas phase is ideal and that all sites are equivalent.

The rate determining step is the dissociation of adsorbed N_2, depicted in step (2). The rate of ammonia synthesis is, therefore, the net rate of step (2)

$$r_2 = k_2 \, \theta_{N_2*} \, \theta_* - k_{-2} (\theta_{N*})^2. \tag{6.4.19}$$

The net rate rather than just the forward rate $k_2 \theta_{N_2*} \theta_*$ must be used because the step is assumed to be reversible. The rate constants can be written in Arrhenius form, for instance,

$$k_2 = A_2 \exp(-\Delta^{\ddagger} H_2 / RT). \tag{6.4.20}$$

Next, we need expressions that relate all of the surface species concentrations so that we can substitute into Eq. (6.4.19) to obtain a final rate expression. The final rate expression relates the experimental parameters – $p_{N_2}, p_{H_2}, p_{NH_3}$, and T – and thermodynamic parameters to the ammonia synthesis rate. As each reaction step is assumed to be equilibrated, we can write the complete set of equilibrium constant expressions to obtain relationships for the coverages for the various species.

$$K_1 = \frac{\theta_{N_2*}}{(p_{N_2}/p_0)\theta_*} \tag{6.4.21}$$

$$K_3 = \frac{\theta_{NH*}\theta_*}{\theta_{N*}\theta_{H*}} \tag{6.4.22}$$

$$K_4 = \frac{\theta_{NH_2*}\theta_*}{\theta_{NH*}\theta_{H*}} \tag{6.4.23}$$

$$K_5 = \frac{\theta_{NH_3*}\theta_*}{\theta_{NH_2*}\theta_{H*}} \tag{6.4.24}$$

$$K_6 = \frac{(p_{NH_3}/p_0)\theta_*}{\theta_{NH_3*}} \tag{6.4.25}$$

$$K_7 = \frac{(\theta_{H*})^2}{(p_{H_2}/p_0)(\theta_*)^2} \tag{6.4.26}$$

where p_0 is the standard state pressure of 1 bar. The equilibrium constant for the overall reaction, Rxn (6.3.1), is

$$K_g = K_1 K_2^2 K_3^2 K_4^2 K_5^2 K_6^2 K_7^3 \tag{6.4.27}$$

From statistical mechanics, each equilibrium constant can be calculated from the molecular partition functions of the reactants and products via Eq. (4.4.7). The molecular partition function is calculated from the product of the translational, vibrational, rotational and electronic partition functions as in Eq. (4.4.9). Finally, we need an expression for the conservation of sites. In terms of fractional coverages of the adspecies and the fractional coverage of free sites, θ_*, this yields

$$\theta_{N_2*} + \theta_{N*} + \theta_{NH*} + \theta_{NH_2*} + \theta_{NH_3*} + \theta_{H*} + \theta_* = 1. \tag{6.4.28}$$

From Eqs (6.4.21)–(6.4.26) and (6.4.28), expressions for the fractional coverages of each surface intermediate can be calculated (see Exercise 6.3). Substitution of the expressions for θ_{N_2*}, θ_* and θ_{N*} into Eq. (6.4.19) yields

$$r_2 = 2k_2 K_1 \left(\frac{p_{N_2}}{p_0} - \frac{p_{NH_3}^2 p_0}{K_g p_{H_2}^3} \right) (\theta_*)^2. \tag{6.4.29}$$

The rate at which NH_3 is formed is $2r_2$ molecules per second per surface site. The coverage of empty sites is given by

$$\theta_* = \left(1 + K_1 \frac{p_{N_2}}{p_0} + \frac{p_{NH_3}p_0^{0.5}}{K_3 K_4 K_5 K_6 K_7^{1.5} p_{H_2}^{1.5}} + \frac{p_{NH_3}}{K_4 K_5 K_6 K_7 p_{H_2}} + \frac{p_{NH_3}}{K_5 K_6 K_7^{0.5} p_0^{0.5}} + \frac{p_{NH_3}}{K_6} + K_7^{0.5} \frac{p_{H_2}^{0.5}}{p_0^{0.5}} \right) \tag{6.4.30}$$

Eqs (6.4.29) and (6.4.30) contain thermodynamic and extensive variables that can be determined by experiment and theory. Thermodynamic data for the gas-phase species is readily available in the JANAF tables [58]. The equilibrium constants are calculated from the partition functions for the vibrational properties of the adsorbates. Measurements of the initial sticking coefficient of N_2 into $2N*$ and its activation energy are used to determine A_2 and $\Delta^{\ddagger}H_2$ [53]. To compare with a working catalyst, also the surface area of the catalyst is required. The calculations of Stoltze and Nørskov [5] show that apart from the pressures, temperature, surface area and K_g, the only critical parameters in the model are $A_2, \Delta^{\ddagger}H_2 + \varepsilon_{elec, N_2*}$ and $\varepsilon_{elec, N*}$, where the $\varepsilon_{elec, x}$ terms are the electronic ground state energies as in Eq. (4.4.10). All of these can be determined rather directly from experiments.

Figure 6.4 compares the calculated and measured outputs from a working catalytic reactor. As is clear to see the agreement is remarkably good. The calculations are based on values derived from UHV

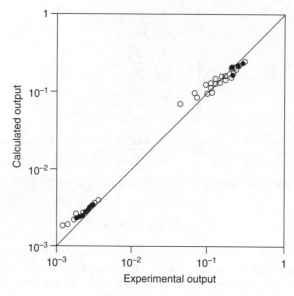

Figure 6.4   *The measured NH₃ mole fraction at the reactor outlet (data points) is compared to the calculation of Stoltze and Nørskov (solid line). The Topsøe KM1R catalyst was operated at 1–300 atm and 375–500°C. Reproduced from P. Stoltze, J. K. Nørskov, J. Catal., 110, 1. © 1988 with permission from Academic Press.*

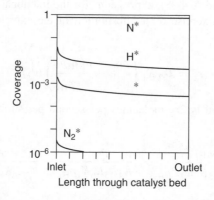

Figure 6.5   *Coverages calculated by Stoltze and Nørskov for a K-promoted Fe catalyst operating at 400°C with an initial mixture (at inlet) of 25% N₂, 75% H₂ and 0% NH₃. Percentage of NH₃ at outlet: 20.6%. The curves are for adsorbed nitrogen (N*), hydrogen (H*), free sites (*) and molecular nitrogen (N₂*). Reproduced from P. Stoltze, J. K. Nørskov, J. Catal., 110, 1. © 1988 with permission from Academic Press.*

single-crystal surface studies. The experiments were performed at 1–300 atm on a practical catalyst. The extremely good agreement demonstrates that the mechanism of ammonia synthesis is understood and quantitatively describable on a molecular scale.

Figure 6.5 depicts the state of catalyst under operating conditions. The reactor has an inlet at one end, a catalyst bed, and an outlet at the other end. The gas-phase composition changes from a 1:3 (stoichiometric) mixture of N_2:H_2 to an equilibrium mixture at the end. Nonetheless, the surface coverage trend is uniform over the length of the catalyst. The surface of the catalyst is essentially covered with adsorbed N. N* is the most abundant reactive intermediate or MARI [44]. The number of free sites is extremely small, on the order of 10^{-3}.

6.5   Fischer-Tropsch synthesis and related chemistry

Several catalytic processes are built around the use of synthesis gas or syngas, a mixture of $CO + H_2$. Fischer-Tropsch (FT) synthesis [59] is the production of hydrocarbons and oxygenated hydrocarbons (oxygenates) from synthesis gas. Closely related are the methanation reaction (production of CH_4 from syngas), methanol synthesis, and the Mobil process, which converts methanol (often produced from syngas) into transportation fuels.

Synthesis gas is produced from oil, natural gas, coal or other carbonaceous mineralogical deposits by steam reforming [60]. Carbon can also be supplied in a renewable form by the use of biomass. Steam reforming is the reaction of hydrocarbons with water to form CO and H_2. Specifically for natural gas (which is primarily methane), this is written

$$CH_4(g) + H_2O(g) \rightleftharpoons CO(g) + 3H_2(g) \quad \Delta H_{298}^0 = -207 \, \text{kJ mol}^{-1} \tag{6.5.1}$$

The first challenge of steam reforming is activity and, thus, a K_2O promoted Ni catalyst is used at 700–830 °C, 20–40 bar pressure, on an alumina or calcium aluminate substrate [61]. The reverse reaction is strongly favoured in the low temperature range of 250–350 °C because the entropy term of the forward reaction is favourable. The second challenge is S poisoning, but the catalyst is also deactivated by As, halogens, P, Pb and Cu. The latter are easily removed from the feedstock prior to exposure to the catalyst. S presents the biggest problem; however, it is much more easily removed from natural gas than, for instance, from coal, tar sands or oil shale. Two other potential forms of catalyst deactivation are carbon formation, also known as coking [62], and sintering.

The water gas shift reaction is the reaction of water gas ($CO + H_2O$) to form CO_2 and H_2

$$CO(g) + H_2O(g) \rightleftharpoons CO_2(g) + H_2(g) \quad \Delta H_{298}^0 = -42 \, \text{kJ mol}^{-1}. \tag{6.5.2}$$

This reaction is used to increase the H_2 content of synthesis gas, and is important in automotive catalysis. An Fe_3O_4 catalyst supported on Cr_2O_3 is used as a high-temperature shift catalyst (400–500 °C). This catalyst is rather robust with respect to S poisoning; indeed, the sulfide, Fe_3S_4, also acts as a catalyst for the reaction, albeit with a lower activity. If the feedstock has a low S content then a low temperature shift catalyst can be used. These catalysts consist of oxides of Cu + Zn or Cu + Zn + Al, and operate at 190–260 °C. From the stoichiometry of the reaction, the change in the moles of gas molecules is $\Delta n_{gas} = 0$. Therefore, the entropy term is not significant, a factor favourable for a low temperature catalyst. However, the selectivity of the low temperature catalyst then becomes a paramount property, because the production of methane and higher hydrocarbons is thermodynamically favoured. The presence of Cu makes low temperature shift catalysts highly susceptible to S poisoning.

Methanol is produced industrially by the ICI low-pressure methanol process. Low pressure is in the eye of the beholder as the process is normally run at 50–100 bar. As an exothermic reaction with a highly unfavourable entropy factor,

$$CO(g) + 2H_2(g) \rightleftharpoons CH_3OH(g) \quad \Delta H_{298}^0 = -92 \, \text{kJ mol}^{-1}, \tag{6.5.3}$$

high pressure and low temperatures favour the process. The temperature must not be too low as the catalyst is deactivated if, for example, the methanol, which is a liquid at room temperature (bp 64.7 °C at 1 bar), does not rapidly desorb from the catalyst. The use of $CuO+ZnO+Al_2O_3$ or $CuO+ZnO$ catalyst allows the process to be run at the relatively mild temperature of $230-270\,°C$. Again, the Cu-based catalyst is highly susceptible to poisoning from S as well as Cl. The selectivity of the catalyst is remarkably high. Although higher alcohols, ethers and alkanes are thermodynamically preferred, the selectivity for methanol is often >99%. In these catalysts the Cu acts as the active site. The ZnO provides a matrix into which it can dissolve, and the Al_2O_3 ensures a high surface area (dispersion) for the catalyst. The reaction is strongly structure-sensitive. This fact is essential for accurate modelling of the kinetics observed under industrial conditions as the structure of the catalyst depends on the reactant gas composition [63].

Fischer-Tropsch chemistry proceeds via a complex set of reactions, which consume CO and H_2, and produce alkanes (C_nH_{2n+2}), alkenes (C_nH_{2n}), alcohols ($C_nH_{2n+1}OH$) and other oxygenated compounds, aromatics as well as CO_2 and H_2O. Some examples of these reactions are

$$nCO + (2n+1)H_2 \rightarrow C_nH_{2n+2} + nH_2O \tag{6.5.4}$$

$$nCO + 2nH_2 \rightarrow C_nH_{2n} + H_2O \tag{6.5.5}$$

$$nCO + 2nH_2 \rightarrow C_nH_{2n+1}OH + (n-1)H_2O \tag{6.5.6}$$

$$2nCO + (n+1)H_2 \rightarrow C_nH_{2n+2} + nCO_2 \tag{6.5.7}$$

$$2nCO + nH_2 \rightarrow C_nH_{2n} + nCO_2 \tag{6.5.8}$$

$$(2n-1)CO + (n-1)H_2 \rightarrow C_nH_{2n+1}OH + (n-1)CO_2. \tag{6.5.9}$$

All of these reactions are accompanied by negative free energy changes and are exothermic [64]. It is important, however, that the catalyst does not produce an equilibrium mixture of products. If this were to happen, the reaction products would be an unwieldy molasses, which would be of little economic value. Therefore, selectivity is perhaps the most important characteristic of a Fischer-Tropsch catalyst. Most group 8, 9 and 10 metals are good catalysts for these reactions; however, the product distributions obtained depend on the metal. A Co or Fe catalyst is best for producing the low molecular weight hydrocarbons preferred as liquid fuels (both petrol and diesel), whereas a Ru catalyst is selective for producing high molecular weight waxy hydrocarbons. Rh-based catalysts give increased yields of oxygenates, particularly methanol and ethanol. A Ni catalyst is the best for formation of methane. The SASOL process uses K as a promoter. In addition, a small amount of Cu added to the catalyst aids in the activation of the catalyst, and likely helps to maintain a high surface area.

Two environmentally attractive aspects of Fischer-Tropsch chemistry are that a renewable source of carbon (biomass) can be used as the source of synthesis gas, and that syngas contains no S, P or N. The liquid fuels obtained by this method are, therefore, very clean burning. The challenge in Fischer-Tropsch chemistry remains an increase in the selectivity. A breakthrough in catalytic chemistry that would allow for the selective formation of specific alkenes or oxygenates would be of tremendous impact. To achieve this, a greater understanding of the mechanisms involved in Fischer-Tropsch synthesis is required [65].

The mechanisms involved in Rxns (6.5.4)-(6.5.9) are obviously complex. Further obscuring the matter, numerous reactions are occurring along parallel paths. Nonetheless, progress has been made in understanding some mechanistic aspects of Fischer-Tropsch chemistry. First, one needs to realize that the product distribution, and therefore the reaction mechanisms, depends sensitively on the surface temperature and composition of the catalyst. Not one unique mechanism operates at all temperatures and for all catalysts. This is to be expected as a result of the interwoven nature of the numerous parallel reactions that occur.

There are at least two general mechanisms for chain propagation and termination: the alkyl mechanism of Brady and Pettit [66] and the alkenyl mechanism of Maitlis. Both have strengths and weaknesses and neither accounts well for the formation of oxygenates [59]. The balance between them may be influenced

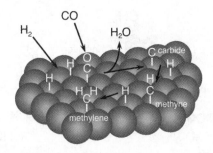

Figure 6.6 *The dissociation of CO followed by hydrogenation to form an adsorbed methylene (CH₂) species. The formation of CH₂ is an essential step in Fischer-Tropsch synthesis. CH₂ is the product of sequential H addition steps that follow the dissociative adsorption of H₂ and CO. Oxygen is removed from the surface via H₂O formation.*

by surface polarity effects [65]. Here we examine in some detail the alkenyl mechanism. Maitlis and co-workers studied FT synthesis over Rh/CeO$_2$/SiO$_2$ and Ru/SiO$_2$ catalysts at 433–473 K and a CO:H$_2$ ratio of 1:2 [67, 68]. The primary products are 1-alkenes, with propene (CH$_2$=CH–CH$_3$) the most abundant. They have established two important pathways that are components of FT chemistry by tracking the incorporation of ^{13}C containing reactants. The first is the dissociation of CO with subsequent hydrogenation to form an adsorbed methylene species (CH$_2$). The second series of reactions corresponds to the polymerization of CH$_2$ moieties with adsorbed alkenyl fragments. These two reaction cycles are depicted in Figs 6.6 and 6.7.

Figure 6.7 depicts a series of non-stoichiometric schematic reactions [67, 68]. The initiating steps of FT synthesis are dissociative H$_2$ adsorption and the chemisorption of CO. Adsorbed CO then dissociates, and the resulting carbide is sequentially hydrogenated to CH, CH$_2$ and CH$_3$, the presence of which has been confirmed in surface science experiments. There is general agreement that what ensues is a stepwise polymerization of methylene (CH$_2$ groups). The beginning of the hydrocarbon formation cycle shown in Fig. 6.7 is the reaction of CH + CH$_2$ to form an adsorbed vinyl species. A CH$_2$ species then adds to this unit, followed by an isomerization to reproduce a structure similar to that of the adsorbed vinyl group. Subsequent additions of CH$_2$ groups then compete with the addition of H. When H is added, the cycle terminates and a 1-alkene desorbs. That H$_2$^{13}C=^{13}CHBr is an efficient initiator of the reaction confirms this mechanism. H$_2$^{13}C=^{13}CHBr adsorbs as H$_2$^{13}C=^{13}CH as required by the mechanism. On the other hand, H$_3$^{13}C–^{13}CH$_2$Br, which does not readily form this species, is not an efficient initiator of reaction.

Several comments must be made about these mechanistic steps. Note that the exact structure of the adsorbed intermediates is unknown. Nonetheless, we know that the ^{13}C=^{13}C bond is not severed completely, and that the two isotopically labelled C atoms remain neighbours throughout, regardless of how many CH$_2$ additions occur to the molecule that contains them. This supports the isomerization/addition mechanism of Fig. 6.7. In addition, desorption of the 1-alkene after the addition of a hydrogen atom to the adsorbed alkenyl species (hydrogenolysis) competes at each step with the addition of a further CH$_2$ group. This type of kinetic competition makes the product distribution particularly sensitive to temperature and catalyst activity. Importantly, the catalyst treats adsorbed double bonds gently. At 433 K the C=C found in step one is not completely broken, and the two ^{13}C atoms remain bound to each other.

When the conditions of the reaction are changed, the reaction product distribution can change. The product distribution for the Ru/SiO$_2$ catalyst is much different at 463 K compared to 433 K. At 463 K significant ^{13}C=^{13}C bond scission occurs, and the two ^{13}C do not often remain neighbours. Surprisingly, a Rh/CeO$_x$/SiO$_2$ at 463 K exhibits similar behaviour to the Ru/SiO$_2$ catalyst at 433 K.

It seems probable that the C–C bond-forming steps are rate-determining. Thus, the promotion of effective FT activity is enhanced by a catalyst that meets the following criteria [65]. The catalyst must comprise

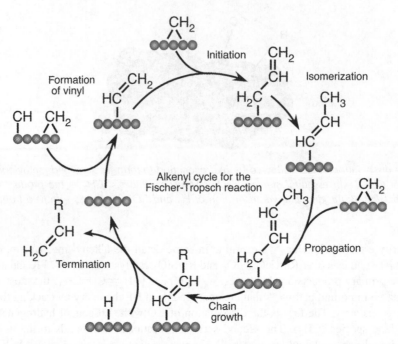

Figure 6.7   *The alkenyl carrier cycle. The cycle begins in the top left with the formation of adsorbed vinyl ($HC=CH_2$) from CH and CH_2. Chain growth ensues, initiated by the addition of CH_2. Isomerization forms an adsorbed allyl ($H_2C-CH=CH_2$). Subsequently, further addition of CH_2 (propagation and chain growth) competes with addition of H (termination). Reproduced from P. M. Maitlis, H. C. Long, R. Quyoum, M. L. Turner, Z. Q. Wang, Chem. Commun., 1. © 1996 with permission from the Royal Society of Chemistry.*

a transition metal. Fe, Co, Ru, and Rh are among the best as they bond strongly to CO and promote its dissociation. This metal must also dissociatively adsorb hydrogen, and bind it strongly enough to maintain a significant coverage. Reaction is strongly enhanced by the presences of defects; hence, the metal should be in a highly disperse nanoparticulate form. The nature of the support is also important, and it should provide islands of an oxide associated with the metal nanoparticles. The oxide needs to function as a good Lewis acid.

6.6   The three-way automotive catalyst

The goal of achieving clean burning internal combustion engines represents both a challenge to and a triumph of catalytic chemistry [69]. Beginning in the 1970s in the United States, governments have imposed progressively stricter regulations on the release of hydrocarbons (HC), nitrogen oxides (NO_x) and CO in automotive emissions.

The strategy of designing an automotive catalyst is very different from that of the industrial processes that we have considered so far. The reaction conditions of temperature (673–773 K) and pressure (~1 bar) are determined, within narrow limits, by the operation of the engine and the desire for high fuel economy. The automotive catalyst aims to take the products of a near stoichiometric mixture of air and fuel after it has been combusted in the engine, and ensure that the emissions consist of an equilibrium mixture containing only CO_2, H_2O and N_2. The reactant mixture that reaches the catalyst consists of H_2, H_2O, O_2,

N_2, NO_x, CO, CO_2 and HC (alkanes, alkenes and aromatics containing 1–8 C atoms). A certain level of impurities, such as S (as SO_2), P and Pb (as tetraethyl lead), is derived from the fuel and engine oil. Fuel and lubrication additives may also include potential poisons. For example, a common form of antiwear agent added to lubricants is a zinc dialkyldithiophosphate. This forms a tough, cross-linked zinc phosphate film on the surface of the engine but also provides a source of Zn, S and P. The use of alcohol in fuels leads to the introduction of aldehydes into the exhaust stream.

As summarized by Taylor [70], the pertinent chemical reactions to be considered are

$$CO + 1/2\ O_2 \rightarrow CO_2 \tag{6.6.1}$$

$$\text{hydrocarbons} + O_2 \rightarrow H_2O + CO_2 \tag{6.6.2}$$

$$H_2 + 1/2\ O_2 \rightarrow H_2O \tag{6.6.3}$$

$$NO + CO \rightarrow 1/2\ N_2 + CO_2 \tag{6.6.4}$$

$$NO + H_2 \rightarrow 1/2\ N_2 + H_2O \tag{6.6.5}$$

$$\text{hydrocarbons} + NO \rightarrow N_2 + H_2O + CO_2 \tag{6.6.6}$$

$$NO + 5/2\ H_2 \rightarrow NH_3 + H_2O \tag{6.6.7}$$

$$CO + H_2O \rightarrow CO_2 + H_2 \tag{6.6.8}$$

$$\text{hydrocarbons} + H_2O \rightarrow CO + CO_2 + H_2 \tag{6.6.9}$$

$$3\ NO + 2\ NH_3 \rightarrow 5/2\ N_2 + 3\ H_2O \tag{6.6.10}$$

$$2\ NO + H_2 \rightarrow N_2O + H_2O \tag{6.6.11}$$

$$2\ N_2O \rightarrow 2\ N_2 + O_2 \tag{6.6.12}$$

$$2\ NH_3 \rightarrow N_2 + 3\ H_2. \tag{6.6.13}$$

These are grouped more or less in order of importance. Both *activity* and *selectivity* are important for the operation of the catalyst. The importance of activity is obvious. The catalyst must efficiently oxidize CO and HC, and reduce NO regardless of whether the exhaust mixture is rich in oxidizing agents or reducing agents. The selectivity of the catalyst is exemplified by the desire to produce N_2, not NH_3, from NO. Furthermore, O_2 should be used for complete oxidation of HC and CO to CO_2 rather than for the production of H_2O from H_2. Given these requirements and constraints we can now set upon the search for a perfect catalyst.

The three-way automotive catalyst, so named because it removes the three unwanted products CO, HC and NO_x, is the result of a multifaceted scientific approach that has spanned from high pressure engineering studies on high surface area supported metal catalysts to UHV surface science on single crystal model catalysts. The general composition consists of rhodium, platinum and palladium dispersed on Al_2O_3 with CeO_2 added as a type of promoter.

Rhodium is almost the perfect material for an automotive catalyst. It is the best catalyst for the reduction of NO to N_2. Rh is particularly active in promoting the reactions of NO + CO and NO + H_2, even at low temperatures. Rh also acts as an efficient steam reforming catalyst, though it does have a tendency to oxidize HCs only partially to CO. While increasing the need for CO oxidation, this is not a severe constraint as Rh functions as an efficient CO oxidation catalyst as well, even at relatively low temperatures. The one great drawback of Rh is its price. It is estimated to comprise only 0.001 ppm of the Earth's crust (compared to 0.005 ppm for Pt and 0.01 ppm for Pd) [71]. Major ore bearing deposits are located only in South Africa and Russia and there is a lower concentration reserve in Canada. Hence Rh is quite expensive – its price per ounce spiked to over $10,000 in 2008, falling to about $1650 in 2011. In addition, its supply has been subject to political volatility. Research continues to find replacements for it.

Platinum is not normally considered an inexpensive metal but at only $1516 per oz it compares quite favourably to rhodium. Nonetheless, this is still more than twice the 2001 price of $616 per oz and it rose to ~$2300 in 2008. Pt is added to the three-way catalyst for its ability to oxidize CO and HC. The warm-up time of an engine represents by far the dirtiest phase of combustion, and presents the most severe problems for the catalytic conversions of CO and HC. Pt has relatively high activity during this phase. Pt makes little contribution to the reduction of NO_x. Though it can reduce NO in isolation, in the presence of SO_2 and high concentrations of CO, the activity of Pt for NO reduction is quite low.

Palladium has regained its status as the least expensive of the three primary platinum group metals used in the three-way catalyst. Dropping from $1094 per oz in 2001 to only $325 per oz in 2007, it stood at ~$800 in 2011. This compares with $160 per oz for Ru and $420 per oz for Ir, two other platinum group metals commonly used in catalysis. Palladium's primary role is as an oxidation catalyst for CO and HC. It is frequently used in a separate part of the catalyst, away from the Rh, so that oxidation and NO reduction occur in separate parts of the overall automotive catalyst. The trend from 2001 for decreased Pd loadings and increased Pt loadings in automotive catalysts has reversed because of the prices of the metals. This demonstrates that catalyst composition cannot be based solely on performance, but is also sensitive to cost.

Alumina is the preferred support material. It combines high surface area, favourable pore structure and strong mechanical stability. Furthermore, Al_2O_3 is an inexpensive and readily available material.

Ceria (CeO_2) is added at the level of ~1% as a promoter. Ceria acts as a structural promoter. It stabilizes the substrate with respect to surface area loss (sintering), as well as enhancing the dispersion of the catalyst. Ceria also promotes the water gas shift reaction, Rxn (6.6.8). It is unclear whether this is due to reaction on the CeO_2 itself or whether this results from some type of electronic promotion. In addition, ceria acts as a reservoir for oxygen. The lattice of ceria is unusually forgiving with respect to the amount of oxygen contained in it, much more so than alumina, because it can exist as a mixed valence oxide. O_2 can dissociate on the metal particles, then diffuse into the surrounding ceria. This effect is known as spillover. The oxygen is not permanently trapped in the ceria; rather, it remains available for future reaction. Oxygen storage capacity is an important characteristic because the air/fuel ratio is not constant in a working engine. The presence of ceria allows the catalyst/substrate system to store oxygen when it is too plentiful, and release it for reaction when the oxygen concentration is too low.

Poisoning is an ever-present problem for catalysts, and the three-way catalyst is no exception. Fortunately for the environment and its inhabitants, the automotive catalyst is poisoned by Pb. Tetraethyl lead has been added to automotive fuels as an octane enhancer as well as for its beneficial tribological properties. Researchers quickly realized, however, that Pb had a detrimental effect on the performance of the three-way catalyst. Removing the lead from fuel easily solved the problem, though this has necessitated some changes in engine design. This solution was rapidly adopted in the United States, but took significantly longer to implement in Europe. The reduced lead burden on the environment is an unintended beneficial aspect of the catalytic converter.

6.7   Promoters

The aim of catalysis is to enhance the rate of formation of a desired product. A catalyst does so by lowering the appropriate activation barriers along the reactive path. In the case of ammonia synthesis, the problematic activation barrier is that of breaking the $N\equiv N$ bond. An Fe surface enhances the reactivity of N_2 by accommodating it into a molecularly chemisorbed state that has a relatively low barrier toward dissociation into the atomic fragments. The barrier to dissociation out of the molecularly chemisorbed states is lower than the barrier to dissociation of the gas-phase molecule because chemisorption has weakened the N–N bond, making it more susceptible to scission.

Nonetheless, a clean Fe surface is still not as reactive as we would like for an effective catalyst. The industrial catalyst also contains K on the surface of the Fe particles. K significantly increases the activity of the catalyst, and is therefore called a promoter. The effect of an electronic promoter is clear: it lowers the barrier to reaction. The mechanism by which a promoter achieves this is less unambiguous.

In Chapter 3 we investigated the dissociative adsorption of H_2 (§3.5) and the molecular chemisorption of CO (§3.4.2). In both these cases, the filling of an antibonding electronic state led to an increased binding of the molecular species to the surface and a concomitant increase of the intramolecular bond strength. This suggests one mechanism by which a promoter can act. If the promoter facilitates charge transfer from the substrate into antibonding electronic states of the adsorbate, the adsorbate is more likely to dissociate. In other words, increased substrate to adsorbate charge transfer can reduce activation barriers. Molecular states near the Fermi energy are the most sensitive to changes in the electronic structure of the substrate because slight shifts in the positions of these states relative to E_F changes their occupation and the extent of mixing with substrate levels. The donation of electrons from an electropositive adsorbate, such as an alkali metal, to the substrate significantly perturbs the surface density of states near E_F. The greater availability of electrons near E_F leads to an enhanced ability of the substrate to donate electrons into the adsorbate, which lowers the barrier to dissociation. This mechanism of promotion has been developed theoretically by Feibelman and Hamann [72].

Nørskov, Holloway and Lang [73] have investigated an alternative mechanism for promotion with a theoretical treatment of the transition state (TS) to dissociation. If the promoter stabilizes the transition state relative to the initial state of the reactants, it lowers the activation energy. Because of the extended bonds and distorted electronic structure of transitions states compared to stable molecules, transition states tend to exhibit larger dipole moments than ground state molecules. This makes them susceptible to the influence of electrostatic fields. Once an alkali metal has donated an electron to the substrate, it is effectively an ion adsorbed on the surface. This ion is strongly screened on a metal; nonetheless, a local electrostatic field is associated with the alkali metal. The interaction of this field with the TS can reduce the energy of the TS, and thereby reduce the activation barrier to reaction through this TS. Whether this effect is attractive or repulsive depends on the relative sign of the electrostatic fields on the two adsorbates. For most molecular adsorbates this field is related to electron transfer into the antibonding molecular orbitals.

It is still not clear which one of these promotion mechanisms predominates. In all likelihood, both are possible, and the mechanism of promotion is a system specific property rather than being represented by only one global mechanism. Calculations suggest that the electrostatic stabilization of the transition state is responsible for the alkali–metal promotion of N_2 dissociation on Ru surfaces [74]. On the other hand, stabilization of chemisorbed $N_2(a)$ appears to account for a large fraction of the promotion effect on K-promoted Fe [75]. Therefore, the mechanism of promotion must be considered on a case-by-case basis, and there may even be circumstances in which both models contribute to the overall effect.

While promotion is most commonly thought of in terms of the electronic promotion that is described above, it is important to remember that an effective catalyst is one that stably and selectively produces the desired product. Thus, promoters are added to a catalyst to enhance the stability and selectivity of the catalyst, not just the activity. In the ammonia synthesis catalyst, Al_2O_3 and CaO are added as structural promoters. They enhance the surface area of the dispersed Fe particles, and work against the sintering of these into larger particles. In Fischer-Tropsch synthesis, K is added to the catalyst when an increased selectivity for higher molecular weight products is desired.

6.8   Poisons

Poisons are adsorbed species that lower the activity of a catalyst. The above case studies show that poisoning of catalysts is a ubiquitous phenomenon. Poisons can sometimes be avoided simply by changing the reactive feed, but this is not always possible. Poisons can sometimes be removed by side reactions. This represents a type of self-cleaning catalyst. Oxygen can often be removed from a catalyst if the reactive mixture is strongly reducing, as in the ammonia synthesis. Carbon can be removed under strongly oxidizing conditions. It is the formation of tenacious deposits, which cannot be removed chemically, that must be avoided.

Poisons can reduce the rate of a surface reaction by two distinctly different mechanisms. The first is an electronic mechanism, which is an indirect, long-range effect. This is essentially the inverse of the process described above for promotion in which the poison acts to change activation energies in an unfavourable manner. The presence of a poison can destabilize either a chemisorbed reactive intermediate or the transition state of an elementary step. Again, electron donation or electrostatic effects may be involved in this process. Reaction rates are suppressed either because an activation barrier is increased or because the concentration of one of the reactants is reduced. If a poison acts to reduce the chemisorption bond strength, the affected adsorbate is more likely to desorb than to proceed in the reaction. The second mechanism is called site blocking. This is a direct, geometric effect. If a poison occupies all of the adsorption sites, there are none left over to take part in the catalytic reaction. For instance, when a three-way catalyst is covered with Pb, it is inactivated because Pb is not an effective catalyst for a catalytic converter.

Site blocking is a localized effect that can be utilized in the nanoscale modification of surfaces. H adsorption on Si surfaces greatly reduces the reactivity of these surfaces toward oxidation by O_2 and H_2O. If the H(a) is removed from the surface, the uncovered sites become reactive and are readily oxidized in the presence of O_2 or H_2O. Lyding, Avouris, Quate and co-workers have taken advantage of this poisoning of Si oxidation to pattern Si surfaces [76–78]. Using a scanning probe tip to desorb hydrogen, it is possible to modify the Si surface with near atomic resolution, as shown in Fig. I.4(a).

Poisons are not always undesirable. They are sometimes intentionally added to a catalyst to stabilize their performance. Thus, a calculated reduction of the initial reactivity of a catalyst can pay off if the long-term activity is higher on a specifically poisoned catalyst than on an unintentionally poisoned one. Another reason for intentionally poisoning a catalyst is that the poison may be specific for a certain reaction. Step sites can be particularly effective at breaking bonds. Depending on the reaction, we may want to suppress this function. By decorating the steps with a poison, the reactivity of the steps can be turned off so that no reaction occurs there. Hence, poisons are sometimes added to catalyst to enhance their selectivity.

In Fischer-Tropsch chemistry, coking, the formation of graphite on the surface, must be avoided. One solution is the intentional partial poisoning of the catalyst with S by adding H_2S to the feed. An ingeniously novel suggestion to avoid graphite formation comes from Besenbacher et al. [79]. The addition of a small amount of Au to the Ni catalyst leads to the formation of a surface alloy. This alloyed surface has a slightly lower activity than the pure Ni surface. The formation of graphite, on the other hand, is completely suppressed. Therefore, while the initial reactivity is somewhat lower for the Au/Ni catalyst, no long-term degradation of the catalytic activity occurs. This strategic implementation of poisoning represents a great victory for the fundamental surface science approach to heterogeneous catalysis. The poisoning effect of Au was predicted based on an understanding of the reactivity of metal surfaces gleaned from theoretical studies of dissociative hydrogen adsorption (§3.5). The successful implementation of this knowledge in a practical catalyst demonstrates that the principles governing the interactions of small molecules under UHV conditions can, indeed, be used to develop an understanding of real world catalysts.

A catalyst can be deactivated in a number of other ways apart from poisoning. If the metal clusters in the active catalyst agglomerate into larger particles, a net reduction of surface area occurs. If the reaction were structure-sensitive, the loss of reactivity may be larger than the loss of surface area would indicate. Some catalysts have materials added to them that hinder the sintering of the active phase. Al_2O_3 and CaO perform this function in the Haber-Bosch process and are known as structural promoters. The catalyst may actually be lost if it is volatile, reacts to form volatile products, diffuses into the support, or reacts with the support to form non-active phases. The oxidation of K on the ammonia synthesis catalyst reduces the volatility of K. The choice of substrate and control of the reaction temperature are critical to avoid volatilization and reactions between the catalyst and support. Most heterogeneous catalysts exist in porous matrices. Shape selective catalysis in zeolites depends on molecular scale pores controlling the transport of reactants and products to and from the catalyst. If the pores become blocked by reaction products or by-products, the reactivity suffers.

6.9   Bimetallic and bifunctional catalysts

Consider the mixture of two metals in a catalyst and the types of structures depicted in Fig. 1.6. In light of our discussion of the ability of co-adsorbates to promote or poison catalysts, it should be clear that these structures present very different reactivity as compared to a monometallic catalyst. The added metal may act either as a poison or promoter. Poisoning and promotion need to be discussed with respect to three effects: site blocking, bifunctionality and the ligand effect The effects of poisoning and promotion can affect both the activity and selectivity of a catalyst.

As seen in §6.8, poisoning is not always deleterious. Addition of Au to Ni attenuates the formation of carbonaceous deposits. The surface Au/Ni alloy has a lower reactivity than pure Ni due to an electronic effect akin to that discussed above. Au deposited on Ru does not form an alloy. Instead, Au segregates to the steps. The step sites are exceedingly more reactive than terrace sites toward the dissociation of nitrogen [80]. Therefore, the presence of a sufficient amount of Au to completely decorate the steps shuts down the reactivity of Ru(0001) for NH_3 production because the rate limiting step is the dissociation of N_2. Au poisons the reaction by blocking the active sites.

Promotion can occur because the added metal changes the electronic structure of the mixture as compared to the pure metals. This is also known as the ligand effect. This is the same effect as that discussed above in conjunction with the coadsorption of electronegative and electropositive atoms on pure metals. Shifts in the *d* bands lead to changes in binding energies and activation energies, which are reflected in catalytic activity and selectivity. The ligand effect is most likely to occur when an alloy is formed, that is, when one metal is dispersed in the other either randomly or in an ordered fashion.

If islands of one metal form on top of the other metal, then the possibility exists that each metal (apart perhaps from the atoms at the edges of the islands) acts much like the pure metals. Thus, if one metal is particularly good at dissociating one molecule while the other has a particularly well suited binding energy for the other reactant, the combination of the favourable properties of both pure metals can lead to a catalyst that is more effective than the sum of its parts. This is known as bifunctionality: each metal contributes a function to the overall chemical mechanism, which is the sum of at least two parallel steps that occur at different sites. For example, in the electrocatalytic oxidation of CO in an aqueous solution on a Pt electrode promoted with Ru [81], CO is bound at sites on the Pt surface, whereas OH is formed from the dissociation of water on Ru islands. Oxidation occurs at an appropriate voltage when CO diffuses from the Pt sites to the edge sites of the Ru islands.

Not always is the substrate inert and a spectator in the reaction. This is especially true in the case of so-called solid-acid catalysts and solid-base catalysts. Consider the aqueous phase reforming (APR) and aqueous phase dehydration/hydrogenation (APD/H) of oxygenated hydrocarbons, in particular sorbitol ($C_6O_6H_4$) [82]. On a Pt/Al_2O_3 catalyst, the polyol adsorbs dissociatively on Pt sites, releasing H_2. The metal catalyzes the breaking of C–C bonds and leads to the formation of CO(a). The water gas shift

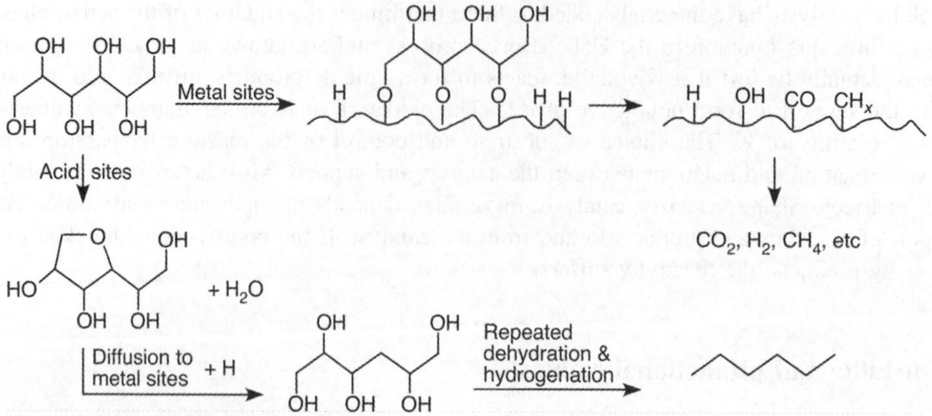

Figure 6.8 *A schematic representation of the function of a bifunctional catalyst containing supported metal particles and acid sites on the support.*

reactions then converts the adsorbed CO to $CO_2 + H_2$. If SiO_2 is mixed into the Al_2O_3 carrier, then acid sites are introduced into the catalyst. These acid sites induce the dehydration of sorbitol upon absorption to form ring compounds. The products of dissociative absorption on the acid sites then migrate to the metal sites where they are hydrogenated. Repeated hydrogenation leads to the formation of alkanes. The selectivity to form more alkanes and less H_2, that is to favour hydrogenation rather than recombinative desorption of H_2 at metal sites, is enhanced by increasing the amount of SiO_2 added to the catalyst (i.e. by increasing the number of acid sites). The overall process is depicted schematically in Fig. 6.8.

In this context it is important to note that bifunctional catalysts can either be composed of two distinct metal sites or metal sites juxtaposed with acid or base sites. In addition, there is no reason for the ligand effect and bifunctionality to be mutually exclusive. In any bimetallic system both may contribute to a greater or lesser degree.

6.10   Rate oscillations and spatiotemporal pattern formation

Non-linear dynamicss is a term used to describe a broad range of phenomena that includes not only oscillating chemical reactions and pattern formation, but also chaos and turbulence. Non-linear dynamics has been applied to the study of phenomena as diverse as traffic jams and heart arrhythmia. Oscillating chemical reactions have been known for a long time with the most celebrated example being the Belousov-Zhabotinskii reaction. The first observation of oscillating kinetics in heterogeneous catalysis was made in the group of Wicke [83]. Imbihl and Ertl [84, 85] have reviewed the field with particular emphasis on studies carried out on single crystal surfaces. They point out that, compared to pattern formation in homogeneous systems, surface reactions provide three unique aspects. First, anisotropic diffusion is possible on surfaces. Second, mass transport in the gas phase provides a means of communication across the surface that can lead to global synchronization. Finally, since the reactions occur on a surface, we can use a library of techniques that have high spatial resolution to characterize the spatial and temporal concentration gradients that are formed.

Imbihl and Ertl [84] list 16 surface reactions that exhibit oscillatory kinetics. Single crystal studies have concentrated on CO oxidation on Pt and Pd, and NO reduction by CO, H_2 and NH_3 on Pt(100) and Rh(110). In general, oscillating reactions are accompanied by spatiotemporal pattern formation and surface reactions are no exception. We concentrate on CO oxidation on Pt as it exhibits particularly rich chemical behaviour.

CO oxidation follows the Langmuir-Hinshelwood mechanism described in §3.14 and §6.2. Particularly important for the observation of oscillations is that CO oxidation exhibits asymmetric inhibition. Inhibition occurs when a reactant poisons the reaction. In this case, CO forms a densely packed layer upon which O_2 cannot dissociatively adsorb. Conversely, O atoms form an open adlayer into which CO readily adsorbs. Therefore, the reaction is only poisoned by high coverages of CO, and the reaction rate exhibits two branches. On the high-rate branch at low p_{CO}, the CO_2 production rate increases linearly with p_{CO}. On the low-rate branch at high p_{CO}, the reaction rate decreases with increasing p_{CO}. In other words, the effective reaction order relative to p_{CO} changes from $n = 1$ to $n < 0$ even though there is no change in the reaction mechanism. This highlights the need to understand what is occurring on the surface in order to understand effective reaction orders whereas the converse cannot be done unambiguously.

The reaction dynamics described above are sufficient for explaining the bistability of the reaction rate, but not yet sufficient to explain oscillations. A further mechanistic element is required to introduce a non-linearity into the reaction kinetics. The first is the adsorbate-induced reconstruction of the Pt(100) and Pt(110) surfaces. The second is that the sticking coefficient of O_2 changes dramatically on the different reconstructions.

Rate oscillations are observed under conditions where O_2 adsorption is rate limiting. To illustrate, consider a CO covered (but not saturated) Pt(110)–(1 × 1) surface. O_2 adsorbs readily on this surface, and the reaction rate increases as more O_2 adsorbs. Eventually the rate becomes too high for adsorption of CO to replenish the surface. Consequently, the CO coverage starts to drop and the rate passes through a maximum. When the CO coverage drops below $\theta_{CO} \approx 0.2$ ML, the surface reconstructs into the (1 × 2) phase. O_2 does not stick well on this surface, but CO adsorption is much less structure-sensitive. The reaction rate drops to a minimum, and the CO coverage begins to build up on the (1 × 2) surface until the critical coverage is exceeded. The surface reconstructs, and the system returns to the initial conditions so that the next oscillation can begin.

Reconstructions are not a required element of oscillating surface reactions. A more general statement is that periodic site blocking leads to oscillations. Thermokinetic oscillations can be related to a periodic site blocking mechanism. In this mechanism, the heat of reaction can lead to either heating or cooling of the surface depending on whether the reaction is exo- or endothermic. Because reaction rates depend exponentially on temperature, a change in the surface temperature dramatically changes the balance between adsorption and desorption. A decreasing temperature caused by an endothermic reaction leads to the build up of site blocking adsorbates, killing off reaction and forming a deactivated phase. After deactivation, the temperature rises, the site blocker desorbs, the activated state is restored, and oscillations ensue. Alternatively, an exothermic reaction can lead to a heating of the surface that accelerates either desorption or reaction involving the blocking species. If the reaction rate exceeds the rate of replenishment of reactants, the catalyst is eventually cleaned off at which point the temperature drops and the site blocker again begins to build up. Autocatalysis can also provide the mechanism for oscillations. In the reaction of NO + CO, NO dissociation requires two empty sites. The formation of N_2 and CO_2 leads to the formation of four empty sites. Under conditions in which NO dissociation is rate limiting, the formation of products increases the rate of NO dissociation. The rate can accelerate so rapidly that a surface explosion occurs in which products desorb in extremely narrow peaks only 2–5 K wide.

Rate oscillations are closely coupled to some mechanism of synchronization. If there were none, different reaction rates across the surface would simply average out to a uniform rate. Synchronization can occur via heat transfer, partial pressure variations or surface diffusion. Only the latter two are relevant in low-pressure ($p < 10^{-3}$ mbar) single crystal studies.

These couplings can result in pattern formation as well. Imbihl, Ertl and co-workers [86, 87] were the first to identify propagating waves on single crystal surfaces. Several classes of patterns can form solely on the basis of diffusional coupling both in fluid phases and on surfaces. These include target patterns,

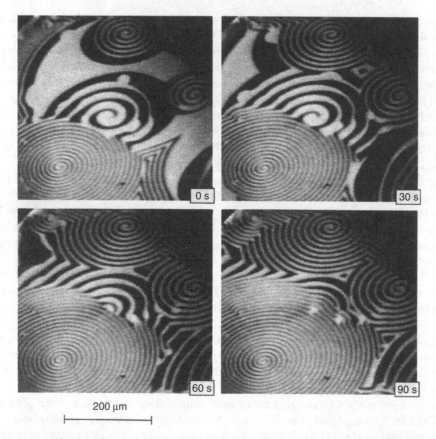

200 μm

Figure 6.9 *Spatiotemporal pattern formation in CO oxidation over Pt. Reproduced from S. Nettesheim, A. von Oertzen, H. H. Rotermund and G. Ertl, J. Chem. Phys. 98 (1993) 9977. © 1993 with permission from the American Institute of Physics.*

spiral waves and pulses. In addition to these travelling waves, stationary waves known as Turing structures can form. Anisotropic diffusion and lateral interactions lead to deformations in patterns, altering what would have been circular patterns into elliptical patterns. Gas-phase coupling leads to either stabilization or destabilization of a uniformly oscillating surface depending on conditions. An example of travelling spiral waves observed for CO oxidation on Pt(110) is shown in Fig. 6.9.

Advanced Topic: Cluster assembled catalysts

Techniques now exist for the production of beams of atomic and molecular clusters [88–90]. Furthermore, the cluster can be size selected, in favourable circumstances, for clusters ranging from two to thousands of atoms. These clusters can be deposited onto a suitable substrate for further study. Alternatively, organometallic chemistry can be used to create multicentre cluster compounds of metals [91] even inside of zeolite pores by "ship in a bottle synthesis" [92]. The use of size-selected, deposited metallic clusters represents an alternative route to the formation of model catalysts [91, 93].

Gas-phase clusters are known to exhibit size-dependent reactivity in specific reactions, for example H_2 dissociation on metals [89] or O_2 on Si [94]. Thus, we might also expect supported clusters to exhibit size-dependent reactivity. Size-dependent reactivity, of course, ties into the concepts of structure-sensitive and structure-insensitive reactions that were mentioned in §6.2. First, we note that gas-phase reactivity has little predictive relevance for reactivity of supported clusters. The interaction of the cluster with the support not only changes the electronic and geometric structure of a cluster, it also opens up new energy relaxation pathways that are not available to gas-phase clusters, such as loss of thermal energy to substrate phonons. Therefore, the reactivity of clusters is greatly modified by interactions with the substrate. For instance, the dissociation of O_2 on Si clusters exhibits strong size-dependence in the gas phase, but virtually none for supported clusters [94]. Nonetheless, size-dependent reactivity has been observed for supported clusters as in the case of CO oxidation on small gold clusters [93] and the question of why one of the most noble of metals changes from a catalytically useless bulk material into an extremely effective catalysis when present as supported nanoscale clusters is a very active issue [95, 96].

The study of size-selected supported clusters promises to bring new insight to catalytic studies. When coupled with structural probes such as STM, it addresses important questions regarding the interactions between catalyst particles and substrates. Supported metal clusters with diameters <10 Å are smaller than those that have generally been used in practical catalysts [91]. The study of this neglected size range may lead to developments in catalytic chemistry.

6.11 Sabatier analysis and optimal catalyst selection

It has long been assumed that there is a correlation between the activation energy for a reaction E_a and the energy released during reaction (the enthalpy of reaction, which we denote here as ΔE as it could be either an adsorption enthalpy or the enthalpy of a surface reaction). Calculations based on density functional theory (DFT) have shown that for bond breaking on different transition metal surfaces, such a relationship exists. Specifically, Pallassana and Neurock [97] for C–H bond breaking, Liu and Hu [98] for CO dissociation, and Logadóttir et al. [39] for N_2 dissociation have all demonstrated this relationship. Indeed, Nørskov and co-workers [99, 100] have shown that not only do many surface reactions exhibit this correlation, there are classes of similar reactions that follow the same "universal" relationship. Dissociation of N_2, CO, NO, and O_2 on transition metals obeys the relationship

$$E_a = \alpha_1 \Delta E_1 + \beta_1 \tag{6.11.1}$$

with $\alpha_1 = 0.87$ and $\beta_1 = 1.34\,\text{eV}$. This relationship between barrier height and enthalpy is known as a Brønsted-Evans-Polanyi (BEP) relationship, and is illustrated in Fig. 6.10. Furthermore, it has been shown that the linear BEP relationship of Eq. (6.11.1) is directly responsible for the phenomenon known as a volcano plot, in which, classically, the reaction rate is plotted against a parameter such as the group number of the transition metal used to catalyze the reaction or the exothermicity of oxide formation. The reactivity is found to increase up to a plateau, then decrease in a fashion similar to the profile of a volcano.

A good catalyst is one that has a low activation energy, but it must also not bind the intermediates or products so tightly that they cannot go on to react or desorb from the surface. The BEP relationship shows that low activation barriers are correlated with strong binding and visa versa. Because of this it has been recognized that the best catalyst is one with intermediately strong interactions with the species involved in the reaction – a concept known as the Sabatier principle [101, 102]. Furthermore, if we can identify

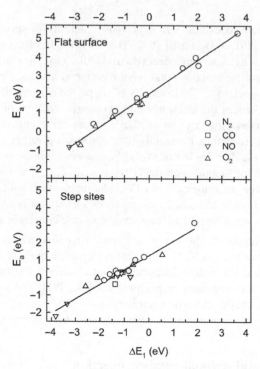

Figure 6.10   *Calculated transition state energies (E_a) and dissociative chemisorption energies (ΔE_1) for N_2, CO, NO, and O_2 on a number of transition metal surfaces. Results for close packed as well as stepped surfaces are shown. Reproduced from T. Bligaard, J. K. Nørskov, S. Dahl, J. Matthiesen, C. H. Christensen, J. Sehested, J. Catal., 224, 206. © 2004 with permission from Elsevier.*

reactions for which the dissociation of a small molecules such as N_2, CO, NO, and O_2 is rate determining, Eq. (6.11.1) allows us to identify which material is the optimal catalyst based on energetics and rates of reactions because we can use Eq. (6.11.1) to help us determine which material has the most favourable rate for the reaction. Examples include not only the ammonia synthesis reaction, which is rate limited by N_2 dissociation, but also the methanation reaction, which is rate limited by CO dissociation [100].

6.12   Summary of important concepts

- The pressure gap and materials gap are not intrinsic barriers to understanding catalysis on the basis of UHV surface studies.
- Some catalytic reactions exhibit kinetics that depends strongly on the presence of certain types of sites or crystallographic planes. These are structure-sensitive reactions. Particularly reactive sites are known as active sites.
- Structure-insensitive reactions have kinetic parameters that are similar across a range of sites.
- The kinetics of a series of reactions is often controlled by a single reaction – the rate determining step – which has the largest degree of rate control.
- The rate of an elementary step (or an irreversible one-step reaction) is given by

$$r = k[\theta_A^n \theta_B^m \ldots] \tag{6.12.1}$$

whereas the rate of a reversible elementary reaction is given by the net rate of reaction

$$r = k_1[\theta_A^m \theta_B^n \ldots] - k_{-1}[\theta_A^{m'} \theta_B^{n'} \ldots] \tag{6.12.2}$$

where k_1 is the rate constant of the forward reaction and k_{-1} that of the reverse reaction.

The net rate can also be written using De Donder relations and the affinity

$$r_i = r_{i,\,f}[1 - \exp(-A_i/RT)] \tag{6.12.3}$$

- An accurate count of empty sites must be taken into account in the kinetics of surface reactions.
- Both activity and selectivity are important characteristics of industrial catalysts.
- A practical catalyst is usually composed of an active phase and a support that helps to maintain the catalytic material in the active phase.
- A promoter enhances the reactivity and/or selectivity of the active phase by either electronic effects or direct interactions with the reactants.
- Poisons reduce activity either through site blocking or electronic effects.
- Poisons are sometimes strategically added to practical catalysts.
- Bifunctional catalysts incorporate different activities as different sites, for example, by combining two different metal sites or a metal site with an acid or base site on the support.
- Non-linearities in surface kinetics can lead to rate oscillations and spatiotemporal pattern formation.
- An optimal catalyst exhibits balanced binding interactions so that the appropriate bonds are weakened or broken to lower activation energies, while not binding reactants and products too tightly so that site blocking occurs.

6.13 Frontiers and challenges

- What is the structure of a catalyst under reaction conditions?
- Understanding and controlling sintering.
- Reducing or removing precious metals and substituting base metals or reactive porous solids, particularly in fuel cell catalysis [103–105].
- Formulation of catalysts to refine biomass and pyrolysis oils economically, in particular, catalysis associated with the chemistry of aqueous phase sugar solutions, levulinic acid and γ-valeroactone reactions, hydrogen production from aqueous phase reforming, and the production of alkanes from aqueous phase dehydration/hydrogenation [10, 82].
- Catalysts that make useful products from glycerol (a by-product of biodiesel synthesis) and formic acid (a by-product of carbohydrate to levulinic acid transformation).
- More selective Fischer-Tropsch synthesis.
- Beyond Fischer-Tropsch: Direct conversion of CH_4 to feedstocks, particularly easily transportable liquids.
- Short contact time catalysis [106]. Can it be usefully controlled and can it be scaled up to industrially relevant quantities?
- Making homogeneous catalysis heterogeneous. Homogeneous catalysts often exhibit desirable qualities especially in the realm of chiral catalysis. The need to separate them from the products, however, often complicates scaling them up for commercial purposes. Can centres that exhibit desirable catalytic activity be attached to substrates such that they retain their activity/selectivity while obviating the need for separations?
- Gold catalysis and its implications [95, 96]. How can it be that we have only recently learned that making nanoscale Au particle leads to interesting catalytic properties? What is the underlying mechanism, and are there more such surprises to be found?

6.14   Further reading

M. Boudart and G. Djéga-Mariadassou, *Kinetics of Heterogeneous Catalytic Reactions* (Princeton University Press, Princeton, NJ, 1984).

I. Chorkendorff and J. W. Niemantsverdriet, *Concepts of Modern Catalysis and Kinetics* (Wiley-VCH, Weinheim, 2007).

G. Ertl, *Reactions at Solid Surfaces* (John Wiley & Sons, Hoboken, NJ, 2009).

G. Ertl, H. Knözinger, F. Schüth and J. Weitkamp (Eds), *Handbook of Heterogeneous Catalysis* (Wiley-VCH, Weinheim, 2008).

B. C. Gates and H. Knözinger (Eds), *Impact of Surface Science on Catalysis* (Academic Press, Boston, 2001).

D. A. King and D. P. Woodruff (Eds), *The Chemical Physics of Solid Surfaces and Heterogeneous Catalysis: Fundamental Studies of Heterogeneous Catalysis*, Vol. 4 (Elsevier, Amsterdam, 1982).

G. W. Huber and J. A. Dumesic, *An overview of aqueous-phase catalytic processes for production of hydrogen and alkanes in a biorefinery*, Catal. Today **111**, 119–132 (2006).

R. Imbihl, Non-linear dynamics in catalytic reactions, in *Handbook of Surface Science: Dynamics*, edited by E. Hasselbrink and I. Lundqvist (Elsevier, Amsterdam, 2008, Vol. **3**), pp. 341–428.

J. H. Larsen and I. Chorkendorff, *From fundamental studies of reactivity on single crystals to the design of catalysts*, Surf. Sci. Rep. **35**, 163–222 (1999).

J. V. Lauritsen, R. T. Vang, and F. Besenbacher, *From atom-resolved scanning tunneling microscopy (STM) studies to the design of new catalysts*, Catal. Today **111**, 34–43 (2006).

T. Mallat, E. Orglmeister, and A. Baiker, *Asymmetric catalysis at chiral metal surfaces*, Chem. Rev. **107**, 4863–4890 (2007).

B. E. Nieuwenhuys, Toward understanding automotive exhaust conversion catalysis at the atomic level, in *Advances in Catalysis*, Vol. **44**, edited by W. O. Haag, B. C. Gates, and H. Knözinger (Academic Press, Boston, 1999), p. 260.

J. K. Nørskov, *Electronic factors in catalysis*, Prog. Surf. Sci. **38**, 103–144 (1991).

P. Stoltze, *Microkinetic simulation of catalytic reactions*, Prog. Surf. Sci. **65**, 65–150 (2000).

J. M. Thomas and W. J. Thomas, *Principles and Practice of Heterogeneous Catalysis* (VCH, Weinheim, 1996).

A. Wieckowski and M. Koper, *Fuel Cell Catalysis: A Surface Science Approach* (Wiley Interscience, New York, 2007).

F. Zaera, *Probing catalytic reactions at surfaces*, Prog. Surf. Sci. **69**, 1–98 (2001).

6.15   Exercises

6.1   Confirm Eq. (6.1.1).

6.2   Explain with the aid of a one-dimensional potential energy diagram why the Haber-Bosch synthesis of NH_3 is less energy intensive and more rapid than the homogeneous process.

6.3   Show that the following expressions [5] hold for the equilibrium surface coverages during ammonia synthesis.

$$\theta_{N_2*} = K_1 \frac{p_{N_2}}{p_0} \theta_* \tag{6.15.1}$$

$$\theta_{N*} = \frac{p_{NH_3} p_0^{0.5}}{K_3 K_4 K_5 K_6 K_7^{0.5} p_{H_2}^{1.5}} \theta_* \tag{6.15.2}$$

$$\theta_{NH*} = \frac{p_{NH_3}}{K_4 K_5 K_6 p_{H_2}} \theta_* \tag{6.15.3}$$

$$\theta_{NH_2*} = \frac{p_{NH_3}}{K_5 K_6 K_7^{0.5} p_{H_2}^{0.5} p_0^{0.5}} \theta_* \tag{6.15.4}$$

$$\theta_{NH_3*} = \frac{p_{NH_3}}{K_6 p_0} \theta_*$$

(6.15.5)

$$\theta_{H*} = K_7^{0.5} \frac{p_{H_2}^{0.5}}{p_0^{0.5}} \theta_*$$

(6.15.6)

6.4 Consider the dissociative adsorption of N_2 through a molecularly bound state

$$N_2(g) + {}^* \rightleftharpoons N_2{}^*$$

(i)

$$N_2{}^* + {}^* \rightleftharpoons 2N^*$$

(ii)

Show that the rate of formation of adsorbed N atoms is

$$\frac{d\theta_{N*}}{dt} = 2k_2 K_1 \frac{p_{N_2}}{p_0} \left(\frac{1 - \theta_{N*}}{1 + K_1 \frac{p_{N_2}}{p_0}}\right)^2 - \frac{2k_2}{K_2}\theta_{N*}^2$$

(6.15.7)

Assume that the system is at equilibrium.

6.5 A number of heterogeneous reactions, including NH_3 formation, exhibit a maximum in the rate in the middle of the row of transition metals [39]. Plots of reactivity versus Periodic Chart group number with such a maximum are called volcano plots because of their shape [44]. Discuss the origin of this trend for ammonia synthesis.

6.6 The following results are observed in co-adsorption studies of CO, C_6H_6 or PF_3 with K on Pt(111) [107]. (i) CO + K/Pt(111): the CO-Pt bond is significantly strengthened. (ii) C_6H_6, which adsorb molecularly at low T_s, is less likely to dissociate upon heating in the presence of co-adsorbed K. (iii) The amount of PF_3 that can adsorb drops roughly linearly with θ_K. Explain these observations. If the rate determining step of a catalytic reaction is (a) CO dissociation or (b) C_6H_6 desorption, how does the presence of K affect the reaction rate?

6.7 The three-way catalytic converter used in automobiles catalyzes the oxidation of unburned hydrocarbons and CO while reducing simultaneously NO to N_2. Consider only the reactions of H_2 and NO, Rxns (6.5.5) and (6.5.7). The desired products are N_2 and H_2O. NH_3 formation must be suppressed. Write out a complete set of elementary reactions for which H_2O, N_2 and NH_3 are the final products. Pinpoint particular elementary steps that are decisive for determining the selectivity of the catalyst for N_2 over NH_3.

6.8 A reaction is carried out on a single crystal Pt(111) sample. The reaction requires the scission of a C–H bond. The reaction is completely poisoned by just a few percent of a monolayer of pre-adsorbed oxygen at low temperature but that at high temperature, oxygen coverage up about a tenth of a monolayer has little effect on the reaction rate. Discuss the poisoning behaviour of pre-adsorbed oxygen.

6.9 The NO + CO $\rightarrow \frac{1}{2} N_2 + CO_2$ reaction is a useful reaction in an automotive catalyst. It follows Langmuir-Hinshelwood kinetics. Consider the following to be a complete set of reactions for this system.

$$NO(g) \rightleftharpoons NO(a)$$

$$CO(g) \rightleftharpoons CO(a)$$

$$NO(a) \rightleftharpoons N(a) + O(a)$$

$$2\,N(a) \rightleftharpoons N_2(g)$$

$$CO(a) + O(a) \rightarrow CO_2(g)$$

(a) A trace of S weakens the NO–surface bond. If S co-adsorption substantially affects only NO(a), explain what the effect of the presence of a trace of S(a) has on the rate of production of N_2 and CO_2.

(b) Assume that K co-adsorption promotes N_2 dissociation but affects no other reaction. Does K co-adsorption change the rate of CO_2 production? Explain.

(c) NO only dissociates on the steps of Pt(111). Assume that the resulting O atoms are immobile at 770 K – the reaction temperature in a typical automatic catalytic converter. Given that CO desorption is first order and characterized by $E_{des} = 135 \, \text{kJ mol}^{-1}$, $A = 3.00 \times 10^{13} \, \text{s}^{-1}$ and that CO diffusion is characterized by $E_{dif} = 19.8 \, \text{kJ mol}^{-1}$, $D_0 = 5.00 \times 10^{-6} \, \text{cm}^2 \, \text{s}^{-1}$, would it be wise to use Pt(111) single crystals with terrace widths of 3500 Å to eliminate CO from the exhaust gas? Justify your answer with calculations.

6.10 For methanation of CO over Ni, CO dissociation is not the rate determining step. K co-adsorption increases the CO dissociation rate. Explain how in this case K acts as a poison. Hint: Consider the net effect on H_2 adsorption [108].

6.11 Consider the electrocatalytic oxidation of CO with water over a Ru/Pt bimetallic catalyst to form $CO_2 + H^+ + e^-$. Write out a set of elementary steps in which the water dissociates on Ru sites, CO adsorbs on Pt sites and the reaction of OH + CO to form $CO_2 + H^+ + e^-$ occurs on a Pt cluster.

6.12 3-nitrostyrene ($NO_2\text{–}C_6H_4\text{–}CH\text{=}CH_2$, 3 NS) was treated with H_2 at 120 °C in the presence of a gold nanoparticle catalyst supported on titania (1.5 wt % Au/TiO$_2$) for 6 h [109]. The primary product was 3-vinylaniline ($NH_2\text{–}C_6H_4\text{–}CH\text{=}CH_2$, 3VA). Gas chromatography with mass spectrometric (GC-MS) analysis of the product stream reveals that the peak corresponding to 3 NS has been reduced from its initial value (normalized to 1.00) to a peak that is 50 times smaller. The peak corresponding to 3VA is 48 times larger than the final 3 NS peak. Assuming that the peak areas have been corrected for sensitivity factors and that the peak areas are, therefore, proportional to the molar ratio (or equivalently the concentrations), calculate the % selectivity and % conversion. That is, calculate the % of 3 NS that is selectively hydrogenated to the desired product (3VA) and calculate the % of 3 NS that reacts to form any and all products.

6.13 A chemical reaction is found to have a well-defined rate determining step which involves the breaking of a N–O bond in NO. Explain how a transition metal surface can effectively act as a heterogeneous catalyst.

References

[1] G. A. Somorjai, *Chem. Rev.*, **96** (1996) 1223.

[2] T. Engel, G. Ertl, Oxidation of carbon monoxide, in *The Chemical Physics of Solid Surfaces and Heterogeneous Catalysis, Vol. 4* (Eds: D. A. King, D. P. Woodruff), Elsevier, Amsterdam, 1982, p. 73.

[3] T. Engel, G. Ertl, *Adv. Catal.*, **28** (1979) 1.

[4] M. Grunze, Synthesis and decomposition of ammonia, in *The Chemical Physics of Solid Surfaces and Heterogeneous Catalysis, Vol. 4* (Eds: D. A. King, D. P. Woodruff), Elsevier, Amsterdam, 1982, p. 143.

[5] P. Stoltze, J. K. Nørskov, *J. Catal.*, **110** (1988) 1.

[6] D. W. Goodman, *Chem. Rev.*, **95** (1995) 523.

[7] P. Stoltze, J. K. Nørskov, *Phys. Rev. Lett.*, **55** (1985) 2502.

[8] D. W. Goodman, J. E. Houston, *Science*, **236** (1987) 403.

[9] S. M. Davis, G. A. Somorjai, Hydrocarbon conversion over metal catalysts, in *The Chemical Physics of Solid Surfaces and Heterogeneous Catalysis: Fundamental Studies of Heterogeneous Catalysis, Vol. 4* (Eds: D. A. King, D. P. Woodruff), Elsevier, Amsterdam, 1982, p. 217.

[10] H. H. Szmant, *Organic Building Blocks of the Chemical Industry*, John Wiley & Sons, Inc., New York, 1989.

[11] C. Marcilly, *J. Catal.*, **216** (2003) 47.

[12] R. J. Davis, *J. Catal.*, **216** (2003) 396.

[13] A. Corma, *J. Catal.*, **216** (2003) 298.

[14] K. J. Laidler, *Chemical Kinetics*, HarperCollins, New York, 1987.

[15] K. W. Kolasinski, F. Cemič, A. de Meijere, E. Hasselbrink, *Surf. Sci.*, **334** (1995) 19.

[16] K. W. Kolasinski, F. Cemič, E. Hasselbrink, *Chem. Phys. Lett.*, **219** (1994) 113.

[17] J. T. Yates, Jr., K. Kolasinski, *J. Chem. Phys.*, **79** (1983) 1026.

[18] H. S. Taylor, *Proc. R. Soc. London, A*, **108** (1925) 105.

[19] G. A. Somorjai, *Adv. Catal.*, **26** (1977) 1.

[20] T. Zambelli, J. Wintterlin, J. Trost, G. Ertl, *Science*, **273** (1996) 1688.

[21] J. V. Lauritsen, R. T. Vang, F. Besenbacher, *Catal. Today*, **111** (2006) 34.

[22] Á. Logadóttir, P. G. Moses, B. Hinnemann, N. Y. Topsøe, K. G. Knudsen, H. Topsøe, J. K. Nørskov, *Catal. Today*, **111** (2006) 44.

[23] H. Over, Y. D. Kim, A. P. Seitsonen, S. Wendt, E. Lundgren, M. Schmid, P. Varga, A. Morgante, G. Ertl, *Science*, **287** (2000) 1474.

[24] Z.-P. Liu, P. Hu, A. Alavi, *J. Chem. Phys.*, **114** (2001) 5956.

[25] J. A. Schwarz, R. J. Madix, *Surf. Sci.*, **46** (1974) 317.

[26] M. P. D'Evelyn, R. J. Madix, *Surf. Sci. Rep.*, **3** (1983) 413.

[27] C. R. Arumainayagam, R. J. Madix, *Prog. Surf. Sci.*, **38** (1991) 1.

[28] D. F. Padowitz, S. J. Sibener, *J. Vac. Sci. Technol., A*, **9** (1991) 2289.

[29] D. F. Padowitz, S. J. Sibener, *Surf. Sci.*, **254** (1991) 125.

[30] R. H. Jones, D. R. Olander, W. J. Siekhaus, J. A. Schwarz, *J. Vac. Sci. Technol.*, **9** (1972) 1429.

[31] D. R. Olander, A. Ullman, *Int. J. Chem. Kinet.*, **8** (1976) 625.

[32] D. R. Olander, *J. Colloid Interface Sci.*, **58** (1977) 169.

[33] H. H. Sawin, R. P. Merrill, *J. Vac. Sci. Technol.*, **19** (1981) 40.

[34] USGS, minerals.usgs.gov/minerals/pubs/commodity/nitrogen/, 2011.

[35] C. J. H. Jacobsen, *Chem. Commun.* (2000) 1057.

[36] G. Ertl, Reaction mechanisms in catalysis by metals, in *Critical Reviews in Solid State and Materials Science, Vol. 10*, CRC Press, Boca Raton, 1982, p. 349.

[37] N. Brønsted, *Chem. Rev.*, **5** (1928) 231.

[38] M. G. Evans, N. P. Polanyi, *Trans. Faraday Soc.*, **34** (1938) 11.

[39] Á. Logadóttir, T. H. Rod, J. K. Nørskov, B. Hammer, S. Dahl, C. J. H. Jacobsen, *J. Catal.*, **197** (2001) 229.

[40] F. Bozso, G. Ertl, M. Weiss, *J. Catal.*, **50** (1977) 519.

[41] T. H. Rod, Á. Logadóttir, J. K. Nørskov, *J. Chem. Phys.*, **112** (2000) 5343.

[42] G. Ertl, *Angew. Chem., Int. Ed. Engl.*, **29** (1990) 1219.

[43] M. Boudart, Catalysis by supported metals, in *Advanced Catalysis., Vol. 20* (Eds: D. D. Eley, H. Pines, P. B. Weisz), Academic Press, New York, 1969, pp. 153.

[44] M. Boudart, G. Djéga-Mariadassou, *Kinetics of Heterogeneous Catalytic Reactions*, Princeton University Press, Princeton, NJ, 1984.

[45] D. D. Strongin, J. Carrazza, S. R. Bare, G. A. Somorjai, *J. Catal.*, **103** (1987) 213.

[46] N. D. Spencer, R. C. Schoonmaker, G. A. Somorjai, *J. Catal.*, **74** (1982) 129.

[47] F. Bozso, G. Ertl, M. Grunze, M. Weiss, *J. Catal.*, **49** (1977) 18.

[48] G. Ertl, S. B. Lee, M. Weiss, *Surf. Sci.*, **114** (1982) 515.

[49] C. T. Rettner, H. Stein, *J. Chem. Phys.*, **87** (1987) 770.

[50] C. T. Rettner, H. E. Pfnür, H. Stein, D. J. Auerbach, *J. Vac. Sci. Technol., A*, **6** (1988) 899.

[51] C. T. Rettner, H. Stein, *Phys. Rev. Lett.*, **59** (1987) 2768.

[52] J. J. Mortensen, L. B. Hansen, B. Hammer, J. K. Nørskov, *J. Catal.*, **182** (1999) 479.

[53] G. Ertl, S. B. Lee, M. Weiss, *Surf. Sci.*, **114** (1982) 527.

[54] J. A. Dumesic, *J. Catal.*, **185** (1999) 496.

[55] T. De Donder, *L'Affinité*, Gauthier-Villers, Paris, 1927.

[56] M. Boudart, K. Tamaru, *Catal. Lett.*, **9** (1991) 15.

[57] C. T. Campbell, *J. Catal.*, **204** (2001) 520.

[58] M. W. Chase (Ed.), *JANAF Thermochemical Tables*, 4th ed., National Institute of Standards and Technology, Gaithersburg, MD, 1998.

[59] M. J. Overett, R. O. Hill, J. R. Moss, *Coord. Chem. Rev.*, **206–207** (2000) 581.

[60] J. Sehested, *Catal. Today*, **111** (2006) 103.

[61] R. Pearce, M. V. Twigg, Coal- and natural gas-based chemistry, in *Catalysis and Chemical Processes* (Eds: R. Pearce, W. R. Patterson), Blackie & Sons, Glasgow, 1981, p. 114.

[62] K. P. De Jong, J. W. Geus, *Catal. Rev.*, **42** (2000) 481.

[63] H. Topsøe, C. V. Ovesen, G. S. Clausen, N.-Y. Topsøe, P. E. Højlung Nielsen, E. Törnqvist, J. K. Nørskov, Importance of dynamics in real catalyst systems, in *Dynamics of Surfaces and Reaction Kinetics in Heterogeneous Catalysis* (Eds: G. F. Froment, K. C. Waugh), Elsevier, Amsterdam, 1997, p. 121.

[64] J. M. Thomas, W. J. Thomas, *Principles and Practice of Heterogeneous Catalysis*, VCH, Weinheim, 1996.

[65] P. M. Maitlis, V. Zanotti, *Chem. Commun.* (2009) 1619.

[66] R. C. Brady, R. Pettit, *J. Am. Chem. Soc.*, **102** (1980) 6181.

[67] P. M. Maitlis, H. C. Long, R. Quyoum, M. L. Turner, Z.-Q. Wang, *Chem. Commun.*, **13**, (1996) 1.

[68] B. E. Mann, M. L. Turner, R. Quyoum, N. Marsih, P. M. Maitlis, *J. Am. Chem. Soc.*, **121** (1999) 6497.

[69] H. S. Gandhi, G. W. Graham, R. W. McCabe, *J. Catal.*, **216** (2003) 433.

[70] K. C. Taylor, Automobile catalytic converters, in *Catalysis: Science and Technology, Vol. 5* (Eds: J. R. Anderson, M. Boudart), Springer-Verlag, Berlin, 1984, p. 119.

[71] C. A. Hampel, G. G. Hawley (Eds), *The Encyclopedia of Chemistry*, 3rd ed., Van Nostrand Reinhold, New York, 1973.

[72] P. J. Feibelman, D. R. Hamann, *Surf. Sci.*, **149** (1985) 48.

[73] J. K. Nørskov, S. Holloway, N. D. Lang, *Surf. Sci.*, **137** (1984) 65.

[74] J. J. Mortensen, B. Hammer, J. K. Nørskov, *Phys. Rev. Lett.*, **80** (1998) 4333.

[75] J. K. Nørskov, P. Stoltze, *Surf. Sci.*, **189/190** (1987) 91.

[76] S. C. Minne, S. R. Manalis, A. Atalar, C. F. Quate, *J. Vac. Sci. Technol. B*, **14** (1996) 2456.

[77] T.-C. Shen, C. Wang, G. C. Abeln, J. R. Tucker, J. W. Lyding, P. Avouris, R. E. Walkup, *Science*, **268** (1995) 1590.

[78] J. W. Lyding, T.-C. Shen, J. S. Hubacek, J. R. Tucker, G. C. Abeln, *Appl. Phys. Lett.*, **64** (1994) 2010.

[79] F. Besenbacher, I. Chorkendorff, B. S. Clausen, B. Hammer, A. M. Molenbroek, J. K. Nørskov, I. Stensgaard, *Science*, **279** (1998) 1913.

[80] S. Dahl, Á. Logadóttir, R. C. Egeberg, J. H. Larson, I. Chorkendorff, E. Törnqvist, J. K. Nørskov, *Phys. Rev. Lett.*, **83** (1999) 1814.

[81] J. S. Spendelow, P. K. Babu, A. Wieckowski, *Curr. Opin. Solid State Mater. Sci.*, **9** (2005), 37–48.

[82] G. W. Huber, J. A. Dumesic, *Catal. Today*, **111** (2006) 119.

[83] P. Hugo, *Ber. Bunsen-Ges. Phys. Chem.*, **74** (1970) 121.

[84] R. Imbihl, G. Ertl, *Chem. Rev.*, **95** (1995) 697.

[85] G. Ertl, *Reactions at Solid Surfaces*, John Wiley & Sons, Inc., Hoboken, NJ, 2009.

[86] R. Imbihl, M. P. Cox, G. Ertl, G. Müller, W. Brenig, *J. Chem. Phys.*, **83** (1985) 1578.

[87] M. P. Cox, G. Ertl, R. Imbihl, *Phys. Rev. Lett.*, **54** (1985) 1725.

[88] H. Haberland (Ed.), *Clusters of Atoms and Molecules, Vol. 52*, Springer-Verlag, Berlin, 1993.

[89] H. Haberland (Ed.), *Clusters of Atoms and Molecules II*, *Vol. 56*, Springer-Verlag, Berlin, 1994.

[90] G. Scoles (Ed.), *Atomic and Molecular Beam Methods, Vol. 1*, Oxford University Press, New York, 1988.

[91] B. C. Gates, *Chem. Rev.*, **95** (1995) 511.

[92] T. Beutel, S. Kawi, S. K. Purnell, H. Knözinger, B. C. Gates, *J. Phys. Chem.*, **97** (1993) 7284.

[93] U. Heiz, W.-D. Schneider, *J. Phys. D: Appl. Phys.*, **33** (2000) R85.

[94] M. F. Jarrold, *Science*, **252** (1991) 1085.

[95] M. S. Chen, D. W. Goodman, *Catal. Today*, **111** (2006) 22.

[96] S. Chrétien, S. K. Buratto, H. Metiu, *Curr. Opin. Solid State Mater. Sci.*, **11** (2007) 62.

[97] V. Pallassana, M. Neurock, *J. Catal.*, **191** (2000) 301.

[98] Z.-P. Liu, P. Hu, *J. Chem. Phys.*, **114** (2001) 8244.

[99] J. K. Nørskov, T. Bligaard, Á. Logadóttir, S. Bahn, L. B. Hansen, M. Bollinger, H. Bengaard, B. Hammer, Z. Sljivancanin, M. Mavrikakis, Y. Xu, S. Dahl, C. J. H. Jacobsen, *J. Catal.*, **209** (2002) 275.

[100] T. Bligaard, J. K. Nørskov, S. Dahl, J. Matthiesen, C. H. Christensen, J. Sehested, *J. Catal.*, **224** (2004) 206.

[101] A. A. Balandin, Modern state of the multiplet theory of heterogeneous catalysis, in *Advanced Catalysis, Vol. 19* (Eds: D. D. Eley, H. Pines, P. B. Weisz), Academic Press, New York, 1969, p. 1.

[102] P. Sabatier, *La Catalyse en Chimie Organique*, Bérange, Paris, 1920.

[103] J. P. Dodelet, M. Lefevre, E. Proietti, F. Jaouen, *Science*, **324** (2009) 71.

[104] G. Wu, K. L. More, C. M. Johnston, P. Zelenay, *Science*, **332** (2011) 443.

[105] F. Jaouen, E. Proietti, M. Lefevre, R. Chenitz, J. P. Dodelet, G. Wu, H. T. Chung, C. M. Johnston, P. Zelenay, *Energy & Environmental Science*, **4** (2011) 114.

[106] J. R. Salge, B. J. Dreyer, P. J. Dauenhauer, L. D. Schmidt, *Science*, **314** (2006) 801.

[107] E. L. Garfunkel, J. J. Maj, J. C. Frost, M. H. Farias, G. A. Somorjai, *J. Phys. Chem.*, **87** (1983) 3629.

[108] C. T. Campbell, D. W. Goodman, *Surf. Sci.*, **123** (1982) 413.

[109] A. Corma, P. Serna, *Science*, **313** (2006) 332.

7

Growth and Epitaxy

Up to this point we have mainly concentrated on adsorption at the level of one monolayer or less, but now we need to go beyond thinking about only a single monolayer. What happens as the coverage approaches, then exceeds one monolayer? We have seen that strong chemisorption often leads to ordered overlayers. Particularly at high coverage, chemisorbates often take on structures that are related to the structure of the substrate. The factors that influence the order of a monolayer are the strength of the adsorption interaction, the strength of lateral interactions, the relative strength of adsorption versus lateral interactions, corrugation in the adsorbate/substrate interaction potential, and mobility in the layer. In short, we need to know how the adsorbate fits into the template of the substrate and how it gets there. When we consider the adsorption of a second layer on top of the first, we again need to ask how the subsequent layers fit onto the layers below them. In order to quantify this fit and to understand the different modes of layer growth, we need to understand strain and surface tension.

Growth and etching are two of the most important techniques of surface modification. They are the basic tools of integrated circuit device and solar cell fabrication. They also are the means by which lab-on-a-chip devices are created. Anisotropic wet etching of Si is used to created low reflectivity surfaces to improve the performance of photovoltaic devices. Understanding growth and etching is fundamental to understanding the manner in which numerous nanostructured materials are produced.

7.1 Stress and strain

Consider two separate single crystal solids, A and B, taken to be two Si(111) surfaces. If two unreconstructed Si(111) surfaces are brought in contact, the dangling bonds match up perfectly and once the new bonds between the two surfaces are formed, the two crystals form an interface that is indistinguishable from the rest of the crystal. The interface is perfectly matched geometrically. If we were to bring two unreconstructed (111) surfaces of Si and Ge in contact, they would not match. The lattice constant of Ge is larger than that of Si, 5.658 Å versus 5.43 Å. If we force the two surfaces to form bonds, these bonds are highly perturbed, and this perturbation must be relaxed in some way if the crystal is to lower its energy. Far from the interface, the new material has the structures of the pure materials, but what does the new interface look like? Is the perturbation to be relaxed gradually? Does the interface assume the geometry of Si, Ge or something completely new?

The examples on the matching of Si with Si and Ge with Si demonstrate that we need to distinguish two different types of growth. In homoepitaxy, the growing layer has the same atomic dimensions and lattice

Surface Science: Foundations of Catalysis and Nanoscience, Third Edition. Kurt W. Kolasinski.
© 2012 John Wiley & Sons, Ltd. Published 2012 by John Wiley & Sons, Ltd.

structure as the substrate. This greatly simplifies the growth process and the types of interfaces that can be formed. In heteroepitaxy – the growth of chemically dissimilar layers which nonetheless share the same structure as the substrate – the difference in lattices and chemical bonding between the two materials gives rise to a number of important phenomena that need to be controlled and understood to obtain multilayer structures with desired properties. Strain is not only important in determining growth characteristics, but also in defining the electrical characteristics of multilayer structures. For instance, strain reduces the band gap of GeSi layers grown on Si and this property makes them useful for fabrication of high-speed switches. Similarly, the lasers fabricated from InGaAs/GaAs would not operate without the band structure modifications imposed by strain. A layer that approximately assumes the structure of the substrate rather than exhibiting the structure found in the pure bulk material is known as a pseudomorphic layer.

Stress and strain are illustrated in Fig. 7.1. The difference between the ideal and actual positions of the surface plane is the surface strain. Most clean surfaces exhibit tensile strain, that is, they contract away from the boundary between the bulk terminated ideal structure and the vacuum. A compressive stress is experienced when the tendency of a material is to expand into a neighbouring region. A one-monolayer (1 ML) pseudomorphic Ge film on a Si substrate is compressively strained because of the larger natural lattice spacing of Ge compared to Si. If a material has a tendency to contract, it is under tensile stress. A pseudomorphic Si monolayer on Ge is correspondingly under tension. Stress is nontrivial to measure but its magnitude is extremely important for understanding reconstructions, epitaxial growth and the formation of structures at surfaces [1–4]. Surface layers tend to contract away from the position of the ideal bulk-terminated material to create the selvage (the near surface region that differs from the bulk). Because, as is shown in §7.2, the surface energy is dependent on strain, at equilibrium the minimum energy does not correspond to zero surface strain.

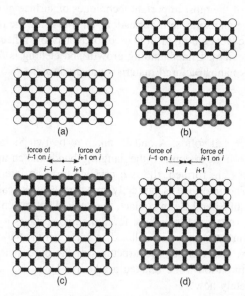

Figure 7.1   *Stress and strain in pseudomorphic layers in cross-sectional view. (a) Before attachment of the film to the substrate, the substrate has a larger lattice spacing than the film. (b) The film has a larger lattice spacing. (c) The film relaxes upon attachment, expanding to the lattice spacing of the substrate and is therefore under tension. (d) The film must contract upon attachment and is under compression. The forces along a line are as indicated. The net force vanishes in both cases but the directions of the components differ.*

The bulk stress tensor σ and the bulk strain tensor ε are straightforward to define. Bulk stress is described by a 3×3 tensor (third-order tensor of rank two) with elements σ_{ij}, the diagonal elements σ_{ii} describe the normal stresses. The off-diagonal elements describe shearing stresses. A full three-dimensional description of strain [5] requires a description of the shape and orientation of the deformed unit cell by a 3×3 matrix (the strain tensor). The 6×1 reduced Voigt strain matrix describes the deformation but not the orientation of the unit cell. The stress–strain relationships are crystal symmetry dependent and can be found elsewhere [6].

What concerns us here is the stress and strain of an interface (i.e. where material A meets material B) or surface (i.e. when material A is either vacuum or an adsorbate), which require a meticulous definition, the details of which can be found, for example, in the work of Müller and Saúl [5]. Here we denote the interfacial (or surface) stress tensor **s** with elements s_{ij} and the interfacial (or surface) strain tensor **e** with elements e_{ij}. The elements s_{ij} are defined as the interfacial excess quantity of the parallel components of the bulk stress tensor with units of energy per unit area. The elements e_{ij} are defined as the interfacial excess quantity of the perpendicular components of the bulk strain tensor with units of length. We write that the difference between the ideal and actual positions of the surface plane is the surface strain on substrate B, e.g.

$$e_{zz}^{\text{surf}} = u(z_0) - u^{\text{B}}(z_0), \tag{7.1.1}$$

or for an interface between overlayer A on substrate B, the interfacial strain is

$$e_{zz}^{\text{int}} = u^{\text{A}}(z_0) - u^{\text{B}}(z_0). \tag{7.1.2}$$

The lattice misfit (or misfit strain), ε_0, is defined in terms of the lattice parameters of the overlayer a_l and the substrate a_s as

$$\varepsilon_0 = \frac{a_l - a_s}{a_s}. \tag{7.1.3}$$

This quantity is, in principle, anisotropic but in cubic systems it is always isotropic. Group IV semiconductors (diamond, Si, Ge) have cubic lattices as do the most common zincblende forms of III-V and II-VI compound semiconductors. For a Ge overlayer on Si, $\varepsilon_0 = 0.042$. For a III-V compound semiconductor such as $In_x Ga_{1-x} As$ grown on GaAs, ε_0 depends on the stoichiometry (the value of x) of the overlayer. Hence, the lattice misfit can be tuned by controlling the chemistry of the overlayer as embodied in the Vegard law [7], which states that there is a linear variation of lattice constant with alloy composition. For example, consider $In_x Ga_{1-x} As$ grown on GaAs [6] for which it can be shown, see Exercise 7.1, that the linear dependence of the lattice constant also leads to a linear dependence of the misfit strain on the composition according to

$$\varepsilon_0(x) = 0.0726 x. \tag{7.1.4}$$

The strain ε is defined by

$$\varepsilon = \frac{a_l - a_s}{a_l} = \left(\frac{a_s}{a_l}\right) \varepsilon_0 = \frac{\varepsilon_0}{1 + \varepsilon_0}. \tag{7.1.5}$$

It is the extent to which the layer is deformed compared to its natural lattice constant a_l.

Strain is a *deformation*. Bulk *stress* is a *force* per unit area as opposed to a force per unit length for surface stress [1, 2]. The interfacial forces arising from stress lead to the deformations known as strained layers. Strain depends both on the nature and magnitude of stress. When the stress is not too large, the strain in a material responds linearly to stress, that is, Hooke's law applies. The difference between misfit strain and strain is only significant in the regime where non-linear response to stress must be used. Conventions differ in different realms of physics, but generally in surface physics tensile stress is chosen to be positive and compressive stress negative [5].

The amount of strain at the interface of two materials, A and B, depends on several factors. As discussed above, the relative sizes of the two materials, that is the lattice constants, are important factors. The lattice symmetry of the two materials must be taken into account as well. Something obviously has to give if one tries to fit an hcp lattice onto a diamond lattice. The relative strengths of the A–A, A–B and B–B interactions, as well as the temperature and deposition conditions, are important because they affect the type of interface that is formed. If two materials have different coefficients of thermal expansion, the stress experienced at the interface is temperature dependent.

The presence of adsorbates causes changes in surface stress and strain and, thereby, the surface energy. The electronegativity of the adsorbate, the work function change upon adsorption, the degree of filling of bonding and antibonding orbitals, and the spatial extent of filled electronic states all make contributions to the surface stress changes. At present we do not possess a complete understanding of the electronic origin of adsorbate-induced surface stress changes [8]; however, these changes are quite important for an understanding of adsorbate-induced reconstructions, adsorbate-induced changes in crystallite shape during growth, and for understanding the role that elastic interactions between stress domains play in the formation of ordered arrays of islands during growth on semiconductor surfaces [3].

7.2   Types of interfaces

Before considering the dynamics of growth, let us consider how strain relaxation affects the types of interfaces that are formed. These are depicted in Fig. 7.2. If there is no mixing of the two materials, a sharp interface is formed. When mixing occurs, a non-abrupt interface is formed. The mixing may take two different forms. If one of the materials is soluble in the other, then it can diffuse into the other material and create a region of variable composition. The dissolved material may act as a substitutional impurity or reside in interstitial sites. If A and B can form a stable compound, a reactive interface is formed. In this case, a new compound is formed between the two pure phases; hence, two interfaces are created by the growth process, rather than just one. When growth occurs on a crystalline substrate, a type of growth that is of particular interest for semiconductor processing, the overlayer, may either grow in a crystalline or amorphous state. A special case of growth is epitaxial growth, in which the overlayer takes on the same

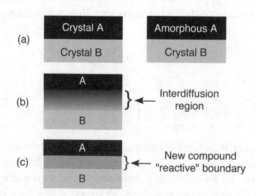

Figure 7.2   *The variety of interfaces that can be formed at the interface of two materials: (a) sharp interface; (b) non-abrupt interface; (c) reactive interface.*

structure as the substrate and grows a pseudomorphic layer in registry with the substrate. Such films are called epitaxial layers or epilayers.

7.2.1 Strain relief

The situation shown in Fig. 7.2 is often oversimplified in that the transition from one material may be abrupt from the standpoint of composition; nonetheless, from the standpoint of structure, interfaces are quite often found to be more slowly varying beasts. By this I mean that when making the transition from A to B, the composition may change from pure A to pure B within one or two atomic layers. The structure, on the other hand, changes from the structure of pure A to the structure of pure B over a much longer length scale. This is frequently the case for materials of similar structure but with different lattice constants. The long-range change of structure allows for a gradual easing of the strain associated with the formation of an interface between two materials with lattices that do not match exactly.

The type of interface obtained not only depends on the materials involved but also on the *manner in which they are grown*, in other words, the morphology and composition of layered structures depend on the balance between kinetics and thermodynamics. Many important structures are intentionally grown to be non-equilibrium or quasi-equilibrium structures. The reaction conditions, in particular the temperature, the rate of deposition and the composition of the gas phase, are regulated to grow interfaces of the desired structure. The desire to form a particular structure with the correct composition often leads to a competitive balance of processing conditions. For instance, high temperature tends to favour surface diffusion that facilitates the formation of well-ordered structures; however, it also can promote interdiffusion between layers and the loss of abrupt interfaces. Because of the importance of the manner in which growth is performed, below we investigate both the dynamics of growth and the techniques used to create designer films of controlled composition and thickness.

An important parameter that affects layer structure is the layer thickness. One mechanism of strain relief is the introduction of defects (misfit dislocations) at the interface. This generally only occurs after the overlayer has grown to a value greater than the critical thickness h_c. Alternatively, the upper surface of the overlayer may experience roughening to relax the interfacial strain. These mechanisms of strain relief, along with interdiffusion, may either appear cooperatively or competitively. The generation of misfit dislocations is typically confronted with a barrier that is much greater than the thermal energy available at the growth temperature. Since misfit dislocations are only generated after the layer exceeds the critical thickness, this type of defect may need to be annealed *into* the system. In other words, the system with defects actually represents the relaxed lower energy state, and the interface without defects represents a metastable state.

Misfit dislocations only form once the overlayer has reached and surpassed a critical thickness. This is the thickness beyond which the fully strained layer is no longer stable, a concept first defined by Frank and van der Merwe [9]. Sufficiently thin strained layers are thermodynamically stable because the strain energy is linearly proportional to the film thickness and vanishes at zero thickness. The energy of the defects, in contrast, has a non-vanishing lower limit and depends more weakly on thickness.

While the concept of critical thickness is straightforward, the quantitative calculation of this quantity in terms of materials properties has enjoyed a lifetime of over 50 years that is pockmarked with proclamations of the problem being solved. The Matthews equation [10], which purports to describe this relationship, has many forms and derivations [11] that have the ability to deliver a disturbingly divergent range of predictions [6].

Efforts continue to solve the problem quantitatively but a simple geometric argument by Dunstan et al. [12] leads to a useful rule of thumb that can be used in device design. The starting point is to consider the simplest geometric rather than energetic criterion for relaxation: the misfit dislocation should not generate more strain than the strain it is meant to relieve. This can be used to derive the result that h_c expressed in monolayers is inversely proportional to strain. For CdTe grown on CdZnTe(100) with lattice misfits from 0.15 to 6%, the relationship

$$h_c \simeq \frac{1}{\varepsilon_0} \qquad (7.2.1)$$

is found to hold within a factor of two (except at the very highest strain) even though h_c varies from 1000 ML to 5 ML over this range.

7.3   Surface energy, surface tension and strain energy

While developing the thermodynamics of interfaces, Gibbs [13] introduced the concept of surface stress and how it relates to surface energy and other thermodynamic quantities. This was expanded upon later by, for instance, Brillouin [14] and Shuttleworth [15]. A general formulation of the thermodynamics of solid/solid and fluid/solid interfaces came later in the work of Ericksson [16], Halsey [17], Andreev and Kosevich [18], and Nozières and Wolf [19, 20], who introduced the concept of surface strain. For a liquid, the creation and the stretching of the surface are equivalent processes. The energetics are therefore governed by one parameter, γ, and surface energy (associated with surface creation) is identical to surface tension (associated with surface deformation). For a solid, these two processes are inequivalent because increased solid surface area can be achieved not only by adding atoms from the bulk but also by increasing the distance between surface atoms. Furthermore, deformations of a solid surface change the intrinsic properties of the surface, and this leads to a further differentiation between solid and liquid surfaces.

The surface energy of a solid, see for example [1, 2, 5, 21] for more detail, is defined as the energy required to create a unit of surface area at constant strain. This is a scalar quantity, but for solids it depends on the surface orientation, which is defined by a unit vector **m** along the surface normal. The surface energy in the absence of strain could thus be denoted $\gamma(\mathbf{m})$ but the **m** is usually suppressed. The surface area of a solid can be increased either by deformations or adding material. Since the bonding in a solid is not isotropic, the energy required to deform the solid depends on the direction in which the atom is moved. The product of the elements of the intrinsic surface stress tensor s_{ij} and those of the strain tensor ε_{ij} describes the elastic strain energy E_{strain} of the layer according to

$$E_{\text{strain}} = \int \frac{1}{2} \sum_{ij} s_{ij} \varepsilon_{ij} \, dV \qquad (7.3.1)$$

This is valid in the limit of small deformations ε_{ij} for which the change of surface energy is linear in strain. Hence, in the presence of strain, an additional contribution (neglecting non-linear terms) must be added to the surface energy to account for the deformation, which in Lagrangian co-ordinates (see below) is

$$\gamma(\varepsilon_{ij}) = \gamma_0 + s_{ij} \varepsilon_{ij} \qquad (7.3.2)$$

For more on the tensor properties of the intrinsic surface stress tensor, see the review of Koch [21]. Equation (7.3.1) can be rewritten for pseudomorphic layers to express the elastic strain energy per unit area as

$$E_{\text{strain}} = \tfrac{1}{2} h \, s_0 \, \varepsilon_0 = \tfrac{1}{2} h \, M \varepsilon_0^2 \qquad (7.3.3)$$

where M is the modulus for biaxial strain, a two-dimensional equivalent of the uniaxial Young modulus, and h is the thickness.

In calculations of the surface energy, in particular how the surface energy depends on deformation, it is important to note that there are two different co-ordinate systems that can be chosen: the Lagrangian and the Eulerian. In Lagrangian co-ordinates, the reference state is always the undeformed one, and the Gibbs surface is attached to a given piece of matter [19]. In Eulerian co-ordinates, the Gibbs surface is fixed.

While the difference may seem to be obscure and academic, it leads to two different forms of Eq. (7.3.2) or, equivalently, two different forms of the Shuttleworth equation [15], which relates strain and surface energy as discussed by Müller and Saúl [5]:

$$\text{Euler} \quad s^E(0) = s_0 = \gamma_0 + \left. \frac{\delta\gamma^E}{\delta\varepsilon} \right|_{\varepsilon=0} \tag{7.3.4}$$

$$\text{Lagrange} \quad s^L(0) = s_0 = \left. \frac{\delta\gamma^L}{\delta\varepsilon} \right|_{\varepsilon=0} \tag{7.3.5}$$

Crystalline materials spontaneously form particles of particular shapes that reflect the symmetry of the lattice and expose the surfaces which exhibit the lowest surface energies. Hence, cubic materials tend to form cubic crystallites whereas hexagonal materials have a tendency to pack themselves into crystallites with angled faces. The Wulff construction [22] describes the equilibrium crystal shape (ECS), which is spontaneously assumed by a crystalline material if it is grown at equilibrium to minimize the surface energy and, thereby, the total energy of the particle. Its exact thermodynamic formulation has been given, for example, by Rottman and Wortis [23] and Desjonquères and Spanjaard [24].

Small crystallites supported on a substrate are particularly important in catalysis. In many cases, only small energy differences exist for crystallites that expose different crystallographic planes. Thus, a distribution of shapes and exposed crystal faces exists at equilibrium. The distribution of shapes affects the reactivity of a catalyst for structure sensitive reactions such as methanol formation on Cu/ZnO_2. The Wulff construction was conceived to describe the shape of large free-standing particles, but it can also be used to describe the distribution of crystallite sizes observed, for instance, for metal clusters formed on oxide surfaces [25]. If the clusters are unstrained, the Wulff construction predicts that cluster shape is independent of size. Daruka, Tersoff and Barabási [26], however, have shown that size dependent cluster shapes can occur for strained islands. It must also be emphasized that adsorbates can change surface energies with concomitant changes in surface reconstructions and equilibrium crystal shapes in the presence of sufficiently strongly bound adsorbates.

The ECS can only be expected if the crystallites are grown under thermodynamic control. Growth of structures on patterned substrates is complicated by diffusion between different crystal facets (intersurface diffusion). Nishinaga and co-workers [27] demonstrated that the *direction* of intersurface diffusion can be reversed by varying the As pressure in GaAs molecular beam epitaxy (defined in §7.7.1). Growth at equilibrium leads to the ECS according to the Wulff construction, subject to some modifications introduced by strain. However, for growth under kinetic control, it is intersurface diffusion that determines the profile of facets. When the substrate on which growth begins is convex, the slowest growth rate facet dominates the final structure, whereas if the initial structure is concave, the fastest growth rate structure dominates. As an aside, we shall see below that this is the opposite pattern to that followed by etching, i.e. a convex object etches to reveal fast etching planes while a concave object etches to reveal the slow etching planes. By controlling the As pressure, either pyramids or flat-topped pyramids can be grown. Furthermore, control of diffusion by temperature, pressure and composition can be used to control the profile of structures grown on mesas [28, 29].

7.4 Growth modes

7.4.1 Solid-on-solid growth

Life without stress would be easy. The same applies to a description of growth phenomena at equilibrium. Growth in the absence of stress is a rather simple process. Such is the case for the growth of thin liquid

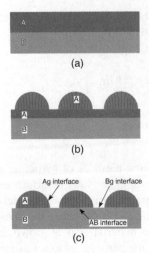

Figure 7.3   *Three modes of thermodynamically controlled growth of a solid overlayer A on solid substrate B in the presence of a gas (or, more generally, a fluid or vacuum) g. (a) Layer-by-layer growth (Frank-van der Merwe, FM) of two lattice matched ($\varepsilon_0 = 0$) materials. (b) Layer-plus-island growth (Stranski-Krastanov, SK). The strained wetting layer does not exhibit dislocations at either of its interfaces; however, the islands continuously relax with a lattice distortion in the growth direction. (c) Island growth (Volmer-Weber, VW) of lattice mismatched ($\varepsilon_o \neq 0$) materials with dislocations at the interface.*

layers on a solid substrate, in which gravity can be neglected and we need only consider the balance of forces at the liquid/gas (lg), solid/liquid (sl) and solid/gas interfaces (sg) that lead to the Young equation, Eq. (5.3.1). For liquid-on-solid growth there are only two modes of growth: wetting (2D layer-by-layer) and nonwetting (3D island formation). Investigations of solid-on-solid growth reveal the three growth modes outlined in Fig. 7.3. The appearance of strain, resulting from lattice mismatch, is the reason why solid-on-solid growth is more complicated than liquid-on-solid growth.

As we shall see below, growth is often performed in a vacuum chamber in which the pressure of the gas phase does not correspond to the equilibrium vapour pressure of the depositing material, p^*. In this case, a correction to the change in the Gibbs energy of the system must be made

$$\Delta G = n k_B T \ln(p/p^*). \tag{7.4.1}$$

The ratio

$$\zeta = p/p^* \tag{7.4.2}$$

is known as the degree of supersaturation. This leads to a further parameter that determines the thermodynamically controlled structure of the growing interface. Thus, by controlling p and T, the range of structures that can be grown as well as their growth mode can, in principle, be changed.

Equation (5.2.4) allows us to identify γ as the Gibbs surface energy per unit area; therefore, we can use Eq. (7.4.1) to write an effective Young's equation for solid-on-solid growth under the assumption that the structure formed has been equilibrated, i.e. that there was sufficient mobility during growth that a metastable structure has not been formed,

$$\gamma_{Bg} = \gamma_{Ag} \cos \psi + \gamma_{AB} + C k_B T \ln \zeta. \tag{7.4.3}$$

The constant C corrects the units of the supersaturation term. The surface energy terms are for the substrate-gas (Bg), overlayer-gas (Ag) and overlayer/substrate (AB) interfaces, respectively.

7.4.2 Strain in solid-on-solid growth

Even using the correction to the Young equation provided by Eq. (7.4.3) we still cannot explain the occurrence of the Stranski-Krastanov growth mode, Fig. 7.3(b). The observation of equilibrium structures formed by this mode is only explicable by means of the influence of strain. The three growth modes can be summarized as follows:

Layer by layer growth = Frank-van der Merwe (FM) [30]

- $\psi = 0 \Rightarrow \gamma_{Bg} \geq \gamma_{AB} + \gamma_{Ag} + Ck_B T \ln(p^*/p)$.
- Requires precise lattice matching ($|\varepsilon_0| < 1.5\%$) and balanced interactions ($\Delta = \Phi_{AA} - \Phi_{AB} \approx 0$, as defined in Eq. (5.3.4)).
- Overlayer A wets the substrate B, flat layers are expected.

Layer + island growth = Stranski-Krastanov (SK) [31]

- $|\varepsilon_0| \gtrsim 2\%$.
- After the growth of a wetting layer, often 1–4 atomic layers thick, the overlayer cannot continue to grow in the strained structure and instead forms less strained three-dimensional islands.
- Adsorbate experiences a strong interaction with the surface but the overlayer experiences significant strain.

Three-dimensional island growth = Volmer-Weber (VW) [32]

- $\psi > 0 \Rightarrow \gamma_{Bg} < \gamma_{AB} + \gamma_{Ag} + Ck_B T \ln(p^*/p)$.
- Non-wetting adsorbate, $\varepsilon_0 \neq 0$.
- Adsorbate–adsorbate interactions far stronger than adsorbate–substrate interactions, coupled with lattice mismatch that is too great to be relieved by subtle structural relaxations.

At equilibrium $\zeta = 1$ and the supersaturation term vanishes. For finely balanced systems ($\Delta \approx 0$), this term acts as a switch that can be used under nonequilibrium conditions to push the system in and out of the Frank-van der Merwe growth regime, even though the system is still under thermodynamic control, i.e. that the system and the structure assumed are not limited by restriction in the transport of material. The growth window represents the range of temperatures and compositions over which layer-by-layer growth is followed. The MBE growth window is particularly large for the $In_x Ga_{1-x} As/GaAs$ system. Substrate temperatures of 670–870 K are used. The V:III flux ratio can be varied between 2.5:1 and 25:1. The In content can extend from $0 \leq x \leq 0.25$ [6].

Stranski-Krastanov growth is the most commonly observed growth mode. It is also an intriguing growth mode as it is sometimes accompanied by the observation of islands with a small size distribution, i.e. an optimum size, which can form ordered arrays [3]. The spontaneous formation of islands represents a type of self-assembly process for quantum dot formation. The interplay between kinetics and thermodynamics in determining the shape and size of the islands remains a topic of intense study. Island formation in semiconductor epitaxy, especially of II-VI and III-V compounds as well as the Si/Ge system, has attracted great interest because of the unique optical and electronic properties of semiconductor quantum dots [33, 34]. Quantum dots get their name from the observation of size-dependent quantum mechanical effects in nanoscale objects.

An example of Stranski-Krastanov growth is the deposition of Ge onto Si(100). After a wetting layer of ~3 ML coats the Si substrate, Ge islands begin to form, which are dislocation free and known as coherent islands. Both square-based pyramidal and dome-shaped islands are formed. Similar morphologies are found in Ge/Si alloys. The evolution of these islands has been investigated by Ross, Tromp and Reuter [35] during the chemical vapour deposition growth of Ge_xSi_{1-x} films from disilane/digermane (Si_2H_6/Ge_2H_6) mixtures at 923–973 K. Islands initially take on pyramidal shapes, and these grow in size until a critical size is achieved at which point they transform into domes. A transitional truncated pyramidal structure is also observed. The pyramids do not appear to be stable. They either grow large enough so that they can convert into domes or else they shrink and disappear. The growth of larger islands at the expense of smaller ones is reminiscent of Ostwald ripening; however, the establishment of an optimum island size does not accompany conventional Ostwald ripening. The relative proportions of the three structures depend sensitively on the conditions, in particular, the temperature.

Apart from temperature and pressure, one other variable available to the experimentalist to influence the growth mode is the presence of a surfactant [36, 37]. A surfactant can lower the surface energy of interfaces and islands and thereby change the strain profile. The change in strain can also lead to a change in growth mode. Si grows on Ge(100) in a Volmer-Weber mode. This inversion of growth mode is a general rule and is a direct result of Ge possessing a lower surface energy than Si and, therefore, that Δ in Eq. (5.3.4) changes sign when the roles of substrate and overlayer are switched. Both As and Sb can be used to change the growth mode of Ge on Si. They do so by acting as a surfactant that segregates to the top surface. At the surface, they lower the energy of a flat film and inhibit the formation of islands. Consequently, they promote layer-by-layer growth at the expense of Stranski-Krastanov growth.

A general understanding of the *equilibrium* structure and growth modes of epitaxial systems is developed further below. From the above discussion we have learned that the geometric structure of overlayers depends critically on several parameters:

- Relative strengths of adsorbate–adsorbate and adsorbate–substrate interactions.
- Lattice matching – how the lattice of the substrate fits with the lattice of the overlayer.
- Strain fields – how strain relaxes in the substrate, islands and the area around the islands.
- Temperature – surface energies and, therefore, reconstructions change with temperature.
- The amount of material deposited and whether the critical thickness has been exceeded.

7.4.3 Ostwald ripening

Ostwald ripening has already been mentioned several times and deserves a more quantitative treatment. Imagine the following *Gedanken* experiment. Cover a hydrophobic surface, e.g. the H-terminated Si surface, with a thin mobile polar layer such as water. Now release the system and allow it to approach equilibrium. Because this is a nonwetting adsorbate, we expect the system to follow Volmer-Weber growth, that is, to form islands. The water overlayer contracts into 3D islands. The absolute minimum of energy, as surmised by Eqs (5.2.5) and (5.7.12), is achieved when the contact area between the overlayer and the substrate is minimized. The surface area is minimized when one large island forms rather than any greater number of smaller islands. In other words, small islands are unstable with respect to larger islands. In the limit of infinite mobility, one large hemispherical island forms.

We now express what drives this result quantitatively. The chemical potential μ_i of an unstrained island of radius r_i can be written [38]

$$\mu_i(r_i) = v\sigma/r_i \qquad (7.4.4)$$

where v is the surface area per atom in the island and σ is the island edge (step) energy. In the absence of strain, Eq. (7.4.4) shows that the chemical potential of an island is inversely proportional to its radius. Therefore as expected, a thin film that does not wet a surface breaks up into a distribution of 3D islands. If the system evolves without constraint and in the absence of strain, coarsening of the initial layer/island distribution will tend to form one large island rather than any number of smaller islands.

In any real finite system, the communication distance between islands is limited. Once the distance between the contracting islands exceeds the diffusion length, the islands become independent of one another and their accretion is arrested. During accretion, small islands feed material to larger islands. Since the position of islands is random, the number and size of initial neighbouring islands is random, and the amount of material that can ultimately be accreted into the largest island varies randomly across the surface. There is no optimum island size and the size distribution of islands is broad. Ostwald ripening was first observed and described for the growth of grains in solution [39]. As shown below, even at equilibrium and with no constraint on diffusion, ripening can be observed under appropriate conditions in the presence of strain.

The chemical potential of a strained island can be written [38]

$$\mu_i(r_i) = v \left[\frac{\sigma - \alpha}{r_i} - \frac{\alpha}{r_i} \ln \frac{r_i}{a_0} \right] \tag{7.4.5}$$

where

$$\alpha = \frac{4\pi F^2 (1 - v^2)}{\mu}. \tag{7.4.6}$$

F is the misfit strain induced elastic force monopole along the island edge, v the Poisson ratio, μ the Young modulus and a_0 is a cutoff length on the order of the surface lattice constant. The strained system exhibits a thermodynamically stable island size that is resistant to further coarsening. A system of many islands evolves until they reach a radius given by

$$r_0 = a_0 \exp(\sigma / \alpha). \tag{7.4.7}$$

At finite temperatures, entropic effects broaden the size distribution into a Gaussian distribution about a mean value that is somewhat different than r_0. Lagally and co-workers [38] have shown that further considering the influence of island–island interactions, strain leads not only to the establishment of a preferred size, but also that it can result in self-organization and a narrowing of the size distribution.

7.4.4 Equilibrium overlayer structure and growth mode

As presented above, all three of the growth modes can be observed at *equilibrium*. The development of a theory to describe the formation of different equilibrium layer structures remains an active pursuit, and whether these structures are truly equilibrium structures or relics of a kinetically controlled process is still debated. A theory to describe the role of strain in growth at equilibrium has been derived by Daruka and Barabási [40] from the theoretical framework of Shchukin and Bimberg [3].

Daruka and Barabási have calculated the phase diagram that appears in Fig. 7.4. This phase diagram confirms that FM, SK and VW can each represent the *equilibrium* growth mode for the appropriate combinations of deposited material and lattice mismatch. Two SK phases are found, differing as to whether the wetting layer forms before (SK$_1$) or after (SK$_2$) the islands form. Also found in Fig. 7.4 are three distinct ripening phases. R$_1$ corresponds to classic Ostwald ripening in the presence of a wetting layer. R$_2$ is a modified ripening phase with a wetting layer and stable small islands. The islands are formed during the SK stage of growth. Subsequently, their growth is arrested but not all of them are lost to the ripening islands. The R$_3$ phase is similar to R$_2$ but lacks the wetting layer.

In the theory presented above, allowance is made for the island–island interactions. Just as for atomic and molecular adsorbates, these lateral interactions become progressively more important as the coverage increases. If the strain fields related to these interactions are anisotropic, they have the potential to lead to

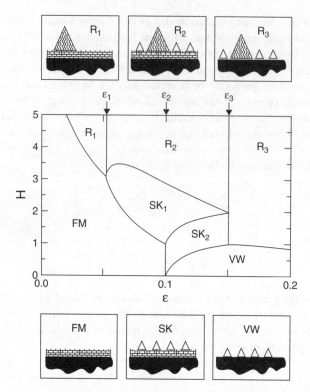

Figure 7.4 *Phase diagram calculated by Daruka and Barabási as a function of coverage H and misfit strain ε. The small panels at the top and bottom schematically represent the growth modes observed. The small empty islands indicate the presence of stable islands, while the large shaded ones refer to ripened islands. FM, Frank-van der Merwe growth; SK, Stranski-Krastanov growth (SK$_1$, wetting layer forms before islands; SK$_2$, wetting layer forms after islands); VW, Volmer-Weber growth. Ripening phases: R$_1$, Ostwald ripening in the presence of a wetting layer; R$_2$, a modified ripening phase with a wetting layer and small stable islands; R$_3$, a modified ripening phase with small stable islands but without a wetting layer. Reproduced from I. Daruka, A.-L. Barabási, Phys. Rev. Lett., 79, 3708. © 1997 with permission from the American Physical Society.*

ordered arrays of islands. Shchukin and Bimberg [3] have investigated these interactions theoretically and have shown that under appropriate condition, islands coalesce into an ordered square array.

One factor that has not been fully incorporated into this model is entropy. Strictly speaking, the phase diagram of Fig. 7.4 is valid only at $T = 0$ K. The extension of the theory outlined in Ref. [40] to include entropic effects remains an outstanding challenge. This is required to describe the temperature dependence of growth systems. Temperature plays a vital role because surface energies are temperature dependent. The most obvious consequence of this is that the most stable reconstruction of a surface is temperature dependent. Furthermore, some surfaces are known to undergo reversible roughening transitions at a specific temperature. Additional temperature effects can arise because the two materials may have different coefficients of thermal expansion, which results in a temperature dependence to ε. This last effect, assuming all other material dependent parameters are constant, is accounted for within the model.

Whether islands are truly equilibrium structures has been questioned by Scheffler and co-workers [41–43]. They showed that any equilibrium theory that includes energetic contributions only from island surface energy and elastic relaxation does not predict a finite equilibrium size distribution. In this case, strain relaxation dominates and at high coverage small islands are unstable with respect to large islands, and

Ostwald ripening occurs. They offer a theory of so-called constrained equilibrium in which island size is determined by island density and coverage. Nuclei grow to a size determined by material transport between the wetting layer and the islands, which is governed by energy balance. They calculate the elastic energy of islands and substrate within continuum elasticity theory and use density functional theory to calculate the surface energies of the wetting layer as well as the islands' facets. The result is a three phase process: (i) a nucleation phase that determines the island density; (ii) island growth at the expense of the wetting layer; and (iii) Ostwald ripening. Joyce and Vvedensky, however, point out that this theory, in spite of several appealing aspects, fails to predict the correct form of the dependence of critical thickness on the growth conditions of temperature and amount deposited [44]. Critical thickness increases with increasing surface temperature T_s. The theory of Wang et al. predicts a decrease. Critical thickness is observed experimentally to be independent of number density at constant T_s. Contrarily, the theory predicts an increase in critical thickness with increasing number density. Developments in theory to address equilibrium structure in heteroepitaxy remain a challenging frontier.

7.5 Nucleation theory

Nucleation of a new phase is naturally resisted by a pure system. A condensing vapour generally does not form droplets unless the vapour is supersaturated. A liquid does not solidify unless supercooled in the absence of a heterogeneous interface that catalyzes the process. As discussed in relation to the Kelvin equation Eq. (5.2.13), the resistance to nucleation is associated with the variation in surface energy with the size of small particles. The surface energy, or change in the Gibbs energy, is unfavourable for addition of matter to small clusters up to a critical size. After attaining the critical nucleus size, further addition of material to the cluster is energetically favourable. The formulation of this size dependence on the change in Gibbs energy is the basis of classical nucleation theory.

The change in Gibbs energy upon forming a spherical cluster or droplet of radius r from N atoms (or molecules) is

$$\Delta G = -N \, \Delta \mu + 4\pi r^2 \gamma. \tag{7.5.1}$$

The first term is the chemical potential change brought about by the phase transformation of N atoms into a sphere with a surface energy given by the second term. The number of atoms in the cluster and the radius are related by

$$N = \frac{4\pi}{3} \frac{r^3 N_A \rho}{M} = \frac{4\pi}{3} r^3 \overline{\rho}, \tag{7.5.2}$$

where N_A is the Avogadro constant, ρ the mass density, M the molar mass, and $\overline{\rho}$ the number density. Combining Eqs (7.5.1) and (7.5.2) and differentiating with respect to either N or r yields values for the size of the critical classical nucleus

$$N_c = \frac{32 \, \pi \gamma^3}{3 \overline{\rho}^2 \Delta \mu^3} \tag{7.5.3}$$

and

$$r_c = \frac{2\gamma}{\overline{\rho} \Delta \mu}. \tag{7.5.4}$$

A cluster with the critical radius r_c is the smallest structure for which the probability of growth is greater than that of decay. Substitution of the critical radius into Eq. (7.5.1) gives the energy barrier to nucleation.

$$\Delta_{\max} G = \frac{4\pi r_c^2 \gamma}{3} \tag{7.5.5}$$

Assuming that there is no additional activation barrier to formation of the critical nucleus, that is, that the maximum energy difference along the reaction path is equal to the Gibbs energy change required to form the critical nucleus, then $E_{act} = \Delta_{max}G$ and the rate of nucleation depends exponentially on the nucleation barrier. Nonspherical shapes, such as the equilibrium crystal shape, lead to slight modifications of the energy terms but the same principles are followed. The presence of a substrate, rather than nucleation of droplets out of the gas phase, leads to further complications. On a semiconductor surface, the substrate must reconstruct under a growing island, which modifies the energetic requirements for nucleation. The energy required to reconstruct the substrate must be added [45] to Eq. (7.5.1) and this may lead to a change in the size of critical nucleus.

The Burton-Cabrera-Frank (BCF) theory of surface nucleation [46] uses an analogy to droplets in contact with a vapour phase of adatoms to estimate the conditions under which islands nucleate and form stable entities. This model considers a stepped surface with terraces of width l on which atoms diffuse with a mean diffusion length λ. If the atoms, which are deposited with flux J, "re-evaporate" from the step with a lifetime τ_s, then the equilibrium concentration of terrace and step atoms can be evaluated. This model shows that for a given set of flux and diffusion parameters, a critical temperature exists below which nucleation occurs. Above this temperature, the atoms accumulate at the steps. This is the basis of step-flow growth, which is discussed in more detail in §7.6.2.

The difficulty of nucleating the new phase is expressed by the positive value of $\Delta_{max}G$. It is unfavourable for small clusters to grow until the critical size is surpassed. The accretion and release of atoms from small particles leads to some distribution of small particles. At thermal equilibrium the population of each size cluster conforms to a Boltzmann distribution. Attachment and evaporation of atoms from the clusters lead to constant interconversion of clusters of different sizes. Random fluctuations in attachment occasionally push small clusters past the critical size, at which point they rapidly grow larger. The rate of formation of clusters with the critical size determines the rate of formation of the new phase. On a surface, this rate is related to the rate at which atoms are deposited as well as the rate of surface diffusion to and from steps and between the islands. In addition to energetic factors, nucleation can be affected by structural factors. Steps and/or elastic strain interactions can lead to preferential sites for nucleation, which can have dramatic implications for structure formation [47]. Numerous metal-on-metal growth systems show evidence for nucleation at preferred sites [48]. Below we examine in more detail how preferential nucleation can be used to influence the structure of multilayer structures in the III-V family.

Transport processes as well as the critical nucleus size affect the evolution of layer morphology as material is added to the system, in particular, the evolution of the island size distribution if the system is not following step flow growth. In the nucleation regime, addition of material to the substrate leads to the formation of islands, which share the same mean size. The number of islands changes but not their size distribution. In the growth regime, the island density no longer changes but the size of the islands increases. The mobility and coverage of atoms added to the substrate must be such that they are more likely to encounter and add to an island than to find another adatom and form a nucleus. The extent of the nucleation phase is, therefore, determined by the density of islands.

Because of the importance of transport processes – the interplay between deposition rate, diffusion rate and the distance between islands–the critical nucleus size can be determined not only by thermodynamics as outlined above, but by kinetic phenomena. Venables [49–52] developed a kinetic framework to describe critical nucleus size, cluster size and the density of clusters during growth. The single atom population density, n_1 (assuming atomic deposition), determines the critical nucleus density, n_i. In turn, n_1 is determined by the rate of atom impingement (J) and the characteristic times for evaporation (τ_a), nucleation (τ_n) and diffusion capture (τ_c) by stable clusters. At high coverage direct incorporation of impinging atoms into the stable clusters can also become significant. Venables has shown [51] that when re-evaporation is negligible, taking σ_0 to be the substrate atom density and $\beta = (k_BT_s)^{-1}$, the relative coverage of stable clusters n_x can be written

$$\left(\frac{n_x}{\sigma_0}\right)^{i+2} = f(Z_0, i) \left(\frac{J}{\sigma_0^2 D}\right)^i \exp(\beta E_i)$$ (7.5.6)

where $f(Z_0, i)$ is nominally a constant that depends on the maximum cluster density Z_0 and the number of atoms in the critical nucleus i. The diffusion coefficient D is given by

$$D = D_0 \exp(-\beta E_{\mathrm{dif}})$$ (7.5.7)

where E_i is the critical cluster binding energy, and E_{dif} the diffusion activation energy. Equations (7.5.6) and (7.5.7) demonstrate explicitly how the cluster coverage depends on kinetic parameters such as the impingement rate, which is directly proportional to the growth rate, and the temperature dependence of diffusion and detachment from the critical cluster.

One consequence of the importance of transport dynamics is that the critical cluster size as well as the island density can depend on the flux of incident atoms. Furthermore, anisotropic diffusion, strain fields or anisotropic accommodation of adatoms to islands can lead to the formation of anisotropic islands, such as elongated chains on Si(100). On Si(100), the critical cluster size is simply a dimer ($i = 2$). Lagally and co-workers have shown that elongated chains result from anisotropic accommodation [38, 53–56]. Of course, any distribution of islands that is created using kinetic control is subject to Ostwald ripening once the temperature is raised sufficiently to facilitate diffusion in all directions.

7.6 Growth away from equilibrium

7.6.1 Thermodynamics versus dynamics

Measurements on the InAs/GaAs system have shown that Frank-van der Merwe growth gives way to Stranski-Krastanov growth at a critical coverage of ~1.7 ML [57]. This is a true phase transition; thus, the driving force of the growth mode transition is thermodynamic rather than kinetic. Other systems exhibit clear evidence of the influence of non-equilibrium processes upon layer morphology [58]. Two important parameters – temperature and pressure – are at the disposal of the experimentalist to drive a system away from equilibrium and into a regime in which dynamical and kinetic parameters determine the layer morphology. The temperature of the substrate controls the diffusion and desorption rates of adsorbates. Sticking coefficients as well as surface energies may also be temperature dependent. The pressure (and relative composition if more than one species is involved) of the gas phase controls the impingement rate of adsorbing atoms and/or molecules. Consequently, whether growth is thermodynamically or dynamically controlled depends upon these experimental conditions.

If the temperature is low enough for islands to nucleate and if the diffusion rate is anisotropic, highly non-equilibrium structures can be formed. In this regime, especially for large amounts of deposited material, the size and shape of islands is determined by the interplay between the kinetics of adsorption and diffusion. Often pyramidal islands are the predominant feature of a generally rough surface profile. Atomistic understanding of the morphology in this regime is difficult, but attempts have been made to understand the morphology in terms of phenomenological models and scaling relationships [59].

The spontaneous formation and ordering of islands is an example of self assembly. Kern, Comsa and co-workers [60] have found periodic stripes of adsorbed oxygen atoms on Cu(110). Triangular islands of Ag grown on Pt demonstrate that self-ordered growth can also occur in metal-on-metal systems [61]. Strain relief, which is involved in Stranski-Krastanov growth, also forms the basis of this phenomenon. The lattice mismatch between Pt and Ag is large. The second monolayer of Ag on Pt assumes a structure in which a trigonal network of dislocations partially relaxes the compressive strain. Careful preparation of this layer leads to long-range ordering of this strain relief structure. Further deposition of Ag leads to the formation of triangular Ag islands. These islands preferentially nucleate away from the dislocation and, therefore, they form an ordered array with a narrow size distribution. This example illustrates not only the importance of strain relief, but also how control of the nucleation process can be used to create self-organized structures.

We also need to distinguish between the dynamics that is operative during growth and after. During growth, we must treat an open system. Matter is being added to the surface. During kinetic control, equilibrium on the surface and between the surface and the gas phase is not established. After growth, the film morphology is subject to change due to the kinetics of diffusion as well as desorption. In addition, the chemical potential of surface species is affected by the change in pressure. The morphology of films is often studied at temperatures lower than the growth temperature. The phase of a film is temperature dependent; therefore, it is always important to establish whether the observed film structure is the equilibrium structure corresponding to the observation temperature (and cooling conditions), or whether it represents the structure characteristic of the growth conditions. The latter may be observed if growth is interrupted and the temperature is decreased so rapidly that diffusion does not have the opportunity to equilibrate the film to its new conditions.

Because of the unstable nature of growth in this regime and since the system evolves after the flux of gases is turned off and while the system cools, *in situ* experimental techniques, which can probe the system during growth, are required to reveal unambiguously the characteristics of the system. Alternatively, the system can be rapidly quenched. If the temperature drops rapidly enough, diffusion and the development of the system can be arrested before the equilibrium structure is achieved. *Ex situ* imaging and spectroscopic techniques can then be applied to probe the system.

7.6.2   Non-equilibrium growth modes

When dynamics rather than thermodynamics determines the structure of growing films, the kinetic balance of various surface processes must be analyzed. These processes are explicated in Fig. 7.5. Under kinetic control the relative rates of terrace diffusion, accommodation of atoms at steps, nucleation, diffusion across steps, and deposition determine the growth mode and film morphology. As found in §3.2, diffusion across a terrace is generally more facile than transport across a step. In addition, atoms are generally bound more tightly at the bottom of a step. As a result, the barrier for step-up diffusion is substantially greater than that for step-down diffusion, and only step-down diffusion is kinetically significant unless the temperature is very high. The added barrier for step-down diffusion compared to terrace diffusion is known as E_s, the Ehrlich-Schwoebel barrier [62, 63]. In Fig. 7.5, desorption is neglected because it is usually not important for homoepitaxy of metals and semiconductors.

Proof that kinetic control of growth can occur is easily demonstrated by studying homoepitaxial growth. Under thermodynamic control, homoepitaxial growth leads to layer-by-layer growth because there is

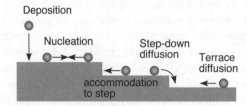

Figure 7.5   *Surface processes involved in film growth.*

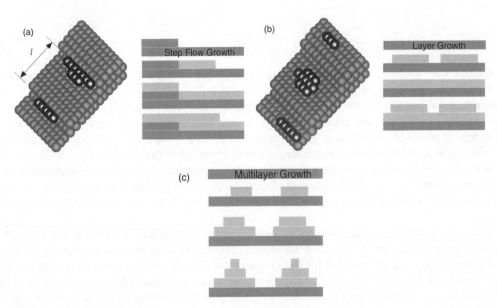

Figure 7.6 *Non-equilibrium growth modes: (a) step-flow growth; (b) layer growth by island coalescence; (c) multilayer growth.*

necessarily no strain. Nonetheless, three distinct growth modes, shown in Fig. 7.6, are observed. Close to equilibrium, step-flow growth is observed. In step-flow growth, no interlayer transport occurs and terrace diffusion is so rapid that all atoms reach a step before nucleation of islands can occur.

Far from equilibrium, the nucleation of islands is rapid. The growth mode is decided by the extent of interlayer transport. In the limit of high interlayer transport, no islands nucleate on a layer until it is complete. This is ideal layer growth (2D growth). In the limit of no interlayer transport, islands nucleate on top of incomplete layers. The higher the amount of deposition, the greater the number of islands on islands. This is multilayer growth (3D growth). For $E_{\mathrm{dif}}^{\mathrm{step}} \to \infty$, ideal multilayer growth occurs. However, in the opposite limit, $E_{\mathrm{dif}}^{\mathrm{step}} \to 0$, ideal layer growth does not occur as long as there is a finite probability for island nucleation on an incomplete layer. Hence, ideal layer growth is never observed, though some systems do approach this behaviour. Nonetheless, after the deposition of many layers, any far-from-equilibrium system grows rough and, as noted above, tends toward pyramidal structures.

The difference between step-flow growth and layer growth is best illustrated by their different behaviours when observed by reflection high energy electron diffraction (RHEED). In step-flow growth, all atoms are added at the step edge, and the surface roughness (i.e. step density) does not change during growth. In layer growth, in contrast, the density of islands rises until they begin to coalesce and eventually form one uniform layer. Therefore, the roughness goes through a maximum when the island density is highest, decreases at the onset of coalescence, and reaches a minimum when the layer is complete. As a consequence, step-flow growth leads to no variation in the intensity of RHEED reflexes, whereas distinct oscillations in reflex intensity are observed during layer growth.

The distinction between layer and multilayer growth can be made quantitative as follows [64]. Let θ_{c} be the critical coverage at which nucleation occurs on top of the growing islands and θ_{coal} the coverage at which islands coalesce to form a connected layer. Thus, the conditions defining the growth modes are

$$\theta_{\mathrm{c}} > \theta_{\mathrm{coal}} \Rightarrow \text{layer growth} \tag{7.6.1}$$

and

$$\theta_c < \theta_{coal} \Rightarrow \text{multilayer growth.} \tag{7.6.2}$$

Typically, $0.5 \le \theta_c \le 0.8$. θ_c is not a constant; however, it can only be increased by increasing the temperature or decreasing the deposition rate.

Temperature and pressure are the two defining parameters that determine the growth mode. Temperature controls the rate of diffusion, and pressure controls the rate of deposition, r_{dep}. As $T_s \to \infty$ and $r_{dep} \to 0$, equilibrium is approached and step-flow growth is observed. As $T_s \to 0$ and $r_{dep} \to \infty$, interlayer transport stops while nucleation still occurs. Multilayer growth is observed in this limit. If $E_s = 0$, i.e. terrace and step-down diffusion have the same barrier height, then for any r_{dep} transitions from step-flow to layer to multilayer growth occur as the temperature decreases. If $E_s > 0$, the transitions step-flow $\to$ layer $\to$ multilayer occur only for higher values of r_{dep}, while a direct step-flow $\to$ multilayer transition occurs for small r_{dep}.

We are now in a position to make predictions about growth modes in real systems. On non-reconstructing fcc(100) transition metal surface, E_s is on the order of 0.05 eV or even smaller. In contrast, E_s is on the order of a few tenths of an electron volt for fcc(111) surfaces. Consequently, growth on fcc(100) surfaces is generally much smoother than on fcc(111). Higher temperatures are required for the observation of smooth fcc(111), not because step flow growth occurs at a significantly higher temperature on these surfaces, rather because these surfaces do not usually exhibit the layer growth mode. For example, the onset of step-flow growth for both Ag(100) and Ag(111) is ~500 K. However, Ag(100) exhibits layer growth down to 77 K whereas Ag(111) switches immediately to multilayer growth.

As mentioned above, the transition from 2D nucleation (layer growth) to step-flow growth can be observed experimentally using RHEED intensity oscillations. This can be used to investigate the properties of the surface and the energetics of growth. The terrace diffusion length λ increases with increasing T_s, decreases with increasing J (the incident flux of growth material). Step-flow growth occurs when $\lambda \ge l$, whereas 2D nucleation occurs for $\lambda < l$. The transition from layer growth to step-flow growth allows for the determination of λ.

Such experiments allow determination of the diffusion rate parameters [65]. The mean displacement distance for random isotropic diffusion in the one dimension across the terrace is

$$\langle x^2 \rangle = 2D\tau, \tag{7.6.3}$$

where D is the diffusion constant, and τ is the average time of arrival of atoms at a specific site

$$\tau = \sigma_0/J, \tag{7.6.4}$$

where σ_0 is the areal density of surface atoms. By varying J and observing the temperature at which RHEED oscillations disappear (i.e. where step-flow growth occurs and $l = \lambda$), D and E_{dif} can be determined from

$$D = l^2\sigma_0/2J \tag{7.6.5}$$

and

$$D = D_0 \exp(-E_{dif}/k_B T_s). \tag{7.6.6}$$

7.7   Techniques for growing layers

We have shown that the structure of interfaces is controlled not only by thermodynamics, but also by the kinetics of adsorption, desorption, diffusion across the interface and diffusion into the substrate. Therefore, the techniques used to grow interfaces play an important role in determining the types of interfaces that are formed. Below we introduce several growth techniques that are commonly used.

7.7.1 Molecular beam epitaxy (MBE)

Molecular beam epitaxy grew out of fundamental studies on the sticking coefficients of metal atoms performed at Bell Laboratories. In MBE, a substrate is placed inside a vacuum chamber. Often this substrate is a single crystal sample of a semiconductor, most commonly GaAs or another III-V compound, but other materials such as a Si or CdTe can be used. As shown in Fig. 7.7, the chamber is fitted with facilities for heating the substrate and probing the surface structure and composition. The defining elements of the chamber are several Knudsen cells. A Knudsen cell is a source from which a *thermal* beam emanates after the evaporation of a solid or a liquid. It is composed of a heated crucible placed behind a small aperture from which a directional but highly divergent beam propagates. A shutter is placed in front of

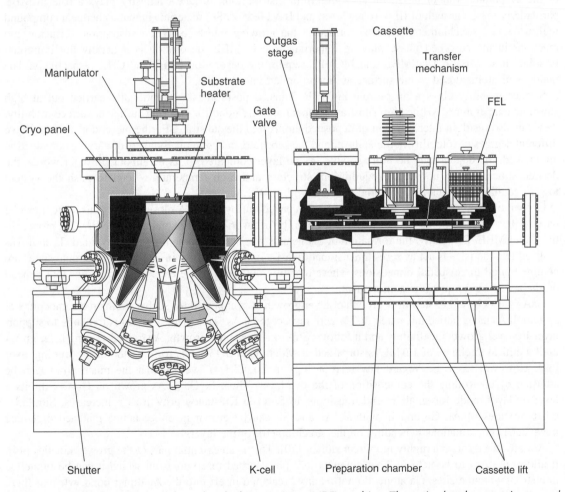

Figure 7.7 *Schematic diagram of a molecular beam epitaxy (MBE) machine. The main chamber contains several Knudsen cells (K-cells) for the evaporation of material onto the substrate. Reproduced from the VG Semicon manual with permission.*

the aperture so that the flow of gas can be rapidly turned on and off. This allows for the precise control of the dose provided by the Knudsen cell such that the amount of deposited material can be controlled at the submonolayer level. This is important not only for the control of the layer structure but also so that precise amount of dopants can be added to the layers. The control of dopant concentrations allows us to control the electrical properties of the semiconductor films.

Solid-source MBE uses evaporation of, for example, In, Ga and As from high purity ingots held within the Knudsen cells. Gas source MBE (GSMBE) uses volatile reagents such as AsH_3 and PH_3 to deliver the Group V element. Metallorganic MBE (also called metal-organic vapour phase epitaxy MOVPE) delivers the Group III element with molecules such as $Ga(CH_3)_3$ (trimethylgallium, TMGa), triethylgallium (TEGa) or trimethyleindium (TMIn). MBE allows for exquisite control of the composition and thickness of the deposited layers, which makes possible the growth of binary, ternary and quaternary compounds in superlattice structures of defined thickness.

MBE generally involves the deposition of atoms, though dimers, trimers, etc can constitute some fraction of the evaporated vapour based on an equilibrium distribution. Surface chemistry plays a role in MBE particularly for the growth of III-V (e.g. GaAs) and II-VI (e.g. ZnSe) materials in that a chemical compound with a defined stoichiometry must be formed at the growing interface. While adsorption, diffusion and reactions in the adsorbed phase play an important role in MBE, desorption is generally not important in solid phase MBE but GSMBE and MOVPE require the desorption of H_2 and CH_3, respectively. The majority of atoms dosed to the surface are incorporated into the film.

Surface mobility plays a major role in MBE. Thus, deposition is conventionally carried out at high substrate temperatures. While high substrate temperatures are good for ensuring sufficient surface mobility, they can also result in interdiffusion of deposited atoms into the substrate. Furthermore, dopants may have different degrees of solubility in the epilayer and the substrate, or if several epilayers of varying composition are created, the dopant may be more soluble in one layer compared to its neighbour. In other words, the desired structure may not be the equilibrium structure and high temperatures tend to push the system toward equilibrium. This is a potential problem with MBE.

Growth of III-V compounds is most commonly carried out by MBE. A closer look at this system can be used to illustrate several aspects of MBE [44, 66, 67]. Arthur [68] and Foxon and Joyce [69, 70] pioneered the use of MBE. Originally, the formation of 3D islands was seen as something to be avoided. Then it was realized that the islands act as zero-dimensional (quasi-atomic) structures in which the electronic states are confined in all three spatial dimensions. These quantum dots (QD) have electronic properties that depend on their size due to quantum confinement.

GaAs grows by either step-flow growth or layer growth as is expected for such a strain-free homoepitaxial system. InAs has a lattice mismatch of 7% with respect to GaAs. This level of strain is too large to support layer-by-layer growth indefinitely and it follows Stranski-Krastanov growth. A strained wetting layer grows until a critical thickness of 1.6 ML is surpassed at which point 3D islands form on top of the wetting layer [67]. However, from the Vegard law relationships in Eq. (7.1.4), we see that the misfit strain can be adjusted by controlling the composition of the overlayer. Thus, $In_x Ga_{1-x} As$ grown on GaAs exhibits a layer-by-layer mode for small x, and transitions to Stranski-Krastanov growth as x increases. Similarly, other mixtures of Al, Ga and In with P, As and Sb can be grown in either lattice matched or lattice mismatched combinations depending on the stoichiometry of the layers.

As_2 adsorbs through a highly mobile precursor [70]. This is an essential part of the growth kinetics [44]. It allows the As_2 to be held in what amounts to a physisorbed reservoir from which it can be funnelled into strongly bound sites. Ga atoms do not readily break and insert into the As dimer bond, whereas there is no barrier to adding another As dimer (the addimer) on top of a dimer pair. A pair of Ga atoms then binds to the addimer. The addimer decomposes, releasing As_2 back into a mobile physisorbed state while the Ga atoms incorporate into the surface lattice, each bridging one of the original dimers. Another Ga

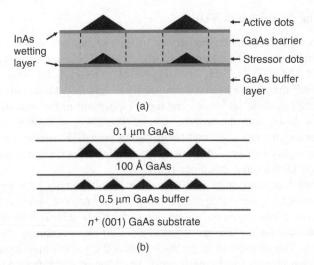

Figure 7.8 *Mechanism of quantum dot stacking. (a) Schematic of a bi-layer quantum dot heterostructure that shows perfect vertical coupling of the active (second) quantum dot layer due to the strain field generated by the stressor quantum dots. The dashed lines indicate the lateral extent of the tensile strain field in the GaAs barrier. (b) Schematic layout of a quantum dot heterostructure grown by molecular beam epitaxy. Reproduced from Zetian Mi and Pallab Bhattacharya, J. Appl. Phys., 98, 023510. © 2005, with permission from the American Institute of Physics.*

atom (or two) then binds into the long bridge site next to and in the same row as the dimer that have already accommodated the first two Ga atoms. The stable nucleus is completed when As_2 forms a dimer atop the four Ga atoms.

In a device structure, the quantum dots are capped with an overgrown layer. For instance, InAs QDs grown on GaAs might be capped with GaAs. The capping layer must also relax. This sets up a strain field in the layer unless its stoichiometry is chosen to lattice match the QDs and their wetting layer. The strain field in the capping layer influences the growth characteristics of subsequent layers as shown in Fig. 7.8.

Figure 7.8 demonstrates that the location of the subsequently grown QDs is affected by the strain field [71]. Vertical stacking, confirmed by TEM, is observed in many cases, not only for III-V materials [72, 73] but also for SiGe superlattices [71]. Hence, this can be considered a method of strain patterning to align as well as to make the QDs more uniform [71]. Not only are the QDs aligned and larger with a narrower size distribution, they also nucleate at a reduced critical coverage compared to the unstrained substrate [74]. Due to the strain fields produced by the dots in the first layer (stressor dots), the lattice constant of the GaAs layer is expanded directly above the buried dots, that is, it exhibits tensile strain above the dots. However, the island and the substrate around the island are under compressive strain.

Scheffler and co-workers [75, 76] have investigated the effects of strain on the diffusion and binding of In on GaAs(001). Changes from the unstrained lattice alter both the prefactor and the activation energy for diffusion, the latter being perhaps the more important factor. The binding energy of In increases with tensile strain. Therefore the equilibrium concentration of In at steady state is higher above the underlying stressor dots, and this enhances nucleation at these sites. E_{dif} displays a maximum at compressive strain and is a decreasing function for tensile strain. This has implications for the growth kinetics of InAs islands and can lead to a self-limiting nature of island growth because at a certain size a repulsive interaction between the quantum dot and a diffusing In atom occurs. In accord with these theoretical results, Jones and co-workers [77, 78] reported a narrower size distribution for second layer QDs grown on a 10 nm capping layer.

7.7.2   Chemical vapour deposition (CVD)

In chemical vapour deposition, molecular precursors to film growth are dosed onto a surface from the gas phase or using a supersonic or Knudsen molecular beam. These molecular precursors include, for example, $AsH_3 + Ga(CH_3)_3$ for GaAs, SiH_4 or $Si_2H_6 + GeH_4$ or Ge_2H_6 for SiGe alloys. The wide range of metallorganic compounds that can be used to fabricate III–V and II–VI materials can be found elsewhere [79]. From the compositions of these molecules and the compositions of the desired epilayers, it is obvious that not all of the atoms dosed onto the surface remain. Therefore, in contrast to MBE, CVD requires facile desorption to be among the reactions that occur during growth. Fundamental surface science studies have shed light on the interplay between adsorption, the stability of surface intermediates, and desorption in determining the growth kinetics and outcomes [80].

For instance in the growth of homoepitaxial Si layers or heteroepitaxial SiGe layers on Si from silanes and germanes, the desorption of molecular hydrogen plays a vital role. The H-terminated surface is virtually inert and neither silanes nor germanes dissociate on this surface [80–83]. Therefore, at low temperature as H(a) builds up on the surface, the surface growth is choked off and eventually stops. This is an example of a self-limiting reaction. The desorption of hydrogen limits the growth kinetics and sets a lower bound to the temperature at which growth can be performed. Hydrogen desorbs from Si at about 800 K and at about 650 K from Ge. At higher temperatures, the dissociative adsorption of silanes and germanes limits the growth. The presence of H(a) also affects the diffusion of Si and Ge atoms, which leads to distinct differences in layer morphology in CVD-grown as compared to MBE-grown layers.

Two growth strategies that exploit self-limiting chemistry to control layer thickness and stoichiometry are atomic layer epitaxy (ALE) and atomic layer deposition (ALD), as it is more generally called when referring to growth of chemically dissimilar layers that are not in registry with one another. These have been extended from relatively simple systems such as Si and SiGe to systems as complex as YBCO [84], and are of great interest for growing dielectric layers such as hafnium and aluminium oxide [85]. Alumina films are grown by alternately exposing a surface to trimethylaluminum and water. The $Al(CH_3)_3$ decomposes on hydroxide groups that have been previously formed on the surface to form a single $O–Al(CH_3)_2$ layer that is unreactive toward further $Al(CH_3)_3$ decomposition. This layer is then exposed to water, which displaces the methyl groups, and reforms surface hydroxides. The process can then be repeated as often as required. The thickness of the layer can be controlled with single layer precision. The substrate need not be planar; indeed, the surfaces of trenches, nanowires [86] or even spider silk [87] can be coated with oxide in this fashion. The range of materials that can be deposited with ALD can be further extended by using not only thermal processes, but also plasma or radical assisted chemistry, or to polymers by using molecular precursors [86].

High surface temperatures again bring about potential problems with the interdiffusion of dopants and other atoms. However, thermal desorption need not be the only means of removing adsorbates from the surface. Photon or electron irradiation can be combined with CVD. This opens the possibility of directly writing structures during CVD. If one of the reactants has a low sticking coefficient, energy can be put into the system to increase the sticking coefficient. This can be provided by the high translational energy of a supersonic beam. Sticking and desorption can also be enhanced by igniting a plasma in front of the surface.

The poster child of CVD is the growth of Si layers from silane, SiH_4, which is used in IC fabrication to make polycrystalline silicon (poly-Si) interconnects. There is much uncertainty in the dissociative adsorption dynamics of SiH_4. The process is direct and activated with an activation energy of $\sim$0.2 eV but there are differences in the dissociation kinetics between Si(100) and Si(111). It is unclear whether one or two silicon atoms are directly involved and on the dimers of Si(100)–(2 × 1) it is unclear whether a single dimer or two adjacent dimers are involved. On both Si(100)–(2 × 1) and Si(111)–(7 × 7), the sticking coefficient of SiH_4 is a complex function of H coverage [88] and does not follow simple Langmuirian kinetics. Breaking a Si–H bond is involved in the first step which leads to adsorbed H atoms and SiH_3.

The moderate activation barrier is consistent with the observed weak surface temperature dependence of the sticking coefficient s_0 when adsorption proceeds on the clean surface. s_0 is a strong function of T_s in the region dominated by H_2 desorption but at high T_s, i.e. on clean surfaces, s_0 only increases weakly with T_s. The sticking coefficient of SiH_4 depends nearly exponentially on the incident translational energy E_i [89, 90]. The dissociative sticking coefficient of SiH_4 on Si(100)–(2 × 1) and Si(111)–(7 × 7) at $T_s = 673$ K is 3×10^{-5} for both surfaces when dosed from a room temperature gas [88].

In order to increase the dissociation probability of SiH_4 to useful levels, the surface and/or gas temperature must be raised to considerable levels. Because high temperatures above $T_s = 770$ K are generally used to dissociate SiH_4, H_2 desorption is rapid, db sites are available, and unless the SiH_4 flux is extraordinarily high, the deposition rate is proportional to the SiH_4 sticking coefficient on the clean surface. Below this temperature, H_2 desorption, i.e. the creation of empty sites, represents the rate limiting event. High temperatures during growth are not only important for the removal of H(a) to free up adsorption sites, but also to maintain high crystallinity in the homoepitaxial layer.

Dissociation according to Rxn. (7.7.1), which requires two empty dangling bond (db) sites,

$$SiH_4(g) + 2\,db \rightarrow SiH_3(a) + H(a), \tag{7.7.1}$$

is followed by successive hydride decomposition, Si incorporation and H_2 desorption as long as T_s is sufficiently high, as described in the following four reaction steps

$$SiH_3(a) + db \longrightarrow SiH_2(a) + H(a) \tag{7.7.2}$$

$$2\,SiH_2(a) \longrightarrow 2\,SiH(a) + H_2(g) \tag{7.7.3}$$

$$SiH_2(a) + db \longrightarrow 2\,SiH(a) \tag{7.7.4}$$

$$2SiH(a) \longrightarrow 2\,Si(a) + H_2(g). \tag{7.7.5}$$

7.7.3 Ablation techniques

Some materials are not easily evaporated or form compounds that are not well suited to chemical vapour deposition. Fortunately, other methods exist to volatilize refractory materials and allow them to be deposited on a substrate. Collectively, these are known as ablation techniques. Ablation (see §8.3.3) occurs when high-energy particles (or high intensity photons) encounter a surface and dislodge surface and possibly bulk atoms. When high-energy ions are used, the technique is called sputtering. Sputtering can be achieved with lower-energy ions when a plasma, as in magnetron sputtering, assists the process. A high-power laser can also ablate atoms from a target. For the ejection of atoms from a target under laser irradiation we usually differentiate between true laser ablation, the explosive release of atoms from a target as the result of a non-equilibrium process, and laser vaporization, the evaporative release of atoms due to a thermal process. Details of the processes occurring during laser ablation, its use in pulsed laser deposition [91], and structure formation as a result of laser ablation [92] can be found elsewhere. Alternatively, high-energy electron irradiation can be used to induce vaporization of a target.

7.8 Catalytic growth of nanotubes and nanowires

Growth does not occur only on flat surfaces or lead to layer-like structures. This is particularly true in the case of catalytic growth, for instance, for the formation of semiconductor nanowires (solid core structures with diameters below ∼100 nm) [93–95], carbon nanofibres [96] and carbon nanotubes (single

Figure 7.9 *Scanning electron micrographs of (a) a surface covered by flower-like silica nanostructures generated by heating SiC + Co under an Ar-CO atmosphere, (b) a typical 3D feature radiated from a central spherical Co-rich particle, and (c) a uniform nanoflower film formed from SiC + Co heated in pure CO. Reproduced from M. Terrones, N. Grobert, W. K. Hsu, Y. Q. Zhu, W. B. Hu, H. Terrones, J. P. Hare, H. W. Kroto, D. R. M. Walton, MRS Bulletin, 24(8), 43. © 1999, with permission from the Material Research Society.*

or multi-walled hollow core structures with diameters below ~100 nm) [97, 98] and whiskers (larger solid core structures). Fullerenes such as C_{60} are formed in any number of reactive systems in which carbon is volatilized and allowed to condense. Single-walled and multi-walled carbon nanotubes (SWCNT and MWCNT, respectively) are formed when certain metals are added to graphite that is subsequently vaporized [99, 100]. Smalley and co-workers [101] have shown that extremely high yields of SWCNT are obtained from a mixture of ~1 at% Co/Ni or Co/Pt and graphite. This mixture is formed into a rod that is used as a laser ablation target. The metals form clusters in the gas phase. The clusters act as catalytic agents upon which C atoms adsorb and subsequently react to form nanotubes. The nanotubes grow outward from the catalytic metal cluster. The size of the cluster during the initial stages of nanotube formation is responsible for determining whether SWCNT or MWCNT are formed.

Terrones, Kroto, Walton and co-workers [102] have used a catalyzed solid state reaction to form SiO_x nanoflowers (see Fig. 7.9), SiC and Si nanotubes, and SiC nanotubes wrapped in a-SiO_2. A powder of SiC is heated to ~1800 K in CO. The product distribution is dependent upon the catalyst that is mixed into the powder. One dimensional SiC and Si nanowires are obtained from an Fe catalyzed reaction. Co catalyzes the production of three-dimensional SiO_x nanoflowers. When Fe and Co are both combined with SiC, the result is the formation of elongated single crystal of β-SiC wrapped in a sheath of amorphous SiO_x. These nanowires have a SiC core diameter of ~6–40 nm and an outer diameter of ~10–60 nm. Exceptionally long nanowires of up to 100 µm are sometimes found. The strongly reducing reaction conditions maintain the working catalyst in the metallic state. The metallic particles appear to be acting as catalytic sites for the dissociation of CO. They also act as tethers for the growing nanowires. The reactants (Si, C and O) diffuse across the surface of the metallic clusters and, again, the nanotubes grow out from the metal. Volatile SiO may play a role in the reaction as well.

A wide range of materials has been found to form nanowires as a result of catalytic growth [103]. Figure 7.10 displays GaP nanowires grown in Lund by the catalytic action of a gold nanoparticle that can clearly be seen to ride atop the nanowires. Growth of this type was first observed by Wagner and Ellis, who deposited Au particles on Si, then exposed them to a mixture of $SiCl_4$ and H_2 at 1220 K. Under these simple conditions, Wagner and Ellis [104] demonstrated the growth of crystalline Si columns grown along the $\langle 111 \rangle$ direction with diameters from 100 nm to 0.2 mm. This work was followed by that of Givargizov [105] but it was not until Lieber and co-workers [95, 106] rediscovered it that the power of catalytic growth for nanowire formation was appreciated. The question now is whether we can explain mechanistically how some or all of this growth occurs and, fortunately, significant progress has been made in this respect.

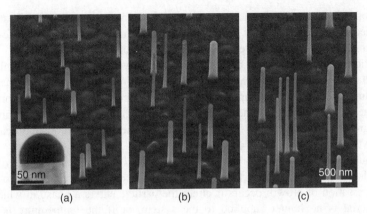

Figure 7.10 *Scanning electron micrograph of GaP nanowires grown via the VLS mechanism. The inset shows a hemispherical Au particle atop the GaP nanowire. Reproduced from J. Johansson, C. P. T. Svensson, T. Mårtensson, L. Samuelson and W. Seifert, J. Phys. Chem. B, 109, 13567. © 2005, with permission from the American Chemical Society.*

Catalytic growth of nanotubes and nanowires has been observed under a wide variety of conditions and in a number of phases such that catalytic growth is known by a number of names including vapour-liquid-solid (VLS), vapour-solid-solid (VSS), supercritical fluid-liquid-solid (SFLS), solution-liquid-solid (SLS) and solid-liquid-solid growth (unfortunately also SLS). Nanowires produced by catalytic growth are often found to have a uniform diameter. The wires are not always round but might also exhibit other crystallographically defined shapes, such as hexagonal for ZnO or rectangular for In_2O_3. Under some conditions, particularly for long growth times, tapering of diameter to smaller (or less commonly larger) values is found. Often growth requires a bit of an induction period before uniform nanowires begin to grow. These considerations are represented schematically in Fig. 7.11.

The initial period before uniform growth commences is associated with any of a number of processes. In some cases, the catalytic particles of radius r_p must be formed ($dr_p/dt > 0$) by vapour phase and/or surface diffusion or else their surfaces have to be cleansed of impurities (oxides or terminating thiols). The particles may be deposited directly, for instance from the evaporation of a colloidal solution with a well-defined size. Alternatively, a thin film of metal can be evaporated directly onto a substrate or

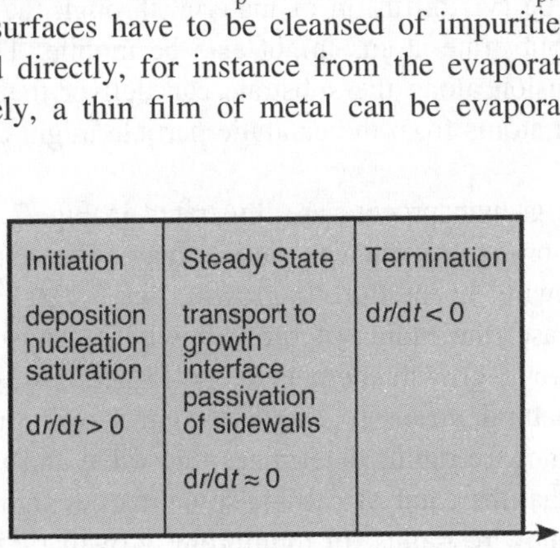

Initiation	Steady State	Termination
deposition nucleation saturation	transport to growth interface passivation of sidewalls	$dr/dt < 0$
$dr/dt > 0$	$dr/dt \approx 0$	

time

Figure 7.11 *General considerations on the different regimes that occur during catalytic growth of nanowires and nanotubes. Reproduced from K. W. Kolasinski, Curr. Opin. Solid State Mater. Sci. 10, 182. © 2006, with permission from Elsevier.*

as the byproduct of carbothermal reduction and if the metal does not wet the substrate, it balls up into islands either immediately as the result of Volmer-Weber growth, or else subsequently when the system is annealed, the onset of Ostwald ripening leads to a distribution of island sizes.

Once the catalytic particles are formed or deposited, they may still need to be primed for the growth of nanowires. For instance, the pure metal catalytic particle might not be that active for nanowire formation. Instead, an admixture of the growth compound and the metal might be required to form an (unstable or stable) alloy, a true eutectic or some other solid/liquid solution. In this case, saturation of the catalytic particle with the growth material or the formation of the proper composition may lead to an induction period before growth.

The growth of nanowires with a uniform radius is associated with a steady state growth ($dr_p/dt \approx 0$) in which material is transported to the particle/nanowire interface. Finally, a tapering to smaller diameters ($dr_p/dt < 0$) and cessation of growth occurs if either the particle enters a phase in which it is consumed, if the growth material is no longer supplied to the system, or if the temperature is reduced below a critical value.

Figures 7.9 and 7.10 illustrate that wildly different structures can be formed as a result of catalytic growth, and that we need to distinguish between the differences that are depicted schematically in Fig. 7.12. Is the nanowire produced from root growth in which the catalyst particle is found at the base of the nanowire, Fig. 7.12(a), or by float growth, in which the particle is located at the tip of the nanowire? Does multiple prong growth ensue, Fig. 7.12(c), in which more than one nanowire emanates from each particle, or does single-prong growth occur, Fig. 7.12(d)?

Figures 7.12(a) and 7.12(b) also illustrate several dynamical processes that can affect catalytic growth. Adsorption can occur from the fluid (whether gaseous, liquid or supercritical) phase. Adsorption might be molecular or dissociative and may either occur (vii) on the nanowire (viii) on the particle, (ix) on the substrate. A natural way for the catalytic particle to direct material to the growth interface is if the sticking coefficient is higher on the particle and vanishingly small elsewhere. Diffusion of adatoms occurs (i) across the substrate (if the sticking probability is not negligible), (ii) across the particle and (iii) along the sidewalls. Diffusion across the substrate and along the sidewalls must be rapid and cannot lead to nucleation events. Nucleation of the nanowire anywhere other than on the particle must be suppressed so that growth only occurs at the particle/nanowire interface, and so that sidewalls do not grow independently of the axial growth. There may be (vi) diffusion of material through the catalytic particle in addition to (ii) diffusion along its surface. Substrate atoms might also be mobile. They might (v) enter the particle directly or else (iv) surface diffusion along the substrate can deliver them to the surface of the particle. Not shown in the diagram is that atoms from the catalytic particle might also be mobile and diffuse along the sidewalls and the substrate.

Four major distinctions in the growth process are illustrated in Fig. 7.12: root growth vs float growth and multiprong growth vs single-prong growth. The particle may either end up at the bottom (root growth) or top (float growth) of the nanowire. In multiprong growth, Fig. 7.12(c), more than one nanowire grows from a single particle. In this case, the radius of the nanowire r_w must be less than the radius of the catalytic particle r_p. In single-prong growth there is a one-to-one correspondence between the number of particles and nanowires. A natural means to exercise control over the nanowire diameter in single-prong growth would be if the nanowire radius determines this value and $r_w \approx r_p$. In single-prong growth, $r_w \approx r_p$ is usually observed but that the catalyst particle sometimes is significantly larger and occasionally is somewhat smaller than the nanowire radius. In multiprong growth r_w is not determined directly by r_p but must be related to other structural factors, such as the curvature of the growth interface and lattice matching between the catalytic particle and the nanowire.

There are circumstances under which the catalyst can be liquid, and others in which it can be solid and the nanowires that are formed do not appear to be substantially affected. As far as the dynamics of

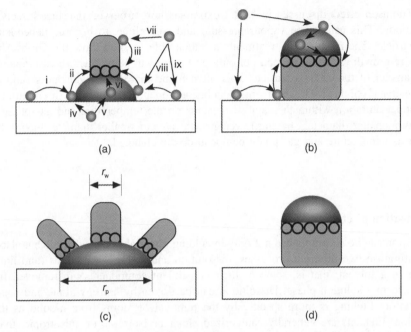

Figure 7.12   *The processes that occur during catalytic growth. (a) In root growth, the particle stays at the bottom of the nanowire. (b) In float growth, the particle remains at the top of the nanowire. (c) In multiple prong growth, more than one nanowire grows from one particle and the nanowires must necessarily have a smaller radius than the particle. (d) In single-prong growth, one nanowire corresponds to one particle. One of the surest signs of this mode is that the particle and nanowire have very similar radii. Reproduced from K. W. Kolasinski,* Curr. Opin. Solid State Mater. Sci. *10, 182. © 2006, with permission from Elsevier.*

nanowire and nanotube formation are concerned, it does not appear to matter whether the catalytic particle is liquid or solid. Likewise, the phase from which the growth material is taken is of little consequence. The growth material may come from a gas that is unreactive on the substrate and only reactive on the catalyst surface, such as in the case of silicon nanowires (SiNW) growth from silane. It may come from an atomic vapour that has unit sticking probability on both the substrate and the catalyst, as in MBE growth of SiNW or III-V compounds. It may come from a plasma, a solution or even supercritical fluids. What is required is that the growth material is mobile and can readily reach the growth interface with a low probability of nucleating a crystallite (alternate growth front) anywhere other than at the nanowire/catalyst interface.

The essential role of the catalyst is lowering the activation energy of nucleation at the axial growth interface. Equation (7.5.5) shows that, according to classical nucleation theory, a substantial barrier is associated with the formation of the critical nucleation cluster at a random position on the substrate or nanowire. If the catalyst can lower the nucleation barrier at the particle/nanowire interface, then growth only occurs there. Any of a number of processes may be the rate determining step depending on the exact conditions, but the most important role of the catalyst particle is to ensure that material is preferentially incorporated at the growth interface.

Virtually all of the theoretical work on catalytic growth has treated the VLS mechanism explicitly. Therefore, there is little information on why root growth and multiprong growth occur under some circumstances. Theoretical approaches to nanowire growth can be separated into at least three different categories: molecular dynamics [107–109], thermodynamics [105, 110–118] and kinetics [119–125].

The Gibbs-Thomson effect discussed in §5.2.2 expresses how a curved interface affects the chemical potential of a body. This causes the vapour pressure and solubilities to become dependent on the size of a catalyst particle. Thermodynamic treatments are then able to show how the Gibbs-Thomson effect leads to nanowire growth rates that depend not only on the growth parameters (pressure and temperature) but also the diameter of the catalyst particle. It is often found that the growth rate should decrease with decreasing diameter [105, 114]. However, this conclusion depends on the growth conditions [125] since the extent of supersaturation within the catalyst depends on the temperature and gas-phase composition. A transition from smaller diameters having lower growth rates to smaller diameter having higher growth rates can occur as temperature and gas-phase composition are changed.

7.9   Etching

7.9.1   Classification of etching

Structure formation can be accomplished not only by adding material to the substrate but also by removing it. Here we distinguish two different processes: dissolution and etching. All solids (and liquids) dissolve to some extent in a solvent; that is, some of their constituent components are removed from the solid and solvated to form a solution phase. Dissolution can be driven by entropy alone and need not involve a chemical reaction. Etching is more specifically the removal of atoms from a solid as the result of a chemical reaction. Etchants are generally categorized either as isotropic or anisotropic, that is, they are categorized based on whether the etch rate does not or does depend on the crystallography of the etching surface. Etching is also classified as dry, in which surface atoms are volatilized by chemical reaction, i.e. the products are gases, or wet, in which reaction products are released into a solution phase above the surface. Two other processes that remove material but are not to be confused with etching are evaporation (desorption from a liquid) and sublimation (desorption from a solid). Much of the literature on etching is technical rather than fundamental in nature, and the delineations between dissolution, etching, evaporation and sublimation are not always clearly stated.

The concept of anisotropic etching is illustrated in Fig. 7.13. In Fig. 7.13(a) we see a Si pillar that has been made by laser ablation with a nanosecond pulsed excimer laser of Si in the presence of $SF_6(g)$. The tip of the pillar is rounded with a radius of $\sim 2\,\mu m$ and the length is roughly $50\,\mu m$, resulting in an aspect ratio of $\sim 50\,\mu m/2\,\mu m \approx 25$. If we were to etch this pillar isotropically, the top would recede as quickly as the sides, the pillar would get smaller but its shape would not change. In Fig. 7.13(b) we see the result of etching the pillar in 40 wt% KOH at 80° C for ~ 2 min. The pillar has taken on a pyramidal shape with a sharp tip. Indeed, the tip is approximately $10\,nm$ across. With a length of over $50\,\mu m$, the aspect ratio is now > 5000. Figures 7.13(c) and (d) display pillar-covered surfaces in plan view. When etching in KOH to remove the pillars, the final structure observed depends on whether the substrate is a Si(100) crystal, Fig. 7.13(c), or a Si(111) crystal, Fig. 7.13(d). An etched Si(100) substrate presents rectangular pores whereas an etched Si(111) substrate presents triangular pores. Note that a pore is a structure that is deeper than it is wide, whereas a pit is not. It is obvious that the final structure resulting from etching depends on the initial structure, crystallography of the substrate, and the etching conditions.

Adsorption and desorption are related by microscopic reversibility so we might ask whether growth and etching are similarly related. Growth is a much more general phenomenon than etching; consequently, etching is not the reverse of growth. Growth is the result of adsorption, and can be accomplished by the reverse of etching, dissolution, evaporation or sublimation. Furthermore, while adsorption and desorption are related by reversing a single step in a chemical process, growth necessarily involves several steps including adsorption, diffusion, and incorporation of the adsorbate into a growing island or step. To use microscopic reversibility it is always important to compare the forward and reverse processes such that

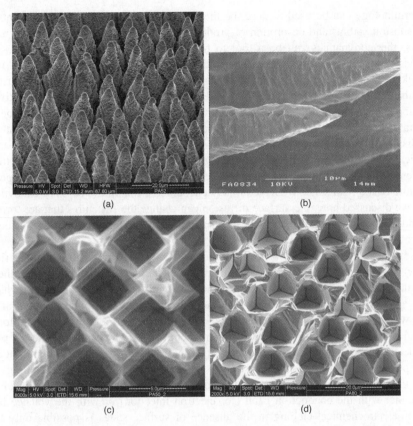

Figure 7.13 *(a) A Si pillar created by laser ablation. (b) A lightly etched Si pillar. (c) A Si(100) substrate, initially pillar-covered and then heavily etched in KOH. (d) A Si(111) substrate, initially pillar-covered and then heavily etched in KOH. (b) D. Mills, K. W. Kolasinski, J. Phys. D: Appl. Phys., 38, 632. © 2005, with permission from the Institute of Physics. (a), (c) and (d) Reproduced from M. E. Dudley, K. W. Kolasinski, J. Electrochem. Soc., 155, H164. © 2008, with permission from the Electrochemical Society.*

they are run under the same conditions. When several steps are involved in structure formation, a following step and its possible outcomes are contingent on the outcome of previous steps. Etching of a pit wall along a network of dislocations into a crystal cannot occur unless the etch pit is first formed, and only if the dislocation network was there in the first place. The only way etching and growth can possibly be treated as if they are the reverse of one another is if both are considered while proceeding at equilibrium.

While growth is often observed and carried out at equilibrium (or at least under thermodynamic control at pressures for which the chemical potential is not strongly perturbed from equilibrium and/or at temperatures high enough that activation barriers are not blocking thermodynamically favoured outcomes), the reverse of equilibrium homoepitaxial growth is simple evaporation (or dissolution in solution). Most commonly, etching is carried out under kinetic control, sometimes in steady state, but rarely at equilibrium. This distinguishes etching from dissolution of a mineral phase in contact with a saturated (or nearly saturated) solution phase, for which the reaction dynamics can be addressed in terms of growth models [126, 127]. Etching modes can resemble non-equilibrium growth modes in reverse. Step flow modes are observed for both. Etch pits play an analogous role in etching to that played by islands in growth and, therefore, some of the same terminology can be used to describe the two processes. Adsorption of etchants, diffusion of reactants and substrate atoms and desorption of products all have roles to play in etching reactions.

Nevertheless, thermodynamics can be a tool to relate growth and removal processes. The power of thermodynamics is that it can be used to relate initial and final states in a manner that is independent of the path taken. Thus, if etching, dissolution, or, more generally, desorption are carried out either close to equilibrium or else at temperatures that are sufficiently high so that kinetic activation barriers are not directing the course of the process, then only the final state energetics determine the structure of the final state. In such cases, purely thermodynamic treatments can be used to relate growth with etching and dissolution. Therefore, we expect that it should be easier to relate growth and dissolution thermodynamically, but that connections between growth and etching are more likely to be found in kinetic arguments. For both growth and etching, the structure of the surface is very important and can change the nature of the final state. We have already encountered this in the cases of catalytic growth, and in §7.3 where we discussed how intersurface diffusion can change the structures that are formed. Similarly for etching, and as shown in Fig. 7.13, the initial structure of the surface is important for determining the final structures that are formed by etching.

There are several types of etching processes. Etching can be either electrochemical (involving free charge transfer) or chemical, dry or wet. Dry etching is chemical but wet etching can be either chemical or electrochemical. Electrochemical etching can be further distinguished as either anodic, in which the sample and a counter electrode are connected to a power supply and a bias is supplied to control current flow; electroless, in which no electrodes or power supply is used but instead a redox couple is formed between the sample and a species in solution; or photoelectrochemical, in which photons act to provide charge carriers. Photoelectrochemical etching can be performed either with or without the presence of a counter electrode and power supply. In the latter case, it is sometimes called contactless etching. As discussed in Chapter 8, charge injection only occurs if the solution species has an electronic level that overlaps energetically either with the conduction band (for electron injection) or with the valence band (for hole injection). Photoelectrochemical etching in the absence of surface states is possible only for irradiation with above band gap radiation to create an electron-hole pair. In almost all cases, semiconductors exhibit band bending in the space charge region and this causes charge separation under illumination. For instance, in n-type Si, the bands are bent upward such that holes are forced to the surface, whereas for p-type Si, the bands bend downward and electrons are driven to the surface. Thus, control of the doping type allows for unequivocal identification of the charge species that is responsible for initiating etching or other reactions at the surface of a semiconductor.

There have been numerous attempts to propose general theories of dissolution and etching [128]. There are kinematic theories along the lines of the Burton-Cabrera-Frank (BCF) theory of surface nucleation [46], which do not specifically account for crystal structure. Molecular-kinetic theories such as those developed by Heimann and co-workers [129] explicitly include the lattice of the solid. Topochemical theories such as those developed by Kleber [130] and Knacke and Stranski [131], attempt to describe etching by using, for example, a Langmuir isotherm to model the adsorption of reactive etchants with an explicit inclusion of sites (terrace, step, kink) with different binding energies or activation energies for etching. All of these models have advantages and limitations. They struggle with a theme that also unites growth with etching: to what extent does thermodynamics versus kinetics control the outcome? If the energetics of the final structure that is formed – an equilibrium crystal structure (ECS) or an ECS that is modified by adsorption of surfactants – is the most important element in determining the structural outcome because, for instance, the system is allowed to evolve to equilibrium (or at least a steady state very close to equilibrium), then thermodynamics approaches in which growth and etching act very much like mirrors of each other can be used. In this case, much of what we have learned above concerning growth can be used to describe etching with the signs reversed to indicate the removal rather than the addition of material. However, if the reactions are kinetically constrained and occur far from equilibrium or are determined by reactions that only occur at defects (such as dislocation chains), then the results are completely different than what might be inferred from growth models.

7.9.2   Etch morphologies

Etching results in one of three general classes of substrate morphology: flat surfaces, rough surfaces or porous films. Under the rubric porous films we can also add nanowire forests and bundles of nanotubes. Starting with a rough surface, if material is preferentially removed from asperities compared to flat regions, the surface becomes progressive flatter as it is etched. This is the basic mechanism of electropolishing in which, for example, electrical field enhancement at asperities increases the rate of an electrochemical reaction and ultimately leads to smoother surfaces. To determine the final surface structure at the atomic level, we similarly need to consider the relative rates of etching at different sites on the surface and how these affect the morphology.

A real single-crystal surface contains terraces separated by one-atom-high steps that are randomly spaced across the surface and which have kink sites randomly distributed along the step. The steps are randomly shaped and there may also be pits scattered randomly on the terraces. Now consider the relative rate of removal of a step atom compared to a terrace atom. The removal of a terrace atom corresponds to the nucleation of an etch pit. The formation of a pit also transforms the atoms that ring the pit from terrace atoms to step atoms. If the removal rate is different for step and terrace atoms, then etch pit formation also changes the etch rate for these atoms.

If the rate of terrace atom etching (etch pit nucleation) is small compared to the step atom etch rate, the surface etches by removal of steps, and the morphology of the terraces is unaffected. Atomically flat terraces separated by steps result from a mechanism that is analogous to step-flow growth. The terrace atom etch rate does not need to be zero, simply low enough that the rate of pit nucleation is low compared to the rate at which an entire terrace is removed by step flow. Any pits that initially populated the surface are removed as they expand until they encounter and merge with receding steps. Importantly, since terrace atom removal is slow, it is extremely unlikely that a pit is nucleated within a pit; hence, no new steps are created and surface roughening is avoided.

Now consider the step morphology. If kink atoms etch faster than normal step atoms, kink sites are preferentially removed and the step straightens. If the kink atoms are removed more slowly than step atoms, the steps are roughened. If by some mechanism the steps repel one another – for instance, by a dipole interaction resulting from Smoluchowski smoothing – the etch rate slows as the steps approach each other, and the steps tend toward regular spacing. On the other hand, if they tend to attract each other – for instance, if a strain field near the step accelerates the etching – then the steps tend to bunch. Step bunching can lead to an increase in the step height, a reduction in the number of steps and an increase in the mean terrace width. If the steps do not merge, step bunching leads to packs of closely spaced steps (short terraces) separated by broad atomically flat regions. It is also important to note that the etchant, particularly for etching in solution, provides a means of communication and coupling between different regions of the crystal. If inhomogeneities in concentration or temperature arise in the etchant, they can couple to the surface reactions and lead to step bunching, as shown by Hines and co-workers [132] for KOH etching of Si(100). Couplings in reaction–diffusion systems can lead to autocatalytic phenomena that affect the rate and nature of pitting corrosion [133].

At the opposite extreme, terrace atoms etch more rapidly than step atoms and pits form readily inside pits. The pits increase in density and depth until the surface is composed of nothing but steps. The result is a rough surface with a morphology much like what might be observed from multilayer growth. If the steps sites have zero etch rate, perhaps because they are passivated by selective adsorption, and only the terrace atoms etch, a rough surface profile composed of pyramidal structures is the ultimate morphology of the surface.

7.9.3   Porous solid formation

If etching is sufficiently anisotropic, a transition occurs from a rough surface to the formation of a porous solid. Electrochemical etching has been used to form porous semiconductors from, e.g. Si [134], Ge, SiC, GaAs, InP, GaP and GaN [135] as well as Ta_2O_5 [136], TiO_2 [137], WO_3 [138], ZrO_2 [139] and Al_2O_3 [140]. To describe porous solid formation we adopt a coarse-grained or continuum model rather than a molecular description. Three rates become important: pit nucleation, pit wall etching and pit bottom etching. If pit nucleation and pit wall etching are both rapid and faster than pit bottom etching, pits merge frequently and more rapidly than they grow in depth. If pit wall etching is much faster than pit nucleation, flat surfaces result (step-flow etching). If pit bottom etching is much more rapid than wall etching, a rough surface results if new pits are continually nucleated at the tops of the pit walls. However, if the pits do not move, that is, if nucleation can be arrested after the first wave of nucleation events, then stable pore growth results.

Perhaps the most celebrated example of well defined pore nucleation and propagation is shown in Fig. 7.14. Lehmann and co-workers [141–147] used lithography and anisotropic wet etching of Si in KOH to define nucleation sites to initiate pore formation. The etching for pore propagation was performed photoelectrochemically in acidic fluoride solution. Both KOH and fluoride solution etching are highly anisotropic for much different reasons. The etch rate of Si in aqueous KOH is highly dependent on the exposed crystallographic plane [148], and is used to create an inverted pyramidal tip. Etching in fluoride, as discussed further in §7.9.4, is enhanced at the pyramidal tips because electric field lines preferentially

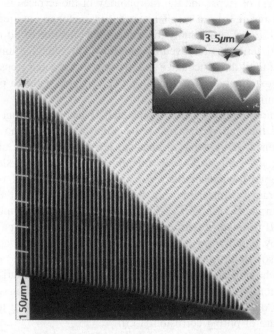

Figure 7.14   *Surface, cross section, and a 45° bevel of an n-type silicon sample showing a predetermined pattern of macropores. Pore growth was induced by a regular pattern of etch pits produced by standard photolithography as shown in the inset. Reproduced from S. Ottow, V. Lehmann and H. Föll, J. Electrochem. Soc. 143, 385. © 1996, with permission from the Electrochemical Society.*

direct holes generated by photon absorption at the back of the Si wafer to the tips. The result is that the pit bottom etches rapidly while the pit walls are inert and the formation of a regular array of rectangular macropores, which can extend hundreds of μm deep, indeed, through the entire depth of the wafer. The lithographically produced array of nucleation sites is not required for the formation of straight walled macropores; however, it does lead to the most regular arrays.

The formation of a porous solid usually results from a combination of nucleation, growth, branching and coalescence of pores [134, 149–152]. An in-depth review of, in particular, macropore formation in silicon has been given by Föll et al. [153]. Passivation, either chemical or electrical passivation, must play an important role because if pore wall etching is not somehow arrested, the pores coalesce into ever larger structures and eventually disappear. The relative rates (as well as any crystallographic dependence) of nucleation, wall, bottom and top etching determine not only whether a porous solid forms, but also the pore morphology and pore size distribution.

Anisotropy in etch rates is important for pore formation. Isotropic etching can only lead to flat or rough surfaces. In anisotropic etching, the initial shape of the surface being etched determines whether fast or slow etching planes are survivors. A general rule of thumb is that a convex object, e.g. a sphere, etches to reveal fast etching planes. Conversely, a concave object such as a hollow, etches to reveal the slow etching planes [154]. However, as Fig. 7.13 shows, the etching of a complex initial surface profile can lead to complex crystallographically defined structures. Masking as well as patterning of the substrate prior to etching can be used to structure the initial surface. Any factor that affects the rate of reaction can affect the degree of anisotropy of an etchant, including temperature, concentration, solvent, surfactant addition, addition of an electrochemically active species or applying a bias to the surface.

Anisotropy is a rather elastic word. There are two senses in which it is used with respect to etching. One is that when comparing one crystallographic plane to another, there are differences in etch rates. This type of anisotropy is essential for describing why Si(100) etches into square pores with inverted pyramid bottoms, Fig. 7.13(c), as opposed to Si(111) which etches to expose triangular pores with sloped walls terminating in a flat triangular bottom, Fig. 7.13(d). The second sense is that within a single plane, different sites can etch at different rates. Thus, the surface morphology that results from etching depends not only on the crystallographic orientation of the initial surface but also on the relative etch rates of kinks versus steps versus terraces. This type of anisotropy is essential for the formation of atomically smooth Si(111)–(1 × 1) by wet etching in alkaline fluoride solutions [155]. While it is impossible to predict *a priori* what morphologies will result from an arbitrary combination of rate parameters, simple rule-based simulations can be run, for instance, based on a kinetic Monte Carlo scheme to predict expected steady-state morphologies [156] and to extract relative rates.

7.9.4 Silicon etching in aqueous fluoride solutions

It is often claimed that the wet etching of semiconductors proceeds by the oxidation of the semiconductor followed by the dissolution of the oxide. This is not always the case [157], and its widespread dissemination is derived partly from the ambiguity of the term oxidation. Take Si etching as an example. When Si is anodically etched by aqueous HF in the electropolishing regime, a surface oxide forms and is then chemically etched by the fluoride solution. However, at less positive bias, no oxide is formed while Si is still etched and forms porous silicon. Si is oxidized in the electrochemical sense, that is, its oxidation state changes from zero to a more positive value, but a surface oxide need not play a role in the etching mechanism.

The etching of Si surfaces in aqueous fluoride solutions is of great scientific and technological interest [134, 149, 158, 159] because depending on the reaction conditions it can produce either a perfectly flat, ideally terminated surface or else exceedingly complex, high-surface-area porous silicon (por-Si). Solutions of HF are one of the few that can etch glass. Glass is largely composed of a-SiO$_2$. A native oxide layer – SiO$_2$ of ~4 nm thickness – covers a Si wafer because of air oxidation of the Si single crystal after cutting and polishing. HF(aq) is used in semiconductor processing to remove the native oxide [160]. There is great interest in obtaining the flattest possible Si surface after cleaning as surface roughness influences device performance [161, 162]. Por-Si is an inexpensive means for production of nanostructured Si that emits visible light because of quantum confinement effects [163]. Por-Si has numerous other characteristics that make it interesting for a variety of other applications in optics and drug delivery, as a biomaterial and for chemical sensing [134, 164, 165].

The Si–F bond is significantly stronger than the Si–H bond. Therefore, it was assumed for years that the surface obtained by etching in HF(aq) (or other solutions obtained from NH$_4$F etc) was terminated by chemisorbed F atoms. Chabal and co-workers [166] demonstrated with IR spectroscopy that the surface is actually terminated with H atoms. Furthermore, a nearly perfect Si(111)–(1 × 1) surface can be formed under appropriate conditions. The Si(100) surfaces produced by fluoride etching can be quite flat, but do not exhibit the same degree of perfection [155, 158, 167–169].

The mechanism of Si etching in fluoride solutions has received a great deal of attention and while there are still open questions, many of the details are understood. [170–172]. A description of the reaction dynamics illustrates a number of important concepts, and shows how the surface science approach can be used to help understand complex reactions that occur at the liquid/solid interface. We use the model of Gerischer, Allongue and Costa-Kieling [171] (known as the Gerischer model) as the basis of our discussion. The mechanism is not complete in every detail nor for all reaction conditions. Its simplicity and intuitive nature, however, make it a good pedagogical tool and foundation for understanding this system. This mechanism was proposed to explain the electrochemical etching of the surface, in which a hole is captured by the Si–H bond to initiate the reaction. Nevertheless, we can use it to illustrate the chemical reactions that occur after initiation.

The first question to be addressed is why the surface is H-terminated. The answer is related to the electronegativities of F and H. While the Si–F bond is much stronger and therefore more stable on thermodynamics grounds, it is also highly electron withdrawing. The adsorption of F leads to an induction effect whereby the Si–Si bonds, in particular the backbonds of surface atoms to the underlying Si atoms, are polarized. The induction effect weakens the Si–Si bonds and makes them susceptible to further chemical attack. Thus, the monofluoride-covered surface is unstable on kinetic grounds as the addition of a F atom makes the surface more reactive. The addition of a second F atom increases the inductive effect, making the addition of a third F atom that much easier, Fig. 7.15. The resulting HSiF$_3$ is liberated as the final etch product. In solution, this is hydrolyzed to H$_2$SiF$_6$ as silicon (hydrogen) fluorides are unstable in water.

From a purely thermodynamic viewpoint the system Si + HF is unstable. The reaction of Si + 6HF to make 2H$_2$ + H$_2$SiF$_6$ (the ultimate etch product) is exothermic by 733 kJ mol^{-1} [157]. We might then wonder why the etching of Si in HF stops instead of proceeding completely to H$_2$SiF$_6$ until either all of the Si or all of the F is consumed. The answer is that a kinetic constraint stops the reaction. The kinetic constraint arises from the H-termination of the surface. The origin of the H-termination can be observed by inspection of Fig. 7.15. In step one, a surface hydrogen is replaced by a F atom. This occurs preferentially at a defect site occupied by a dihydride species. In the two subsequent steps, HF adds to the Si surface. The F atom attaches to the Si atom that already has F bound to it. The H atom adds to the Si atom that has no F atoms bound to it. The F atom preferentially attaches to the Si with other F atoms attached because this Si atom is deficient of electrons. Since F is a strong nucleophile it prefers to attach to this Si atom rather than the one bound in the lattice. The result is that the surface is left H-terminated after the etched Si atom departs.

The Si–H bond is close to non-polar. The electrons are shared nearly equally, and no polarization of backbonds is induced by H(a). The non-polar bond results in a hydrophobic surface. The surface relaxes in

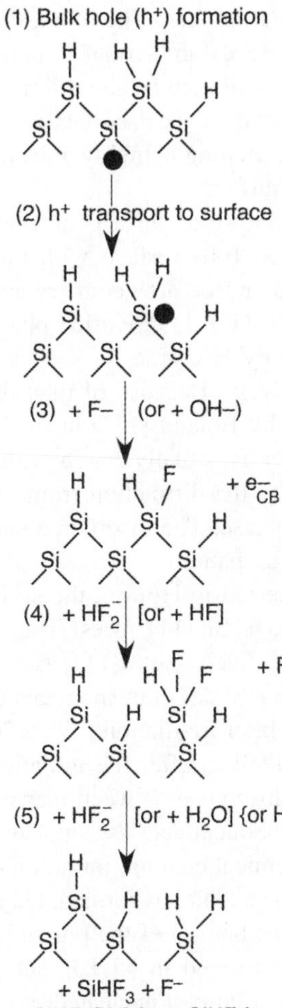

Figure 7.15 *The Gerischer mechanism of Si etching in fluoride solutions as proposed by Gerischer, Allongue and Costa-Kieling [171] modified by Kolasinski [172]. Electrochemical or photo-assisted etching is initiated by hole capture at the surface in a Si–Si back bond. This is followed by rapid replacement of H(a) by F(a), and slow addition of HF or HF_2^-. A final HF addition completes the etch cycle and leaves the surface H terminated.*

the presence of chemisorbed H such that the Si atoms approach the bulk-terminated structure. Because of the lack of strain in the Si lattice, lack of dangling bonds and the non-polar nature of the Si–H bond, the H-terminated Si surface is chemically passivated. Exposed to the ambient of a normal vacuum chamber, the surface remains clean indefinitely. Even when exposed to the atmosphere, the surface resists oxidation. While hydrocarbons physisorb readily on the H-terminated surface, water and O_2 take weeks to oxidize completely the H-terminated surface even at atmospheric pressure. It is this passivation effect that protects the Si crystal, and the etch rate of Si in HF(aq) in the dark is $<1\,\text{Å}\,\text{min}^{-1}$ even though the system is thermodynamically unstable.

The extreme passivation of perfect Si(111)–(1 × 1):H surface also explains why HF(aq) solutions of the proper pH can lead to the formation of nearly perfect H-terminated Si(111)–(1 × 1):H surfaces. The step sites have a higher reactivity than the terraces. Therefore, etching occurs preferentially at the steps. The steps are stripped away while the terraces are virtually immune to pitting. Expansive ideal terraces are formed, the size of which are ultimately limited by the miscut of the crystal from the true (111) plane. Two essential characteristics of etching lead to ideal surfaces: (1) the initiation step is not random but preferentially starts at the step, and (2) the etching is highly anisotropic, starting at the step and proceeding laterally across instead of down into the surface.

The $Si(111)-(1 \times 1)$:H surface is a strain-free surface in which the surface atoms occupy positions close to the bulk-termination values. A $Si(100)$ surface with the same degree of perfection cannot be produced. This can be understood based on adsorbate geometry and strain. Contrary to repeated assertions in the literature, there is no stable $Si(100)-(1 \times 1)$ dihydride phase that corresponds to the bulk structure with the two dangling bonds terminated by H atoms. In such a structure, the H atoms are so close to their neighbours that they interact repulsively. Because of this, the $Si(100)-(1 \times 1)$ dihydride surface is strained and reactive. UHV STM studies by Boland [173] have shown that while the LEED pattern from such a surface appears (1×1), the surface is actually rough with both monohydride and dihydride units present. The (1×1) pattern corresponds to the diffraction from the ordered subsurface layers rather than to diffraction from an ordered dihydride phase. The disordered surface causes only diffuse scattering and an increase in the background of the LEED pattern.

If a Si crystal is left for an extended period in HF(aq), the surface spontaneously roughens. A $Si(100)$ surface develops pyramidal facets that have $Si(111)$ faces. This is because the $Si(111)$ surface is more stable than the $Si(100)$ versus etching, effectively the $Si(111)$ is a survivor while the $Si(100)$ is etched at a higher rate and disappears. Nonetheless, por-Si does not spontaneously form. Por-Si can be produced under appropriate anodic conditions [149, 174], laser irradiation [175, 176], or by electroless etching [177–179]. Essentially the same etch chemistry is involved after the initiation step, but now an extremely complex and non-planar structure is formed. The difference between planar etching and por-Si formation is a consequence of the initiation step. In electrochemical and laser-assisted etching, a hole is driven to the surface and captured there. Unlike the case of chemical etching, initiation occurs randomly across the surface, and pitting of the terraces occurs. Moreover, once a pit has formed, the electronic structure of the Si responds in a way that preferentially directs holes to the bottom of the pits rather than to the pit walls [157, 180–182], which creates the electrical passivation discussed in §7.9.3. Subsequent etching is at the bottom of the pore rather than the sidewalls. Thus, once a pore is nucleated, it continues to propagate into the surface.

7.9.5   Coal gasification and graphite etching

Coal gasification is the reaction of an inhomogeneous carbonaceous solid with H_2O to form H_2, CO, CO_2 and CH_4. The water gas shift and methanation reactions discussed in §6.4 and §6.5 are of obvious importance to this class of reactions. Conditions can be optimized so that either methane or synthesis gas is the primary product. Impurities within the coal itself may act as catalysts for gasification. The demands put upon a catalyst are extremely severe due to poisoning, and because it is inherently more difficult to catalyze a reaction involving the etching of a solid. One way to avoid mass transport problems is to use a catalyst that is liquid under the reaction conditions. K_2CO_3 has been used to catalyze coal gasification in a process developed by Exxon [183]. The reaction is run at 1000 K where K_2CO_3 is liquid. The reactions are by no means understood. Various surface oxides are formed and these appear to exert control over the reaction [184]. In addition, numerous hydrocarbon decomposition reactions occur simultaneously with the reactions of solid C.

The reaction of graphite with H_2O [185], O_2 [186, 187] and H_2 [188] have been studied under well-defined conditions. Low temperature adsorption of H_2O is non-dissociative and does not lead to etching of graphite. The addition of K to the surface catalyzes the formation of CO_2 at a temperature of just

750 K, consistent with its use in the Exxon process. The reaction of modulated molecular beams of O_2 with pyrolytic graphite depends strongly on the surface structure. CO is the primary product with much less CO_2 being formed. The reaction has an appreciable rate only above 1000 K, then increases slowly with temperature. As etching proceeds the surface is roughened significantly. Nonetheless, the reaction probability remains low ($\leq 10^{-2}$). This implies that only certain defect sites act as active sites for the reactions. The reaction probability is insensitive to changes in the O_2 energy, which indicates that the reaction is not activated in the co-ordinates of the molecule. Activation only occurs by increasing the surface temperature. Graphite is not etched by molecular hydrogen. Even a beam of atomic hydrogen is only able to etch pyrolytic graphite with a reaction probability of $<10^{-2}$. Methane is produced below 800 K, whereas acetylene (C_2H_2) is the major etch product above 1000 K. Between these temperatures, the H atoms recombine to form H_2 without etching the surface. The basal plane is much less reactive and remains relatively smooth after etching. The prism plane is more reactive and forms ridges parallel to the basal plane much as is found after etching in O_2.

The etching of highly oriented pyrolytic graphite (HOPG) in O_2 or air can lead to the formation of circular etch pits [189], which have been used as molecular corrals for the study of self assembled monolayers (see §5.6). When etched for short periods at ~920 K, flat-bottomed, one monolayer deep etch pits with diameters of 50 to 5000 Å are formed. Again, defects are of great importance in the etching process. The defects have much higher reactivity than the perfect terraces; therefore, they act as nucleation sites from which the circular pits grow. Etching in air can be accompanied by the introduction of particulates onto the surface. Particles that land on the surface can act as further nucleation centres that lead to enhanced etching via a catalytic process. The particles often appear to be mobile. This leads to the formation of channels formed by etching in the wake of the diffusing particle.

7.9.6 Selective area growth and etching

In the formation of integrated circuits (ICs), solar cells as well as micro- or nano-electromechanical systems (MEMS or NEMS) manufacturing [190], the substrate must be etched in a precise manner to create two- and three-dimensional structures. To achieve this, the surface must be patterned in some way so that its resistance to etching can be manipulated in a controlled manner.

Lithography [191], as opposed to relief printing which uses a raised edge or intaglio in which ink resides in a crevice, is a planar printing process that was invented by Alois Senefelder in Bohemia in 1798 [192]. In photolithography for manufacturing [193, 194], etching plays a central role. It is the prototypical top-down process for manufacturing and nanostructure formation although schemes exist for integrating it with bottom-up methods to produce hierarchical structures [195]. A typical process involves spin coating of a resist onto the substrate. The resist is a polymer [196, 197] whose solubility (etch rate) depends on whether it has been exposed to electron, photon, or ion beam irradiation. This exposure is usually performed through a stencil mask in photolithography. However, it is also possible to perform maskless lithography in which the irradiating beam is moved across the surface to expose the resist in a pre-determined pattern. In §6.8 we briefly mentioned how hydrogen can be used as a resist for maskless lithography using electron beam irradiation from an STM tip. After transfer of the pattern into the resist, the next step may either involve etching or growth. Successive combinations of etching and deposition are then used to create either raised or sunken features in materials that are either similar or dissimilar to the substrate.

In conventional lithography, the resist is a thin polymer film that is applied from a solution onto the substrate by spin coating. Irradiation of the resist leads to chemical changes in the polymer, which lead to a change in the solubility between the irradiated and unirradiated regions. In a negative resist such as poly(methylmethacrylate) (PMMA) or cyclized poly(*cis*-isoprene) with bisazide cross-linkers, irradiation leads to cross-linking of the polymer chains. The irradiated area is rendered insoluble in the next step, which is development in a solvent. A positive resist such as a diazonaphthoquinone (DNQ)/novolak resins is insoluble. Irradiation transforms the hydrophobic DNQ into indenecarboxylic acid. This renders the exposed regions soluble in the development step. Various compounds have been used as resists. These and the issues involved in resist performance are treated by Gutmann [196] and Ito [197]. Advanced lithographic techniques involving wavelengths below 250 nm require chemically amplified (CA) resists. Irradiation does not change the solubility of CA resists directly, but it does release a chemical component into the thin film of resist. In a subsequent baking step, this component reacts with the thin film to form a region of altered solubility [191].

In the development step, the latent image from lithography is transformed into a real physical image. Development usually consists of introducing a solvent to the exposed resist. The chemical identity of the solvent depends on the resist. The solvent may be an alkaline solution such as KOH or $N(CH_3)_4OH$ (trimethyl ammonium hydroxide, TMAH) or a polar or non-polar organic solvent such as butylacetate, xylene or ethylpyruvate. The most important feature is that the solvent must selectively dissolve either only the exposed region (positive tone resist) or the unexposed region (negative tone resist). The process of lithography and development is illustrated in Fig. 7.16.

The substrate is now ready for selective area modification. One type of modification is selective area growth. For instance, if the patterned and unpatterned regions alternate between hydrophobic and hydrophilic tendencies, then Langmuir-Blodgett films and self-assembled monolayers (Chapter 5) can be used to transfer patterns of assorted organic molecules onto the surface. The grafting of controlled patches of biomolecules (proteins, peptides, DNA, etc) unto Si surfaces forms the basis of biochips used in immunoassays and the study of genomics and proteomics.

Chemical vapour deposition can be used to transfer inorganic films into the patterned regions. To be effective, the patterned region must have a different reactivity than the unpatterned region. This is the basis of a technique for the selective area growth of carbon nanotubes via catalytic growth. In a resistless

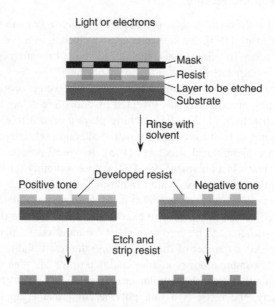

Figure 7.16 *A schematic illustration of the steps involved in lithographic pattern transfer. Irradiation through a mask transfers a pattern into the resist. Development removes the irradiated area of a positive tone resist, whereas the unirradiated region is removed from a negative tone resist. Etching transfers the pattern into the layer beneath the resist.*

approach, Fe is evaporated through a mask to transfer a pattern of catalytically active regions onto a Si substrate. Alternatively, one can pattern a resist, develop it, then fill the developed regions with a solution containing an Fe compound [198]. The Fe compound is activated by annealing to form patterned regions that are catalytically active for the production of carbon nanotubes.

After development, the resist may be called upon to function as a mask. There are two distinct ways in which the resist may act as a mask: (1) as a chemical mask, and (2) as an etch mask. In the first mode, the developed substrate is exposed to a reactive chemical environment. This could be oxygen, ammonia, an evaporated metal or a laser ablated insulator such as SiO_2. The mask does not react with the chemicals but the bare substrate does. After reaction, the resist is removed with an appropriate solvent leaving behind a substrate that is patterned with, e.g. an oxide, nitride or metal. In the second mode, the substrate is exposed to a reactive chemical environment that can etch the surface, such as KOH (for Si), HF (for SiO_2) ion radiation or a halogen-containing plasma. The resist is chosen such that its etch behaviour is opposite to that of the substrate. After removal of the resist, the substrate is left with a pattern of troughs. By successive alternation of various etching and growth steps, the intricate structures of ICs and MEMS are constructed. Selective area growth and etching allow for the incorporation of materials of various compositions into the final structure.

Advanced Topic: Si Pillar Formation

To illustrate these principles, we refer to Fig. 7.17. First, a Si wafer surface is coated with an unconventional "resist" [199]. This resist consists of polystyrene spheres with a diameter of 500 nm dispersed in water. Upon drying of the solvent, the polystyrene spheres form a self-assembled closed packed hexagonal array, as shown in Fig. 7.17(a). This layer is then irradiated with evaporated metal atoms, e.g. Ag or Au. The metal atoms fill the interstices. Rinsing with chloroform, toluene or acetone then develops the film. This dissolves the polystyrene spheres leaving an ordered array of metal clusters. More conventionally, the polystyrene spheres would be called a shadow mask, but the analogy to classical lithography should not be lost. The rinsing step develops a negative tone image.

Figure 7.17 *The creation of nanoscale Si pillars via reactive ion etching. (a) Polystyrene spheres deposited from solution self-assemble into a close-packed lattice. Various vacancy and domain boundary defects are visible in the lattice. (b) After irradiation with metal vapour and reactive ion etching, the Si pillars are formed. (c) A close up demonstrates that the pillars are not only ordered, but also exhibit a small size distribution.*

The array of metal clusters can now be used as an etch mask. Reactive ion etching (RIE) using a mixture of SF_6 and CF_4 preferentially etches the bare Si surface whereas the clusters inhibit the etching of the Si beneath them. The result is the formation of an ordered array of Si nanopillars. The height and width of the pillars depend on the details of the lithographic and etching procedures. Heights of >100 nm and widths of <50 nm are readily achieved with this method as seen in Figs 7.17(b) and 7.17(c). Alternatively, the metal clusters can be exposed to HF + H_2O_2 solutions to create forests of higher aspect ratio silicon nanowires over large areas [200, 201].

7.10 Summary of important concepts

- A stress applied to a lattice causes a deformation called strain.
- Strain relief is critical in the determination of interface structure. Heteroepitaxial layers often relax by the introduction of defects at the interface after they grow thicker than the critical thickness.
- The term surface tension should only be applied to liquids. For solids the proper term is surface energy.
- Surface energy can be thought of as being analogous to a two-dimensional pressure, compare Eqs (5.2.1) and (5.2.2), or else as the surface Gibbs energy, Eq. (5.2.4).
- The multilayer growth of an adsorbate on a substrate is characterized by one of three growth modes: Frank-van der Merwe (layer-by-layer), Stranski-Krastanov (layer + island) or Volmer-Weber (island).
- At equilibrium the growth mode is determined by the effects of strain and the relative strength of the A–A, A–B and B–B interactions (A = adsorbate atom and B = substrate atom).
- Under non-equilibrium conditions the dynamics of adsorption, desorption and diffusion can lead to deviations from the equilibrium growth mode (kinetically controlled growth).
- Self-limiting chemistry is used to grow layers with atomic composition and thickness control.
- Etching involves reactions that consume the surface.
- Anisotropy (with regard to crystallography and surface site) in the etch process can lead to a range of final structures ranging from sharp tips to smooth surfaces to porous solids.
- The region of the surface that is etched can be controlled by the use of a mask.
- In a positive tone resist, the exposed portion of the resist is removed during development, whereas in a negative resist the unexposed region is removed.

7.11 Frontiers and challenges

- Formation and characterization of free-standing 2D films, graphene being the most popular example [202], but many other layered compounds (BN, MoS_2, WS_2, etc) can also be turned into 2D films [203].
- Calculation and measurement of surface energies as a function of nanoparticle size and composition. Relative values of surface energies are a primary driving force for determining phase stability in nanoscale objects and can lead to changes in phase stability compared to bulk materials [204].
- Surface functionalization of nanoparticles and the inhibition of nanoparticle aggregation. Both of these are essentially growth problems, and both have a direct impact on the incorporation of nanoparticles into nanoparticle/polymer composites. Nanoparticle/polymer composites can have desirable materials properties. Common examples include Bakelite and tires but the greatest hurdle to the routine production of engineered composites is the lack of cost-effective methods for controlling the dispersion of nanoparticles within the polymer matrix [205].
- Growth of nanoparticles and nanoparticle aggregates including colloidal crystals and nucleation of ice particles.
- Dislocation networks formed during wafer bonding as a method of nanostructure formation.
- Development of theory to describe equilibrium structures in heteroepitaxy.
- Development of models to explain the mechanism of catalytic growth, and determination of the limits of catalytic growth to produce nanowires and nanotubes from different materials with controlled sizes and placement.
- Whereas unified kinetic models can be proposed to explain growth and the dissolution of minerals [126, 127] in aqueous solutions, how far can a similar unification of underlying principles be achieved to relate growth and etching more generally?
- How does the first chemical step in the etching of Si in acidic fluoride solution occur? Does F^- replace H(a) in a concerted manner or does it abstract H(a) with the resulting dangling bond capped by F^-

(with injection of an electron into the conduction band) in a second step? Does abstraction play a role generally at the liquid/solid interface? Is abstraction involved in the activation of H-terminated Si surface in its reactions with organic molecule such as alkene, alkynes, alcohols, etc?

- Thin films for integrated circuits, photovoltaics and optoelectronics: diffusion barriers, high *k* dielectrics, quantum dots, (nanocrystalline) silicon structures for optoelectronics, nanostructures for photovoltaics.
- Atomic layer epitaxy: How versatile is it, and how many materials properties can be controlled with it?
- Biomineralization and templated growth. Do these systems generally not follow classical nucleation theory [206]?
- Producing porous materials with controlled composition, pore size and pore morphology – either from the bottom up with growth or from the top down with etching.

7.12 Further reading

D. Bimberg (Ed.), *Semiconductor Nanostructures* (Springer, Berlin, 2010).

D. A. King and D. P. Woodruff (Eds), *The Chemical Physics of Solid Surfaces and Heterogeneous Catalysis: Surface Properties of Electronic Materials*, Vol. 5 (Elsevier, Amsterdam, 1988).

D. A. King and D. P. Woodruff (Eds), *The Chemical Physics of Solid Surfaces and Heterogeneous Catalysis: Growth and Properties of Ultrathin Epitaxial Layers*, Vol. 8 (Elsevier, Amsterdam, 1997).

H. Ibach, *The role of surface stress in reconstruction, epitaxial growth and stabilization of mesoscopic structures*, Surf. Sci. Rep. **29** (1997) 193.

P. Jensen, *Growth of nanostructures by cluster deposition: Experiments and simple models*, Rev. Mod. Phys. **71** (1999) 1695.

H. Lüth, *Solid Surfaces, Interfaces and Thin Films*, 4th ed. (Springer-Verlag, Berlin, 2001).

S. M. Prokes and K. L. Wang (Guest Eds), *Novel methods of nanoscale wire formation*, MRS Bulletin, Vol. 24 (Aug. 1999).

V. A. Shchukin and D. Bimberg, *Spontaneous ordering of nanostructures on crystal surfaces*, Rev. Mod. Phys. **71** (1999) 1125.

V. Shchukin, N. N. Ledentsov, and D. Bimberg, *Epitaxy of Nanostructures* (Springer, Berlin, 2010).

G. M. Wallraff and W. D. Hinsberg, *Lithographic imaging techniques for the formation of nanoscopic features*, Chem. Rev. **99**, (1999), 1801.

J. A. Venables, *Atomic processes in crystal growth*, Surf. Sci. **299/300**, (1994), 798–817.

J. A. Venables, G. D. T. Spiller, and M. Hanbucken, *Nucleation and growth of thin-films*, Rep. Prog. Phys. **47**, (1984), 399–459.

7.13 Exercises

7.1 Dunstan [6] has shown that there is a linear dependence of the $In_x Ga_{1-x}As$ lattice constant on the lattice constants of its constituents according to

$$a_{In_x Ga_{1-x}As} = x\, a_{InAs} + (1 - x)\, a_{GaAs}. \tag{7.16.1}$$

Substitute this dependence into Eq. (7.1.3) and derive Eq. (7.1.4).

7.2 Consider a system that for a given set of conditions exhibits step-flow growth. Discuss the effects that the adsorption of heteroatoms can have on homoepitaxial growth. Consider two low heteroatom coverage cases: (a) the heteroatoms decorate the steps, and (b) the heteroatoms occupy isolated terrace sites.

7.3   Si is the most important semiconductor for electronic applications. GaAs and its III-V sister compounds are better suited than Si as building blocks for optical devices such as light emitting diodes (LEDs) and lasers. The integration of optical components with electronics is a highly desirable manufacturing goal for improved communications, computing and display devices. Discuss fundamental physical reasons why it is difficult to integrate GaAs circuitry with Si.

7.4   Discuss how Auger electron spectroscopy or XPS can be used to distinguish Frank-van der Merwe from Volmer-Weber growth. Hint: Look at Fig. 7.3 and consider how the substrate signal varies.

7.5   (a) Consider the epitaxial growth by MBE of $In_{0.67}Al_{0.33}P$ layer on an InGaAs substrate. What must the relative fluxes of In, Al and P be in order to maintain this composition? What influence does the substrate temperature have on epitaxy and the required fluxes?

(b) Consider the CVD growth of P-doped (at a concentration of 10^{16} cm^{-3}) $Si_{(1-x)}Ge_x$ with $x = 0.05$ from the respective hydrides. Discuss the influence of surface temperature on epitaxy and the fluxes required to maintain this composition.

7.6   The dimensionless formation energy $E(V)$ of a single-facetted quantum dot as a function of its dimensionless volume is given by [4]

$$E(V) = -\alpha V + \frac{2\beta V^{2/3}}{e^{1/2}} - 2V^{1/3}\ln(e^{1/2}V^{1/3}) \qquad (7.16.1)$$

The chemical potential of an island is given by

$$\mu(V) = \frac{dE(V)}{dV} \qquad (7.16.2)$$

Assuming that $\alpha = 0$, predict the most probable island volume for $\beta = 1.4$, 0.5 and -0.5.

7.7   The incident flux can be used to tune the chemical potential of a system of islands on a surface. Predict what occurs to the island size distribution when $\beta = -0.7$ and the flux is turned off for a system with a chemical potential of (a) $+1$, (b) -3, (c) -4 given the functional form of μ vs V given in Fig. 7.18. How does the size distribution evolve when the flux is turned off for $\beta = 2$ and $\mu = 1$?

7.8   Determine the orientation of the pore walls formed on a Si(100) wafer given that they are straight and that their orientation with the respect to the {110} planes is as shown in Fig. 7.19.

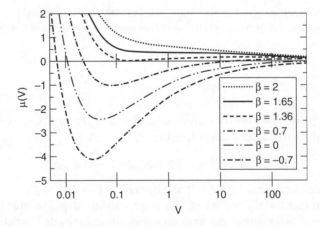

Figure 7.18 *Graph of μ vs V. Reproduced from T. P. Munt, D. E. Jesson, V. A. Shchukin, D. Bimberg, Appl. Phys. Lett., 85, 1784. © 2004, with permission from the American Institute of Physics.*

Figure 7.19 *Si(100) pore orientation. Reproduced from M. E. Dudley, K. W. Kolasinski,* J. Electrochem. Soc., *155, H164.* © *2008, with permission from the Electrochemical Society.*

References

[1]   H. Ibach, *Surf. Sci. Rep.*, **29** (1997) 193.

[2]   H. Ibach, *Surf. Sci. Rep.*, **35** (1999) 71.

[3]   V. A. Shchukin, D. Bimberg, *Rev. Mod. Phys.*, **71** (1999) 1125.

[4]   T. P. Munt, D. E. Jesson, V. A. Shchukin, D. Bimberg, *Appl. Phys. Lett.*, **85** (2004) 1784.

[5]   P. Müller, A. Saúl, *Surf. Sci. Rep.*, **54** (2004) 157.

[6]   D. J. Dunstan, *J. Mater. Sci.*, **8** (1997) 337.

[7]   L. Vegard, *Z. Phys.*, **5** (1921) 17.

[8]   D. Sander, *Curr. Opin. Solid State Mater. Sci.*, **7** (2003) 51.

[9]   F. C. Frank, J. H. van der Merwe, *Proc. R. Soc. London, A*, **198** (1949) 216.

[10] J. W. Matthews, J. L. Crawford, *Thin Solid Films*, **5** (1970) 187.

[11] E. A. Fitzgerald, *Mater. Sci. Reports*, **7** (1991) 87.

[12] D. J. Dunstan, S. Young, R. H. Dixon, *J. Appl. Phys.*, **70** (1991) 3038.

[13] J. W. Gibbs, *Collected Works, Vol. 1, Thermodynamics*, Longmans, London, 1928.

[14] L. Brillouin, *J. Phys. Radium*, **9** (1938) 462.

[15] R. Shuttleworth, *Proc. Phys. Soc. A*, **63** (1950) 444.

[16] J. C. Eriksson, *Surf. Sci.*, **14** (1969) 221.

[17] G. D. Halsey, *Surf. Sci.*, **72** (1978) 1.

[18] A. F. Andreev, Y. A. Kosevich, *JETP*, **54** (1981) 761 [*Zh. Eksp. Teor. Fiz.* **81**, 1435 (1981)].

[19] P. Nozières, D. E. Wolf, *Z. Phys. B: Condens. Matter*, **70** (1988) 399.

[20] D. E. Wolf, P. Nozières, *Z. Phys. B: Condens. Matter*, **70** (1988) 507.

[21] R. Koch, Intrinsic stress of epitaxial thin films and surface layers, in *The Chemical Physics of Solid Surfaces and Heterogeneous Catalysis: Growth and Properties of Ultrathin Epitaxial Layers, Vol. 8* (Eds: D. A. King, D. P. Woodruff), Elsevier, Amsterdam, 1997, p. 448.

[22] G. Wulff, *Z. Kristallogr. Mineral.*, **34** (1901) 449.

[23] C. Rottman, M. Wortis, *Phys. Rep.*, **103** (1984) 59.

[24] M. C. Desjonquères, D. Spanjaard, *Concepts in Surface Physics*, 2nd ed., Springer-Verlag, Berlin, 2002.

[25] B. S. Clausen, J. Schiøtz, L. Gråkæk, C. V. Ouesen, K. W. Jacobsen, J. K. Nørskov, H. Topsøe, *Topics in Catalysis*, **1** (1994) 367.

[26]   I. Daruka, J. Tersoff, A. L. Barabási, *Phys. Rev. Lett.*, **83** (1999) 2753.

[27]   X. Q. Shen, D. Kishimoto, T. Nishinaga, *Jpn. J. Appl. Phys., Part 1* , **33** (1994) 11.

[28]   R. S. Williams, M. J. Ashwin, T. S. Jones, J. H. Neave, *J. Appl. Phys.*, **97** (2005) 044905.

[29]   R. S. Williams, M. J. Ashwin, T. S. Jones, J. H. Neave, *J. Appl. Phys.*, **95** (2004) 6112.

[30]   F. C. Frank, J. H. van der Merwe, *Proc. R. Soc. London, A*, **198** (1949) 205.

[31]   I. N. Stranski, L. Kr'stanov, *Sitzungsber. Akad. Wiss. Wien*, **146** (1938) 797.

[32]   M. Volmer, A. Weber, *Z. Phys. Chem.*, **119** (1926) 277.

[33]   A. P. Alivisatos, *Science*, **271** (1996) 933.

[34]   A. D. Yoffe, *Adv. Phys.*, **42** (1993) 173.

[35]   F. M. Ross, R. M. Tromp, M. C. Reuter, *Science*, **286** (1999) 1931.

[36]   M. Horn-von Hoegen, F. K. LeGoues, M. Copel, M. C. Reuter, R. M. Tromp, *Phys. Rev. Lett.*, **67** (1991) 1130.

[37]   M. Copel, M. C. Reuter, E. Kaxiras, R. M. Tromp, *Phys. Rev. Lett.*, **63** (1989) 632.

[38]   F. Liu, A. H. Li, M. G. Lagally, *Phys. Rev. Lett.*, **87** (2001) 126103.

[39]   W. Ostwald, *Z. Phys. Chem.*, **34** (1900) 495.

[40]   I. Daruka, A.-L. Barabási, *Phys. Rev. Lett.*, **79** (1997) 3708.

[41]   L. G. Wang, P. Kratzer, M. Scheffler, *Jpn. J. Appl. Phys.*, **39** (2000) 4298.

[42]   L. G. Wang, P. Kratzer, N. Moll, M. Scheffler, *Phys. Rev. B*, **62** (2000) 1897.

[43]   L. G. Wang, P. Kratzer, M. Scheffler, N. Moll, *Phys. Rev. Lett.*, **82** (1999) 4042.

[44]   B. A. Joyce, D. D. Vvedensky, *Mater. Sci. Eng., R*, **46** (2004) 127.

[45]   R. G. S. Pala, F. Liu, *Phys. Rev. Lett.*, **95** (2005) 136106.

[46]   W. K. Burton, N. Cabrera, F. C. Frank, *Philos. Trans. R. Soc. London, A*, **243** (1951) 299.

[47]   H. Röder, E. Hahn, H. Brune, J.-P. Bucher, K. Kern, *Nature (London)*, **366** (1993) 141.

[48]   H. Brune, *Surf. Sci. Rep.*, **31** (1998) 125.

[49]   J. A. Venables, *Philos. Mag.*, **27** (1973) 693.

[50]   J. A. Venables, G. D. T. Spiller, M. Hanbucken, *Rep. Prog. Phys.*, **47** (1984) 399.

[51]   J. A. Venables, *Phys. Rev. B*, **36** (1987) 4153.

[52]   J. A. Venables, *Surf. Sci.*, **299/300** (1994) 798.

[53]   R. J. Hamers, U. K. Köhler, J. E. Demuth, *J. Vac. Sci. Technol., A*, **8** (1990) 195.

[54]   Y.-W. Mo, R. Kariotis, D. E. Savage, M. G. Lagally, *Surf. Sci.*, **219** (1989) L551.

[55]   Y.-W. Mo, R. Kariotis, B. S. Swartzentruber, M. B. Webb, M. G. Lagally, *J. Vac. Sci. Technol., A*, **8** (1990) 201.

[56]   Y. W. Mo, R. Kariotis, B. S. Swartzentruber, M. B. Webb, M. G. Lagally, *J. Vac. Sci. Technol. B*, **8** (1990) 232.

[57]   I. Daruka, A.-L. Barabási, *Phys. Rev. Lett.*, **78** (1997) 3027.

[58]   D. Jesson, *Phys. Rev. Lett.*, **77** (1996) 1330.

[59]   W. M. Tong, R. S. Williams, *Annu. Rev. Phys. Chem.*, **45** (1994) 401.

[60]   K. Kern, H. Niehus, A. Schatz, P. Zeppenfeld, J. George, G. Comsa, *Phys. Rev. Lett.*, **67** (1991) 855.

[61]   H. Brune, M. Giovannini, K. Bromann, K. Kern, *Nature (London)*, **394** (1998) 451.

[62]   G. Ehrlich, F. G. Hudda, *J. Chem. Phys.*, **44** (1966) 1039.

[63]   R. L. Schwoebel, E. J. Shipsey, *J. Appl. Phys.*, **37** (1966) 3682.

[64]   G. Rosenfeld, B. Poelsema, G. Comsa, Epitaxial growth modes far from equilibrium, in *The Chemical Physics of Solid Surfaces and Heterogeneous Catalysis: Growth and Properties of Ultrathin Epitaxial Layers, Vol. 8* (Eds: D. A. King, D. P. Woodruff), Elsevier, Amsterdam, 1997, p. 66.

[65]   J. H. Neave, P. J. Dobson, B. A. Joyce, J. Zhang, *Appl. Phys. Lett.*, **47** (1985) 100.

[66]   T. Nishinaga, *Prog. Cryst. Growth Characterization Mater.*, **48/49** (2004) 104.

[67]   S. Franchi, G. Trevisi, L. Seravalli, P. Frigeri, *Prog. Cryst. Growth Characterization Mater.*, **47** (2003) 166.

[68]   J. R. Arthur, Jr., *J. Appl. Phys.*, **39** (1968) 4032.

[69]   C. T. Foxon, B. A. Joyce, *Surf. Sci.*, **50** (1975) 434.

[70]   C. T. Foxon, B. A. Joyce, *Surf. Sci.*, **64** (1977) 293.

[71]   J. Tersoff, C. Teichert, M. G. Lagally, *Phys. Rev. Lett.*, **76** (1996) 1675.

[72]   Q. Xie, A. Madhukar, P. Chen, N. P. Kobayashi, *Phys. Rev. Lett.*, **75** (1995) 2542.

[73]   Z. Mi, P. Bhattacharya, *J. Appl. Phys.*, **98** (2005) 023510.

[74]   O. G. Schmidt, O. Kienzle, Y. Hao, K. Eberl, F. Ernst, *Appl. Phys. Lett.*, **74** (1999) 1272.

[75]   E. Penev, P. Kratzer, M. Scheffler, *Phys. Rev. B*, **64** (2001) 085401.

[76]   P. Kratzer, E. Penev, M. Scheffler, *Appl. Phys. A*, **75** (2002) 79.

[77]   P. Howe, B. Abbey, E. C. Le Ru, R. Murray, T. S. Jones, *Thin Solid Films*, **464–465** (2004) 225.

[78]   E. C. Le Ru, P. Howe, T. S. Jones, R. Murray, *Phys. Rev. B*, **67** (2003) 165303.

[79]   A. C. Jones, *J. Cryst. Growth*, **129** (1993) 728.

[80]   J. M. Jasinski, S. M. Gates, *Acc. Chem. Res.*, **24** (1991) 9.

[81]   J. M. Jasinski, B. S. Meyerson, B. A. Scott, *Annu. Rev. Phys. Chem.*, **38** (1987) 109.

[82]   J. Wintterlin, P. Avouris, *Surf. Sci.*, **286** (1993) L529.

[83]   J. Wintterlin, P. Avouris, *J. Chem. Phys.*, **100** (1994) 687.

[84]   L. Niinistö, *Curr. Opin. Solid State Mater. Sci.*, **3** (1998) 147.

[85]   M. M. Frank, G. D. Wilk, D. Starodub, T. Gustafsson, E. Garfunkel, Y. J. Chabal, J. Grazul, D. A. Muller, *Appl. Phys. Lett.*, **86** (2005) 152904.

[86]   S. M. George, *Chem. Rev.*, **110** (2010) 111.

[87]   S. M. Lee, E. Pippel, U. Gosele, C. Dresbach, Y. Qin, C. V. Chandran, T. Brauniger, G. Hause, M. Knez, *Science*, **324** (2009) 488.

[88]   S. M. Gates, C. M. Greenlief, D. B. Beach, P. A. Holbert, *J. Chem. Phys.*, **92** (1990) 3144.

[89]   M. E. Jones, L. Q. Xia, N. Maity, J. R. Engstrom, *Chem. Phys. Lett.*, **229** (1994) 401.

[90]   L. Q. Xia, M. E. Jones, N. Maity, J. R. Engstrom, *J. Vac. Sci. Technol., A*, **13** (1995) 2651.

[91]   H.-G. Rubahn, *Laser Applications in Surface Science and Technology*, John Wiley & Sons Ltd, Chichester, 1999.

[92]   K. W. Kolasinski, *Curr. Opin. Solid State Mater. Sci.*, **11** (2007) 76.

[93]   F. D. Wang, A. G. Dong, J. W. Sun, R. Tang, H. Yu, W. E. Buhro, *Inorg. Chem.*, **45** (2006) 7511.

[94]   H. J. Fan, P. Werner, M. Zacharias, *Small*, **2** (2006) 700.

[95]   W. Lu, C. M. Lieber, *J. Phys. D: Appl. Phys.*, **39** (2006) R387.

[96]   K. P. De Jong, J. W. Geus, *Catal. Rev.*, **42** (2000) 481.

[97]   P. M. Ajayan, *Chem. Rev.*, **99** (1999) 1787.

[98]   H. Dai, *Surf. Sci.*, **500** (2002) 218.

[99]   D. S. Bethune, C. H. Klang, M. S. de Vries, G. Gorman, R. Savoy, J. Vazquez, R. Beyers, *Nature (London)*, **363** (1993) 605.

[100]   S. Iijima, T. Ichihashi, *Nature (London)*, **363** (1993) 603.

[101]   T. Guo, P. Nikolaev, A. Thess, D. T. Colbert, R. E. Smalley, *Chem. Phys. Lett.*, **243** (1995) 49.

[102]   Y. Q. Zhu, W. B. Hu, W. K. Hsu, M. Terrones, N. Grobert, J. P. Hare, H. W. Kroto, D. R. M. Walton, H. Terrones, *J. Mater. Chem.*, **9** (1999) 3173.

[103]   K. W. Kolasinski, *Curr. Opin. Solid State Mater. Sci.*, **10** (2006) 182.

[104]   R. S. Wagner, W. C. Ellis, *Appl. Phys. Lett.*, **4** (1964) 89.

[105]   E. I. Givargizov, *J. Cryst. Growth*, **31** (1975) 20.

[106]   A. M. Morales, C. M. Lieber, *Science*, **279** (1998) 208.

[107]   F. Ding, K. Bolton, A. Rosén, *Computational Materials Science*, **35** (2006) 243.

[108]   F. Ding, A. Rosen, K. Bolton, *J. Chem. Phys.*, **121** (2004) 2775.

[109]   F. Ding, K. Bolton, A. Rosen, *J. Phys. Chem. B*, **108** (2004) 17369.

[110]   J. M. Blakely, K. A. Jackson, *J. Chem. Phys.*, **37** (1962) 428.

[111]   S. J. Kwon, J.-G. Park, *J. Phys.: Cond. Matter*, **18** (2006) 3875.

[112]   C. X. Wang, M. Hirano, H. Hosono, *Nano Lett.*, **6** (2006) 1552.

[113]   H. Chandrasekaran, G. U. Sumanasekara, M. K. Sunkara, *J. Phys. Chem. B*, **110** (2006) 18351.

[114]   Z. Chen, C. B. Cao, *Appl. Phys. Lett.*, **88** (2006) 143118.

[115]   S. N. Mohammad, *J. Chem. Phys.*, **125** (2006) 094705.

[116]   T. Y. Tan, N. Li, U. Gösele, *Appl. Phys. A*, **78** (2004) 519.

[117]   T. Y. Tan, N. Li, U. Gösele, *Appl. Phys. Lett.*, **83** (2003) 1199.

[118] N. Li, T. Y. Tan, U. Gösele, *Appl. Phys. A*, **86** (2007) 433.
[119] M. A. Verheijen, G. Immink, T. de Smet, M. T. Borgström, E. P. A. M. Bakkers, *J. Am. Chem. Soc.*, **128** (2006) 1353.
[120] V. G. Dubrovskii, N. V. Sibirev, G. E. Cirlin, J. C. Harmand, V. M. Ustinov, *Phys. Rev. E*, **73** (2006) 021603.
[121] S. Kodambaka, J. Tersoff, M. C. Reuter, F. M. Ross, *Phys. Rev. Lett.*, **96** (2006) 096105.
[122] F. M. Ross, J. Tersoff, M. C. Reuter, *Phys. Rev. Lett.*, **95** (2005) 146104.
[123] A. I. Persson, M. W. Larsson, S. Stenström, B. J. Ohlsson, L. Samuelson, L. R. Wallenberg, *Nature Mater.*, **3** (2004) 677.
[124] J. Johansson, B. A. Wacaser, K. A. Dick, W. Seifert, *Nanotechnology*, **17** (2006) S355.
[125] J. Johansson, C. P. T. Svensson, T. Mårtensson, L. Samuelson, W. Seifert, *J. Phys. Chem. B*, **109** (2005) 13567.
[126] P. M. Dove, N. Han, J. J. De Yoreo, *Proc. Natl. Acad. Sci. U. S. A.*, **102** (2005) 15357.
[127] A. C. Lasaga, A. Luttge, *Science*, **291** (2001) 2400.
[128] K. Sangwal, *Etching of Crystals, Vol. 15*, North-Holland, Amsterdam, 1987.
[129] R. B. Heimann, Principles of chemical etching – The art and science of etching crystals, in *Silicon Chemical Etching* (Ed.: J. Grabmaier), Springer-Verlag, Berlin, 1982, p. 173.
[130] W. Kleber, *Z. Elektrochem.*, **62** (1958) 587.
[131] O. Knacke, I. N. Stranski, *Z. Elektrochem.*, **60** (1956) 816.
[132] S. P. Garcia, H. Bao, M. A. Hines, *Phys. Rev. Lett.*, **93** (2004) 166102.
[133] C. Punckt, M. Bölscher, H. H. Rotermund, A. S. Mikhailov, L. Organ, N. Budiansky, J. R. Scully, J. L. Hudson, *Science*, **305** (2004) 1133.
[134] A. G. Cullis, L. T. Canham, P. D. J. Calcott, *J. Appl. Phys.*, **82** (1997) 909.
[135] S. Langa, J. Carstensen, M. Christophersen, K. Steen, S. Frey, I. M. Tiginyanu, H. Föll, *J. Electrochem. Soc.*, **152** (2005) C525.
[136] I. V. Sieber, P. Schmuki, *J. Electrochem. Soc.*, **152** (2005) C639.
[137] R. Beranek, H. Hildebrand, P. Schmuki, *Electrochem. Solid State Lett.*, **6** (2003) B12.
[138] H. Tsuchiya, J. M. Macak, I. Sieber, L. Taveira, A. Ghicov, K. Sirotna, P. Schmuki, *Electrochem. Commun.*, **7** (2005) 295.
[139] H. Tsuchiya, J. M. Macak, L. Taveira, P. Schmuki, *Chem. Phys. Lett.*, **410** (2005) 188.
[140] H. Masuda, K. Fukuda, *Science*, **268** (1995) 1466.
[141] F. Müller, A. Birner, U. Gösele, V. Lehmann, S. Ottow, H. Föll, *J. Porous Mater.*, **7** (2000) 201.
[142] S. W. Leonard, J. P. Mondia, H. M. van Driel, O. Toader, S. John, K. Busch, A. Birner, U. Gösele, V. Lehmann, *Phys. Rev. B*, **61** (2000) R2389.
[143] S. W. Leonard, H. M. van Driel, K. Busch, S. John, A. Birner, A.-P. Li, F. Müller, U. Gösele, V. Lehmann, *Appl. Phys. Lett.*, **75** (1999) 3063.
[144] V. Lehmann, S. Rönnebeck, *J. Electrochem. Soc.*, **146** (1999) 2968.
[145] V. Lehmann, U. Grünning, *Thin Solid Films*, **297** (1997) 13.
[146] S. Ottow, V. Lehmann, H. Föll, *J. Electrochem. Soc.*, **143** (1996) 385.
[147] V. Lehmann, *J. Electrochem. Soc.*, **140** (1993) 2836.
[148] R. A. Wind, H. Jones, M. J. Little, M. A. Hines, *J. Phys. Chem. B*, **106** (2002) 1557.
[149] R. L. Smith, S. D. Collins, *J. Appl. Phys.*, **71** (1992) R1.
[150] G. C. John, V. A. Singh, *Phys. Rep.*, **263** (1995) 93.
[151] R. L. Smith, S.-F. Chuang, S. D. Collins, *J. Electron. Mater.*, **17** (1988) 533.
[152] V. Parkhutik, *Solid-State Electron.*, **43** (1999) 1121.
[153] H. Föll, M. Christophersen, J. Carstensen, G. Hasse, *Mater. Sci. Eng., R*, **39** (2002) 93.
[154] R. A. Wind, M. A. Hines, *Surf. Sci.*, **460** (2000) 21.
[155] P. Jakob, Y. J. Chabal, K. Raghavachari, R. S. Becker, A. J. Becker, *Surf. Sci.*, **275** (1992) 407.
[156] M. A. Hines, *Annu. Rev. Phys. Chem.*, **54** (2003) 29.
[157] K. W. Kolasinski, *J. Phys. Chem. C*, **114** (2010) 22098.
[158] Y. J. Chabal, A. L. Harris, K. Raghavachari, J. C. Tully, *Internat. J. Mod. Phys. B*, **7** (1993) 1031.
[159] L. A. Jones, G. M. Taylor, F.-X. Wei, D. F. Thomas, *Prog. Surf. Sci.*, **50** (1995) 283.

[160] G. J. Pietsch, *Appl. Phys. A*, **60** (1995) 347.

[161] R. I. Hegde, M. A. Chonko, P. J. Tobin, *J. Vac. Sci. Technol. B*, **14** (1996) 3299.

[162] T. Ohmi, M. Miyashita, M. Itano, T. Imaoka, I. Kawanabe, *IEEE Trans. Electron Devices*, **39** (1992) 537.

[163] L. T. Canham, *Appl. Phys. Lett.*, **57** (1990) 1046.

[164] O. Bisi, S. Ossicini, L. Pavesi, *Surf. Sci. Rep.*, **38** (2000) 1.

[165] S. Ozdemir, J. L. Gole, *Curr. Opin. Solid State Mater. Sci.*, **11** (2007) 92.

[166] G. S. Higashi, Y. J. Chabal, G. W. Trucks, K. Raghavachari, *Appl. Phys. Lett.*, **56** (1990) 656.

[167] P. Jakob, Y. J. Chabal, *J. Chem. Phys.*, **95** (1991) 2897.

[168] P. Jakob, P. Dumas, Y. J. Chabal, *Appl. Phys. Lett.*, **59** (1991) 2968.

[169] P. Dumas, Y. J. Chabal, P. Jakob, *Surf. Sci.*, **269/270** (1992) 867.

[170] G. W. Trucks, K. Raghavachari, G. S. Higashi, Y. J. Chabal, *Phys. Rev. Lett.*, **65** (1990) 504.

[171] H. Gerischer, P. Allongue, V. Costa-Kieling, *Ber. Bunsen-Ges. Phys. Chem.*, **97** (1993) 753.

[172] K. W. Kolasinski, *Phys. Chem. Chem. Phys.*, **5** (2003) 1270.

[173] J. J. Boland, *Adv. Phys.*, **42** (1993) 129.

[174] A. Uhlir, *Bell Syst. Tech. J.*, **35** (1956) 333.

[175] N. Noguchi, I. Suemune, *Appl. Phys. Lett.*, **62** (1993) 1429.

[176] L. Koker, K. W. Kolasinski, *Phys. Chem. Chem. Phys.*, **2** (2000) 277.

[177] K. W. Kolasinski, *Curr. Opin. Solid State Mater. Sci.*, **9** (2005) 73.

[178] M. E. Dudley, K. W. Kolasinski, *Electrochem. Solid State Lett.*, **12** (2009) D22.

[179] K. W. Kolasinski, New approaches to the production of porous silicon by stain etching, in *Nanostructured Semiconductors: From Basic Research to Applications* (Eds: P. Granitzer, K. Rumpf), Pan Stanford Publishing, Singapore, 2012 (in press).

[180] M. I. J. Beale, J. D. Benjamin, M. J. Uren, N. G. Chew, A. G. Cullis, *J. Cryst. Growth*, **73** (1985) 622.

[181] R. T. Collins, M. A. Tischler, J. H. Stathis, *Appl. Phys. Lett.*, **61** (1992) 1649.

[182] S. Frohnhoff, M. Marso, M. G. Berger, M. Thönissen, H. Lüth, H. Münder, *J. Electrochem. Soc.*, **142** (1995) 615.

[183] R. Pearce, M. V. Twigg, Coal- and natural gas-based chemistry, in *Catalysis and Chemical Processes* (Eds: R. Pearce, W. R. Patterson), Blackie & Sons, Glasgow, 1981, p. 114.

[184] M. L. Gorbaty, F. J. Wright, R. K. Lyon, R. B. Long, R. H. Schlosberg, Z. Baset, R. Liotta, B. G. Silbernagel, D. R. Neskora, *Science*, **206** (1979) 1029.

[185] D. V. Chakarov, L. Österlund, B. Kasemo, *Langmuir*, **11** (1995) 1201.

[186] D. R. Olander, W. Siekhaus, R. Jones, J. A. Schwarz, *J. Chem. Phys.*, **57** (1972) 408.

[187] D. R. Olander, R. H. Jones, J. A. Schwarz, W. J. Siekhaus, *J. Chem. Phys.*, **57** (1972) 421.

[188] M. Balooch, D. R. Olander, *J. Chem. Phys.*, **63** (1975) 4772.

[189] H. Chang, A. J. Bard, *J. Am. Chem. Soc.*, **113** (1991) 5588.

[190] M. Esashi, T. Ono, *J. Phys. D: Appl. Phys.*, **38** (2005) R223.

[191] G. M. Wallraff, W. D. Hinsberg, *Chem. Rev.*, **99** (1999) 1801.

[192] Wikipedia, **2006**.

[193] J. N. Helbert (Ed.), *Handbook of VLSI Microlithography*, 2nd ed., Noyes Publications, Norwich, NY, 2001.

[194] R. K. Watts, Lithography, in *VLSI Technology* (Ed.: S. M. Sze), McGraw Hill, Boston, 1988.

[195] P. M. Mendes, J. A. Preece, *Curr. Opin. Colloids Interface Sci.*, **9** (2004) 236.

[196] A. Gutmann, Photolithography, in *Electronic Materials Chemistry* (Ed.: H. B. Pogge), Marcel Dekker, New York, 1996, pp. 199.

[197] H. Ito, *Adv. Polymer Sci.*, **172** (2005) 37.

[198] H. T. Soh, C. F. Quate, A. F. Morpurgo, C. M. Marcus, J. Kong, H. Dai, *Appl. Phys. Lett.*, **75** (1999) 627.

[199] K. Seeger, R. E. Palmer, *J. Phys. D: Appl. Phys.*, **32** (1999) L129.

[200] K. Q. Peng, M. L. Zhang, A. J. Lu, N. B. Wong, R. Q. Zhang, S. T. Lee, *Appl. Phys. Lett.*, **90** (2007) 163123.

[201] M.-L. Zhang, K.-Q. Peng, X. Fan, J.-S. Jie, R.-Q. Zhang, S.-T. Lee, N.-B. Wong, *J. Phys. Chem. C*, **112** (2008) 4444.

[202] M. J. Allen, V. C. Tung, R. B. Kaner, *Chem. Rev.*, **110** (2010) 132.

[203]   J. N. Coleman, M. Lotya, A. O'Neill, S. D. Bergin, P. J. King, U. Khan, K. Young, A. Gaucher, S. De, R. J. Smith, I. V. Shvets, S. K. Arora, G. Stanton, H.-Y. Kim, K. Lee, G. T. Kim, G. S. Duesberg, T. Hallam, J. J. Boland, J. J. Wang, J. F. Donegan, J. C. Grunlan, G. Moriarty, A. Shmeliov, R. J. Nicholls, J. M. Perkins, E. M. Grieveson, K. Theuwissen, D. W. McComb, P. D. Nellist, V. Nicolosi, *Science*, **331** (2011) 568.

[204]   A. Navrotsky, C. C. Ma, K. Lilova, N. Birkner, *Science*, **330** (2010) 199.

[205]   A. C. Balazs, T. Emrick, T. P. Russell, *Science*, **314** (2006) 1107.

[206]   E. M. Pouget, P. H. H. Bomans, J. A. C. M. Goos, P. M. Frederik, G. de With, N. A. J. M. Sommerdijk, *Science*, **323** (2009) 1455.

8

Laser and Non-Thermal Chemistry: Photon and Electron Stimulated Chemistry and Atom Manipulation

Most of the processes that we have discussed in previous chapters have dealt with thermal events, that is, with chemical and physical changes that occur at or close to equilibrium as the result of the addition of heat (random energy) to the system. Exceptions included the many spectroscopic techniques in Chapter 2 that involved photon or electron irradiation, as well as the changes that are engendered by the tip of a scanning probe microscope. These nonthermal or stimulated processes are the types of chemical and physical transformations that we now wish to examine in detail. In general, we focus on electronic excitations since these are the most likely to lead to reaction and desorption. Vibrational excitation can also lead to reaction or, for very weakly adsorbed systems, desorption in special cases, and a few of these are also discussed.

Electron and photon irradiation can bring about many of the same excitations. As we discussed in the section on electron energy loss spectroscopy, however, there are some differences [1–3] particularly for electron scattering at surfaces [4]. In addition to electric dipole transitions of the type that are commonly associated with photon excitation, electrons can also lead to excitations involving resonant capture and direct momentum transfer (impact scattering, momentum transfer from photons is only important in the MeV or greater energy range). Electrons have a spin of $1/2$ whereas photons have an angular momentum of 1; therefore, electrons are efficient at exciting singlet to triplet transitions (or other transitions in which the total spin of the electronic state changes) while photons are inefficient at causing electron spin changes. As the electron energy increases, the importance of electric dipole transitions increases and electrons can induce transitions in a manner that is increasingly like that of photons. Consequently, we can think of electron and photon stimulated processes in much the same way. Once an excitation is made, the excited state dynamics does not depend on how the system was excited. The evolution of the system is the same regardless of whether a laser or an electron beam was responsible for the initial excitation.

Tip induced processes also bear some resemblance to other nonthermal pathways. The electron beam emanating from an STM tip can lead to excitations just like any other electron beam. However, the close proximity of the tip to the species involved in the transformations can lead to complications. Hence, we start by considering photon excitation. Photon sources, as discussed in Chapter 2, are the most versatile and controllable of all excitation sources. We then add to the discussion those extras that are required to understand electron excitation. Finally, tip induced processes are discussed.

Surface Science: Foundations of Catalysis and Nanoscience, Third Edition. Kurt W. Kolasinski.
© 2012 John Wiley & Sons, Ltd. Published 2012 by John Wiley & Sons, Ltd.

8.1   Photon excitation of surfaces

Both lasers and lamps can be used to excite photochemistry. Lasers operate at a well-defined wavelength, though the bandwidth of the light becomes appreciable for ultrafast pulses in the femtosecond regime. Some atomic lamps also have narrow linewidths but often lamps emit a broad continuum form of radiation. While this means that narrow linewidths from continuum lamps come at the cost of vastly diminished radiant power, continuum lamps are also readily tuneable. An extremely useful property of lasers is that they generally have a well-defined collimated beam. Such a beam is easily transported over large distances and can either be focused tightly on a substrate or expanded to irradiate an entire sample.

8.1.1   Light absorption by condensed matter

As with any type of matter, the excitation caused by photon absorption depends on the energy (frequency, wavenumber, wavelength) of the light. When considering if and to what extent absorption occurs, we need to consider the band structure of the sample. Absorption cannot occur if the initial state and final state do not fulfil the resonance condition. Thus absorption is not allowed into a band gap regardless of whether the band gap is in the electron or phonon band structure. The strength of absorption depends on the matrix element that connects the initial and final states. A set of selection rules, as discussed in Chapter 2, arises from these matrix elements. The absorption and reflection coefficients depend not only on the chemical identity of the substrate and the wavelength of light, but also on the angle of incidence and the polarization of the light.

Excluding for the moment multiphoton absorption, the absorption of microwave and far infrared light leads to direct excitation of phonons. In a metal, microwave and infrared light can also produce electron-hole pairs near E_F. In semiconductors there is a range of IR photons (or even visible photons for a wide band gap semiconductor) that cannot be absorbed at all because their energy is less than the band gap energy. However, once the photon energy exceeds the band gap energy, photon absorption results. In a direct band gap semiconductor such as GaAs, the absorption coefficient exhibits a sudden increase as the photon energy exceeds the band gap. For an indirect band gap semiconductor such as Si, the absorption coefficient rises only slowly as the photon energy exceeds the indirect band gap energy. This is because photon absorption must be accompanied by phonon excitation in an indirect gap material, and this is a low probability event. The absorption coefficient in Si increases markedly once the photon energy exceeds the energy at which the first direct excitations are allowed. Figure 8.1 displays the absorption coefficient and normal incidence reflectivity for polished Si.

Electronic excitation by light absorption occurs whenever the photon energy is sufficient to excite electrons from filled states in the valence band to empty states in the conduction band. The minimum energy required for such a process is obviously dependent on the nature of the material. Electron-hole pair excitations represent a continuum of excitations for metals since there is no gap between the valence and conduction bands. Hence electron-hole pairs can be made with photons of any energy. For semiconductors, the minimum energy for electron-hole pair formation corresponds to the band gap energy. This ranges from the IR to the visible for most semiconductors, and even into the UV for GaN and diamond. Insulators have band gaps that extend from the UV to the VUV, with LiF exhibiting the largest band gap at 13.6 eV.

The statement above that the minimum energy for electron-hole pair formation is the band gap energy is a slight oversimplification. This is true for the production of free electron-hole pairs, that is, for the case in which the electron is widely separated from the hole. However, much like image charge states or Rydberg states, bound electron-hole pair states can be formed. A bound electron-hole pair state is known as an exciton. Excitons exist as states with a well-defined energy just below the conduction band of semiconductors and insulators. As long as there are empty states in the valence band, excitons as well as electrons excited to the conduction band have a finite lifetime and are unstable with respect to recombination. In other words, excited electrons tend to fall back down into the holes left behind in the valence band.

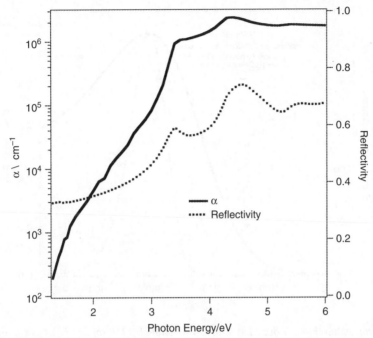

Figure 8.1   *The absorption coefficient and reflectivity for Si. Adapted from the data of Aspnes et al. [5, 6].*

As the photon energy exceeds the work function, photoemission occurs. The threshold for photoemission is a material dependent quantity as discussed previously and lies somewhere in the visible or UV. Since we have already encountered the concept of mean free path for the inelastic scattering of electrons, we should anticipate that a primary excitation of an electron could lead to secondary excitation of electrons. Secondary electrons are electrons that have been excited by collisions with excited electrons following primary excitation. Electrons can also collide with the lattice and exchange energy. The time scales for these various processes are most important in determining the characteristics of nonthermal chemistry. This is explored below.

8.1.2   Lattice heating

One way to think of a laser is simply as a source of energy, in which case it acts like a very elaborate heating element. If the irradiated substrate reacts instantaneously to convert the energy of the photons into lattice motions (phonons), then the result of irradiation is to increase the temperature of the substrate. For cw irradiation with photons with an energy below the work function, this is pretty much what we expect. The temperature rise depends on the intensity of the laser as well as the absorption coefficient, heat capacity and conductivity of the substrate.

As we shall see shortly, the electron–phonon coupling time in a metal is less than a picosecond. Therefore, even for lasers with pulse durations roughly several hundred ps or longer and particularly for IR, visible and not too deep UV light, we expect the response of the substrate to be instantaneous to a first approximation. This means that the vast majority of photon energy is immediately converted into heat, and a pulsed laser with a nanosecond or longer pulse can be used to engender extremely rapid temperature jumps in the irradiated area. Particularly for 100 ps to few ns pulses, the pulse duration is more rapid than thermal diffusion outside of the irradiated area. Therefore, pulses of these durations deposit all of their

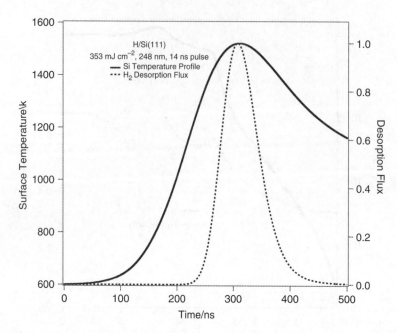

Figure 8.2 *The temperature jump induced in a silicon substrate by 353 mJ cm^{-2}, 14 ns excimer laser irradiation at 248 nm (solid line) from a Si(111) surface with an initial H atom coverage of 1 ML. The thermal desorption of H$_2$ from Si(100) induced by this laser pulse (dashed line).*

energy into a volume defined by the beam's cross sectional area and the penetration depth of the light, which is but a few nm for metals. In such cases the substrate's response is predominantly (though not entirely) thermal. Nonetheless, as shown in Fig. 8.2, the heating rates that can be attained by mJ pulse energy ns duration pulses can exceed 10^{10} K s^{-1}.

Because of the rapid response of the substrate, any process that averages over a time scale comparable to or longer than the pulse length is described by essentially equilibrium thermal kinetics. This is the basis of the technique known as laser induced thermal desorption (LITD) [7, 8]. LITD is a method that allows us to probe surface coverage under conditions where the desorption rate can be arbitrarily small or large. The extremely high heating rates facilitate excitation of a small spot that effectively removes all of the adsorbates in that spot. Under favourable conditions, the adsorbates are removed without any cracking (dissociation) and detection of the desorbed flux in a mass spectrometer can quantify not only how much is desorbed but also chemically identify the desorbates. Since the desorbed flux is proportional to the initial coverage of the species, the intensity of the LITD signal can be calibrated and used to determine the coverage of the adsorbate. Because it provides a snapshot of the contents of the surface, coverage determination by LITD is particularly useful for measuring equilibrium coverage under reaction conditions, and for measuring coverage decay under isothermal desorption conditions.

Figure 8.2 demonstrates the response of an adsorbed layer of H atoms on a Si(100) surface with an initial coverage of 1 ML. Figure 8.2 is obtained simply by using the known thermal desorption parameters for H$_2$ desorption from Si(100) and the temperature profile shown in Fig. 8.2. Note that desorption occurs over a very short time span, and that desorption occurs predominantly at the maximum temperature reached by the surface. Therefore, LITD provides a method to measure the kinetic energy of thermally desorbed molecules by the application of time-of-flight (TOF) techniques.

Time-of-flight determination of velocity is very simple in practice. Here we follow the notation of Auerbach [9]. The narrowness of the desorption laser pulse defines time zero, t_0, quite accurately. The path length L from the surface to point of detection can also be defined accurately. The velocity v is then simply the ratio of the two

$$v = L/t. \tag{8.1.1}$$

However, the molecules leave the surface with a distribution of velocities, $f(v)$. For instance, $f(v)$ may be described by a Maxwellian velocity distribution, i.e. a thermal distribution, which can then be converted into a kinetic energy distribution $f(E)$. The distribution of velocities leads to a distribution of flight times $g(t)$. Experimentally, the situation is complicated slightly by the fact that what we measure directly is a detector time response function $I(t)$. Under the assumption that t_0 is well defined (set equal to 0 and spread out negligibly over time), and that the detector response time is infinitely fast (or at least fast compared to the flight time), we can calculate the TOF function $g(t)$ for a flux-weighted velocity distribution $f(v) \, dv$. The flux-weighted velocity distribution is a measure of the number of particles that pass through a unit of area per unit time (e.g. molecules $m^{-2} \, s^{-1}$) with velocities in the range v to $v + dv$. The transformation from velocity space to time space is governed by the Jacobian derived by differentiating Eq. (8.1.1)

$$dv = -(L/t^2) \, dt \tag{8.1.2}$$

The TOF flux distribution is given by

$$g(t) \, dt = f(v) \, dv = -(L/t^2)f(v) \, dt. \tag{8.1.3}$$

Conventionally, the negative sign is disregarded. The *measured* TOF signal $I(t)$ depends on how the signal is measured and the nature of the detector. In the case at hand, we do not need to consider a convolution of the signal with either the detector response (since it is infinitely fast) or the source function (as we would have to if the particles passed through a chopper). Therefore, $g(t) = I(t)$. For details on how to carry out such a convolution see Ref [9].

A detector can be sensitive to particle density, particle flux or energy flux. In many cases laser ionization (resonance enhanced multiphoton ionization or REMPI) and laser induced fluorescence (LIF) roughly approximate density detectors in which all of the molecules in the focal volume are detected before any additional molecules can enter the volume, and before any of the molecules initially in the focal volume are able to leave. Slow molecules, small detection volumes, high detection efficiencies (independent of velocity), and short detection times (short laser pulses) all push a detector toward the limit of a density detector. A particle flux detector has a detection probability that depends on the amount of time the particles spend in the detection region. In this case, detection probability is inversely proportional to the velocity of the particles. Electron bombardment ionization, as commonly used in a quadrupole mass spectrometer, is an example of a flux detection system. A bolometer measures the energy released when particles hit a surface and is an example of an energy flux detector.

Equation (8.1.3) is given in terms of flux; therefore, a flux detector measures a signal $I_{\text{flux}}(t)$ directly proportional to this

$$I_{\text{flux}}(t) \propto (1/t^2)f(v) \tag{8.1.4}$$

To obtain the flux distribution associated with the signal measured by a density detector $I_{\text{den}}(t)$, we must divide $g(t)$ by $v = L/t$

$$I_{\text{den}}(t) \propto t \, I_{\text{flux}}(t) \propto (1/t)f(v). \tag{8.1.5}$$

To calculate the surface heating produced by irradiation, we start with the heat diffusion equation [10–13]

$$C_p \, \rho \frac{\delta T}{\delta t} = K \nabla^2 T + G \tag{8.1.6}$$

where C_p is the heat capacity, ρ the density, K the thermal conductivity, and G is the energy source term. With the z-axis chosen along the surface normal and the surface defined as the x, y-plane, G is defined by the spatiotemporal intensity profile of the laser and the absorbed irradiance

$$G(x, y, t) = (1 - R) I_0 \, \alpha \, \exp(-\alpha z) f(x, y) h(t) \tag{8.1.7}$$

where R is the reflectivity, I_0 is the peak irradiance (W cm^{-2}), α the absorption coefficient, $f(x, y)$ defines the spatial pattern, and $h(t)$ defines the pulse shape and duration. In principle, all of the materials parameters are temperature dependent, and this complicates the solution of Eq. (8.1.6) for high power irradiation. Analytic solutions of Eq. (8.1.6) can be found [12, 14] under the assumption of temperature independent materials constants. However, all of these properties are temperature dependent, and for large temperature jumps (high power irradiance), numerical calculations that treat all of the non-linearities must be carried out [15].

There are several limits in which solutions to Eq. (8.1.6) provide convenient analytical expressions [13]. In all three of these cases, the irradiance I is sufficiently low that we can neglect the temperature dependence of the materials constants. Consider first a uniformly irradiated surface at $z = 0$ of thickness H that is strongly coupled to a thermal bath on the back surface (at $z = H$). The temperature rise induced by cw irradiation is given by

$$\Delta T \big|_{z=0} = I(1 - R)H/K. \tag{8.1.8}$$

For a cw laser beam that is focussed to a Gaussian profile described by

$$f(x, y) = I_0 \exp\left(-r^2/w^2\right), \tag{8.1.9}$$

heat diffuses out from the irradiated spot to form a heated hemispherical region with a depth of order w [16]. The surface temperature rise at the centre of the beam is given by

$$\Delta T \big|_{x=y=z=0} = \frac{I_0(1 - R)\pi^{1/2} \, w}{2K}. \tag{8.1.10}$$

Pulsed laser irradiation leads to a temporally as well as spatially variant temperature profile, and the analytical solution of the distribution depends on the temporal shape of the laser beam [17]. The energy of the laser is deposited in a region (the thermal diffusion length d) with a characteristic depth of

$$d = (2\kappa t_p)^{1/2}, \tag{8.1.11}$$

where κ is the thermal diffusivity, and t_p is the pulse duration. For nanosecond lasers the solution to Eq. (8.1.6) reduces to a one-dimensional problem because the thermal diffusion length is large compared to the optical penetration depth, and much smaller than the radial extent of the beam unless the beam is extremely tightly focussed. The surface temperature temporal response is given by

$$T(t) = T_0 + \frac{I_0 (1 - R)}{K} \left(\frac{\kappa}{\pi}\right)^{1/2} \int_0^t \frac{h(t - \tau)}{\tau^{1/2}} \, d\tau. \tag{8.1.12}$$

The peak surface temperature at the centre of the beam is approximated by

$$\Delta T \big|_{x=y=z=0} = \frac{I_0(1 - R)t_p}{C_p \rho} \left(\frac{1}{\alpha^{-1} + (\kappa t_p)}\right). \tag{8.1.13}$$

The peak temperature for pulsed irradiation is generally much higher than for cw irradiation both because higher peak irradiances are achievable, and because the energy is deposited over a much reduced volume for pulsed irradiation. In these calculations we have assumed that the laser energy is transferred instantaneously to the lattice, and that one temperature is capable of characterizing the entire system. While this is true for ns and longer illumination, deviations occur for ultrafast pulse lengths, and we discuss this in the next section. If this is your first encounter with ultrafast excitation, you may want to skip to the summary in §8.1.4

Advanced Topic: Temporal evolution of electronic excitations

In Chapter 1, we were introduced to the Fermi-Dirac distribution, which describes a thermal distribution of electrons in a solid. Let us now consider how the electron distribution evolves after laser excitation. Near IR, visible or near UV photons are absorbed by the electrons in the substrate. We known that once equilibration has occurred, the electron distribution will be Fermi-Dirac, but how does it look immediately after excitation? First consider an infinitely narrow pulse incident on a metal. A metal simplifies the discussion since we do not need to consider the complications brought about by a band gap. To further simplify the system, assume the metal is initially at 0 K as in Fig. 8.3(a).

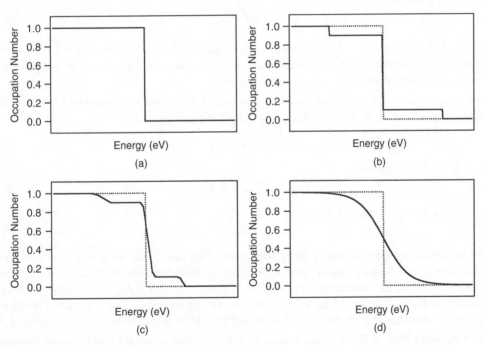

Figure 8.3 *A Fermi-Dirac distribution describes the electron distribution of a metal held at 0 K prior to irradiation with an infinitely narrow laser pulse. (b) The instantaneous absorption of photons with energy hν excites a portion of the distribution above the Fermi level. (c) Electron–electron collisions smooth out the initial excitation, though it is still non-Fermi-Dirac. (d) Eventually, the electrons equilibrate among themselves, and their distribution is described by a Fermi-Dirac distribution at a temperature significantly above the initial temperature.*

The simplest approximation of the electrons in a metal is to consider them to be a gas of noninteracting fermions. This is known as a Fermi gas. However, to understand how electrons exchange energy with themselves and the lattice, we need to include interactions between particles. Landau [8] developed a theory that accounts for electron–electron interactions that is known as Fermi liquid theory.

To begin with, we need to answer the question: what is responsible for the long mean free path for conduction band electrons? As can be inferred from Fig. 2.13, electrons with energies near E_F have extremely long mean free paths. Even though electrons are spaced by less than 1 nm (comparable to and less than the spacing of the atoms in the lattice), at room temperature the mean free path is on the order of 10^3 nm and increases rapidly with decreasing temperature. The Pauli exclusion principle, in combination with the conservation of momentum and energy, is responsible for such a large value of the mean free path because it greatly restricts the number of allowed orbitals into which colliding electrons can scatter. The result is a large reduction in the effective cross section for electron–electron scattering σ by a factor of $(k_B T / E_F)^2$,

$$\sigma \approx (k_B T / E_F)^2 \, \sigma_0, \tag{8.1.14}$$

where $\sigma_0 \approx 10^{-19}$ m^2 is the cross section for the scattering of free electrons. Recall that E_F / k_B is roughly 5×10^4 K; thus at room temperature in a typical metal, $k_B T / E_F$ is on the order of 10^{-2}. Therefore, $\sigma \sim 10^{-4}$ and $\sigma_0 \sim 10^{-23}$ m^2. The mean free path λ is related to the effective scattering cross section and the electron concentration n_e by

$$\lambda = 1 \bigg/ \sqrt{2} n_e \sigma . \tag{8.1.15}$$

Since $n_e \approx 5 \times 10^{28}$ m^{-3}, $\lambda \approx 1 \times 10^{-6}$ m at room temperature, and it increases rapidly with decreasing temperature. We expect the mean free path of excited electrons, ones with energy in excess of E_F, to decrease rapidly with increasing energy.

In some cases it is more convenient to think in terms of the mean time between electron–electron collisions for electrons at the Fermi energy $\tau_{ee,0}$, which is related to λ by

$$\lambda = v_F \tau_{ee,\,0}, \tag{8.1.16}$$

where v_F is the Fermi speed

$$v_F = \left(2 E_F / m_e \right)^{1/2} . \tag{8.1.17}$$

From these equations we can see that

$$\tau_{ee,\,0} \propto T^{-2}. \tag{8.1.18}$$

In order to consider how excitations in the system evolve after irradiation, we need to know how photon absorption changes the initial distribution, how the electrons equilibrate among themselves, and how they equilibrate with the lattice. Instantaneous absorption of photons of energy $E_{ph} = h\nu$ excites electrons that were initially at E_F to an energy of $E_F + h\nu$. Electrons with an energy $E_F - h\nu$ are excited to E_F, and all electrons with an energy in between are excited accordingly to an intermediate energy as in Fig. 8.3(b).

Based on our discussion of Fermi liquid theory, we can now predict that the collision time for electrons excited above E_F is given by

$$\tau_{ee} = \tau_{ee,\,0} \left(E_F / \delta E \right)^2 \tag{8.1.19}$$

where $\tau_{ee,\,0}$ is a few fs, for instance, 5 fs for Au, and $\delta E = E - E_F$ is the electron energy above the Fermi energy. The electron–electron collision time is on the order of a few fs and decreases for increasingly large excitations above E_F. This result also gives us a conception of how the excited electron distribution relaxes. The highest energy electrons have the shortest lifetimes. The greatest density of electrons is that of the unexcited electrons near E_F, and the excited state density is generally much smaller than this. Hence, we can think of the electrons relaxing in the form of electron cascades [18], in which an excited electron at energy $E_F + \delta E$ collides with an electron near E_F. When properly averaged over the Fermi–Dirac distribution of thermal excitations, the collision leads to two electrons with a mean energy $E_F + \frac{1}{3}\delta E$. They go on to collide with two more electrons near E_F to form four electrons at $E_F + \frac{1}{9}\delta E$, etc. Eventually, the density of excited state electrons is such that they also collide with other excited electrons, and the distribution thermalizes to that of a Fermi–Dirac distribution at an elevated temperature.

Electrons scatter from the lattice by the absorption and emission of phonons. Only longitudinal phonons interact with electrons, and since longitudinal acoustic (LA) phonons have lower energies than longitudinal optical (LO) phonons, LA phonons tend to dominate electron–phonon coupling. The temperature dependence of electron–phonon scattering is complicated by the quantum nature of the oscillator. The number of phonons increases with increasing temperature; therefore, electron–phonon scattering increases with increasing temperature. At temperatures above the Debye temperature $T \gg \theta_D = \hbar\omega_D/k_B$, the number of thermal phonons is roughly proportional to the temperature, and the electron–phonon scattering relaxation time τ_{ph} is inversely proportional to temperature,

$$\tau_{ph} \propto T^{-1}. \tag{8.1.20}$$

For $T \ll \theta_D$, the temperature dependence is stronger

$$\tau_{ph} \propto T^{-3}. \tag{8.1.21}$$

To estimate τ_{ph} consider the following. The electron mean free path in Au is $\sim$300 Å. The Fermi speed is $\sim 10^6\,\mathrm{m\,s^{-1}}$. Consequently, there is a spacing of about 30 fs between collisions or roughly 30 electron–phonon collisions per 1 ps. However, because of the large mass difference between electrons and lattice atoms, the energy transfer from electrons to phonons is rather inefficient, and it takes many collisions to transfer a substantial amount of energy. Because of the relative rates of electron–electron energy transfer and electron–phonon energy transfer, photoexcited electrons first equilibrate among themselves, then on a much longer timescale, transfer their energy to the lattice.

Now consider that the heat capacity of N electrons in a free electron gas is given by

$$C_e = \frac{1}{2}\pi^2 N k_B (T/T_F). \tag{8.1.22}$$

The high temperature limit of the heat capacity of $3N$ oscillators in a lattice is approximately

$$C_{ph} = 3N k_B. \tag{8.1.23}$$

The temperature rise engendered by the deposition of energy q is directly proportional to the heat capacity,

$$\Delta T = q/C, \tag{8.1.24}$$

and the ratio of electron and phonon heat capacities is

$$C_{ph}/C_e \approx T_F / T. \tag{8.1.25}$$

The value of T_F is $\sim 10^4$ K; hence, the heat capacity associated with the phonons is generally orders of magnitude higher than the heat capacity of the electrons. The same amount of energy deposited into each system (if the energy is deposited exclusively in one or the other) leads to a much higher temperature jump for the electrons than it does for the phonons.

So how do the energy distributions of electrons and phonons respond to pulsed excitation? To answer this question we need to consider the time scale of the excitation in comparison to the processes that we have outlined above. Here again, recall that we are specifically discussing the interaction of near IR, visible and UV light with a metal. Some modification must be made if we consider direct phonon absorption further in the IR, or if we need to consider how electron relaxation is affected by the presence of a band gap in semiconductors and insulators. In addition, the electron–lattice interaction in insulators (polar solids) is much stronger because the ion–electron interaction is stronger than the neutral–electron interaction encountered in covalent solids and metals.

Slow pulse excitation (>100 ps)

While it may seem rapid to us, because of the ultrafast nature of electron–electron and electron–phonon relaxation, pulses much longer than 100 ps are considered slow compared to the response of the substrate. Therefore, our first-order expectation for long pulse lasers is that the electrons relax so rapidly with the lattice that such pulses act as a heat source, and that the electrons and the lattice are in equilibrium at all times. As was done for the curves in Fig. 8.2, the substrate temperature can be calculated by considering how much energy is deposited, the penetration depth of the light, and the heat capacity of the substrate. For the most accurate calculations, it is important to consider the thermal conductivity of the substrate; however, as the pulse length approaches the ns regime, the pulse length becomes competitive with thermal diffusion.

Any process that averages over the length of these long pulses responds in a thermal fashion. However excitation up to the photon energy does occur, and for the briefest of moments in a metal, excited carriers with energy as high as $E_F + h\nu$ are created consistent with the electron cascade model outlined above. Therefore, if a process, such as electronic excitation of an adsorbate, can respond sufficiently rapidly, it is able to take advantage of these excited carriers and exhibit a nonthermal response. In a semiconductor, relaxation of electrons excited into the conduction band to the conduction band minimum (or a local minimum in the band structure) is an ultrafast process. Relaxation across the band gap back into the valence band is a much slower process governed by the kinetics of radiative and nonradiative process that are often in the nanosecond (for a direct band gap) or longer (for indirect) range. Similarly, hot holes relax to the top of the valence band and wait there for recombination on a similar time scale. The lifetimes of excited carriers at band edges in a semiconductor are, therefore, much longer than the excited carriers in metals and they should correspondingly be much more available for inducing photochemistry.

Ultrafast pulse excitation (1–10 ps)

In this regime, the electrons are thermalizing among themselves (subject to the caveat about electron cascades) and maintaining a Fermi-Dirac distribution throughout the pulse. I say thermalize because their distribution is described by a temperature, but they are not in equilibrium with the lattice. The lattice takes several hundred femtoseconds to begin to heat up and will not approach the same temperature as the electrons until several picoseconds have elapsed. In this regime, the laser pulse is over before the electrons have fully equilibrated with the lattice. Furthermore, because of the extreme difference in heat capacities and because the energy is initially deposited exclusively in the electrons, the temperature of the electrons soars far above the initial temperature and then relaxes back down to the same temperature as the lattice – a final temperature that is above the initial temperature by an amount determined by the pulse energy and the heat capacity of the substrate.

Even faster (<1 ps)

For pulses on the order of 100 fs, the electron cascading process is not complete before the pulse is over. The energy distribution can be pushed out of the thermal limit and no longer conforms completely to a Fermi-Dirac distribution. The exact time that it takes to achieve a Fermi-Dirac distribution depends on the material and the laser characteristics but for a typical metal and pulses from a garden variety Ti:sapphire laser (800 nm, 100 fs, several mJ cm^{-2}), it takes approximately 300–700 fs to be established. The relaxation rate depends not only on the materials properties (electron–electron and electron–phonon coupling times), but also the excitation density produced by the laser and the temperature.

Clearly, in the ultrafast regime (<10 ps) it is no longer adequate to describe the system with a single temperature. At times comparable to the electron–electron coupling time, the electrons are not described by a Fermi-Dirac distribution, and the lattice has yet to react to the excitation. Once the electrons have thermalized, the lattice still has not caught up with them. Therefore, we are led to a two-temperature description of the substrate to describe the temporal regime after electron thermalization but before equilibration. If an adsorbate is present, coupled to the substrate by an adsorption bond, we need to introduce a third temperature to describe the adsorbate since it does not couple to the excitation in the same way as the substrate atoms do.

The two-temperature model (TTM) [19] assumes that the electron and phonon subsystems are each maintained in a thermalized state by Coulombic interactions, in the case of the electrons, and anharmonic interactions, in the case of the phonons. The time evolution of the temperature of the electrons and lattice is given by a set of two coupled differential equations

$$C_e \frac{dT_e}{dt} = \frac{\delta}{\delta z}\left(\kappa_e \frac{\delta}{\delta z} T_e\right) - H\left(T_e, T_{ph}\right) + P(t) \tag{8.1.26}$$

$$C_{ph} \frac{dT_{ph}}{dt} = H\left(T_e, T_{ph}\right). \tag{8.1.27}$$

where $P(t)$ is the absorbed laser power density (W m^{-3}). H is the energy transfer rate in W m^{-3} between electrons and lattice. The form of Eqs (8.1.26 and 8.1.27) assumes that the energy is deposited directly and exclusively into the electrons. The term $\frac{\delta}{\delta z}\left(\kappa_e \frac{\delta}{\delta z} T_e\right)$ describes the transport of heat resulting from electron diffusion out of the excitation volume. The characteristic times for diffusive and ballistic transport in Au are approximately 30 fs and 20 fs, respectively [20]; hence, we take the electronic temperature to be uniform in the irradiated zone. However, since lattice heat transport occurs on the nanosecond time scale, it can be neglected. Frequently, a simplified version of the expression in Eq. (8.1.22) is used

$$C_e = \gamma T_e \tag{8.1.28}$$

to account for the dependence of the electronic heat capacity on temperature. However, this may be an oversimplification for intensities that lead to large electron temperature jumps [21].

The energy transfer rate has been calculated by Kaganov et al. [22] and has a particularly simple form. The total transfer of energy per unit volume per second from electrons to phonons is

$$H\left(T_e, T_{ph}\right) = f(T_e) - f(T_{ph}). \tag{8.1.29}$$

where

$$f(T) = 4g_\infty \left(\frac{T}{\theta_D}\right)^5 \int_0^{\theta_D} \frac{x^4}{e^x - 1} \, dx, \tag{8.1.30}$$

θ_D is the Debye temperature, and g_∞ is the electron–phonon coupling constant. These equations show that the response of the electronic and lattice temperatures are dependent not only on the electron–phonon coupling constant and the respective heat capacities, but also the Debye temperature and the initial temperature of the system. In particular, when the lattice temperature $T_{ph} \ll \theta_D$, df/dt varies as T^4 but when $T_{ph} > \theta_D$, df/dt approaches a constant value of g_∞.

For t_p short compared to the effective electron–phonon relaxation time – in the regime of 100 fs pulses – the peak electron temperature is easily estimated from the initial temperature T_0, the deposited laser energy density U_1 and the constant γ from Eq. (8.1.28) according to [20]

$$T_{e,\,max} = (T_0^2 + 2U_1/\gamma)^{1/2}. \tag{8.1.31}$$

The coupled differential equations that describe the two-temperature model must be solved numerically to obtain time dependent descriptions of T_e and T_{ph}. The results of such a calculation are shown in Fig. 8.4 for the case of two 400 fs duration pulses delayed with respect to each other by a time τ. The TTM provides a good approximation to the behaviour of the distribution of electron and phonon energies. Discrepancies are expected at short times much less than the pulse duration before the electrons thermalize. Discrepancies can also arise from the laser fluence and temperature dependence of the materials properties that go into the coupled differential equations. For a more complete description of these distributions, not only must the dependencies of the parameters be considered, but also diffusion of hot carriers must be accounted for [23].

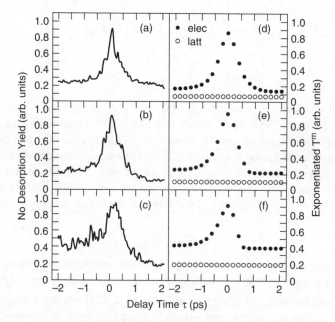

Figure 8.4 *Two-pulse correlation data as a function of pulse delay τ. (a)–(c) Experimental desorption yields of NO from Pd(111) for different pulse-pair fluences: (a) 1.7 and 1.8 mJ cm^{-2}; (b) 1.6 and 2.2 mJ cm^{-2}; (c) 1.3 and 3.0 mJ cm^{-2}. (d)–(f) Calculated values of the peak electronic (solid symbols) and lattice temperature (open symbols, multiplied by 5000) for the conditions corresponding to (a)–(c). (g)–(i) The electron and lattice temperatures resulting from irradiation of Pd with two laser pulses I(t) with fluences of 1.3 and 3.0 mJ cm^{-2}, pulse duration 400 fs and wavelength 620 nm. Note the electronic temperature response $T_e(t)$ depends on the delay τ between the pulses whereas the lattice temperature response $T_{ph}(t)$ is insensitive to this. The dashed line corresponds to the laser intensity profile. Reproduced with permission from F. Budde, T. F. Heinz, M. M. T. Loy, J. A. Misewich, F. de Rougemont, H. Zacharias, Phys. Rev. Lett., 66 (1991) 3024. © 1991 by the American Physical Society.*

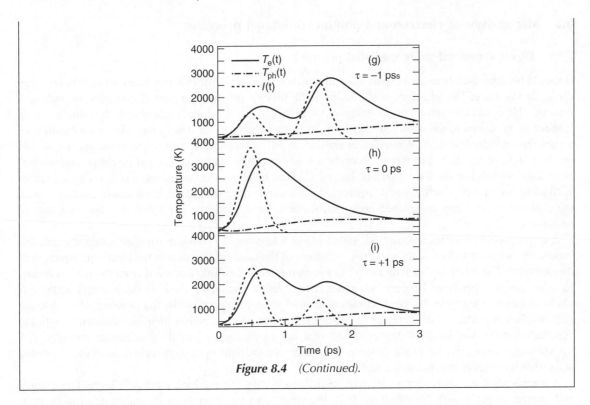

Figure 8.4 (Continued).

8.1.3 Summary of laser excitations

The previous section may contain more information than some want to know, so it is important here to summarize the most critical concepts encountered so far in this chapter.

- Lasers (and other light sources) can lead to resonant and nonresonant excitation of surfaces and the adsorbates bound to them. The type of excitation that occurs (vibrational, electronic, etc) depends on the energy of the photon involved. The optical response of the solid depends on whether or not an electronic band gap exists, in other words, metals act very differently than semiconductors and insulators.
- When a photon is absorbed by the electrons of a solid, hot carriers (excited electrons and holes) are created. These hot carriers relax among themselves and with the lattice extremely rapidly.
- Any process that averages over a timescale of ~ 1 ns or longer is essentially going to experience that the response of the substrate is thermal. A laser can be used to produce extremely fast heating rates.
- If we look with much higher temporal resolution, we see that the electrons and phonon react to photon absorption on different time scales. In the first few tens of femtoseconds after photon absorption, the electron distribution is nonthermal, and the lattice has not yet begun to feel the influence of these excited electrons. The excited electrons thermalize with themselves and establish an electronic temperature that is much higher than the lattice temperature. The lattice acquires energy from the hot electrons on a longer timescale, and eventually after several picoseconds the lattice equilibrates with the electrons. This separation of time scales is the basis of the so-called two-temperature model, which describes the temporal response of the electrons and the lattice.

8.2   Mechanisms of electron and photon stimulated processes

8.2.1   Direct versus substrate mediated processes

It should be clear that there are several ways in which a laser beam (or electron beam for that matter) can excite an adsorbate. The adsorbate could absorb light directly and undergo photodissociation or photodesorption. This is what we term a direct process. Alternatively, the substrate might absorb the radiation, then transfer an excitation to the adsorbate that leads to further chemistry or desorption. This is an example of a substrate mediated process. While the discussion in §8.1 concentrated on excited electrons, we should not lose sight of the fact that photon absorption leads to the formation of excited electrons *and* excited holes (also called hot electrons and hot holes). Excited holes are just as capable of causing adsorbate excitations as excited electrons and, particularly on semiconductors where the excited electrons roam the conduction band and the excited holes patrol the valence band, the hot holes may be the agents of excitation.

A direct process requires resonant excitation of the adsorbate. To consider whether resonant excitation is possible, we need to look at the electronic structure of the adsorbate as well as the electronic structure of the substrate. The electronic structure of a physisorbed molecule is little perturbed from the free molecule, and the energy separations between adsorbate energy levels should be close to those found in the gas phase molecule. We expect that the lifetimes of excited states are decreased by the presence of a substrate, particularly a metallic or semiconducting substrate that has energy levels which are degenerate with the adsorbate levels. Any levels of the adsorbate that fall in an energy gap do not interact strongly with the substrate. Therefore, we might consider the absorption of light by a physisorbed molecule to behave somewhat like that of the gas phase molecule.

A chemisorbed molecule shares electrons with the substrate. It must have electronic levels that overlap and interact strongly with the substrate. It is, therefore, coupled strongly to the substrate. The energies and lifetimes of the levels are perturbed strongly from the gas phase values. In fact, the orbitals that bind the molecule to the substrate are rehybridized into new sets of orbitals. Therefore, it may be impossible to identify these as exclusively adsorbate orbitals, since they also take on the character of the substrate orbitals involved in the chemisorption bond. All electronic states interact with the substrate as long as they do not fall in a band gap and should thus have their lifetimes reduced. Reduced lifetimes lead to broadened absorption peaks as well as reduced oscillator strengths.

How to tell substrate mediated from direct processes is not always easy. A direct process requires resonant behaviour. However, the resonance may be very broad (>1 eV) and strongly shifted from any feature associated with the gas phase molecule. A substrate mediated process should have a wavelength dependence that follows the absorption coefficient of the solid. The dependence of the reaction cross section on the polarization and angle of incidence of the light used to excite photochemistry can also be used as a probe. Richter et al. [24] have shown that whether any useful information can be obtained in this way depends on the orientation of the transition dipole moment responsible for excitation.

For a process that involves direct linear absorption by an adsorbate with coverage θ, the reaction rate should follow the form

$$\frac{d\theta}{dt} \propto \theta \, |\boldsymbol{\mu} \cdot \boldsymbol{E}|^2 , \tag{8.2.1}$$

where $\boldsymbol{\mu}$ is the transition dipole moment for the excitation and $\boldsymbol{E}$ is the electric field at the adsorption complex. Both $\boldsymbol{\mu}$ and $\boldsymbol{E}$ are vectors. The transition dipole moment can be decomposed into three Cartesian components directed along the x, y and z axes. The direction of the electric field vector is fixed by the angle of incidence and whether the laser is s or p polarized (assuming linear polarization).

For a substrate mediated process, the reaction rate is given by

$$\frac{d\theta}{dt} \propto \theta \sigma_r F_{ex} \tag{8.2.2}$$

where σ_r is the cross section for the reaction and F_{ex} is the flux of excitations (hot electrons or hot holes) incident on the adsorbate. F_{ex} is determined by the amount of light *absorbed* by the substrate as well as the *transport* of the excited species to the adsorption complex. The amount of light adsorbed depends on the intensity of the optical field, which depends on the angle of incidence ϑ measured from the surface normal, as well as the substrate absorbance $1 - R_i$, where R_i is the reflectivity for i-polarized light as determined by the angle of incidence dependent complex Fresnel reflection coefficients. Rather than being directly proportional to the incident irradiance of the laser (power per unit area in the beam), F_{ex} is proportional to $\cos \vartheta$ times the irradiance. Note that in both cases we expect the reaction rate to be linear in laser (or electron beam) fluence F,

$$\frac{d\theta}{dt} \propto F^n, \tag{8.2.3}$$

where $n = 1$ because a single photon leads to a single reactive event.

Richter et al. have shown that if the transition dipole moment is in the plane of the surface, a direct adsorbate excitation cannot be distinguished from a substrate mediated process on the basis of their angle of incidence dependence for s vs p polarized light. Unambiguous identification of a direct excitation process dominated by μ_z, that is, for a transition dipole moment oriented along the surface normal, can be made on the basis of such data.

8.2.2 Gas-phase photochemistry

Our explanation of photochemistry is couched within the language of the PESs, or stationary states, that are derived from the solutions to the time independent Schrödinger equation for the molecule. The time dependent behaviour is described by the evolution of a wavepacket that is deposited on the excited state as a result of the irradiation. In the gas phase, electronic excitation is followed by radiative decay (the wavepacket falling back to the ground electronic state accompanied by emission of a photon), nonradiative decay (return to the ground state without photon emission) or photochemistry. Photochemistry can be thought of in very simple terms as being composed of the following steps: (i) excitation, (ii) excited state propagation, and (iii) competition between radiative decay, photochemistry and nonradiative quenching. Note that the excited state may be neutral or ionic (formed by photoionization accompanied by electronic excitation) and, insofar as gas phase molecules are isolated, once a cationic state is formed it cannot be neutralized and there is no essential difference in the propagation dynamics because of the charge of the excited state.

A Franck-Condon transition takes the molecule from the ground electronic state to an excited state. Electronic excitation is often accompanied by vibrational excitation, the extent of which is based on the similarity of the ground and electronic state potential energy curves. Next the excited state wavepacket propagates on the excited state potential. The excited PES has one of three characters as shown in Fig. 8.5: bound, predissociative or dissociative.

Excitation to a bound state is followed by fluorescence back to the ground state unless nonradiative processes, usually involving collisions or spin-orbit coupling, transfer the excitation by intersystem crossing to a PES of different spin multiplicity or by internal conversion to an excited vibrational state of the ground PES. Typical radiative lifetimes for allowed electronic transitions are of the order $10-100$ ns. Excitation to a bound state can only lead to dissociation if the excitation occurs to an energy above the dissociation limit of

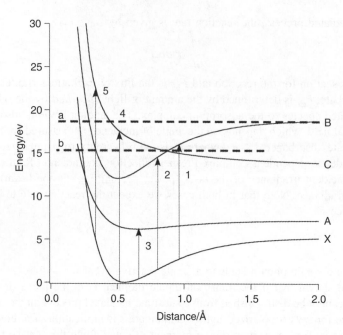

Figure 8.5 *Electronic excited states characterized as either bound, predissociative or dissociative (repulsive). Both the X and A states are bound. The B state curve is crossed by dissociative C state curve. Curve crossing from B to C leads to predissociation. The threshold for normal dissociation of the B state is labelled (a), whereas the predissociation threshold (b) occurs at the curve crossing between the B and C states. (1) A transition to the B state for which curve crossing and predissociation are allowed. (2) X to B bound-bound transition. (3) X to A bound-bound transition. (4) X to C bound-free transition, which leads to dissociation as long as curve crossing to the B state does not occur. (5) X to C bound-free transition, which always leads to dissociation.*

the state. Nonradiative quenching processes affect the quantum yield of fluorescence or photodissociation. The quantum yield of process j is defined as the fraction of molecules that undergo process j subsequent to excitation divided by the total number of excitations N_{ex}, which must be equal to the sum over all of the k different processes that are allowed. Equivalently, the quantum yield can be defined in terms of the rates of the processes involved

$$\phi_j = \frac{N_j}{N_{ex}} = \frac{N_j}{\sum_k N_k} = \frac{r_j}{\sum_k r_k}. \tag{8.2.4}$$

At low energy, excitation to a predissociative state is equivalent to excitation to a bound state. However, at high energy a predissociative state is crossed by a second state that is dissociative. For excitation at an energy above the crossing point, propagation of the wavepacket on the predissociative PES leads to a competitive process in which the wavepacket can either relax to the bottom of the well or else curve cross and hop onto the dissociative PES. Dissociation by curve crossing occurs even though the initial excitation did not place the wavepacket above the dissociative asymptote of the excited state, hence the name predissociation.

A dissociative state is purely repulsive and supports no bound vibrational levels. Excitation to such a PES leads to acceleration of the atoms involved in the bond away from one another at a rate that depends on the slope of the PES. The repulsive force experienced by the atoms is given by the gradient of the potential energy function, which in one dimension is given by

$$F = -\frac{dV}{dr}.$$

(8.2.5)

Newton's second law relates the acceleration to the force

$$F = ma = m\frac{d^2r}{dt^2},$$

(8.2.6)

from which the kinetic energy increase caused by the repulsive force can be calculated. The kinetic energy increase depends not only on the force but, just as importantly, on the length of time τ that the force acts. Indeed, it is the square of the impulse, the product of F and τ, that determines the kinetic energy change according to

$$\Delta E_K = F^2\tau^2/2m.$$

(8.2.7)

Note that because the energy gain depends on the square of the impulse, the sign of the force is unimportant in determining energy transfer.

8.2.3   Gas phase electron stimulated chemistry

Once the excited state has been attained, how the excitation occurred does not influence the propagation dynamics of the wavepacket. Therefore, it is inconsequential whether photon or electron excitation is responsible for the initial excitation and all of the discussion above is equally applicable to an electric dipole transition induced by electron irradiation.

Nonetheless, electron irradiation does introduce some differences compared to photon irradiation. As mentioned in Chapter 2, electrons can act as waves, particles or uniquely as electrons. The wavelike character of electrons allows them to induce electric dipole transitions just like photons, including transitions accompanied by ionization. Because they are massive particles, they can also facilitate direct momentum transfer much more efficiently than photons. Finally as electrons, they can be captured into resonant states. The captured state might have a long lifetime and correspond to a stable anion. Or it may be a short-lived state known as Feshbach resonances in which the electron's angular momentum forms a centrifugal barrier that temporarily traps the electron close to the molecule. The capture probability depends on the molecule and the electron energy. Electron capture cross section tends to decrease with increasing energy; however, superimposed on top of this structure are resonant features that correspond to bound electronic states. The one other factor that differentiates electron irradiation from photon irradiation is that because they are spin $1/2$ particles (Fermions), electrons can readily induce transitions between states of different spin multiplicity such as singlet to triplet transitions.

8.2.4   MGR and Antoniewicz models of DIET

How does the presence of a surface modify excited state dynamics? The short answer is: quite dramatically. Adsorbate levels that interact with the surface are shifted and broadened. The electronic states of the substrate open up numerous pathways for the quenching of excited states. The extent of quenching and perturbation of excited state dynamics from the gas phase limit depends on whether the adsorbate is physisorbed or chemisorbed and whether or not the excited electronic state interacts strongly with the substrate or whether it falls in a band gap and acts essentially independently of the substrate. The first and most ubiquitous change brought about by the surface is the opening of numerous quenching channels. The second and more chemically dependent factor is that if an adsorbate is strongly bound at the surface, the electronic states involved in bonding no longer belong exclusively to the adsorbate. Rather, they belong to an adsorption complex – a larger moiety that is composed of the adsorbate as well as the orbitals of the surface atoms involved in the chemisorption bond.

Physisorbed species, especially those on insulator surfaces or separated from metal or semiconductor surfaces with spacer layers such as a physisorbed layer of noble gas, are slightly perturbed by the substrate. The excited state lifetimes in such species are reduced but little from their gas phase values. Chemisorbed species, on the other hand, are significantly perturbed. The excited state lifetime of chemisorbed adsorbates on metal surface are expected to be on the order of 1 fs. Similar values are expected for chemisorbates on semiconductor surface so long as the electronic state in question does not fall within a band gap. This is a reduction of 7 or 8 orders of magnitude compared to typical radiative lifetimes in the gas phase. Quenching this efficient leads to profound changes in the probability and dynamics of stimulated processes at surfaces. The simplest surface chemical reaction is desorption, and as a generic term for nonthermal chemical change induced by electronic excitation, we discuss desorption induced by electronic transitions (DIET), or more generally dynamics induced by electronic transitions.

A general scheme for stimulated chemistry, which we refer to as DIET for simplification, is described in Fig. 8.6. The first major change compared to gas phase chemistry is that both direct and substrate mediated processes are possible. Second, quenching is of much greater significance. Whereas in the gas phase excited state propagation is predominantly responsible for photochemistry, and quenching is generally the end of photochemistry, at surfaces quenching is so rapid that it is very uncommon that propagation on the excited state potential leads to desorption or reaction. This means that cross sections for DIET of chemisorbed species on metal surfaces (of order 10^{-20} cm^{-2} plus or minus several orders) are significantly reduced compared to gas phase cross sections (of order 10^{-16} cm^{-2} or so). In addition, usually the system has to propagate to its final state along the ground state potential, or at least a lower lying electronic state than the one initially excited.

The principles embodied in this general scheme have been more specifically summarized in the two fundamental mechanisms of the DIET. The first model, depicted in Fig. 8.7(a) was independently proposed by Menzel and Gomer [25] and Redhead [26]. This is known as the Menzel-Gomer-Redhead or MGR model. In this model, excitation to a repulsive state leads to the desorption of either neutral or ionic species.

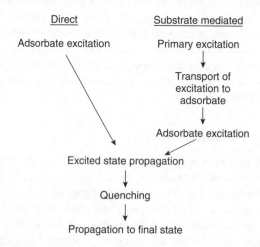

Figure 8.6 *A general scheme for direct and substrate mediated stimulated chemistry and dynamics induced by electronic transitions (DIET).*

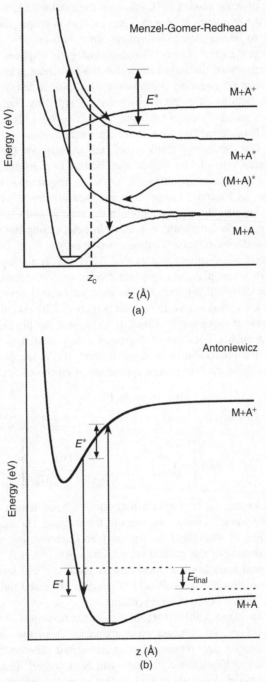

Figure 8.7 *Diagrams of the (a) Menzel-Gomer-Redhead and (b) Antoniewicz models of dynamics induced by electronic transitions (DIET).*

An alternative model is that of Antoniewicz [27], in which the excitation is to an attractive state, as shown in Fig. 8.7(b). In both cases the dynamics follow the form excitation-propagation-quenching-desorption and we expect the cross section to be very low. *Desorption can only occur if sufficient energy E* is acquired during the propagation step in the excited state*. Sufficient energy is acquired if the kinetic energy transfer is greater than the binding energy of the state into which the adsorbate is quenched. This energy transfer is accomplished if excited state propagation continues until a critical distance z_c is passed. A theoretical approach to calculate the probability of a reactive event is provided by Gadzuk's "jumping wavepacket" method [28, 29]. For certain materials other desorption mechanisms [30] are possible, for instance, the Knotek-Feibelman model [31] of desorption from transition metal oxides involving core-hole Auger decay.

Equation (8.2.7) tells us that the energy transfer depends strongly on the magnitude but not the sign of the slope of the excited state potential as well as the lifetime of the state. Therefore it is irrelevant on one level if the excited state is repulsive or attractive. What is important is the degree to which the state is attractive or repulsive and its lifetime. *Longer lifetimes and greater slopes lead to greater desorption probabilities*. We shall also see that differences between the ground and excited states regarding bonding configuration and equilibrium bond length are important in determining the internal state distribution, as these can lead to rotational and vibrational excitation, respectively.

Figure 8.7 describes several other salient features of DIET desorption dynamics. Curve crossings, as shown explicitly in Fig. 8.7(a), can play an important role in the desorption dynamics. Curve crossings can change the nature of the desorbed product, for instance, an excited neutral species can cross onto an excited ionic potential (or vice versa) before relaxation. Figure 8.7(b) also shows that the final energy is not equal to E^* because some of this energy is used to overcome the binding energy that remains when the excited state species is quenched back onto the ground state potential.

Within the MGR model, the photodesorption cross section, σ_d, is the product of the photoexcitation cross section, σ_{ex}, and the probability P to escape quenching until the critical distance, z_c, is achieved

$$\sigma_d = \sigma_{ex}P. \tag{8.2.8}$$

P is given by [32]

$$P = \exp\left\{ -\left(\frac{m}{2}\right)^{1/2} \int_{z_0}^{z_c} \frac{R(z)\,dz}{\sqrt{V(z_0) - V(z_0)}} \right\} \tag{8.2.9}$$

where R is the rate of quenching, z_0, is the adsorbate-surface bond length (z points along the surface normal), V is the excited state potential and m the mass of the particle. The mass dependence in Eq. (8.2.9) leads to a strong isotope effect in stimulated desorption (see Exercises 8.8 and 8.9). Furthermore, since the rate of quenching is the inverse of the excited state lifetime $R(z) = \tau(z)^{-1}$, the desorption probability increases greatly as the excited state lifetime increases. Because excited state lifetimes, particularly for chemisorbates on metals, are extremely short, $P \ll 1$ is generally true and the cross section for stimulated desorption is much less than the probability for excitation.

If the excited electronic state is ionic, the desorbate may either be an ion or a neutral. Neutral desorption can only occur if the quenching also involves charge transfer from the substrate. During this charge transfer/quenching step, the neutral may return either to an excited state or the ground state; hence, both neutrals and metastable (electronically excited neutrals) can be desorbed. The desorption of an ion within the MGR directly from the excited ionic state implies that $z_c = \infty$. For primary excitation to the same excited intermediate state this means that desorbed neutrals spend a shorter time on the excited state potential than desorbed ions. Therefore, we should expect desorbed ions to have a higher kinetic energy than desorbed neutrals. In the Antoniewicz model, desorption of an ionic species only occurs after a curve crossing or relaxation to another ionic state that does not have a sufficiently strong adsorbate-surface bond.

8.2.5   Desorption induced by ultrafast excitation

The MGR and Antoniewicz models of stimulated chemistry are linear models. One photon or electron leads to one excitation. One excitation is followed by a single propagation step, quenching and a single decision as to whether quenching is followed by further relaxation to the ground state or to desorption (or reaction). The desorption dynamics follows only one cycle of excitation-propagation-quenching-relaxation to final state, one EPQR cycle, because it is assumed that the excitation rate is slow compared to the rates of quenching and relaxation. For conventional light sources with pulse lengths in the nanosecond or longer range, this is an accurate assumption. As stated above, excited electronic states have lifetimes of only a few femtoseconds and, as shown in Chapter 3, the lifetimes of vibrational excited states on metals lie in the $\sim$1 ps range. On semiconductor surfaces vibrational lifetimes in systems such as H/Si [33] and CO/Si [34] are significantly longer (in the nanosecond range) as a result of the lack of electron-hole pair mediated relaxation methods. If we push the pulse duration into the femtosecond regime, we might be able to push the system away from the linear regime and into a non-linear desorption dynamics if we can increase the excitation rate such that it is competitive with quenching and relaxation. In other words, we can push the dynamics away from a regime in which a single EPQR cycle is responsible for desorption or reaction and into a regime in which multiple EPQR cycles act on the adsorbate.

Misewich, Heinz and Newns [35] were the first to demonstrate the process that has come to be known as desorption induced by multiple electronic transitions or DIMET [21, 36–38]. There are two different limiting cases to describe high excitation rate ultrafast dynamics. The first is facilitated by direct excitation to an intermediate electronic state and can be thought of as multiple EPQR cycles occurring faster than relaxation can return the adsorbate to the ground vibrational level of the ground electronic state. This is the limit known as DIMET. The second framework is called the electronic friction model, in which the swarm of hot electrons created by the femtosecond pulse induces multiple vibrational transitions at a rate faster than vibrational relaxation to the ground state; however, the EPQR dynamics evolves exclusively on the ground electronic state PES.

In both cases we expect the desorption yield to be non-linear in the fluence of the incident laser

$$\frac{d\theta}{dt} \propto F^n \tag{8.2.10}$$

where $n > 1$ is a signature of a non-linear process. However, in contrast to gas phase non-linear events such as multiphoton ionization, the value of n should *not necessarily be interpreted as the number of photons required to bring about the photochemical change*. If the excitation results from direct resonant excitation, n can be associated directly with the number of photons involved in the excitation; however, if the excitation is substrate mediated and because the electron distribution produced by the femtosecond pulse (the hot electron distribution responsible for excitation of the adsorbate) is itself dependent on the laser fluence in a non-linear manner, n is not directly related to the number of photons (or excitations) involved in the process.

When it is recognized [38] that the valence states of a chemisorbed molecule such as NO or O_2 lie close to the Fermi level, that their width Δ and energy position are coupled to their distance from the surface and that the sea of hot electrons responsible for excitation populates a broad distribution of states at and above E_F corresponding to a temperature on the order of 5000 K, it is possible to construct a theoretical framework in which these two apparently disparate models of femtosecond laser induced desorption can be united. The width Δ of the electronic state (or resonance as it is often referred as) turns out to be a crucial parameter to differentiate the two limits. When Δ is small compared to the excitation energy (the energy of the resonance above E_F), the result can be interpreted in terms of multiple excitations between the ground and excited electronic states. In this case, the excited electronic state is an anionic adsorbate state in which one of the hot electrons has attached to the adsorbate. In the opposite limit in which Δ is large compared to the peak electronic temperature and comparable to the energy separation between the resonance and E_F, low-energy electron-hole pairs dominate the frictional energy transfer and the EPQR cycles play out on the ground electronic state PES.

8.3   Photon and electron induced chemistry at surfaces

8.3.1   Thermal desorption, reaction and diffusion

From the above discussion, we expect that long pulse (nanosecond and longer) lasers should be able to excite thermal dissociation on and desorption from surfaces. How can we prove this? The answer is to probe the molecules desorbed from the surface and see if they behave as though they have desorbed by means of thermal desorption. To accomplish this, we probe the response of the desorption cross section to changes in wavelength and laser intensity. A photothermal (pyrolytic) process follows an Arrhenius rate law

$$\frac{d\theta}{dt} = \theta^n A \exp(-E_{des}/k_B T(t)) \tag{8.3.1}$$

where n is the appropriate order of the desorption (or dissociation) kinetics, and $T(t)$ is the surface temperature profile induced by irradiation. The rate increases exponentially with increasing temperature. According to Eq. (8.1.13), the temperature is linearly related to the incident laser irradiance. The wavelength dependence of the temperature change (at constant irradiance) is determined by the wavelength dependence of the reflectivity and the absorption coefficient (properties of the substrate). Therefore, the rate of a thermal process is an exponential function of the incident irradiance, and the wavelength dependence is related to the properties of the substrate.

We also expect the desorbed atoms or molecules to have the properties of thermally desorbed particles. Their rovibrational and translational energy distributions are roughly Maxwell-Boltzmann in form; however, there will be deviations from equilibrium distributions at T_s just as we would expect for any thermally desorbed distribution of molecules. For instance, a system that exhibits non-activated adsorption can be expected to experience some degree of translational and rotational cooling in desorption. For activated adsorption we expect to see excessive heating of the vibrational and translational degrees of freedom.

Once we have established that laser desorption is thermal for a specific system, we can use it to examine the behaviour of the system. As shown in Fig. 8.2, nanosecond laser irradiation of H atoms adsorbed on Si(100) can induce thermal desorption of H_2 molecules by laser induced thermal desorption. LITD facilitates dynamical studies of thermal desorption by providing pulses of desorbates, whose properties can then be probed with laser spectroscopy or TOF techniques. By using one laser for desorption and one for state resolved detection, the two can be combined to provided rovibrationally state resolved translational energy distributions [39]. Since the laser desorption signal is proportional to the coverage, LITD can be used to probe coverage as a function of T_s and gas phase pressure of reactants to perform surface kinetics studies [8].

If the laser desorbed flux is meant to accurately reflect the properties of thermally desorbed molecules, collisions between desorbing molecules must be avoided. If a full monolayer is removed by a nanosecond pulsed layer, this condition is not fulfilled [40, 41]. Instead the molecules collide with each other and form an expansion in a manner analogous to that discussed in Chapter 2 in the context of molecular beams. Both the translational energy and angular distributions are altered by the expansion. Desorption per laser pulse must be kept significantly below 1 ML per pulse, even as low as 10^{-3} ML per pulse, to ensure collision-free conditions.

Surface diffusion can be very difficult to study. Laser desorption provides several methods to investigate diffusion quantitatively by cleaning off a region of the sample by LITD and then probing the coverage after some time delay τ_d to determine how much material has diffused back into the clean region. A particularly ingenious application of this sort was demonstrated by Shen and co-workers [42]. First, a laser beam is split into two beams, which then interfere with each other when they intersect on the surface. The resulting interference pattern burns stripes in the coverage of an adsorbate initially present on the surface. Where the intensity is high, LITD occurs, but at the nodes no desorption occurs. The stripes act like a diffraction grating that can be used to modulate the optical second harmonic generation signal of a probe laser beam that is incident on the surface after a delay of τ_d. Subsequent analysis of the redistribution of surface coverage as a function of τ_d and T_s can be used to derive the diffusion parameters E_{dif} and D_0.

8.3.2   Stimulated desorption/reaction

Laser and electron irradiation can also lead to direct or substrate mediated stimulated (nonthermal) events. When not using ultrafast pulses, the hallmark of photochemical events is that they normally depend linearly on the incident excitation density. For electron radiation, the primary electron current i_p is easily measured. The response of coverage to irradiation for electron stimulated desorption (ESD) is

$$-\frac{d\theta}{dt} = \sigma_d \frac{i_p}{Ae}\theta, \tag{8.3.2}$$

where e is the elementary charge, A is the area of incidence, and σ_d the desorption cross section as defined in Eq. (8.2.8). Similarly for photon stimulated desorption (PSD)

$$-\frac{d\theta}{dt} = \sigma_d \frac{F_0}{h\nu}\theta, \tag{8.3.3}$$

where $h\nu$ is the photon energy and F_0 the light source (laser, lamp or synchrotron) fluence (e.g. J cm^{-2}). Integration of these equations leads to an expression for the coverage as a function of time, for example,

$$\theta(t) = \theta_0 \exp\left(-\frac{i_p\sigma_d}{Ae}t\right). \tag{8.3.4}$$

As long as the cross section is independent of coverage, which is not true if there are strong lateral interactions, a plot of $\ln(\theta/\theta_0)$ vs t yields a straight line with a slope $(-i_p\sigma_d/Ae)$. Frequently, rather than the coverage, the desorption signal itself is monitored. In this case, an analogous expression can be used to evaluate the cross section. However, the cross section measured is not necessarily the same. If coverage is monitored directly, analysis according to Eq. (8.3.4) yields the cross section for *all* channels that lead to desorption regardless of whether the desorbing species is a positive ion, neutral or negative ion. If the current of positive ions i^+ is measured versus time and then analyzed according to

$$i^+(t) = i_0^+ \exp(-(i_p\sigma_d^+/Ae)t), \tag{8.3.5}$$

then the cross section measured is the cross section for the desorption of positive ions *alone*. Similar measurements for neutrals and anions would determine the cross sections for the respective desorption processes.

8.3.2.1   *High-energy radiation*

The vacuum ultraviolet (VUV) extends from 100 nm $\leq \lambda \leq$ 200 nm, correspondingly 6.2–12.4 eV while the extreme ultraviolet (XUV) lies in the range 10 nm $\leq \lambda \leq$ 100 nm and 12.4–124 eV. Higher energy photons (up to about 10,000 eV) comprise x-rays. The early era of scientific studies of stimulated reactions was dominated by studies performed with electron beams with energies of tens to hundreds of eV. As discussed in Chapter 2, electron beam sources are easily constructed and adapted to UHV conditions. They provide high currents and are broadly tuneable. Synchrotrons and now high harmonic generation [43] can provide photons in the same energy range but studies with high energy photon sources are not nearly so extensive as high energy electron beams. Many of the early studies investigated the production of positive ions, even if cross sections for producing them are low, because detection efficiencies for these are exceptionally high. Neutrals are much more difficult to detect since they normally need to be ionized with electron bombardment or laser irradiation before they can be sent into a quadrupole or time-of-flight mass spectrometer to be identified and counted. Surface photochemistry in the VUV and XUV is of particular importance to the evolution of the interstellar medium and protoplanetary clouds [44, 45] as well as elsewhere in our solar system [46]. Photochemistry and photodesorption are both important processes for interstellar chemistry because photochemistry may lead to the production of new chemical species on dust grains that populate interstellar space and photodesorption provides a pathway to expel these species into the interstellar gas phase, even though the surface temperature may well be below 10 K.

In the energy range where it is possible to ionize valence levels, irradiation of a molecule such as CO adsorbed on a transition metal surface can lead to several different products including CO, metastable electronically excited CO^*, CO^+, C^+ and O^+ as well as negative ion states. Each of these products has its own distinct cross section that depends on electron energy according to which excitations are most effective at producing that species. Deeper analysis of the specific case of CO on Ru(001), which has been extensively studied by Menzel and co-workers [32], gives us insight into the types of electronic excitations and reaction dynamics that are expected for high energy irradiation of chemisorbed systems. A very weak threshold for the formation of desorbed CO^+ is found at 14 eV, with a much stronger threshold at 18 eV. Neutral CO is also found in this energy range though detection sensitivity issues make it more difficult to identify the primary threshold. These thresholds correspond to $5\sigma^{-1}$ and $5\sigma^{-2}$ excitations, that is, removing 1 or 2 e^- from the 5σ molecular orbital of adsorbed CO. For 30–100 eV excitation several other resonant threshold features are observed and can be assigned with the aid of UPS to valence level excitations. The nature of the orbitals excited is correlated with the types of species that are formed. The $6\sigma^*$ orbital, for instance, is strongly antibonding with respect to the C–O bond and a $3\sigma^{-2}6\sigma^{*+2}$ excitation (removal of 2 e^- from the 3σ and addition of 2 e^- to the $6\sigma^*$) is associated with an increase in O^+ desorption yield while the CO and CO^+ yields are little affected.

For higher energies, ESD and PSD cross sections for ions (but not neutrals) often exhibit threshold behaviour near the energies of core level excitations; however, the threshold is delayed slightly compared to the absorption edge. This is related to the importance of the excited state lifetime in the desorption dynamics. The longer the lifetime, the greater the desorption cross section. Multiple electron excitations lead to complex excited states with longer lifetimes than simple one-electron excitations. An excited ionic state corresponds to a multiple-electron excitation of the neutral. If the excited ion is neutralized by filling the hole left behind by photoemission, an excited state of the neutral is formed. This state itself may also be repulsive, which enhances the desorption cross section. Refilling of the initial hole may also involve an Auger process, which leads to a different ionic excited state being populated.

The MGR model is found to be very effective at describing the dynamics of ESD for CO following valence and core level excitation. Consider the consequences of a repulsive state being responsible for desorption. Acceleration occurs in the direction of the repulsive force. The particle that desorbs receives a kick that is directed along the bond that was switched to a repulsive state by the excitation. Therefore, the desorption trajectory follows a path that is determined by the initial bond orientation, and measurement of the angular distribution of the ions desorbed in this manner gives us information about the orientation of the bonds in the adsorbed state. This is the basis of the technique called electron stimulated desorption ion angular distribution (ESDIAD) that was developed by Madey and Yates [47, 48].

For CO bound on Pt(111) [30], it was found that CO^+, O^+ and electronically excited $a^3\pi$ CO* desorbed after bombardment with electrons of energy 200–900 eV but no C^+ was observed. The CO^+ and CO* desorption cross sections are found not to be correlated, which means that CO* results from a separate excitation path rather being produced by simple neutralization of the ionic CO^+ as it attempts to desorb from the surface. At coverages up to 0.5 ML all three species desorbed along the surface normal. This is completely consistent with our expectation from the Blyholder model that CO is bound by the C atom in an upright geometry in which the O points away from the surface along the surface normal.

However, at a coverage of 0.67 ML, six beams directed along the 6 equivalent [110] directions of the hexagonal (111) surface are observed to grow while the intensity of the normal beam disappears. It was proposed that these six beams correspond to CO molecules that are tilted by ~6° with respect to the surface normal because of lateral interactions at high coverage. When T_s is raised from 90 K to 230 K, the pattern blurs out into a halo because vibrational motion increases and leads to a spread of initial orientations of the adsorbed CO.

An important consideration is the interaction of the desorbing ion with its image charge [49, 50]. The image charge causes an attractive force between the desorbing ion and the surface. If the ion desorbs at an angle from the normal, then the force associated with the interaction of the ionic charge with the image charge attracts the desorbing ion toward the surface. The greater the inclination of the initial angle, the greater the attraction, and the more the ion is pulled toward the surface. At some critical angle, the ion is unable to escape the surface and is recaptured.

The properties of the desorbed species also reflect the nature of the changing, highly repulsive excited states as the electron or photon energy is changed. Take, for example, the case of water adsorbed in mono- and multilayers on Ni(111). Stulen and co-workers [51] found a threshold for H^+ ESD at an incident energy of 20–21 eV. This is significantly above the gas-phase threshold for dissociative photoionization (DPI) 18.76 eV. The H^+ kinetic energy distributions measured at an incident energy of 45 eV extends beyond 10 eV, Fig. 8.8. At 25 eV incident energy, however, the desorbed H^+ has a distribution that is substantially less energetic, peaking below 1 eV, Fig. 8.8.

The distributions shown in Fig. 8.8 are representative of the kinetic energies distributions (KED) that are often observed in this energy range. They are very broad and have a roughly Gaussian shape that is skewed to high energy. The high coverage KED for 45 eV excitation may be composed of more than one component, which suggests that more than one channel leads to desorption for excitation at this energy. The

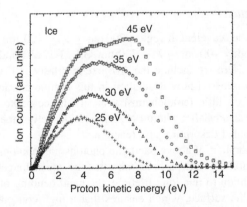

Figure 8.8 *H^+ kinetic energy distributions for ESD from H_2O ($\theta = 16$ ML) adsorbed on Ni(111) for several different excitation energies. Reproduced from J.O. Noell, C.F. Melius and R.H. Stuhlen,* Surf. Sci. *157, 119. © 1985 with permission from Elsevier.*

distributions do not correspond to simple Maxwell-Boltzmann distributions; however, both distributions are very broad. The mean kinetic energy is significantly less than the energy of the exciting photon. Even after subtracting the energy required to break the O–H bond and ionize the H atom, it is clear that much of the 45 eV excitation energy has been lost to quenching.

The first PSD study using ultrafast VUV pulses was performed by Riedel, Kolasinski, Palmer and co-workers [52]. The intensity of these ultrafast pulses was still low enough for the induced photochemistry to be in the linear regime. The system studied was multilayers of O_2 physisorbed on graphite, a system in which the O_2 adsorbs standing up with its bond axis along the surface normal. Because of its physisorbed character, the photoexcitation dynamics that lead to the formation of O^+ by irradiation of O_2 with photons in the 17–38 eV range closely mimics the dissociative photoionization dynamics of gas phase O_2. Comparison of the kinetic energy distributions observed in the gas phase to those in the physisorbed phase reveals that the high-energy side of the distributions is little changed. The low-energy side, however, is drastically transformed by adsorption. In the gas phase, the predominant pathway for dissociative photoionization leads to low-energy O^+ for both 21 eV and 41 eV irradiation. This channel is shut down in the dissociative photodesorption of physisorbed O_2. The departure dynamics of the O^+ fragment is significantly different when comparing gas-phase DPI and adsorbed-phase DIET. In the adsorbed phase, the departing O^+, even though it points along the surface normal, must overcome the image potential it experiences due to the surface. Zero kinetic energy and slow O^+ ions are captured by the image potential and cannot desorb. On the basis of energetics and the similarity of the kinetic energy distributions, the photodesorption of O^+ is assigned to a direct rather than a substrate mediated process.

The presence of a surface can change photodissociation dynamics not only because of image charge effects but also as a result of changes in electronic states. The adsorbate geometry and the possibility of electrons being excited out of the solid and into the adsorbate can also have an influence. On Si(111)–(7 × 7), O_2 chemisorbs with its molecular axis *parallel* to the surface rather than perpendicular as for O_2/graphite. Dujardin et al. [53] used synchrotron radiation at 10–150 eV to study the desorption of O^+ in this system. A threshold energy for desorption of O^+ is found at 33 eV, significantly higher than in the physisorbed system, and was assigned to the direct excitation of the $2\sigma_g$ electron of chemisorbed O_2. Other resonances and thresholds were also observed. The kinetic energy distribution of photodesorbed O^+ demonstrates again that KEDs are sensitive to the nature of the excited states, because both the peak and breadth of the KEDs change with photon energy.

8.3.2.2   IR-visible-UV radiation

The ultraviolet is defined as the wavelength region 200 nm $\leq \lambda \leq$ 400 nm, with corresponding energies 3.1–6.2 eV; the visible is roughly 400 nm $\leq \lambda \leq$ 700 nm, 3.1–1.77 eV; and the IR extends from 700 nm to 1 mm, 1.77 eV–1.24 meV. Here we include the ArF excimer laser (193 nm, 6.4 eV) in our discussion of the UV. We have already discussed how irradiation in this regime can lead to thermal chemistry. But to paraphrase Gadzuk, "there's a little femtochemistry in all surface photochemistry." Direct excitations or the electrons excited immediately after photon absorption and before the electron cascades have fully thermalized can lead to stimulated desorption and reaction.

On semiconductors, the absorption of above band gap photons, i.e. photons with energies larger than the band gap, leads to the formation of electron-hole pairs that are much longer lived than the electron-hole pairs formed in metals. In addition to this long lifetime, the band bending of the space charge region also forces one type of carrier to the surface, which ensures that a high concentration of at least one carrier type is present at the surface after photoirradiation. The carrier of the opposite sign is transported either to the other face of the crystal (the back face) or to the unilluminated part of the front face [54]. The excited carriers can then partake in what is effectively photoelectrochemistry at the surface of the excited semiconductor. Photodesorption by means of direct optical excitation of adsorbed H on Si requires VUV photons [33]. Nonetheless, it is possible for IR photons to have a dramatic effect on the reactivity of H-covered Si surface. Kolasinski [55], showed that in an aqueous environment, the sticking coefficient of F^- on H/Si is no greater than $\sim 5 \times 10^{-11}$. Irradiation of H/Si with a garden variety HeNe laser operating at 633 nm changes this value to ~ 1. In other words, irradiation changes the sticking coefficient of F^- on hydrogen-terminated Si by over 10 orders of magnitude. This occurs because the holes created by photon absorption of n-type doped Si are forced to the surface where they occupy a bulk electronic state. The presence of this hole lowers the barrier for the replacement of adsorbed H atoms with a F^- ion from solution.

Stimulated desorption and reaction with nanosecond to cw irradiation of adsorbate/metal systems is generally only observed at wavelengths closer to the UV. A linear dependence of coverage change on photon flux as required by Eq. (8.3.3) is one of the characteristics of photon stimulated chemistry. The characteristics of the desorbed molecules can also be used to identify the process as stimulated in nature and to investigate the dynamics of the process. NO is a poster child of state-resolved studies because of the high detection sensitivity associated with LIF and REMPI schemes for this molecule. Much of our insight into photochemistry dynamics at surface has been developed by studies of NO [28, 39, 56–60].

Consider, for example, the desorption of NO following laser photodissociation of N_2O_4 on Pd(111) as studied by Hasselbrink and co-workers [61]. The translational energy distribution is found to be roughly Maxwell-Boltzmann in nature; however, it is characterized by a translational temperature $\langle E_{trans} \rangle / 2k_B = 1560$ K. Since the surface is held at 95 K and is heated by only 10 K during the laser pulse, this high temperature is clearly indicative of the nonthermal origin of the NO. When excited with 6.4 eV photons, an excess energy of 3.4 eV is available to be deposited in the translational and internal degrees of freedom of the products. The translational energy of the photodesorbed NO corresponds to 269 meV and the internal energy is only 127 meV; therefore, only a fraction of the excitation energy ends up in the desorbed product. Much of the excitation energy must be lost during quenching or to other substrate and adsorbate degrees of freedom – just as was found for ESD and discussed above.

Why is the velocity distribution roughly described by a Maxwell-Boltzmann distribution? Consider the MGR dynamics depicted in Fig. 8.7 [39]. For desorption of a neutral molecule we do not have to account for any deceleration by the image charge potential. The energy gained by the desorbed molecule in the excited state is given by $V_{ex}(z_0) - V_{ex}(z_{jump})$, where V_{ex} is the excited state potential, z_0 is the initial position of the molecule, and z_{jump} is the position of the molecule when it is quenched back to the ground state potential. A portion of this energy, $V_g(\infty) - V_g(z_{jump})$, is lost to overcome the remainder of the adsorption energy that is encountered when the molecule returns to the ground state potential V_g. Defining the energy scale such that $V_g(\infty) = 0$, the kinetic energy of the desorbed molecule is then

$$E(z_{jump}) = V_{ex}(z_0) - V_{ex}(z_{jump}) + V_g(z_{jump}). \qquad (8.3.6)$$

The longer a particle stays on the excited state potential, the further it propagates and the more energy it acquires. However, the excited state population decays roughly exponentially in time and, therefore, the probability of attaining a given amount of kinetic energy also decreases exponentially. The presence of the absorption well cuts off the low end of the distribution – those molecules that have gained too little energy are decelerated by the adsorption potential to negative kinetic energy and remain on the surface. Those that have gained more than the critical energy are decelerated but some are still able to escape. The kinetic energy probability distribution thus looks like an exponentially decreasing function that peaks at $E = 0$. Once such a distribution is flux weighted, and one takes account of angular detection effects and the initial velocity distribution of the molecule, the broad distribution looks very much like the broad Maxwell-Boltzmann distribution. Such distributions generally fit what is called a modified Maxwell-Boltzmann distribution

$$I_{\text{flux}}(t)\, dt = 2b^2 L^4 t^{-5} \exp(-b(v - v_0)^2)\, dt, \tag{8.3.7}$$

where v_0 is an offset or stream velocity (see Exercise 8.3).

The positive correlation between time spent on the excited state and the amount of energy gained has a rather striking effect on the internal energy of the desorbed molecule. For an equilibrium distribution (a true Maxwell-Boltzmann distribution), there should be no correlation between the internal energy and the translational energy as a function of, for instance, the rotational state. However, Buntin et al. [57] and Budde et al. [60] both observed a positive correlation between translational energy and internal energy, that is, that translational energy increases with increasing rotational state. Hasselbrink [62] has shown that this positive correlation can be understood on the basis of excited state lifetime effects. In the excited state, the molecule not only experiences a repulsion but also a torque. For instance, if the excited state corresponds to a negative ion state, the excited state may have an equilibrium binding geometry that is different from the ground state, and the excited state attempts to relax into its preferred geometry. The longer the torque acts, the greater the rotational excitation, and since we have already shown that longer propagation times in the excited state correspond to higher kinetic energies, we conclude that faster moving molecules should also on average have more rotational energy. Formation of a temporary negative ion state also leads to enhanced vibrational excitation in the desorbed molecules since the negative ion likely has a different bond length than the adsorbed molecule. This was confirmed, for instance, by Freund and co-workers [59], who found that NO photodesorption from a Ni surface leads to a vibrational temperature of $T_{\text{vib}} = 1890 \pm 50\,\text{K}$.

8.3.2.3   *Ultrafast IR-visible-UV radiation*

Irradiation of an adsorbate with ultrafast laser pulses causes such a high density of excitations that an odd thing happens to the desorption dynamics: they return to a regime that is quasi-thermal [63]. What distinguishes thermal processes is that we can describe them as stochastic random walks through energy space. A molecule experiences many collisions, gaining and losing energy until it finally collides with a partner with the right energy and orientation, and reaction ensues. The ultrafast dynamics discussed above also involve repeated excitation and de-excitation steps, which means that the reaction of the system resembles a thermal response in some respects as long as we remember that the electrons, phonons and adsorbates respond with their own characteristic time scales. The degrees of freedom of the adsorbate each couple to substrate excitations with their corresponding time scales, none of which have to be the same. In the end, then, the properties of the desorbed molecules may all look somewhat Maxwell-Boltzmann, but the temperatures describing each degree of freedom need not have anything to do with each other since each degree of freedom responds to the ultrafast substrate-mediated excitations in its own characteristic way.

Desorption with femtosecond pulses is a highly non-linear process. Heinz and co-workers [64] used $\sim$3 mJ, 200 fs pulses at 620 nm to desorb NO from Pd(111). Such pulses lead to a peak electronic temperature of $T_e \approx 3000\,\text{K}$. The desorption cross section for these pulses ($\sim 10^{-18}\,\text{cm}^{-2}$) is much larger than the desorption cross section associated with nanosecond pulses ($< 10^{-21}\,\text{cm}^{-2}$). The desorption yield was found to increase with the absorbed laser fluence raised to the third power. For O_2 on Pd(111) under similar conditions [65], the yield increased as fluence raised to $n \approx 6$. The ratio of population in $v = 1$ to $v = 0$ of NO indicates vibrational excitation in excess of 2000 K, while the rotational population distribution was decidedly non-Boltzmann. A moderately good fit to this distribution could be obtained by overlapping two thermal distributions, one with $T_{\text{rot}} = 400\,\text{K}$ and the other with $T_{\text{rot}} = 2600\,\text{K}$. The translation distribution was described by a temperature of $T_{\text{trans}} \approx 600$ K. Only a moderate correlation between internal state and translational energy was observed.

Desorption can unequivocally be assigned to ultrafast coupling of the excited electron distribution to the adsorbate by observing the time response of the system [66] and showing that desorption has a subpicosecond response time. The time response of the system is recorded by measuring a cross correlation function of the desorption signal. The cross correlation function is measured by recording the desorption yield as a function of the delay between two ultrafast pulses. Figure 8.4 demonstrates how cross correlation reveals the different time responses of the electronic and phonon temperatures. Analogous measurement can also reveal the characteristic response time of the excitation that leads to desorption.

Wolf and co-workers [63, 67] showed that recombinative H_2 desorption from Ru(001) can be driven by 130 fs, 800 nm pulses with pulse energies up to 4.5 mJ. At an absorbed fluence of 60 J m^{-2}, the translational energy distributions are described fairly well by a modified Maxwell-Boltzmann distribution with a mean energy corresponding to $\sim$2000 K. The time response was investigated by measuring the cross correlation response of H_2 and D_2 desorption. The coupling times were found to be 180 fs and 360 fs for H_2 and D_2, respectively, again indicating that ultrafast electronic excitations are responsible for stimulating the reaction. The highly non-linear response to laser fluence ($n \sim 3$) and strong isotope effects are consistent with dynamics described by the electronic friction model outlined above.

8.3.3 Ablation

If we continue to turn up the intensity of the radiation source, we leave the regime of laser desorption and enter the regime of laser ablation. Ablation is a general approach to removing material by photon or particle irradiation in which a substantial amount of the substrate is removed. The dividing line between desorption and ablation is somewhat arbitrary. Ablation can be thought of as explosive desorption. In some cases, the mechanism of ablation is predominantly thermal; however, there can also be a stimulated component. Consequently, it is difficult to define ablation mechanistically. Instead, it may be best to define it in terms of the rate of material removal and the intent to remove a significant portion of the *substrate* rather than just an adsorbed layer. As we have seen above, when the desorption flux during a single pulse exceeds a certain threshold, the rate of desorption becomes so high that the properties of the desorbing flux change due to a process similar to molecular beam expansion. Around this same desorption rate, we can draw the dividing line that separates ablation from desorption. For the sake of having a round number, here we choose to define ablation as the process that is occurring when roughly 1 ML per pulse is removed from the substrate.

Laser ablation leads to the formation of an ablation plume of ejected material [68]. Whereas etched material is often forgotten about once it has left the surface, ablation is used for transferring material from the target to a substrate as well as for texturing the target. Hence, plume interactions (plume/laser, plume/surface and plume/ambient gas) are very important in laser ablation. For ejection into vacuum, collisions among the ablated atoms lead to an expansion phenomenon much like that experienced in a supersonic molecular beam expansion. For ejection into an ambient gas, a shock wave forms. Pressures in excess of 100 atm can be exerted on the surface by the plume. Furthermore, free expansion into vacuum means that redeposition out of the plume is unimportant, whereas redeposition is extremely important when an ambient gas is present to impede plume expansion. Redeposition has a measurable effect on ablation rates [69] and can decrease them by a factor of 3 or more.

The presence of gas above the surface prior to the onset of ablation influences more than just the formation of the plume [70]. Collisions can lead to the formation of clusters [71]. Indeed, laser ablation into a buffer gas creates clusters from the ejecta very efficiently [72]. The clusters are often observed to form aerosol dispersions when ablation is carried out at pressures on the order of 100 mbar or more. These aerosol particles can scatter a significant amount of laser light. Considerable chemistry can also result from the interactions of the ambient gas, plume and surface. Whereas redeposition lowers the ablation rate, the presence of a reactive gas significantly increases the ablation rate. For instance, when a Si surface is irradiated in air with roughly 3 J cm^{-2} of 248 nm light from an KrF excimer laser, the ablation rate is $\sim$24 greater than when the air is replaced by Ar [73]. Si removal is faster yet in the presence of SF$_6$. SF$_6$ does not react with the surface but F, released either by photodissociation or the interaction of SF$_6$ with the plume plasma, does. The enhanced ablation can occur either because of direct etching by the fluorine or because adsorbed species such as SiF, SiF$_2$ and SiF$_3$ are more volatile and have a higher desorption rate than Si. Plume chemistry can also influence redeposition. Redeposition can occur by way of clusters that precipitate on the surface, by the condensation of atoms previously ejected from the surface or by the adsorption of compounds that have formed in the plume and then subsequently encountered the surface.

Because ablation involves the removal of a substantial amount of bulk material rather than just an adsorbed layer, for ablation to occur the laser fluence must be high enough for the absorbed energy to exceed not just the desorption energy of a single atom but the binding energy of the solid. When the absorbed fluence per atom exceeds the heat of fusion of the solid, the solid melts. Many materials have surprisingly low vapour pressures in the liquid phase; hence, this phase transition often is not sufficient to produce a very large desorption flux per pulse. If the laser fluence is increased further such that the absorbed fluence exceeds the heat of vaporization, the solid is heated to its boiling point and vaporization occurs. At this point there is a very significant increase in the rate of material removal.

Nonetheless, ablation is frequently performed at even higher absorbed fluences at the point where the lattice is heated beyond the critical temperature. The material is initially superheated but remains solid until regions of liquid or gas nucleate. At this point an inhomogeneous phase of vapour bubbles and liquid is formed. The volume of this inhomogeneous phase expands in the material, and explosive evaporation ensues. Just above threshold, disordering takes tens of picoseconds consistent with a thermal melting process, and ablation is well characterized as a quasi-thermal process. Nonetheless, ablation always encompasses some degree of both photolytic and photothermal processes. Particularly for molecular solids and polymers irradiated in the deep UV, photochemistry plays a role in ablation. When ultrafast lasers are used with pulse widths of 1 ps or less, a large fraction of electrons can be excited to the conduction band, which is antibonding, and if a sufficient fraction of the electrons are excited (exceeding $\sim$10%), a highly repulsive state is achieved. The resulting forces cause nonthermal mechanisms of melting and ablation [74–77]. In Si and GaAs nonthermal disordering has been observed by optical measurements at excitation fluences >1.5 times the damage threshold.

Ablation can occur in a single laser shot. However, it can also occur via multiple shots. Repeated laser shots lead to rapid heating, melting and resolidification, which can incorporate a growing number of defects into the lattice with each successive shot. The lattice defects correspond to the build up of energy in the lattice and make it progressively easier to ablate [78].

The pulse width of the laser is very important in determining the processes that occur during laser ablation. The comparison of femto- to nanosecond lasers is particularly illustrative [76, 79], as shown schematically in Fig. 8.9. For near IR, visible or UV light absorption by a solid, the electrons absorb the radiation. They equilibrate among themselves within a couple hundred femtoseconds. The hot electrons then equilibrate with the lattice over the course of many picoseconds. Thermal diffusion is important during nanosecond irradiation, but does not play a role in removing energy from the irradiated region when using femtosecond pulses. Because of these timescales, the interaction of femtosecond pulses in the ablation regime is much different than that of ns pulses [71]. For the same energy deposited, the peak temperature is higher for fs versus ns pulses, it is achieved sooner in time and the melt depth (approximately 1μm) is slightly shorter. For a 1 ps pulse incident on a metal at a fluence of 0.15 J cm^{-2}, the electron temperature

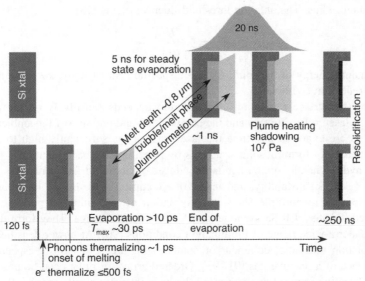

Figure 8.9   *Femtosecond versus nanosecond laser ablation.*

peaks after ~1.8 ps, the lattice temperature at 27.2 ps. The lifetime of the melt, roughly 200–300 ns, is still the same as in nanosecond irradiation since resolidification is determined by the same materials properties.

Femtosecond pulses are absorbed before the lattice has reached its peak temperature and the onset of ablation; hence, they do not interact with the plume, which begins to form at around 10 ps. Before 1 ns has elapsed, ablation ceases. Some molten material may still exist at the bottom of the ablation pit but most of the melted material is removed. Nanosecond pulses heat the sample comparatively slowly (even though heating rates can exceed 10^{10} K s^{-1}). The onset of steady-state ablation is at roughly 5 ns and continues for many ns after the pulse is over; therefore considerable irradiation of the plume occurs. This significantly heats the plume, induces photochemistry and increases the plume pressure [80]. The plasma formed in the plume also shadows the surface and decreases the laser power that can be coupled into the substrate. The plume has a much longer lifetime during ns ablation, and is therefore able to interact more extensively with the molten surface. A decrease in the ablation threshold of two orders of magnitude was observed by Preuss et al. [81] for thin films of Ni when the pulse length was decreased from 14 ns to 500 fs. The difference can be attributed to decreased shadowing and thermal diffusion for fs pulses.

Stuke and co-workers [69] have developed a general formula to describe the ablation depth per pulse d under the following assumptions: (i) absorption is described by the Beer-Lambert law; (ii) reflectivity is constant; (iii) ablation occurs after the laser pulse is over, hence shadowing can be neglected; (iv) thermal conduction is negligible; (v) the laser fluence exceeds the ablation threshold; and (vi) redeposition is negligible. For a Gaussian laser beam the ablation depth is given by

$$d = \frac{1}{\alpha} \ln \left(\frac{\phi_0}{\phi_{th}} \right) \tag{8.3.8}$$

where α is the absorption coefficient, ϕ_0 is the laser fluence at beam centre and ϕ_{th} is the ablation threshold fluence. The diameter of the ablated spot, D, is given by [82]

$$D^2 = 2 w_0^2 \ln \left(\frac{\phi_0}{\phi_{th}} \right) \tag{8.3.9}$$

where w_0 is the beam radius. The ablation threshold fluence, ϕ_{th}, is [78]

$$\phi_{th} = \frac{UC_v}{k_B\,\alpha(1-R)} \tag{8.3.10}$$

where U is the binding energy of the solid, C_v is the specific heat capacity, k_B is the Boltzmann constant and R the optical reflection coefficient.

The topology of the surface remaining after ablation depends sensitively on the power density, the number of shots, the laser pulse duration and the composition and pressure of the ambient gas [79, 83–85]. Topological changes are expected any time the laser intensity is sufficiently high to cause melting, and six different mechanisms of forming solid structures by freezing have been identified: solidification driven extrusion (SDE), hydrodynamic sputtering, laser-induced periodic surface structures (LIPSS), capillary waves, the Mullins-Sekerka instability, and laser zone texturing [86]. Below the ablation threshold and particularly for exposure to multiple shots, the formation of laser-induced periodic surface structures (LIPSS) is observed [87–89]. LIPSS are a general form of laser damage. They form through a variety of mechanisms and under a wide range of illumination conditions [90]. At power densities at and just below the single shot ablation threshold, columnar or conical structures often form, especially when ablation occurs in the presence of a reactive gas [91–93]. Ordered arrays of pillars can be produced either by the use of masks or diffraction-induced modulation of the laser intensity profile [83, 84, 94, 95]. At intensities significantly above the single shot ablation threshold and particularly for femtosecond irradiation, ablation leads to the formation of pits with vertical walls and flat bottoms. Irradiation in this regime can be used for direct writing of pits, trenches and patterns as small as a few hundred nm across.

8.4   Charge transfer and electrochemistry

Electron transfer to an adsorbate can be induced in a number of ways. The simplest way is that an adsorbate has an affinity level that drops below the Fermi level of the substrate when it is bound. Electrons from the substrate then populate this affinity level either fully or partially depending on its energetic position. An electron beam, whether external or emitted by an STM tip in field emission mode (see next section), can supply electrons that attach to an adsorbate. Electrons can tunnel from a tip across a barrier and into an unoccupied state. Electron or photon irradiation can create hot electrons, which then attach to an adsorbate in a substrate mediated process. Charge transfer is, of course, the hallmark of electrochemistry.

Electrochemistry is defined by chemical transformations that proceed with the participation of free charges (not just shared electrons) as shown in Fig. 8.10. Charge is conserved; therefore, one species must be reduced (accept electrons) while another is oxidized (donate electrons). Whether we follow the flow of

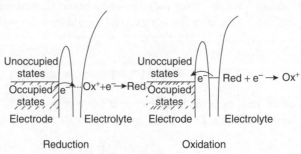

Figure 8.10   *Oxidation and reduction.*

electrons or holes, charge transfer must always proceed as follows: electrons move from occupied to empty states and holes move from unoccupied to occupied states. Furthermore, in order for charge transfer to occur, the states involved must overlap spatially and energetically at least partially within an energy range allowed by thermal fluctuations. The greater the overlap and the smaller any barrier is (in both spatial width and energetic height), then the greater the probability (or rate) of exchange.

8.4.1 Homogeneous electron transfer

A discussion of electron transfer reactions generally begins by framing the process within the parameters of a displaced oscillator theory known as Marcus theory. Rudolph A. Marcus was awarded the 1992 Nobel Prize in Chemistry for his contributions to electron transfer theory. The remarkable thing about Marcus theory is that it allows a simple classical picture to be drawn to explain an inherently quantum phenomenon. Furthermore, this simple picture can explain broad trends – predicting a majority of rate constants for electron transfer to an acceptable degree of accuracy – while also predicting previously unknown phenomena such as the existence of a rate constant maximum and the "inverted region". This predictive capacity is generally accepted as one of the quintessential aspects of a ground-breaking theory. It should be kept in mind that the simplest form of Marcus theory does not predict rate constants with what is considered chemical accuracy by modern standards. Nor is the predicted parabolic shape of rate constants in the inverted region well reproduced by experimental data. Nonetheless, the theory clearly and cogently describes the process, and lays bare those assumptions that can be optimized by quantum mechanical corrections. It also points to chemical processes that clearly lie outside of the model parameters. Thus a description of Marcus theory provides a basis for any discussion of electron transfer kinetics.

Consider a reversible one-electron transfer from a reduced donor species D to an oxidized acceptor species A in an aqueous solution. Both D and A exist in solution with a tightly bound and highly organized inner solvation shell of water molecules. Looser outer shells surround these. The inner shells have much different configurations depending on whether they surround D or A, and it takes a significant amount of energy to reorganize the equilibrium solvent shell surrounding D into the configuration that would surround A at equilibrium. Although there is a high degree of order in the inner shell, it is subject to the normal fluctuations expected from a thermal Boltzmann distribution at equilibrium. Counterions are constantly in flux in the outer solvation shell. Therefore, *while on average a system at equilibrium finds itself at the bottom of its Gibbs energy well, instantaneously a system can have virtually any energy from zero to infinity*. The probability that the system is found at any particular Gibbs energy G is given by the Boltzmann distribution. It depends, among other things, on the energy difference from the ground state and the excited state, and the temperature in an exponential fashion.

Because the electron has a mass that is at least a thousand times less than the nuclei involved in electron transfer, its velocity is significantly greater than that of the nuclei, and it is transferred in a Franck-Condon-like transition. That is, the electron can transfer so quickly that *the nuclei do not have time to adjust their positions during electron transfer*. This means that if an electron were to hop from D to A, the solvation shells of D and A would be fixed in their initial configurations while D is transformed into D^+ and A is transformed into A^-. These solvation atmospheres do not correspond to the equilibrium configurations about D^+ and A^-. Therefore, the system finds itself in an excited state, which violates the conservation of energy during the electron transition. Recall that in a Franck-Condon transition between two electronic states engendered by electromagnetic radiation, the absorption or emission of a photon ensures energy conservation.

Marcus became aware of this state of affairs and realized that the Franck-Condon nature of the transition could only be maintained if the electron transfer were made between degenerate states. He proposed that the solvent shell *first* reconfigures itself. Once the solvent shell of the donor D reconfigures itself so that

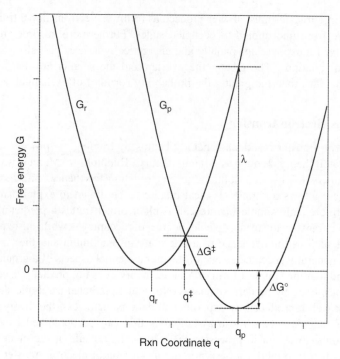

Figure 8.11   *The displaced oscillators used to define the parameters required to calculate electron transfer rate constants within Marcus theory. The Gibbs energy curves of the reactants and products (G_r and G_p respectively) intersect at $q^{\ddagger}$ to form the transition state, which lies $\Delta G^{\ddagger}$ above the ground state of the reactants. The reorganization energy λ is also shown.*

the system has the same energy regardless of whether the electron is on D or A, electron transfer can occur degenerately. The transition state for electron transfer is, therefore, this configuration of the solvent shell, and the activation energy for electron transfer is the Gibbs energy change required to bring about this reorganization.

This simple assertion would have been greeted with approval, and likely relegated to the closet of good but not great ideas if Marcus has not followed up this assertion with a series of theoretical papers [96] that justified the simple picture shown in Fig. 8.11. This body of work also sketched out many of the corrections that any successor theory must address. Let G_r represent the Gibbs energy of the reactants and G_p the Gibbs energy of the products. These are connected by a reaction coordinate q. The nature of the reaction coordinate is quite woolly, and we discuss that more below. At equilibrium the reactants would find themselves at $q = q_r$ and the products at $q = q_p$. We are free to place the zero of Gibbs energy wherever we choose, and for this exercise we set $G_r(q_r) = 0$. In other words, the Gibbs energy origin is placed at the equilibrium position of the reactants. The products are then located ΔG° below this value at equilibrium, $\Delta G^{\circ} = G_p(q_p) - G_r(q_r) = G_p(q_p)$. The transition state occurs at $q = q^{\ddagger}$, and the activation Gibbs energy $\Delta G^{\ddagger}$ is the energy at this point referenced to the Gibbs energy origin.

If we assume a linear response, which one might term a harmonic approximation, between the Gibbs energy and the reaction co-ordinate (and indeed this is generally a good approximation in aqueous solutions [97–99]), the Gibbs energy exhibits a parabolic dependence on the reaction co-ordinate. Furthermore, with our choice of origin and by defining the reorganization energy λ as the energy difference between D when it is surrounded by its equilibrium solvent shell configuration and D when it is surrounded by the solvent shell associated with A at equilibrium, we find that the model reduces to only two parameters (see Exercise 8.17), namely ΔG° and λ according to

$$G_r = \lambda q^2 \qquad (8.4.1)$$

$$G_p = \lambda(q - 1)^2 + \Delta G^\circ \qquad (8.4.2)$$

ΔG° is readily available from tables and is directly related to the difference in E° values by $\Delta G^\circ = -nFE^\circ$ with n the valence of electron transfer, and F the Faraday constant. λ can be determined experimentally. There are explicit molecular and implicit continuum models for calculation of λ. The relative merits of these computational approaches are discussed by Voorhis et al. [100]. It can easily be shown (see Exercise 8.17) that the transition state occurs at

$$q^\ddagger = 0.5 + \Delta G^\circ/2\lambda. \qquad (8.4.3)$$

The activation energy, which is given by the intersection point $\Delta G^\ddagger = G_r(q^\ddagger) = G_p(q^\ddagger)$, is

$$\Delta G^\ddagger = (\lambda + \Delta G^\circ)/4\lambda. \qquad (8.4.4)$$

Within transition state theory, the rate constant for electron transfer can be written

$$k = \kappa A \exp(-\Delta G^\ddagger/RT), \qquad (8.4.5)$$

where κ is the transmission coefficient, A the preexponential factor, and R and T are as usual the gas constant and absolute temperature. Marcus has argued that in many instances $\kappa \approx 1$ for outer shell electron transfer in solution [101], and assertion that needs to be tested at surfaces.

The TST expression for the rate constant combined with Eq. (8.4.4) makes for an extraordinary prediction, which is displayed graphically in Fig. 8.12. As the driving force for the reaction, represented by $-\Delta G^\circ$, increases, the rate constant increases up until the point where $\Delta G^\circ = -\lambda$. *At this point k reaches a maximum and decreases as the reaction becomes increasingly exergonic*, i.e. more thermodynamically favoured. This is known as the inverted region. The inverted region is intimately related to the efficiency of natural and artificial photosynthetic systems. Photosynthesis is shut down if the electron initially excited by photon absorption immediately relaxes back from the acceptor state back into the donor state from which it was excited. However, for sufficiently large donor-acceptor energy differences, the greater the energy stored by electron transfer from the donor to the acceptor, the slower back electron transfer will be. Molecular chemiluminescence also relies on the inverted region. Chemiluminescence occurs when an electronic excited state intervenes between the transition state and the product ground state. If the excited state couples more strongly to the transition state than the ground state, the excited state can be populated as a result of the charge transfer reaction. Radiative relaxation then leads to the conversion of chemical energy (effectively the exothermicity of the reaction) into light as the system drops down to the ground state.

8.4.2 Corrections to and improvements on Marcus theory

Marcus theory provides such a nice picture and is predictive of interesting physics, what could possibly go wrong [99, 102]? To answer this question, let us redefine the rate constant in a quantum mechanical form:

$$k = \frac{2\pi}{\hbar}|H_{DA}|^2(4\pi\lambda k_B T)^{1/2}\exp(-\Delta G^\ddagger/k_B T). \qquad (8.4.6)$$

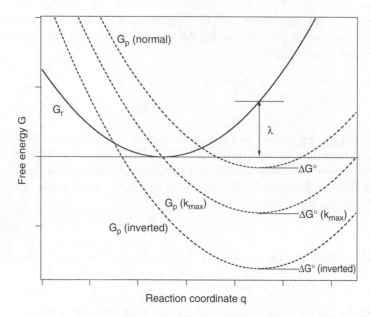

Figure 8.12   *The rate constant for electron transfer passes through a maximum at $\Delta G° = -\lambda$ as the relative values of λ and $\Delta G°$ change, and the system passes from the normal region ($\Delta G° < |\lambda|$) to the inverted region ($\Delta G° > |\lambda|$).*

The coupling matrix element depends on orbital overlap and symmetry. Thus, it varies exponentially with distance according to

$$H_{DA} = \langle D|H|A \rangle^2 = V_0 \exp[-\beta(R - R_0)/2] \tag{8.4.7}$$

where V_0 is the donor/acceptor coupling matrix element at the van der Waals separation R_0, and β is a constant that typically lies in the range 0.8–1.2 Å^{-1} [102]. Furthermore, for certain symmetry combinations, it may vanish or only be allowed due to vibronic mixing of non-totally-symmetric modes.

First, note that the intersection point drawn in Fig. 8.11 is reached only in cases where the coupling between the reactant and product state is weak, $H_{DA} \ll \lambda$. As H_{DA} grows, the intersection takes on the shape of an avoided crossing, the reaction surface rounds off, and is better described as one single adiabatic potential energy surface rather than a curve crossing. Second, by introducing quantum mechanical behaviour, we have opened up the possibility for the involvement of tunnelling. Tunnelling becomes important at low temperatures and causes temperature-independent electron transfer rates.

The transition state approach behind Marcus theory relies on the establishment of a quasi-equilibrium between the reactants and the transition state. This will hold as long as electron transfer is not too fast. The quasi-equilibrium cannot be maintained when electron transfer enters the ultrafast regime. This depletes the high-energy portion of the distribution of states and leads to departures from Boltzmann statistics known as Kramers dilemma.

A breakdown of the linear response of the solvent would lead to potential energy surfaces with nonparabolic shapes or, at the very least, different curvatures for the reactant and product curves. The relationship found in Eq. (8.4.4) was derived from a continuum approximation for the modelling of the solvent. If solvent molecule dynamics become directly involved in the dynamics of electron transfer, deviations are expected, and the value of $\Delta G^{\ddagger}$ would be modified.

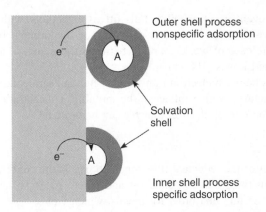

Figure 8.13 *Schematic representations of inner shell and outer shell heterogeneous electron transfer at a solid electrode.*

The inverted region is not well described by a parabolically decreasing rate constant as predicted by Marcus theory. This may result from the involvement of a number of vibronic states that couple to the transition state [103]. Alternatively, it may have something to do with the woolly nature of the reaction co-ordinate mentioned above.

Finally, if electron transfer occurs heterogeneously at a metal or semiconductor interface, new considerations arise. These complications are dealt with in the next section.

8.4.3 Heterogeneous electron transfer

We now turn to charge transfer between a solute and a solid electrode. We can imagine two limiting cases for electron transfer with an electrode occurring either with specific or nonspecific adsorption. Here I denote the specific adsorption case "inner shell" and the nonspecific adsorption case "outer shell" electron transfer in analogy with homogeneous phase reactions. These are illustrated in Fig. 8.13. The inner shell electron transfer case involves direct specific adsorption of A on the surface of the electrode. Specific adsorption proceeds with A losing part or all of its solvation shell, and leads to the formation of a strong chemical interaction (chemisorption) of A with the substrate. Effectively, the chemisorption bond and a surface atom act as the bridge in the heterogeneous inner shell process. Our discussion is framed in terms of electron transfer from a donor level $|D\rangle$ in the electrode to an acceptor level $|A\rangle$ on the solute.

Since Eq. (8.4.7) shows that electron transfer decreases exponentially with distance and that the range parameter is of order $\beta \approx 1\ \text{Å}^{-1}$, then essentially all ($> 90\%$) electron transfer occurs within $\sim 3\text{Å}$ of the surface. Since the width of a water molecule is roughly this size (diameter $= 2.76$ Å), this means that essentially all electron transfer occurs when the acceptor is adsorbed on the surface. Or, put more precisely, electron transfer occurs in the distance range between the closest approach of A with the surface (which may include its inner solvation shell) and this distance plus the width of one water molecule.

Outer shell electron transfer involves the close approach to the surface of the solute with its full inner sphere of solvation, which may, for example, contain six or so water molecules. The electron transfer occurs by tunnelling from an occupied state at or below E_F into the not-fully-occupied acceptor level $|A\rangle$. In the case of a metal, we expect the electron to come from a state close to E_F. From a semiconductor with conventional doping and temperatures compatible with aqueous solutions, the conduction band has a very low density of occupied states, and the band gap is devoid of states (with the possible exception of surface states). Thus, the electron is most likely to come from the top of the valence band at energy E_V.

Because this is a tunnelling event, its probability falls off exponentially with distance. It also requires favourable orbital overlap, which means that the symmetry of the orbitals $|D\rangle$ and $|A\rangle$ can influence the tunnelling probability. The case of weak coupling between these states has been treated by Marcus [104–106], Gerischer [107] and Lewis [108–110].

There are major differences between electron transfer at metal and semiconductor electrodes. These are due to the different band structures, which influence the mobility, concentration and energy of electrons at their surfaces. In general, the rate of electron transfer can be written

$$r_{et} = c_{red}c_{ox}\,k_{et} \tag{8.4.8}$$

where c_{red} is the concentration of the reductant (the donor D), c_{ox} is the concentration of the oxidant (the acceptor A). Because the density of electron states at and below E_F is so large and constant, electron transfer kinetics is effectively pseudo first order for metals, which yields

$$r_{metal} = c_{ox}k_{metal} \tag{8.4.9}$$

However, at semiconductors, the electron density in the conduction band or the hole density in the valence band is very small and, therefore, second-order kinetics is expected

$$r_{sc} = n_s c_{ox}\,k_{sc} \tag{8.4.10}$$

where n_s is the density of occupied states at the band edge (E_V for a semiconductor). This means that the units of the rate constants k_{metal} and k_{sc} differ [108, 109]. In order to get the units correct, Eq. (8.4.6), which was developed for homogeneous reactions, has to be modified slightly to account for the density of states and the distribution of the acceptor.

A second major difference between electron transfer kinetics at metal electrodes and semiconductors is that *the voltage drop near a metal electrode occurs completely in the solution* because the electrons in the metal are so highly polarizable. Therefore, changing the voltage on the metal electrode changes the current exponentially (as shown explicitly below), because the relative position of the Fermi level with respect to the donor/acceptor level in the solution varies. Since this energy difference influences the activation energy for charge transfer, the current changes exponentially. On the other hand, *the voltage drop at a semiconductor electrode occurs completely in the space charge region*. Therefore, changing the voltage does not change the relative positions of the band edges at the surface compared to the donor/acceptor level in solution. Instead, the current varies exponentially with applied voltage because the surface density of states changes with bias.

In other words, bias voltage changes the rate of electron flow through a metal electrode because it changes the rate constant for electron transfer. In contrast, bias voltage changes the rate of electron flow through a semiconductor electrode because it changes one of the concentrations in the rate equation.

To obtain a feeling for the qualitative implications of these differences, consider, for example [108], the ferrocenium/ferrocene ($Fe^{+/0}$) redox system. The difference in dielectric constant between a metal electrode such as Pt and a semiconductor such as Si is quite large. Consequently, the image charge effects in the two systems are much different [106, 111] with efficient screening at a metal surface. This reduces λ_0, the inner shell contribution to λ, at a Pt electrode to about half the value found for a homogeneous electron transfer, but leaves λ_0 little changed at a Si electrode. Thus $\lambda_0 = 0.5\,eV$ for a Pt electrode versus $1.0\,eV$ for a Si electrode. Using $n_{s0} = N_c\exp[(-0.9\,eV)/k_BT]$, with $N_c = 10^{19}$ cm^{-3} for the effective density of states in the Si conduction band, n_{s0} is 10^4 cm^{-3}, which yields $k_{sc}^{\circ} = 10^{-12} - 10^{-13}$ cm s^{-1}, and exchange current densities of 10^{-11} A cm^{-2} at $[Fe^+] = 1.0\,M$. In contrast $k_{metal}^{\circ} = 400$ cm s^{-1} for $Fe^{+/0}$ at Pt, which would lead to an exchange current density of $J_0 = 4 \times 10^4$ A cm^{-2} at $[Fe^+] = 1.0\,M$. The latter prediction cannot be measured directly because the current density is restricted by diffusion before it reaches this ideal limiting value. The 15 orders of magnitude difference is related primarily to the reduction in the electron density in the semiconductor electrode relative to that at the metal, not to a change in mechanism.

Within a transition state theory framework, [101, 108, 110] the rate constant of electron transfer can be written in terms of a pre-exponential factor and a Gibbs energy of activation as usual

$$k_{\mathrm{et}} = A \exp(-\Delta G^{\ddagger}/k_{\mathrm{B}}T). \tag{8.4.11}$$

The units of A are different for a metal (s^{-1}) and for a semiconductor ($m^4\ s^{-1}$) as discussed above. Due to the differences in band structure, the expression for the activation energy is also different. For a metal electrode held at a potential E

$$\Delta G^{\ddagger} = (E - E_{\mathrm{ox}} + \lambda)^2/4\lambda. \tag{8.4.12}$$

For a semiconductor

$$\Delta G^{\ddagger} = (E_{\mathrm{V}} - E_{\mathrm{ox}} + \lambda)^2/4\lambda. \tag{8.4.13}$$

E_{ox} is the Nernst potential of the oxidant,

$$E_{\mathrm{ox}} = E^{\circ} - (RT/nF)\ln Q \tag{8.4.14}$$

λ is the reorganization energy, and E_{F} or E_{V} is referenced to a common vacuum level along with E_{ox} (see §5.8.2).

After electron transfer has occurred, the acceptor molecule with an additional electron A$^-$ can undergo further reactions that may be either heterogeneous or homogeneous. The rate determining step for any given reaction may be either the electron transfer step or one of these chemical reactions.

As long as the acceptor level $|A\rangle$ approaches the donor level $|D\rangle$ from above, that is, from a higher energy, it can be assumed that electron transfer occurs either at the Fermi level for a metal or else the appropriate band edge for a semiconductor after the acceptor level has fluctuated to an appropriate energy to facilitate degenerate electron transfer [97, 98, 101]. In this case, all holes are injected into the electrode in a narrow energy range at E_{F}, the valence band maximum or conduction band minimum regardless of the ground state energy of the acceptor level on the oxidant. However, once the acceptor level lies below the energy of the occupied band, there are always states that are degenerate with respect to electronic energy available for electron transfer from the electrode. The solvation shell still needs to reorganize; therefore, there is still an activation energy.

8.4.4   Current flow at a metal electrode

The above theory is refined in this section to derive expressions for the current flow at a solid electrode in contact with a redox couple composed of Ox$^+$ and Red in solution, following closely the treatment of Gerischer [107]. Each of these species is solvated and the equilibrium solvation shell about Ox$^+$ differs in structure from the equilibration solvation shell about Red. This means that if we put Red into the equilibrium solvation shell associated with Ox$^+$, the system is not in its equilibrium configuration; rather, it is in a configuration that is higher in energy by an amount λ_{red} known as the reorganization energy. Similarly, if Ox$^+$ is placed in the equilibrium solvation shell of Red, the system differs in energy by an amount λ_{ox}. Here we assume that $\lambda_{\mathrm{red}} = \lambda_{\mathrm{ox}} = \lambda$, as was done implicitly above.

The cathodic process occurs from an occupied state in the electrode (the solid surface) to an empty state in the electrolyte, Ox$^+$. The anodic process involves electron transfer from an occupied state on the electrolyte species Red to an empty state in the electrode. Conservation of energy demands that the levels involved in electron transfer need to be degenerate. Thus the transition shown in Fig. 8.10 is a horizontal transition. At equilibrium, where there is no thermodynamic driving force, the anodic and cathodic current densities, j^- and j^+ respectively, are equal in magnitude (no net charge flow) to a quantity known as the exchange current j_0. The anodic and cathodic currents are related to the concentrations of the Ox^+ and $Red - N_{ox}(z)$ and $N_{red}(z)$ where z is the distance from the electrode surface – in terms of ions (or molecules) per cm^3; the concentrations of occupied or empty states in the electrode at an energy $E - D_{occ}(E)$ and $D_{emp}(E)$ – expressed in states cm^{-3} eV^{-1}; the tunnelling probability for electron transfer as a function of E and z – denoted $\kappa(E, z)$; and the probabilities – $W_{ox}(E)$ and $W_{red}(E)$ – that the electronic energy level in the respective redox species reaches by thermal fluctuations an energy level E. The resulting expressions are

$$j^- = \iint N_{ox}(z)\kappa(E,z)D_{occ}(E)W_{ox}(E) \; dE \; dx \qquad (8.4.15)$$

$$j^+ = \iint N_{red}(z)\kappa(E,z)D_{vac}(E)W_{red}(E) \; dE \; dx. \qquad (8.4.16)$$

We now specifically deal with electron transfer at a metal electrode. While Eqs. (8.4.15) and (8.4.16) are also valid for semiconductors, the presence of a band gap in a semiconductor means that both of these expressions have to be split into two integrals covering the energy range of the valence band and the conduction band, respectively. Thereby, one also has to distinguish whether the charge transfer occurs via the conduction band or valence band.

At equilibrium, the Fermi energy of the metal and the free energy of the electron in the redox couple must be equal. Equivalently, this can be stated that the chemical potentials of an electron in the solid must be equal to the electrochemical potential of an electron in the redox couple. As argued by Gerischer [112], Gomer [113] and Reiss [114, 115], this equivalence allows us to identify the Fermi energy with the redox potential. Charging of the double layer in the solution causes a shift in the electronic states on both sides of the interface relative to each other such that the Fermi energy of the metal and the free energy of the electrons coincide. The Galvani potential difference $\Delta\varphi_0$ quantifies the effect of double layer charging such that at equilibrium

$$E_{F, \; metal} - e\Delta\varphi_0 = E_{F, \; redox}^0. \qquad (8.4.17)$$

At low temperatures compatible with aqueous solutions, the sharp separation between filled and empty states at E_F in the metal means that electron transfer can only occur in a narrow energy range close to E_F.

With suitable approximations, the probability of reaching the energy level at $E_{F, \; redox}^0$ requiring $\Delta E = \lambda$ is, therefore,

$$W(\Delta E = \lambda) = (4\pi\lambda k_B T)^{-1/2} \exp\left(\frac{-\lambda}{4k_B T}\right), \qquad (8.4.18)$$

for both Ox^+ and Red. The exchange current density in A cm^{-2} assuming unequal concentrations $N_{ox}(z) \neq N_{red}(z)$ is approximated by

$$j_0 = k_0 \, e(N_{ox, \; 0}N_{red, \; 0})^{1/2} \exp\left(\frac{-\lambda}{4k_B T}\right), \qquad (8.4.19)$$

where k_0 is a rate constant with dimensions cm s^{-1}. Note that the density of states at the Fermi energy does not appear in this expression as it is incorporated into k_0 consistent with the kinetics discussed above.

Current flow is induced by placing a bias on the metal electrode. The overpotential η is defined as the potential of the metal relative to the electrolyte compared to the equilibrium value

$$\eta = \Delta\varphi_0 - \Delta\varphi. \qquad (8.4.20)$$

The effect of the overpotential is to shift the position of the metal's electronic states relative to the electrolyte by an amount $-e\eta$. At anodic ($\eta > 0$) or cathodic ($\eta < 0$) polarization, the position of $E_{F,\,metal}$ relative to the reference state in the electrolyte is given by

$$E_{F,\,metal}(\eta) = E_{F,\,redox}^0 - e\eta. \tag{8.4.21}$$

The resulting anodic and cathodic current densities are then

$$j^+ = j_0 \exp\left(\frac{1}{2}\frac{e\eta}{k_B T}\right) \tag{8.4.22}$$

$$j^- = j_0 \exp\left(-\frac{1}{2}\frac{e\eta}{k_B T}\right) \tag{8.4.23}$$

The symmetrical factors of $1/2$ are only approximate and hold for a symmetrical energy level distribution of the levels of Ox^+ and Red about $E_{F,\,redox}^0$. More generally, for a one-electron transfer they are replaced by the (apparent) charge transfer coefficients α and β subject to the constraint

$$\alpha + \beta = 1. \tag{8.4.24}$$

If we also allow for the concentrations N_{red} and N_{ox} to diverge from the starting values $N_{red,\,0}$ and $N_{ox,\,0}$, the total current density is given by

$$j = j_0 \left[\frac{N_{red}}{N_{red,\,0}} \exp\left(\frac{\alpha e\eta}{k_B T}\right) - \frac{N_{ox}}{N_{ox,\,0}} \exp\left(-\frac{\beta e\eta}{k_B T}\right)\right]. \tag{8.4.25}$$

Equations of this form are known as Butler-Volmer equations [116, 117].

Advanced Topic: Semiconductor photoelectrodes and the Grätzel photovoltaic cell

The rate of charge transfer at a semiconductor electrode responds to an applied voltage much differently than at a metal electrode [118]. The electron concentration at the surface of a semiconductor n_s (cm^{-3}) is related to the potential applied to the electrode E, and the flat band potential E_{fb} by

$$n_s = n_{s0} \exp(q(E_{fb} - E)/k_B T) \tag{8.4.12}$$

where n_{s0} is defined as the value of n_s at E equal to $E(Ox^+/Red)$, the Nernst potential of the redox couple in solution. At forward bias, n_s increases exponentially with $-E$, and the net current across the liquid/semiconductor interface increases. The net flux of electrons $J(E)$ in A cm^{-2} from the conduction band to the acceptor of concentration $[Ox^+]$ (cm^{-3}) in solution is given by

$$J(E) = -ek_{et}[Ox^+]n_s \tag{8.4.13}$$

where k_{et} is the electron transfer rate constant ($cm^4\ s^{-1}$). Unlike the case of a metal electrode, this is a second-order rate law that explicitly contains the electron concentration. The rate of charge transfer changes in response to an applied potential because the potential changes the electron concentration rather than changing the rate constant or the energetics.

A photovoltaic cell converts light into electricity and is the basic technology behind a solar cell. Light is absorbed in a semiconductor to create an electron-hole pair. Two biased electrodes are required to separate the charges and collect the electrons and holes. The voltage produced by the cell is related to the separation between energy levels, while the current is proportional to the amount of light absorbed minus the number of carriers lost to recombination of electron-hole pairs. The external efficiency of a solar cell is defined as the ratio of the incident power to the electrical power produce

$$\Phi_{ext} = P_{in}/P_{out}. \tag{8.4.14}$$

The external efficiency can be related to specific materials properties. First, the light harvesting efficiency *LHE* at a particular wavelength λ is defined as

$$LHE(\lambda) = 1 - 10^{-\alpha d}, \tag{8.4.15}$$

where α is the absorption coefficient and d the thickness of the cell. The external quantum efficiency *EQE* (also called incident photon to charge carrier conversion efficiency, IPCE) relates the number of electrons produced in the external circuit at a particular wavelength to the number of photons at that wavelength incident on the device

$$EQE(\lambda) = LHE(\lambda)\,\phi_{inj}\,\eta_{coll} \tag{8.4.16}$$

where ϕ_{inj} is the quantum yield of electron injection from the excited sensitizer to the conduction band and η_{coll} is collection efficiency of the external circuitry. The quantum yield of injection is related to the competition between the rate of injection (characterized by the rate constant k_{inj}) compared to all radiative and nonradiative deactivation processes (characterized by k_{deact})

$$\phi_{inj} = \frac{k_{inj}}{k_{inj} + k_{deact}}. \tag{8.4.17}$$

Defects in the bulk crystal structure and recombination centres on the surface, known as traps, provide sites for the recombination and deactivation of free carriers. Removal of traps (passivation) often involves a chemical process such as the termination of dangling bonds with a suitable adsorbate. Both oxygen and hydrogen can accomplish this for silicon. Extremely low coverages of surface traps are required for optimal photovoltaic cell operation.

Dye molecules act as the sensitizer – the absorber of the radiation – in a Grätzel cell. Dyes in the solution phase would efficiently lose their electronic excitation by radiative decay with almost unit quantum efficiency. However, here we take advantage of the strongly quenched nature of electronic states in chemisorbates. In the case of an adsorbate that is strongly coupled to a semiconductor surface, the excited state rapidly loses its energy by injecting the excited electron into the conduction band of the substrate. Injection lifetimes on the order of 1 ps or less must be attained such that injection competes favourably compared to deactivation by the electrolyte. With properly engineered molecules, injection efficiencies exceeding 90% have been achieved.

The external efficiency is reduced by recombination and reflection since both reduce the number of carriers that are extracted. Recombination occurs either when conduction band electrons are injected back into the sensitizer or if they are used to reduce triiodide

$$I_3^- + 2e^-(TiO_{2.cb}) \longrightarrow 3I^-. \tag{8.4.18}$$

As shown below, the iodide/triiodide redox couple is used to complete the circuit in the Grätzel cell.

External efficiency (but not the external quantum efficiency) is also reduced by the relaxation of hot electrons and holes. Above we saw that excited electrons are subject to ultrafast relaxation. Electrons excited above the conduction band minimum rapidly relax to E_c, then reside there waiting to recombine unless they are extracted under the influence of an electric field. The difference between the initial electron energy and E_c is lost to the cell as heat. Therefore, the most efficient photovoltaic cell is created by matching the absorption characteristics of the semiconductor to the light source such that the band gap of the semiconductor is only slightly smaller than the photon energy. For a broadband source such as solar radiation, which is 5% UV, 46% visible and 49% near IR, this is impossible with a single material. However a multi-junction solar cell composed of, say, one material to absorb blue light, another to absorb red and a third to absorb IR can approximate this effect rather effectively.

Some of the characteristics of materials used in solar cells are shown in Table 8.1. If solar energy is going to be implemented widely, thousands of square meters of solar panels will have to be installed. By far the most common semiconductor material used in solar cells is silicon in either single crystalline, multicrystalline or amorphous form. Silicon is a great choice of material based on its abundance in the Earth's crust, and because of the vast experience of producing silicon microelectronics. In 2006, for the first time, the amount of Si used in solar cells exceeded the amount used to produce integrated circuits. Abundance does not necessarily translate into inexpensiveness since the Siemens method of producing ultrapure wafer grade silicon is ridiculously expensive and the world capacity has experienced constraints. The market cries out for less expensive methods of producing silicon or alternative materials. Combining Ge with Si in the form of superstructures, alloys or quantum dots is also actively being pursued.

Table 8.1 *The price, world production, economically recoverable world reserves (both in thousand metric ton, kMt) and reserves to production ratio (R/P) for several materials commonly associated with solar cells [119]*[2]

Material	Price/$ kg^{-1}	World production/kMt	World reserves/kMt	R/P
As	2.09	59.2	1 776	30
Al	2.64	33 100	At 8%, 3rd most abundant element in Earth's crust	∞
Au	21 472	2.5	90	36
Cd	2.80	20.9	1 600	77
Cu	6.80	15 300	940 000	61
Ga	500 ~ 2000 as GaAs wafer	0.16	1 000	6250
Ge	880	0.1	no estimate	–
In	855	0.48	6	12.5
Sb	4.95	131	3 900	30
Se	66	1.39	170	122
Si	0.77 metallurgical grade 100–300 as wafer	4 700	At, 28%, 2nd most abundant element in Earth's crust	∞
Sn	11.44	273	11 000	40
Te	220	~ 0.128	47	367
TiO$_2$ (rutile)	4.65/Mt2.57 pigment grade	444	100 000	225
Zn	3.19	10 000	460 000	46

GaAs is an attractive material based on performance, but it is not so easily grown into large wafers. The supply of Ga is limited not by its abundance, but because it is produced as a by-product of Zn and Al smelting. Indium is very useful either in the form of indium tin oxide (ITO), a transparent conducting oxide, or in compound semiconductors. Its supply is severely constrained as it too is a by-product of Zn smelting and is predominantly consumed in the manufacture of LCD flat panel displays and electroluminescent lamps. Copper indium disulfide ($CuInS_2$), known as CIS or CIGS when some Ga is substituted for In, and $CuInSe_2$ are other In containing materials, as well as from $Ga_xIn_{1-x}As$, $Al_xGa_yIn_{1-x-y}P$, $Ga_xIn_{1-x}P$, that are actively being investigated. Another direct bandgap material receiving much attention is CdTe, particularly attractive since it avoids the use of In. Based on abundance and cost analysis, some unconventional photovoltaic materials, such as iron pyrite FeS_2 and zinc phosphide Zn_3P_2, have been proposed as worthy of further development [120].

Titania (TiO_2 in the rutile or anatase forms) is a wide band gap semiconductor that is found naturally in abundant quantities. The ~3 eV band gap of this material means that it is a great scatterer rather than absorber of sunlight and for this reason it is commonly used as a white pigment in paints, etc. That it does not absorb sunlight appears to rule it out as a material of interest for solar cells. However, this is not the case as shown by Grätzel [121, 122].

The idea behind the dye sensitized cell (DSC) or Grätzel cell is illustrated in Fig. 8.14. TiO_2 has a band gap too large to absorb sunlight. However, a suitable dye molecule adsorbed to the surface could potentially absorb solar radiation efficiently (or several different ones can be selected to absorb different parts of the solar spectrum). If a dye molecule can be found that has an excited state located at an energy just above the band gap of TiO_2, charge injection out of the excited state and into the semiconductor should be quite efficient. The TiO_2 anode onto which the dye is adsorbed is biased to transport the electron to a conducting glass such as ITO or fluorine-doped tin oxide (FTO). Apart from the anatase form of TiO_2 also ZnO, SnO_2 and Nb_2O_5 have been investigated as semiconducting substrates. Even though only a monolayer of dye is adsorbed on the semiconductor, photon absorption can be quite high if a high surface area nanocrystalline/porous semiconductor is used. Because of the large band gap of the semiconducting nanoparticles, the skeleton of the porous network, which they comprise, does not absorb light. Only the dye molecules adsorbed on the surface of the nanoparticles absorb light. Ruthenium complexes are commonly employed as the dye.

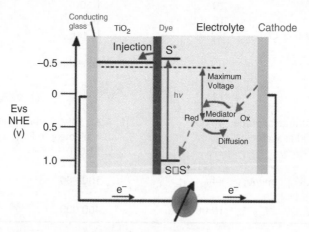

Figure 8.14   *Schematic diagram of a Grätzel cell. Reproduced from M. Grätzel, Inorg. Chem. 44, 6841.* © *2005 with permission from the American Chemical Society.*

The circuit needs to be closed by electron injection from the cathode back to the dye. This does not occur directly because the cathode is physically separated from the dye-covered surface. Instead, a redox couple is used to transport the electron and neutralize the adsorbed dye molecule without reacting with it chemically. The electrolyte usually consists of an organic solvent containing the iodide/triiodide redox couple. Electron transfer from I^- to the dye ensures that the electron injected into the conduction band does not recombine with the dye, thus negating the photoeffect. The I^- is regenerated by reduction of triiodide at the cathode. The cell voltage corresponds to the difference between E_F of the semiconductor and the redox potential of the electrolyte.

Quantum dot-based devices are another area of active research for advanced solar cell device architectures [123]. Quantum dots can take the place of the dye sensitizer. Both CdS and CdSe quantum dots can inject electrons into TiO_2, SnO_2 and ZnO after absorption of light. PbS, InP and InAs quantum dots covering TiO_2 have also attracted attention. One reason why quantum dots are so attractive is that quantum confinement effects allow their absorption spectrum to be tuned by controlling their size so that their adsorption spectrum can be matched to the solar irradiance. A particularly exciting advance is the production of more than one electron per absorbed photon in PbSe nanocrystal quantum dots [124].

8.5   Tip Induced process: mechanisms of atom manipulation

When the tip of a scanning probe microscope (SPM), particularly a scanning tunnelling microscope (STM) is brought close to an adsorbate covered surface there are a number of mechanisms by which the tip can perturb the system. First we list them here and then we look at each in turn in more detail.

- Electric field effects
- Electron (or hole) stimulated desorption
- Vibrational ladder climbing akin to DIMET/electron friction
- Pushing adsorbates with repulsive forces
- Pulling adsorbates with attractive forces
- (Reversible) chemical bonding to tip

We also need to note that there are two very different modes of electron emission from an STM. In field emission mode, the bias on the tip is made sufficiently negative that electrons are excited above the work function barrier and accelerated out of the tip toward the surface. The kinetic energy of the electrons is set by the voltage difference between the tip and surface. The only distance dependence involved in field emission is in the current density rather than the current as the field emission beam is divergent. In tunnelling mode, the electrons are not excited above the barrier that holds them in the metal. Instead, they tunnel through this barrier. The tunnelling rate is dependent on overlap with the state that accepts the electrons. The kinetic energy of the electron is zero because the electron is transported quantum mechanically through the barrier to a level with the same energy on either side of the barrier. When the electron tunnels directly into a localized state, for instance a surface state or an adsorbate related state, the process is termed resonant electron tunnelling.

8.5.1   Electric field effects

In imaging mode an STM tip is brought within ~1 Å of the surface and is biased to a voltage of ~1 V. This corresponds to an enormous electric field of 1 V Å^{-1} and fields of this magnitude represent a significant perturbation to the chemical and physical environment in which adsorbate and substrate atoms find themselves. These fields polarize bonds potentially leading to field induced desorption, can induce new localized modes just below the tip [125], and influence the trajectories of any charged particles. For imaging, the tip-sample distance is often increased by operating at the lowest possible tunnelling current. The increased separation reduces the field and lessens the likelihood of perturbing the sample. If manipulation is the aim of the tip/surface encounter, then the current is cranked up so that the tip approaches the surface as closely as possible without crashing into it. For any biased tip, the electric field must be taken into account for all of the interactions listed below.

8.5.2   Tip induced ESD

An SPM tip can be used as a localized source of electrons. If the bias is properly adjusted, the emitted electrons can cause electronic excitation just as a far-field electron gun can. These electrons can induce electron stimulated desorption and the mechanisms discussed above are just as valid for tip induced ESD as they are for an external electron gun. There are some modifications that need to be mentioned. First is the localized nature of the electron irradiation. A tip acting in field emission mode sprays electrons onto only a few nearby atoms. Only these atoms are excited and, therefore, the removal of atoms by ESD can be a highly localized process. This was demonstrated by Chabal and co-workers [126] for H adsorbed on Si(111). Excitation can also be caused by the injection of holes. The injection of electrons versus holes is controlled by controlling the tip to sample bias denoted V_S, with positive V_S defined as the surface held positive with respect to the tip and $V_S < 0$ when the surface is negative with respect to the tip. Electrons always flow from occupied states to unoccupied. Holes are injected from empty states into filled states.

The various tip induced electronic excitations are illustrated in Fig. 8.15. In panel (a) the bias is set such that the Fermi level lines up with the energy of an unoccupied adsorbate level to affect resonance charge transfer. A temporary negative ion state is formed on the adsorbate (electron attachment). This state may itself be a repulsive (or attractive) state either with respect to the adsorbate–surface bond or one of the intermolecular adsorbate bonds. Decay of this state by loss of the electron into the substrate may lead to vibrational excitation as discussed in §2.4.2. In Fig. 8.15(b) the surface is biased negatively and holes flow from unoccupied states near E_F of the tip into an occupied state of the adsorbate. The evolution of the temporary positive ion adsorbate state by hole attachment is then analogous to the evolution of a temporary anion: it can be neutralized by charge transferred by from the substrate, which can be accompanied by electronic or vibrational excitation of the adsorbate compared to its initial state before hole injection.

The flux of excitations is related to the current. Hence by analogy to Eq. (8.2.10), we should expect the rate of change of coverage for desorption to be proportional to the tunnelling current according to

$$\frac{d\theta}{dt} \propto I^n \tag{8.5.1}$$

The power n can be associated with the number of electrons involved in the process if direct excitation of the adsorbate is responsible for the tip induced process. However, just as in the case of laser excitation, if the excitation is substrate-mediated then the exponent n is not directly interpretable in terms of the number of electronic excitation required to bring about change.

The H/Si system illustrates some of the complications introduced by the tip, in particular, what is the final state of the desorbed product? For ESD with an external electron gun source, the question is fairly easily answered and we expect H$^+$, H and H$^-$ to be formed in various proportions that depend on the exact details of the system including T_s, $\theta(H)$, which Si plane is used and the electron energy. When a

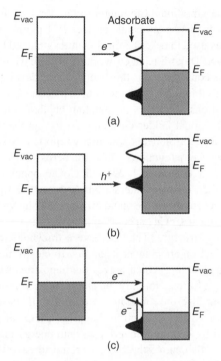

Figure 8.15   *Electronic excitations associated with tip induced processes. (a) Electron injection at $V_S > 0$ into an empty adsorbate state leads to temporary anion formation. (b) Hole injection at $V_S < 0$ leads to temporary cation formation. (c) Inelastic electron scattering with $V_S > 0$ leads to direct electronic excitation. After Comtet et al., Phil Trans. Roy Soc. London Ser. A, 362, 1217. © 2004 with permission from the Royal Society.*

tip is used for excitation, several nearby atoms can be excited simultaneously. The hydrogen may be able to leave in this case by recombinative desorption to form H_2. The tip can also influence the desorption trajectories both because of the electric field and because of its physical presence. If the desorbing particle lands on the tip, the final state of the product is then an adsorbed H atoms or H_2 molecule (or two H atoms if a desorbed molecule adsorbs dissociatively on the tip). The final state for the particles desorbed by an external source was a free gas phase particle; consequently, the energetics of tip induced ESD can be completely different than electron gun induced ESD.

8.5.3   Vibrational ladder climbing

If the current of electrons flowing out of a tip can be increased sufficiently such that the rate of excitation from these electrons rivals the rate of vibrational relaxation for the adsorbate, a vibrational ladder climbing mechanism of desorption similar to that discussed above as DIMET/electronic friction can occur. Both of these limits, one in which direct electronic excitation is involved and one in which only direct vibrational excitation occurs, are possible. The latter mechanism is often thought to be more important; however, again, these two limits are not as distinct as they may at first seem because of the possibility of the adsorbate having a resonance near E_F.

Avouris and co-workers [127] were the first to observe a vibrational ladder climbing mechanism of desorption for H/Si(100) and they demonstrated that single atoms wide lines can be fashioned as a result. This represents the ultimate in lithographic resolution. The H/Si system is particularly well suited for this type of desorption mechanism because the vibrational lifetime is so long ($\sim$1 ns on Si(111) and $\sim$10 ns on Si(100) [33]) compared to, for instance, adsorbates on metal surfaces ($\sim$1 ps). At first an electronic friction model similar to that discussed above was used to describe the dynamics induced by inelastic scattering of the electrons emitted from the tip. However, Dujardin and co-workers [128] have recently suggested that a modified desorption mechanism should be considered. In this mechanism, developed by Persson and co-workers [125], *coherent multiple* quantum vibrational excitations take place rather than incoherent excitations. The roles of these two types of excitations remain an outstanding challenge in dynamics of surface processes.

Both tip induced ESD and vibrational ladder climbing are low probability events, with efficiencies of 10^{-6} desorption events per incident electron being commonly reported. Note that desorption need not be the only outcome of either of these types of excitations. Diffusion across the surface and chemical reactions can also occur. For instance, Stroscio and co-workers [129] built $CoCu_n$ nanostructures using atom manipulation techniques discussed below and then resonant interactions between the tip and the electronic structure of the $CoCu_n$ nanostructure to move atoms in the nanostructure as shown in Fig. 8.16. Inelastic electron scattering vibrationally excites a linear $CoCu_2$ species and causes the Co atom to hop into a neighbouring site, which leads to a bent $CoCu_2$ structure. They observe a threshold for switching configurations at a bias of 15 mV and suggest that electron attachment leads first to electronic excitation (possibly to an ionic intermediate state) followed by quenching, which leaves the nanostructure in a vibrationally excited state that converts from the linear to bent forms. Not only the *energy of excitation* is important for atom resolved manipulation of nanostructures but also the *location of excitation*. If the excitation of a particular orbital is responsible for the manipulation event (desorption, hopping, chemical change), then the probability of the tunnelling event is maximized by maximum overlap in both energy and geometric space. The tip must be not only at the correct bias but also above the correct orbital to have maximum effect.

Ho and co-workers [130] used an STM tip to induce dissociation of O_2 chemisorbed at 50 K on Pt(111) one molecule at a time. Resonant inelastic electron scattering at biases in the range 0.2–0.4 V was used to vibrationally excite the O_2. O_2 is very weakly chemisorbed and was found in an fcc site, a bridge site and at steps. The behaviour of these three species with respect to inelastic electron scattering is distinctly different. O_2 in the fcc site only dissociates whereas the bridge bound O_2 can dissociate or desorb. The step bound O_2 exclusively desorbs. These results show that the dynamics of chemisorbed molecules can be very sensitively dependent on their adsorption configuration.

The adsorption configuration is also very important for inducing chemical bond formation with an STM tip. Rieder and co-workers [131] used an STM tip to cleave the I atoms from two adsorbed iodobenzene molecules by irradiating them with 1.5 V electrons. They then used lateral manipulation (see below) to position two phenyl fragments next to each other along a step on a Cu(111) surface. Because of the low temperature (20 K) and the presence of a reaction barrier, the phenyls do not spontaneously react. When irradiated with 0.5 V electrons, however, they couple to form a biphenyl molecule in a manner consistent with the copper catalyzed Ullmann reaction. The reaction was found to be much easier to induce for two phenyls bound along a step edge than on the terraces.

8.5.4   Pushing

In the macroscopic world, we can easily imagine taking a needle and pushing around small particles. As long as thermal energies are larger than the forces that might hold any of the particles to the needle, the particles do not stick to the needle and are pushed into new positions. This is an effective method of clearing a line but an ineffective method of positioning the particles since it is hard to control the trajectory of the particle. If the needle is idealized as a cylinder and the particle a ball, the difficulty in controlling the trajectory is understood by the high degree of sensitivity of the direction of momentum transfer to the exact angle of impact between two curved surfaces.

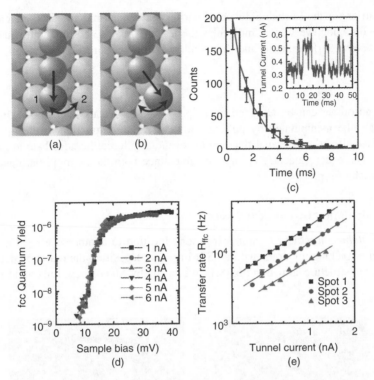

Figure 8.16 *Atomic manipulation of a CoCu$_2$ chain. Model of a CoCu$_2$ molecule on the Cu(111) substrate in (A) linear and (B) canted configurations, corresponding to the Co atom (dark) in the fcc and the hcp sites, respectively. (C) Distribution of residence times of the Co atom in the high current state with a fit to an exponential decay. (Inset) A portion of the tunneling current vs time trace obtained in the left vicinity of the Co atom in the CoCu$_2$ molecule at 15.4 mV sample bias. (D) Fcc quantum yield as a function of sample bias at fixed tip-sample separation. Symbols correspond to different set-point currents for each measurement. The tip-sample separation varied by 0.75 Å when changing the current set point from 1 to 6 nA at 40 mV sample bias. (E) Transfer rate for the Co atom out of the fcc site as a function of tunneling current obtained at a fixed sample bias of 40 mV for the CoCu$_2$ molecule. The three curves were obtained at different locations near the Co atom and are fit to I^n (lines). The average of the three data sets yields n = 1.3 ± 0.1 (lines). Reproduced from J.A. Stroscio, F. Tavazza, J.N. Crain, R.J. Celotta, A. M. Chaka, Science 313, 948. © 2006 with permission from the American Association for the Advancement of Science.*

Many of the same considerations can be brought to bear on the interaction of an inert SPM tip with an unreactive molecule, for instance, the interaction of two closed shell species or if the tip is functionalized with a molecule that does not bind to the adsorbate that is to be manipulated. If the repulsive force generated by the tip–adsorbate interaction can overcome the adsorbate's diffusion barrier, the adsorbate is pushed from one site to the next. The direction of pushing is generally somewhere in the forward half space (the semi-circular region in the plane of the surface in front of the tip) but the exact site into which the adsorbate is pushed can sometimes be hard to control. The height of the tip, scan speed and scan direction must all be controlled to control pushing. The manipulated atom might not follow the path of the tip directly since the diffusion potential is not flat, causing the adsorbate to follow crystallographically preferred paths that deviate from the exact course of the tip. In fact, there may be certain directions in which the diffusion barrier is so high that manipulation by pushing is impossible. Surface defects such as steps may pose insurmountable barriers or the Ehrlich-Schwoebel barrier may mean that an adsorbate can only be pushed over a step in one direction. Very large molecules are less sensitive to corrugation of the surface because they extend over several surface atoms and they tend to follow the direction of the tip more accurately.

8.5.5   Pulling

A weakly bound adsorbate can be attracted to a tip by van der Waals interactions. In this way it may attach to the tip and be manipulated by pulling. However, if the attraction to the tip is comparable to the adsorbate/surface attraction, thermal fluctuations eventually force the adsorbate to fall off the tip at a different location. The exact trajectory is unpredictable since both the hopping onto and the hopping off from the tip are stochastic processes.

8.5.6   Atom manipulation by covalent forces

To facilitate *controllable* placement of atoms from one location to another, we need a more precise and switchable method of picking up atoms with an SPM tip and depositing them at a predetermined location. This is how the corral structure shown in Fig. 8.17 was made by Manoharan, Lutz and Eigler [132]. Eigler

Figure 8.17   *A quantum corral produced by atom manipulation. This STM image shows Co atoms arranged into an ellipse on Cu(111). Under appropriate conditions, a Co atom placed at a focus within the ellipse leads to a mirage signal, which appears as a virtual atom at the other focus. Reprinted from H.C. Manoharan, C.P. Lutz, D.M. Eigler,* Nature (London) *403, 512. © 2000 with permission from Macmillan Magazines Ltd.*

and Schweizer [133] were the first to demonstrate controlled atom manipulation and the ability to make structures by positioning single atoms. They used what is now called lateral manipulation to write IBM in Xe atoms on a Ni(110) surface. Lateral manipulation involves moving the tip above the adsorbate, moving it along the surface to a predetermined location and then retracting the tip while leaving the atom in the new adsorption site. In a sliding mode of manipulation, the atom is trapped between the biased tip and the sample as the tip is moved. The electric field associated with the tip can either strengthen or weaken bonds in the region between it and the surface and this can facilitate the activation of local dynamical phenomena such as sliding of the atom into a new position [125]. Pushing and pulling are also examples of lateral manipulation.

Eigler was also the first to demonstrate vertical manipulation in which the molecule/atom is first adsorbed on the tip, the tip is retracted and moved to a predetermined location at which the molecule/atom is desorbed in a controlled manner. Vertical manipulation is much more controllable and can be extended to producing nanostructures. It can be used not only to move adsorbates but also to extract atoms from a surface and reposition them [134]. Kolb [135] has demonstrated that metal atoms can be placed onto a tip and then deposited onto a substrate in an aqueous electrochemical environment to form reproducibly sized and spaced nanostructures at a high rate. Vertical manipulation is able to transfer atoms and molecules to any position on the surface and is not impeded by defects, such as steps, as is lateral deposition.

Persson and co-workers [136, 137] and Avouris and co-workers [138] independently arrived at very similar explanations of the vertical transfer of Xe atoms. The essential ideas are embodied in Fig. 8.18. The Xe atom can be adsorbed either on the tip or the substrate. The binding, depicted by the wells shown in Fig. 8.18, is roughly equivalent on both surfaces when no bias is applied. Even if the tip is not originally composed of the same metal as the substrate, it can easily be coated with these atoms by ramming the

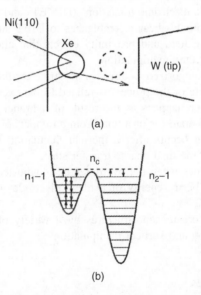

Figure 8.18 *The mechanism of vertical manipulation. (a) Schematic picture of the atomic switch. (b) Double well model for atom transfer based on truncated harmonic oscillators. In the vibrational heating mechanism, the atom transfer results from stepwise vibrational excitation of the adsorbate–substrate bond by inelastic electron tunnelling as depicted by arrows between the bound state levels of the adsorption well 1. Reproduced from S. Gao, M. Persson, B. I. Lundqvist, Phys. Rev. B 55, 4825. © 1997 with permission from the American Physical Society.*

tip into the surface and this is one way in which an equivalency in binding energies between the tip and substrate can be ensured. When a bias is applied to the system, Fig. 8.18(b), the adsorption wells shift such that one is now deeper than the other. In this case, the bias is chosen to increase the binding strength on the tip. At a sufficiently high bias the barrier between the two wells is reduced to the point where current-driven inelastic electron scattering or thermal fluctuations induce vibrational ladder climbing that transfers the atom from one well to the next. The atom then hops onto the tip, which can be retracted and positioned at any arbitrary location. By reversing the bias, the attraction of the atom for the substrate becomes greater than the attraction for the tip and the atom is transferred back to the surface at the new location.

8.6   Summary of important concepts

- Laser irradiation with nanosecond pulses can induce localized temperature increases at rates approaching 10^{10} K s^{-1}.
- Lasers can be used to drive pyrolytic (photothermal) and photolytic (photochemical) process.
- Electronic excitations relax on an ultrafast time scale, and the responses of metals and semiconductors to photoexcitation differ in important ways.
- Electron–electron relaxation processes are much faster than electron–phonon relaxation.
- Excited state lifetimes of adsorbate electronic states that do not fall in a band gap are on the order of 1 fs.
- The MGR and Antoniewicz models, composed of excitation-propagation-quenching-desorption steps, are the basic models to understand desorption induced by electronic transitions (DIET) and surface photochemistry.
- Desorption and reactions stimulated by ultrafast lasers can be understood within the framework of desorption induced by multiple electronic transitions (DIMET) and the electronic friction model.
- Tip induced processes of adsorbates bear resemblances to DIET and DIMET/friction mechanism but we need also consider electric field and proximity effects that allow for the high levels of control associated with atom and molecule manipulation.
- Stimulated chemistry can be categorized as direct or substrate mediated as well as linear or non-linear. The latter refers to how the rate of the process is related to the excitation source fluence.
- Electrochemical charge transfer happens as the result of a Franck-Condon transition of an electron from a filled to an unoccupied state (or of a hole from an unoccupied to a filled state).
- This Franck-Condon transition occurs after a thermal fluctuation reorganizes the solvent shell to a configuration compatible with the final charge transfer state.
- Anodic or cathodic current flows when the overpotential is adjusted so that an electronic state in the electrolyte lines up with the Fermi energy of a metal electrode (or an appropriate band edge of a semiconductor).
- An SPM tip can be used to create nanostructures by a variety of mechanisms including localized desorption, lateral manipulation and vertical manipulation.

8.7   Frontiers and challenges

- Understanding and modelling the nature of excited states at surfaces.
- Using lasers to make sub-wavelength size features.
- Structure formation during laser ablation.
- Further understanding and control of tip induced processes.
- Does coherent multiple quantum vibrational excitation play a role in tip induced desorption?

- The mechanism of charge transfer at solid surfaces, particularly how ions inject or accept charge and what controls the kinetics of reactions such as the oxygen reduction reaction $\frac{1}{2}O_2 + 2H^+ + 2e^- \longrightarrow H_2O$.
- The kinetics of charge exchange between light harvesting adsorbates and semiconductor surfaces [118].
- Photocatalysis with solar radiation, for instance, to create H_2 by means of water splitting [139, 140] or to create other fuels [123].
- The production of multiple electron-hole pairs from one absorbed photon and the creation of ultra-high response photovoltaic and optoelectronic devices either in nanocrystals [124, 141] or carbon nanotubes [142].
- Calculation of the reorganization energy λ with chemical accuracy for electrode reactions.

8.8   Further reading

J. H. Bechtel, *Heating of solid targets with laser pulses*, J. Appl. Phys. **46**, (1975), 1585–1593.

D. Burgess, Jr., P. C. Stair, and E. Weitz, *Calculations of the surface temperature rise and desorption temperature in laser-induced thermal desorption*, J. Vac. Sci. Technol. A **4**, (1986), 1362–1366.

R. R. Cavanagh, D. S. King, J. C. Stephenson, and T. F. Heinz, *Dynamics of nonthermal reactions: Femtosecond surface chemistry*, J. Phys. Chem. **93**, (1993), 786–798.

H.-L. Dai and W. Ho, in *Laser Spectroscopy and Photochemistry on Metal Surfaces: Parts 1 & 2*, Advanced Series in Physical Chemistry, Vol. **5** (World Scientific, Singapore, 1995).

P. M. Echenique, R. Berndt, E. V. Chulkov, T. Fauster, A. Goldmann, and U. Höfer, *Decay of electronic excitations at metal surfaces*, Surf. Sci. Rep. 2004, **52**, 219–317.

S. Fletcher, *The theory of electron transfer*, J. Solid State Electrochem., **14** (2010) 705.

J. Gudde, W. Berthold, and U. Höfer, *Dynamics of electronic transfer processes at metal/insulator interfaces*, Chem. Rev. 2006, **106**, 4261–4280.

J. Gudde and U. Höfer, *Femtosecond time-resolved studies of image-potential states at surfaces and interfaces of rare-gas adlayers*, Prog. Surf. Sci. **80**, 49–91 (2005).

R. B. Hall, *Pulsed-laser-induced desorption studies of the kinetics of surface reactions*, J. Phys. Chem. **91**, (1987), 1007–1015.

E. Hasselbrink, *Photon driven chemistry at surfaces*, in Handbook of Surface Science: Dynamics, Vol. **3**, edited by E. Hasselbrink and I. Lundqvist (Elsevier, Amsterdam, 2008), pp. 621–679.

P. V. Kamat, *Meeting the clean energy demand: Nanostructure architectures for solar energy conversion*, J. Phys. Chem. C **111**, (2007), 2834–2860.

D. M. Kolb and F. C. Simeone, *Nanostructure formation at the solid/liquid interface*, Curr. Opin. Solid State Mater. Sci. **9**, (2005), 91–97.

N. S. Lewis, An analysis of charge-transfer rate constants for semiconductor liquid interfaces, Annu. Rev. Phys. Chem. **42**, (1991), 543–580.

M. Lisowski, P. A. Loukakos, U. Bovensiepen, J. Stahler, C. Gahl, and M. Wolf, *Ultrafast dynamics of electron thermalization, cooling and transport effects in Ru(001)*, Appl. Phys. A **78**, (2004), 165–176.

L. J. Richter, S. A. Buntin, D. S. King, and R. R. Cavanagh, *Constraints on the use of polarization and angle-of-incidence to characterize surface photoreactions*, Chem. Phys. Lett. **186**, (1991), 423–426.

8.9   Exercises

8.1   Calculate the peak power, time averaged power, fluence and number of photons per pulse for (i) a Nd:YAG laser operating at 532 nm with a pulse energy of 100 mJ, a pulse length of 7 ns and a repetition rate of 10 Hz, and (ii) a Ti:sapphire laser operating at 800 nm with a pulse energy of 10 nJ, a pulse length of 80 fs and a repetition rate of 1 kHz. Assume that the laser beam is focussed into a spot with a diameter of 500 µm.

8.2   A 1 nW cw HeNe laser operating at 633 nm is incident on a surface. Describe qualitatively the distribution of excited carries near the surface for (a) an n-type doped Si substrate with a 100 nm space charge layer, (b) a p-type doped Si substrate with a 100 nm space charge layer, and (c) a metal substrate.

8.3   Given that the Maxwell number density speed distribution is

$$N(v)\,dv = \frac{4b^{3/2}}{\pi^{1/2}} v^2 \exp(-bv^2)\,dv \tag{8.9.1}$$

and that the Maxwell flux distribution of velocities is obtained by multiplying $N(v)$ by v

$$f(v)\,dv = vN(v)\,dv, \tag{8.9.2}$$

show that the TOF distribution measured by a flux detector is given by

$$I_{\text{flux}}(t)\,dt = a_f \frac{L^4}{t^5} \exp(-b(L/t)^2)\,dt \tag{8.9.3}$$

and that the TOF distribution measured by a density detector is given by

$$I_{\text{den}}(t)\,dt = a_d \frac{L^4}{t^4} \exp(-b(L/t)^2)\,dt. \tag{8.9.4}$$

Note that $b = m/2k_{\text{B}}T$ and a_f and a_d are normalization constants.

8.4   Explain why the temperature of the electrons in a metal excited by a 10 fs pulse significantly exceeds the phonon temperature in the subpicosecond time scale.

8.5   Describe the difference between a pyrolytic and a photolytic process.

8.6   CO was dosed onto Pt(111) at 80 K as in Exercise 3.18. TPD clearly demonstrates that CO adsorbs with near unit sticking coefficient. Using Auger spectroscopy, however, only a small signal near the detection limit is measured for all exposures. Explain this result and the discrepancy between the two methods of CO coverage determination.

8.7   With the assumption that the excitation probabilities are the same for both isotopologues because isotopic substitution does not lead to a change in the structure of the electronic potential energy curves, the excitation probabilities in Eq. (8.2.8) are the same. Show that the ratio of the desorption cross section of one isotope to another is given by

$$\frac{\sigma_d(1)}{\sigma_d(2)} = \left(\frac{1}{P}\right)^{\sqrt{m_1/m_2}-1}. \tag{8.9.5}$$

8.8   Consider (a) H_2O and (b) D_2O adsorbed on Pd(111) and then irradiated with 6.4 eV light [143]. Given the data in the table below, calculate σ_{ex} and P.

	$\sigma_d(H_2O)/ \times 10^{-18}$ cm^2	$\sigma_d(D_2O)/ \times 10^{-18}$ cm^2
Desorption	0.046	0.033
Dissociation	1.35	0.600

8.9   Consider a CO molecule adsorbed on a metal surface such that its 5σ state lies far below E_{F}, its $2\pi^*$ state is very close to but slightly above E_{F} and the $6\sigma^*$ state lies far above E_{F}. Describe the electron dynamics that lead to two different $5\sigma^{-1}$ state after absorption by the absorbed CO of a photon with an energy that is resonant with the $5\sigma \to 6\sigma^*$ transition.

8.10   (a) Describe the desorption dynamics and the desorbate formed for the process shown in Fig. 8.7(a).
(b) Consider now the case in which a curve crossing is made to the M+A$^+$ potential curve. Describe the desorption dynamics and the desorbate formed.

8.11   Describe why vertical manipulation is more versatile than lateral manipulation.

8.12   Describe how the presence of a band gap changes the absorption of light and relaxation of charge carriers in a semiconductor compared to a metal.

8.13   Describe how carrier recombination and ultrafast relaxation of electrons affect the efficiency of a solar cell.

8.14   The bonding orbital associated with the Si–H bond is a 5σ orbital at -5.3 eV below E_F. It has an intrinsic width of $\Gamma = 0.6$ eV [144]. Assuming that direct excitation of this orbital can be used to form a repulsive state, what is the minimum force and the minimum excited state potential slope required to desorb H from the surface by direct electronic excitation of this orbital. Take the Si–H bond strength to be 3.05 eV [145].

8.15   Hydrogen-terminated Si surfaces can be made with a level of perfection that exceeds all other surfaces. When free carriers are made they recombine at defects in the bulk or at the surface in traps. Si crystals can be made with such perfection that for a $d = 200\,\mu$m thick sample, recombination of electrons and holes only occurs appreciably at the surface. The surface recombination velocity S is related to the coverage of surface traps σ_{ss}, the recombination cross section σ_{rec} and the electron velocity v by

$$S = \sigma_{ss}\sigma_{rec}v. \tag{8.9.6}$$

The recombination cross section is $\sigma_{rec} \approx 10^{-15}$ cm^2 for nonresonant capture. The velocity is reflective of a thermal value. The recombination velocity of H-terminated Si can be as low as 0.25 cm s^{-1} [146]. Use the relationship between hot carrier lifetime τ and S

$$\frac{1}{\tau} = \frac{1}{\tau_b} + \frac{2S}{d}, \tag{8.9.7}$$

where τ_b is the bulk lifetime, to estimate the coverage of surface traps σ_{ss} and the lifetime of hot carriers τ. How does σ_{ss} compare to the number of surface atoms?

8.16   Direct vibrational excitation with IR lasers is possible [56, 147, 148]. Consider the behaviour of two very different systems. (i) H$_2$ and HD coadsorbed on LiF. The energy of the $v = 0 \rightarrow v = 1$ transition is greater than the energy of the physisorption bond. Absorption of one photon resonant with the HD($v = 0 \rightarrow v = 1$) desorbs HD but not H$_2$. (ii) NH$_3$ and ND$_3$ coadsorbed on Cu. The energy of the $v = 0 \rightarrow v = 1$ transition is less than the energy of the chemisorption bond. Absorption of photons resonant with the N–H($v = 0 \rightarrow v = 1$) stretching mode of NH$_3$ desorbs both NH$_3$ and ND$_3$. Discuss two different types of desorption dynamics that explain these results.

8.17   (a) Show that for electron transfer in which the solvent has a linear response described by the force constant k_0, the transition state occurs at the reaction co-ordinate

$$q^{\ddagger} = \frac{q_p^2 - q_r^2 + \Delta G^\circ / k_0}{2q_p - 2q_r}. \tag{8.9.8}$$

(b) Show that for such a system the activation Gibbs energy is given by Eq. (8.4.4).

(c) Show that for a one electron transfer occurring with a cell potential E and for which we take the reaction co-ordinate to be the fractional charge transfer (or fluctuation), the transition state is located at

$$q^{\ddagger} = \tfrac{1}{2} - FE/2\lambda. \tag{8.9.9}$$

8.18   Show that if the probability of electron transfer from a level in an electrode at energy ε to an acceptor level in solution depends on ε according to

$$P(\varepsilon) \propto n(\varepsilon) f(\varepsilon) \exp(\varepsilon/2k_B T), \tag{8.9.10}$$

where $f(\varepsilon)$ is the Fermi-Dirac function, then the maximum transfer probability occurs at $\varepsilon = \overline{\mu}_e$. Note that density of states $n(\varepsilon)$ is a weak function of ε.

8.19   Show that the Tafel equation [149]

$$\eta = A + b \log |i| \tag{8.9.11}$$

can be derived from the Butler-Volmer equation Eq. (8.4.12). To do so consider a cathodic reaction for which

$$\eta \ll -\frac{RT}{nF}. \tag{8.9.12}$$

8.20   The rate constant for a proton transfer reaction as a function of applied potential U can be written

$$k(U) = k_0 \exp(-\Delta G^{\ddagger}(U)/k_B T) \tag{8.9.13}$$

where the Gibbs energy of activation as a function of applied potential U is written $\Delta G^{\ddagger}(U)$. In terms of current density this is written

$$i_k(U) = 2e \, \sigma_{\text{site}} \, k(U) = i_0 \exp(-\Delta G^{\ddagger}(U)/k_B T) \tag{8.9.14}$$

where σ_{site} is the number of sites per unit area and e is the elementary charge.

(a) If a current under no driving force, that is the exchange current measured at $U = U_0$ ($U_0 =$ the equilibrium potential of the cell), is measured to be $i_0 = 100 \, \text{mA cm}^{-2}$, estimate the value of k_0 in units of $\text{s}^{-1} \, \text{site}^{-1}$.

(b) If the Gibbs energy of activation depends on the overpotential $\eta = U - U_0$ according to

$$\Delta G^{\ddagger}(U) = \Delta G^{\ddagger}(U_0) - e\eta, \tag{8.9.15}$$

show that the Butler-Volmer relation is given by

$$U = U_0 - b \log_{10}\left(\frac{i_k}{i_k^0}\right) \tag{8.9.16}$$

and calculate the value of the Tafel slope b at 300 K.

References

[1]   P. G. Burke, C. J. Noble, Comm. At. Mol. Phys., **18** (1986) 181.
[2]   W. Domcke, J. Phys. B: At., Mol. Opt. Phys., **14** (1981) 4889.
[3]   G. J. Schultz, Rev. Mod. Phys., **45** (1973) 423.
[4]   R. E. Palmer, Prog. Surf. Sci., **41** (1992) 51.
[5]   D. E. Aspnes, A. A. Studna, Phys. Rev. B, **27** (1983) 985.
[6]   D. E. Aspnes, Optical functions of intrinsic Si, in *EMIS Datareview Series No. 4: Properties of Silicon*, INSPEC, London, 1988, p. 61.
[7]   G. Ertl, M. Neumann, Z. Naturforsch., A: Phys. Sci., **27** (1972) 1607.
[8]   R. B. Hall, J. Phys. Chem., **91** (1987) 1007.

[9] D. J. Auerbach, Velocity measurements by time-of-flight methods, in *Atomic and Molecular Beam Methods, Vol. 1* (Ed.: G. Scoles), Oxford University Press, Oxford, 1988, p. 362.

[10] J. F. Ready, *Effects of High Power Laser Radiation*, Academic Press, New York, 1971.

[11] D. Burgess, Jr., P. C. Stair, E. Weitz, J. Vac. Sci. Technol., A, **4** (1986) 1362.

[12] J. H. Bechtel, J. Appl. Phys., **46** (1975) 1585.

[13] L. J. Richter, R. R. Cavanagh, Prog. Surf. Sci., **39** (1992) 155.

[14] H. S. Carslaw, J. C. Jaeger, *Conduction of Heat in Solids*, Clarendon Press, Oxford, 1959.

[15] P. Baeri, S. U. Campisano, G. Foti, E. Rimini, J. Appl. Phys., **50** (1979) 788.

[16] M. Lax, J. Appl. Phys., **48** (1977) 3919.

[17] J. L. Brand, S. M. George, Surf. Sci., **167** (1986) 341.

[18] F. Weik, A. de Meijere, E. Hasselbrink, J. Chem. Phys., **99** (1993) 682.

[19] S. I. Anisimov, B. L. Kapeliovitch, T. L. Perel'man, Sov. Phys. JETP, **39** (1974) 375.

[20] R. H. M. Groeneveld, R. Sprik, A. Lagendijk, Phys. Rev. B, **51** (1995) 11433.

[21] J. A. Misewich, T. F. Heinz, P. Weigand, A. Kalamarides, Femtosecond surface science: The dynamics of desorption, in *Laser Spectroscopy and Photochemistry on Metal Surfaces* (Eds: H. L. Dai, W. Ho), World Scientific, Singapore, 1995, p. 764.

[22] M. I. Kaganov, I. M. Lifshitz, L. V. Tanatarov, Sov. Phys. JETP, **4** (1957) 173.

[23] M. Lisowski, P. A. Loukakos, U. Bovensiepen, J. Stahler, C. Gahl, M. Wolf, Appl. Phys. A, **78** (2004) 165.

[24] L. J. Richter, S. A. Buntin, D. S. King, R. R. Cavanagh, Chem. Phys. Lett., **186** (1991) 423.

[25] D. Menzel, R. Gomer, J. Chem. Phys., **41** (1964) 3311.

[26] P. A. Redhead, Can. J. Phys., **42** (1964) 886.

[27] P. R. Antoniewicz, Phys. Rev. B, **21** (1980) 3811.

[28] J. W. Gadzuk, L. J. Richter, S. A. Buntin, D. S. King, R. R. Cavanagh, Surf. Sci., **235** (1990) 317.

[29] J. W. Gadzuk, Surf. Sci., **342** (1995) 345.

[30] R. D. Ramsier, J. T. Yates, Jr., Surf. Sci. Rep., **12** (1991) 243.

[31] M. L. Knotek, P. J. Feibelman, Phys. Rev. Lett., **40** (1978) 964.

[32] P. Feulner, D. Menzel, Electronically stimulated desorption of neutrals and ions from adsorbed and condensed layers, in *Laser Spectroscopy and Photochemistry on Metal Surfaces* (Eds: H. L. Dai, W. Ho), World Scientific, Singapore, 1995, p. 627.

[33] K. W. Kolasinski, Curr. Opin. Solid State Mater. Sci., **8** (2004) 353.

[34] K. Lass, X. Han, E. Hasselbrink, J. Chem. Phys., **123** (2005) 051102.

[35] J. A. Misewich, T. F. Heinz, D. M. Newns, Phys. Rev. Lett., **68** (1992) 3737.

[36] P. Saalfrank, Curr. Opin. Solid State Mater. Sci., **8** (2004) 334.

[37] J. W. Gadzuk, Chem. Phys., **251** (2000) 87.

[38] M. Brandbyge, P. Hedegård, T. F. Heinz, J. A. Misewich, D. M. Newns, Phys. Rev. B, **52** (1995) 6042.

[39] E. Hasselbrink, State-resolved probes of molecular desorption dynamics induced by short-lived electronic excitations, in *Laser Spectroscopy and Photochemistry on Metal Surfaces, Vol. 2* (Eds: H. L. Dai, W. Ho), World Scientific, Singapore, 1995, p. 685.

[40] J. P. Cowin, D. J. Auerbach, C. Becker, L. Wharton, Surf. Sci., **78** (1978) 545.

[41] J. P. Cowin, Phys. Rev. Lett., **54** (1985) 368.

[42] X. D. Zhu, T. Rasing, Y. R. Shen, Phys. Rev. Lett., **61** (1988) 2883.

[43] K. W. Kolasinski, J. Phys.: Cond. Matter, **18** (2006) S1655.

[44] D. P. Ruffle, E. Herbst, Mon. Not. R. Astron. Soc., **322** (2000) 770.

[45] K. Willacy, W. D. Langer, Astrophys. J., **544** (2000) 903.

[46] T. E. Madey, R. E. Johnson, T. M. Orlando, Surf. Sci., **500** (2002) 838.

[47] T. E. Madey, J. J. Czyzewski, J. T. Yates, Jr., Surf. Sci., **57** (1976) 580.

[48] T. E. Madey, J. T. Yates, Jr., Surf. Sci., **63** (1977) 203.

[49] W. L. Clinton, M. Esrick, H. Ruf, W. Sacks, Physica Review B, **31** (1985) 722.

[50] W. L. Clinton, M. A. Esrick, W. S. Sacks, Phys. Rev. B, **31** (1985) 7550.

[51] J. O. Noell, C. F. Melius, R. H. Stulen, Surf. Sci., **157** (1985) 119.

[52] D. Riedel, L. M. A. Perdigão, J. L. Hernández-Pozos, Q. Guo, R. E. Palmer, J. S. Foord, K. W. Kolasinski, Phys. Rev. B, **66** (2002) 233405.

[53] G. Dujardin, G. Comtet, L. Hellner, T. Hirayama, M. Rose, L. Philippe, M. J. Besnardramage, Phys. Rev. Lett., **73** (1994) 1727.

[54] L. Koker, K. W. Kolasinski, Phys. Chem. Chem. Phys., **2** (2000) 277.

[55] K. W. Kolasinski, Phys. Chem. Chem. Phys., **5** (2003) 1270.

[56] W. Ho, Surface photochemistry, in *Laser Spectroscopy and Photochemistry on Metal Surfaces, Vol. 2* (Eds: H. L. Dai, W. Ho), World Scientific, Singapore, 1995, p. 1047.

[57] S. A. Buntin, L. J. Richter, R. R. Cavanagh, D. S. King, Phys. Rev. Lett., **61** (1988) 1321.

[58] J. W. Gadzuk, Phys. Rev. Lett., **76** (1996) 4234.

[59] T. Mull, B. Baumeister, M. Menges, H.-J. Freund, D. Weide, F. C, P. Andresen, J. Chem. Phys., **96** (1992) 7108.

[60] F. Budde, A. V. Hamza, P. M. Ferm, G. Ertl, D. Weide, P. Andresen, H.-J. Freund, Phys. Rev. Lett., **60** (1988) 1518.

[61] E. Hasselbrink, S. Jakubith, S. Nettesheim, M. Wolf, A. Cassuto, G. Ertl, J. Chem. Phys., **92** (1990) 3154.

[62] E. Hasselbrink, Chem. Phys. Lett., **170** (1990) 329.

[63] A. C. Luntz, M. Persson, S. Wagner, C. Frischkorn, M. Wolf, J. Chem. Phys., **124** (2007) 244702.

[64] J. A. Prybyla, T. F. Heinz, J. A. Misewich, M. M. T. Loy, J. H. Glowina, Phys. Rev. Lett., **64** (1990) 1537.

[65] J. A. Misewich, A. Kalamarides, T. F. Heinz, U. Höfer, M. M. T. Loy, J. Chem. Phys., **100** (1994) 736.

[66] F. Budde, T. F. Heinz, M. M. T. Loy, J. A. Misewich, F. de Rougemont, H. Zacharias, Phys. Rev. Lett., **66** (1991) 3024.

[67] D. N. Denzler, C. Frischkorn, M. Wolf, G. Ertl, J. Phys. Chem. B, **108** (2004) 14503.

[68] D. B. Geohegan, Diagnostics and characteristics of pulsed laser deposition laser plasmas, in *Pulsed Laser Deposition of Thin Films* (Eds: D. B. Chrisey, G. K. Hubler), John Wiley & Sons, Inc., New York, 1994, p. 115.

[69] S. Preuss, A. Demchuk, M. Stuke, Appl. Phys. A, **61** (1995) 33.

[70] L. V. Zhigilei, Appl. Phys. A, **76** (2003) 339.

[71] S. I. Anisimov, B. S. Luk'yanchuk, Phys. Usp., **45** (2002) 293.

[72] T. Makimura, T. Mizuta, K. Murakami, Appl. Phys. Lett., **76** (2000) 1401.

[73] A. J. Pedraza, J. D. Fowlkes, D. H. Lowndes, Appl. Phys. A, **69** (1999) S731.

[74] C. V. Shank, R. Yen, C. Hirlimann, Phys. Rev. Lett., **51** (1983) 900.

[75] P. Stampfli, K. H. Bennemann, Appl. Phys. A, **60** (1995) 191.

[76] S. K. Sundaram, E. Mazur, Nature Mater., **1** (2002) 217.

[77] K. Sokolowski-Tinten, C. Blome, J. Blums, A. Cavalleri, C. Dietrich, A. Tarasevitch, I. Uschmann, E. Förster, M. Kammler, M. Horn-von-Hoegen, D. von der Linde, Nature (London), **422** (2003) 287.

[78] V. I. Emel'yanov, D. V. Babak, Appl. Phys. A, **74** (2002) 797.

[79] K. W. Kolasinski, Laser-assisted restructuring of silicon over nano-, meso- and macro-scales, in *Recent Research Advances in Applied Physics, Vol. 7* (Ed.: S. G. Pandalai), Transworld Research Network, Kerala, India, 2004, p. 267.

[80] F. Claeyssens, S. J. Henley, M. N. R. Ashfold, J. Appl. Phys., **94** (2003) 2203.

[81] S. Preuss, E. Matthias, M. Stuke, Appl. Phys. A, **59** (1994) 79.

[82] J. Jandeleit, G. Urbasch, H. D. Hoffmann, H. G. Treusch, E. W. Kreutz, Appl. Phys. A, **63** (1996) 117.

[83] D. Riedel, J. L. Hernández-Pozos, K. W. Kolasinski, R. E. Palmer, Appl. Phys. A, **78** (2004) 381.

[84] D. Mills, K. W. Kolasinski, J. Vac. Sci. Technol., A, **22** (2004) 1647.

[85] C. H. Crouch, J. E. Carey, J. M. Warrender, M. J. Aziz, E. Mazur, F. Y. Génin, Appl. Phys. Lett., **84** (2004) 1850.

[86] K. W. Kolasinski, Curr. Opin. Solid State Mater. Sci., **11** (2007) 76.

[87] J. F. Young, J. S. Preston, H. M. Van Driel, J. E. Sipe, Phys. Rev. B, **27** (1983) 1155.

[88] J. F. Young, J. E. Sipe, H. M. Van Driel, Phys. Rev. B, **30** (1984) 2001.

[89] J. E. Sipe, J. F. Young, J. S. Preston, H. M. Van Driel, Phys. Rev. B, **27** (1983) 1141.

[90] A. E. Siegman, P. M. Fauchet, IEEE J. Quantum Electron., **22** (1986) 1384.

[91]   S. R. Foltyn, Surface modification of materials by cumulative laser irradiation, in *Pulsed Laser Deposition of Thin Films* (Eds: D. B. Chrisey, G. K. Hubler), John Wiley & Sons, Inc., New York, 1994, p. 89.

[92]   T.-H. Her, R. J. Finlay, C. Wu, S. Deliwala, E. Mazur, Appl. Phys. Lett., **73** (1998) 1673.

[93]   A. J. Pedraza, J. D. Fowlkes, D. H. Lowndes, Appl. Phys. Lett., **74** (1999) 2322.

[94]   M. Y. Shen, C. H. Crouch, J. E. Carey, R. Younkin, E. Mazur, M. Sheehy, C. M. Friend, Appl. Phys. Lett., **82** (2003) 1715.

[95]   D. Mills, K. W. Kolasinski, J. Phys. D: Appl. Phys., **38** (2005) 632.

[96]   R. A. Marcus, Rev. Mod. Phys., **65** (1993) 599.

[97]   H. Gerischer, Z. Phys. Chem. N. F., **26** (1960) 233.

[98]   H. Gerischer, Z. Phys. Chem., **27** (1961) 40.

[99]   S. Fletcher, J. Solid State Electrochem., **14** (2010) 705.

[100]   T. V. Voorhis, T. Kowalczyk, B. Kaduk, L. P. Wang, C. L. Cheng, Q. Wu, Annu. Rev. Phys. Chem., **61** (2010) 149.

[101]   R. A. Marcus, J. Chem. Phys., **43** (1965) 679.

[102]   P. F. Barbara, T. J. Meyer, M. A. Ratner, J. Phys. Chem., **100** (1996) 13148.

[103]   J. K. Hwang, A. Warshel, J. Am. Chem. Soc., **109** (1987) 715.

[104]   R. A. Marcus, Annual Reviews of Physical Chemistry, **15** (1964) 155.

[105]   R. A. Marcus, N. Sutin, Biochim. Biophys. Acta, **811** (1985) 265.

[106]   R. A. Marcus, J. Phys. Chem., **94** (1990) 1050.

[107]   H. Gerischer, Principles of electrochemistry, in *The CRC Handbook of Solid State Electrochemistry* (Eds: P. Gellings, H. Bouwmeester), CRC Press, Boca Raton, 1997, p. 9.

[108]   N. S. Lewis, Annu. Rev. Phys. Chem., **42** (1991) 543.

[109]   W. J. Royea, A. M. Fajardo, N. S. Lewis, J. Phys. Chem. B, **101** (1997) 11 152.

[110]   N. S. Lewis, J. Phys. Chem. B, **102** (1998) 4843.

[111]   B. B. Smith, C. A. Koval, J. Electroanal. Chem., **277** (1990) 43.

[112]   H. Gerischer, W. Ekardt, Appl. Phys. Lett., **43** (1983) 393.

[113]   R. Gomer, G. Tryson, J. Chem. Phys., **66** (1977) 4413.

[114]   H. Reiss, A. Heller, J. Phys. Chem., **89** (1985) 4207.

[115]   H. Reiss, J. Phys. Chem., **89** (1985) 3783.

[116]   T. Erdey Gruz, M. Volmer, Z. Phys. Chem., **150** (1930) 203.

[117]   J. A. V. Butler, Trans. Faraday Soc., **18** (1924) 729.

[118]   N. S. Lewis, Inorg. Chem., **44** (2005) 6900.

[119]   U. S. Geological Survey, 2007.

[120]   C. Wadia, A. P. Alivisatos, D. M. Kammen, Environmental Science & Technology, **43** (2009) 2072.

[121]   M. Grätzel, Inorg. Chem., **44** (2005) 6841.

[122]   B. O'Regan, M. Grätzel, Nature (London), **353** (1991) 737.

[123]   P. V. Kamat, J. Phys. Chem. C, **111** (2007) 2834.

[124]   R. D. Schaller, V. I. Klimov, Phys. Rev. Lett., **92** (2004) 186601.

[125]   B. N. J. Persson, P. Avouris, Chem. Phys. Lett., **242** (1995) 483.

[126]   R. S. Becker, G. S. Higashi, Y. J. Chabal, A. J. Becker, Phys. Rev. Lett., **65** (1990) 1917.

[127]   T.-C. Shen, C. Wang, G. C. Abeln, J. R. Tucker, J. W. Lyding, P. Avouris, R. E. Walkup, Science, **268** (1995) 1590.

[128]   L. Soukiassian, A. J. Mayne, M. Carbone, G. Dujardin, Phys. Rev. B, **68** (2003) 035303.

[129]   J. A. Stroscio, F. Tavazza, J. N. Crain, R. J. Celotta, A. M. Chaka, Science, **313** (2006) 948.

[130]   B. C. Stipe, M. A. Rezaei, W. Ho, S. Gao, M. Persson, B. I. Lundqvist, Phys. Rev. Lett., **78** (1997) 4410.

[131]   S.-W. Hla, L. Bartels, G. Meyer, K.-H. Rieder, Phys. Rev. Lett., **85** (2000) 2777.

[132]   H. C. Manoharan, C. P. Lutz, D. M. Eigler, Nature (London), **403** (2000) 512.

[133]   D. M. Eigler, E. K. Schweizer, Nature (London), **344** (1990) 524.

[134]   G. Dujardin, A. Mayne, O. Robert, F. Rose, C. Joachim, H. Tang, Phys. Rev. Lett., **80** (1998) 3085.

[135]   D. M. Kolb, F. C. Simeone, Curr. Opin. Solid State Mater. Sci., **9** (2006) 91.

[136]   S. Gao, M. Persson, B. I. Lundqvist, Phys. Rev. B, **55** (1997) 4825.

[137]   S. Gao, M. Persson, B. I. Lundqvist, Solid State Commun., **84** (1992) 271.

[138]   R. E. Walkup, D. M. Newns, P. Avouris, Vibrational heating and atom transfer with the STM, in *Atomic and Nanometer-Scale Modification of Materials: Fundamentals and Applications* (Ed.: P. Avouris), Kluwer, Dordrecht, 1993, p. 97.

[139]   J. A. Turner, Science, **305** (2004) 972.

[140]   J. H. Park, A. J. Bard, Electrochem. Solid State Lett., **9** (2006) E5.

[141]   S. J. Kim, W. J. Kim, Y. Sahoo, A. N. Cartwright, P. N. Prasad, Appl. Phys. Lett., **92** (2008) 031107.

[142]   N. M. Gabor, Z. H. Zhong, K. Bosnick, J. Park, P. L. McEuen, Science, **325** (2009) 1367.

[143]   M. Wolf, S. Nettesheim, J. M. White, E. Hasselbrink, G. Ertl, J. Chem. Phys., **94** (1991) 4609.

[144]   K. Stokbro, C. Thirstrup, M. Sakurai, U. Quaade, B. Y.-K. Hu, F. Perez-Murano, F. Grey, Phys. Rev. Lett., **80** (1998) 2618.

[145]   M. B. Raschke, U. Höfer, Phys. Rev. B, **59** (1999) 2783.

[146]   E. Yablonovitch, D. L. Allara, C. C. Chang, T. Gmitter, T. B. Bright, Phys. Rev. Lett., **57** (1986) 249.

[147]   P. M. Ferm, S. R. Kurtz, K. A. Pearlstine, G. M. McClelland, Phys. Rev. Lett., **58** (1987) 2602.

[148]   I. Hussla, H. Seki, T. J. Chuang, Z. W. Gortel, H. J. Kreutzer, P. Piercy, Phys. Rev. B, **32** (1985) 3489.

[149]   J. Tafel, Z. Phys. Chem., **50** (1905) 641.

Answers to Exercises

9

Answers to Exercises from Chapter 1. Surface and Adsorbate Structure

1.1 (a) Determine the surface atom density of Ag(221).

Ag packs as an fcc crystal with a lattice parameter of 0.408 nm. Therefore according to Eq. (1.1.1)

$$\sigma_0 = \frac{4}{2(0.408 \times 10^{-7}\,\text{cm})^2 (2^2 + 2^2 + 1^2)^{1/2}} = 4.00 \times 10^{14}\,\text{cm}^{-2}$$

(b) For the basal (cleavage) plane of graphite, determine the unit cell length, a, the included angle between sides of the unit cell, γ, and the density of surface C atoms, σ_0, given that the C–C nearest neighbour distance is 1.415 Å. A representation of the graphite surface is given in Fig. 1.20. First make a few geometrical identifications

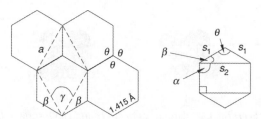

Figure Exercise 1.1 *Graphite basal plane.*

$$\theta_1 = \theta_2 = \theta_3 = \theta; \quad 3\theta = 360° \therefore \theta = 120°$$

$$\alpha = 90°,\ \alpha + \beta = \theta \therefore \beta = 30°$$

$$\gamma + 2\beta = \theta \therefore \gamma = 60°$$

There are two ways to calculate σ_0.

Surface Science: Foundations of Catalysis and Nanoscience, Third Edition. Kurt W. Kolasinski.
© 2012 John Wiley & Sons, Ltd. Published 2012 by John Wiley & Sons, Ltd.

(i) Calculate the area of the hexagon. Note that each C atom is shared between three equivalent hexagons and therefore contributes 1/3 of an atom to any given hexagon. $6(1/3) = 2$ atoms per hexagon. Dividing 2 atoms by the area of the hexagon will yield the density of surface atoms.

Divide the hexagon into two equivalent triangles with side lengths s_1, s_1 and s_2 and one rectangle of side lengths s_1 and s_2. The area of the hexagon is the sum of the areas of the two triangles and the rectangle.

$$\frac{s_1}{\sin \beta} = \frac{s_2}{\sin \theta} \Rightarrow s_2 = s_1 \frac{\sin \theta}{\sin \beta} = 2.451 \, \text{Å}$$

$$A_{\text{hex}} = 2\left(\frac{s_1 s_2 \sin \beta}{2}\right) + s_1 s_2 = s_1 s_2 (\sin \beta + 1) = s_1^2 \frac{\sin \theta}{\sin \beta}(\sin \beta + 1) = 5.202 \, \text{Å}^2$$

$$\sigma_0 = \frac{2}{A_{\text{hex}}} = \frac{2}{5.202 \, \text{Å} \left(\frac{1 \, \text{cm}}{10^8 \, \text{Å}}\right)^2} = 3.845 \times 10^{15} \, \text{cm}^{-2}$$

$$a = s_2 = 2.451 \, \text{Å}.$$

(ii) Calculate the area of the unit cell. Note that there are two atoms per unit cell (the one in the center of the cell and the one derived from the four corner atoms, which are shared between four unit cells.

$$A_{\text{unit cell}} = a^2 \sin \gamma = s_2^2 \sin \gamma = 5.202 \, \text{Å}^2$$

$$\sigma_0 = \frac{2}{A_{\text{unit cell}}} = \frac{2}{5.202 \, \text{Å} \, (1 \, \text{cm}/10^8 \, \text{Å})^2} = 3.845 \times 10^{15} \, \text{cm}^{-2}$$

The answers are, as expected, equal.

1.2   The Pt(111) surface has a surface atom density of $\sigma_0 = 1.503 \times 10^{15} \, \text{cm}^{-2}$. (a) Calculate the diameter of a Pt atom. (b) Calculate the atom density of the Pt(100) surface.

(a)   (b)

Figure Exercise 1.2   *Unit cells of (111) and (100) surface lattices.*

There is one atom in the (111) unit cell. The side of the unit cell is equal to the diameter of the Pt atom. The area of a parallelogram is equal to the length of the base times the height.

$$A_{\text{unit cell}} = ah.$$

The length of the base, a, is also equal to the diameter of a Pt atom d_{Pt}. The line segment that defines the height bisects the base into two equal portions of length $a/2$ to form a right triangle. Thus the height is $a^2 = h^2 + (a/2)^2$

$$h = \frac{\sqrt{3}}{2}a$$

and the area of the unit cell is

$$A_{\text{unit cell}} = \frac{\sqrt{3}}{2}a^2 = \frac{\sqrt{3}}{2}d_{\text{Pt}}^2$$

The areal density is then

$$\sigma_0(111) = \frac{1}{A_{\text{unit cell}}} = \frac{1}{\frac{\sqrt{3}}{2}d_{\text{Pt}}^2}$$

From which the Pt atom diameter can be calculated.

$$d_{\text{Pt}} = \sqrt{\frac{2}{\sqrt{3}\sigma_0}} = \sqrt{\frac{2}{1.732(1.503 \times 10^{15})}} = 2.772 \times 10^{-8} \text{ cm}.$$

(a) In the square lattice of the Pt(100) surface, there is one atom per unit cell and the area of the unit cell is simple a^2.

$$\sigma_0(100) = \frac{1}{A_{\text{unit cell}}} = \frac{1}{d_{\text{Pt}}^2} = \frac{1}{(2.772 \times 10^{-8} \text{ cm})^2} = 1.302 \times 10^{15} \text{ cm}^{-2}$$

1.3 (a) Derive a general expression for the step density, ρ_{step}, of an fcc crystal with single-atom-height steps induced by a miscut angle χ from the ideal surface plane.

(b) Make a plot of step density versus miscut angle for an fcc crystal.

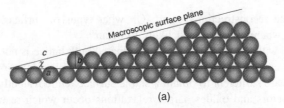

(a)

Figure Exercise 1.3(a) *Side view of a stepped surface.*

(a) Steps occur each time an atom is intersected by the macroscopic surface plane. In other words, you cannot have a fraction of an atom. A step occurs each time the side b of the triangle is equal to the diameter of a metal atom. The terrace width is the length a.

$$\tan \chi = \frac{a}{b} \Rightarrow a = \frac{b}{\tan \chi} = n_{\text{terrace}}d_M,$$

where n_{terrace} is the number of terrace atoms and d_M is the metal atom diameter. There is one step for every $n_{terrace}$ surface atoms; hence the step density is

$$\rho_{\text{step}} = \frac{1}{n_{\text{terrace}}} = \tan \chi$$

(b) As shown in the figure below, a plot of $\tan \chi$ is essentially linear at the small angles of interest.

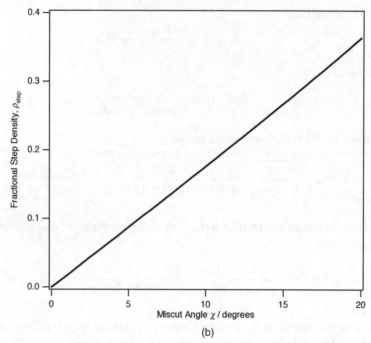

Figure Exercise 1.3(b)   *Step density as a function of miscut angle.*

1.4   Discuss why surface reconstructions occur. On what types of surfaces are reconstructions most likely to occur and where are they least likely to occur?

A surface created with the structure of the terminated bulk lattice is not the configuration with the lowest total surface energy. Generally, on metal surfaces simple relaxations (small changes in bond lengths and angles) are sufficient to obtain a stable surface. On covalently bonded surfaces, such as those of semiconductors and oxides, larger relaxations occur which result in a new periodicity of the surface atoms with respect to the bulk-truncated structure. The formation of a structure with a new periodicity is a reconstruction.

1.5   The Fermi energies of Cs, Ag and Al are 1.59, 5.49 and 11.7 eV, respectively. Calculate the density of the Fermi electron gas in each of these metals as well as the Fermi temperature. Calculate the difference between the chemical potential and the Fermi energy for each of these metals at their respective melting points.

From Eq. (1.3.8)

$$E_F = \frac{\hbar^2}{2m_e}(3\pi^2\rho)^{2/3}$$

and Eq. (1.3.7)

$$\mu(T) \approx E_F\left[1 - \frac{\pi^2}{12}\left(\frac{k_B T}{E_F}\right)^2\right]$$

with the Fermi temperature

$$T_F = \frac{E_F}{k_B}.$$

Therefore,

$$\rho = \frac{1}{3\pi^2} \left(\frac{2m_e}{\hbar^2}\right)^{3/2} E_F^{3/2}$$

and

$$\mu(T) - E_F \approx E_F\left[1 - \frac{\pi^2}{12}\left(\frac{k_B T}{E_F}\right)^2\right] - E_F = -E_F\left[\frac{\pi^2 k_B^2}{12}\left(\frac{T}{E_F}\right)^2\right].$$

This can be used to complete the following table.

	Cs	Ag	Al
E_F/ev	1.59	5.49	11.7
$E_F/10^{-19}$ J	2.547	8.796	18.75
T_F/K	18 460	63 740	135 900
T_f/K	300.55	1235.08	933.52
$\rho/10^{28}$ m^{-3}	0.922	5.92	18.4
$\mu - E_F$/eV	-3.46×10^{-4}	-1.70×10^{-3}	-4.55×10^{-4}

Note that the shift in the chemical potential is negative and very small.

1.6   A clean Ag(111) surface has a work function of 4.7 eV. As a submonolayer coverage of Ba is dosed onto the surface, the work function drops and reaches a minimum of 2.35 eV. Calculate the surface potential associated with the clean and Ba covered surfaces and explain the effect of Ba adsorption on the work function of Ag(111).

The Fermi energy of Ag is 5.49 eV. Thus, by rearranging Eq. (1.3.4) and with both Φ and E_F expressed in eV, we find that the surface potential is

$$\chi = -\left(\frac{\Phi + E_F}{e}\right).$$

Hence for the clean and Ba covered surfaces

$$\chi\{Ag(111)\} = -\left(\frac{4.7\,eV + 5.49\,eV}{e}\right) = -10.19\,V.$$

$$\chi\{Ba/Ag(111)\} = -\left(\frac{2.35\,eV + 5.49\,eV}{e}\right) = -7.84\,V.$$

Ba adsorption lowers the work function of the surface by reducing the magnitude of the surface dipole. Ba is more electropositive than Ag, i.e. it has a lower work function than Ag; therefore it donates charge to the Ag substrate.

1.7   The work function of clean Al(111) is 4.24 eV [1] (a free electron *sp* metal), clean Ag(111) is 4.7 eV (a coinage metal), and for bulk polycrystalline Cs is 2.14 eV (an alkali metal). Use the value of the Fermi energy given in Exercise 1.5 to calculate the surface potential for these three different types of metals.

From Eq. (1.3.4), the surface potential is

$$\chi = -\left(\frac{\Phi + E_F}{e}\right).$$

The Fermi energies of Al, Ag and Cs are 11.7, 5.49 and 1.59 eV, respectively. Hence the surface potentials are -15.94, -10.19 and -3.78 V, respectively. Note that the surface potential is over 4 times greater on Al(111) as compared to Cs and over 2.5 higher on Ag(111) than on Cs.

1.8   The magnitude of an electric dipole μ is

$$\mu = 2qR \tag{1.8.1}$$

for a charges $+q$ and $-q$ separated by a distance $2R$. The work function change $\Delta\Phi$ expressed in V associated with adsorption of a species with charge q, located a distance R from the surface (hence its image charge is at a distance $2R$ from the adsorbate and $-R$ below the plane of the surface) at a coverage σ is given by the Helmholtz equation

$$\Delta\Phi = \sigma\mu/2\varepsilon_0 = \sigma qR/\varepsilon_0, \tag{1.8.2}$$

when the dipole moment μ and coverage are expressed in SI units.

Calculate the work function changes expected for 0.1 ML of either peroxo (O_2^{2-}) or superoxo (O_2^-) species bound on Pd(111) given that their bond distance is 2 Å. The measured work function change is only on the order of 1 eV. Explain the difference between your estimates and the measured value.

From Table 1.1, we find $\sigma_0 = 1.53 \times 10^{19}$ m^{-2} for Pd(111). For the superoxo species, we substitute $q = 1.602 \times 10^{-19}$ C and $R = 2 \times 10^{-10}$ m. Hence

$$\Delta\Phi = \frac{0.1(1.53 \times 10^{19} \text{ m}^{-2})(1.602 \times 10^{-19} \text{ C})(2 \times 10^{-10} \text{ m})}{8.85 \times 10^{-12} \text{ C}^2 \text{ J}^{-1} \text{ m}^{-1}} = 5.74 \text{ JC}^{-1} = 5.74 \text{ V},$$

Since the charge q is twice as large on the peroxo species, the predicted work function change is $\Delta\Phi = 11.5$ V.

The differences can be ascribed to polarization effects within the adsorbed layer as well as to screening effects, and the breakdown in assuming that the transferred charged is localized as a point charge rather than a charge distribution about the adsorbed molecule [2].

1.9   The work function of Pt(111) is 5.93 eV. A Ru film has a work function of 4.71 eV. If Ru islands are deposited on a Pt(111) surface, in which direction does electron transfer occur?

Charge transfer occurs from the low work function metal to the high work function metal; thus from Ru to Pt. Consider this: when not in contact, the vacuum levels are at the same position. The Fermi levels are at two different energies, with the Pt Fermi level significantly below that of the Ru in order to achieve the larger work function of Pt. When the two metals are brought in contact, the Fermi levels must align, and the vacuum levels will be realigned by the change in the surface potential. Electron transfer must occur to achieve this realignment. Since electron flow from high energy to low, electrons will flow out of the Ru toward the Pt.

1.10   Redraw Fig. 1.17 for a *p*-type semiconductor.

The energy difference *in the bulk* between E_C, E_F and E_V is a constant. The energy difference between E_C and E_V is constant throughout the semiconductor. In *n*-type material, E_F is just below E_C whereas in *p*-type material, E_F is just above E_V. At equilibrium the Fermi energies of the metal and semiconductor must be equal. In order to accommodate these conditions, the position of E_C and E_V (and E_{vac}) relative to E_F must change at the surface and in the space charge region. The sense of the change is opposite for *n*-type and *p*-type materials.

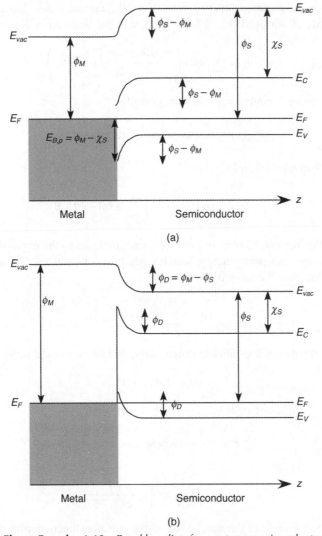

Figure Exercise 1.10 *Band bending for a p-type semiconductor.*

1.11 Given that the partition function, q, is defined by a summation over all states according to

$$q = \sum_{i=1}^{\infty} \exp(-E_i/k_B T) \tag{1.8.3}$$

where E_i is the energy of the ith state. Use Eq. (1.4.4) to show that the mean vibrational energy of a solid at equilibrium is given by Eq. (1.4.6). Hint: The mean energy is given by

$$\langle E \rangle = k_B T^2 \frac{\partial(\ln q)}{\partial T} \tag{1.8.4}$$

From Eq. (1.4.4), the energy of the nth state is given by

$$E(\mathbf{k}, p) = \left[n(\mathbf{k}, p) + \tfrac{1}{2} \right] \hbar \omega_k(p)$$

The summation in the partition function runs over all quantum states. Since the quantum number of the ground state of our oscillator is for $n(\mathbf{k}, p) = 0$, the summation is given by

$$q = \sum_{n=0}^{\infty} \exp\left(\frac{-[n(\mathbf{k}, p) + \frac{1}{2}]\hbar\omega_k(p)}{k_B T}\right).$$

To simplify, we define a vibrational quantum given by

$$\theta_{\text{vib}}^k(p) = \theta_{\text{vib}} = \hbar\omega_k(p), \text{ introduce } \beta = \frac{1}{k_B T},$$

and drop the branch label, therefore

$$q = \exp\left(\frac{-\theta_{\text{vib}}\beta}{2}\right) \sum_{n=0}^{\infty} \exp(-n\theta_{\text{vib}}\beta).$$

Note, however, that the $\hbar\omega_k/2$ term in Eq. (1.4.4), which leads to the $\exp(-\theta_{\text{vib}}\beta/2)$ term above, is simply related to the zero point energy, which leads to a constant $\hbar\omega_k/2$ that we can add into the mean energy at the end. We are left to evaluate

$$q = \sum_{n=0}^{\infty} \exp(-n\theta_{\text{vib}}\beta).$$

This summation represents a geometric progression, that is, an infinite series

$$s = 1 + x + x^2 \ldots$$

which, because $x^2 < 1$, rearranges to

$$s = \frac{1}{1-x}, \text{ where } x = \exp(-\theta_{\text{vib}}\beta)$$

In our case

$$q = \frac{1}{1-x} = \frac{1}{1 - \exp(-\theta_{\text{vib}}\beta)}.$$

We now take the derivative of the natural log of the partition function with respect to T, or rather, we must convert to a derivative with respect to β.

$$\frac{\partial \ln q}{\partial \beta} = \frac{\partial \ln q}{\partial(1/k_B T)} = k_B \frac{\partial \ln q}{\partial(1/T)} = -k_B T^2 \frac{\partial \ln q}{\partial T}$$

Thus

$$\langle E \rangle = k_B T^2 \frac{\partial \ln q}{\partial T} = -\frac{\partial \ln q}{\partial \beta} = -\frac{\partial}{\partial \beta} \ln\left(\frac{1}{1 - \exp(-\theta_{\text{vib}}\beta)}\right) = \frac{\theta_{\text{vib}} \exp(-\theta_{\text{vib}}\beta)}{1 - \exp(-\theta_{\text{vib}}\beta)}.$$

We need to add in the zero-point contribution and, upon rearranging

$$\langle E \rangle = \frac{\theta_{\text{vib}}}{2} + \frac{\theta_{\text{vib}} \exp(-\theta_{\text{vib}}\beta)}{1 - \exp(-\theta_{\text{vib}}\beta)}\left(\frac{\exp(\theta_{\text{vib}}\beta)}{\exp(\theta_{\text{vib}}\beta)}\right) = \frac{\theta_{\text{vib}}}{2} + \frac{\theta_{\text{vib}}}{\exp(\theta_{\text{vib}}\beta) - 1}$$

as was meant to be shown.

1.12   The Debye temperature,

$$\theta_D = \hbar\omega_D/k_B, \tag{1.8.5}$$

is more commonly tabulated and determined than is the Debye frequency because of its relationship to the thermodynamic properties of solids.

(a) Calculate the Debye frequencies of the elemental solids listed below in Hz, meV and cm^{-1}.

(b) Calculate the mean phonon occupation number at the Debye frequency and room temperature for each of these materials at 100, 300 and 1000 K.

	Ag	Au	diamond	graphite	Pt	Si	W
θ_D (K)	225	165	2230	760	240	645	400

(a) Rearranging Eq. (1.8.5) and converting appropriately, we have

$$\omega_D = \left(\frac{k_B}{\hbar}\right)\theta_D = (1.309216 \times 10^{11}\,\text{s}^{-1}\,\text{K})\theta_D$$

$$\bar{\nu}_D = \frac{\omega_D}{2\pi c} \quad (c \text{ in cm s}^{-1} \text{ then } \bar{\nu}_D \text{ in cm}^{-1})$$

For meV $\hbar\omega_D = \dfrac{\nu_D}{8.065\,\text{cm}^{-1}\,\text{meV}^{-1}}$

$$n(\mathbf{k},p) \equiv \langle n(\omega_k(p),T)\rangle = \frac{1}{\exp(\hbar\omega_k(p)/k_B T) - 1}$$

	Ag	Au	diamond	graphite	Pt	Si	W
θ_D (K)	225	165	2230	760	240	645	400
ω_D (/10^{13} Hz)	2.94	2.16	29.2	9.95	3.14	8.44	5.24
$\bar{\nu}_D$ (cm^{-1})	156	115	1549	528	167	448	278
meV	19.4	14.2	192	65.5	20.7	55.5	34.4
$\langle n(\omega_D, 100\text{ K})\rangle$	0.118	0.238	2×10^{-7}	5×10^{-4}	0.0998	1.6×10^{-3}	0.0187
$\langle n(\omega_D, 300\text{ K})\rangle$	0.895	1.36	6×10^{-4}	0.0862	0.821	0.132	0.358
$\langle n(\omega_D, 1000\text{ K})\rangle$	3.96	5.57	0.12	0.879	3.69	1.1	2.03

1.13   The Debye model can be used to calculate the mean square displacement of an oscillator in a solid. In the high-temperature limit this is given by

$$\langle u^2 \rangle = \frac{3N_A \hbar^2 T}{M k_B \theta_D^2} \tag{1.8.6}$$

(a) Compare the root-mean-square displacements of Pt at 300 K to that at its melting point (2045 K). What is the fractional displacement of the metal atoms relative to the interatomic distance at the melting temperature?

(b) Compare this to the root-mean-square displacement of the C atoms at the surface of diamond at the same two temperatures.

According to the Debye model the root-mean-square displacement of Pt is given by

$$\langle u^2 \rangle^{1/2} = \sqrt{\frac{3N_A \hbar^2 T}{M k_B \theta_D^2}} = 3.815 \times 10^{-11} \text{ m kg}^{1/2} \text{ K}^{1/2} \sqrt{\frac{T}{M \theta_D^2}} = 0.3815 \text{ Å kg}^{1/2} \text{ K}^{1/2} \sqrt{\frac{T}{M \theta_D^2}}$$

with $M = 0.19508$ kg mol^{-1} and $\Theta_D = 240$ K.

Thus $\langle u^2 \rangle^{1/2}|_{300 \text{ K}} = 0.0623$ Å and $\langle u^2 \rangle^{1/2}|_{2045 \text{ K}} = 0.163$ Å.

For the fractional displacement we need to calculate the equilibrium Pt–Pt distance. Since Pt is an fcc metal, the nearest neighbour distance is given by $a/\sqrt{2}$ (where a is the lattice constant) $= 3.92$ Å/$\sqrt{2} = 2.77$ Å. Therefore,

$$\frac{\langle u^2 \rangle^{1/2}|_{300 \text{ K}}}{2.77 \text{Å}} = 0.0225 \text{ and } \frac{\langle u^2 \rangle^{1/2}|_{2045 \text{ K}}}{2.77 \text{ Å}} = 0.0588.$$

(a) For diamond $M = 0.0120$ kg mol^{-1} and $\theta_D = 2230$ K.

Thus $\langle u^2 \rangle^{1/2}|_{300 \text{ K}} = 0.0271$ Å and $\langle u^2 \rangle^{1/2}|_{2045 \text{ K}} = 0.0706$ Å.

The absolute motion of a C(diamond) atom at these temperatures corresponds to only about 43% of that of a Pt atom.

1.14   The surface Debye temperature of Pt(100) is 110 K. Take the definition of melting to be the point at which the fractional displacement relative to the lattice constant is equal to $\sim 8.3\%$ (Lindemann criterion). What is the surface melting temperature of Pt(100)? What is the implication of a surface that melts at a lower temperature than the bulk?

Solving the root mean square displacement equation, Eq. (1.8.6), for T yields

$$T_f = M \theta_D^2 (0.083 x/c)^2$$

where $x = \sqrt{2}a$ is the nearest neighbour distance and $c = 0.3815$ Å kg$^{-1/2}$ K$^{-1/2}$. The surface melting temperature is found by substituting $\theta_s = 110$ K for θ_D and $x = 2.77$ Å. T_f(surface) $= 857$ K, significantly below the bulk melting point of 2045 K.

A liquid layer covers the bulk solid. The layer grows in thickness as each successively deeper solid layer becomes the surface layer of the solid. Thus melting starts at the surface and proceeds back into the bulk.

1.15   The bulk terminated Si(100)–(1 × 1) surface has two dangling bonds per surface atom and is, therefore, unstable toward reconstruction. Approximate the dangling bonds as effectively being half-filled sp^3 orbitals. The driving force of reconstruction is the removal of dangling bonds. (a) The stable room temperature surface reconstructs into a (2 × 1) unit cell in which the surface atoms move closer to each other in one direction but the distance is not changed in the perpendicular direction. Discuss how the loss of one dangling bond on each Si atom leads to the formation of a (2 × 1) unit cell. Hint: The nearest neighbour surface Si atoms are called dimers. (b) This leaves one dangling bond per surface atom. Describe the nature of the interaction of these dangling bonds that leads to (i) symmetric dimers and (ii) tilted dimers. (c) Predict the effect of hydrogen adsorption on the symmetry of these two types of dimers. Hint: Consider first the types of bonds that sp^3 orbitals can make. Second, two equivalent dangling bonds represent two degenerate electronic states.

There are two ways in which the sp^3 like dangling bond orbitals can interact. (a) π bond formation. This leads to symmetric dimers as one would expect based on the symmetry of the bonding interaction and the equivalence of the two bonding orbitals. (b) The two dangling bonds represent two degenerate half-filled electronic states when they are in a symmetric configuration. Such a system is unstable with respect to a Jahn-Teller distortion unless some other interaction overrides the Jahn-Teller effect. Therefore, the system will spontaneously break symmetry by tilting the dimer. This leads to a splitting of the electronic states into two states, with a higher population of electrons in the lower energy state than in the higher energy state.

H atom adsorption breaks the π bond and negates the Jahn-Teller effect. Thus, regardless of which interaction was first present, the structure of the dimer with one H atom on it will be the same in either case. It is unclear whether this should be symmetric or not based on simple arguments but it will be less symmetric than a dimer with two H atoms adsorbed and less tilted than a dimer with no H atoms adsorbed. A dimer with two adsorbed H atoms will be symmetric.

1.16 Describe the features a, b, c, and d in Fig. 1.19.

(a) is an occupied surface resonance. It is a resonance since it falls in an allowed part of the projected bulk band structure, i.e. it overlaps bulk bands.

(b) is an occupied surface state. It is a surface state because it appears in part of the band gap (a region where bulk states are forbidden). Both a and b are occupied because they are located below the Fermi energy, E_F.

(c) is a normally unoccupied surface state.

(d) is a normally unoccupied surface resonance. Both c and d are normally unoccupied because they lie above E_F.

1.17 What is the significance of a band gap? (b) What differentiates a partial band gap from a full band gap?

(a) A band gap is a region of k space in which no bulk bands exist. Only defect states or surface states can have the combination of energy and momentum that lies in a band gap. Any state that lies in a band gap will be weakly coupled to bulk states and the farther away it is from a band edge, the more weakly it will be coupled (all other things such as symmetry being equal). The presence and size of a band gap also determines whether a material is a metal, semimetal, semiconductor or insulator.

(b) A partial band gap is a small region of k space in which states of a given energy are not allowed. A full band gap exists in a certain energy range for all values of crystal momentum.

1.18 What are E_g, E_F, E_C, E_V and E_{vac} as shown, for instance, in Fig. 1.19?

E_g is the magnitude of the band gap. A band gap can be either direct (the minimum vertical energy difference in a diagram such as Fig. 1.19) or indirect (the minimum energy difference, which occurs between points with different values of **k**).

E_F is the Fermi energy. It is the highest allowed energy for electrons at 0 K. However, in a perfect intrinsic (undoped) semiconductor, there are no states at E_F because it lies midway in the band gap.

E_C is the conduction band minimum, the lowest energy point in the conduction band.

E_V is the valence band maximum, the highest energy point in the valence band.

E_{vac} is the vacuum energy, above which electrons are no longer bound to the solid. The difference between E_{vac} and E_C is the electron affinity. The difference between E_{vac} and E_F is the work function.

References

[1]   A. Hohlfeld, M. Sunjic, K. Horn, *J. Vac. Sci. Technol. A*, **5** (1987) 679.
[2]   L. Schmidt, R. Gomer, *J. Chem. Phys.*, **42** (1965) 3573.

10
Answers to Exercises from Chapter 2. Experimental Probes and Techniques

2.1 Derive equation (2.1.2).

$$Z_W = \frac{p}{(2\pi m k_B T)^{1/2}} = \left(2\pi \frac{0.028 \, \text{kg mol}^{-1}}{6.02 \times 10^{23} \, \text{mol}^{-1}} (1.38 \times 10^{23} \, \text{JK}^{-1} \text{mol}^{-1})(300\text{K})\right)^{-1} p$$

$$Z_W = (2.875 \times 10^{25} \, \text{m}^{-2} \, \text{s}^{-1}) p$$

This is for the pressure in Pa. Now make the identification that 1 ML corresponds to $1 \times 10^{19} \, \text{m}^{-2}$ surface atoms and convert the pressure to torr

$$Z_W = \frac{(2.875 \times 10^{25} \, \text{m}^{-2} \, \text{s}^{-1})}{1 \times 10^{19} \, \text{ML m}^{-2}} p \frac{101325 \, \text{Pa}}{760 \, \text{torr}} = (3.8 \times 10^5 \, \text{ML s}^{-1}) p$$

2.2 Show that the pressure in a chamber of volume V, at initial pressure p_0, pumped with a pumping speed S changes as a function of time according to

$$p(t) = p_0 \exp(-t/\tau), \qquad\qquad 2.12.1$$

where $\tau = V/S$ is the time constant of the chamber/pump combination.
From Eq. (2.1.5)

$$Q = k_B T \frac{dN}{dt}$$

Now we substitute to transform N into p and for Q we introduce the pumping speed from Eq.(2.1.4)

$$-Sp = k_B T \frac{d}{dt} \left(\frac{pV}{k_B T}\right).$$

Surface Science: Foundations of Catalysis and Nanoscience, Third Edition. Kurt W. Kolasinski.
© 2012 John Wiley & Sons, Ltd. Published 2012 by John Wiley & Sons, Ltd.

Note the sign on S, which is defined as a positive number, because the chamber is being pumped out and, therefore, dN/dt must be negative. V and T are constant and can be pulled out of the derivative

$$\frac{-Sp}{V} = \frac{dp}{dt}.$$

$$\frac{-S}{V} dt = \frac{dp}{p}.$$

$$\frac{-S}{V} \int_0^t dt = \int_{p_0}^p \frac{dp}{p}.$$

$$-\frac{S}{V}t = \ln p - \ln p_0 = \ln(p/p_0)$$

$$p = p_0 \exp\left(-\frac{S}{V}t\right) = p_0 \exp(-t/\tau)$$

where $\tau = V/S$ as was to be shown.

2.3 (a) What is the gas flux striking a surface in air at 1 atm and 300 K?

(b) Calculate the pressure necessary to keep a $1\,cm^2$ Pt(100) surface clean for 1 hr at 300 K, assuming a sticking coefficient of 1, no dissociation of the gas upon adsorption and that "clean" means < 0.01 ML of adsorbed impurities.

(a) Using the equation

$$Z_w = \frac{p}{\sqrt{2\pi m k_B T}} = 2.63 \times 10^{24} m^{-2} s^{-1} \frac{p/Pa}{\sqrt{MT}} = 3.51 \times 10^{22}\ cm^{-2}\ s^{-1} \frac{p/Torr}{\sqrt{MT}}$$

where p is pressure in either Pa or Torr, appropriately, T is the temperature in K and M is the molecular weight of the gas in g mol^{-1}, we obtain $Z_w \approx 2.9 \times 10^{23}$ molecules cm^{-2} s^{-1}, assuming $M_{air} \approx 29$ g mol^{-1}.

(b) With a constant sticking coefficient, the coverage is equal to the product of the sticking coefficient, s, and the exposure, ε,

$$\theta = s\varepsilon = sZ_w t.$$

With $s = 1$ (given) and making the standard assumption of 1×10^{15} Pt atoms cm^{-2} and since the coverage on the surface must be less than 0.01 ML, we have

$$Z_w t < (0.01\ ML)(1 \times 10^{15}\ cm^{-2})$$

$$3.51 \times 10^{22} \frac{p}{\sqrt{MT}}t < 1 \times 10^{13}\ cm^{-2},\ t = 3600\ s$$

$$p < 7.4 \times 10^{-12}\ Torr$$

Using the correct areal density of Pt(100), $\sigma_0 = 1.30 \times 10^{15}$ cm^{-2}, one obtains $p < 9.6 \times 10^{-12}$ Torr.

2.4 (a) Estimate the maximum exposure delivered per pulse (in units of ML/pulse) if N_2 is expanded through a chopped Knudsen source with a skimmer radius of $r_s = 50\,\mu m$, an effective pressure of $p_0 = 20$ Pa at the skimmer, a pulse length of 5 ms, and a skimmer to surface distance of $x = 15$ cm. Take the surface to be Si(111) with a density of surface atoms $\sigma_0 = 7.8 \times 10^{18}$ m^{-2}. (b) Using the same parameters, make an estimate for a hard expansion through a supersonic source with an effective Mach number of $M_{eff} = 20$.

(a) Equation (2.3.11) can be used with an appropriate value of the density at the skimmer. However, we can use the fact that an equilibrium gas distribution is incident upon the skimmer in a Knudsen source, a fact that is used to derive this equation. Assuming that the pulse train of the chopped beam has a rectangular waveform with infinitely fast rise and fall times, the flux through the skimmer during the pulse is

$$Z_w(\text{open}) = \frac{p}{\sqrt{2\pi m k_B T}} = \frac{2.63 \times 10^{24} \text{ m}^{-2}\text{s}^{-1}}{7.8 \times 10^{18} \text{ m}^{-2}} \frac{20 \text{ Pa}}{\sqrt{(30 \text{ g mol}^{-1})(298 \text{ K})}} = 7.1 \times 10^4 \text{ ML s}^{-1}$$

Note that dividing Z_w through by the number of surface atoms converts the flux from $\text{m}^{-2}\text{s}^{-1}$ to ML s^{-1}. Since the chopper is only open for 5 ms per pulse, the flux incident on the skimmer is

$$Z_w = (5 \times 10^{-3} \text{ s pulse}^{-1})(7.1 \times 10^4 \text{ ML s}^{-1}) = 360 \text{ ML pulse}^{-1}.$$

According to Eq. (2.3.11) the centreline intensity, which corresponds to the maximum flux, from a Knudsen source scales as a function of skimmer radius and distance from the skimmer as r_s^2/x^2. Hence,

$$\varepsilon_{\text{Knudsen}}(\text{max}) = \varepsilon(\text{centerline}) = Z_w \frac{r_s^2}{x^2} = 360 \text{ ML s}^{-1} \frac{(0.005 \text{ cm})^2}{(15 \text{ cm})^2} = 4.0 \times 10^{-5} \text{ ML pulse}^{-1}.$$

The flux at the centre of the beam on the surface of the crystal is reduced to only 4×10^{-5} ML per pulse due to the spread of the beam. Note also that the flux drops radially from the centre of the beam to the edge of the crystal due to the $\cos\theta$ angular distribution of the flux from a Knudsen source.

(b) For the supersonic source, the note under Eq. (2.3.15) tells us that the intensity ratio of a supersonic N_2 beam with $M_{\text{eff}} = 20$ to a Knudsen beam is 1860. Therefore,

$$\varepsilon_{\text{supersonic}} = 1860\varepsilon_{\text{Knudsen}} = 7.4 \times 10^{-2} \text{ ML pulse}^{-1}.$$

The centreline intensity is significantly higher. Furthermore, since the intensity is greater because the supersonic beam has an angular divergence that is significantly narrower than the $\cos\theta$ angular distribution of the flux from a Knudsen source, the beam size is significantly smaller. Packing more molecules into a smaller angular profile is, after all, why the intensity is greater. The supersonic beam is focussed compared to the Knudsen beam.

2.5 Lieber and co-workers [1] used AFM to measure the adhesion force arising from the contact of CH_3 groups. The adhesion force was 1.0 nN. If the tip/surface contact area was 3.1 nm^2 and the radius of a CH_3 group is $(0.2)^{1/2}$ nm, calculate the interaction force resulting from the contact of two individual CH_3 groups.

The area of a CH_3 group is

$$A_{CH_3} = \pi\, r_{CH_3}^2 = \pi\,(0.2).$$

The number of CH_3 groups on the tip, assuming close packing, is the ratio of the area of the tip to the area of a single CH_3 group

$$N(CH_3) = \frac{A_{\text{tip}}}{A_{CH_3}} = \frac{3.1}{0.2\pi} = 4.93 \quad \therefore 5CH_3 \text{ groups.}$$

As the total forces was 1.0 nN, the force per CH_3 group is 1.0 nN/5 = 0.2 nN.

2.6   In an STM image, does an adsorbate sitting on top of a surface always look like a raised bump compared to the substrate?

No. Whether an adsorbate looks like a bump or a hollow depends on the specific system, the imaging voltage and the composition of the tip. An STM image depends not only on the topography of the surface but also on the electronic structure of the surface (and tip). It is electron density not atoms that are imaged. If the tip is biased positive with respect to the surface, electrons tunnel from the occupied states of the surface into the tip. If the tip is biased negatively, electrons tunnel from the tip into the unoccupied states of the surface. Therefore, the voltage applied to the tip can affect the image that is produced if the occupied and unoccupied states have different structures.

2.7   Some adsorbates can be imaged in STM at low temperatures but disappear at higher temperatures even though they have not desorbed from the surface. Explain.

Thermal energy leads to diffusion. If the adsorbate diffuses too rapidly it will either be imaged as a streak or not at all. The tip itself can also play a role in enhancing diffusion. The enhanced diffusion can be tip voltage and tunnelling current dependent.

2.8   The dimer unit on a Si(100) surface has a bonding orbital just below E_F and an antibonding orbital just above E_F. Make a prediction about STM images that are taken at positive compared with negative voltages. Do the images look the same and, if not, how do they differ?

A bonding orbital has enhanced electron density between the nuclei involved in the bond. An antibonding orbital has a node. When imaged below E_F (positive tip bias) the bonding orbital is imaged and an electron cloud extending over both nuclei is expected to have a maximum in the middle. Above E_F (imaging with a negative tip), the antibonding state with a node between the dimer atoms is imaged. See Fig. 1 in the Introduction.

2.9   Use Eqs (2.5.19) and (2.5.20) to show that the diffraction reflexes appear at

$$\Delta \mathbf{s}/\lambda = h_1 \mathbf{a}_1^* + h_2 \mathbf{a}_2^* \tag{2.12.2}$$

Assume normal incidence of the incoming electrons and $\Delta \mathbf{s} = \mathbf{s} - \mathbf{s}_0$.

From Eq. (2.5.19) $\mathbf{a}_1 \cdot (\mathbf{s} - \mathbf{s}_0) = h_1 \lambda$.

Thus $\mathbf{a}_1 \cdot (\mathbf{s} - \mathbf{s}_0) = \lambda \mathbf{a}_1 \cdot \dfrac{(\mathbf{s} - \mathbf{s}_0)}{\lambda} = \lambda \mathbf{a}_1 \cdot \dfrac{\Delta \mathbf{s}}{\lambda}$.

Now substitute from Eq.(2.12.2)

$$\mathbf{a}_1 \cdot (\mathbf{s} - \mathbf{s}_0) = \lambda \mathbf{a}_1 \cdot (h_1 \mathbf{a}_1^* + h_2 \mathbf{a}_2^*).$$

From the Kronecker delta relation in Eq. (2.5.5) $\mathbf{a}_1 \cdot \mathbf{a}_1^* = 1$ and $\mathbf{a}_1 \cdot \mathbf{a}_2^* = 0$, which yields

$$\mathbf{a}_1 \cdot (\mathbf{s} - \mathbf{s}_0) = \lambda h_1.$$

Similarly $\mathbf{a}_2 \cdot (\mathbf{s} - \mathbf{s}_0) = \lambda \mathbf{a}_1 \cdot \dfrac{\Delta \mathbf{s}}{\lambda} = \lambda h_2.$

Both of these results can only be true if condition (2.12.2) holds. Thus the directions of the interference maxima are determined by the vectors of the reciprocal lattice. Also the spacings between the spots in the reciprocal lattice, i.e. the spots in the diffraction pattern, are dependent on the incident energy of the electron beam, which determines λ through the de Broglie relationship Eq. (2.5.1).

2.10   Describe how the spots in a LEED pattern would evolve if:
(a) incident molecules adsorbed randomly onto a surface forming an ordered overlayer only when one quarter of the substrate atoms are covered (a quarter of a monolayer);
(b) the incident molecules formed ordered islands that continually grow in size until they reach a saturation coverage of a quarter of a monolayer;

(c) an overlayer is ordered in one direction but not in the orthogonal direction (either due to random adsorption or diffusion in the orthogonal direction).

(a) The clean substrate will produce its characteristic pattern. A disordered overlayer gives rise to diffuse background scattering. As the number of disordered scatterers increases, the intensity of the background will increase. At one quarter of a monolayer there is a sharp transition to an ordered phase. Hence, there will be a sharp change in the LEED pattern from substrate spots + diffuse background to substrate spots + spots indicative of the overlayer over a narrow coverage range. As the substrate and overlayer structures are not specified, we cannot assign them unambiguously; however, the following (2×2) structure is consistent with the above.

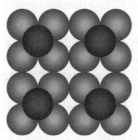

Figure Exercise 2.10(a)   *Absorbate overlayer with a (2 × 2) structure corresponding to $\theta = 0.25$ ML.*

(b) "Extra" LEED spots, that is, LEED spots associated with the overlayer, will begin to appear even at the lowest coverages due to the presence of ordered overlayer islands. As the number of scatterers increases, the intensity of these extra spots increases. As the size of the islands grows, the LEED spots become sharper (narrower) until the islands approach the size of the coherence length of the electron beam. For the above example, the LEED pattern would develop as shown in Fig. Ex.2.10(b).

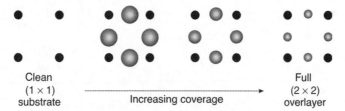

Figure Exercise 2.10(b)   *An illustrator of how a LEED pattern develops during island information for an absorbate overlayer with a (2 × 2) stucture corresponding to a local coverage of $\theta = 0.25$ ML.*

(c) Streaky spots appear. That is, the "spots" are narrow in one direction but long in the other. The spots turn into lines in the limit of complete delocalization in one dimention.

2.11 Consider clean fcc(100), fcc(110) and fcc(111) surfaces (fcc = face-centred cubic). Draw the unit cell and include the primitive lattice vectors $\mathbf{a}_1$ and $\mathbf{a}_2$. Calculate the reciprocal lattice vectors $\mathbf{a}_1*$ and $\mathbf{a}_2*$ and draw the LEED pattern including the reciprocal lattice vectors.

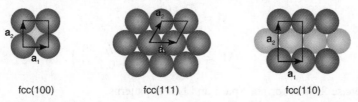

Figure Exercise 2.11(a)   *Lattice vectors and unit cells of three fcc low index planes.*

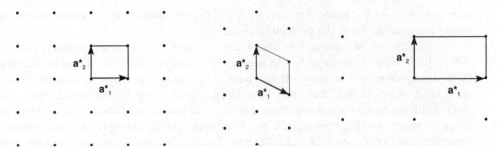

Figure Exercise 2.11(b)   *LEED patterns for three fcc low index planes.*

2.12   For structures (a)–(i) in Fig. 2.30 determine the associated LEED patterns. Classify the structures in both Wood's and matrix notation.

(a) In Wood's notation (3 × 4). We have to determine the M matrix for matrix notation

$$\mathbf{M} = \begin{pmatrix} 3 & 0 \\ 0 & 4 \end{pmatrix}.$$

To determine the LEED pattern we need the M* matrix.

$$\mathbf{M}^* = \frac{1}{(3)(4)} \begin{pmatrix} 4 & 0 \\ 0 & 3 \end{pmatrix} = \begin{pmatrix} 1/3 & 0 \\ 0 & 1/4 \end{pmatrix}.$$

Figure Exercise 2.12(a)   *Real space and reciprocal space depictions of a (3 × 4) structure.*

(b) In Wood's notation p(2 × 2).

$$\mathbf{M} = \begin{pmatrix} 2 & 0 \\ 0 & 2 \end{pmatrix}.$$

$$\mathbf{M}^* = \frac{1}{4} \begin{pmatrix} 2 & 0 \\ 0 & 2 \end{pmatrix} = \begin{pmatrix} 1/2 & 0 \\ 0 & 1/2 \end{pmatrix}.$$

See Exercise 2.14(b) for real space and LEED patterns.

(c) In Wood's notation c(2 × 2) or $p\left(\sqrt{2} \times \sqrt{2}\right)R45°$. The $p\left(\sqrt{2} \times \sqrt{2}\right)R45°$ nomenclature reflects the true geometry of the overlayer unit cell relative to the substrate unit cell and can be worked out from the relationship of the **a** vectors to the **b** vectors as done in analogy to (g) below.

$$\mathbf{M} = \begin{pmatrix} 1 & -1 \\ 1 & 1 \end{pmatrix}.$$

$$\mathbf{M}^* = \frac{1}{2}\begin{pmatrix} 1 & -1 \\ 1 & 1 \end{pmatrix} = \begin{pmatrix} 1/2 & -1/2 \\ 1/2 & 1/2 \end{pmatrix}.$$

See Exercise 2.14(a) for real space and LEED patterns.

(d) In Wood's notation p(2 × 2).

$$\mathbf{M} = \begin{pmatrix} 2 & 0 \\ 0 & 2 \end{pmatrix}.$$

$$\mathbf{M}^* = \frac{1}{4}\begin{pmatrix} 2 & 0 \\ 0 & 2 \end{pmatrix} = \begin{pmatrix} 1/2 & 0 \\ 0 & 1/2 \end{pmatrix}.$$

Figure Exercise 2.12(d) *Real space and reciprocal space depictions of a p(2 × 2) structure.*

(e) In Wood's notation p(2 × 2).

$$\mathbf{M} = \begin{pmatrix} 2 & 0 \\ 0 & 2 \end{pmatrix}.$$

This is equivalent to (d) as far as the LEED pattern is concerned. A LEED pattern can only determine the lattice of the surface unit cell, not its position relative to the substrate. Further information such as a dynamical LEED calculation of the spot intensities as a function of incident electron energy, infrared adsorption spectra, STM imaging, or x-ray absorption fine structure would be required to determine whether the adsorbate occupies the hollow or on-top site.

(f) In Wood's notation c(2 × 2).

$$M = \begin{pmatrix} 1 & -1 \\ 1 & 1 \end{pmatrix}.$$

$$M^* = \begin{pmatrix} 1/2 & -1/2 \\ 1/2 & 1/2 \end{pmatrix}.$$

Whereas the notation and matrices look equivalent to the (100) case, the LEED pattern looks different because of the different orientation of the unit vectors.

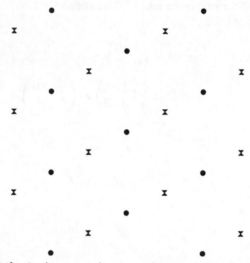

Figure Exercise 2.12(f)   *Real space and reciprocal space depictions of a c(2 × 2) structure.*

(g) This is the $\left(\sqrt{3} \times \sqrt{3} \right)$ R30° structure. To determine this, we first determine the real space transformation matrix

$$M = \begin{pmatrix} 1 & 1 \\ -1 & 2 \end{pmatrix}.$$

Therefore

$$b_1 = a_1 + a_2.$$

To work out the length of b_1 and the angle between a_1 and b_1, we need the real space co-ordinates of a_2 and b_1. The real space components of a_1 are trivially (1 0). The others are found from geometry by noting that the angle between a_1 and a_2 is 60° and both have unit length. Thus the tip of a_2 lies at $(x\ y) = (\cos 60°\ \sin 60°) = \left(\frac{1}{2} 0.866 \right)$ and the real space components are $b_1 = \left(1 + \frac{1}{2} 0 + 0.866 \right) = \left(\frac{3}{2} 0.866 \right)$. The length of the real space overlayer unit cell is the length of b_1 given by

$$\|b_1\| = \left(\left(\frac{3}{2} \right)^2 + (0.866)^2 \right)^{1/2} = \sqrt{3}.$$

The angle between a_1 and b_1 is given by the scalar product

$$\mathbf{a_1} \cdot \mathbf{b_1} = (10) \cdot \left(\frac{3}{2}0.866\right) = \frac{3}{2} = \|\mathbf{a_1}\|\|\mathbf{b_1}\| \cos\theta = (1)\left(\sqrt{3}\right)\cos\theta$$

$$\theta = 30°$$

$$\mathbf{M}^* = \frac{1}{2+1}\begin{pmatrix} 2 & 1 \\ -1 & 1 \end{pmatrix} = \begin{pmatrix} ^2/_3 & ^1/_3 \\ -^1/_3 & ^1/_3 \end{pmatrix}.$$

Figure Exercise 2.12(g)   *Real space and reciprocal space depictions of a* $(\sqrt{3} \times \sqrt{3})R30°$ *structure.*

(h) In Wood's notation p(2 × 1).

$$\mathbf{M} = \begin{pmatrix} 2 & 0 \\ 0 & 1 \end{pmatrix}.$$

$$\mathbf{M}^* = \frac{1}{2}\begin{pmatrix} 1 & 0 \\ 0 & 2 \end{pmatrix} = \begin{pmatrix} ^1/_2 & 0 \\ 0 & 1 \end{pmatrix}.$$

See Exercise 2.14(e) for real space and LEED patterns, which are related by a 90° rotation.

(i) In Wood's notation p(4 × 1).

$$\mathbf{M} = \begin{pmatrix} 4 & 0 \\ 0 & 1 \end{pmatrix}.$$

$$\mathbf{M}^* = \frac{1}{4}\begin{pmatrix} 1 & 0 \\ 0 & 4 \end{pmatrix} = \begin{pmatrix} ^1/_4 & 0 \\ 0 & 1 \end{pmatrix}.$$

Figure Exercise 2.12(i)   *Real space and reciprocal space depictions of a p(4 × 1) structure.*

2.13   Fractional coverage can be defined as the number of adsorbates divided by the number of surface atoms:

$$\theta = N_{ads} / N_0. \tag{2.12.3}$$

For each of the structures in Exercise 2.12, calculate the coverage. Note any correlations between coverage and the LEED patterns.

The area of a unit cell is proportional to the product of the unit vectors that describe the unit cell. The coverage is given by the number of adsorbates per unit cell, n_{unit}, divided by the area of the adsorbate unit cell, divided by the area of the substrate unit cell. In Wood's notation the unit cell has a (1×1) dimension and an area of 1 unit. Thus the fractional coverage is simply

$$\theta = n_{unit}/(nm) = 1/\det \mathbf{M}$$

where n and m are the indicies used in Wood's notation. Obviously, this relationship only holds if Wood's notation can be used to describe the structure. It must also be kept in mind that an ordered array of vacancies gives the same diffraction pattern as the analogous ordered array of filled sites.

(a) $\theta = 1/12\,\text{ML}$. (b) $\theta = 0.25\,\text{ML}$. (c) $\theta = 0.5\,\text{ML}$. (d) $\theta = 0.25\,\text{ML}$. (e) $\theta = 0.25\,\text{ML}$. (f) $\theta = 0.5\,\text{ML}$. (g) $\theta = 0.33\,\text{ML}$. (h) $\theta = 0.5\,\text{ML}$. (i) $\theta = 0.25\,\text{ML}$.

2.14   Given LEED patterns (a)–(g) in Fig. 2.30 obtained from adsorbate-covered face-centred cubic (fcc) substrates, determine the surface structures. Substrate reflexes are marked while the additional adsorbate induced reflexes are marked ×. Assume no reconstruction of the surface.

(a) $\mathbf{M}^* = \begin{pmatrix} \frac{1}{2} & -\frac{1}{2} \\ \frac{1}{2} & \frac{1}{2} \end{pmatrix}$.

$$\mathbf{M} = \frac{1}{\frac{1}{4} + \frac{1}{4}} \begin{pmatrix} \frac{1}{2} & -\frac{1}{2} \\ \frac{1}{2} & \frac{1}{2} \end{pmatrix} = \begin{pmatrix} 1 & -1 \\ 1 & 1 \end{pmatrix}.$$

$\theta = 0.5\,\text{ML}$. In Wood's notation c(2×2) or p$\left(\sqrt{2} \times \sqrt{2}\right)$ R45°.

Figure Exercise 2.14(a)   *Real space and reciprocal space depictions of a c(2 × 2)p$\left(\sqrt{2} \times \sqrt{2}\right)$ R45° structure.*

(b) $\mathbf{M}^* = \begin{pmatrix} \frac{1}{2} & 0 \\ 0 & \frac{1}{2} \end{pmatrix}.$

    $\mathbf{M} = \dfrac{1}{\frac{1}{4}} \begin{pmatrix} \frac{1}{2} & 0 \\ 0 & \frac{1}{2} \end{pmatrix} = \begin{pmatrix} 2 & 0 \\ 0 & 2 \end{pmatrix}.$

The p(2 × 2) structure can either correspond to $\frac{1}{4}$ ML of ordered filled sites or $\frac{1}{4}$ ML of ordered vacancies, that is, $\frac{3}{4}$ ML of ordered filled sites. Care must be taken when performing the inverse problem of relating a diffraction pattern, which is not unique, to a real space structure, which is unique.

Figure Exercise 2.14(b)   *Real space and reciprocal space depictions of a c(2 × 2) structure.*

(c) $\mathbf{M}^* = \begin{pmatrix} \frac{1}{4} & -\frac{1}{2} \\ \frac{1}{4} & \frac{1}{2} \end{pmatrix}.$

    $\mathbf{M} = \dfrac{1}{\frac{1}{4}} \begin{pmatrix} \frac{1}{2} & -\frac{1}{4} \\ \frac{1}{2} & \frac{1}{4} \end{pmatrix} = \begin{pmatrix} 2 & -1 \\ 2 & 1 \end{pmatrix}.$

Figure Exercise 2.14(c)   *Real space and reciprocal space depictions of a c(4 × 2) structure.*

The c(4 × 2) structure can either correspond to $\frac{1}{4}$ ML of ordered filled sites or $\frac{1}{4}$ ML of ordered vacancies, that is, $\frac{3}{4}$ ML of ordered filled sites.

(d) $\mathbf{M}^* = \begin{pmatrix} \frac{1}{4} & -\frac{1}{2} \\ \frac{1}{4} & \frac{1}{2} \end{pmatrix}.$

$\mathbf{M} = \frac{1}{\frac{1}{4}} \begin{pmatrix} \frac{1}{2} & -\frac{1}{4} \\ \frac{1}{2} & \frac{1}{4} \end{pmatrix} = \begin{pmatrix} 2 & -1 \\ 2 & 1 \end{pmatrix}.$

This apparent c(4 × 2) structure can either correspond to $\frac{1}{4}$ ML of ordered filled sites or $\frac{1}{4}$ ML of ordered vacancies, that is, $\frac{3}{4}$ ML of ordered filled sites. The choice of site is arbitrary as the pattern cannot differentiate between long bridge, short bridge or on-top site. However, contrary to the c(4 × 2) on the square lattice (100), c(4 × 2) can be distinguished from c(2 × 4) because of the rectangular lattice on (110).

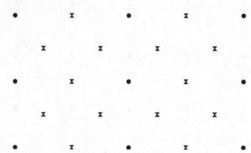

Figure Exercise 2.14(d)   *Real space and reciprocal space depictions of a c(4 × 2) structure.*

(e) $\mathbf{M}^* = \begin{pmatrix} 1 & 0 \\ 0 & \frac{1}{2} \end{pmatrix}.$

$\mathbf{M} = \frac{1}{\frac{1}{2}} \begin{pmatrix} \frac{1}{2} & 0 \\ 0 & 1 \end{pmatrix} = \begin{pmatrix} 1 & 0 \\ 0 & 2 \end{pmatrix}.$

The p(1 × 2) structure, 0.5 ML.

Figure Exercise 2.14(e)   *Real space and reciprocal space depictions of a c(4 × 2) structure.*

(f) $\mathbf{M}^* = \begin{pmatrix} \frac{1}{3} & 0 \\ 0 & \frac{1}{2} \end{pmatrix}$.

$$\mathbf{M} = \frac{1}{\frac{1}{6}} \begin{pmatrix} \frac{1}{2} & 0 \\ 0 & \frac{1}{3} \end{pmatrix} = \begin{pmatrix} 3 & 0 \\ 0 & 2 \end{pmatrix}.$$

The p(3 × 2) structure shown with $\theta = 1/6\,\mathrm{ML}$ as opposed to the $\theta = 5/6\,\mathrm{ML}$ structure.

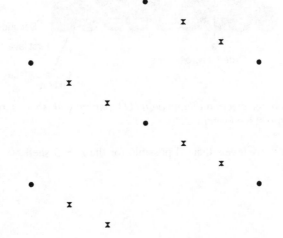

Figure Exercise 2.14(f)   *Real space and reciprocal space depictions of a p(3 × 2) structure.*

(g) $\mathbf{M}^* = \begin{pmatrix} \frac{1}{3} & 0 \\ 0 & 1 \end{pmatrix}$.

$$\mathbf{M} = \frac{1}{\frac{1}{3}} \begin{pmatrix} 1 & 0 \\ 0 & \frac{1}{3} \end{pmatrix} = \begin{pmatrix} 3 & 0 \\ 0 & 1 \end{pmatrix}.$$

Figure Exercise 2.14(g)   *Real space and reciprocal space depictions of a p(3 × 1) structure.*

The p(3 × 1) structure shown with $\theta = 1/3$ ML and $\theta = 2/3$ ML structures. Note that because of the three-fold symmetry of the substrate, it is likely that a real diffraction pattern would appear as a superposition of three (3 × 1) patterns related by rotation.

2.15 When O atoms adsorb on Ag(110), a series of streaky LEED patterns is observed. The LEED patterns are elongated in one direction as shown in Fig. 2.32. Describe the structure of the overlayer and what might be causing this phenomenon. Refer directly to the structure of the overlayer with respect to the Ag(110) surface.

The LEED pattern shows a (2 × 1) overlayer structure that is disordered along the long reciprocal space dimension, i.e. the short real space dimension. An explanation is that the oxygen atoms occupy half of the sites in the trough. Adsorption in the trough is more likely than on top because atoms tend toward highest co-ordination. This leads to a well-defined spacing between adsorbed O atoms in one direction. Occupying half of the trough sites leads to the doubling of the overlayer unit cell with respect to the substrate. Disorder along the direction parallel to the troughs leads to the streaking. The disorder may be static or may be due to diffusion.

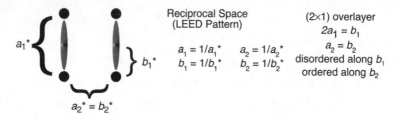

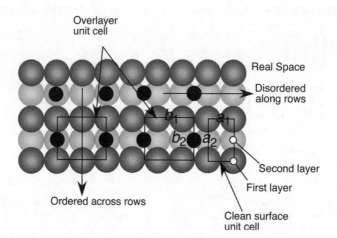

Figure Exercise 2.15 *Low-energy electron diffraction (LEED) pattern with sharp substrate spots (•) and streaky adsorbate overlayer related spots: see Exercise 2.15.*

2.16 Determine all of the x-ray levels that all possible for the $n = 3$ shell.

n	l	j	X-ray Level	Electron level
3	0	1/2	M_1	$3s$
3	1	1/2	M_2	$3p_{1/2}$
3	1	3/2	M_3	$3p_{3/2}$
3	2	3/2	M_4	$3d_{3/2}$
3	2	5/2	M_5	$3d_{5/2}$

2.17   Show that XPS can be made more surface sensitive by detecting the photoelectrons in the off normal direction.

Any electrons that are inelastically scattered will not be detected in the XPS peak associated with the atoms in the surface. The probability of escaping the film without inelastic scattering decreases exponentially with the path length traversed by the electron. An electron emitted normal to the substrate surface travels a distance d, where d is the depth at which the electron was generated. An electron produced at d but emitted at angle ϑ travels $d\,/\cos\vartheta$ before leaving the film. Thus, if only electrons that travel a path $d\,/\cos\vartheta \leq \lambda_M$ contribute significantly to the XPS signal, then only electrons generated at a depth $d \leq \lambda_M \cos\vartheta$ contribute, and the XPS signal becomes increasingly surface sensitive, i.e. d becomes smaller, as the angle increases.

2.18   Explain why Auger electron spectroscopy is surface sensitive. Are Auger peaks recorded at 90, 350 and 1500 eV equally surface sensitive?

Auger electron spectroscopy is surface sensitive because of the short mean free path of electrons with kinetic energies of roughly ~ 50–500 eV. In this kinetic energy range, the mean free path is ≤ 10 Å, that is, the mean free path is only as long as 2–5 atom layers. Therefore, electrons ejected with these kinetic energies are created in the near-surface region of the sample. There are also secondary electrons throughout this energy range, particularly at lower kinetic energy; however, the electrons detected in the peaks associated with particular transitions have not been subjected to inelastic scattering and, therefore, must come from the surface and near-surface region.

The mean free path changes with kinetic energy as shown in Fig. 2.13. The mean free path increases from ~ 5 Å at 90 eV to ~ 10 Å at 350 eV to ~ 20 Å at 1500 eV. Therefore, the peaks at those energies are not equally surface sensitive. Instead, they sample progressively more and more of the subsurface region.

2.19   The Auger data for the intensity of the Pt transition at 267 eV and the C transition at 272 eV found in Table 2.3 were taken for adsorption of CH_4 on Pt(111) at $T_s = 800$ K, $T_{gas} = 298$ K. At this temperature CH_4 dissociates and H_2 desorbs leaving C(a) on the surface. Use the data to make a plot of θ_{Pt} and θ_C versus the exposure, ε, in other words make a plot of the uncovered and covered surface sites as a function of ε. The AES sensitivity factors are $s_{Pt} = 0.030$ and $s_C = 0.18$.

The relative coverage from Auger data is given by

$$\theta_A = \frac{I_A\,/\,s_A}{I_A\,/\,s_A + I_B\,/\,s_B}.$$

Table 2.3   *Integrated Auger peak areas measured for Pt and C after dosing with* CH_4

$\varepsilon/10^{20}$ cm^{-2}	I_{Pt}	I_C	θ_{Pt}/ML	θ_C/ML
0	11350.8	166.2	0.998	0.002
0.176	10248.4	664.7	0.989	0.011
1.59	11693.5	1871.3	0.974	0.026
2.48	11360.1	2382.5	0.967	0.033
3.45	10967.6	3178.5	0.954	0.046
4.29	11610.2	4464.4	0.940	0.060
5.07	9626.0	4277.8	0.932	0.068
6.19	11313.7	5876.1	0.921	0.079
9.96	11575.4	8358.7	0.893	0.107
14.2	10964.3	8730.4	0.884	0.116
17.4	10864.1	10292.9	0.865	0.135
21.5	11850.8	11689.7	0.860	0.140
25.7	12912.7	13086.6	0.857	0.143
30.5	11529.1	12753.0	0.846	0.154
48.8	10791.4	10713.9	0.859	0.141

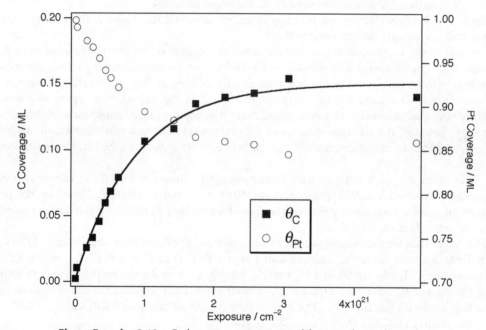

Figure Exercise 2.19   *Carbon coverage extracted from analysis of AES data.*

2.20   Consider the spectrum of adsorbed CO in Fig. 2.24. CO is adsorbed as a gem-dicarbonyl on the Rh atoms present on the Al_2O_3 substrate as shown in Figure 2.31.

(a) If the CO molecules in the gem-dicarbonyl were independent oscillators, only one CO stretch peak would be observed. Explain why there are two peaks in the spectrum [77].

(b) Explain why substitution of ^{18}O for ^{16}O changes the positions of the bands.

(a) The two oscillators are coupled. Two coupled oscillators produce symmetric and antisymmetric composite modes. These two modes have slightly different frequencies and, consequently, two peaks are observed. Similarly, the two monohydride H atoms bound to a Si dimer on the Si(100)–(2 × 1) surface are coupled and lead to two peaks in the Si–H stretch region.

(b) The wavenumber of an oscillator,$\tilde{v}$, depends on the mass of the oscillator according to

$$\tilde{v} = \frac{1}{2\pi c} \left(\frac{k}{\mu}\right)^{1/2}$$

where k is the force constant and μ the effective mass,

$$\mu = \frac{m_1 m_2}{m_1 + m_2}.$$

If c is given in cm s^{-1}, then $\tilde{v}$ is in cm^{-1}. To a first approximation, k is independent of mass. In this system, for instance, the force constant changes by $\leq 0.2\%$ for complete substitution of ^{18}O for ^{16}O. The different isotopes have different masses and, therefore, different stretch frequencies. The isotope shift is expected to go as the inverse of the square root of the effective masses. The predicted shift for an uncoupled CO stretch is

$$\frac{\tilde{v}_{16}}{\tilde{v}_{18}} = \left(\frac{\mu_{18}}{\mu_{16}}\right)^{1/2} = \left(\frac{\frac{(12)(18)}{12+18}}{\frac{(12)(16)}{12+16}}\right)^{1/2} = 1.025.$$

In other words, we expect a 2.5% larger wavenumber for the lighter isotope.

References

[1]   C. D. Frisbie, L. F. Rozsnyai, A. Noy, M. S. Wrighton, C. M. Lieber, *Science*, **265** (1994) 2071.

11

Answers to Exercises from Chapter 3. Chemisorption, Physisorption and Dynamics

3.1 Draw the structure of the fcc(100), fcc(111) and fcc(110) surfaces (top view). Indicate the unit cell and identify all possible surface adsorption sites expected for chemisorption.

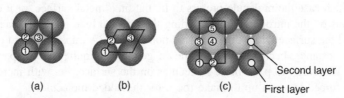

Figure Exercise 3.1 *Top view of fcc(100), fcc(111) and fcc(110) surfaces.*

(a) On the (100) there are three unique high-symmetry sites: (1) on-top, (2) 2-fold bridge, and (3) 4-fold hollow.

(b) On (111): (1) on-top, (2) 2-fold bridge, and (3) 3-fold hollow.

(c) On (110) there are seven possible sites: (1) on-top, (2) 2-fold (short) bridge between two first-layer atoms in the same row, (3a) 2-fold (long) bridge between two first layer atoms in adjacent rows, (3b) 2-fold bridge between two second-layer atoms in the same row, (4a) on top on a second-layer atom, (4b) 4-fold site between four first layer atoms, and (5) 3-fold site between two first layer and one second layer atom (much like a (111) site).

3.2 Given that the mean lifetime of an adsorbate is

$$\tau = \frac{1}{A} \exp\left(E_{\text{des}}/RT_{\text{s}}\right) \tag{3.19.1}$$

where A is the pre-exponential factor for desorption and E_{des} is the desorption activation energy, show that the mean random walk distance travelled by an adsorbate is

$$\langle x^2 \rangle^{1/2} = \left(\frac{4D_0}{A} \exp\left(\frac{E_{\text{des}} - E_{\text{dif}}}{RT}\right)\right)^{1/2}. \tag{3.19.2}$$

Surface Science: Foundations of Catalysis and Nanoscience, Third Edition. Kurt W. Kolasinski.
© 2012 John Wiley & Sons, Ltd. Published 2012 by John Wiley & Sons, Ltd.

The average lifetime on the surface is given by

$$\tau = \frac{1}{A} \exp\left(E_{des}/RT\right)$$

The average distance travelled on the surface due to diffusion, d, is given by

$$d = \langle x^2 \rangle^{1/2} = (4D\tau)^{1/2}$$

where $D = D_0 \exp(-E_{dif}/RT)$.

Therefore

$$\langle x^2 \rangle^{1/2} = \left(\frac{1}{A} \exp(E_d/RT) \times 4D_0 \exp(-E_{dif}/RT)\right)$$

$$\langle x^2 \rangle^{1/2} = \left[\frac{4D_0}{A} \exp\left(\frac{E_{des} - E_{dif}}{RT}\right)\right]^{1/2}$$

3.3   When pyridine adsorbs on various metal surfaces, it changes its orientation as a function of coverage. Describe the bonding interactions that pyridine can experience and how this affects the orientation of adsorbed pyridine.

   The bonding of pyridine to metal surfaces can involve either the system of π electrons parallel to the ring, in which case the molecule prefers to lie flat on a metal surface, or it can involve the lone pair of electrons on the nitrogen atom. Bonding through the lone pair will lead to the N end being pointed toward the surface while the other end points away. A transition from one type to the other can occur, for instance, because the lying down geometry requires more space than the standing up geometry. Therefore, to pack more molecules on the surface, i.e. with increasing coverage, the molecules are forced to stand up to make room for the added molecules.

   Such a transition as a function of coverage has been observed on Ag(111), Ni(100), and Pt(110). Temperature also changes the bonding, as is seen on Pd(111). Tilted pyridine has been observed on Ir(111) and Cu(110).

3.4   Cyclopentene, C_5H_8, is chemisorbed very weakly on Ag(111). Given that the double bond in C_5H_8 leads to a dipole that is oriented as shown in the figure below, suggest a configuration for the molecule bound at low coverage on a stepped Ag surface with (111) terraces.

   Since the chemisorption is weak, the dipole–dipole interaction between the molecule and a step on the surface will strongly influence the adsorbate geometry. The lowest energy configuration is one in which the dipoles align anti-parallel. Thus the top of the pentagon will point in the downstairs direction. The double bond will be toward the top of the step.

3.5   When CO binds in sites of progressively higher co-ordination number (on-top → two-fold bridge → three-fold hollow → four-fold hollow), both the π and σ contributions to bonding increase in magnitude.

   (a) Predict the trends that are expected in the CO stretching frequency and chemisorption bond energy with change of site.

   (b) When the π bonding interaction with the surface is weak, which adsorption site is preferred?

   The bonding of CO to metal atoms occurs via charge donation from the 5σ orbital and back-donation from the metal into the $2\pi^*$ antibonding orbital. The occupied 1π orbital is also involved in the interactions of the CO π system with the metal. The interaction of the σ orbitals (the occupied 4σ also being involved to a lesser extent) is more or less non-bonding (perhaps even repulsive), whereas the interactions with the π system are primarily responsible for the metal –CO

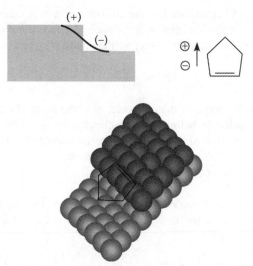

Figure Exercise 3.4   *Schematic drawing of cyclopentene adsorbed on a stepped Ag surface.*

chemisorption bond. This back-donation weakens the C $\equiv$O bond. At higher coordination, there is an increased back-donation of charge from more than one atom. As a result, the C $\equiv$O bond becomes weaker and the force constant decreases accordingly. Thus the ν_{CO} stretching frequency shifts to lower wavenumbers. It is generally accepted that $\nu_{CO} \approx 2000$ cm^{-1} corresponds to linear bonding, $1850 \leq \tilde{\nu} \leq 2000$ cm^{-1} indicates bridge bonding (two-fold symmetry) and higher co-ordination can be assigned to frequencies below 1850 cm^{-1}.

In adsorption systems where the π interaction is weaker, such as CO on Cu or N$_2$ on Ni(100), the σ repulsion will become too dominant at higher co-ordination, leading to population of only on-top sites. Strong π interactions favour higher co-ordination sites.

3.6   CO bound to Pt(111) submerged in 0.1 M HClO$_4$ exhibits an FTIR peak associated with a linearly bound on-top species at 2070 cm^{-1} [1]. 0.6 ML of Ru is deposited on the Pt(111) electrode to form islands of Ru. When CO is adsorbed on the resulting surface the peak at 2070 cm^{-1} shifts by -10 cm^{-1}, and decreases in intensity while a new peak appears at 1999 cm^{-1}. The new peak is shifted by $+6$ cm^{-1} compared to the peak associated with CO bound in an on-top site on a clean Ru electrode. Interpret the data as to where and how the CO is bound.

According to the Blyholder model of CO adsorption, the higher the frequency of the C–O stretch, the weaker the M–CO chemisorption bond. Accordingly, CO with a vibrational wavenumber of 1993 cm^{-1} is more strongly bound on clean Ru than on clean Pt, which has a wavenumber of 2070 cm^{-1}. The shift of -10 cm^{-1} as Ru is added to the Pt surface indicates an electronic effect in which coadsorbed Ru atoms strengthen the Pt–CO bond. The $+6$ cm^{-1} shift on Ru islands, compared to clean Ru, indicates that an electronic effect caused by the adsorption on Ru on Pt leads to a weakening of the Ru–CO chemisorption bond compared to chemisorption on the clean Ru surface. The electronic effect arises from charge transfer between Pt and Ru.

Since the CO is more strongly bound on Ru than on Pt, if CO were initially adsorbed on Pt and then Ru is added to the surface, the CO will migrate from Pt adsorption sites to Ru sites. Thus, the 2070 cm^{-1} peak will decrease in intensity after the deposition of Ru as CO migrates to Ru islands, while the accompanying 1999 cm^{-1} peak will grow in intensity.

3.7   The amount of energy, δE, transferred in the collision of a molecule with a chain of atoms in the limit of a fast, impulsive collision (that is, a collision that is fast compared to the time that it takes the struck atom to recoil and transfer energy to the chain) is given by the Baule formula,

$$\delta E = \frac{4\mu}{(1 + \mu)^2}(E_\mathrm{i} + q_\mathrm{ads}) \tag{3.19.3}$$

where $\mu = M/m$, M = the mass of the molecule, m is the mass of one chain atom, E_i is the initial kinetic energy of the molecule before it is accelerated by q_ads (the depth of the attractive well, effectively the heat of adsorption). Estimate the energy transfer for H_2, CH_4 and O_2 incident upon copper or platinum chains. Take the incident energy to be
(a) the mean kinetic energy at 300 K;
(b) $E_K = 1.0\,\mathrm{eV}$.
    Take the well depths to be 20 meV, 50 meV and 200 meV for H_2, CH_4 and O_2, respectively.
    The mean kinetic energy of the flux onto the surface is $2k_\mathrm{B}T$ (see Exercise 3.16). This is the same for all of the molecules.

$$2k_\mathrm{B}T = 2(1.38 \times 10^{-23}\,\mathrm{J\,K^{-1}})(300\,\mathrm{K})\left(\frac{1\,\mathrm{J}}{1.602 \times 10^{-19}\,\mathrm{eV}}\right) = 0.0517\,\mathrm{eV}$$

The appropriate masses in amu are $H_2 = 2$, $CH_4 = 16$, $O_2 = 32$, Cu = 63.5, Pt = 195. Thus

μ	H_2	CH_4	O_2
Cu	0.0315	0.252	0.504
Pt	0.0103	0.0821	0.164

$$\delta E = \frac{4(0.0315)}{(1 + 0.0315)^2}(0.0517 + 0.02) = 0.00849\,\mathrm{eV}\ H_2/\mathrm{Cu}$$

$$\delta E = \frac{4(0.252)}{(1 + 0.252)^2}(0.0517 + 0.050) = 0.0654\,\mathrm{eV}\ CH_4/\mathrm{Cu}$$

$$\delta E = \frac{4(0.504)}{(1 + 0.504)^2}(0.0517 + 0.2) = 0.224\,\mathrm{eV}\ O_2/\mathrm{Cu}$$

	H_2		CH_4		O_2	
E_i	300 K	1 eV	300 K	1 eV	300 K	1 eV
Cu	0.00849	0.121	0.0654	0.675	0.224	1.07
Pt	0.00289	0.0412	0.0285	0.294	0.122	0.581

Note the increase in energy transfer for increasing mass of the incident particle, decreasing mass of the chain atom, increasing incident energy and increasing heat of adsorption.

3.8   When a molecule strikes a surface it loses on average an amount of energy $\langle \Delta E \rangle$ given by

$$\langle \Delta E \rangle = -\gamma \alpha_{dsp} E \qquad (3.19.4)$$

where γ is a constant characteristic of the potential energy surface, α_{dsp} is a constant that depends on the collision partners and E is the kinetic energy upon collision. For H/Cu(111), $\alpha_{dsp} = 0.0024$, $\gamma = 4.0$ and the binding energy chemisorbed H is 2.5 eV.

(a) For an H atom with an initial $E_K = 0.1$ eV, 10 Å away from the surface, calculate the energy transfer on the first bounce.

(b) Assuming the same amount of energy transfer on each subsequent collision, how many collisions are required for the H atom to reach the bottom of the well?

(c) Given that α_{dsp} changes from one molecule to the next, in a manner analogous to the Baule formula, we write

$$\alpha_{dsp} = k \frac{4\mu}{(1+\mu)^2} \qquad (3.19.5)$$

where μ is calculated assuming one surface atom participates in the collision, calculate α_{dsp} for CO assuming the same proportionality factor as for H. Then make a rough estimate of the number of collisions CO with an initial kinetic energy of 0.1 eV requires to reach the bottom of a 1.2 eV chemisorption well with $\gamma = 4.0$.

   A hydrogen atom 10 Å away from the surface does not feel the attraction of the chemisorption well; however, as it approaches the surface, the chemisorption interaction increases in strength. The incident energy is equal to the initial kinetic energy plus the kinetic energy gained by acceleration due to the chemisorption potential.

$$E_i = E_K + q_{ads}$$

(a) $\langle \Delta E \rangle = -\gamma \alpha_{dsp} E = -4.0(0.0024)(0.1 + 2.5) = -0.0250 \text{eV}$

(b) The well depth is 2.5 eV and this amount of energy is transferred in

$$2.5/0.025 = 100 \text{ bounces.}$$

(c) For CO, we first calculate the value of k from the data in part (a).

$$k = \frac{\alpha_{dsp}(1+\mu)^2}{4\mu} = \frac{0.0024(1 + 1/63.5)}{4(1/63.5)} = 0.03931$$

Then we calculate α_{dsp} for CO.

$$\alpha_{dsp} = k \frac{4\mu}{(1+\mu)^2} = 0.3931 \frac{4(28/63.5)}{(1 + (28/63.5))^2} = 0.03339$$

The energy lost per bounce is

$$\langle \Delta E \rangle = -\gamma \alpha_{dsp} E = -4.0(0.03339)(0.1 + 1.2) = -0.1736 \text{ eV}$$

and the number of bounces to the bottom is

$$1.2/0.1736 = 7 \text{ bounces.}$$

3.9   Classically, a chemical reaction cannot occur if the collision partners do not have sufficient energy to overcome the activation barrier. This and the thermal distribution of energy are the basis of the Arrhenius formulation of reaction rate constants. For an atom, the thermal energy is distributed among the translational degrees of freedom. The velocity distribution is governed by Maxwell distribution

$$f(v) = 4\pi \left(\frac{M}{2\pi RT}\right)^{3/2} v^2 \exp\left(\frac{-Mv^2}{2RT}\right), \tag{3.19.6}$$

where M is the molar mass and v the speed. Assuming that there is no steric requirement for sticking, i.e. that energy is the only determining factor, calculate the sticking coefficient for adsorption activation barriers of $E_{ads} = 0$, 0.1, 0.5 and 1.0 eV for an atomic gas held at (a) 300 K and (b) 1000 K.

$$\text{Use} \quad s = s_0 \exp\left(-E_{ads}/k_B T_S\right) \tag{3.E9.1}$$

with $s_0 = 1$ (no steric hindrance) and $k_B = 1.3807 \times 10^{-23} \, \text{J K}^{-1}/1.6022 \times 10^{-23} \, \text{eV J}^{-1} = 8.617 \times 10^{-5} \, \text{eV K}^{-1}$.

	s	
E_{ads} (eV)	300 K	1000 K
0	1	1
0.1	2.1×10^{-2}	3.1×10^{-1}
0.5	4.0×10^{-9}	3.0×10^{-3}
1.0	1.6×10^{-17}	9.2×10^{-6}

Note that temperature has no effect if the barrier is absent and a progressively greater relative effect for higher barriers. A barrier of just 0.1 eV $\approx$ 10 kJ mol^{-1} is sufficient to reduce the reactivity at room temperature by almost 2 orders of magnitude.

3.10   A real molecule has quantized rotational and vibration energy levels. The Maxwell-Boltzmann distribution law describes the occupation of these levels. The distribution among rotational levels is given by

$$N_{vJ} = N_v (hc/k_B T)(2J + 1) \exp(-E_{rot}/k_B T), \tag{3.19.7}$$

where N_{vJ} is the number of molecules in the rotational state with quantum numbers v and J and N_v is the total number of molecules in the vibrational state v. The energy of rigid rotor levels is given by

$$E_{rot} = hcB_v J (J+1) \tag{3.19.8}$$

where B_v is the rotational constant of the appropriate vibrational state. The vibrational population is distributed according to

$$N_v = N \exp(-hcG_0(v)/k_B T) \tag{3.19.9}$$

where N is the total number of molecules and $G_0(v)$ is the wavenumber of the vibrational level v above the ground vibrational level. At thermal equilibrium $T = T_{trans} = T_{vib} = T_{rot}$ and the mean energy is distributed according to

$$\langle E \rangle = \langle E_{trans} \rangle + \langle E_{rot} \rangle + \langle E_{vib} \rangle \tag{3.19.10}$$

where for a diatomic molecule

$$\langle E_{\text{trans}} \rangle = 2k_B T \tag{3.19.11}$$

$$\langle E_{\text{rot}} \rangle = k_B T \tag{3.19.12}$$

$$\langle E_{\text{vib}} \rangle = \sum_{n > 0} \frac{h\nu_n}{\exp(h\nu_n / k_B T) - 1}. \tag{3.19.13}$$

Note that Eq. (3.19.13) neglects the contribution of zero-point energy to the vibrational energy. Assume that the sticking coefficient exhibits Arrhenius behaviour when the temperature of the gas is varied.

For the same barrier heights as in Exercise 3.9, calculate the sticking coefficient for molecules with mean total energies of 0.1, 0.5 and 1.0 eV. Use NO as the molecule and assume that only $v = 1$ contributes to the vibrational energy for which $G_0 = 1904 \, \text{cm}^{-1}$. Again, we should first assume that the Arrhenius model is valid; thus,

$$s = s_0 \exp(-E_{\text{ads}} / k_B T_{\text{gas}}).$$

We assume thermal equilibrium between all of the degrees of freedom of the molecule. The problem now is that we need to determine the temperature that corresponds to the given mean energies. That is, we need to solve for T in

$$\langle E \rangle = \langle E_{\text{trans}} \rangle + \langle E_{\text{rot}} \rangle + \langle E_{\text{vib}} \rangle = 2k_B T + k_B T + \frac{hcG_0}{\exp(G_0 / k_B T) - 1}$$

$$\langle E \rangle = 3k_B T + \frac{hcG_0}{\exp(G_0 / k_B T) - 1} \approx 4k_B T$$

The approximate answer, $4k_B T$, follows from the classical approximation in which $k_B T \gg h\nu_n$. Here

$$h\nu_n = \frac{G_0}{8065.5 \frac{\text{cm}^{-1}}{\text{eV}}} = 0.2361 \, \text{eV}$$

and the temperature above which we can make the classical approximation is

$$T = \frac{h\nu_n}{k_B} = \frac{0.2361}{8.617 \times 10^{-5}} = 2740 \, \text{K}.$$

Let us first make an estimation of the temperature based on the classical result.

$$T = \frac{\langle E_{\text{tot}} \rangle}{4k_B} = \frac{0.1, 0.5, 1.0 \, \text{eV}}{4(8.617 \times 10^{-5} \text{eV K}^{-1})} = 290, 1450, 2900 \, \text{K}$$

		s	
E_{ads} (eV)	290 K	1450 K	2900 K
0	1	1	1
0.1	1.8×10^{-2}	0.45	0.67
0.5	2.0×10^{-9}	1.8×10^{-2}	0.14
1.0	4.2×10^{-18}	3.3×10^{-4}	1.8×10^{-2}

Note, however, that the vibrational spacing is very large compared to thermal energy for all but the very highest temperature.

$$\langle E_{\text{vib}} \rangle = \frac{h\nu}{\exp(h\nu/k_{\text{B}}T) - 1} = \frac{0.2361\,\text{eV}}{\exp(2740/T) - 1} = 1.9 \times 10^{-5}, 0.042, 0.15\,\text{eV}$$

for 290, 1450 and 2900 K, respectively. Therefore, the contribution of vibrational energy to the total energy is negligible for 290 K. We need to solve

$$\langle E \rangle = 2k_{\text{B}}T + k_{\text{B}}T + \frac{hcG_0}{\exp(G_0/k_{\text{B}}T) - 1} = 3k_{\text{B}}T + \frac{0.2361\,\text{eV}}{\exp(2740/T) - 1}.$$

for T. At low temperature, we make the approximation

$$\langle E \rangle = 3k_{\text{B}}T$$

$$T = \frac{0.1\,\text{eV}}{3k_{\text{B}}} = 390\,\text{K}.$$

Near 2900 K, we expand e^x according to

$$e^x = 1 + x + \frac{x^2}{2!} + \frac{x^3}{3!} + \dots$$

For $x = 2740/2900$, $e^x = 1 + 0.9445 + 0.4461 + 0.1404 \dots \approx 2$, from which we find

$$\langle E \rangle \approx 3k_{\text{B}}T + \frac{G_0}{2 - 1} \approx 3k_{\text{B}}T + G_0$$

and

$$T = \frac{\langle E \rangle - G_0}{3k_{\text{B}}} = 2960\,\text{K}.$$

The 0.5 eV case demands a numerical solution. We make a first approximation using $T = 1450\,\text{K}$, calculate the appropriate vibrational energy, then use this to calculate a new T from which a new vibrational energy is calculated, and a new estimate of T is made. This process can be repeated until the answers converge. Ideally, this procedure would be done by computer but I'll do it by hand here for pedagogical clarity.

Initial estimate: $T = 1450\,\text{K}$ from classical approximation

$$\langle E_{\text{vib}} \rangle = \frac{0.2361\,\text{eV}}{\exp(2740/1450) - 1} = 0.042\,\text{eV}$$

Second estimate using total energy expression

$$\langle E \rangle = 0.5 \approx 3k_{\text{B}}T + \langle E_{\text{vib}} \rangle \approx 3k_{\text{B}}T + 0.042$$

$$T = \frac{0.5 - \langle E_{\text{vib}} \rangle}{3k_{\text{B}}} = \frac{0.5 - 0.042}{3(8.617 \times 10^{-5})} = 1770\,\text{K}$$

Recalculate the vibrational energy

$$\langle E_{\text{vib}} \rangle = \frac{0.2361\,\text{eV}}{\exp(2740/1770) - 1} = 0.0638\,\text{eV}$$

Third estimate of the temperature

$$T = \frac{0.5 - 0.0637}{3(8.617 \times 10^{-5})} = 1690\,\text{K}$$

Recalculate the vibrational energy

$$\langle E_{vib} \rangle = \frac{0.2361 \, eV}{\exp(2740/1690) - 1} = 0.0582 \, eV$$

Fourth estimate of the temperature

$$T = \frac{0.5 - 0.0582}{3(8.617 \times 10^{-5})} = 1710 \, K$$

Recalculate the vibrational energy

$$\langle E_{vib} \rangle = \frac{0.2361 \, eV}{\exp(2740/1710) - 1} = 0.0596 \, eV$$

Fifth estimate of the temperature

$$T = \frac{0.5 - 0.0596}{3(8.617 \times 10^{-5})} = 1700 \, K$$

I'll call this converged to our degree of accuracy. Now you can see why we use computers to perform iterative, numerical solutions.

The final answer then is

		s	
	$\langle E \rangle = 0.1 \, eV$	$\langle E \rangle = 0.5 \, eV$	$\langle E \rangle = 1 \, eV$
E_{ads} (eV)	$T = 390 \, K$	$T = 1700 \, K$	$T = 2960 \, K$
0	1	1	1
0.1	5.1×10^{-2}	0.51	0.68
0.5	3.5×10^{-7}	3.3×10^{-2}	0.14
1.0	1.2×10^{-13}	1.1×10^{-3}	2.0×10^{-2}

3.11   Consider an extremely late barrier in which translational energy plays no role, vibrational energy is 100% effective at overcoming the barrier and rotational energy is 50% efficient. Calculate the classical sticking coefficient of H_2 and D_2 as a function of rovibrational state for the first three vibrational levels and an adsorption barrier of 0.5 eV. Assume that zero point energy plays no role and that molecules can be described as rigid rotors. The vibrational energy spacings and rotational constants of H_2 and D_2 are given below

	H_2		D_2	
v	B_v/cm^{-1}	$G_0(v)$/cm^{-1}	B_v/cm^{-1}	$G_0(v)$/cm^{-1}
0	59.3	0.0	29.9	0.0
1	56.4	4161.1	28.8	2994.0
2	53.5	8087.1	28.0	5868.8

Assume a classical system with no tunnelling. Therefore, if the energy in the reaction co-ordinate exceeds the barrier height, then the sticking coefficient is unity. If the energy is less than the barrier height, the sticking coefficient is zero. According to the description, the sticking criterion is

$$E_{vib} + 0.5\,E_{rot} > 0.5\,eV \Rightarrow s = 1$$

$$E_{vib} + 0.5\,E_{rot} < 0.5\,eV \Rightarrow s = 0$$

First calculate the energy of each vibrational level, then compare this to the barrier height. For those vibrational levels for which the energetic criterion is met, the sticking coefficient is unity. If not, it is zero.

v	H_2			D_2		
	$G_0(v)$/cm^{-1}	$G_0(v)$/eV	$s(v = 0, J = 0)$	$G_0(v)$/cm^{-1}	$G_0(v)$/eV	$s(v = 0, J = 0)$
0	0.0	0.0	0	0.0	0.0	0
1	4161.1	0.52	1	2994.0	0.371	0
2	8087.1	1.00	1	5868.8	0.73	1

Now consider the $D_2(v = 1)$ level in more detail. The energy deficit is 0.1287 eV. However, rotational energy is only 50% effective. Thus we can find the rotational level for which the energy criterion is met by solving

$$2B_v J(J + 1) = 0.1287\,eV.$$

This yields a quadratic equation with solution

$$J = \frac{-b \pm \sqrt{b^2 - 4ac}}{2a} \Rightarrow J = 8.04 \text{ (the second root is negative)}.$$

At $J = 8, E_{vib} + 0.5E_{rot} = 0.499\,eV$, which is slightly below the energetic cutoff. Thus, for $D_2(v = 1, J \geq 9)$, as long as the molecule has sufficient kinetic energy to get it to the surface, it will stick.

3.12   Consider the adsorption of D_2. Assuming that normal translational energy is 100% effective and vibrational energy is 60% effective at overcoming the adsorption barrier, calculate the sticking coefficient of the first three vibrational levels as a function of normal translational energy. Neglect the effects of rotation.

Again assume a classical barrier with no tunnelling. The sticking criterion is

$$E_{trans} + 0.6E_{vib} > E_{ads} \Rightarrow s = 1$$

$$E_{trans} + 0.6E_{vib} < E_{ads} \Rightarrow s = 0$$

We could then introduce the concept of a vibrational state dependent sticking coefficient, each of which has an activation energy that decreases with increasing vibrational excitation. Within the context of the Arrhenius model, then, we write

$$s_0(v) = A_s(v)\exp\left(\frac{-E_{ads}(v)}{k_B T_{trans}}\right) = A_s(v)\exp\left(\frac{-(E_{ads}(v = 0)) - 0.6E_v}{k_B T_{trans}}\right)$$

where $A_s(v)$ is the saturation value of the sticking coefficient for the vibrational state v, $E_{ads}(v = 0)$ is the activation energy in the $v = 0$ state and E_v is the vibrational energy above the ground state for the vibrational state v.

If the model introduced in Exercise 3.16 is used, the functional form of the sticking coefficient versus normal translational energy would be

$$s_0(E_n) = \frac{A_s}{2}\left[1 + \mathrm{erf}\left(\frac{E_n + 0.6\,E_{vib} - E_0}{W}\right)\right]$$

We might also need to allow for both A_s and W being dependent on v.

3.13   The flux of molecules striking a surface follows a cosine distribution, $\cos\vartheta$ where ϑ is the angle from surface normal. If the perpendicular component of translational energy is effective at overcoming the adsorption barrier and the parallel component is not, the angular distribution of the flux that sticks is tightly constrained about the surface normal. The desorbing flux is similarly peaked about the surface normal. It is often observed that the desorbing flux can be described by a $\cos^n\vartheta$ distribution in which $n > 1$, the greater the value of n, the more peaked the distribution. The angular distribution of D_2 desorbing from Cu(100) has been measured by Comsa and David. They found the following relationship between the normalized desorption intensity, $N(\vartheta)/N(0°)$, and the desorption angle measured from the surface normal, ϑ:

ϑ	0°	5°	10°	15°	20°	25°	30°	35°	45°
$\dfrac{N(\vartheta)}{N(0°)}$	1.00	0.99	0.98	0.77	0.63	0.48	0.38	0.21	0.06

Determine n.

Fitting the data to

$$N(\theta)/N(0°) = a + \cos^n\theta$$

we obtain $n = 3.4 \pm 0.4$ and the following graph of the data and fitted function.

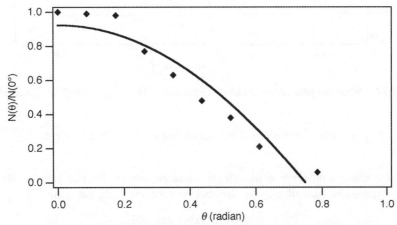

Figure Exercise 3.13   *Fitted angular distribution of D2 desorbed from Cu(100).*

3.14   The sticking of molecular hydrogen on Si is highly activated in the surface co-ordinates. Bratu and Höfer have determined the sticking coefficient of H_2 on $Si(111)-(7 \times 7)$ as a function of surface temperature and have recorded the following data:

T_s/K	587	613	637	667	719	766	826	891	946	1000	1058
s_0	2.8×10^{-9}	6.5×10^{-9}	1.3×10^{-8}	2.0×10^{-8}	5.3×10^{-8}	1.7×10^{-7}	5.4×10^{-7}	1.3×10^{-6}	2.3×10^{-6}	2.7×10^{-6}	5.0×10^{-6}

Using an Arrhenius formulation (Eq. 3.19.7), determine the pre-exponential factor and the activation barrier height.

This is a straightforward application of the Arrhenius equation.

$$s_0 = A \exp\left(-E_{ads}/RT\right)$$

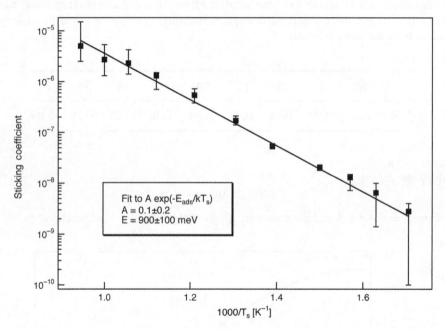

Fit to $A \exp(-E_{ads}/kT_s)$
$A = 0.1 \pm 0.2$
$E = 900 \pm 100$ meV

Figure Exercise 3.14   *Arrhenius plot of the sticking coefficient of H_2 on $Si(111)-(7 \times 7)$ from the data of Bratu and Höfer.*

$$E_{ads} = 900 \pm 100\,\text{meV} = 0.9 \pm 0.1\,\text{eV/molecule} = 78 \pm 9\,\text{kJ mol}^{-1}.$$

3.15   If a chemical reaction proceeds with a single activation barrier, the rate constant should follow the Arrhenius expression. Accordingly for the sticking coefficient, s_0, we write

$$s_0 = A_s \exp(-E_{ads}/RT), \tag{3.19.14}$$

where A is a constant, E_{ads} is the adsorption activation energy, R is the gas constant and T is the temperature. However, if a distribution of barriers rather than a single barrier participates in the

Table 3.6 *Initial sticking coefficient, s_0, with normal kinetic energy, E_n; see Exercise 3.15. Source: H. A. Michelsen, C. T. Rettner, D. J. Auerbach, and R. N. Zare,* J. Chem. Phys. *98, 8294 (1993)*

$E_\perp/\text{eV}$	0.1	0.2	0.3	0.4	0.5	0.6	0.7	0.8	0.9	1.0
s_0	4×10^{-6}	1×10^{-4}	0.0021	0.0167	0.0670	0.151	0.219	0.245	0.250	0.250

reaction, a different form is followed. In the case of H_2 sticking on Cu(100), the sticking coefficient as a function of kinetic energy is found to follow

$$s_0(E_n) = \frac{A_s}{2}\left[1 + \text{erf}\left(\frac{E_n - E_0}{W}\right)\right] \tag{3.19.15}$$

where E_0 is the mean position of a distribution of barriers that has a width W. Given the data in Table 3.6 of s_0 for $H_2(v = 0)$ versus E_n, determine E_0 and W. Make plots of s_0 vs E_n with different values of E_0 and W to observe the effects these have on the shape of the sticking curve.

A fit leads to the values

$A_s = 0.250, E_0 = 0.57\,\text{eV}, \ W = 0.16\,\text{eV}$. A_s is the saturation value of the sticking coefficient. The low value of A_s found in the fit remains an unsolved puzzle.

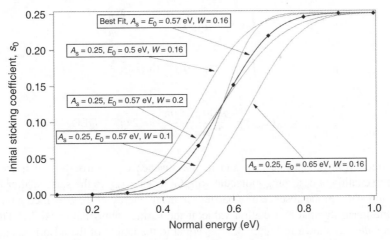

Figure Exercise 3.15 *Fits using the error function form of the sticking coefficient from the data of Michelsen, Rettner and Auerbach.*

Note how a change in the mean barrier height, E_0, changes the position of the curve. A narrower barrier distribution – smaller value of W – leads to a steeper curve, whereas a broad distribution leads to a sticking curve that has a gentler slope.

3.16 Classically we assign $\frac{1}{2}k_B T$ of energy to each active degree of freedom and, therefore, we assign a value of $\frac{3}{2}k_B T$ to the kinetic energy. This is true for a *volume* sample of a gas. For a *flux* of gas, such as that desorbing from a surface, the answer is different. Use the Maxwell velocity distribution to show that the equilibrium mean kinetic energy of a flux of gas emanating from (or passing through) a surface is

$$\langle E_{\text{trans}} \rangle = 2k_B T_s. \tag{3.19.16}$$

The mean kinetic energy is defined by the moments of the velocity distribution according to

$$\langle E_{\text{trans}} \rangle = \frac{1}{2} \frac{m M_3}{M_1} \tag{3.19.17}$$

where the moments are calculated according to

$$M_i = \int_0^\infty v^i f(v) \, dv. \tag{3.19.18}$$

The Maxwell velocity distribution is taken from Eq. (3.19.6)

$$f(v) = 4\pi \left(\frac{M}{2\pi RT} \right)^{3/2} v^2 \exp \left(\frac{-Mv^2}{2RT} \right),$$

Thus

$$\langle E_{\text{trans}} \rangle = \frac{m}{2} \frac{\int_0^\infty v^5 \exp(-\beta v^2) \, dv}{\int_0^\infty v^3 \exp(-\beta v^2) \, dv}$$

where $\beta = m/2k_{\text{B}}T$. Note that all the other constants cancel out.

$$\int_0^\infty v^3 \exp(-\beta v^2) \, dv = \frac{2k_{\text{B}}^2 T^2}{m^2}$$

$$\int_0^\infty v^5 \exp(-\beta v^2) \, dv = \frac{8k_{\text{B}}^3 T^3}{m^3}$$

Therefore

$$\langle E_{\text{trans}} \rangle = \frac{m}{2} \frac{8k_{\text{B}}^3 T^3 / m^3}{2k_{\text{B}}^2 T^2 / m^2} = 2k_{\text{B}}T \quad \text{QED}$$

3.17   For desorption from a rigid surface and in the absence of electron-hole pair formation or other electronic excitation, a desorbing molecule will not lose energy to the surface after it passes through the transition state. In the absence of a barrier the mean energy of the desorbed molecules is roughly equal to the mean thermal expectation value at the surface temperature $\langle k_{\text{B}}T_{\text{s}} \rangle$. The excess energy above this value, as shown in Figure 3.27, is equal to the height of the adsorption activation energy. Therefore a measurement of the mean total energy of the desorbed molecules, $\langle E \rangle$, can be used to estimate the height of the adsorption barrier. The mean total energy is given by

$$\langle E \rangle = \langle E_{\text{trans}} \rangle + \langle E_{\text{rot}} \rangle + \langle E_{\text{vib}} \rangle. \tag{3.19.19}$$

(a) Given the data for $D_2/Cu(111)$ given in Table 3.7, show that the above approximations hold and, therefore, to a first approximation we need not consider surface atom motions in the ad/desorption dynamics.

(b) Given the data for $D_2/Si(100)-(2 \times 1)$ given in Table 3.7 show that the above approximations do not hold and, therefore, the static potential picture cannot be used to interpret the ad/desorption dynamics.

In both cases assume that the vibrational distribution is thermal and is describe by a temperature T_{vib}.

Table 3.7 *Data for D_2/Cu(111) and D_2/Si(100)–(2 × 1); see Exercise 3.16. Source: H. A. Michelsen, C. T. Rettner, and D. J. Auerbach, in* Surface Reactions: Springer Series in Surface Sciences, *edited by R. J. Madix (Springer-Verlag, Berlin, 1994), Vol. 34, p. 185., and K. W. Kolasinski, W. Nessler, A. de Meijere, and E. Hasselbrink,* Phys. Rev. Lett. *72, 1356 (1994)*

T_s (K)	$\langle E_{rot} \rangle$ (K)	T_{vib} (K)	$\langle E_{trans} \rangle$ (K)	E_{ads} (eV)
D_2/Cu(111): 925	1020	1820	3360	0.5
D_2/Si(100): 780	330	1700	960	0.8

The proper way to calculate the vibrational energy, accounting for the contribution of all vibrational levels, is to use

$$\langle E_{vib} \rangle = \frac{\Theta_v}{\exp(\Theta_v / T_{vib}) - 1}$$

where $\Theta_v(D_2) = 4294$ K is the characteristic vibrational temperature of D_2. The adsorption barrier is then estimated from the difference between the measured mean molecular energy and the expected mean molecular energy based on thermal equilibrium

$$E_{ads} = \langle E \rangle_{measured} - \langle E \rangle_{equilibrium}$$

$$E_{ads} = 2k_B T_{trans} + k_B T_{rot} + \frac{k_B \Theta_v}{\exp(\Theta_v / T_{vib})} - \left[2k_B T_s + k_B T_s + \frac{k_B \Theta_v}{\exp(\Theta_v / T_s)} \right]$$

$$\frac{E_{ads}}{k_B} = 2(T_{trans} - T_s) + (T_{rot} - T_s) + \frac{\Theta_v}{\exp(\Theta_v / T_{vib})} - \frac{\Theta_v}{\exp(\Theta_v / T_s)}$$

In the last line we have clearly separated the comparative translational, rotational and vibrational contributions to the energy excess that is related to the barrier height.

(a) D_2/Cu

$$\frac{E_{ads}}{k_B} = 2(3360 - 925) + (1020 - 925) + \frac{\Theta_v}{\exp(\Theta_v / 1820) - 1} - \frac{\Theta_v}{\exp(\Theta_v / 925) - 1}$$

$$\frac{E_{ads}}{k_B} 7 = 4870 + 95 + 393 = 5358 \text{ K}$$

This compares to $0.5 \text{ eV}/8.617 \times 10^{-5} \text{ eV/K} = 5802$ K. Thus only $(5802 - 5358)8.617 \times 10^{-5} = 0.038$ eV is missing from the energy balance and can be presumed to be deposited in the substrate.

(b) H_2/Si

$$\frac{E_{ads}}{k_B} = 2(960 - 780) + (330 - 780) + \frac{\Theta_v}{\exp(\Theta_v / 1700) - 1} - \frac{\Theta_v}{\exp(\Theta_v / 780) - 1}$$

$$\frac{E_{ads}}{k_B} = 360 - 450 + 343 = 253 \text{ K}$$

This compares to $0.8 \text{ eV}/8.617 \times 10^{-5} \text{ eV/K} = 9284$ K. Thus $(9284 - 253) \, 8.617 \times 10^{-5} = 0.78$ eV is missing from the energy balance and can be presumed to be deposited in the substrate. The energy must be deposited in the form of either lattice excitations (phonons) or electronic excitations. Clearly there is something decidedly different about the dynamics of H_2 adsorption on and desorption from Si compared to Cu [2].

3.18   A Pt(997) is a stepped surface that contains 8 times more terrace atoms than step atoms. When CO is exposed to a Pt(997) surface held at $T_s = 11$ K, the ratio of CO molecules adsorbed at the atop terrace sites to those adsorbed in atop step sites is 3.6:1. Discuss what the expected coverage ratio is in terms of simple Langmuirian adsorption. Propose an explanation for why the Langmuirian result is not obtained.

In simple Langmuirian adsorption, molecules stick where they land. At $T_s = 11$ K diffusion should be frozen out. Since the incidence of molecules from the gas phase is random, the coverage would be expected to follow the probability of striking a terrace atom versus a step atom. Assuming that the sticking coefficient on terrace and step atoms is the same, the expected ratio would then be 8:1, that is, the coverage of CO in atop terrace sites should exceed the coverage of CO in atop step sites by a factor of 8. Two possible explanations of the observed overabundance of CO at step sites are: (1) either the sticking coefficient of CO at step sites is roughly twice as great as the sticking coefficient on terrace sites, or (2) the CO molecules exhibit transient mobility due to the slow relaxation and dissipation of the adsorption energy to electron-hole pairs and phonons (as in Exercise 3.8).

3.19   CO from a bottle held at 298K was exposed to Pt(111) held at $T_s = 80$ K. Thermal desorption detected by a mass spectrometer (temperature programmed desorption or TPD, see Chapter 4) was then used to quantify how much CO stuck to the surface after each defined exposure. The data obtained is given in Table 3.8. Interpret the data in terms of the adsorption dynamics of CO on Pt(111). Note that the saturation coverage of CO is 0.50 ML measured with respect to the Pt(111) surface atom density. To interpret your data, first make a plot of TPD peak area vs exposure and then convert this into a plot of CO coverage vs exposure (both in ML), from which you can infer the CO sticking coefficient and its dependence on coverage.

Table 3.8   *TPD peak areas as a function of CO exposure on to Pt(111) at 80 K*

Exposure/Langmuir	0.198	0.350	0.733	1.082	1.393	1.755	2.033	2.481	2.843	3.172
Integrated TPD peak area	1.46	2.62	5.70	8.48	11.3	13.8	16.8	15.9	16.1	16.6

First plot TPD area versus exposure. From the plot found in Fig. Ex. 3.19(a) it is clear that a linear relationship is found until a saturation value is reached. Therefore, the sticking coefficient is not a function of coverage and precursor-mediated adsorption is occurring.

Next, covert the TPD area into coverage. This requires two normalization factors. First, average the last four data points to arrive at a mean value for the TPD peak area that corresponds to saturation coverage, from which we obtain 16.35 ± 0.42. Thus the standard deviation is less than 3% of the absolute value, which corresponds to an error bar that is comparable in size to the size of the markers chosen in the figure. Divide each area by 16.35 to convert to coverage relative to saturation, then multiply by 0.5 to convert to coverage relative to the number of Pt surface atoms. In other words, the saturation coverage now corresponds to 0.5 ML.

The exposure in monolayer equivalents is obtained by plugging the exposure in Langmuir into the Hertz-Knudsen equation and then dividing by the surface density of Pt atoms in cm^{-2}:

$$\varepsilon(\text{in ML}) = 3.51 \times 10^{22} \varepsilon(\text{in L}) \times 10^{-6}/(MT)^{1/2}\sigma_0$$

The calculations yield the following results

Exposure/L	Exposure/ML	TPD peak area	θ/ML
0.000	0.000	0.00	0.000
0.198	0.051	1.46	0.045
0.350	0.089	2.62	0.080
0.733	0.187	5.70	0.174
1.082	0.277	8.48	0.259
1.393	0.356	11.30	0.346
1.755	0.449	13.80	0.422
2.034	0.520	16.80	0.514
2.481	0.634	15.90	0.486
2.843	0.727	16.10	0.492
3.172	0.811	16.60	0.508

Adding a point at zero exposure is justified as long as the surface is clean before each TPD run. A fit that is constrained to go through the origin yields a sticking coefficient over the linear portion of the graph of $s = 0.96 \pm 0.01$.

3.20 Explain why multilayer absorption can occur for physisorption but not for chemisorption.

Chemisorption occurs on a fixed number of adsorption sites. The only way for the chemisorption coverage to exceed the initial number of adsorption sites is for the surface to reconstruct or facet. The one exception to this is if we consider a growth reaction (for instance, the evaporation of the atoms of a metal or semiconductor onto a cold substrate) in which the adsorbing atoms are incorporated into the substrate. Usually, though, when we are thinking of chemisorption, we think in terms of an adsorbate that sticks to a chemically distinct substrate without creating more substrate.

Physisorption can occur both on the bare surface and on a physisorbed layer. For instance, in the condensation of a liquid on a cold surface the adsorbing atoms or molecules of the liquid stick on top of the physisorbed multilayers of the liquid.

3.21 A metal single crystal sample is dosed with a Knudsen source, i.e. a nozzle that produces an equilibrium flux rather than a supersonic jet. The gas expands out of the nozzle with a cosine angular distribution and then intercepts the crystal. The flux intercepted by the crystal F_N depends on the temperature in the nozzle T_N, the distance of the crystal away from the nozzle d, the axial distance from the centre of the crystal x, the area of the hole in the nozzle A_N, and the pressure in the gas manifold behind the nozzle p_m according to

$$F_N = \frac{p_m A_N}{\sqrt{2\pi m k_B T_N}} \frac{\cos^4(\tan^{-1}(x/d))}{\pi d^2}. \tag{3.19.20}$$

The effective sticking coefficient is defined according to

$$\theta_{\text{total}} = s_{\text{eff}}(F_N t/\sigma_0), \tag{3.19.21}$$

where θ_{total} is the total coverage accumulated after a time t on the crystal, which has σ_0 surface atoms per unit area. The total coverage, however, is made up of contributions from adsorption directly from the nozzle as well as adsorption from background gas. Assuming that the background is made up from molecules that originate from the nozzle and miss the crystal (or do not stick on the first bounce), then the background pressure is proportional to the nozzle pressure, that is

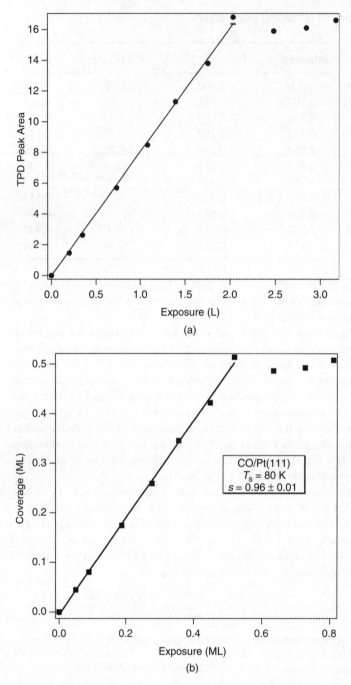

Figure Exercise 3.19 *(a) CO TPD area versus exposure in Langmuir. (b) CO coverage versus exposure in monolayer equivalents.*

$p_{bkg} = c_0 p_m$, show that the true sticking coefficient of molecules emanating directly from the nozzle is given by

$$s_N = s_{eff} - s_{bkg} \left(\frac{c_0 \sqrt{T_N} \pi d^2}{A_N \sqrt{T} \cos^4(\tan^{-1}(x/d))} \right). \tag{3.19.22}$$

There are two contributions to the total C coverage, one from molecules coming directly out of the nozzle, $\theta_{C,N}$, and one from molecules that first equilibrate with the chamber and then adsorb on the surface. The latter contribution is the background contribution $\theta_{C,bkg}$. Hence,

$$\theta_{C,total} = \theta_{C,N} + \theta_{C,bkg}$$

with $\theta_{C,N} = \dfrac{s_N F_N t}{\sigma_{Pt}}$ and $\theta_{C,bkg} = \dfrac{s_{bkg} F_{bkg} t}{\sigma_{Pt}}$,

where s_{bkg} is the sticking coefficient of molecules equilibrated to the chamber temperature (300 K) and the flux from the background at 300 K is given by

$$F_{bkg} = \frac{p_{bkg}}{\sqrt{2\pi m k_B T}} = \frac{c_0 p_m}{\sqrt{2\pi m k_B T}}.$$

In the last expression, we use the fact that the background pressure p_{bkg} is related simply by a proportionality constant c_0 to p_m. Independent experiments are used to determine the values of c_0 and s_{bkg}.

Consequently, we can relate the effective sticking coefficient to the value that we want to determine (s_N) and experimentally determined quantities.

$$\theta_{C,total} = \frac{s_{eff} F_N t}{\sigma_{Pt}} = \frac{s_N F_N t}{\sigma_{Pt}} + \frac{s_{bkg} F_{bkg} t}{\sigma_{Pt}}$$

$$s_{eff} F_N = s_N F_N + s_{bkg} F_{bkg}$$

$$(s_{eff} - s_N) F_N = s_{bkg} F_{bkg}$$

$$(s_{eff} - s_N) = s_{bkg} (F_{bkg}/F_N)$$

$$s_{eff} - s_{bkg} (F_{bkg}/F_N) = s_N$$

$$s_{eff} - s_{bkg} \left(\frac{\dfrac{c_0 p_m}{\sqrt{2\pi m k_B T}}}{\dfrac{p_m A_N}{\sqrt{2\pi m k_B T_N}} \dfrac{\cos^4(\tan^{-1}(x/d))}{\pi d^2}} \right) = s_N$$

$$s_N = s_{eff} - s_{bkg} \left(\frac{c_0 \sqrt{T_N} \pi d^2}{A_N \sqrt{T} \cos^4(\tan^{-1}(x/d))} \right)$$

3.22   Consider the bonding of NO on Pt(111).

(a) Describe a likely binding geometry for adsorbed NO and justify your answer.

(b) The adsorption of NO on Pt(111) is non-activated. The sticking of NO is less likely for molecules that have high translational energy compared to low translational energy. Molecules with high rotational quantum number (high J states) also stick less effectively than low J states. Discuss the magnitude of the sticking coefficient and how it will change as the temperature of the NO is changed.

(c) Real Pt(111) surfaces often have a step density of about 1%. On these surfaces NO is found to dissociate at room temperature but only at very low coverages. Typically, the first 1% of a monolayer (ML) dissociates but after this all further adsorption is molecular. If the NO is dosed onto the clean surface at 90 K (a temperature at which there is no surface diffusion of NO), virtually no dissociation is observed. Discuss the dissociation of NO. Explain a possible dissociation mechanism.

(a) Since it has one more electron than CO, to a first approximation NO binds analogously to CO. Thus, N end down and O pointing away from the surface.

(b) Since sticking is non-activated, the sticking coefficient should be high, approaching unity. However, thermal excitation of translational and rotational degrees of freedom will reduce sticking coefficient because we are told that the high-energy states do not stick as well. Therefore, the sticking coefficient should be close to 1 at low temperature, and drop slowly as the temperature of the molecule is raised.

(c) Any real surface will have some defects. The most common kind is steps. NO is dissociating at the steps, which must be present at about 0.01 ML. At room temperature, NO is diffusing to the steps, dissociating and the dissociation products stay at the step. After the step fills with dissociation products, there is no more room for further NO to dissociate. At low temperature, the NO cannot diffuse to the step so the dissociation is shut down.

3.23 When 0.25 L (L = Langmuir) of O_2 is dosed onto a Pd(111) surface at 30 K, a peak can be seen in the vibrational spectrum. The wavenumber of this peak is 1585 cm^{-1}. This value is virtually the same as the vibrational frequency of a gas-phase O_2 molecule. When the crystal is warmed to above 50 K, the 1585 cm^{-1} peak disappears and a peak at 850 cm^{-1} appears. No O_2 is observed to desorb from the surface below 100 K. Interpret these results and describe what is occurring in the adsorbed layer.

At 30 K, O_2 is physisorbed. There is no charge transfer into antibonding orbitals, and the intramolecular bonding is very similar to that found in the gas phase. This is mirrored in the lack of a shift in the vibrational frequency. Heating to above 50 K is sufficient to overcome the small activation barrier that separates the physisorbed state from a chemisorbed state. The chemisorbed state is more strongly bound (strong enough that it does not desorb) and this binding comes about through charge transfer from the surface into antibonding molecular orbitals. The added population in the $O_2 \pi^*$ states reduces the O–O bond strength and, therefore, the O–O vibrational wavenumber.

References

[1] W. F. Lin, M. S. Zei, M. Eiswirth, G. Ertl, T. Iwasita, W. Vielstich, *J. Phys. Chem. B*, **103** (1999) 6969.
[2] M. Dürr, U. Höfer, *Surf. Sci. Rep.*, **61** (2006) 465.

12

Answers to Exercises from Chapter 4. Thermodynamics and Kinetics of Adsorption and Desorption

4.1 The canonical partition function appropriate for the Langmuir model is

$$Q_{ads} = C_{N_0}^{N_{ads}} \exp\left(\frac{N_{ads}\varepsilon}{k_B T}\right) \qquad (4.11.1)$$

and the grand canonical partition function is

$$\Xi = \sum_{N_{ads}}^{N_0} Q_{ads} \exp\left(\frac{N_{ads}\mu_{ads}}{k_B T}\right) \qquad (4.11.2)$$

where $C_{N_0}^{N_{ads}}$ is the number of configurations and μ_{ads} is the chemical potential of the adsorbate phase. Derive Eq. (4.2.1) by calculating

$$\theta = \langle N_{ads}\rangle/N_0 \qquad (4.11.3)$$

where $\langle N_{ads}\rangle$ is the average number of adsorbed atoms.

For any given state of the system at thermal equilibrium, the total number of adsorbed particles follows the Boltzmann distribution and is given by:

$$N_{ads} = Q_{ads} \exp\left(\frac{N_{ads}\,\mu_{ads}}{k_B\,T}\right)$$

The total number of particles in that state is given by the grand canonical partition function. (which we take to be N_0 since this would be the total number of particles for a full monolayer for which $\theta = 1$.) Therefore the probability of finding a given number of particles is:

$$P(N_{ads}) = \frac{N_{ads}}{\Xi} = \frac{Q_{ads} \exp\left(\dfrac{N_{ads}\,\mu_{ads}}{k_B\,T}\right)}{\Xi}$$

which demonstrates relation (4.2.1).

Surface Science: Foundations of Catalysis and Nanoscience, Third Edition. Kurt W. Kolasinski.
© 2012 John Wiley & Sons, Ltd. Published 2012 by John Wiley & Sons, Ltd.

4.2   (a) If the sticking coefficient of NO is 0.85 on Pt(111), what is the initial rate of adsorption of NO onto Pt(111) at 250 K if the NO pressure is 1.00×10^{-6} Torr? (b) If the dissociative sticking coefficient of O_2 is constant at a value of 0.25, how many oxygen atoms will be adsorbed on a Pd(111) surface after it has been exposed to 1.00×10^{-5} Pa of O_2 for 1 s. The system is held at 100 K.

(a)

$$r_{ads} = s_0 Z_w = \frac{s_0 N_A p}{(2\pi MRT)^{1/2}}; p = 1.00 \times 10^{-6} \text{ Torr} \left(\frac{101325 \text{ Pa}}{760 \text{Torr}} \right) = 1.333 \times 10^{-4} \text{ Pa}$$

$$r_{ads} = \frac{0.85(6.02 \times 10^{23} \text{ mol}^{-1})(1.333 \times 10^{-4} \text{ Pa})}{(2\pi (0.030 \text{ kg mol}^{-1})(8.314 \text{ J mol}^{-1}\text{K}^{-1})(250 \text{ K}))^{1/2}}$$

$$r_{ads} = 4.32 \times 10^{18} \text{m}^{-2} \text{ s}^{-1}$$

(b) Since the O_2 molecule dissociates to form $2 O$ atoms, we need to take this into account to find the coverage.

$$\sigma = 2s_0 Z_w t = \frac{2s_0 N_A p t}{(2\pi MRT)^{1/2}}$$

$$\sigma = \frac{2(0.25)(6.02 \times 10^{23} \text{ mol}^{-1})(1.00 \times 10^{-5} \text{ Pa})}{(2\pi (0.032 \text{ kg mol}^{-1})(8.314 \text{ J mol}^{-1}\text{K}^{-1})(100 \text{ K}))^{1/2}}$$

$$\sigma = 2.33 \times 10^{17} \text{ m}^{-2}$$

4.3   Consider non-dissociative adsorption. Assuming both adsorption and desorption are first-order processes, write down an expression for the coverage of an adsorbate as a function of time. The system is at a temperature T, there is only one component in the gas phase above the surface at pressure p and the saturation surface coverage is given by θ_{max}. Assume adsorption is non-activated and that the sticking is direct.

In general, the coverage at time t, $\theta(t)$ can be found by solving the differential equation

$$\frac{d\sigma(t)}{dt} = \sigma_0 \frac{d\theta(t)}{dt} = r_{ads} - r_{des}$$

where r_{ads} is the rate of adsorption, r_{des} the rate of desorption, and the absolute coverage σ is given by the product of the areal density of surface atoms σ_0 and the fractional coverage θ.

$$r_{ads} = s_0(T_s, T_g) f(\theta(t)) Z_w$$

$$Z_w = \frac{p}{\sqrt{2\pi m k_B T}}$$

where $s_0(T_s, T_g)$ is a surface and gas temperature dependent initial sticking coefficient, $f(\theta(t))$ represents the coverage dependence of the sticking coefficient and Z_w is the flux of molecules onto the surface. For simplicity, we assume that s_0 is independent of temperature (non-activated adsorption).

The rate of desorption follows the Polanyi-Wigner expression

$$r_{des} = \sigma_0 \theta(t)^n k_d = \sigma_0 \theta(t)^n A_d \exp(-E_{des}/RT_s)$$

where n is the desorption order, k_d the desorption rate constant, A_d the desorption Arrhenius prefactor, E_{des} the desorption activation energy, and R the gas constant.

In order to proceed further we must now assume kinetic orders for the adsorption and desorption process. That is, we have to set the function $f(\theta(t))$ and the value of n. *First-Order Adsorption, First-Order Desorption, Langmuirian Sticking*

In this case

$$f(\theta(t)) = \frac{\theta_{\max} - \theta(t)}{\theta_{\max}} = 1 - \theta(t)/\theta_{\max}$$

and, therefore,

$$\frac{d\theta(t)}{dt} = \frac{s_0 Z_w}{\sigma_0}\left(1 - \frac{\theta(t)}{\theta_{\max}}\right) - \theta(t)A_d \exp\left(\frac{-E_d}{RT_s}\right)$$

Using the shorthand notation s_0 for $s_0(T_s, T_n)$, k_d for the desorption rate constant and

$$r_0 = s_0 Z_w/\sigma_0$$

for the rate of adsorption at zero coverage, we obtain

$$\frac{d\theta(t)}{dt} = r_0 - r_0\frac{\theta(t)}{\theta_{\max}} - \theta(t)k_d$$

and thus

$$\int_0^{\theta(t)} \frac{d\theta}{r_0 - \left(\frac{r_0}{\theta_{\max}} + K_d\right)\theta} = \int_0^t dt$$

Integrating yields

$$\frac{\theta_{ax}\ln\left[r_0 - \theta\left(k_d - \frac{r_0}{\theta_{\max}}\right)\right]}{r_0 + k_d\theta_{\max}}\bigg|_0^\theta = t.$$

Taking $a = r_0/\theta_{\max} + k_d$ leads to an expression for the coverage as a function of time

$$\theta(t) = \frac{r_0}{a}(1 - \exp(-at))$$

Note that if $k_d = 0$

$$\theta(t) = \theta_{\max}(1 - \exp\left(-r_0 t/\theta_{\max}\right)) = \theta_{\max}\left(1 - \exp\left(-\frac{s_0 Z_w t}{\theta_{\max}\sigma_0}\right)\right).$$

This function is plotted in Fig. 4.7.

4.4 Consider sticking that requires an ensemble of two adjacent sites as is often true in dissociative adsorption. Assuming that there is no desorption, write down an expression for the coverage of an adsorbate as a function of time. The system is at a temperature T, there is only one component in the gas phase above the surface at pressure p and the saturation surface coverage is given by $\theta_{\max}$. Assume adsorption is non-activated and that the sticking is direct.

For second-order Langmuirian adsorption, i.e. adsorption requiring two adjacent sites, the sticking coefficient varies as

$$s = s_0(1 - \theta/\theta_{\max})^2,$$

and the rate of change of coverage is equal to the rate of adsorption

$$R_{ads} = \frac{d\theta(t)}{dt} = r_0\left(1 - \frac{\theta(t)}{\theta_{\max}}\right)^2,$$

where now

$$r_0 = 2s_0 Z_w / \sigma_0$$

and the factor of 2 arises from two sites being filled after dissociation. We need to solve the following integral to obtain the uptake curve,

$$\int_0^{\theta(t)} \frac{d\theta}{(1 - \theta/\theta_{max})^2} = r_0 \int_0^t dt.$$

Integrating yields

$$\left. \frac{\theta_{max}}{(1 - \theta/\theta_{max})} \right|_0^\theta = r_0 t.$$

The uptake curve is given by

$$\theta(t) = \frac{r_0 t \theta_{max}}{r_0 t + \theta_{max}}.$$

This curve is plotted in Fig. 4.7. Note that Fig. 4.7 has been drawn for *equal initial rates* of adsorption for the first- and second-order processes to emphasize the change in the sticking coefficient as a function of exposure.

4.5   Consider sticking that requires an ensemble of three adjacent sites. Assuming that there is no desorption, write down an expression for the coverage of an adsorbate as a function of time. Compare the results to what is expected for adsorption that requires only one site or two sites for the case of A_3 sticking on a Pt(111) surface. Take $\sigma_0 = 1.503 \times 10^{19}$ m^{-2}, $\theta_{max} = 0.25$ ML, $T = 800$ K and the initial sticking coefficient is 2.5×10^{-7}. You will have to consider exposures up to 1×10^{26} m^{-2}.

For third-order Langmuirian adsorption, i.e. adsorption requires three adjacent sites according to

$$A_3(g) \rightarrow 3A(a),$$

the sticking coefficient varies as

$$s = s_0(1 - \theta/\theta_{max})^3$$

and the rate of change of coverage is

$$\frac{d\theta(t)}{dt} = r_0 \left(1 - \frac{\theta(t)}{\theta_{sat}} \right)^3.$$

We need to solve the following integral to obtain the uptake curve,

$$\int_0^{\theta(t)} \frac{d\theta}{r_0 \left(1 - \dfrac{\theta}{\theta_{max}} \right)^3} = \int_0^t dt$$

Integrating yields

$$\left. \frac{\theta_{max}^3}{2r_0(\theta_{max} - \theta)^2} \right|_0^\theta = t.$$

The uptake curve is given by

$$\theta(t) = \theta_{max} - \sqrt{\frac{\theta_{max}^3}{2r_0 t + \theta_{max}}}.$$

Assuming that the molecule breaks up into three adsorbates (Note: termolecular processes are highly unlikely; therefore, by microscopic reversibility it is unlikely that a true third-order process occurs except under unusual conditions or else that the adsorption is irreversible)

$$r_0 = \frac{3s_0 Z_w}{\sigma_0}$$

and

$$\varepsilon = Z_w t = \frac{pt}{(2\pi m k_B T)^{1/2}} = 3.51 \times 10^{22} \frac{pt}{(MT)^{1/2}}$$

where p is in torr, T in Kelvin and M in g mol^{-1}.

The resulting curves look like Fig. Ex. 4.5.

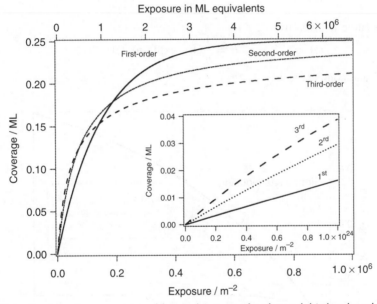

Figure Exercise 4.5 *A comparison of first-order, second-order and third-order adsorption.*

Note that initially the rate of adsorption is equal to r_0. Since in second-order two adsorbates per incident molecule and in third-order adsorption three adsorbates per incident molecule are formed, the rate of adsorption in terms of $d\theta/dt$ for equal flux onto the surface and equal s_0 values is highest for third-order and lowest for first-order adsorption. Eventually, as the coverage increases and more sites are blocked, the rate of second-order adsorption exceeds that of third-order, and then the rate of first-order adsorption exceeds both second and third order.

4.6 Consider sticking that requires an ensemble of only one surface atom but that poisons n sites. In other words, each adsorbing molecule appears to occupy n sites. Assuming that there is no desorption, write down an expression for the coverage of the adsorbate as a function of time. Compare the result for $n = 7$ to what is expected for adsorption that requires only one site for the case of methane sticking on a Pt(111) surface. Take $\sigma_0 = 1.503 \times 10^{15}$ cm^{-2}, $\theta_{max} = 1$, $T = 800$ K and the initial sticking coefficient is 2.5×10^{-7}. You will have to consider exposures up to 1×10^{22} cm^{-2}.

Use the same set up as for the first-order model. The sticking coefficient and rate of adsorption then behave as

$$s = s_0(1 - n\theta)$$

$$\frac{d\theta(t)}{dt} = r_0(1 - n\theta(t)).$$

We need to solve the following integral to obtain the uptake curve,

$$\int_0^{\theta(t)} \frac{d\theta}{r_0(1 - n\theta)} = \int_0^t dt$$

Integrating yields

$$\frac{\theta_{max} \ln[r_0(n\theta - \theta_{max})]}{nr_0} \Big|_0^\theta = t$$

The uptake curve is given by

$$\theta = \frac{1}{n}(1 - \exp(-nr_0 t)).$$

Take $n = 7$ and compare to $n = 1$.

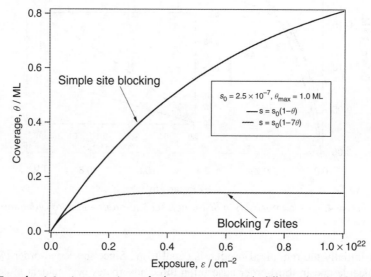

Figure Exercise 4.6 *A comparison of adsorption rates with different levels of site blocking.*

Note that the $n = 1$ case has a saturation coverage of 1 ML, whereas the $n = 7$ case has a saturation coverage of just 0.14 ML. Thus a reason for a saturation coverage below 1 ML can be that the adsorption of a species is self-limiting, that is, that it poisons its own adsorption more strongly than simple site blocking by excluding adsorption at neighbouring sites.

4.7   Given that the kinetic parameters for diffusion are $D_0 = 5 \times 10^{-6} \, cm^2 \, s^{-1}$ and $E_{dif} = 20.5 \, kJ \, mol^{-1}$ and those for desorption are $A = 10^{12} \, s^{-1}$ and $E_{des} = 110 \, kJ \, mol^{-1}$ (first order) for CO on Ni(100), how far does the average CO molecule roam across the surface at $T_s = 480$ K (the top of the temperature-programmed desorption peak).

The lifetime of an adsorbate is given by the inverse of the rate of desorption, otherwise known as the Frenkel equation

$$\tau = \tau_0 \exp\left(\frac{E_d}{RT}\right) = 10^{-12}\,\text{s} \times \exp\left(\frac{110000\,\text{J mol}^{-1}}{8.31\,\text{J mol}^{-1}\text{K}^{-1} \times 480\,\text{K}}\right)$$

$$\tau = 0.948\,\text{s}$$

where τ_0, the characteristic lifetime, is equal to the inverse of the Arrhenius pre-exponential factor.

The mean square displacement due to diffusion of the adsorbate is given by

$$\langle x^2 \rangle^{1/2} = (4\,Dt)^{1/2}$$

where D is the diffusion constant given by

$$D = D_0 \exp\left(-\frac{E_{dif}}{RT}\right) = 5 \times 10^{-6}\,\text{cm}^2\text{s}^{-1} \exp\left(-\frac{20500\,\text{J mol}^{-1}}{8.31\,\text{J mol}^{-1}\text{K}^{-1} \times 480\,\text{K}}\right)$$

$$D = 2.93 \times 10^{-8}\,\text{cm}^2\text{s}^{-1}.$$

Thus,

$$\langle x^2 \rangle^{1/2} = [4\left(2.93 \times 10^{-8}\,\text{cm}^2\text{s}^{-1}\right) 0.948\,\text{s}]^{1/2} = 3.33 \times 10^{-4}\,\text{cm}$$

That is, the average CO molecule diffuses roughly 30,000 Å before desorbing. Since adsorption sites are roughly 3 Å apart, this means that each CO visits over 10,000 sites before it bids the surface adieu.

4.8   The experimentally measured desorption prefactor for NO desorption from Pt(111) is $A = 10^{16\pm0.5}\,\text{s}^{-1}$ with a TPD peak temperature of 340 K and an initial sticking coefficient $s_0 = 0.9$. Take the partition function of the transition state as equal to the partition function of the free molecule and determine whether the experimental value is consistent with the transition state theory prediction. In the gas phase: $\tilde{\upsilon}(NO) = \omega_e = 1904\,\text{cm}^{-1}, B_0 = 1.705\,\text{cm}^{-1}$. In the adsorbed phase: $\tilde{\upsilon}(NO) = 1710\,\text{cm}^{-1}, \tilde{\upsilon}(NO\text{–}Pt) = 306\,\text{cm}^{-1}, \tilde{\upsilon}(\text{frust rot}) = 230\,\text{cm}^{-1}, \tilde{\upsilon}(\text{frust trans}) = 60\,\text{cm}^{-1}$.

According to TST, the pre-exponential factor is written

$$A = \kappa \frac{k_B T}{h} \frac{q_{\ddagger}}{q_a} = s_0 \frac{k_B T}{h} \frac{q_{\ddagger}}{q_a}$$

The adsorbed phase partition function is straightforwardly written

$$q_a = \prod_i \frac{1}{1 - \exp(-h\nu_i/k_B T)}.$$

Note that both ν_3 (frust rot) and ν_4 (frust trans) are two-fold degenerate and must be squared in the product. Using $\nu = c\tilde{\upsilon}$ with c in cm s^{-1}, we obtain

$$q_a = (1.00069)(1.37727)(1.6073)^2(4.45969)^2 = 70.8151.$$

The transition state partition function with one degree of freedom taken out as a $k_B T/h$ term is

$$q_{\ddagger} = \frac{(2\pi m k_B T)}{h^2} A_s \frac{k_B T}{hcB_0} \frac{1}{1 - \exp(-h\nu/k_B T)}$$

Note the presence of the area A_s over which the transition state is delocalized. In this case we will take the area to be that of the unit cell. This is easily calculated according to Eq. (1.1.1) from the inverse of σ_0 given in Table 1.1 to be $6.653 \times 10^{-20}\,\mathrm{m}^2$. Combining A_s and q_{trans} into one term we have

$$q_{\ddagger} = (222.661)(138.6)(1.00032) = 30870.6.$$

From this we see that the partition function of the transition state is significantly greater as a result of the delocalization in both the translational and rotational degrees of freedom.

The value of the pre-exponential factor predicted by TST is thus

$$A = s_0 \frac{k_B T}{h} \frac{q_{\ddagger}}{q_a} = 0.9 \left(7.0845 \times 10^{12}\,\mathrm{s}^{-1}\right) \frac{24029.2}{46.5686} = 2.8 \times 10^{15}\,\mathrm{s}^{-1}$$

The experimental value is $3.2 \times 10^{15}\,\mathrm{s}^{-1} \le A \le 3.2 \times 10^{16}\,\mathrm{s}^{-1}$. Hence the prediction of TST is quite close to the experimental value within the uncertainty of the approximations made.

4.9   Estimate the evaporative cooling rate for a hemispherical droplet of Si with a radius of 900 nm held at its melting point.

We use values from Lange's *Handbook of Chemistry* in the following.

Evaporation is simply the phase transition from liquid to gas, that is, the desorption of Si from the surface of the liquid into the gas phase

$$Si(l) \rightarrow Si(g)$$

The energy change associated with this is the enthalpy of vaporization and since adsorption of Si onto Si is non-activated, the enthalpy of vaporization is also the activation energy for the desorption of Si from the surface of liquid Si, $\Delta_{\mathrm{vap}}H = E_{\mathrm{des}} = 359\,\mathrm{kJ\,mol}^{-1}$. Evaporative cooling is caused by loss of Si atoms from the surface and the energy consumed by this event. We calculate the rate of desorption and then estimate the power loss to evaporation, P_{evap}, from the product of energy lost per desorption event times the rate of desorption,

$$P_{\mathrm{evap}} = \frac{\Delta_{\mathrm{vap}}H}{N_A} r_{\mathrm{evap}}\, \sigma_0\, A.$$

The area of the hemisphere is

$$A = \tfrac{1}{2}(4\pi r^2) = 2\pi (900\,\mathrm{nm})^2 = 5.09 \times 10^{-12}\,\mathrm{m}^2,$$

the rate of desorption is

$$r_{\mathrm{evap}} = A_{\mathrm{evap}} \exp\left(-\frac{E_{\mathrm{des}}}{RT}\right) = 10^{13}\,\mathrm{s}^{-1} \exp\left(-\frac{359000\,\mathrm{J\,mol}^{-1}}{8.314\,\mathrm{J\,mol}^{-1}\mathrm{K}^{-1}(1688\,\mathrm{K})}\right) = 77.7\,\mathrm{s}^{-1},$$

and σ_0 is the surface density of atoms which we estimate with the canonical $1 \times 10^{19}\,\mathrm{m}^{-2}$, thus,

$$P_{\mathrm{evap}} = \frac{359000\,\mathrm{J\,mol}^{-1}}{6.02 \times 10^{23}\,\mathrm{mol}^{-1}} (77.7\,\mathrm{s}^{-1})(1 \times 10^{19}\,\mathrm{m}^{-2})(5.09 \times 10^{-12}\,\mathrm{m}^2) = 2.36 \times 10^{-9}\,\mathrm{W}.$$

4.10   Show that for precursor mediated adsorption described by the Kisliuk model, a plot of $\ln[(\alpha/s_0) - 1]$ versus $1/T_s$ is linear with a slope of $-(E_{\mathrm{des}} - E_a)/R$, where E_{des} and E_a are the activation energies for desorption and adsorption out of the precursor state, respectively.

$$s_0 = \alpha \left(1 + \frac{f_{\mathrm{ads}}}{f_{\mathrm{des}}}\right)^{-1} = \alpha \left(1 + \frac{r_{\mathrm{ads}}}{r_{\mathrm{des}}}\right)^{-1}$$

$$s_0 \left(1 + \frac{r_{\mathrm{des}}}{r_{\mathrm{ads}}}\right) = \alpha$$

$$\frac{r_{\text{des}}}{r_{\text{ads}}} = \frac{\alpha}{s_0} - 1$$

$$\frac{\theta A \ \exp(-E_{\text{des}}/RT)}{\exp(-E_{\text{a}}/RT)} = \frac{\alpha}{s_0} - 1$$

$$\theta A \ \exp\left(\frac{-(E_{\text{des}} - E_{\text{a}})}{RT}\right) = \frac{\alpha}{s_0} - 1$$

$$\ln \theta A - \frac{E_{\text{des}} - E_{\text{a}}}{RT} = \ln\left(\frac{\alpha}{s_0} - 1\right) = a + mx$$

where a is the intercept, $m = -(E_{\text{des}} - E_{\text{a}})/R$ is the slope and $x = 1/T$, as was to be shown.

4.11 Consider precursor mediated adsorption through an equilibrated precursor state. The activation barrier to desorption out of the precursor is E_{des} and the activation barrier separating the precursor from the chemisorbed state is E_{a}. Prove mathematically that in precursor mediated adsorption, if $E_{\text{des}} > E_{\text{a}}$, increasing the surface temperature decreases the sticking coefficient and if $E_{\text{des}} < E_{\text{a}}$, increasing T_{s} favours sticking.

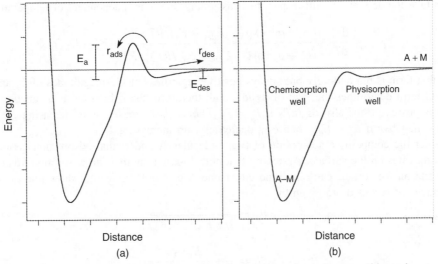

Figure Exercise 4.11 *(a) Activated or (b) non-activated adsorption through an equilibrated precursor state.*

The rate of conversion out of the precursor into the chemisorbed state is

$$r_{\text{ads}} = \sigma_{\text{pre}} A_{\text{ads}} \exp(-E_{\text{a}}/RT).$$

The rate of desorption out of the precursor is

$$r_{\text{des}} = \sigma_{\text{pre}} A_{\text{des}} \exp(-E_{\text{des}}/RT).$$

σ_{pre} is the coverage in the precursor. Differentiate each rate with respect to T_{s} and take the ratio.

$$\frac{\mathrm{d}r_{\text{ads}}/\mathrm{d}T}{\mathrm{d}r_{\text{des}}/\mathrm{d}T} = \frac{A_{\text{ads}}}{A_{\text{des}}} \frac{E_{\text{a}}}{E_{\text{des}}} \exp(-(E_{\text{a}} - E_{\text{des}})/RT)$$

If $E_{des} > E_a$, then $E_a - E_{des} < 0$

$$\therefore \exp(-(E_a - E_{des})/RT) = \exp(X/RT)$$

where X is a positive number. Therefore, as T increase, $\exp(X/RT)$ decreases and the ratio decreases. In other words, as the temperature increases the rate of desorption increases more rapidly than the rate of adsorption. Sticking is disfavoured by the increase in temperature.

If $E_{des} < E_a$, then $E_a - E_{des} > 0$

$$\therefore \exp(-(E_a - E_{des})/RT) = \exp(-X/RT)$$

where X is a positive number. Therefore, as T increase, $\exp(-X/RT)$ increases and the ratio increases. In other words, as the temperature increases the rate of adsorption increases more rapidly than the rate of desorption. Sticking is favoured by the increase in temperature.

More elegantly, rearrange Eq (4.5.21) to

$$s_0 = \frac{\alpha}{1 + a\,\exp\left(-(E_{des} - E_a)/RT\right)}, a = \frac{A_{des}}{A_{ads}}$$

Then take the derivative with respect to T assuming that α is independent of temperature.

$$\frac{ds_0}{dT} = \frac{a\alpha\,\exp\left(-(E_{des} - E_a)/RT\right)}{\left(1 + a\,\exp\left(-(E_{des} - E_a)/RT\right)\right)^2 RT^2}(E_a - E_{des}).$$

The first term is big and ugly but always positive regardless of the magnitudes of E_a and E_{des}. The second term determines the sign of ds_0/dT and, therefore, the behaviour as a function of T. This term is positive (thus also ds_0/dT) if $E_a > E_{des}$, hence s_0 increases with increasing T. The second term is negative if $E_a < E_{des}$, hence s_0 decreases with increasing T.

4.12   Consider the competitive adsorption of two molecules A and B that adsorb non-dissociatively on the same sites on the surface. Assuming that both follow Langmuirian adsorption kinetics and that each site on the surface can only bind either one A or one B molecule, show that the equilibrium coverages of A and B are given by

$$\theta_A = \frac{K_A p_A}{1 + K_A p_A + K_B p_B} \tag{4.11.4}$$

$$\theta_B = \frac{K_B p_B}{1 + K_A p_A + K_B p_B} \tag{4.11.5}$$

The fraction of the surface covered by A is θ_A, that covered by B is θ_B. The coverage of empty sites is $1 - \theta_A - \theta_B$. Since empty sites are required for adsorption, the rates of adsorption can be written

$$r_{ads}^A = k_{ads}^A p_A (1 - \theta_A - \theta_B)$$
$$r_{ads}^B = k_{ads}^B p_B (1 - \theta_A - \theta_B).$$

The rates of desorption are $r_{des}^A = k_{des}^A \theta_A$

$$r_{des}^B = k_{des}^B \theta_B$$

At equilibrium, the rates of the adsorption and desorption of A are equal and the rates of adsorption and desorption of B are equal. Setting these equal, we arrive at two equations that are similar to the one component system

$$\frac{k_{ads}^A}{k_{des}^A} p_A = K_A p_A = \frac{\theta_A}{1 - \theta_A - \theta_B}$$

$$\frac{k_{ads}^B}{k_{des}^B} p_B = K_B p_B = \frac{\theta_B}{1 - \theta_A - \theta_B}$$

These represent two simultaneous equations that can be solved for θ_A and θ_B to yield

$$\theta_A = \frac{K_A p_A}{1 + K_A p_A + K_B p_B}$$

$$\theta_B = \frac{K_B p_B}{1 + K_A p_A + K_B p_B}$$

as was to be shown.

4.13 Many different pressure units are encountered in surface science and conversions are inevitable. Show that

$$Z_w = 3.51 \times 10^{22} \, \text{cm}^{-2} \, \text{s}^{-1} p/(MT)^{1/2} \qquad (12.0.1)$$

where p is given in torr, M in g mol^{-1} and T in Kelvin. Derive a similar expression that relates the SI units of pressure (Pa), molar mass (kg mol^{-1}) to the flux in m^{-2} s^{-1}.

Note that kg mol^{-1} is the SI unit of molar mass whereas the molecular weight in g mol^{-1} is used in the first expression.

The original expression for the impingement rate is

$$Z_w = \frac{p}{(2\pi m k_B T)^{1/2}}$$

where all SI units are used, p is in Pa, m is the particle mass in kg and T is in Kelvin. Now substitute

$$R = N_A k_B, N_A = \text{Avogadro constant}$$

$$M = N_A m \text{ (but beware of units, this is molar mass in kg mol}^{-1})$$

$$Z_w = \frac{p}{\left(2\pi \dfrac{M}{N_A} \dfrac{R}{N_A} T\right)^{1/2}} = \frac{N_A p}{(2\pi MRT)^{1/2}} = 2.63 \times 10^{24} \text{m}^{-2} \, \text{s}^{-1} \frac{p}{(MT)^{1/2}}$$

where p is in Pa and M is in kg mol^{-1}.

$$Z_w = (2.63 \times 10^{24} \, \text{m}^{-2} \, \text{s}^{-1}) \left(\frac{1 \, \text{m}^2}{10^4 \, \text{cm}^2}\right) \frac{p}{(M(1000 \, \text{g}/1 \, \text{kg})T)^{1/2}} \left(\frac{760 \, \text{torr}}{101325 \, \text{Pa}}\right)$$

$$Z_w = 3.51 \times 10^{22} \, \text{cm}^{-2} \, \text{s}^{-1} \, p/(MT)^{1/2}$$

Now p is in torr and M is in g mol^{-1}.

4.14 Show that for dissociative adsorption within the Langmuir model, the isotherm equation is given by

$$\theta = \frac{(pK)^{1/2}}{1 + (pK)^{1/2}} \qquad (12.0.2)$$

In dissociative Langmuirian adsorption, two adjacent sites are required. The probability of finding two adjacent sites is given by $(1 - \theta)^2$. Therefore the rate of adsorption is

$$r_a = s_0(1 - \theta)^2 Z_w = k_a p(1 - \theta)^2.$$

The rate of desorption is second order in coverage. Thus

$$r_{des} = k_d \theta^2.$$

At equilibrium the two rates are equal, thus,

$$k_a p(1 - \theta)^2 = k_d \theta^2$$

$$\frac{k_a p}{k_d} = Kp = \frac{\theta^2}{(1 - \theta)^2},$$

where K, the ratio of rate constants, is an equilibrium constant, again with the caveats mentioned in Chapter 4.

$$(Kp)^{1/2} = \frac{\theta}{(1 - \theta)}$$

$$(Kp)^{1/2}(1 - \theta) = \theta$$

$$\frac{(Kp)^{1/2}}{1 + (Kp)^{1/2}} = \theta$$

as was to be shown.

4.15   An alternative representation of the Langmuir isotherm for a single component system is

$$c = \frac{\lambda_V p}{\lambda_p + p}, \tag{4.11.8}$$

where c is the surface coverage, λ_v is the Langmuir volume, λ_p is the Langmuir pressure and p the pressure. Note that the dimensions of λ_v set the dimension of c, which is often expressed in some form of concentration units (such as molecules adsorbed per gram of adsorbent) rather than areal density. Use Eq. (4.6.3) to derive an equation of the form of Eq. (4.11.8) and determine the meaning of λ_v and λ_p. Instead of fractional coverage explicitly use the areal density and saturation coverage σ_{sat}.

   Start with

$$\frac{\sigma}{\sigma_{sat} - \sigma} = pK.$$

Rearranging we obtain

$$\sigma = \frac{\sigma_{sat} pK}{1 + pK} = \frac{\sigma_{sat} p}{1/K + p} = \frac{\lambda_v p}{\lambda_p + p}.$$

We now see that the Langmuir volume is equivalent to the saturation coverage and the Langmuir pressure is the inverse of the adsorption/desorption equilibrium constant.

4.16   Methane is often found adsorbed on coal. Consider the reversible, non-dissociative molecular adsorption of CO_2 and CH_4 on coal. The parameters describing adsorption are given below

	λ_v/mol g^{-1}	λ_p/kPa
CO_2	0.0017	2100
CH_4	0.00068	4800

Calculate the equilibrium coverage for CO_2 and CH_4 on coal exposed to a total gas pressure of 3500 kPa for (a) pure CH_4, (b) pure CO_2, and (c) a 50/50 mixture of CH_4 and CO_2. Which species is bound more strongly, and what is the implication of pumping CO_2 into a coal seam saturated with CH_4?

The equilibrium coverage is given by

(a)

$$c(CH_4) = \frac{\lambda_v p}{\lambda_p + p} = \frac{(0.00068)(3500)}{4800 + 3500} = 2.87 \times 10^{-4}\,mol\,g^{-1}.$$

(b)

$$c(CO_2) = \frac{(0.0017)(3500)}{2100 + 3500} = 1.06 \times 10^{-3}\,mol\,g^{-1}.$$

The equilibrium coverage of CO_2 is almost 4 times greater than that of CH_4 indicative of a stronger adsorption for CO_2 compared to CH_4.

(c) From Eqs (4.11.4) and (4.11.5)

$$\theta_{CO_2} = \frac{K_{CO_2} p_{CO_2}}{1 + K_{CO_2} p_{CO_2} + K_{CH_4} p_{CH_4}} = \frac{p_{CO_2}}{\frac{1}{K_{CO_2}} + p_{CO_2} + \frac{K_{CH_4}}{K_{CO_2}} p_{CH_4}}$$

$$\theta_{CO_2} = \frac{p_{CO_2}}{\lambda_{CO_2} + p_{CO_2} + \frac{\lambda_{CO_2}}{\lambda_{CH_4}} p_{CH_4}} = \frac{1750}{2100 + 1750 + \frac{2100}{4800}(1750)} = 0.38\,ML$$

$$\theta_{CH_4} = \frac{p_{CH_4}}{\lambda_{p,CH_4} + p_{CH_4} + \frac{\lambda_{p,CH_4}}{\lambda_{p,CO_2}} p_{CH_4}} = \frac{1750}{4800 + 1750 + \frac{4800}{2100}(1750)} = 0.17\,ML$$

With equal partial pressures, the coverage of CO_2 is higher. In other words, CO_2 is able to displace CH_4 from the surface of the coal. Therefore, by pumping CO_2 into the coal, CH_4 can be displaced and CO_2 will remain in its place. This is a method of extracting CH_4 (so-called coal seam methane) that might otherwise not leave the coal, and of sequestering CO_2 in a geological reserve.

4.17   Given that A adsorbs in an on-top site on a fcc(111) lattice (fcc, face-centred cubic) with an initial heat of adsorption of $q_0 = 100\,kJ\,mol^{-1}$, calculate the heat of adsorption as a function of coverage for repulsive lateral interactions of 0, 2, 5 and $10\,kJ\,mol^{-1}$. Make a series of plots to demonstrate the effects of repulsive interactions at $T = 80\,K$ and $300\,K$.

For the fcc(111) lattice $z = 6$. Recall that a repulsive interaction has a negative values; thus $\omega = 0, -2, -5$ and $-10\,kJ\,mol^{-1}$. Using these values, Figs. Ex. 4.17(a) and 4.17(b) were generated for $T = 80\,K$ and $300\,K$, respectively. Note the near linear behaviour at $(\omega = -2\,kJ\,mol^{-1}, T = 300\,K)$ as opposed to the step-like and near step-like behaviour at $(\omega = -5\,kJ\,mol^{-1}, T = 80\,K)$ and $(\omega = -10\,kJ\,mol^{-1}, T = 80\,K)$, respectively, whereas $(\omega = -2\,kJ\,mol^{-1}, T = 80\,K)$ and $(\omega = -5\,kJ\,mol^{-1}, T = 300\,K)$ yield more smoothly varying sigmoidal behaviour.

4.18   On Pt(111) straight-chain alkanes exhibit non-activated molecular adsorption at low T_s. The desorption activation energy as determined by TPD is 18.6, 36.8, 42.4 and $54.0\,kJ\,mol^{-1}$ for CH_4, C_2H_6, C_3H_8 and C_4H_{10}, respectively. Explain this trend.

The low binding energies reported are indicative of physisorption. The primary interactions are related to the dispersion forces of the van der Waals interactions. These straight chain molecules can maximize their interactions by lying flat on the surface, and in this configuration, as the chain length increases, more CH_2 (or CH_3) groups interact with the surface. With each successive group, the overall attraction to the surface increases.

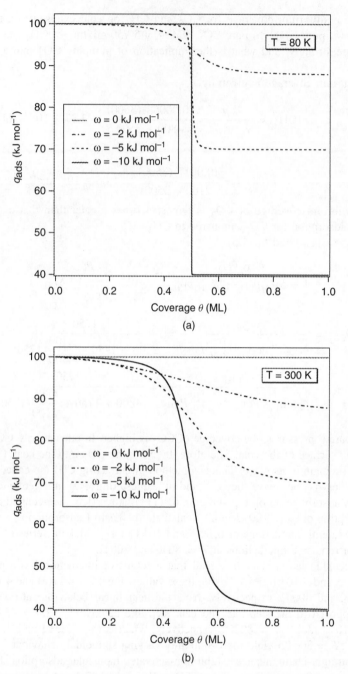

Figure Exercise 4.17   *The behaviour of the heat of adsorption q_{ads} as a function of coverage for (a) 80 K and (b) 300 K for the conditions described in Ex. 4.17.*

4.19   Derive Eqs (4.5.16–18).

The coverage dependence of the sticking coefficient is given by

$$s = s_0 \theta_{\text{req}}$$

where

$$\theta_{\text{rec}} = \theta_{\text{OO}} = 1 - \theta - \frac{2\theta(1-\theta)}{[1 - 4\theta(1-\theta)(1 - \exp(\omega/k_B T_s))]^{1/2} + 1}$$

and the strength of the lateral interactions is characterized by the term

$$B = 1 - \exp(\omega/k_B T_s).$$

There are three limiting cases for the value of ω and, thus, three corresponding values of θ_{rec} and three expressions for s:

No interactions: $\omega = 0, B = 1 - \exp(0) \therefore B = 0$

$$\theta_{\text{rec}} = 1 - \theta - \frac{2\theta(1-\theta)}{[1 - 4\theta(1-\theta)(0)]^{1/2} + 1} = 1 - \theta - \frac{2\theta(1-\theta)}{[1]^{1/2} + 1} = 1 - \theta - \theta(1-\theta)$$

$$\theta_{\text{rec}} = 1 - 2\theta + \theta^2 = (1-\theta)^2$$

$$s = s_0(1-\theta)^2$$

Strong repulsive interactions: $\omega \ll 0, B = 1 - \exp(-\infty) \therefore B = 1$

$$\theta_{\text{rec}} = 1 - \theta - \frac{2\theta(1-\theta)}{[1 - 4\theta(1-\theta)(1)]^{1/2} + 1} = 1 - \theta - \frac{2\theta(1-\theta)}{[1 - 4\theta + 4\theta^2]^{1/2} + 1}$$

$$= 1 - \theta - \frac{2\theta(1-\theta)}{[(1 - 2\theta)^2]^{1/2} + 1}$$

$$\theta_{\text{rec}} = 1 - \theta - \frac{\theta(2 - 2\theta)}{2 - 2\theta} = 1 - 2\theta$$

$$s = s_0(1 - 2\theta)$$

Since the sticking coefficient is defined as a positive number, this equation must hold in the region $\theta \leq 0.5$. At that point the sticking coefficient falls to zero and the coverage does not increase above 0.5.

Strong attractive interactions: $(\omega \gg 0, B = 1 - \exp(\infty) \therefore B = -\infty)$

$$\theta_{\text{rec}} = 1 - \theta - \frac{2\theta(1-\theta)}{[1 - 4\theta(1-\theta)\infty]^{1/2} + 1} = 1 - \theta - \frac{2\theta(1-\theta)}{\infty} = 1 - \theta$$

$$s = s_0(1-\theta)$$

4.20   Write out expressions of Eq. (4.5.23) in the limits of

(a) large desorption rate from the precursor,
(b) large adsorption rate into the chemisorbed state,
(c) large desorption rate from the chemisorbed state.

Explain the answers for (b) and (c) with recourse to the value of s_0.

$$\frac{s(\theta)}{s_0} = \left[1 + K\left(\frac{1}{\theta_{\text{req}}} - 1\right)\right]^{-1} \Rightarrow s(\theta) = \frac{s_0}{1 + \dfrac{f'_{\text{des}}}{f_{\text{ads}} + f_{\text{des}}}a} = \frac{\alpha}{1 + \dfrac{f'_{\text{des}}}{f_{\text{ads}} + f_{\text{des}}}a\left(1 + \dfrac{f_{\text{des}}}{f_{\text{ads}}}\right)}$$

where a is a constant.

(a) For a large desorption rate from the precursor

$$f'_{\text{des}} \to \infty$$

and

$$s(\theta) = \frac{\alpha}{1 + \dfrac{\infty}{f_{\text{ads}} + f_{\text{des}}} a \left(1 + \dfrac{f_{\text{des}}}{f_{\text{ads}}}\right)} = \frac{\alpha}{\infty} = 0.$$

Hence, the sticking coefficient tends toward zero since there is no conversion from the precursor state into the adsorbed state.

(b) For large adsorption rate into the chemisorbed state

$$f_{\text{ads}} \to \infty$$

and

$$s(\theta) = \frac{\alpha}{1 + \dfrac{f'_{\text{des}}}{f_{\text{ads}} + f_{\text{des}}} a \left(1 + \dfrac{f_{\text{des}}}{f_{\text{ads}}}\right)} = \frac{\alpha}{1 + (0)a(1 + 0)} = \alpha$$

Hence, the sticking coefficient remains constant as the coverage increases and equal to the probability of entering the precursor state.

(c) For large desorption rate from the chemisorbed state

$$f_{\text{des}} \to \infty$$

and

$$s(\theta) = \frac{\alpha}{1 + \dfrac{f'_{\text{des}}}{f_{\text{des}}} a \left(\dfrac{f_{\text{des}}}{f_{\text{ads}}}\right)} = \frac{\alpha}{1 + a \left(\dfrac{f'_{\text{des}}}{f_{\text{ads}}}\right)}$$

Hence, the sticking coefficient is again constant. Note, however, that

$$s_0 = \alpha \left(1 + \frac{f_{\text{des}}}{f_{\text{ads}}}\right)^{-1} = \frac{\alpha}{\infty} = 0.$$

That is, the initial sticking probability is equal to zero and remains that way since the rate of desorption is too high to allow for adsorption.

4.21 In recombinative desorption, the pre-exponential factor can be written

$$A = D\bar{v} \tag{4.11.9}$$

where D is the mean molecular diameter and $\bar{v}$ is the mean speed of surface diffusion,

$$\bar{v} = \frac{\lambda}{\tau} \tag{4.11.10}$$

where λ is the mean hopping distance (the distance between sites) and τ is the mean time between hops,

$$\tau = \tau_0 \exp(E_{\text{dif}}/RT). \tag{4.11.11}$$

where τ_0 is the characteristic hopping time. Use this information and the expression for the desorption rate constant to show that a compensation effect could be observed in a system that exhibits a coverage dependent diffusion activation energy, E_{dif}.

Substituting from (4.11.11) and (4.11.12) into (4.11.10), we obtain

$$A = D^2 \tau_0^{-1} \exp(-E_{\text{dif}}/RT).$$

Experimentally, we measure the rate constant for desorption according to

$$k_{\text{des}} = A \exp(-E_{\text{des}}/RT).$$

Therefore,

$$k_{\text{des}} = D^2 \tau_0^{-1} \exp[-(E_{\text{dif}} + E_{\text{des}})/RT].$$

For a compensation effect to occur, the pre-exponential factor $D^2 \tau_0^{-1}$ must increase in concert with the activation energy $E_{\text{dif}} + E_{\text{des}}$. D is the mean molecular diameter, a quantity that cannot be coverage dependent. τ_0, the diffusion pre-exponential or characteristic hopping time, must therefore increase as E_{dif} increases or decrease as E_{dif} decreases to facilitate a compensation effect.

4.22   When 0.5 L of H_2 is dosed onto Ir(110)–(2 × 1) at 100 K, it adsorbs dissociatively. Its TPD spectrum exhibits one peak at 400 K. When propane is dosed to saturation at 100 K on initially clean Ir(110)–(2 × 1), the H_2 TPD spectrum exhibits two peaks, one at 400 K and one at 550 K. Interpret the origin of these two H_2 TPD peaks from C_3H_8 adsorption.

The H_2 TPD peaks result from the recombination of H atoms adsorbed on the surface. Recombination normally occurs at 400 K for chemisorbed atoms. If chemisorbed atoms are present on the surface below 400 K, they will contribute to the 400 K TPD feature. Therefore, propane is either dissociating during adsorption at 100 K or during heating of the surface from 100 K to 400 K. This produces chemisorbed H atoms that desorb in a so-called desorption limited peak at 400 K. The initial dissociation product must be a hydrocarbon species that dissociates around 550 K. The dissociation of this species leads to a reaction limited H_2 desorption peak at 550 K. Since chemisorbed H atoms are not stable on the surface at this elevated temperature (they desorb already at 400 K), they will recombine rapidly and desorb promptly after the dissociation reaction. Hence, it is the kinetics of the dissociation reaction, not recombinative desorption, that determines the peak temperature of this TPD feature.

4.23   Show that for the simple adsorption reaction

$$A + {}^* \rightleftharpoons A^*$$

the following relationship holds:

$$\theta_* = \frac{1}{1 + (\theta_A/\theta_*)}. \tag{4.11.12}$$

Assuming that the surface does not reconstruct and that there is only one type of adsorption site, then we can write a conservation of total adsorption sites expression such that

$$\theta_A + \theta_* = 1.$$

Dividing through by θ_*,

$$\frac{\theta_A}{\theta_*} + 1 = \frac{1}{\theta_*}.$$

Now take the reciprocal of both sides

$$\frac{1}{\theta_A/\theta_* + 1} = \theta_*$$

as was to be shown.

4.24   Using transition state theory and the lattice-gas approximation the rate constants for adsorption and desorption can be written

$$k_a = (1 - \theta)k_a^0 \sum_i P_{0,i} \exp(-\varepsilon_i^*/k_B T) \tag{4.11.13}$$

$$k_d = \frac{(1 - \theta)}{\theta} k_d^0 \exp(\mu_a/k_B T) \sum_i P_{0,i} \exp(-\varepsilon_i^*/k_B T), \tag{4.11.14}$$

where k_a^0 and k_d^0 are the rate constants in the limit of low coverage, $P_{0,i}$ is the probability that a vacant site has the environment denoted by index i and ε_i^* describes the lateral interactions in the activated state. The units of these rate constants are such that the net rate of coverage change can be written

$$\frac{d\theta}{dt} = k_a p - k_d \theta. \tag{4.11.15}$$

Demonstrate that Eqs (4.11.14) and (4.11.15) obey microscopic reversibility by showing that at equilibrium

$$\mu_a = \mu_d. \tag{4.11.16}$$

You will need to use the standard expression for the chemical potential of gas phase particles

$$\mu_g = \mu_g{}^\circ + k_B T \ln p. \tag{4.11.17}$$

At equilibrium

$$\frac{d\theta}{dt} = k_a p - k_d \theta = 0.$$

Therefore

$$\frac{\theta}{p} = \frac{k_a}{k_d} = \frac{(1 - \theta)k_a^0 \sum_i P_{0,i} \exp(-\varepsilon_i^*/k_B T)}{\frac{(1 - \theta)}{\theta} k_d^0 \exp(\mu_a/k_B T) \sum_i P_{0,i} \exp(-\varepsilon_i^*/k_B T)} = \frac{(1 - \theta)k_a^0}{\frac{(1 - \theta)}{\theta} k_d^0 \exp(\mu_a/k_B T)}$$

$$k_d^0(1 - \theta) \exp(\mu_a/k_B T) = (1 - \theta)p k_a^0$$

$$k_d^0 \exp(\mu_a/k_B T) = p k_a^0$$

$$\mu_a/k_B T = \ln(p k_a^0/k_d^0)$$

$$k_B T \ln p + k_B T \ln(k_a^0/k_d^0) = \mu_a$$

Since

$$\mu_g = \mu_g{}^\circ + k_B T \ln p.$$

If we define

$$\mu_g{}^\circ = k_B T \ln(k_a^0/k_d^0),$$

then

$$\mu_a = k_B T \ln p + \mu_g^0 = \mu_g. \qquad \text{QED}$$

4.25   The initial sticking coefficient at 500 K for CH_4 on a terrace site on Ni(111) is 2.1×10^{-9} whereas on a step site it is 2.8×10^{-7}. Assuming an Arrhenius form to describe the temperature dependence of the initial sticking coefficient and that the entropy of activation is the same for both terrace and step sites, calculate the difference in the barrier to dissociation between a terrace and step site.

By assuming an Arrhenius form, we write

$$s_{step} = s_0 \exp(-E_{ads}^{step})$$

$$s_{terr} = s_0 \exp(-E_{ads}^{terr}),$$

where s_0 is equal in the two cases because we assume that the entropy of activation is the same. Taking the natural logarithm of both equations and subtracting the second from the first yields

$$\ln s_{step} - \ln s_{terr} = \frac{E_{ads}^{terr}}{RT} - \frac{E_{ads}^{step}}{RT}$$

$$RT \ln \left(\frac{s_{step}}{s_{terr}} \right) = E_{ads}^{terr} - E_{ads}^{step} = \Delta E$$

$$\Delta E = (8.314\,\text{J mol}^{-1}\,\text{K}^{-1})(500\,\text{K}) \ln \left(\frac{2.8 \times 10^{-7}}{2.1 \times 10^{-9}} \right) = 20\,\text{kJ mol}^{-1},$$

4.26   The dissociative sticking coefficient of H_2 on Si(100) is $\sim 1 \times 10^{-11}$ at room temperature. There are 6.8×10^{18} Si atoms per m^2 on the Si(100) surface. Estimate the coverage of H atoms that results from exposing a Si(100) surface to 1×10^{-5} Pa of H_2 for 3 min. Justify the use of a constant sticking coefficient.

Remember the factor of 2, which arises because of dissociation,

$$H_2(g) \rightarrow 2H(a).$$

Therefore with $s = 1 \times 10^{-11}$ (assumed constant), $p = 1 \times 10^{-5}$ Pa, $t = 3(60) = 180$ s, $T = 300$ K, $M = 0.002$ kg mol^{-1}, and $\sigma_0 = 6.8 \times 10^{18}$ m^{-2} we have

$$\sigma = 2s\varepsilon = 2sZ_w t = 2s \frac{N_A p}{\sqrt{2\pi MRT}} t = 3.88 \times 10^9\,\text{m}^{-2}$$

$$\theta = \frac{\sigma}{\sigma_0} = \frac{3.88 \times 10^9\,\text{m}^{-2}}{6.8 \times 10^{18}\,\text{m}^{-2}} = 5.7 \times 10^{-10}\,\text{ML}$$

Since the coverage is so low $\theta \ll 1$, there is no need to account for a $(1 - \theta)$ (or similar) term, and the assumption of a constant sticking coefficient is justified.

4.27   Determine the steady-state saturation coverage that results if atomic H is exposed to a surface. Assume that the sticking coefficient of H atoms on an empty site is s_{ad} and the abstraction probability is s_{ab} if the H atom strikes a filled site. Assume that one and only one H atom can adsorb per surface atom and that both s_{ad} and s_{ab} are independent of coverage. Can the coverage ever reach 1 ML (as defined by one adsorbate per surface atom)? Assume that T_s is low enough that desorption can be neglected.

At steady state, the rate of adsorption

$$r_{ad} = \frac{Z_w}{\sigma_0} s_{ad}(1 - \theta)$$

equals the rate of abstraction

$$r_{ab} = \frac{Z_w}{\sigma_0} s_{ab}\,\theta$$

and there is no net change in coverage

$$\frac{d\theta}{dt} = r_{ad} - r_{ab} = 0.$$

Here Z_w is the flux of H atoms onto the surface, θ equals the probability of hitting a filled site, $(1 - \theta)$ equals the probability of hitting an empty site and σ_0 is the areal density of surface atoms. Substituting the expressions for the rates into the differential equation yields

$$\frac{Z_w}{\sigma_0} s_{ad}(1 - \theta) = \frac{Z_w}{\sigma_0} s_{ab} \theta$$

$$\frac{1 - \theta}{\theta} = \frac{s_{ab}}{s_{ad}}.$$

This result indicates that $\theta = 1 \, \text{ML}$ implies that $s_{ab} = 0$ or at least that it is much less than s_{ad}. As long as there is a finite probability of abstraction that is comparable to the sticking coefficient, the coverage cannot reach 1 ML.

4.28   Under the same assumptions found in problem 4.29, calculate how the H atom coverage θ changes with time.

We now have

$$\frac{d\theta}{dt} = r_{ad} - r_{ab} = \frac{Z_w}{\sigma_0} s_{ad}(1 - \theta) - \frac{Z_w}{\sigma_0} s_{ab} \theta.$$

To obtain an expression for θ as a function of time, we move all t terms to one side and all θ terms to the other. Then we integrate from 0 to t and 0 to θ, respectively.

$$\frac{d\theta}{\dfrac{Z_w}{\sigma_0} s_{ad}(1 - \theta) - \dfrac{Z_w}{\sigma_0} s_{ab} \theta} = dt.$$

If we collect the θ terms together

$$\int_0^\theta \frac{d\theta}{\dfrac{Z_w s_{ad}}{\sigma_0} - \left(\dfrac{Z_w s_{ad}}{\sigma_0} + \dfrac{Z_w s_{ab}}{\sigma_0} \right) \theta} = \int_0^t dt$$

we see that we obtain an integral of the form

$$\int \frac{d\theta}{X} = \frac{1}{a} \ln X, \quad \text{with } X = b + a\,\theta,$$

where

$$a = -\frac{Z_w}{\sigma_0}(s_{ad} + s_{ab}) \quad \text{and} \quad b = \frac{Z_w s_{ad}}{\sigma_0}.$$

Therefore, the integral becomes

$$\int_0^\theta \frac{d\theta}{X} = \int_0^t dt$$

$$\frac{1}{a} \ln(a\theta + b) \Big|_0^\theta = t \big|_0^\theta$$

$$\frac{1}{a} \ln(a\theta + b) - \frac{1}{a} \ln(b) = t$$

$$\ln\left(\frac{a\theta + b}{b}\right) = at$$

$$\theta = \frac{b}{a}(e^{at} - 1)$$

$$\theta = \frac{s_{ad}}{s_{ad} + s_{ab}}\left(1 - \exp\left(-\frac{Z_w}{N_0}(s_{ad} + s_{ab})t\right)\right)$$

Note that the answer is equivalent to what we found in Exercise 4.3, in which the opposing reactions were first-order adsorption and first-order desorption. The kinetics is the same even though the processes are different.

4.29 Derive Equation (4.6.6).

From Eq. (4.6.1)

$$\theta = pK - pK\theta$$

$$\sigma_0\theta = \sigma_0 pK - pK\theta\sigma_0$$

$$\sigma = \sigma_0 pK - pK\sigma$$

$$\frac{\sigma}{\sigma_0} = pK - pK\frac{\sigma}{\sigma_0} = \left(1 - \frac{\sigma}{\sigma_0}\right)pK$$

$$\sigma = \left(1 - \frac{\sigma}{\sigma_0}\right)pK\sigma_0 = \left(1 - \frac{\sigma}{\sigma_0}\right)pb$$

where

$$b = K\sigma_0 = \frac{\sigma_0 k_{ads}}{k_{ads}} = \frac{\sigma_0(2\pi mk_B T)^{1/2}e^{-E_{ads}/RT}}{\sigma_0 A e^{-E_{des}/RT}}$$

$$b = (2\pi mk_B T)^{1/2}A^{-1}e^{-(E_{ads}-E_{des})/RT}.$$

4.30 The equilibrium vapour pressure of water in Pa is given as a function of absolute temperature T by

$$\ln p = 23.195 - \frac{3.814 \times 10^3}{T - 46.29}.$$

(a) Assuming that the desorption activation energy is given by the enthalpy of vaporization of pure water $40.016\,\mathrm{kJ\,mol^{-1}}$, calculate the coverage of water expected on a surface exposed to the equilibrium vapour pressure of water at 298 K.

(b) What is the value of K in Eq. (4.6.4)?

(a) Inserting $T = 298$ K into the vapour pressure equation yields a value of 3111 Pa. The coverage is calculated from Eq. (4.2.4)

$$\theta = \frac{p}{p + p_0(T)}$$

using Eq. (4.2.5)

$$p_0(T) = \left(\frac{2\pi mk_B T}{h^2}\right)^{3/2}k_B T \exp\left(\frac{-\varepsilon}{k_B T}\right) = 5737\,\mathrm{Pa}$$

with $m = \dfrac{0.180\,\mathrm{kg\,mol^{-1}}}{6.02 \times 10^{23}\mathrm{mol^{-1}}} = 2.99 \times 10^{-26}\,\mathrm{kg}$ and $\varepsilon = \dfrac{44016\,\mathrm{Jmol^{-1}}}{6.02 \times 10^{23}\,\mathrm{mol^{-1}}} = 7.312 \times 10^{-20}\,\mathrm{J}.$

$$\text{Thus, } \theta = \frac{3111}{3111 + 5737} = 0.352\,\text{ML}.$$

In other words, any surface exposed to air will be covered with a thin film of water, the amount of water depends on the temperature and relative humidity but a few tenths of a monolayer can be expected.

(b) Since

$$\theta = \frac{p}{p + p_0(T)} = \frac{pK}{1 + pK} = \frac{p}{1/K + p},$$

then by inspection

$$K = 1/p_0(T) = 1.743 \times 10^{-4}\text{Pa}^{-1}.$$

13

Answers to Exercises from Chapter 5. Liquid Interfaces

5.1 Derive the Young equation, Eq. (5.3.1).

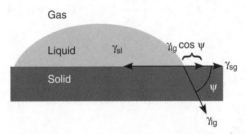

Figure Exercise 5.1 *Balance of forces when a liquid droplet forms on a solid surface.*

Surface tension acts parallel to the surface, exerting a force in this direction. At equilibrium, the drop maintains a constant shape because of the balance of forces exerted at the three interfaces. While γ_{sl} opposes γ_{sg}, the contribution of γ_{lg} depends on the component of γ_{lg} in the plane of the surface. By simple trigonometry, this component can be worked out.

$$\gamma_{sl} + \gamma_{lg} \cos \psi - \gamma_{sg} = 0$$

$$\gamma_{sl} + \gamma_{lg} \cos \psi = \gamma_{sg}$$

5.2 Consider a hemispherical liquid island of radius r with surface energies $\gamma_{sl} > \gamma_{lg}$ in equilibrium with its vapour. Calculate the island surface energy as a function of r and demonstrate that small islands are unstable with respect to large islands. Assume the substrate to be rigid and that the island energy is composed only of island-substrate and island-vapour terms.

Surface Science: Foundations of Catalysis and Nanoscience, Third Edition. Kurt W. Kolasinski.
© 2012 John Wiley & Sons, Ltd. Published 2012 by John Wiley & Sons, Ltd.

The total island surface energy is equal to the sum of all of the contributions of all of the islands. For an individual island i and a system for which the total is the sum of the contributions of all islands, we define the area, volume and surface energy according to

$$A_{sl}^i = \pi r_i^2$$

$$A_{lg}^i = 4\pi r_i^2$$

$$V_i = \tfrac{1}{2}(\tfrac{4}{3})\pi r_i^3$$

$$V = \sum_i V_i = \frac{2\pi}{3} \sum_i r_i^3$$

$$G = \sum_i \gamma_{sl} A_{ls}^i + \sum_i \gamma_{lg} A_{lg}^i = \pi \gamma_{ls} \sum_i r_i^2 + 4\pi \gamma_{lg} \sum_i r_i^2 = (\pi \gamma_{sl} + 4\pi \gamma_{lg}) \sum_i r_i^2$$

Note that regardless of the relative magnitude of the surface tensions, to minimize G, the island energy, we need to minimize the summation over r_i^2.

Consider the case of two islands, with island energy G_2, and compare this to the case of one island with island energy G_1. Note that V, the total island volume, is a constant.

One island

$$V = \frac{2\pi}{3} r^3, \quad r = \left(\frac{3V}{2\pi}\right)^{1/3}, \quad G_1 = \pi(\gamma_{sl} + 4\gamma_{lg}) r^3$$

Two islands

$$V = \frac{2\pi}{3} \left(r_1^3 + r_2^3\right)$$

$$r_2 = \left(\frac{3V}{2\pi} - r_1^3\right)^{1/3}$$

$$G_2 = \pi \left(\gamma_{sl} + 4\gamma_{lg}\right) \left(r_1^2 + r_2^2\right) = \pi(\gamma_{sl} + 4\gamma_{lg}) \left(r_1^2 + \left(\frac{3V}{2\pi} - r_1^3\right)^{2/3}\right)$$

Now take the first and second derivatives utilizing the chain rule

$$\frac{dy}{dx} = \frac{dy}{du}\frac{du}{dx}$$

$$G_2 = c\left(r_1^2 + u^{2/3}\right), u = \frac{3V}{2\pi} - r_1^3$$

$$\frac{dG_2}{dr_1} = c\left(2r_1 - 2r_1^2\left(\frac{3V}{2\pi} - r_1^3\right)^{-1/3}\right)$$

$$\frac{d^2G_2}{dr_1} = c\left(2 - \frac{2r_1}{(b + r_1^3 \text{ term})} - \frac{2r_1^2}{(d + r_1^3 \text{ term})}\right)$$

The first derivative goes to zero at $r_1 = 0$. The second derivative is positive at $r_1 = 0$, therefore, a minimum in the $G_2(r_1)$ is found at $r_1 = 0$. In other words, the minimum energy occurs when only one island is formed.

Alternatively, if the energy of one large island is lower than the energy of two smaller islands of the same total volume, then

$$V = V_1 + V_2 \text{ and } \frac{G_2}{G_1} > 1$$

We now try to prove the second relation.

$$\frac{G_2}{G_1} = \frac{\left(\frac{3V_1}{2\pi}\right)^{2/3} + \left(\frac{3V_2}{2\pi}\right)^{3/2}}{\left(\frac{3V}{2\pi}\right)^{3/2}} = \frac{V_1^{3/2} + V_2^{3/2}}{V^{3/2}} > 1$$

Thus $G_2 > G_1$ if

$$V_1^{2/3} + V_2^{2/3} > V^{2/3}$$

which is, indeed, true as shown below.

$$V^2 = \left(V^{2/3}\right)^3 = V_1^2 + 2V_1 V_2 + V_2^2$$

$$\left(V_1^{2/3} + V_2^{2/3}\right)^3 = V_1^2 + 3V_1^{4/3} V_2^{2/3} + 3V_1^{2/3} V_2^{4/3} + V_2^2$$

$$\left(V_1^{2/3} + V_2^{2/3}\right)^3 - V^2 = 3V_1^{4/3} V_2^{2/3} + 3V_1^{2/3} V_2^{4/3} - 2V_1 V_2$$

$$= 3V_1 V_2 ((V_1/V_2)^{1/3} + (V_2/V_1)^{1/3}) - 2V_1 V_2$$

If $V_1 > V_2$ then the first term of the bit in brackets > 1, if $V_2 > V_1$ then the second term is greater than 1. Either way, the factor in brackets is > 1, so the difference is $> V_1 V_2 > 0$.

$$\text{Therefore } \left(V_1^{2/3} + V_2^{2/3}\right)^3 > V^2 \Rightarrow V^{2/3} < V_1^{2/3} + V_2^{2/3}.$$

5.3 Consider the dynamics of deposition of X, Y and Z multilayer films. For each case, determine whether deposition occurs on the downstroke (insertion of substrate into the LB trough), upstroke (retraction) or in both directions. Discuss the reasons for these dependencies.

Y film deposition is the easiest to understand. The substrate is hydrophilic. The amphiphile is strongly polar on one side and hydrophobic on the other. Deposition occurs on both the up- and down-strokes as the surface alternates between hydrophobic and hydrophilic. The meniscus alternates between upward (for hydrophilic) and downward (for hydrophobic surfaces).

Z-type film only deposits on upstroke. It starts with a hydrophilic substrate. Neither side of amphiphile is particularly hydrophobic or hydrophilic. The outer surface is always weakly hydrophobic but the dynamic contact angle, i.e. the curvature of the meniscus while the substrate moves, changes based on whether the substrate is being immersed or retracted. Deposition on the downstroke does not occur due to a combination of weak tail–tail interactions, and because the substrate is immersed too rapidly to allow effective deposition.

X film deposition only occurs on the downstroke. The story is similar to that of Z-type films but all of the roles are reversed. The substrate is hydrophobic, and the head-head interactions are weak. Not so hydrophobic and weak polar groups such as –$COOCH_3$ and –NO_2 favour X- and Z-type films.

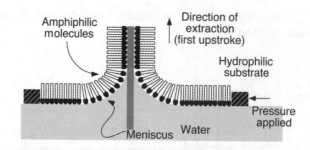

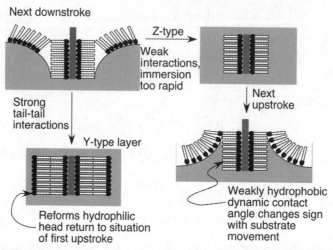

Figure Exercise 5.3   *The formation of X, Y and Z multilayer films.*

5.4   The sticking coefficientis defined as

$$s = r_{ads}/Z_w \qquad (5.12.1)$$

and represents the probability of a successful adsorption event. The collision frequency in solution is given by

$$Z_w = c_{sol} \left(k_B T / 2\pi m \right)^{1/2} \qquad (5.12.2)$$

where c_{sol} is the concentration in molecules per cubic metre. The initial sticking coefficient of $CH_3(CH_2)_7SH$ on a gold film is 9×10^{-8}. Assuming a constant sticking coefficient, which is valid only at low coverage, estimate the time required to achieve a coverage of 0.01 ML for adsorption from a 5×10^{-3} mol l^{-1} solution. Take the surface density of atoms to be 1×10^{19} m^{-2}.

Concentration must be converted

$$c_{sol} = \left(5 \times 10^{-3} \text{ mol dm}^{-3}\right) \left(10^3 \frac{dm^3}{m^3}\right) (6.02 \times 10^{23} \text{ mol}^{-1}) = 3.01 \times 10^{24} \text{ m}^{-3}$$

The molecular weight is $M = 0.1463$ kg mol^{-1}. Then substitute R for k_B and Z_w is

$$Z_w = c_{sol} \left(\frac{RT}{2\pi M} \right)^{1/2} = 3.01 \times 10^{24} \left(\frac{8.3145 \times 298}{2\pi \times 0.1463} \right)^{1/2} = 1.56 \times 10^{26} \text{ m}^{-2} \text{ s}^{-1}$$

The time comes from $\theta\sigma_0 = sZ_wt$

$$t = \frac{\theta\sigma_0}{sZ_w} = \frac{(0.01)(1 \times 10^{19})}{(9 \times 10^{-8})(1.56 \times 10^{26})} = 7.1 \times 10^{-3}\ \text{s}$$

The sticking coefficient of $CH_3(CH_2)_7SH$ is ≈ 1 for gas phase $CH_3(CH_2)_7SH$. From the solution, the sticking coefficient is significantly smaller because (1) solvent molecules must be displaced from the surface and (2) solvent molecules must be displaced from the solvation shell around the $CH_3(CH_2)_7SH$.

5.5   Your lab partner has prepared two Si crystals but has not labelled them. One is H-terminated, the other is terminated with an oxide layer. Propose and explain an experiment you could perform in your kitchen that would distinguish the two.

   The oxide-terminated surface is an OH-terminated surface and is hydrophilic. The H-terminated surface is hydrophobic. They can be identified simply by placing a drop of water on them. The surface that is wetted is OH-terminated whereas the H-terminated surface causes the water to form droplets.

5.6   Explain the observed trend that C_4 straight-chain amphiphile generally do not form LB films or SAMs that exhibit a structure that is as well ordered as that of C_{12} straight-chain amphiphiles.

   Whereas the interaction of the head group anchors the molecule to the substrate, the ordering is in large part due to the non-covalent interactions between the chains. The chain–chain interactions in the C_4 amphiphiles are not sufficiently strong to overcome the disordering effects of thermal excitations. The longer C_{12} chains have stronger chain–chain interactions because of the greater number of CH_2 groups that can interact with each other. These interactions are strong enough to overcome the destabilizing influence of other forces.

5.7   Describe what would occur during vertical deposition of an LB film if the barriers of the trough were stationary and a large surface area substrate was used.

   The trough is a shallow rectangular tray filled with the liquid substrate. The film is prepared by carefully pipetting a dilute solution of the film-forming material onto the substrate. After evaporation of the solvent the sweep is moved towards the float and the force necessary to maintain the position of the float is measured by a torsion balance. Three types of behavior can be described:

- *Gaseous films*: In films with very large surface area per molecule ($>100\ \text{Å}^2$), the film behaves like an ideal gas. The films can be expanded or contracted without phase transitions. The gaseous film can be regarded as a dilute two-dimensional solution of the film-forming material and the substrate.
- *Liquid film*: There exists a certain degree of cooperative interaction between the film-forming molecules. Two types have been observed: liquid expanded and liquid condensed films. The first type of film can be characterized by high compressibility and the absence of islands. These films show a first-order "liquid-gas" phase transition. Condensed films are formed by compressing expanded films.
- *Solid films*: The closest packing of the film-forming material is realized if the surface pressure is further increased.

   As molecules deposit on the substrate, the number of molecules in the Langmuir film decreases. If the sweep is not moved towards the float to decrease the surface area and maintain a constant areal density, the density will drop. As the density drops the film will undergo phase transitions in the reverse order to those listed above, and become progressively less well ordered.

5.8   After a 4 h exposure to pure, deoxygenated H_2O, a H-terminated Si(111) surface is found to have an oxygen atom coverage of 0.6 ML measured with respect to the number of Si atoms in the

Si(111)–(1 × 1) layer. Estimate the sticking coefficient, (i) assuming that all of the oxygen is the result of dissociative H_2O adsorption and (ii) assuming that all of the oxygen results from the adsorption of OH^-.

(a) First calculate the concentration of water in H_2O.

$$c = \left(\frac{1.00\,\text{g cm}^{-3}}{18\,\text{g mol}^{-1}} \right) 6.02 \times 10^{23}\,\text{mol}^{-1} \left(\frac{100\,\text{cm}}{1\,\text{m}} \right)^3 = 3.34 \times 10^{28}\,\text{m}^{-3}$$

Next, calculate the impingement rate of water onto the surface using Eq. (5.12.2)

$$Z_w = c \left(\frac{k_B T}{2\pi m} \right)^{1/2} = (3.34 \times 10^{28}) \left(\frac{1.381 \times 10^{-23}(300)}{2\pi (18)(1.661 \times 10^{-27})} \right)^{1/2}$$

$$Z_w = 3.34 \times 10^{28} (630.1)(18)^{-1/2} = 4.96 \times 10^{30}\,\text{m}^{-2}\,\text{s}^{-1}$$

Convert time to seconds $t = 4(3600) = 14400\,\text{s}$

Finally, estimate the sticking coefficient (assuming it to be independent of coverage) from

$$\theta \sigma_0 = s Z_w t$$

$$s = \frac{\theta \sigma_0}{Z_w t} = \frac{0.6(7.83 \times 10^{18})}{(4.96 \times 10^{30})(14400)} = 6.58 \times 10^{-17}$$

(b) The concentration of OH^- in water at 25°C is found from $pK_w = 13.9965$, which is as good as saying 14 and $[OH^-] = 10^{-7}\,\text{mol l}^{-1}$. Therefore

$$c = 1 \times 10^{-7}(6.02 \times 10^{26}) = 6.02 \times 10^{19}\,\text{m}^{-3}$$

$$Z_w = 6.02 \times 10^{19}(630.1)(17)^{-1/2} = 9.20 \times 10^{21}\,\text{m}^{-2}\,\text{s}^{-1}$$

and

$$s(OH^-) = \frac{Z_w(H_2O)}{Z_w(OH^-)} s(H_2O) = \frac{4.96 \times 10^{30}}{9.20 \times 10^{21}} 6.58 \times 10^{-17} = 3.54 \times 10^{-8}.$$

5.9 Calculate the energy in eV released when H^+ (aq) reacts with a dangling bond to form an Si–H bond on an otherwise hydrogen-terminated surface. You will need the following: the ionization potential of H atoms $IP(H) = 13.61$ eV; the enthalpy of solvation of protons $\Delta_{solv}H(H^+) = -11.92$ eV and the Si–H bond strength $D(Si–H) = 3.05$ eV. Assume that the enthalpy of solvation of the Si dangling bond on an otherwise H-terminated surface is the same as the enthalpy of solvation of the Si–H unit that is formed.

Take as the zero of energy a H atom infinitely separated from a H-terminated Si surface in an aqueous solution. Call this $E_0 = 0$. This surface has one dangling bond.

If the H atom is allowed to bind to this site, 3.05 eV is liberated. Call the energy of this state $E_1 = -3.05$ eV.

Now consider a state where a solvated H^+ atom is placed in solution with the surface considered in defining E_0. This corresponds to leaving the electron of the H atom infinitely far away from the solution while placing the H^+ ion in the solution. The energy, E_2, is then given by E_0 plus the ionization potential of the H atom plus the heat of solvation of the H^+ ion.

$$E_2 = E_0 + IP(H) + \Delta_{solv}H(H^+) = 0 + 13.61 - 11.92 = 1.69\,\text{eV}$$

5.10   In UHV surface science the "usual" first-order pre-exponential factor is 1×10^{13} s^{-1}. In electrochemistry the "usual" pre-exponential factor for electron transfer at a metal electrode surface is 1×10^4 m s^{-1}. Explain how these values are derived.

In transition state theory, the rate constant for a unimolecular reaction is written

$$k = \frac{k_B T}{h} \frac{q_{\ddagger}}{q_A} e^{-E_a/RT} = A e^{-E_a/RT}$$

For first-order (unimolecular) desorption, if the partition function of the activated complex $q_{\ddagger}$ does not deviate much from the partition function of the partition function of the reactant q_A; then,

$$\frac{q_{\ddagger}}{q_A} \approx 1 \quad \text{and} \quad A_{des} = \frac{k_B T}{h} \frac{q_{\ddagger}}{q_A} = 6.2 \times 10^{12} \text{ s}^{-1}.$$

With rounding off, the "usual" pre-exponential factor is found.

Electron transfer at an electrode surface can be thought of as a first-order adsorption process. The reactant is free to translate in three dimensions in the solution whereas in the adsorbed activated complex it only has two degrees of translational freedom. Thus, assuming no other changes in the partition functions

$$\frac{q_{\ddagger}}{q_A} = \frac{h}{(2\pi m k_B T)^{1/2}}.$$

Therefore, if we take $T = 298$ K m = mass of water, and assume a non-activated process,

$$A_{electrode} = \frac{k_B T}{h} \frac{h}{(2\pi m k_B T)^{1/2}} = 1.5 \times 10^4 \text{ m s}^{-1}$$

which is the usual approximation for an "electrochemical" pre-exponential factor.

5.11   When Si is placed in HF(aq) a dark current of about 0.4 µA cm^{-2} is measured, i.e. a current that is measured in the absence of an applied potential and in the absence of illumination. Assuming that this dark current is due to an etching reaction which is initiated by the absorption of F$^-$ (aq) and that one electron is injected into the Si substrate for every one Si atom that is removed by etching, calculate the etch rate of unilluminated Si in HF(aq).

If we assume the injection of one electron from a F$^-$ ion per etched Si atom, the dark current is related to the etch rate (in Si atoms removed cm^{-2} s^{-1}) by the Faraday and Avogadro constants according to

$$R_{dark} = j_{dark} N_A / F.$$

$$R_{dark} = \frac{j_{dark} N_A}{F} = \frac{(0.4 \,\mu\text{A cm}^{-2})(6.022 \times 10^{23} \text{ mol}^{-1})}{96490 \,\text{C mol}^{-1}} = 2.5 \times 10^{12} \text{ cm}^{-2} \text{ s}^{-1}.$$

This etch rate can be converted into a linear etch rate in nm s^{-1} by first calculating the atomic density of Si ρ_A from the mass density ρ according to

$$\rho_A = \frac{N_A \rho}{M} = \frac{(6.022 \times 10^{23} \text{ mol}^{-1})(2.328 \text{ g cm}^{-3})}{28.0855 \text{ g mol}^{-1}} = 4.992 \times 10^{22} \text{ atoms cm}^{-3}$$

Then we convert the etch rate

$$\frac{R_{dark}}{\rho_A} \left(10^7 \frac{\text{nm}}{\text{cm}}\right) = \frac{2.5 \times 10^{12} \text{ cm}^{-2}\text{s}^{-1}}{4.99 \times 10^{22} \text{ cm}^{-3}} \left(10^7 \frac{\text{nm}}{\text{cm}}\right) = 5.0 \times 10^{-4} \text{ nm s}^{-1}.$$

5.12 Derive the Nernst equation for the Fe^{2+}/Fe^{3+} redox couple.

The half-reaction is

$$Fe^{3+} + e^- \rightleftharpoons Fe^{2+}$$

where the electron is in the electrode of this half-cell. The electrochemical potential of the electron $\bar{\mu}$ is given by

$$\bar{\mu} = \mu + qU$$

where μ is the chemical potential of the electron, q is the charge and U the local electrostatic potential. The electrochemical potentials of the iron ions are related to their standard electrochemical potentials and their activities by

$$\bar{\mu}_{Fe^{2+}} = \bar{\mu}^0_{Fe^{2+}} + k_B T \ln a_{Fe^{2+}}$$

$$\bar{\mu}_{Fe^{3+}} = \bar{\mu}^0_{Fe^{3+}} + k_B T \ln a_{Fe^{3+}}.$$

At equilibrium the electrochemical potentials of both sides of the half-reaction are equal

$$\bar{\mu}_{Fe^{3+}} + \bar{\mu} = \bar{\mu}_{Fe^{2+}}.$$

Therefore,

$$\bar{\mu}^0_{Fe^{3+}} + k_B T \ln a_{Fe^{3+}} + \bar{\mu} = \bar{\mu}^0_{Fe^{2+}} + k_B T \ln a_{Fe^{2+}}$$

$$\bar{\mu} = \bar{\mu}^0_{Fe^{2+}} + k_B T \ln a_{Fe^{2+}} - \bar{\mu}^0_{Fe^{3+}} - k_B T \ln a_{Fe^{3+}}$$

$$\bar{\mu} = \bar{\mu}^0_{Fe^{2+}} - \bar{\mu}^0_{Fe^{3+}} + k_B T \ln \frac{a_{Fe^{2+}}}{a_{Fe^{3+}}}.$$

The Nernst equation for this redox couple is

$$U_{redox}(Fe^{2+}/Fe^{3+}) = E^0_{redox}(Fe^{2+}/Fe^{3+}) + \frac{k_B T}{e} \ln \frac{a_{Fe^{3+}}}{a_{Fe^{2+}}}.$$

Now make the identification that

$$\bar{\mu} = e U_{redox}(Fe^{2+}/Fe^{3+}).$$

and

$$\bar{\mu}^0 = \bar{\mu}^0_{Fe^{2+}} - \bar{\mu}^0_{Fe^{3+}} = E^0_{redox}(Fe^{2+}/Fe^{3+}).$$

5.13 The dipole moments of O(a) and OH(a) on Pt(111) are 0.035 and 0.05 e Å, respectively. Calculate the shift in the adsorption energy caused by the interaction of the electric field in the double layer with the dipoles when the surface is biased at 1 V relative to the point of zero charge.

The dipole moment of OH(a) is $D(OH) = 0.05$ e Å, that of O(a) is $D(O) = 0.035$ e Å. The bias potential at the surface is $U = 1$ V. The typical double layer width is $d = 3$ Å. Hence the electric field in the double layer is

$$\mathcal{E} = \frac{U}{d} = \frac{1\,V}{3\,Å} = 0.33\,V\,Å^{-1}$$

The shift in energy of the dipole is given by the product of the electric field and the dipole moment:

$$\text{OH(a): } \Delta E(\text{OH}) = \mathcal{E}D = (0.3\,\text{V Å}^{-1})(0.05\,\text{e Å}) = 0.017\,\text{eV}$$

$$\text{O(a): } \Delta E(\text{O}) = \mathcal{E}D = (0.33\,\text{V Å}^{-1})(0.035\,\text{e Å}) = 0.012\,\text{eV}$$

5.14 Calculate the radius of curvature and discuss capillary condensation in a conical pore that has surfaces with (i) $\psi = 90°$ and (ii) $\psi = 180°$.

For pore walls that exhibit a nonzero contact angle, the radius of curvature of the meniscus of the condensed liquid is decreased as compared to the radius of the pore according to

$$r = r_c / \cos\psi$$

or better for this exercise $r\cos\psi = r_c$.

As $\psi \rightarrow 90°$, $\cos\psi \rightarrow 0$, and $r_c \rightarrow 0$. In other words, at the point where the contact angle is 90°, condensation should no longer occur. As the surface becomes even more hydrophobic, the radius of the bubble again takes on a finite value, but with the opposite sign. The bubble is now convex instead of concave, and is now a droplet. If we can neglect gravity, the spherical droplet would be suspended in the pore. This, of course, begs the question of how the droplet formed in the first place, since its formation would never nucleate on the perfectly hydrophobic pore wall. If, however, a droplet were first formed outside of the pore and then rolled into the pore, it would suspend itself at the point where it fitted snugly into the pore where $r = r_c$.

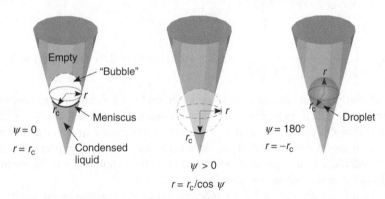

Figure Exercise 5.14   *Capillary condensation for $\psi = 90°$ and $\psi = 180°$.*

5.15 Consider a material with cylindrical pores exposed to air at 25°C with a humidity of 85%. Into pores of what size will water condense?

According to Eq. (5.5.2)

$$r_1 = -\frac{\gamma V_m}{RT\ln(p/p^*)} = -\frac{(0.072\,\text{Jm}^{-2})(18 \times 10^{-6}\,\text{m}^3\text{mol}^{-1})}{(8.31\,\text{JK}^{-1}\text{mol}^{-1})(298\,\text{K})\ln(0.85)}$$

$$r_1 = 3.22\,\text{nm}$$

5.16 Calculate the effective pressure resulting from capillary forces and the critical film thickness for a porous silicon film with a porosity $\varepsilon = 0.90$ when dried in air after rinsing in water or ethanol. The mean pore diameter is $r_p = 5\,\text{nm}$. $\gamma_{\text{EtOH}} = 22.75$ mN m^{-1}, $\gamma_{\text{water}} = 71.99$ mN m^{-1}, $\gamma_{\text{Si}} = 1000$ mN m^{-1}, $E_{\text{Si}} = 1.62 \times 10^{11}$ N m^{-2}.

The effective pressure is given by Eq. (5.5.3)

$$\Delta p = \gamma / r_p.$$

Thus for ethanol and water, respectively, we have

$$\Delta p(\text{EtOH}) = \frac{22.75 \times 10^{-3}\,\text{N}\,\text{m}^{-1}}{5 \times 10^{-9}\,\text{m}} = 4.55 \times 10^6\,\text{Pa}$$

$$\Delta p(\text{H}_2\text{O}) = \frac{71.99 \times 10^{-3}\,\text{N}\,\text{m}^{-1}}{5 \times 10^{-9}\,\text{m}} = 1.44 \times 10^7\,\text{Pa}.$$

Note that in the case of water this corresponds to

$$\Delta p(\text{H}_2\text{O}) = \frac{1.44 \times 10^7\,\text{Pa}}{101\ 325\,\text{Pa/atm}} = 142\,\text{atm},$$

that is, a pressure of over 140 times normal atmospheric pressure!

The critical thickness is given by Eq. (5.5.6)

$$h_c = (r_p / \gamma_L)^2 E_S (1 - \varepsilon)^3 \gamma_S.$$

Ethanol

$$h_c = \left(\frac{5 \times 10^{-9}\,\text{m}}{22.75 \times 10^{-3}\,\text{N}\,\text{m}} \right)^2 (1.62 \times 10^{11}\,\text{N}\,\text{m})(1 - 0.9)^3 (1.00\,\text{N}\,\text{m}) = 7.83\,\mu\text{m}.$$

Water

$$h_c = \left(\frac{5 \times 10^{-9}\,\text{m}}{71.99 \times 10^{-3}\,\text{N}\,\text{m}} \right)^2 (1.62 \times 10^{11}\,\text{N}\,\text{m})(1 - 0.9)^3 (1.00\,\text{N}\,\text{m}) = 0.781\,\mu\text{m}.$$

The significantly higher surface tension of water leads to a much greater pressure being exerted on the pores' walls which, in turn, means that a porous film of a much smaller thickness (one order of magnitude thinner) will be subjected to cracking when rinsed in water and then dried in air as compared to rinsing in ethanol and drying in air. Or put another way, films of equal thickness are subjected to significantly less pressure when rinsed with ethanol, and will be much less susceptible to drying-induced modifications and damage.

5.17   Show that the relative absorption $\Gamma_j^{(1)}$ is independent of the position of the Gibbs surface.

Take component 1 to be the solvent with one or more solutes denoted by the components j. From Eq. (5.7.6), the number of moles of component j is

$$n_j^S = n_j - c_j^A V^A - c_j^B V^B. \tag{5.E7.1}$$

The number of moles of solvent is

$$n_1^S = n_1 - c_1^A V^A - c_1^B V^B. \tag{5.E7.2}$$

Note that

$$V^A = V - V^B. \tag{5.E7.3}$$

Thus, we can eliminate V^A by substitution

$$n_1^S = n_1 - c_1^A(V - V^B) - c_1^B V^B = n_1 - c_1^A V + (c_1^A - c_1^B)V^B \tag{5.E7.4}$$

and similarly for all other components

$$n_j^S = n_j - c_j^A V + (c_j^A - c_j^B)V^B. \tag{5.E7.5}$$

Only V^B depends on the choice of the position of the Gibbs surface. Hence, if we can eliminate V^B, then we should obtain an equation that is invariant upon choice of the Gibbs surface. Multiplying Eq. (5.7.2) by $\left(c_j^A - c_j^B\right)/\left(c_1^A - c_1^B\right)$, we obtain

$$n_1^S \frac{\left(c_j^A - c_j^B\right)}{\left(c_1^A - c_1^B\right)} = \left(n_1 - c_1^A V^A - c_1^B V^B\right) \frac{\left(c_j^A - c_j^B\right)}{\left(c_1^A - c_1^B\right)}, \tag{5.E7.6}$$

which we subtract from Eq. (5.E7.5) to obtain

$$n_j^S - n_1^S \frac{\left(c_j^A - c_j^B\right)}{\left(c_1^A - c_1^B\right)} = n_j - c_j^A V + \left(c_j^A - c_j^B\right) V^B - \left(n_1 - c_1^A V^A - c_1^B V^B\right) \frac{\left(c_j^A - c_j^B\right)}{\left(c_1^A - c_1^B\right)}.$$

This simplifies to

$$n_j^S - n_1^S \frac{\left(c_j^A - c_j^B\right)}{\left(c_1^A - c_1^B\right)} = n_j - c_j^A V - \left(n_1 - c_1^A V\right) \frac{\left(c_j^A - c_j^B\right)}{\left(c_1^A - c_1^B\right)}. \tag{5.E7.7}$$

The right side does not depend on the position of the Gibbs surface, therefore, neither does the left side.

If we now take the left side of Eq. (5.16.7), which we have shown to be invariant, and divide by the surface area A, we will arrive at an expression for the relative absorption of component j with respect to the solvent 1, as was to be shown.

$$\left[n_j^S - n_1^S \frac{\left(c_j^A - c_j^B\right)}{\left(c_1^A - c_1^B\right)}\right] \frac{1}{A} = \frac{n_j^S}{A} - \frac{n_1^S}{A} \frac{\left(c_j^A - c_j^B\right)}{\left(c_1^A - c_1^B\right)} = \Gamma_j^S - \Gamma_j^S \frac{\left(C_j^A - C_j^B\right)}{\left(C_1^A - C_1^B\right)} = \Gamma_j^{(1)} \tag{5.E7.8}$$

5.18 Derive Eq. (5.7.19).

According to Eq. (5.7.14)

$$U^S = TS^S + \sum \mu_j n_j^S + \gamma A$$

Differentiation yields

$$dU^S = T\, dS^S + S^S\, dT + \sum \mu_j\, dn_j^S + \sum n_j^S\, d\mu_j + \gamma\, dA + A\, d\gamma.$$

We now equate this expression to Eq. (5.7.13)

$$dU^S = T\, dS^S + \sum \mu_j\, dn_j^S + \gamma\, dA$$

from which we find

$$T\, dS^S + S^S\, dT + \sum \mu_j\, dn_j^S + \sum n_j^S\, d\mu_j + \gamma\, dA + A\, d\gamma = T\, dS^S + \sum \mu_j\, dn_j^S + \gamma\, dA.$$

and therefore $S^S\, dT + \sum n_j^S\, d\mu_j + A\, d\gamma = 0$.

At constant temperature $dT = 0$ and dividing through by A, we obtain

$$d\gamma = -\sum n_j^S \, d\mu_j.$$

5.19 The Langmuir isotherm also holds for adsorption from solution onto a solid substrate. Recognizing that the pressure p of a gas is related to concentration c, it is straightforward to rewrite the isotherm as

$$\theta = \frac{Kc}{1 + Kc}. \tag{5.12.3}$$

The specific surface area a_s (area per unit mass) is used to characterize high surface area solids such as activated charcoal and porous materials. Therefore, coverage is also conveniently expressed in terms of quantities expressed per unit mass of the substrate. Thus, fractional coverage is equal to the ratio of the number of molecules adsorbed per gram of adsorbent N_s divided by the number of sites per gram of adsorbent ($=$ the number of molecules required to form a monolayer per gram of adsorbent N_m). More conveniently yet, we convert number of molecules to moles (e.g. $n_s = N_s/N_A$) and

$$\theta = n_s/n_m. \tag{5.12.4}$$

Use Eq. (5.12.3) and (5.12.4) to show that

$$\frac{c}{n_s} = \frac{1}{n_m K} + \left(\frac{1}{n_m}\right) c. \tag{5.12.5}$$

$$\theta = \frac{n_s}{n_m} = \frac{Kc}{1 + Kc}$$

$$n_s(1 + Kc) = Kcn_m$$

$$\frac{c}{n_s} = \frac{1}{Kn_m} + \frac{1}{n_m} c.$$

Therefore, a plot of c/n_s vs c should be linear with an intercept of $(1/Kn_m)$ and a slope of $(1/n_m)$. The moles of molecules required to form a monolayer per gram of adsorbent is $(1/\text{slope})$. The temperature dependence of K will be related to the enthalpy of adsorption.

14

Answers to Exercises from Chapter 6. Heterogeneous Catalysis

6.1 Confirm Eq. (6.1.1).

Dispersion is defined as the ratio of the number of surface atoms N_{surf} to the total number of atoms N. The particle is a sphere of surface area A, radius r and volume V. We find N_{surf} by relating it to A, r and the radius of a single atom r_{atom} according to

$$A_{sp} = 4\pi r^2 = N_{surf}\, \pi r_{atom}^2$$

$$N_{surf} = \frac{4\pi r^2}{\pi r_{atom}^2} = \frac{4r^2}{r_{atom}^2}$$

The volume of the sphere is $\quad V = \frac{4}{3}\pi r^3$

The volume of one atom is the volume of the sphere divided by the number of atoms, thus

$$V/N = \frac{4}{3}\pi r_{atom}^3$$

The radii can then be written

$$r = \left(\frac{3V}{4\pi}\right)^{1/3} \text{ and } r_{atom} = \left(\frac{3V}{4\pi N}\right)^{1/3}$$

Substitute into the expression for N_{surf}

$$N_{surf} = \frac{4(3V/4\pi)^{2/3}}{(3V/4\pi N)^{2/3}} = 4N^{2/3}$$

The total number of atoms is

$$N = mN_A/M$$

where m is the mass, N_A is Avogadro's constant and M the molar mass. Thus, from the definition of dispersion

$$D = N_{surf}/N = 4N^{2/3}/N = 4N^{1/3} = 4(M/mN_A)^{1/3}$$

Surface Science: Foundations of Catalysis and Nanoscience, Third Edition. Kurt W. Kolasinski.
© 2012 John Wiley & Sons, Ltd. Published 2012 by John Wiley & Sons, Ltd.

The mass is the density times the volume,

$$m = \frac{4\pi}{3}\rho r^3.$$

Thus, $D = 4\left(\dfrac{3M}{4\pi\rho r^3 N_A}\right)^{1/3} = \left(\dfrac{(4)^3 3M}{4\pi\rho N_A}\right)^{1/3}\dfrac{1}{r} == \left(\dfrac{48M}{\pi\rho N_A}\right)^{1/3}\dfrac{1}{r}.$

6.2   Explain with the aid of a one-dimensional potential energy diagram why the Haber-Bosch synthesis of NH_3 is less energy intensive and more rapid than the homogeneous process.

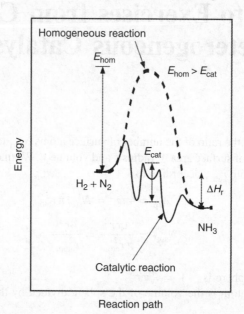

Figure Exercise 6.2   *The effect of a catalyst on a reaction energy profile.*

The catalyst does not change the overall thermodynamics of the reaction; therefore, the heat released (or consumed) overall does not change. However, the activation energy required for the catalytic process is much lower than for the homogeneous reaction. In fact, the activation energy of the industrial process is virtually zero. Therefore, the energy required is not required to break the N−N bond (the rate limiting step); rather, the energy required for the Haber-Bosch process is used to heat the gases and catalyst (required to keep the catalyst clean and active) and to compress the gases (required to increase the rate of formation and increase the equilibrium concentration of the product).

6.3   Show that the following expressions hold for the equilibrium surface coverages during ammonia synthesis.

$$\theta_{N_2*} = K_1\frac{p_{N_2}}{p_0}\theta_* \tag{6.14.1}$$

$$\theta_{N*} = \frac{p_{NH_3}p_0^{0.5}}{K_3 K_4 K_5 K_6 K_7^{0.5} p_{H_2}^{1.5}}\theta_* \tag{6.14.2}$$

$$\theta_{NH^*} = \frac{p_{NH_3}}{K_4 K_5 K_6 p_{H_2}} \theta_* \tag{6.14.3}$$

$$\theta_{NH_2} = \frac{p_{NH_3}}{K_5 K_6 K_7^{0.5} p_{H_2}^{0.5} p_0^{0.5}} \theta_* \tag{614.4}$$

$$\theta_{NH_3} = \frac{p_{NH_3}}{K_6 p_0} \theta_* \tag{6.14.5}$$

$$\theta_{H^*} = K_7^{0.5} \frac{p_{H_2}^{0.5}}{p_0^{0.5}} \theta_* \tag{6.14.6}$$

In the reaction mechanism embodied in seven Rxns (6.3.2)–(6.3.8), we take six of the steps to be equilibrated, while the rate of the rate limiting step Rxn (6.3.3) is

$$r_2 = k_2 \theta_{N_2*} \theta_* - k_{-2} (\theta_{N*})^2 \tag{6.4.19}$$

For the equilibrated steps, we need simply to rearrange Eqs (6.4.21)–(6.4.26), to arrive at the desired equilibrium coverages such that the coverage of each intermediate is expressed in terms only of θ_* and experimentally accessible values (pressures and equilibrium constants). The three equilibrium coverage expressions are easy because they follow directly from the equilibrium constant expressions

$$\theta_{N_2*} = K_1 (p_{N_2}/p_0) \theta_* \text{ from (6.4.21)}$$

$$\theta_{NH_3*} = \frac{(p_{NH_3}/p_0)}{K_6} \theta_* \text{ from (6.4.25)}$$

$$\theta_{H*} = K_7^{0.5} (p_{H_2}/p_0)^{1/2} \theta_* \quad \text{from (6.4.26)}$$

Next, we solve for θ_{NH_2*} because its equilibrium constant expression involves only θ_{NH_3*} and θ_{H*}, for which we already have expression from above.

$$\theta_{NH_2*} = \frac{\theta_{NH_3*}}{K_5 \theta_{H*}} \theta_* \text{ from (6.4.24)}$$

Now substitute for θ_{NH_3*} and θ_{H*}

$$\theta_{NH_2*} = \frac{\dfrac{(p_{NH_3}/p_0)}{K_6} \theta_*}{K_5 K_7^{0.5} (p_{H_2}/p_0)^{1/2} \theta_*} \theta_* = \frac{(p_{NH_3}/p_0)}{K_5 K_6 K_7^{0.5} (p_{H_2}/p_0)^{1/2}} \theta_*$$

Similarly

$$\theta_{NH*} = \frac{\theta_{NH_2*}}{K_4 \theta_{H*}} \theta_* \text{ from (6.4.23)}$$

Then substitute for θ_{NH_2*} and θ_{H*},

$$\theta_{NH*} = \frac{\dfrac{(p_{NH_3}/p_0)}{K_5 K_6 K_7^{0.5} (p_{H_2}/p_0)^{1/2}} \theta_*}{K_4 K_7^{0.5} (p_{H_2}/p_0)^{1/2} \theta_*} \theta_* = \frac{p_{NH_3}}{K_4 K_5 K_6 p_{H_2}} \theta_*$$

And finally

$$\theta_{N*} = \frac{\theta_{NH*}}{K_3 \theta_{H*}} \theta_* \text{ from (6.4.22)}$$

Substituting

$$\theta_{N*} = \frac{\frac{p_{NH_3}}{K_4 K_5 K_6 p_{H_2}}\theta_*}{K_3 K_7^{0.5}(p_{H_2}/p_0)^{1/2}\theta_*}\theta_* = \frac{p_{NH_3}(p_0)^{1/2}}{K_3 K_4 K_5 K_6 K_7^{0.5}(p_{H_2})^{3/2}}\theta_*$$

The above corrects two typographical errors in the original manuscript (Stoltze and Nørskov, *J. Catal.* 110 (1988) 1) in Eqs (6.4.26) and (6.14.2).

6.4   Consider the dissociative adsorption of N_2 through a molecularly bound state

$$N_2(g) + {}^* \rightleftharpoons N_2{-}^* \tag{i}$$

$$N_2^{-*} + {}^* \rightleftharpoons 2N^{-*} \tag{ii}$$

Show that the rate of formation of adsorbed N atoms is

$$\frac{d\theta_{N*}}{dt} = 2k_2 K_1 \frac{p_{N_2}}{p_0}\left(\frac{1-\theta_{N*}}{1+K_1 p_{N_2}/p_0}\right)^2 - \frac{2k_2}{K_2}\theta_{N*}^2 \tag{6.14.7}$$

Assume that the system is at equilibrium.

The aim is to obtain a differential equation for θ_{N*} that contains only constants and θ_{N*} terms. The differential equation can then be integrated to obtain θ_{N*} as a function of time. First write the net rate equation for θ_{N*}

$$\frac{d\theta_{N*}}{dt} = 2k_2\theta_{N_2*}\theta_* - 2k_{-2}\theta_{N*}^2.$$

The θ_{N_2*} term can be eliminated by noting that since the reactions are equilibrated

$$K_1 = \frac{\theta_{N_2*}}{(p_{N_2}/p^\circ)\theta_*}$$

and $\theta_{N_2*} = K_1(p_{N_2}/p^\circ)\theta_*.$

We use conservation of sites to eliminate the θ_* term.

$$\theta_* = 1 - \theta_{N*} - \theta_{N_2*} = 1 - \theta_{N*} - K_1(p_{N_2}/p^\circ)\theta_*$$

$$\theta_* = \frac{1-\theta_{N*}}{1+K_1(p_{N_2}/p^\circ)}$$

Rate constants require knowledge of the transition state, which are difficult under the best of circumstances to address computationally, and at surfaces are even more difficult than for small molecule reactions. Equilibrium constants, on the other hand, depend on ground state properties of the reactants and products, which are much easier to evaluate. Thus, if we want to compare our rate expression to experimentally and computationally accessible quantities, we want to minimize the number of rate constants in the final expression. We can reduce the number of rate constants that need to be determined to one by noting that

$$K_2 = k_2/k_{-2}$$

and substituting for k_{-2}. Hence,

$$\frac{d\theta_{N*}}{dt} = 2k_2 K_1 (p_{N_2}/p^\circ)\theta_* \theta_* - \frac{2k_2}{K_2}\theta_{N*}^2.$$

$$\frac{d\theta_{N*}}{dt} = 2k_2 K_1 \left(\frac{p_{N_2}}{p^\circ}\right)\left(\frac{1-\theta_{N*}}{1+K_1(p_{N_2}/p^\circ)}\right)^2 - \frac{2k_2}{K_2}\theta_{N*}^2.$$

As was to be shown.

6.5   A number of heterogeneous reactions, including NH_3 formation, exhibit a maximum in the rate in the middle of the row of transition metals. Plots of reactivity versus Periodic Chart group number with such a maximum are called volcano plots because of their shape. Discuss the origin of this trend for ammonia synthesis.

The volcano plot of catalytic activity refers to the catalytic efficacy across a row of transition metals in the periodic table. Activity tends to be the highest in the center of a row (for e.g. hydrogenation reactions) and falls off to either side. The reason has to do with the strength of the chemisorption interaction of the reactants on the surface. The chemisorption interactions have to be strong enough to allow the reactants to stick to the surface in a sufficient coverage. This could mean either that the adsorption energy must be high enough (for a molecular reagent) or that the barrier to the dissociation of the gas-phase species is lowered by a strong chemisorption interaction for the fragments. However, if chemisorption of the reactants is too strong, they will not go on to react into the product. They will simply remain chemisorbed. Hence there is a balance between weak and strong interactions, which is best met in the middle of the row.

6.6   The following results are observed in co-adsorption studies of CO, C_6H_6 or PF_3 with K on Pt(111). (i) CO + K/Pt(111): the CO–Pt bond is significantly strengthened. (ii) C_6H_6, which adsorbs molecularly at low T_s, is less likely to dissociate upon heating in the presence of co-adsorbed K. (iii) The amount of PF_3 that can adsorb drops roughly linearly with θ_K. Explain these observations. If the rate determining step of a catalytic reaction is (a) CO dissociation or (b) C_6H_6 desorption, how does the presence of K affect the reaction rate?

The adsorption of K on a transition metal surface is accompanied by substantial electron transfer from K to the metal. This lowers the work function and effectively "frees up" electrons to interact with adsorbates.

(i) The Blyholder model of CO adsorption tells us that the CO chemisorption bond has two components: donation of electrons from the CO into the metal and backdonation of metal electrons into the CO $2\pi*$ antibonding level. K adsorption will effectively add more electrons to the surface, and these electrons will be available for backdonation. Increased backdonation leads to an increased CO-substrate bond strength while weakening the $C\equiv O$. This will facilitate CO dissociation, that is the rate of CO dissociation will increase and since this is the rate-limiting step in (a), the overall reaction rate will increase.

(ii) Molecular benzene tends to donate electrons from the π system. The presence of co-adsorbed K atoms will lower the capacity of the surface to accept electrons and, therefore, decrease the strength of the π bonding interaction. A weaker benzene/surface interaction will favour desorption of molecular benzene over dissociation. If benzene desorption is the rate limiting step, as in (b), then the presence of K acts as a promoter.

(iii) PF_3 is an electron donor. Therefore, we can expect that it chemisorbs by donating electron density from its lone pair to the surface. This interaction is weakened by K because the electrons donated by K tend to saturate the surface's ability to take on more electrons. K coverage reduces the coverage of PF_3 in a linear fashion, that is, a constant number of adsorption sites that could have been used for PF_3 adsorption is being consumed by each adsorbed K atom. The K atoms are acting as site blockers.

6.7   The three-way catalytic converter used in automobiles catalyzes the oxidation of unburned hydrocarbons and CO while reducing simultaneously NO to N_2. Consider only the reactions of H_2 and NO, Eqs (6.5.5) and (6.5.7). The desired products are N_2 and H_2O. NH_3 formation must be suppressed. Write out a complete set of elementary reactions for which H_2O, N_2 and NH_3 are the final products. Pinpoint particular elementary steps that are decisive for determining the selectivity of the catalyst for N_2 over NH_3.

$$NO(g) \rightarrow NO(a)$$

$$H_2(g) \rightarrow 2H(a)$$

$$NO(a) \rightarrow N(a) + O(a)$$

$$2N(a) \rightarrow N_2(g) \quad ***$$

$$O(a) + H(a) \rightarrow OH(a)$$

$$OH(a) + H(a) \rightarrow H_2O(a)$$

$$H_2O(a) \rightarrow H_2O(g)$$

$$N(a) + H(a) \rightarrow NH(a) \quad ***$$

$$NH(a) + H(a) \rightarrow NH_2(a)$$

$$NH_2(a) + H(a) \rightarrow NH_3(a)$$

$$NH_3(a) \rightarrow NH_3(g)$$

The key elementary reactions are labeled ***. The proper catalyst will suppress NH(a) formation in favour of $N_2(g)$ formation.

6.8   A reaction is carried out on a single crystal Pt(111) sample. The reaction requires the scission of a C–H bond. The reaction is completely poisoned by just a few percent of a monolayer of preadsorbed oxygen at low temperature, but at high temperature, oxygen coverage up to about a tenth of a monolayer has little effect on the reaction rate. Discuss the poisoning behaviour of pre-adsorbed oxygen.

Consider the following scenario. O atoms preferentially adsorb at step sites. Step sites are known to be particularly active for C–H bond scission. Diffusion of O atoms from the steps onto the terraces is thermally activated. The surface of a real Pt(111) crystal is likely to have a few percent (or fewer) of step sites. Thus at low temperature, the O blocks the active step sites. At high temperature, the O atoms easily diffuse away from the step sites. This frees up the active sites and reduces the poisoning effect of the pre-adsorbed O atoms.

6.9 The $NO + CO \rightarrow \frac{1}{2}N_2 + CO_2$ reaction is a useful reaction in an automotive catalyst. It follows Langmuir-Hinshelwood kinetics. Consider the following to be a complete set of reactions for this system.

$$NO(g) \rightleftharpoons NO(a)$$

$$CO(g) \rightleftharpoons CO(a)$$

$$NO(a) \rightleftharpoons N(a) + O(a)$$

$$2N(a) \rightleftharpoons \frac{1}{2}N_2(g)$$

$$CO(a) + O(a) \rightarrow CO_2(g)$$

(a) A trace of S weakens the NO–surface bond. If S co-adsorption substantially affects only NO(a), explain what the effect of the presence of a trace of S(a) has on the rate of production of N_2 and CO_2.

(b) Assume that K co-adsorption promotes N_2 dissociation but affects no other reaction. Does K co-adsorption change the rate of CO_2 production? Explain.

(c) NO only dissociates on the steps of Pt(111). Assume that the resulting O atoms are immobile at 770 K – the reaction temperature in a typical automatic catalytic converter. Given that CO desorption is first order and characterized by $E_d = 135\,kJ\,mol^{-1}$, $A = 3.00 \times 10^{13}\,s^{-1}$ and that CO diffusion is characterized by $E_{dif} = 19.8\,kJ\,mol^{-1}$, $D_0 = 5.00 \times 0^{-6}\,cm^2\,s^{-1}$, would it be wise to use Pt(111) single crystals with terrace widths of 3500 Å to eliminate CO from the exhaust gas? Justify your answer with calculations.

The essential requirement for an effective three-way catalyst is high conversions of NO_x (NO and N_2O), CO and hydrocarbons (HC). The overall catalytic reactions that are important for controlling exhaust emissions are given by the following equations (grouped more or less by importance):

$$CO + 1/2\,O_2 \rightarrow CO_2$$

$$\text{hydrocarbons} + O_2 \rightarrow H_2O + CO_2$$

$$H_2 + 1/2\,O_2 \rightarrow H_2O$$

$$NO + CO \rightarrow 1/2\,N_2 + CO_2$$

$$NO + H_2 \rightarrow 1/2\,N_2 + H_2O$$

$$\text{hydrocarbons} + NO \rightarrow N_2 + H_2O + CO_2$$

$$NO + 5/2\,H_2 \rightarrow NH_3 + H_2O$$

$$CO + H_2O \rightarrow CO_2 + H_2$$

$$\text{hydrocarbons} + H_2O \rightarrow CO + CO_2 + H_2$$

$$3\,NO + 2\,NH_3 \rightarrow 5/2\,N_2 + 3\,H_2O$$

$$2\,NO + H_2 \rightarrow N_2O + H_2O$$

$$2\,N_2O \rightarrow 2\,N_2 + O_2$$

$$2\,NH_3 \rightarrow N_2 + 3\,H_2$$

The desired products are N_2, CO_2 and H_2O. Typical reaction (exhaust) temperatures are about 770 K. The selectivity of the three-way catalyst is that it promotes CO and HC oxidation rather than H_2 oxidation and promotes NO reduction rather than H_2 oxidation. Furthermore, NO should be reduced to N_2 rather than NH_3. Hence, high efficiency as well as selectivity are important.

While the composition of the three-way catalyst varies from one manufacturer to the next, the basic formula is a mixture of Rh, Pt and Pd dispersed over an Al_2O_3 substrate. Cerium oxide (CeO_2) is often added as a promoter.

Rh is the catalyst chosen to promote the reduction of NO to N_2. Rh promotes the steam reforming of exhaust HC. Rh also contributes to the oxidation of CO. Unfortunately it is also the most expensive component.

Pt is used for its contribution to CO and HC conversions, especially during the warm-up of converters from a cold start.

Pd is also used for the conversion of CO and HC.

(a) If S weakens the NO–surface bond (1) the N–O bond of NO(a) on the S-covered surface will be stronger than on the clean surface, and (2) the NO adsorption energy is lowered, which lowers the coverage of NO because desorption becomes easier. Hence, NO dissociation is less likely on the S-covered surface. Both effects decrease the rate of production of N(a) and O(a), and decrease the rate of production of $N_2(g)$ and $CO_2(g)$.

(b) If K co-adsorption promotes $N_2(g)$ dissociation, this will increase the amount of N(a) on the surface relative to the clean surface case. This will decrease the rate of $CO_2(g)$ formation because N(a) will block more sites on the K-covered surface.

(c) The average lifetime on the surface is given by

$$\tau = A^{-1} \exp(E_d/RT)$$

The average distance travelled on the surface due to diffusion, d, is given by

$$d = \langle x^2 \rangle^{1/2} = (4D\tau)^{1/2}$$

where

$$D = D_0 \exp(-E_{\text{dif}}/RT).$$

Therefore

$$d = (A^{-1} \exp(E_d/RT) \times 4D_0 \exp(-E_{\text{dif}}/RT))$$

$$d = \left[\frac{4D_0}{A} \exp\left(\frac{E_d - E_{\text{dif}}}{RT} \right) \right]^{1/2}$$

$$d = \left[\frac{4(5 \times 10^{-6}\,\text{cm}^2\text{s}^{-1})}{3 \times 10^{13}\text{s}^{-1}} \exp\left(\frac{135\,000 - 19\,800\,\text{J mol}^{-1}}{8.31\text{JK}^{-1}\text{mol}^{-1}(770\,\text{K})} \right) \right]^{1/2}$$

$$d = 6.62 \times 10^{-6}\,\text{cm} = 662\,\text{Å}$$

Since 662 Å (the mean distance travelled by diffusing CO molecules) is less than 3500 Å/2 = 1750 Å (the mean distance away from a step at which randomly adsorbing CO molecules land), the reaction probability is low and the Pt(111) crystals would not be a good choice for a catalyst.

6.10 For methanation of CO over Ni, CO dissociation is not the rate determining step. K co-adsorption increases the CO dissociation rate. Explain how in this case K acts as a poison. Hint: Consider the net effect on H_2 adsorption.

Methanation of CO is the reaction of CO with H_2 to form methane.

$$CO(g) + 3H_2(g) \rightarrow CH_4(g) + H_2O(g)$$

CO if CO dissociation is too fast compared to H_2 adsorption and H_2 is unable to adsorb on the sites covered by O(a) and C(a), in which case K enhances the production of site blockers. Since CO dissociation is not rate limiting, then a step involving H atoms (either dissociative adsorption of H_2 or a bond formation reaction with adsorbed H atoms) must be rate limiting. By reducing the rate of H_2 adsorption, the rate of the rate limiting step will also be reduced by K co-adsorption.

6.11 Consider the electrocatalytic oxidation of CO with water over a Ru/Pt bimetallic catalyst to form $CO_2 + H^+ + e^-$. Write out a set of elementary steps in which the water dissociates on Ru sites, CO adsorbs on Pt sites and the reaction of OH + CO to form $CO_2 + H^+ + e^-$ occurs on a Pt cluster.

$$2Ru^* + H_2O \rightarrow Ru\text{–}H + Ru\text{–}OH$$

$$^*Pt^* + CO \rightarrow {}^*Pt\text{–}CO$$

$$Ru\text{–}OH + {}^*Pt\text{–}CO \rightarrow OH\text{–}Pt\text{–}CO + Ru^*$$

$$OH\text{–}Pt\text{–}CO \rightarrow H\text{–}Pt^* + CO_2$$

$$Ru\text{–}H \rightarrow Ru^* + H^+ + e^-$$

(alternatively H(a) could migrate to $H\text{–}Pt^*$)

$$H\text{–}Pt^* \rightarrow {}^*Pt^* + H^+ + e^-$$

$$\overline{\hspace{7cm}}$$

$$H_2O + CO \rightarrow CO_2 + 2H^+ + 2e^-$$

6.12 3-nitrostyrene ($NO_2\text{–}C_6H_4\text{–}CH{=}CH_2$, 3NS) was treated with H_2 at 120°C in the presence of a gold nanoparticle catalyst supported on titania (1.5 wt % Au/TiO_2) for 6 h. The primary product was 3-vinylaniline ($NH_2\text{–}C_6H_4\text{–}CH{=}CH_2$, 3VA). Gas chromatography with mass spectrometric (GC-MS) analysis of the product stream reveals that the peak corresponding to 3NS has been reduced from its initial value (normalized to 1.00) to a peak that is 50 times smaller. The peak corresponding to 3VA is 48 times larger than the final 3NS peak. Assuming that the peak areas have been corrected for sensitivity factors and that the peak areas are, therefore, proportional to the molar ratio (or equivalently the concentrations), calculate the % selectivity and % conversion. That is, calculate the % of 3NS that is selectively hydrogenated to the desired product (3VA) and calculate the % of 3NS that reacts to form any and all products.

The % selectivity (%*S*) is defined as the number of moles of desired product n_p (or concentration at constant volume or partial pressure for gases at constant temperature and constant volume) divided by the number of moles of the limiting reagent n_r. In order to calculate this fraction correctly, the stoichiometry of the reaction must also be considered. In this case,

$$NO_2\text{–}C_6H_4\text{–}CH{=}CH_2(l) + 3\,H_2(g) \rightarrow NO_2\text{–}C_6H_4\text{–}CH{=}CH_2(l) + 2\,H_2O(g)$$

The 1:1 mole ratio of reactant (3NS) and product (3VA) in this case makes the calculation simpler as the stoichiometric coefficients of the reactant $v_r = 1$ and of the product $v_p = 1$.

$$\%S = \frac{n_p/v_p}{n_{r,i}/v_r}(100\%)$$

$n_{r,i}$ is the initial number of moles of the reactant as opposed to its final value $n_{r,f}$. % Conversion (%C) is simply related to the initial number of moles of reactant divided by the final number of moles of reactant times 100%

$$\%C = \left(1 - \frac{n_{r,f}}{n_{r,i}}\right)(100\%)$$

The GC-MS analysis reveals that $n_{r,i} = 50\,n_{r,f}$ and $n_p = 48\,n_{r,f}$. Therefore,

$$\%S = \frac{48n_{r,f}}{50n_{r,f}}(100\,\%) = 96\%$$

$$\text{and } \%C = \left(1 - \frac{n_{r,f}}{50n_{r,f}}\right)(100\%) = 98\%.$$

6.13   A chemical reaction is found to have a well-defined rate determining step which involves the breaking of a N–O bond in NO. Explain how a transition metal surface can effectively act as a heterogeneous catalyst.

To speed up the reaction, we need to speed up the rate determining step (RDS). The RDS can be accelerated if its activation energy is lowered. The activation energy is related in this case to the energy required to break a molecular bond. The transition metal surface can do this by chemisorbing the molecule and weakening or breaking this bond. As long as the resulting adsorbates are not bound too tightly, they should have a higher reactivity and the transition metal surface will act as an effective catalyst.

15
Answers to Exercises From Chapter 7. Growth and Epitaxy

7.1 Dunstan has shown that there is a linear dependence of the $In_x Ga_{1-x}As$ lattice constant on the lattice constants of its constituents according to

$$a_{In_x Ga_{1-x}As} = x\, a_{InAs} + (1-x)\, a_{GaAs}. \tag{7.16.1}$$

Substitute this dependence into Eq. (7.1.3) and derive Eq. (7.1.4). From Eq. (7.1.3)

$$\varepsilon_0 = \frac{a_l - a_s}{a_s} = \frac{x\, a_{InAs} + (1-x)\, a_{GaAs} - a_{GaAs}}{a_{GaAs}} = \frac{x\, a_{InAs} + a_{GaAs} - x\, a_{GaAs} - a_{GaAs}}{a_{GaAs}}$$

$$\varepsilon_0 = \frac{x\, a_{InAs} - x a_{GaAs}}{a_{GaAs}} = \left(\frac{a_{InAs}}{a_{GaAs}}\right)x - x = \left(\frac{6.06\,Å}{5.65Å}\right)x - x = 0.0726x$$

7.2 Consider a system that for a given set of conditions exhibits step-flow growth. Discuss the effects that the adsorption of heteroatoms can have on homoepitaxial growth. Consider two low heteroatom coverage cases: (a) the heteroatoms decorate the steps, and (b) the heteroatoms occupy isolated terrace sites.

In ideal step flow growth, the adatoms have sufficient mobility that when they land on the terrace, they diffuse to a step. The adatoms also have a sufficiently strong interaction with the step atoms that they are effectively bound there, and are not released back onto the terrace. The diffusion to the step must occur faster than the nucleation of islands on the terrace.

(a) If the heteroatoms decorate the steps, they do not change the mobility on the terrace, that is, they do not affect the diffusion barrier for motion to the steps. However, they may change the diffusion barrier for release of atoms from the step. If the heteroatoms bind strongly to the incoming adatoms and increase the barrier to release, they will capture incoming diffusing adatoms and become covered. They have no effect on the growth mechanism. However, if they lower the diffusion barrier for adatoms to be released from the step, the presence of heteroatoms at the step will change the growth mode from step flow to island growth once the net rate of addition of atoms to the steps decreases below the rate of island nucleation.

Surface Science: Foundations of Catalysis and Nanoscience, Third Edition. Kurt W. Kolasinski.
© 2012 John Wiley & Sons, Ltd. Published 2012 by John Wiley & Sons, Ltd.

(b) Isolated heteroatoms on the terraces have no effect upon the release rate from the steps. However, they may act as nucleation centres. If the heteroatoms are effective at providing nucleation sites, island growth will be preferred to step flow growth. If the heteroatoms are ineffectual at providing nucleation sites, they will not change the growth mode.

7.3   Si is the most important semiconductor for electronic applications. GaAs and its III-V sister compounds are better suited than Si as building blocks for optical devices such as light emitting diodes (LEDs) and lasers. The integration of optical components with electronics is a highly desirable manufacturing goal for improved communications, computing and display devices. Discuss fundamental physical reasons why it is difficult to integrate GaAs circuitry with Si.

GaAs assumes a different crystal structure than Si because it has two different types of atoms. Forcing GaAs to assume a diamond lattice with the same spacing as Si would result in a great deal of strain because of the different atomic sizes of Ga and As compared to Si. One would expect a growth mode (Volmer-Weber) in which 3D islands form from the start. Therefore, circuitry that requires epitaxial growth of smooth layers will be difficult to achieve.

7.4   Discuss how Auger electron spectroscopy or XPS can be used to distinguish Frank-van der Merwe from Volmer-Weber growth. Hint: Look at Fig. 2 and consider how the substrate signal varies.

In the FM growth mode , the substrate is covered in a layer-by-layer fashion. In VM growth, 3D islands form. For both AES and XPS, the substrate signal is attenuated with increasing growth due to inelastic scattering of the photoelectrons in the adsorbed layer. In FM growth, the signal is the sum of substrate atoms covered by n adsorbate layers plus the signal of substrate atoms covered by $n-1$ atoms. The resulting attenuation is linear in adsorbate coverage for each layer and will show a break in the slope of a plot of the intensity of the substrate signal versus the intensity of the adsorbate signal when the adsorbate coverage reaches 1 ML. Progressively weaker break points are observed as each successive layer is completed.

In WM growth, the substrate signal is composed of a series of terms from bare surface + covered by one layer + covered by two layers + ... The slope of a plot of substrate versus adsorbate signal decreases monotonically (without break points) and does not decrease as rapidly as for FM growth.

7.5   (a) Consider the epitaxial growth by MBE of $In_{0.67}Al_{0.33}P$ layer on an InGaAs substrate. What must the relative fluxes of In, Al and P be in order to maintain this composition? What influence does the substrate temperature have on epitaxy and the required fluxes?

(b) Consider the CVD growth of P-doped (at a concentration of $10^{16}\,cm^{-3}$) $Si_{(1-x)}Ge_x$ with $x = 0.05$ from the respective hydrides. Discuss the influence of surface temperature on epitaxy and the fluxes required to maintain this composition.

(a) In MBE, the pure elements In, Al and P are evaporated from solid sources. The temperature of the evaporators is adjusted to achieve the required flux. In this case, a relative flux of In: Al: P of 0.67: 0.33: 0.75 will yield approximately the correct film composition because the sticking coefficients of the three components are all ≈ 1. Slight corrections to the relative fluxes might have to be made to correct for deviations from unit sticking coefficients.

The temperature of the substrate is adjusted to allow for adequate diffusion. If T_s is too low, the diffusion rate will not be sufficiently high for epitaxy to occur. Higher deposition rates demand greater diffusion and, thus, higher T_s. However, T_s does not have an effect on the required relative fluxes. An exception is if T_s is raised so high that one component, e.g. In, begins to desorb. The temperature must also not be raised so high that interdiffusion (diffusion of adlayers into the substrate) begins to occur.

(b) In CVD of these layers, SiH_4, GeH_4 and PH_3 must adsorb and decompose at the surface. Si, Ge and P then must diffuse sufficiently to grow epitaxially. H atoms must desorb as H_2 from the surface, otherwise they will block sites and effectively poison the decomposition reactions.

The adsorption of SiH_4 and GeH_4 is activated and, therefore, T_s dependent. The growth rate is a complicated function of T_s and the fluxes, since adsorption as well as desorption and diffusion must all be considered.

It is impossible for any arbitrary T_s to say what exactly the relative fluxes must be to achieve the desired composition without a calculation of the kinetics. In the limit of high temperature such that the rate of desorption of H_2 is equal to or greater than the rate of adsorption of any of the hydrides and the sticking coefficients have reached their saturation values (which should be close to unity), then the flux of SiH_4 to GeH_4 should be roughly 0.95: 0.05 (subject to correction for differences in the sticking coefficients) and the flux of PH_3 will be $\approx 10^{16}/10^{23} \approx 10^{-7}$ of the SiH_4 flux. (10^{23} cm^{-3} is the typical order of magnitude for the density of a solid.)

7.6   The dimensionless formation energy $E(V)$ of a single-faceted quantum dot as a function of its dimensionless volume is given by

$$E(V) = -\alpha V + \frac{2\beta V^{2/3}}{e^{1/2}} - 2V^{1/3}\ln(e^{1/2}V^{1/3}) \qquad (7.16.1)$$

The chemical potential of an island is given by

$$\mu(V) = \frac{dE(V)}{dV} \qquad (7.16.2)$$

Assuming that $\alpha = 0$, predict the most probable island volume for $\beta = 1.36$, 0.7 and -0.7.

The most probable island volume is given when the chemical potential is minimized. Taking the derivative yields

$$\mu(V) = \frac{dE(V)}{dV} = -\alpha + \frac{4\beta}{3e^{1/3}}V^{-1/3} - \frac{2}{3}V^{-2/3}\ln(e^{1/2}V^{1/3}) - \frac{2}{3}V^{-2/3}$$

The fourth term is missing from the original reference. This function is plotted in Fig. Ex. 7.6 below and is identical to Fig. 7.18. It shows that the minimum moves to greater values of V as β increases.

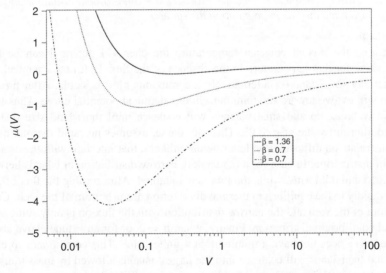

Figure Exercise 7.6   *Plots of chemical potential versus the size of a quantum dot for different values of β.*

7.7   The incident flux can be used to tune the chemical potential of a system of islands on a surface. Predict what occurs to the island size distribution when $\beta = -0.7$ and the flux is turned off for a system with a chemical potential of (a) $+1$, (b)-3, (c)-4 given the functional form of μ vs V given in Fig. 7.18. How does the size distribution evolve when the flux is turned off for $\beta = 2$ and $\mu = 1$?

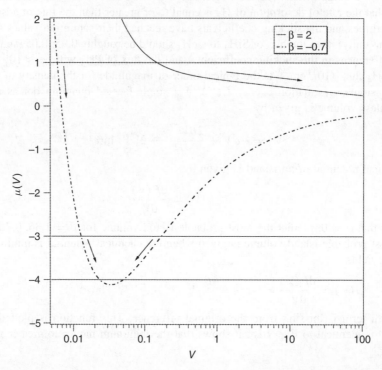

Figure Exercise 7.7 *Plots of chemical potential versus size are used to demonstrate island size distributions change in response to varying the flux incident upon the surface.*

By adjusting the flux at constant temperature the chemical potential can be raised above its minimum value for that temperature in the absence of the flux. In (a) the chemical potential is held so high that any size island (or alternatively a continuous film) is stable. After turning off the flux, the system will evolve toward the minimum in the chemical potential curve. Thus large islands will spontaneously break up and small islands will coalesce until the island size distribution focuses near the equilibrium value of ~ 0.03. This, of course, assumes no rapid thermal quenching and no kinetic constraints on diffusion, the latter being unlikely, that interfere with the establishment of the equilibrium distribution. In case (b) a moderately narrow distribution of islands between volumes of roughly 0.015 and 0.2 forms, while the flux is maintained. After turning the flux off, the distribution will relax rapidly to the equilibrium thermal distribution at the bottom of the well. Case (c) lies close to the bottom of the well and the narrow distribution with the flux on is only somewhat broader than the thermal equilibrium distribution. Finally, when $\beta = 2$, or for any value above about 1.65 for that matter, the curve does not have a minimum at a finite value. This corresponds to classical Ostwald ripening and the islands will coalesce into the largest islands allowed by mass transport constraints.

7.8   Determine the orientation of the pore walls formed on a Si(100) wafer given that they are straight and that their orientation with the respect to the {110} planes is as shown in Fig. 7.20.

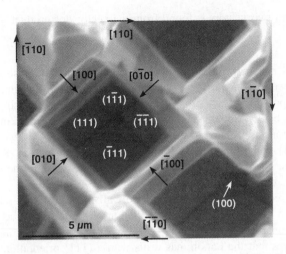

Figure Exercise 7.8 *Crystallographic orientation of macropores obtained by alkaline etching of an initially pillar-covered Si(100) surface.*

Table 7.8 *Angle formed by the intersection of a selection of planes with the (001) and (111) planes 15.1*

Plane	Angle in degrees when intersecting with			Plane	Angle in degrees when intersecting with		
	(001)	(111)	(110)		(001)	(111)	(110)
(001)	–	54.7	90				
(00$\bar{1}$)	180	125.3	90				
(100), (010)	90	54.7	45				
($\bar{1}$00), (0$\bar{1}$0)	90	125.3	135				
(110)	90	35.3		(111)	54.7	–	35.3
(011), (101)	45	35.3		($\bar{1}$11), (1$\bar{1}$1)	54.7	70.5	90
($\bar{1}$10), (1$\bar{1}$0)	90	90		(11$\bar{1}$)	125.3	70.5	35.3
(0$\bar{1}$1), ($\bar{1}$01)	45	90		($\bar{1}$$\bar{1}$1)	54.7	109.5	144.7
(01$\bar{1}$), (10$\bar{1}$)	135	90		($\bar{1}$1$\bar{1}$), (1$\bar{1}$$\bar{1}$)	125.2	109.5	90
($\bar{1}$$\bar{1}$0)	90	144.7		($\bar{1}$$\bar{1}$$\bar{1}$)	125.3	180	144.7
(0$\bar{1}$$\bar{1}$), ($\bar{1}0\bar{1}$)	135	144.7					
(112)	35.3	19.5		(113)	25.2	29.5	
($\bar{1}$12), (1$\bar{1}$2)	35.3	61.9		($\bar{1}$13), (1$\bar{1}$3)	25.2	58.5	
(11$\bar{2}$)	144.7	90		(11$\bar{3}$)	154.8	100	
($\bar{1}$1$\bar{2}$)	35.3	90		($\bar{1}$1$\bar{3}$)	25.2	80	
(1$\bar{1}$$\bar{2}$), ($\bar{1}1\bar{2}$)	144.7	118.1		(1$\bar{1}$$\bar{3}$), ($\bar{1}1\bar{3}$)	154.8	121.5	
(1$\bar{1}$$\bar{2}$)	144.7	160.5		(1$\bar{1}$$\bar{3}$)	154.8	150.5	
(211), (121)	65.9	19.5	30	(311), (131)	72.5	29.5	
($\bar{2}$11), (1$\bar{2}$1)	65.9	90	106.8	($\bar{3}$11), (1$\bar{3}$1)	72.5	100	
(2$\bar{1}$1), ($\bar{1}$21)	65.9	61.9	73.2	(3$\bar{1}$1), ($\bar{1}$31)	72.5	58.5	
(21$\bar{1}$), (12$\bar{1}$)	114.1	61.9	30	(31$\bar{1}$), (13$\bar{1}$)	107.5	58.5	
($\bar{2}$1$\bar{1}$), ($\bar{1}$2$\bar{1}$)	65.9	118.1	150	($\bar{3}$11), ($\bar{1}$31)	72.5	121.5	
(2$\bar{1}$$\bar{1}$), (1$\bar{2}$$\bar{1}$)	114.1	118.1	106.9	(3$\bar{1}$$\bar{1}$), (1$\bar{3}$$\bar{1}$)	107.5	121.5	
(2$\bar{1}$$\bar{1}$), ($\bar{1}2\bar{1}$)	114.1	90	73.2	(3$\bar{1}$$\bar{1}$), ($\bar{1}3\bar{1}$)	107.5	80	
($\bar{2}$1$\bar{1}$), ($\bar{1}$2$\bar{1}$)	114.1	160.5	150	($\bar{3}$1$\bar{1}$), ($\bar{1}$3$\bar{1}$)	107.5	150.5	

The angle between two vectors is found from the vector dot product

$$\mathbf{a} \cdot \mathbf{b} = ab \cos \theta.$$

In the abstract, this can make for a difficult general solution because of the infinite number of (hkl) available. However, once a few relationships between the low index planes are recognized, the combination of possible planes can be greatly limited. The relationships can be recognized from the data in the table below. Squares and 45° angles are associated with the {001} and {011} planes. Triangles and hexagons are related to the {111} system. Look for 54.7° (or its complement) for the intersection of {111} with {001}. Look for 35.3° for the intersection of {111} with {011}. These tend to be the lowest energy planes; thus they are the place to begin the search. If other angles are found, then consider other planes.

In this case, the rectangular shape of the pore wall identifies it as related to the {001} system since it lies at 45° to the [110] direction. Calculation of the dot products confirms the orientations given. A cross sectional image of the pore reveals that the bottom is comprised of walls slanted at 54.7°, which indicates that the bottom has walls with a {111} orientation.

16
Answers to Exercises from Chapter 8. Laser and Nonthermal Chemistry

8.1 Calculate the peak power, time averaged power, fluence and number of photons per pulse for (i) a Nd:YAG laser operating at 532 nm with a pulse energy of 100 mJ, a pulse length of 7 ns and a repetition rate of 10 Hz, and (ii) a Ti:sapphire laser operating at 800 nm with a pulse energy of 10 nJ, a pulse length of 80 fs and a repetition rate of 1 kHz. Assume that the laser beam is focused into a spot with a diameter of 500 μm.

(i) Nd:YAG wavelength 532 nm, pulse energy 100 mJ, pulse length 7 ns, repetition rate 10 Hz.

The photon energy is given by

$$E_{photon} = \frac{hc}{\lambda} = \frac{6.63 \times 10^{-34}\,J\,s(3.00 \times 10^{-34}\,m\,s^{-1})}{532 \times 10^{-9}\,m} = \frac{1.99 \times 10^{-25}\,J\,m}{532 \times 10^{-9}\,m} = 3.74 \times 10^{-19}\,J.$$

The number of photons per pulse is given by

$$N_{photon} = \frac{pulse\ energy}{photon\ energy} = \frac{0.100\,J}{3.74 \times 10^{-19}\,J} = 2.67 \times 10^{17}\ photons\ per\ pulse.$$

The peak power is given by

$$P_{peak} = \frac{pulse\ energy}{pulse\ length} = \frac{0.100\,J}{7 \times 10^{-9}\,s} = 1.43 \times 10^{7}\,J\,s^{-1} = 14.3\,MW.$$

The average power is given by

$$P_{avg} = pulse\ energy(repetition\ rate) = 0.100\,J(10\,s^{-1}) = 1.00\,W.$$

If you look up radiometric quantities, you will generally find the irradiance

$$I_{peak} = peak\ power\ per\ unit\ area = \frac{peak\ power}{\frac{1}{4}\pi d^2}$$

$$I_{peak} = \frac{1.43 \times 10^{7}\,J\,s^{-1}}{\frac{1}{4}\pi(500 \times 10^{-6}\,m)^2} = \frac{1.43 \times 10^{7}\,W}{1.96 \times 10^{-7}\,m^2} = 7.30 \times 10^{13}\,W\,m^{-2}$$

Surface Science: Foundations of Catalysis and Nanoscience, Third Edition. Kurt W. Kolasinski.
© 2012 John Wiley & Sons, Ltd. Published 2012 by John Wiley & Sons, Ltd.

and photon flux

$$\Phi_{peak} = \text{photons per unit area} = \frac{2.67 \times 10^{17} \text{ photons}}{1.96 \times 10^{-7} \text{ m}^2} = 1.36 \times 10^{24} \text{ photons m}^{-2}$$

defined either for peak (as done here) or time averaged values. However, in the literature of pulsed laser, you will usually find the energy fluence referenced. In this case, the fluence is

$$F_{peak} = \text{pulse energy per unit area} = \frac{0.100 \text{ J}}{1.96 \times 10^{-7} \text{ m}^2} = 5.10 \times 10^5 \text{ J m}^{-2}.$$

Note also that a laser beam has an intensity profile that can vary across the beam. An excimer laser has a (close to) top hat profile, that is, the intensity is even across the width of the beam and then falls abruptly to zero. A Nd:YAG laser will usually have a Gaussian intensity profile. Most ultrafast lasers aspire to Gaussian profiles but usually deviate substantially on a pulse-to-pulse basis. There may also be a significant pedestal of lower intensity and longer pulse width on which the femtosecond pulses ride. In the case of nonconstant intensity profiles, all of the per unit area quantities represent either some sort of average over the laser spot profile or else are quoted for the maximum value at the centre of the beam.

(ii) Ti:sapphire wavelength 800 nm, pulse energy 10 nJ, pulse length 80 fs, repetition rate 1000 Hz. The photon energy is given by

$$E_{photon} = \frac{1.99 \times 10^{-25} \text{ J m}}{800 \times 10^{-9} \text{ m}} = 2.49 \times 10^{-19} \text{ J}.$$

The number of photons per pulse is given by

$$N_{photon} = \frac{10 \times 10^{-9} \text{ J}}{2.49 \times 10^{-19} \text{ J}} = 4.02 \times 10^{10} \text{ photons per pulse.}$$

The peak power is given by

$$P_{peak} = \frac{10 \times 10^{-9} \text{ J}}{80 \times 10^{-15} \text{ s}} = 1.25 \times 10^5 \text{ J s}^{-1} = 0.125 \text{ MW.}$$

The average power is given by

$$P_{avg} = 10 \times 10^{-9} \text{ J}(1000 \text{ s}^{-1}) = 10 \text{ μW.}$$

The fluence is

$$F_{peak} = \frac{10 \times 10^{-9} \text{ J}}{1.96 \times 10^{-7} \text{ m}^2} = 5.10 \times 10^{-2} \text{ J m}^{-2}.$$

8.2   A 1 nW cw HeNe laser operating at 633 nm is incident on a surface. Describe qualitatively the distribution of excited carriers near the surface for (a) an *n*-type doped Si substrate with a 100 nm space charge layer, (b) a *p*-type doped Si substrate with a 100 nm space charge layer, and (c) a metal substrate.

(a) A 1 nW cw beam is sufficiently low in intensity that it hardly heats the sample. From Fig. 8.1 it is clear that the penetration depth of the light ($= 1/\alpha$) is much deeper than the width of the space charge layer. The electrons absorb the visible (red light) laser irradiation. Since the photon energy is greater than the band gap of Si, electrons are excited from the valence band to the conduction band, leaving behind excited holes in the valence band. The minimum excited electron energy corresponds to the band gap energy. The maximum excited electron energy corresponds to the energy of the valence band maximum plus the energy of the photon, which is somewhat above the conduction band minimum.

(b) The difference between (a) and (b) is that the bands are bent in opposite directions for *n*-type and *p*-type doped Si. Immediately after the excited electrons and holes are created in the space charge layer, they are forced to move by the electric field associated with band bending . In *n*-type material, the holes are forced to the surface and the electrons are forced into the bulk of the Si sample. This leads to an excess concentration of holes at the surface. In *p*-type material, the electrons are forced to the surface and the holes are forced into the bulk. The *p*-type material will have an excess population of electrons at the surface.

The imbalance of charge leads to an electric field that counterbalances the electric field associated with the space charge region. If the laser intensity is high enough (much higher than in this example), the electric field created by the excited carriers will totally counteract the electric field in the absence of laser excitation. This is known as band flattening.

(c) The penetration depth of the light into the metal is only a few nanometres. The metal does not experience band bending. Therefore, the excited carriers only move according to diffusion. There will be no imbalance of the free charge created by excitation of the metal. There will be a distribution of hot carriers, i.e. electrons excited above the Fermi energy and holes excited below the Fermi energy; however these excited carriers will have a very short lifetime. The great majority of the Fermi-Dirac distribution will only be modestly distorted from the thermal equilibrium distribution.

8.3   Given that the Maxwell number density speed distribution is

$$N(v)\, dv = \frac{4b^{3/2}}{\pi^{1/2}} v^2 \exp(-bv^2)\, dv \qquad (8.9.1)$$

and that the Maxwell flux distribution of velocities is obtained by multiplying $N(v)$ by v

$$f(v)\, dv = vN(v)\, dv, \qquad (8.9.2)$$

show that the TOF distribution measured by a flux detector is given by

$$I_{flux}(t)\, dt = a_f \frac{L^4}{t^5} \exp(-b(L/t)^2)\, dt \qquad (8.9.3)$$

and that the TOF distribution measured by a density detector is given by

$$I_{den}(t)\, dt = a_d \frac{L^4}{t^4} \exp(-b(L/t)^2)\, dt. \qquad (8.9.4)$$

Note that $b = m/2k_B T$ and a_f and a_d are normalization constants.

The flux TOF distribution is given by

$$g(t) = f(v)\frac{dv}{dt}$$

Since $v = \frac{L}{t}$ and $\frac{dv}{dt} = -\frac{v}{t^2}$,

therefore $dv = -\frac{L}{t^2}\, dt$.

Substituting for $f(v)$ from Eq. (8.9.2), for dv from the line above, and then for $N(v)$ from Eq. (8.9.1) yields

$$g(t)\ dt = vn(v)\left(-\frac{L}{t^2}\right)\ dt = \left(\frac{L}{t}\right)\frac{4b^{3/2}}{\pi^{1/2}}\left(\frac{L}{t}\right)^2\left(-\frac{L}{t^2}\right)\exp(-bv^2)\ dt$$

$$I_{\text{flux}} = g(t) = a_{\text{f}}\left(\frac{L^4}{t^5}\right)\exp(-bv^2)$$

$$I_{\text{den}} = t\ I_{\text{flux}} = a_{\text{d}}\left(\frac{L^4}{t^4}\right)\exp(-bv^2)$$

8.4   Explain why the temperature of the electrons in a metal excited by a 100 fs pulse significantly exceeds the phonon temperature in the subpicosecond time scale.

Photons are absorbed by excitation of the electrons to form electron-hole pairs. The absorption of each photon leads to the formation of one electron-hole pair. This essentially excites a rectangular chunk out of the initial Fermi-Dirac distribution of electrons into excited states.

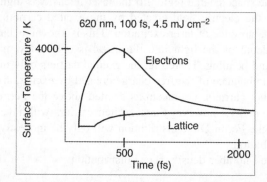

Figure Exercise 8.4   *Approximate temporal profiles of the electron temperature T_e and the phonon temperature T_p at the surface of a metal exposed to a 100 fs visible light pulse with a fluence of 4.5 mJ cm².*

This rectangular distribution relaxes by electron–electron collisions. Most commonly, excited electrons collide with electrons near the Fermi energy, but at very high laser powers, sufficiently high excited state densities may be created so that collisions between excited electrons will also occur. The electron–electron relaxation time is on the order of a few fs, therefore, the rectangular distribution quickly (on the order of 100–500 fs) relaxes to a thermalized distribution. This distribution is characterized by a temperature of typically $> 10^3$ K, but is not in equilibrium with the lattice. The temperature of the electron distribution will greatly exceed that of the phonons (lattice) because the heat capacity of the electrons is much lower than that of the phonon. It is the electrons that initially absorb the energy, and electron–electron collisions are much more frequent and efficient at transferring energy than electron–phonon collisions . The electron–phonon relaxation time is on the order of tens of fs (material dependent). The phonons begin to get heated for times longer than about 100 fs. The electrons and phonons equilibrate on a time scale of several ps. The temperature of the electron distribution goes through a maximum, then approaches the phonon temperature. A rough sketch of the temperature profiles expected for 100 fs pulses with a wavelength of 620 nm and fluence of 4.5 mJ cm^{-2} incident on a metal are given in Fig. Ex. 8.4.

8.5   Describe the difference between a pyrolytic and a photolytic process.

A pyrolytic process is a photothermal process. The photon energy is delivered to the system by resonant absorption but is rapidly transferred to thermal energy by inter- and intramolecular energy transfer.

A photolytic process is a resonant process in which photoreaction occurs before the randomization of the photons' energy by relaxation processes.

8.6 CO was dosed onto Pt(111) at 80 K as in Exercise 3.18. TPD clearly demonstrates that CO adsorbs with near unit sticking coefficient. Using Auger spectroscopy, however, only a small signal near the detection limit is measured for all exposures. Explain this result and the discrepancy between the two methods of CO coverage determination.

CO has a very high cross section for electron stimulated desorption with the >1 keV electrons commonly used for Auger analysis. The electron beam very effectively desorbs CO before it can be detected through AES; hence the surface appears to be clean. One must always pay attention to whether photon or electron beam irradiation is changing the system under observation. For NO on Pt(111), I found that the oxygen AES signal was proportional to the NO coverage. Nevertheless, no N signal was ever apparent in AES because of electron beam-induced dissociation of NO with concurrent desorption of the nitrogen.

TPD analysis yields a signal that is truly proportional to the coverage on the surface (as long as everything desorbs as it does in this case for CO), and delivers the correct result.

8.7 With the assumption that the excitation probabilities are the same for both isotopologues because isotopic substitution does not lead to a change in the structure of the electronic potential energy curves, the excitation probabilities in Eq. (8.2.8) are the same. Show that the ratio of the desorption cross section of one isotope to another is given by

$$\frac{\sigma_d(1)}{\sigma_d(2)} = \left(\frac{1}{P}\right)^{(\sqrt{m_1/m_2}-1)}. \tag{8.9.5}$$

Under the assumption given above, the integral is independent of isotope and can be set equal to a constant.

Therefore,

$$P = \exp\left(-\frac{m^{1/2}}{\sqrt{2}} \int_{z_0}^{z_z} \frac{R(z)\ dz}{\sqrt{V(z_0) - V(z)}}\right) = \exp(-Km^{1/2}).$$

$$\frac{\sigma_1}{\sigma_2} = \frac{\sigma_{ex}P_1}{\sigma_{ex}P_2} = \frac{P_1}{P_2} = \frac{\exp(-K\,m_1^{1/2})}{\exp(-K\,m_2^{1/2})} = \exp(-K\,(m_1^{1/2} - m_2^{1/2}))$$

$$\ln\left(\frac{\sigma_1}{\sigma_2}\right) = -K\,(m_1^{1/2} - m_2^{1/2}) = Km_1^{1/2}\left(\frac{m_2^{1/2}}{m_1^{1/2}} - 1\right)$$

Note that

$$1/P_1 = \exp(Km_1^{1/2}),$$

therefore,

$$\frac{\sigma_1}{\sigma_2} = \exp(Km_1^{1/2})^{(\sqrt{m_2/m_1}-1)} = \left(\frac{1}{P_1}\right)^{(\sqrt{m_2/m_1}-1)}.$$

8.8 Consider (a) H_2O and (b) D_2O adsorbed on Pd(111) and then irradiated with 6.4 eV light. Given the data in the table below, calculate σ_{ex} and P.

	$\sigma_d(H_2O)/\times 10^{-18}$ cm^2	$\sigma_d(D_2O)/\times 10^{-18}$ cm^2
Desorption	0.046	0.033
Dissociation	1.35	0.600

Use the result of Exercise 8.7. For desorption, the effective mass is that of the entire molecule. However, for dissociation of an O–H or O–D bond, the effective mass is that of the H or D atom, respectively. Thus,

$$\frac{\sigma_{des}(H_2O)}{\sigma_{des}(D_2O)} = \left(\frac{1}{P^{des}_{H_2O}}\right)^{(\sqrt{20/18}-1)}$$

$$\text{and } \frac{\sigma_{dis}(H_2O)}{\sigma_{dis}(D_2O)} = \left(\frac{1}{P^{dis}_{H_2O}}\right)^{(\sqrt{2/1}-1)} = \left(\frac{1}{P^{dis}_{H_2O}}\right)^{(\sqrt{2}-1)}$$

To solve for P, you need to take the log of both sides of the equation

$$\log\left(\frac{\sigma_{des}(H_2O)}{\sigma_{des}(D_2O)}\right) = (\sqrt{20/18}-1)\ \log\left(\frac{1}{P^{des}_{H_2O}}\right)$$

$$\frac{-1}{(\sqrt{20/18}-1)}\log\left(\frac{\sigma_{des}(H_2O)}{\sigma_{des}(D_2O)}\right) = \log\left(P^{des}_{H_2O}\right)$$

$$\log(P^{des}_{H_2O}) = \frac{-1}{(\sqrt{20/18}-1)}\log\left(\frac{0.046}{0.033}\right) = -2.66$$

$$P^{des}_{H_2O} = 2.15 \times 10^{-3}$$

$$\log(P^{dis}_{H_2O}) = \frac{-1}{(\sqrt{2}-1)}\log\left(\frac{1.35}{0.6}\right) = -0.850244$$

$$P^{des}_{H_2O} = 0.141$$

The excitation probability is the same for both isotopologues because electronic states are not altered by isotopic substitution. In principle, however, the excitation probabilities may be distinct for the two different photochemical events, because different excitations may be responsible for desorption and dissociation. The excitation probabilities are found to be

$$\sigma^{des}_{ex} = \frac{\sigma_{des}(H_2O)}{P_{des}(H_2O)} = \frac{0.046 \times 10^{-18}\text{ cm}^2}{2.154 \times 10^{-3}} = 21 \times 10^{-18}\text{ cm}^2$$

$$\sigma^{dis}_{ex} = \frac{\sigma_{dis}(H_2O)}{P_{dis}(H_2O)} = \frac{1.35 \times 10^{-18}\text{ cm}^2}{0.141174} = 9.56 \times 10^{-18}\text{ cm}^2.$$

The factor of 2 difference is not large enough to unambiguously indicate whether different excitations are responsible for the two different processes.

The escape probability is, of course, different for H$_2$O versus D$_2$O. Since

$$\frac{\sigma_1}{\sigma_2} = \frac{\sigma_{ex}P_1}{\sigma_{ex}P_2} = \frac{P_1}{P_2},$$

then once we have calculated the escape probability for H_2O, we can calculate that of D_2O from

$$P_2 = \frac{\sigma_2}{\sigma_1} P_1$$

$$P_{dis}(D_2O) = \frac{\sigma_{dis}(D_2O)}{\sigma_{dis}(H_2O)} P_{dis}(H_2O) = \frac{0.6}{1.35}(0.141174) = 6.3 \times 10^{-2}$$

$$P_{des}(D_2O) = \frac{\sigma_{des}(D_2O)}{\sigma_{des}(H_2O)} P_{des}(H_2O) = \frac{0.033}{0.046}(2.154 \times 10^{-3}) = 1.55 \times 10^{-3}.$$

8.9 Consider a CO molecule adsorbed on a metal surface such that its 5σ state lies far below E_F, its $2\pi^*$ state is very close to but slightly above E_F and the $6\sigma^*$ state lies far above E_F. Describe the electron dynamics that lead to two different $5\sigma^{-1}$ state after absorption by the absorbed CO of a photon with an energy that is resonant with the $5\sigma \rightarrow 6\sigma^*$ transition.

There are two scenarios depicted in Fig. Ex. 8.10.

(a) In the first scenario, the initial excitation $5\sigma \rightarrow 6\sigma^*$ takes the system from a $5\sigma\ ^2 2\pi^*\ 6\sigma^{*0}$ configuration to a $5\sigma\ ^1 2\pi^*\ 6\sigma^{*1}$ configuration. The creation of the 5σ hole can pull the $2\pi^*$ level down in energy. If it drops partially below E_F, it will be filled rapidly by an electron from the metal near E_F. Relaxation of the electron from the $2\pi^*$ level into the hole in the 5σ level can eject the other 5σ electron via an Auger-like process in which the electron is lost to the metal.

The system ends in a $5\sigma^1\ 2\pi^*\ 6\sigma^{*1}$ configuration, which corresponds to $5\sigma^{-1} 6\sigma^{*+1}$ in comparison to the original configuration.

(b) In the second scenario, the initial $5\sigma \rightarrow 6\sigma^*$ excitation is followed by relaxation of the $2\pi^*$ partially below E_F and direct filling of the 5σ hole by an electron from the metal near E_F. However, now the simultaneous transition of the other 5σ electron lands in the $2\pi^*$ orbital rather than ionizing the adsorbate. The final configuration corresponds to a $5\sigma^{-1}\ 2\pi^{*+1} 6\sigma^{*+1}$ state.

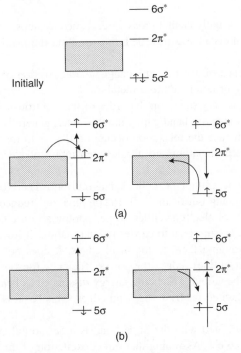

Figure Exercise 8.10 *Electron dynamics that lead to two different $5\sigma^{-1}$ state after resonant photon absorption causes a $5\sigma \rightarrow 6\sigma^*$ transition in CO(a) .*

8.10 (a) Describe the desorption dynamics and the desorbate formed for the process shown in Fig. 8.7(a). (b) Consider now the case in which a curve crossing is made to the M+A$^+$ potential curve. Describe the desorption dynamics and the desorbate formed.

(a) The adsorbate/substrate complex is excited to a neutral excited state. The system then propagates on the excited state potential past the critical distance, which means that the adsorbate has acquired enough energy to allow desorption. The system then relaxes to the ground state potential energy surface (PES). Because the system has returned to the ground electronic state PES with sufficient energy for desorption, the adsorbate A leaves as a neutral particle.

(b) The crossing point to the M+A$^+$ PES occurs after the critical distance, therefore, we might still expect desorption (a caveat below). As the system evolves, the adsorbate will have a certain probability of continuing to propagate along the M+A$^+$ PES, which results in the desorption of an ionic A$^+$ particle. Concurrently, there may also be a certain probability for reneutralization of the particle as it desorbs; hence, a certain fraction of the desorbing particles may appear as neutrals. Note that an ion has an image potential attraction to the surface. This is an additional attraction that the ion must overcome in comparison to a neutral desorbate. Therefore, the critical distance for ion desorption will be slightly larger than for neutral desorption, and we cannot guarantee that the ions will desorb since what is marked in the figure is the critical distance for neutral desorption from the ground state potential. In addition, the time spent close to the surface depends on the component of velocity normal to the surface. Thus, it depends not only on the speed but also on the angle of the velocity vector with respect to the surface normal. Ionic desorbates are deflected toward the surface for non-normal desorption trajectories. Above a critical angle, ionic desorption will not be observed as recapture by the surface becomes overwhelming.

8.11 Describe why vertical manipulation is more versatile than lateral manipulation. Vertical manipulation allows one to move adsorbates arbitrarily long distances in any direction and across any kind of defect.

Lateral manipulation usually cannot cross defects such as steps, and the adsorbate will often hop off the tip after a certain distance is travelled, thus limiting the maximum reliable distance that can be travelled.

8.12 Describe how the presence of a band gap changes the absorption of light and relaxation of charge carriers in a semiconductor compared to a metal.

Intraband relaxation is very rapid, on the order of the electron–electron collision time, which is but a few fs. Relaxation across a band gap is much slower, generally in the ns time scale unless the gap is indirect, in which case the relaxation occurs on the μs to ms time scale.

8.13 Describe how carrier recombination and ultrafast relaxation of electrons affect the efficiency of a solar cell.

Carriers that recombine cannot be collected as current. Therefore, each recombination event decreases the current coming out of the cell. Recombination reduces the quantum efficiency, that is, it reduces the number of electrons collected per photon absorbed.

Ultrafast relaxation does not lead to carrier recombination. It lowers the energy of the electrons and holes, eventually bringing them to the band edges. It does not directly reduce the current or quantum yield. However, it does reduce the power efficiency (power output divided by light power incident), and increase the amount of photon energy dissipated as heat because high-energy photons create carriers that are excited far from the band edges. This "excess" energy is lost to phonons by relaxation.

8.14 The bonding orbital associated with the Si–H bond is a 5σ orbital at -5.3 eV below E_F. It has an intrinsic width of $\Gamma = 0.6$ eV. Assuming that direct excitation of this orbital can be used to form a repulsive state, what is the minimum force and the minimum excited state potential slope required to desorb H from the surface by direct electronic excitation of this orbital. Take the Si–H bond strength to be 3.05 eV.

From Eq. (2.6.11), the lifetime τ is related to the intrinsic width by

$$\tau = h/\Gamma.$$

The energy acquired by the propagation on the repulsive surface is obtained from Eq. (8.2.7), and at a minimum it must equal the bond energy of 3.05 eV.

$$\Delta E = 3.05\,\text{eV} = (F\tau)^2/2m.$$

Therefore, the minimum force required to break the Si–H bond, and desorb the H atom is

$$F = \sqrt{2m\,(3.05\,\text{eV})}(h/\Gamma) = 5.84 \times 10^{-9}\,\text{N}.$$

The minimum required excited state potential gradient is the negative of the force. It can be expressed in the more intuitive units of eV Å^{-1} as

$$\frac{\text{d}V}{\text{d}r} = -F = -5.84 \times 10^{-9}\,\frac{\text{J}}{\text{m}}\left(\frac{1 \times 10^{-10}\,\text{m/Å}}{1.602 \times 10^{-19}\,\text{J/eV}}\right) = -3.65\,\text{eV Å}.$$

8.15   Hydrogen-terminated Si surfaces can be made with a level of perfection that exceeds all other surfaces. When free carriers are made they recombine at defects in the bulk or at the surface in traps. Si crystals can be made with such perfection that for a $d = 200\,\mu\text{m}$ thick sample, recombination of electrons and holes only occurs appreciably at the surface. The surface recombination velocity S is related to the coverage of surface traps σ_{ss}, the recombination cross section σ_{rec} and the electron velocity v by

$$S = \sigma_{ss}\sigma_{rec}v. \tag{8.9.6}$$

The recombination cross section is $\sigma_{rec} \approx 10^{-15}$ cm^2 for nonresonant capture. The velocity is reflective of a thermal value. The recombination velocity of H-terminated Si can be as low as 0.25 cm s^{-1}. Use the relationship between hot carrier lifetime τ and S

$$\frac{1}{\tau} = \frac{1}{\tau_b} + \frac{2S}{d}, \tag{8.9.7}$$

where τ_b is the bulk lifetime, to estimate the coverage of surface traps σ_{ss} and the lifetime of hot carriers τ. How does σ_{ss} compare to the number of surface atoms?

Since we are told that recombination occurs only at the surfaces, bulk recombination is negligible in comparison, and we can assume that $\tau_b \to \infty$. Hence

$$\frac{1}{\tau} \approx \frac{2S}{d} = \frac{2(0.25\,\text{cm s}^{-1})}{0.02\,\text{cm}} = 25\,\text{s}^{-1},$$

and the lifetime is $\tau = 4 \times 10^{-2}$ s.

The thermal velocity of the electrons is approximated from the mean kinetic energy

$$\tfrac{1}{2}mv^2 = \tfrac{3}{2}k_B T$$

hence

$$v = \sqrt{\frac{3k_B T}{m}} = \sqrt{\frac{3(1.38 \times 10^{-23}\,\text{J K}^{-1})(300\,\text{K})}{9.11 \times 10^{-31}\,\text{kg}}} = 1.2 \times 10^5\,\text{m s}^{-1}.$$

The absolute coverage of surface traps is thus

$$\sigma_{ss} = \frac{S}{\sigma_{rec}\,v} = \frac{0.25\,\text{cm s}^{-1}}{(10^{15}\,\text{cm}^{-2})(1.2 \times 10^7\,\text{cm s}^{-1})} = 2.1 \times 10^7\,\text{cm}^{-2}.$$

The fractional coverage is the ratio of the number of surface traps to the number of surface atoms. For the Si(100) surface, there are 6.8×10^{15} cm^{-2}, hence, the ratio of traps to surface atoms, i.e. the coverage in monolayers of traps, is

$$\theta_{ss} = \frac{\sigma_{ss}}{\sigma_0} = \frac{2.1 \times 10^7\,\text{cm}^{-2}}{6.8 \times 10^{15}\,\text{cm}^{-2}} = 3 \times 10^{-8}\,\text{ML}.$$

Note that the lack of surface traps is only part of the answer for superior electrical properties of well-terminated Si surfaces. Banding effects and the accumulation of carriers at the surface in response to the electrochemical potential of a solution in contact with the surface can also play a role [1].

8.16   Direct vibrational excitation with IR lasers is possible. Consider the behaviour of two very different systems. (i) H$_2$ and HD co-adsorbed on LiF. The energy of the $v = 0 \to v = 1$ transition is greater than the energy of the physisorption bond. Absorption of one photon resonant with the HD($v = 0 \to v = 1$) desorbs HD but not H$_2$. (ii) NH$_3$ and ND$_3$ co-adsorbed on Cu. The energy of the $v = 0 \to v = 1$ transition is less than the energy of the chemisorption bond. Absorption of photons resonant with the N–H($v = 0 \to v = 1$) stretching mode of NH$_3$ desorbs both NH$_3$ and ND$_3$. Discuss two different types of desorption dynamics that explain these results.

(i) Resonant absorption of a photon by HD provides the molecule that has absorbed the photon with sufficient energy to overcome the desorption barrier. Therefore, after excitation, if the molecule can transfer the energy from the H–D bond into the HD–LiF chemisorption bond, the molecule will leave the surface via a resonant photodesorption process. Since only HD desorbs, the transfer of energy into the chemisorption bond must be much more efficient than transfer of energy to neighbouring H$_2$ molecules; therefore, none of the H$_2$ molecules desorb.

(ii) Intra-adsorbate energy transfer (from the N–H stretch to the N–Cu stretch) now competes with inter-adsorbate energy transfer (from NH$_3$ to ND$_3$). Several photons need to be adsorbed to cause desorption. The resonance between the photon energy and the vibrational energy creates an efficient pathway to pump thermal energy into the adsorbed layer (both NH$_3$ and ND$_3$), and a resonance-assisted photothermal desorption mechanism causes both types of adsorbates to desorb.

8.17   (a) Show that for electron transfer in which the solvent has a linear response described by the force constant k_0, the transition state occurs at the reaction coordinate

$$q^{\ddagger} = \frac{q_p^2 - q_r^2 + \Delta G / k_0}{2q_p - 2q_r}. \tag{8.9.8}$$

$q^{\ddagger}$ is found by solving for the intersection point between the reactant Gibbs energy curve

$$G_r(q) = k_0(q - q_r)^2.$$

and the product Gibbs energy curve

$$G_p(q) = k_0(q - q_p)^2 + \Delta G.$$

The intersection point is found by equating these two functions when $q = q^{\ddagger}$:

$$k_0(q^{\ddagger} - q_r)^2 = k_0(q^{\ddagger} - q_p)^2 + \Delta G.$$

$$q^{\ddagger 2} - 2q^{\ddagger}q_r + q_r^2 = q^{\ddagger 2} - 2q^{\ddagger}q_p + q_p^2 + \Delta G/k_0$$

Solving for $q^{\ddagger}$ yields the desired result

$$q^{\ddagger} = \frac{q_p^2 - q_r^2 + \Delta G/k_0}{2q_p^- 2q_r}.$$ QED

(b) Show that for such a system the activation Gibbs energy is given by Eq. (8.4.4).
The activation Gibbs energy is found by evaluating either $G_r(q^{\ddagger})$ or $G_p(q^{\ddagger})$.

$$\Delta G^{\ddagger} = G_r(q^{\ddagger}) = k_0(q^{\ddagger} - q_r)^2 = k_0 \left(\frac{q_p^2 - q_r^2 + \Delta G/k_0}{2q_p^- 2q_r} - q_r \right)^2.$$

Putting all the terms over a common denominator we find

$$\Delta G^{\ddagger} = k_0 \left(\frac{k_0(q_p - q_r)^2 + \Delta G}{2k_0(q_p - q_r)} \right)^2.$$

Inspection of Fig. 8.4.4 and the definition of the reorganization energy λ allows the identification

$$\lambda = G_r(q_p) = k_0(q_p - q_r)^2.$$

Thus

$$\Delta G^{\ddagger} = k_0 \left(\frac{\lambda + \Delta G}{2k_0(q_p - q_r)} \right)^2 = \frac{k_0(\lambda + \Delta G)^2}{4k_0^2(q_p - q_r)^2} = \frac{(\lambda + \Delta G)^2}{4\lambda}.$$

(c) Show that for a one electron transfer occurring with a cell potential E and for which we take the reaction co-ordinate to be the fractional charge transfer (or fluctuation), the transition state is located at

$$q^{\ddagger} = \tfrac{1}{2} - FE/2\lambda. \tag{8.9.9}$$

For a one electron transfer with the reaction co-ordinate taken to be the partial charge transfer or fluctuation, the products lie at $q_p = 0$ and a full electron has been transferred to reach the product state $q_p = 1$. Therefore

$$\lambda = k_0(1 - 0)^2 = k_0$$

and $$q^{\ddagger} = \frac{q_p^2 - q_r^2 + \Delta G/k_0}{2q_p - 2q_r} = \frac{1 - 0 + \Delta G/\lambda}{2 - 0} = \tfrac{1}{2} + \frac{\Delta G}{2\lambda}.$$

Recall that the Gibbs energy and the cell potential are related by

$$\Delta G = -nFE = -nF(E - RT \ln Q).$$

If all reactant and products are at unit activity then the reaction quotient $Q = 1$ and $\Delta G^\circ = -nFE$. In this case $n = 1$ and

$$q^{\ddagger} = \frac{1}{2} - FE/2\lambda$$

as was to be shown. This result show that for a thermoneutral reaction for which $\Delta G^\circ = E^\circ = 0$, such as charge transfer between isotopes of the same nucleus, the transition state occurs at the symmetric position $q^{\ddagger} = \frac{1}{2}$. For endothermic reactions with $\Delta G^\circ 0$, the transition state occurs beyond this point (resembles the products more) and for exothermic reactions with $\Delta G^\circ > 0$, the transition state occurs before this point (resembling the reactants more). Finally, when $\Delta G^\circ = -\lambda$ the maximum rate constant occurs, and the transition state occurs at $q^{\ddagger} = 0$, i.e. the products are the transition state and the electron transfer rate is ruled by the distance and symmetry dependence of the orbital overlap.

8.18   Show that if the probability of electron transfer from a level in an electrode at energy ε to an acceptor level in solution depends on ε according to

$$P(\varepsilon) \propto n(\varepsilon)f(\varepsilon)\exp(\varepsilon/2k_BT), \qquad (8.9.10)$$

where $f(\varepsilon)$ is the Fermi-Dirac function, then the maximum transfer probability occurs at $\varepsilon = \overline{\mu}_e$, i.e. from a level near the Fermi level of the metal. Note that density of states $n(\varepsilon)$ is a weak function of ε.

That the density of states is a weak function of ε means that we can take its derivative as a constant. Therefore, we find the maximum transfer probability by taking the derivative of $P(\varepsilon)$, equating it to zero, and solving for the value of ε that makes this equation true.

$$\frac{dP(\varepsilon)}{d\varepsilon} = \frac{d}{d\varepsilon}[n(\varepsilon)f(\varepsilon)\exp(\varepsilon/2k_BT)] \approx \frac{d}{d\varepsilon}[f(\varepsilon)\exp(\varepsilon/2k_BT)]$$

$$\frac{dP(\varepsilon)}{d\varepsilon} = \frac{df(\varepsilon)}{d\varepsilon}[\exp(\varepsilon/2k_BT)] + \frac{d\exp(\varepsilon/2k_BT)}{d\varepsilon}[f(\varepsilon)] = 0$$

$$\frac{e^{\varepsilon/2k_BT}}{2k_BT(1 + e^{(\varepsilon-\overline{\mu}_e)/k_BT})} - \frac{e^{(\varepsilon-\overline{\mu}_e)/k_BT}e^{\varepsilon/2k_BT}}{k_BT(1 + e^{(\varepsilon-\overline{\mu}_e)/k_BT})^2} = 0$$

$$\frac{e^{\varepsilon/2k_BT}}{k_BT(1 + e^{(\varepsilon-\overline{\mu}_e)/k_BT})}\left[\frac{1}{2} - \frac{e^{(\varepsilon-\overline{\mu}_e)/k_BT}}{(1 + e^{(\varepsilon-\overline{\mu}_e)/k_BT})}\right] = 0$$

$$\frac{1}{2} = \frac{e^{(\varepsilon-\overline{\mu}_e)/k_BT}}{(1 + e^{(\varepsilon-\overline{\mu}_e)/k_BT})}$$

Since $\exp[(\overline{\mu}_e - \overline{\mu}_e)/k_BT] = 1$, the maximum transfer probability occurs at $\varepsilon = \overline{\mu}_e$. $\overline{\mu}_e$ represents the position of the Fermi level in the metal; therefore, the electron transfer is most likely to occur from energy levels in a narrow range close to E_F.

Note that $n(\varepsilon)$ is essentially zero at E_F for a semiconductor, then suddenly jumps to a large and comparatively slowly varying number at the conduction band and valence band edges. Therefore, electron transfer involving a semiconductor electrode primarily occurs at the conduction band edge (when an electron is transferred to the semiconductor) or at the valence band edge (when an electron is transferred from the semiconductor to the solution species). All of this, of course, being for a normally doped (nondegenerate) semiconductor at normal temperatures, and for when an acceptor level approaches the metal or semiconductor "from above". When the level is energetically far below the band edge, then occupied states below E_F (in the metal) or E_V (in the semiconductor) will make horizontal transitions into the acceptor level.

8.18   Show that the Tafel equation

$$\eta = A + b \log |j| \tag{8.9.11}$$

can be derived from the Butler-Volmer equation Eq. (8.4.11). To do so, consider a cathodic reaction for which

$$\eta \ll -\frac{RT}{nF} \tag{8.9.12}$$

Equation (8.4.11) is

$$j = j_0 \left[\frac{N_{\text{red}}}{N_{\text{red},0}} \exp\left(\frac{\alpha e \eta}{k_B T} \right) - \frac{N_{\text{ox}}}{N_{\text{ox},0}} \exp\left(-\frac{\beta e \eta}{k_B T} \right) \right]$$

For a cathodic reaction, for which $\eta < 0$, the first term is negligible and only the second need be considered. Next note that

$$\frac{e}{k_B T} = \frac{N_A e}{N_A k_B T} = \frac{F}{RT}.$$

Therefore,

$$j = -j_0 \frac{N_{\text{ox}}}{N_{\text{ox},0}} \exp\left(-\frac{\beta F \eta}{RT} \right).$$

Take the log base 10 of both sides

$$\log |j| = \log\left(j_0 \frac{N_{\text{ox}}}{N_{\text{ox},0}} \right) - \left(\frac{\beta F}{2.303\, RT} \right) \eta.$$

Solve for η $\quad \eta = \log\left(i_0 \frac{N_{\text{ox}}}{N_{\text{ox},0}} \right) - \left(\frac{2.303\, RT}{\beta F} \right) \log |j| = A + b \log |j|$

$$\text{with} A = \log\left(j_0 \frac{N_{\text{ox}}}{N_{\text{ox},0}} \right) \text{ and } b = -\left(\frac{2.303\, RT}{\beta F} \right)$$

as was to be shown.

8.21   The rate constant for a proton transfer reaction as a function of applied potential U can be written

$$k(U) = k_0 \exp(-\Delta G^{\ddagger}(U)/k_B T) \tag{8.9.13}$$

where the Gibbs energy of activation as a function of applied potential U is written $\Delta G^{\ddagger}(U)$. In terms of current density this is written

$$j_k(U) = 2e\,\sigma_0\,k(U) = j_0 \exp(-\Delta G^{\ddagger}(U)/k_{\mathrm{B}}T) \qquad (8.9.14)$$

where σ_0 is the number of sites per unit area and e is the elementary charge.

(a) If a current under no driving force (that is the exchange current measured at $U = U_0$ ($U_0 =$ the equilibrium potential of the cell), is measured to be $j_0 = 100\,\mathrm{mA\,cm^{-2}}$, estimate the value of k_0 in units of $\mathrm{s^{-1}\,site^{-1}}$.

(b) If the Gibbs energy of activation depends on the overpotential $\eta = U_0 - U$ according to

$$\Delta G^{\ddagger}(U) = \Delta G^{\ddagger}(U_0) - e\eta, \qquad (8.9.15)$$

show that the Butler-Volmer relation is given by

$$U = U_0 - b\log_{10}(j_k/j_k^0) \qquad (8.9.16)$$

and calculate the value of the Tafel slope b at 300 K.

(a) Comparing Eqs (8.9.10) and (8.9.11)

$$j_0 = 2e\sigma_0 k_0$$

and, therefore,

$$k_0 = j_0/2e\sigma_0.$$

Insert the given value of j_0, $e = 1.6 \times 10^{-19}$ C and the typical value $\sigma_{\mathrm{w}} = 1 \times 10^{15}$ cm^2 to find

$$k_0 = \frac{100 \times 10^{-3}\,\mathrm{A\,cm^{-2}}}{2(1.6 \times 10^{-19}\,\mathrm{C})(1 \times 10^{15}\,\mathrm{sites\,cm^{-2}})} \approx 300\,\mathrm{s^{-1}\,site^{-1}}.$$

(b) Since

$$j_k(U) = j_0 \exp(-\Delta G^{\ddagger}(U)/k_{\mathrm{B}}T)$$

If we introduce

$$\Delta G^{\ddagger}(U) = \Delta G^{\ddagger}(U_0) - e\eta = \Delta G^{\ddagger}(U_0) - eU_0 + eU$$

Then we obtain

$$j_k(U) = j_0 \exp(-(\Delta G^{\ddagger}(U_0) - eU_0 + eU)/k_{\mathrm{B}}T) = j_0 e^{-\Delta G^{\ddagger}(U_0)/k_{\mathrm{B}}T} e^{-eU/k_{\mathrm{B}}T} e^{eU_0/k_{\mathrm{B}}T}$$

Defining

$$j_k^0 = j_0 e^{-\Delta G^{\ddagger}(U_0)/k_{\mathrm{B}}T}$$

we have

$$j_k(U) = j_k^0\, e^{-eU/k_{\mathrm{B}}T}\, e^{eU_0/k_{\mathrm{B}}T}$$

$$j_k/j_k^0 = e^{-eU/k_{\mathrm{B}}T}\, e^{eU_0/k_{\mathrm{B}}T}$$

$$\frac{k_{\mathrm{B}}T}{e}\ln(j_k/j_k^0) = -U + U_0$$

$$U = U_0 - \frac{k_{\mathrm{B}}T}{e}\ln\left(\frac{j_k}{j_k^0}\right) = U_0 - b\log_{10}\left(\frac{j_k}{j_k^0}\right) \text{ with } b = \frac{k_{\mathrm{B}}T\ln 10}{e} = 60\,\mathrm{mV}\;@\;300\,\mathrm{K}$$

References

[1]   F. Gstrein, D. J. Michalak, W. J. Royea, N. S. Lewis, *J. Phys. Chem. B* **106** (2002) 2950.

Appendix I

Abbreviations and Prefixes

Prefixes

a	f	p	n	μ	m	c	d	k	M	G	T
atto	femto	pico	nano	micro	milli	centi	deci	kilo	mega	giga	terra
10^{-18}	10^{-15}	10^{-12}	10^{-9}	10^{-6}	10^{-3}	10^{-2}	10^{-1}	10^{3}	10^{6}	10^{9}	10^{12}

$\langle x \rangle$	mean value of variable x
1D, 2D, 3D	one-, two-, and three-dimensional
II-VI	compound semiconductors composed of elements from the old groups II and VI, now groups 12 and 16, respectively
III-V	compound semiconductors composed of elements from the old groups III and V, now groups 13 and 15, respectively
AES	Auger electron spectroscopy
AFM	atomic force microscope or microscopy
ATR	attenuated total reflectance
bcc	body-centred cubic
BET	Brunauer-Emmett-Teller
CA	chemically amplified
CI	configuration interaction
CITS	constant current tunnelling spectroscopy
CTST	conventional transition state theory
CVD	chemical vapour deposition
det **M**	determinant of the matrix **M**
DFT	density functional theory
DIET	desorption induced by electronic transitions
DIMET	desorption induced by multiple electronic excitations
DNA	deoxyribonucleic acid

Surface Science: Foundations of Catalysis and Nanoscience, Third Edition. Kurt W. Kolasinski.
© 2012 John Wiley & Sons, Ltd. Published 2012 by John Wiley & Sons, Ltd.

DNQ	diazonaphthoquinone
DRIFTS	diffuse reflectance infrared Fourier transform spectroscopy
e^-	electron
ECS	equilibrium crystal shape
EELS	electron energy loss spectroscopy
Eq	equation, equilibrium
EQE	external quantum efficiency
ER	Eley-Rideal reaction mechanism
ESCA	electron spectroscopy for chemical analysis
ESD	electron stimulated desorption
ESDIAD	electron stimulated desorption ion angular distribution
fcc	face-centred cubic
FM	Frank-van der Merwe (layer-by-layer) growth
FT	Fischer-Tropsch or Fourier transform
FTIR	Fourier transform infrared spectroscopy
FWHM	full width at half maximum
GGA	generalized gradient approximation
h^+	hole (the absence of an electron)
HA	hot atom reaction mechanism
HC	hydrocarbons
hcp	hexagonal close packed
HHG	high harmonic generation
HOMO	highest occupied molecular orbital
HOPG	highly oriented pyrolytic graphite
HREELS	high resolution electron energy loss spectroscopy
IC	integrated circuit
IPCE	incident photon to charge carrier conversion efficiency
IR	infrared
IRAS	infrared reflection absorption spectroscopy
K-cell	Knudsen cell
KED	kinetic energy distribution
k-space	crystal momentum space. **k** is the wavevector of an electron or phonon in a lattice
L	Langmuir, a unit of exposure
LB	Langmuir-Blodgett
LEED	low energy electron diffraction
LH	Langmuir-Hinshelwood reaction mechanism
LHE	light harvesting efficiency
LID	laser induced desorption
LITD	laser induced thermal desorption
LUMO	lowest unoccupied molecular orbital
MARI	most abundant reaction intermediate
MBE	molecular beam epitaxy
MBRS	molecular beam relaxation spectrometry
MEMS	microelectromechanical systems
MGR	Menzel-Gomer-Redhead model
MIR	multiple internal reflection

ML	monolayer
MWT	multi-walled nanotube
NO_x or NOX	nitrogen oxides
NSOM	near-field scanning optical microscope (also called SNOM)
PEEM	photoemission electron microscope
PES	potential energy (hyper)surface
PMMA	poly(methylmethacrylate)
PDS	photon stimulated desorption
PSTM	photon scanning tunnelling microscope
QMS	quadrupole mass spectrometer
RAIRS	reflection absorption infrared spectroscopy
RSH	alkane thiol
Rxn	reaction
SAM	self-assembled monolayer
SATP	standard ambient temperature and pressure (298 K and 1 bar)
SEM	scanning electron microscope
SFG	sum frequency generation
SHG	second harmonic generation
SK	Stranski-Krastanov (layer-plus-island) growth
SFM	scanning force microscope
SPM	scanning probe microscope
STM	scanning tunnelling microscope
STS	scanning tunnelling spectroscopy
STOM	scanning tunnelling optical microscope
SWT	single-walled nanotube
syn gas	synthesis gas, $H_2 + CO$
TPD	temperature programmed desorption (also called TDS, thermal desorption spectrometry)
TPRS	temperature programmed reaction spectrometry
TS	transition state
TST	transition state theory
UHV	ultrahigh vacuum, pressure range below 10^{-8} torr (10^{-6} Pa)
UPS	ultraviolet photoelectron spectroscopy
VUV	vacuum ultraviolet
VW	Volmer-Weber (island) growth
XPS	x-ray photoelectron spectroscopy
XUV	extreme ultraviolet
α-end	head group of an amphiphile
$\delta, \gamma, \nu, \tau$	when referring to vibrational modes, these stand for the in-plane bends, out-of-plane bends, stretches and torsions, respectively
ω-end	tail group of an amphiphile

Appendix II
Symbols

*	surface site
a	absorptivity
a	basis vector of a substrate lattice
a*	reciprocal lattice vector of a substrate
A	Arrhenius pre-exponential factor for a first order process; absorbance
A_p	peak area
A_s	surface area
b	basis vector of an overlayer
b*	reciprocal lattice vector of an overlayer
B_v	rotational constant for vibrational level v
c	speed of light; concentration
d	depletion layer width; tip-to-surface separation, island diameter
D	diffusion coefficient
D_0	diffusion pre-exponential factor
$D(E)$	transmission probability in tunnelling
$D(\text{M–A})$	dissociation energy of the M–A bond
E	energy
E_0	in EELS, the energy of the incident electron beam (primary energy)
E_a	generalized activation energy
E_{ads}	adsorption activation energy
E_B	electron binding energy
E_C	energy of the conduction band minimum
E_{des}	desorption activation energy
E_{dif}	diffusion activation energy
E_f	final energy
E_F	Fermi energy
E_g	band gap energy
E_i	initial energy; position of E_F in an intrinsic semiconductor
E_K	kinetic energy

Surface Science: Foundations of Catalysis and Nanoscience, Third Edition. Kurt W. Kolasinski.
© 2012 John Wiley & Sons, Ltd. Published 2012 by John Wiley & Sons, Ltd.

E_n	component of kinetic energy along the surface normal (normal energy); energy of an image potential (Rydberg) state of principle quantum number n
E_V	energy of the valence band maximum
E_{vac}	vacuum energy
F_N	force on a cantilever
g	degeneracy
G^s	surface Gibbs energy
H	Hamiltonian operator, enthalpy
I	tunnelling current; intensity; moment of inertia
k	rate constant; crystal momentum
k_1	rate constant of forward reaction 1
k_{-1}	rate constant of reverse reaction 1
$\mathbf{k}$	wavevector
K	equilibrium constant
k_N	cantilever force constant
l	path length
L	desorption rate from chamber walls
m	mass
M	molar mass (also called molecular or atomic weight)
$\mathbf{M}$	transformation matrix between $\mathbf{a}$ and $\mathbf{b}$
n	principle quantum number; real part of the index of refraction; number of moles
$\tilde{n}$	complex index of refraction
$n(\mathbf{k}, p)$	occupation number of the phonon with wavevector $\mathbf{k}$ in branch p
n_i	intrinsic carrier density
N_0	number of surface sites or atoms
N_{ads}	number of adsorbates
N_A	density of electrically active acceptor atoms, Avogadro's constant
N_C	effective density of states in the conduction band
N_{exp}	number of atoms/molecules exposed to (incident upon) the surface
N_D	density of electrically active donor atoms
N_V	effective density of states in the valence band
N_{vj}	number of molecules in vibrational level v and rotational level j
p_b	base (steady-state) pressure
q	partition function
q_{ads}	heat of adsorption for a single molecule/atom
Q_{ads}	canonical partition function for adsorption
r	rate; radial distance; molecular bond length
R_0	reflectivity of a clean surface
R	reflectivity of an adsorbate covered surface
R_∞	reflectivity of an optically thick sample, Rydberg constant
s	sticking coefficient (also called sticking probability)
s_0	initial sticking coefficient
s_A	Auger sensitivity factor for element A
$\mathbf{s}$	direction of diffracted electron beam
$\mathbf{s}_0$	direction of incident electron beam
S	scattering function; vacuum chamber pumping speed

t	time
T	temperature, transmittance
T_p	temperature at which a TPD peak maximum occurs
T_s	surface temperature
T_θ	isokinetic temperature
u	displacement
v	velocity
V	voltage; potential; volume
V_B	tunnelling barrier height
V_{fb}	flat band potential
V_{surf}	magnitude of band bending
W	work
x, y	directions in the plane of the surface
x_A	in quantitative electron spectroscopy, the mole fraction of element A
Y	Young's modulus
z	direction normal to the surface
Z	atomic number
Z_w	collision rate
α	absorption coefficient; polarizability
δ_{ij}	Kronecker delta function
$\delta\varepsilon_{relax}, \ \delta\varepsilon_{rel}, \ \delta\varepsilon_{corr}$	in electron spectroscopy, corrections to the Hartree-Fock energies involving electron relaxation, relativistic effects and electron correlation
Δ	change; the energy difference between substrate bonds, Φ_{AA}, and the interface bonds, Φ_{AB}, within a heterolayer
ΔE_b	chemical shift
ΔG	Gibbs energy change
ΔH	enthalpy change
$\Delta_{ads}H$	integral heat of adsorption
$\Delta_{ads}H^{diff}$	differential heat of adsorption
$\Delta_{ads}H^{st}$	isosteric heat of adsorption
ΔS	enthalpy change
β	heating rate
χ	electron affinity
ε	exposure; lattice mismatch; permittivity, porosity
ε_k	orbital energy
$\varepsilon_{\alpha\beta}$	lattice deformation
$\varepsilon(\omega)$	dielectric function
γ	surface tension or energy; angle between basis vectors
η	electrochemical potential
ϕ	azimuthal angle
ϕ_c	contact potential
Φ	work function
$\Phi_{AA}, \ \Phi_{AB}, \ \Phi_{BB}$	at the interface between two materials, the energies of A–A, A–B and B–B bonds
Γ	intrinsic (homogeneous) peak width
κ	imaginary part of the index of refraction; transmission coefficient
λ	wavelength, reorganization energy

λ_m	mean free path in material m
μ	chemical potential; reduced mass; dipole moment
ν	frequency
ν_n	pre-exponential factor for a process of order n
θ	fractional coverage in monolayers
θ_{00}	number of adjacent sites in the presence of lateral interactions
θ_D	Debye temperature
θ_p	coverage at the maximum of a TPD peak
θ_{req}	functional form of the sticking coefficient dependence on coverage
ϑ	polar angle
$\rho_s(E)$	density of states at energy E
σ	areal density of adsorbates; symmetry number in rotational partition function; differential conductance
σ_0	areal density of surface sites or atoms
σ_*	areal density of empty sites
τ	lifetime; delay between laser pulses
$\tau_{\alpha\beta}(\mathbf{m})$	intrinsic surface stress tensor
ω_0	radial frequency of the fundamental mode of a harmonic oscillator
ω_D	Debye frequency
Ξ	grand canonical partition function
ψ	contact angle; wavefunction
ζ	degree of supersaturation

Kisliuk model of precursor-mediated adsorption

f_c	probability of chemisorption from the intrinsic precursor
f_{des}	probability of desorption from the intrinsic precursor
f_m	probability of migration to next site in the intrinsic precursor
f_c'	probability of chemisorption from the extrinsic precursor
f_{des}'	probability of desorption from the extrinsic precursor
f_m'	probability of migration to next site in the extrinsic precursor
α	trapping coefficient into the precursor

Symbols commonly used in subscripts and superscripts

$\ddagger$	transition state
0 or i	initial
ads	adsorption
des	desorption
dif	diffusion
elec	electronic
f	final
g	gas
in	incident

K	kinetic
max	maximum or saturation value
p	peak
rot	rotational
s	surface
trans	translational
vib	vibrational

Appendix III

Useful Mathematical Expressions

$E =$	$\left(n + \frac{1}{2}\right)\hbar\omega_0$	harmonic oscillator energy levels
$E_K =$	$\frac{1}{2}mv^2$	kinetic energy
$E_J =$	$hcB_v J(J+1)$	rigid rotor energy levels
$f(E) =$	$\{\exp[(E - \mu)/k_B T] + 1\}^{-1}$	Fermi-Dirac distribution
$f(v) =$	$4\pi\left(\dfrac{M}{2\pi RT}\right)^{3/2} v^2 \exp\left(\dfrac{-Mv^2}{2RT}\right)$	Maxwellian velocity distribution
$k =$	$A\exp(E_a/RT)$	Arrhenius expression
$D =$	$D_0 \exp(-E_{dif}/k_B T_s)$	diffusion coefficient
$1L =$	1×10^{-6} torr s	definition of the unit of exposure Langmuir
	1.3332×10^{-6} mbar s	
	1.3332×10^{-4} Pa s	
$\langle n(\omega, T)\rangle =$	$\{\exp[(\hbar\omega)/k_B T] - 1\}^{-1}$	Planck distribution
$N_v =$	$N\exp(-E_v/k_B T) = N\exp(-hcG_0(v)/k_B T)$	vibrational Maxwell-Boltzmann distribution
$N_{vj} =$	$N_v \dfrac{hcB_v}{kT}(2J+1)\exp(-E_J/k_B T)$	rotational Maxwell-Boltzmann distribution
$q =$	$\displaystyle\sum_{i=1}^{\infty} g_i \exp\left(\dfrac{-E_i}{k_B T}\right)$	general partition function
$q_{trans} =$	$\displaystyle\prod_i \dfrac{(2\pi m k_B T)^{1/2}}{h}$	translational partition function
$q_{rot} =$	$\dfrac{8\pi^2 I k_B T}{\sigma h^2} = \dfrac{k_B T}{\sigma hcB}$	rotational partition function, linear molecule
	$\dfrac{8\pi^2(8\pi^3 I_A I_B I_C)^{1/2}(k_B T)^{3/2}}{\sigma h^3}$	non-linear molecule

Surface Science: Foundations of Catalysis and Nanoscience, Third Edition. Kurt W. Kolasinski.
© 2012 John Wiley & Sons, Ltd. Published 2012 by John Wiley & Sons, Ltd.

$q_{\text{vib}} =$ $\displaystyle\prod_i \frac{1}{1 - \exp(-h\upsilon_i/k_B T)}$ vibrational partition function

$\langle x^2 \rangle^{1/2} =$ $\displaystyle\sqrt{4Dt} = \left(\frac{4D_0}{A}\exp\left(\frac{E_{\text{ads}} - E_{\text{dif}}}{RT}\right)\right)^{1/2}$ mean square displacement on a uniform 2D PES

$Z_w =$ $\displaystyle\frac{N_A p}{\sqrt{2\pi MRT}}$ impingement rate = flux

$\displaystyle\frac{p}{\sqrt{2\pi m k_B T}}$ M = molar mass
m = particle mass

$\displaystyle 3.51 \times 10^{22}\,\text{cm}^{-2}\text{s}^{-1}\,\frac{p}{\sqrt{MT}}$ Z_w in cm^{-2} s^{-1}, M in g mol^{-1}, p in torr, T in K

$\displaystyle 2.63 \times 10^{24}\,\text{m}^{-2}\text{s}^{-1}\,\frac{p}{\sqrt{MT}}$ Z_w in m^{-2} s^{-1}, M in g mol^{-1}, p in Pa

$\Delta G =$ $\Delta H - T\Delta S$ definition of change in Gibbs energy

$k =$ $\kappa A \exp(-\Delta G^{\ddagger}/RT)$ rate constant within transition state theory

$k =$ $A \exp(-E_a/RT)$ Arrhenius formulation of rate constant

$\varepsilon =$ $Z_w t$ definition of exposure

$\lambda =$ h/p de Broglie relationship

$\mu =$ $\displaystyle\frac{m_1 m_2}{m_1 + m_2}$ reduced mass

$\theta =$ $\displaystyle\frac{p}{p + p_0(T)}$ thermodynamic expression of the Langmuir isotherm for molecular adsorption

$\displaystyle p_0(T) = \left(\frac{2\pi k_B T}{h^2}\right)^{3/2} k_B T \exp\left(\frac{-q_{\text{ads}}}{k_B T}\right)$

$\theta =$ $\displaystyle\frac{pK}{1 + pK}$ kinetic expression of the Langmuir isotherm for molecular adsorption

$K = k_{\text{ads}}/k_{\text{des}}$

$\displaystyle\frac{p}{n^a(p^{\circ} - p)} = \frac{1}{n_m^a C} + \frac{C-1}{n_m^a C}\frac{p}{p^{\circ}}$ BET isotherm. n^a = amount adsorbed, n_m^a = monolayer capacity

$\sigma =$ $s\varepsilon$ coverage resulting from exposure ε when the sticking coefficient is constant

$r =$ $\displaystyle\frac{s_0 Z_w}{\sigma_0}$ rate of absorption in ML s^{-1} at zero coverage

$\theta =$ $\displaystyle\frac{r_0}{a}(1 - \exp(-at))$ fractional coverage as a function of time resulting from concurrent first-order adsorption and first-order desorption, $a = (r_0/\theta_{\text{max}}) + k_{\text{des}}$

$\theta =$ $\displaystyle\frac{r_0 t \theta_{\text{max}}}{r_0 t + \theta_{\text{max}}}$ fractional coverage resulting from second-order adsorption

$\theta =$ $\quad \dfrac{1}{n}(1 - \exp(-nr_0 t))$ — fractional coverage resulting from first-order adsorption in which the adsorbate deactivates n sites

$\Delta p =$ $\quad 2\gamma/r$ — Young-Laplace equation

$RT\ln(p/p^*) = 2\gamma V_m/r$ — Kelvin equation

$p =$ $\quad p^* \exp(2\gamma V_m/RTr)$

$\gamma_{sg} =$ $\quad \gamma_{lg}\cos\psi + \gamma_{sl}$ — Young equation

$F =$ $\quad 2\pi\gamma R_p$ — capillary force between two particles of radius R_p

$\Gamma_j^{(1)} =$ $\quad \Gamma_j^S - \Gamma_1^S \dfrac{c_j^A - c_j^B}{c_1^A - c_1^A}$ — relative adsorption

$h_c =$ $\quad \left(\dfrac{r_p}{\gamma_L}\right)^2 E_S(1-\varepsilon)^3 \gamma_S$ — critical thickness for cracking of a thin porous film

$h_c =$ $\quad 1/\varepsilon_0$ — critical thickness for strain relief by formation of misfit dislocations

$\varepsilon_0 =$ $\quad \dfrac{a_1 - a_s}{a_s}$ — lattice misfit or misfit strain

$\mu =$ $\quad \mu^\circ + RT\ln(a/a_0)$ — chemical potential

$\overline{\mu}_i =$ $\quad \mu_i + z_i F\varphi$ — electrochemical potential

$E =$ $\quad E^\circ - (RT/nF)\ln Q$ — Nernst equation

$j =$ $\quad j_0\left[\dfrac{N_{red}}{N_{red,0}}\exp\left(\dfrac{\alpha e\eta}{k_B T}\right) - \dfrac{N_{ox}}{N_{ox,0}}\exp\left(-\dfrac{\beta e\eta}{k_B T}\right)\right]$ — Butler-Volmer equation

$r_c =$ $\quad 2\gamma/\overline{\rho}\,\Delta\mu$ — classical critical radius of nucleation

$\Delta_{max}G =$ $\quad \frac{4}{3}\pi r_c^2\gamma$ — classical nucleation barrier

$\phi_j =$ $\quad \dfrac{N_j}{N_{ex}} = \dfrac{N_j}{\sum_k N_k} = \dfrac{r_j}{\sum_k r_k}$ — quantum yield

$F =$ $\quad -\dfrac{dV}{dr} = ma = m\dfrac{d^2 r}{dt^2}$ — force defined in terms of the potential

$\phi_{th} =$ $\quad \dfrac{UC_v}{k_B\,\alpha(1-R)}$ — laser ablation threshold fluence

Thermal energy	4 K	77 K	298 K
$RT/\text{kJ mol}^{-1}$	0.03254	0.6264	2.424
$k_B T/\text{meV}$	0.3447	6.635	25.68
$k_B T/hc/\text{cm}^{-1}$	2.780	53.52	207.1

Index